SUSTAINABLE WATER AND ENVIRONMENTAL MANAGEMENT

Chief Patron

Prof. A. Venugopal Reddy

Vice-Chancellor, J.N.T. University Hyderabad

Patron

Dr. N. Yadaiah

Registrar, J.N.T. University Hyderabad

Chairperson

Dr. A. Jaya Shree

Director, IST, J.N.T.U Hyderabad

Convener

Dr. M.V.S.S. Giridhar

Associate Professor, CWR, IST, J.N.T. University Hyderabad

Organized by

Centre for Water Resources

Institute of Science and Technology

Jawaharlal Nehru Technological University Hyderabad

Kukatpally, Hyderabad - 500 085

Disclaimer

The Publisher and Organisers do not claim any responsibility for the accuracy of the data, statements made, and opinions expressed by authors. The authors are solely responsible for the contents published in the paper.

Published by

BSP BS Publications

A unit of **BSP Books Pvt. Ltd.**

4-4-309/316, Giriraj Lane, Sultan Bazar,
Hyderabad - 500 095
Phone : 040 - 23445688, 23445600
e-mail : info@bspbooks.net
www.bspbooks.net

ISBN: 978-93-87593-23-7

Preface

It is indeed a matter of great concern that our natural resources are getting depleted raising many environmental issues. It is imperative therefore to conserve and protect the environment. Conservation is a sustainable use and management of natural resource including wild life, water, air and earth deposits. Natural resources may be renewable and non renewable. The conservation of renewable resources warrants a balance between consumption and their replacement. The conservation of non-renewable resources, like fossil fuels, involves ensuring that the sufficient quantities are maintained for future generation to utilize. Conservationist accept that development is necessary for the better future, but not at the cost of environmental degradation.

Preservation, in contrast to conservation, attempts to maintain the existing environmental conditions. This is due to the concern that mankind is encroaching on to the environment on such a rate that many untamed landscapes are being taken for farming, industry, housing, tourism and other developments in turn damaging ecosystem.

As the priority and availability of water changes, we must also remember that everyone should have the right to a safe source of drinking water. We have to create a more sustainable approach to water management and identify the key influencing factors. It can be concluded that water sustainability, climate change, ecological health, population changes, land management, urbanisation and economic prosperity are all intrinsically linked. It is therefore important to recognise that the challenges we face on a global scale can only be resolved by scientists, engineers and policy makers working together to create aligned objectives and strategies.

It is in this context and backdrop that the Centre for Water Resources, Institute of Science and Technology, JNTUH felt the need to organize a three day National Conference on Sustainable Water and Environmental Management (SWEM-2017) to take stock of the current status of applications in water resources development and management and also to identify areas most relevant to ensure sustainable development of water resources and environment to benefit the society at large.

Researchers, engineers, site managers, regulatory agents, policy makers, Consultants, NGO's, academicians and vendors will all benefit from the opportunity to exchange information on recent research trends and to examine ongoing research programs in the areas of water and environment. The conference is expected to recommend suitable strategies and policy guidelines to operationalize the initiatives and dovetail them into various watershed development programmes appropriately. Keeping in view the importance and need of the hour, this issue of proceedings is brought out to coincide with the conduct of the national conference. The high value contributions by eminent speakers, Research scholars and participants have been overwhelming and encouraging.

The three day national conference on SWEM will focus its attention on various themes such as

1. Mathematical Modeling in Water & Environmental Management
2. Impact of climate change, mitigation and adaption
3. Floods and droughts and its effects
4. Applications of GIS and remote sensing for water and environmental management.
5. Sustainable Irrigation management
6. Reservoir operation and soil erosion
7. Sustainable rainwater harvesting and recharge methodologies
8. Sustainable storm water management, Reuse, Stakeholder participation
9. Water Resources Planning, modeling and Monitoring
10. Groundwater exploration, development and Modeling
11. Urban water and environmental management
12. Water conservation practices
13. Surface water quality and pollutant control
14. Water, food, energy and health
15. Wetland development and management

I hope the present conference would serve as a link between technology, policy, practice and decision making in the quest for synergetic solutions for sustainable development of water resources and environment.

I wish and expect that the participants will find this conference useful and give their total participation to make it a grand success.

It is with this great pleasure; I extend a warm welcome to all the delegates, speakers and participants to SWEM- 2017.

Dr. M.V.S.S Giridhar
-Editor

Acknowledgement

I would like to express their gratitude to all the people that have helped us during these months for the organization of the conference. The National Conference on Sustainable Water and Environmental Management SWEM -2017 has been made possible with the support of many technical experts, individuals and organizations both in man power and finance. This support is gratefully acknowledged.

I owe a deep sense of gratitude to ***Prof. A. Venugopal Reddy***, Vice-Chancellor, Jawaharlal Nehru Technological University Hyderabad and Chief patron of the conference for his constant encouragement valuable guidance in organizing the conference in most efficient way.

My sincere and special thanks to ***Dr. N. Yadaiah***, Registrar, Jawaharlal Nehru Technological University Hyderabad as the Patron of the conference for his cordial, time to time permissions and support.

I am deeply indebted to ***Dr. A. Jaya Shree***, Director, IST, JNTUH and Chairman of this conference for having taken every responsibility for completing this task through various stages.

I would like to extend my very great appreciation to ***Prof C. Sarala***, Head, Centre for Water Resources for her valuable and constructive suggestions during planning, development and implementation of this task.

I would like to extend my grateful thanks to ***Dr. B. Venkateswar Rao*** and ***Dr. K. Rammohan Reddy***, Professors of Centre for Water Resources for their valuable support throughout the conference.

My sincere thanks to the officials of Technical Education Quality Improvement Program (TEQIP), Phase-III, IST, Science and Engineering Research Board (SERB), ISRO (Indian Space Research Organisation) for sponsoring this event. Without their help organization of this conference would not have been possible.

Further the financial assistance received from research and development fund of National Bank for Agriculture and Rural Development (NABARD) towards publication of journal/ printing of proceeding of the conference is gratefully acknowledged. Without their help organization of this conference would not have been possible.

We have been very fortunate enough to be backed by a team of very motivated and dedicated experts of various committees in guiding us throughout the conference very meticulously. My sincere thanks to all the members of the Scientific and Advisory Committee, Technical Committee and Organizing Committee for their sincere advice and help from time to time.

I profusely thank all the Key note speakers, Chair persons and Co-chair persons of various technical sessions of conference have readily responded to our invitation to conduct the proceedings and to address the gathering and for their kind gesture in the conference.

I thank the research scholars who have assisted in every event of conference.

My thanks are also due to various other Teaching and Non-teaching staff of IST and Engineering Staff of JNTUH who have cooperated on several occasions in organizing this Conference.

I sincerely thank M/s BS Publications for bringing out the souvenir and pre-conference proceedings well in advance.

My sincere thanks to my students Smt. P. Sowmya, Research Scholar and Ms. Shyama Mohan for their continuous day and night support for this conference.

Finally, I thank all the people and organizations who are directly and indirectly involved in organizing the conference, but I could not mention their names due to paucity of space.

I thank one and all.

MVSSGiridhar

M.V.S.S. Giridhar
Convenor

Contents

INAUGURAL KEYNOTE

CHALLENGES AND POSSIBLE SOLUTIONS FOR WATER RESOURCES MANAGEMENT IN THE 21st CENTURY

S. Mohan

Professor, Environmental and Water Resources Engineering,
Department of Civil Engineering, Indian Institute of Technology, Madras

INTRODUCTION

Drought in Ethiopia, national conflicts over the Middle East waters, floods along the Yangze river in China, floods and droughts in alternate years in India, riots over irrigation water in the Punjab, cholera in Peru, subsidence in Mexico City, groundwater pollution in Western Europe, industrial effluents in the Volga and arsenic pollution in the state of West Bengal and in different corners of the world water-related problems take different shapes, mirroring the looming water crisis, which will undoubtedly increase in magnitude during the 21st century.

Because the world is facing extraordinary tough decisions about the finite extent of all its natural resources, the only way is to establish policy is through objective and transparent methods that everyone accepts. This technical meeting, which is pioneering these methods, may have repercussions on other environmental policy areas such as climate change, air quality, biodiversity, etc.

The World Health Organization reports that for the first time ever, the majority of the world's population lives in a city, and this proportion continues to grow with projections of 70 percent by 2050. Currently, around half of all urban dwellers live in cities with populations between 100,000 and 500,000 people, and almost 10 percent of urban dwellers live in megacities, which are defined by UN HABITAT as a city with a population of more than 10 million.

As cities around the world experience this exploding growth, the need to ensure they can expand sustainably, operate efficiently and maintain a high quality of life for residents becomes even greater than it is today. This is where smart cities come into the picture. The term "smart cities" is trending amongst governments, urban planners and even the private sector to address the projected demands of cities in the future. Making cities smarter to support growth is emerging as a key area of focus for governments and the private sector alike. A recent estimate of the investment going to be made in the near future on the smart cities initiative is around $108 billion, especially for the smart city infrastructure, such as smart meters and grids, energy-efficient buildings and data analytics.

This keynote address mainly deals with key issues dealing with major water and environmental challenges, namely, raising awareness, data/information shortage, integration, and the management of environmental resources. We all have specific interests on water and environmental issues and possible solution methods specifically smart water management and a brief accounting of these points are discussed in this paper.

Global Perspective on Water and Environment

Before concentrating on water related problems in India, let us have a look at the global situation. The Second World Water Forum in The Hague in March 2000, the largest freshwater conference ever, showed it very clearly to the world public: Water will be one of the **central**

issues of the 21st **century** of this globe and the life of billions of people will depend on the wise management of this resource. During the Forum, the International Water Management Institute published a study on the world water situation in **2025**. Some of the statements of this study may be quoted here:

*"Nearly one-third of the population of developing countries in 2025, some **2.7 billion** people, will live in regions of **severe water scarcity**. They will have to reduce the amount of water used in irrigation and transfer it to the domestic, industrial and environmental sector" (IWMI 2000).*

In **India**, some 460 million people and in **China** more than 500 million people will live in regions that will face **absolute water scarcity**.

Groundwater reserves will be increasingly **depleted** in large areas of the world. In some instances this will threaten the food security of entire nations, such as India. This will certainly lead to major problems in food security and excess to save water.

Groundwater contamination by human interference, e.g. by industrial effluents, agricultural pollution or domestic sewage water intrusion is another world wide problem, which asks for urgent counteractions.

The world's **primary water supply** will need to increase **by 41%** to meet the needs of all sectors in a sustainable way in 2025. This increase in water demand is largely due to the increase in the **world population**, which is estimated to increase to some 80 million people every year, at least up to 2015. This means **another India for the world to feed every decade** (Falkenmark 1998)!

Another problem which faces mainly the Developing World is the phenomenon of **urbanization**: "Safe water and sanitation close to the home for everyone" as demanded by the Mar del Plata Action Plan, seems to remain a dream when acknowledging that in 2025 nearly **4 billion people** will live in urban areas – and the process is most dramatic in countries with relatively few resources (GWP 2000). **Seawater intrusion** in coastal aquifers due to overdrafting is another urgent problem in most tropical and subtropical countries alike. This anthropogenic problem accelerates in some areas the already existing natural problem of **poor groundwater quality**, e.g. of high **salinity**. In Tunisia, 26% of surface water, 90% of water pumped from shallow and 80% from deep aquifers have a salinity of more than 1.5 g per litre. There are many other water related problems, like frequent floods, soil salinization, acid rain etc. which are encountered worldwide. This is clearly stated in the statement of World Water Commission as follows:

"Every human being, now and in the future, should have enough clean water, appropriate sanitation and enough food and energy at reasonable cost. Providing adequate water to meet these basic needs must be done in a manner that works in harmony with nature."

World Water Commission, quoted from GWP 2000

This is the **global water challenge** and it is well known (theoretically), how to reach this goal, how to solve or least to alleviate the above mentioned problems: " Mobilizing the political will to act", "Making water governance effective", "Generating Water Wisdom", "Tackling urgent water priorities", or "Investing for a secure water future" (GWP 2000).

Among the "**urgent water priorities**", which have to be tackled, some should be mentioned here:

- Increasing productivity of water consumed in agriculture,
- More water storage in reservoirs or underground,
- Better water recycling/reuse practices: "Use each drop of water four times!"
- Reduction in water distribution losses in cities as well as in large irrigation systems
- Regulating groundwater extraction
- Conservation of Available Water
- Effective waste water treatment, less water pollution etc.

The '**Dublin Principles'** shall be the guidelines, among them the highly debated one: "Water shall be regarded as an **economic good**". Will we ever meet the "global water challenge" as mentioned above? Why do the problems persist, in spite of so many declarations, so many 'Visions' and 'Frameworks for Actions'? What are the **underlying causes?.** At least for the Developing World, we can identify some reasons why the situation worsens:

- Pauperisation,
- Lack of social structures,
- Lack of education and training, awareness
- Marked oriented production on a global scale ("Globalisation") with missing or at least insufficient enforcement of environmental laws, etc.

(At least the latter point is also true for many countries in the Industrialized World.)

It is clearly evident that there will be hardly any change to the better during the first decades of the 21st century in this respect, unless the **global economic order** is adjusted. A purely profit-oriented global society will neither go for "sustainable development", nor solve the problems of lower income classes nor of the Developing World in general. There is not only the need for a change in water related paradigms, but a general thinking-over, what paradigms we should accept, what goals we should set. There is **no alternative** than to give a larger share of the global profits to the poor and to the Developing World, allowing them a more stable economic, social and environmentally friendly development (Loucks 1994).

Water and Environmental problems in India

"Our vision is that in two or three decades there will be sufficient, save, clean and healthy water for nature and people living in stable societies in the region". This **statement should be the 'motto' for the management of Water for the 21st Century**. There is definitely a long way to go to reach this goal: **drinking water supply** and **waste water management** are very unsatisfactory in rural areas of many states, where they use **low quality** water resources, such as shallow wells. One reason for the low quality of water in biological terms is the so called "**utility gap"**, which means the percentage of population connected to the water supply minus that of connected to sewerage. This "utility gap" is large in many countries. This situation leads to increase pollution of surface and sub-surface waters and health risks. To tackle this precarious situation on water quality, an **anti-pollution programme that includes the following aspects needs to be implemented**:

- Enforcement of environmental laws and regulations
- Harmonization of laws and regulations within the European Union and beyond

- Development of the knowledge base required by actors (= legislators, regulators, industry, local authorities, public health bodies, consumer organizations etc).
- Knowledge of the state and evolution of the resources in quantitative and qualitative terms
- Understanding of the nature of pollution within water supply systems
- Understanding of the impact of the pollution on human health and the environment
- Research on socio-economic aspects of the fight against pollution with the purpose of understanding better the system of interaction linking different players and developing new approaches for combating pollution
- Improving waste water treatment quantitatively and qualitatively, addressing also the specific socio-technological and economic constrains on operators in suburban and rural areas.

Prevention of pollution diffusion by:

"On site" treatment of effluents e.g. pre-treatment before disposal in public sewers

- Treatment in accordance with the specific re-use /re-cycling
- Rehabilitation of contaminated soils, sediments and aquifers
- Implementation of cropping techniques appropriate for preventing or limiting this diffusion of pesticides and fertilisers to nearby aquifers.

Most of the water bodies in the country (surface water and groundwater bodies) receive excessive amounts of untreated or insufficiently treated municipal waste water.

- Pre-treatment of **industrial effluent** is often insufficient for biological treatment processes in treatment plants; industry accounts for a significant part of the discharge of polluting substances into inland waters in the region.
- Both the **agro-food** and pulp and paper industries discharge substantial amounts of oxygen- consuming, nutrient-rich and slowly degradable substances.
- The production of **synthetic chemicals and emerging contaminants** brings new and more exotic types of production wastes. and waste water treatment installations typically suffer from insufficiency, overloading and poor operation and maintenance.
- **Agricultural activities** contribute substantially to the overall nutrient load in surface and in groundwater. **Eutrophication** of rivers, inland and coastal waters is one of the major problems. The problems created by agricultural production include ammonia volatilisation, nutrient leaching, discharge of farm wastes.

Past strategies of water supply development have generally been based on the development of supply by improving abstraction techniques and exploitation of new "primary" sources (European Commission 1996). Needed are strategies for rationalisation of **demand** through

1. **Use of "secondary" sources**, mainly **re-use /recycling** of waste water or drainage water (Prinz 1999a). Re-use and recycling of water as well in industry as in agriculture is needed.
2. **Reduction of consumption of water** as well in private as in industry or in irrigation. To encourage the various categories of water users to save water, economic, fiscal and statutory instruments have to be designed and applied. In industry, water

can be conserved through finding alternatives to water for cooling, solvent precipitation medium, or by development of better water equipment and management methodologies. In agriculture, water can be conserved either by improved irrigation scheduling (better instrumentation, information and control systems) or by using irrigation methods with a higher water efficiency (sprinkler, drip irrigation).

3. **Minimization of losses:** The losses in **public supply** and in irrigation networks are about **15-30%** of the total water extracted. Therefore, the strong need exists to minimize those losses by (1) leak detection, (2) leak repair and (3) by leak prevention, e.g. by using material which is less subject to corrosion and mechanical fatigue. Adequate pricing policies or subsidies should therefore guarantee sufficient clean water supply **to all sectors of the society**.

For the development and **diversification** of the supply sources the following actions have to be taken:

1. Exploitation of presently under- or unexploited water resources, e.g.
    - Collecting rain / runoff for use as domestic water of for agriculture
    - Collecting runoff for artificial recharge of aquifers, taking into account the potential environment impact
    - Exploiting the karstic aquifers found in many of the Mediterranean coutries, evaluating the impact of such exploitation on the local and regional hydrological equilibrium.
2. Extended application of the **desalinisation technology** with particular attention being paid to the use of renewable energy technologies (e.g. photovoltaic cells).

Future climate change will have strong impact on the hydrologic , especially in areas already characterised by water scarcity. Additionally to impacts on availability of water, extremes and fluctuations may become larger and seasonal distribution of precipitation and / or runoff may alter negatively. Increased **temperatures** will increase the demand of crops as well as of vegetation in regard to water and irrigation will be necessary in areas where it was less needed before. Higher average temperatures will result in higher evaporation rates from reservoirs and additionally stronger winds / storms are expected. As climate change is (to a large extend) caused by anthropogenic actions, e.g. burning of fossil fuels, **counteractions** to reduce the release of relevant trace gases are a necessity.

Extreme **floods** are not only caused by the coincidence of unusual natural factors, but also influenced by land use changes, narrowing of the river bed e.g. by constructions in the flood plain, etc. It is therefore extremely important to look for **organisational and institutional structures** for national as well as international **river basin management**. The establishment of **river basin authorities** is therefore an essential part of the flood management. There is no doubt that water resources management in quality and quantity is possible only within a watershed and therefore this regulation is overdue. The importance of flood protection as one of the major elements of watershed planning is exemplified by the fact, that 85% of all civil protection measures taken by states are concerned with flooding.

Specific actions to avoid extreme events like floods and sudden or exceptional pollution are:

- Improving the capacity to forecast the occurrence (incl. likely amplitude , extent, impact)
- Improving of the knowledge about the causes of "natural" catastrophes

- Establishment of preventive practices (e.g. in the framework of land management)
- Development of management tools to interact rapidly and efficiently
- Development of emergency systems e.g. for supplying water to affected populations.

The principles of Integrated Water Resources Management (IWRM) should be the cornerstones of any implementation strategy in the Indian water sector.

The IWRM principles focus on :

- **Integration**
  - of water quality and quantity management
  - of groundwater and surface water management
  - of freshwater and coastal zone management
  - of water supply and sanitation planning
  - of upstream and downstream demands in regard to water quantity and quality
  - of physical, economical and social aspects of water resources management
- E**cosystem management** in recognizing all values of biodiversity and the integrity of ecosystems
- **Communication** between main actors in water resources management, including water consumers, water specialists, interest groups, etc.
- **Capacity building** by training of professional skills, raising public awareness, etc.,
- **Public participation** in decision making, based on access to water-related information.

SMART CITIES AND SMART WATER

Smart cities encompass six important sectors that need to work in unison to achieve a common goal of making a city more livable, sustainable and efficient for its residents (Fig.1). These sectors are smart energy, smart integration, smart public services, smart mobility, smart buildings, and smart water.

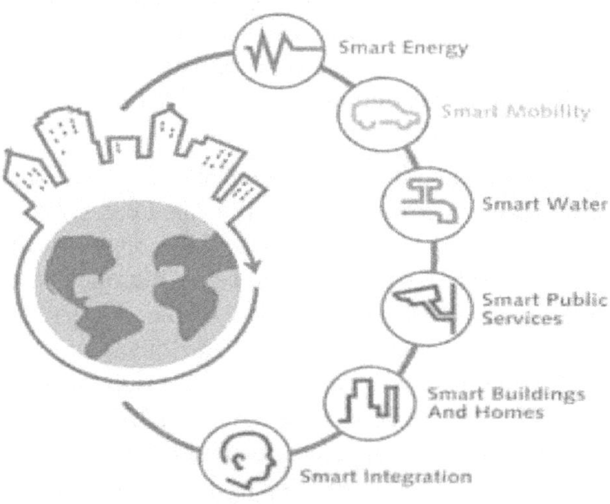

Fig. 1 Conceptual Model for the Smart City

Building smart cities upon the six sectors is crucial for sustainable global growth, but the financial, logistical and political challenges are enormous. The conversations about growth of smart cities have historically been dominated by large IT companies that focus on analyzing "big data" taking a top-down, software-centric approach. However, when it comes to the modernization of hundred-year-old systems like water distribution or the power grid, advanced software and networking capabilities are rarely broad enough in scope to make the necessary impact. Conversely, a bottom-up approach to smart city development is based on the belief that the rapid migration to cities will tax municipal infrastructures beyond their breaking points. The cities that succeed in transitioning to "smart" operations will be those that improve their critical systems and infrastructure at a fundamental level as well as integrate their systems through advanced technology. Lastly, smart cities will apply advanced monitoring and analytics to continuously measure and improve performance.

One of a city's most important pieces of critical infrastructure is its water system. With populations in cities growing, it is inevitable that water consumption will grow as well. The term "smart water" points to water and wastewater infrastructure that ensures this precious resource - and the energy used to transport it - is managed effectively. A smart water system is designed to gather meaningful and actionable data about the flow, pressure and distribution of a city's water. Further, it is critical that that the consumption and forecasting of water use is accurate.

A city's water distribution and management system must be sound and viable in the long term to maintain its growth and should be equipped with the capacity to be monitored and networked with other critical systems to obtain more sophisticated and granular information on how they are performing and affecting each other. Additional efficiencies are gained when departments are able to share relevant, actionable information. One example is that the watershed management team can automatically share storm water modeling information which indicates probable flooding zones and times based on predictive precipitation intelligence. The transportation department can then reroute traffic accordingly and pre-emptively alert the population using mass notification.

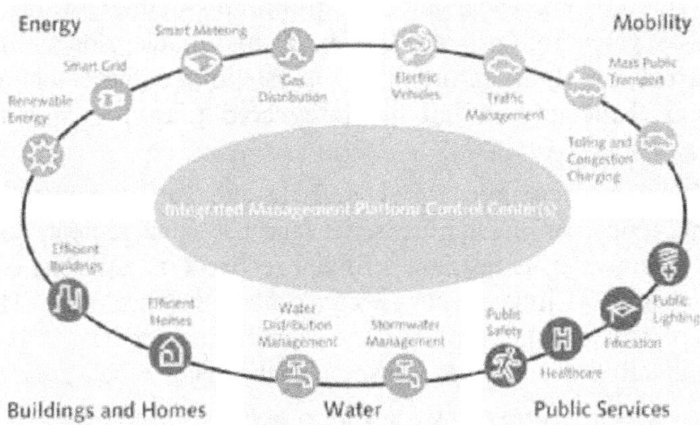

Fig. 2 Integrated Management Platform Control

Water systems are often overlooked yet are critical components of energy management in smart cities, typically comprising 50 percent of a city's total energy spend. Energy is the largest controllable cost in water/wastewater operations, yet optimizing treatment plants and distribution networks has often been overlooked as a source of freeing up operating funds by cash-strapped municipalities. Once facilities are optimized and designed to gather meaningful and actionable data, municipal leaders can make better and faster decisions about their operations, which can result in up to 30 percent energy savings and up to 15 percent reduction of water losses.

Water loss management is becoming increasingly important as supplies are stressed by population growth or water scarcity. Many regions are experiencing record droughts, and others are depleting aquifers faster than they are being replenished. Incorporating smart water technologies allows water providers to minimize non-revenue water (NRW) by finding leaks quickly and even predicatively using real-time SCADA data and comparing that to model network simulations. Reducing NRW also allows municipalities to recover costs incurred in treatment and pumping - this can be significant. A medium-sized city with 100 million gallons per day of produced water that loses 25 percent (not an unusual amount) is incurring over $13 million per year in non-recoverable labor, chemical and energy expenses.

Final Remarks

"Every human being, now and in the future, should have enough clean water, appropriate sanitation and enough food and energy at reasonable cost... "

This 'vision' will never become true, but we should act in India in solidarity with the billions of people of the world whose 'basic needs' are by far not met, - and we should do it "in harmony with nature."

Water in India is a precious, an endangered and often a problem-causing resource. Water is in many places in **short supply** and an increasing standard of living is normally paralleled by an increase in water demand. Additional stress will arise from anthropogenic climate change. The application and adjustment of the various means of **supply and demand management** plus adoption of "smart water" technology are great challenges for the future.

Water is all over the country endangered by **pollution** - either by insufficiently or even totally untreated waste water from private households and industry or by agro-chemicals. Not only **pollution control** based on information, legislation, enforcement of legislation and peoples' participation challenges us, but the **integrated planning of water quantity and quality**, too.

Many states in India face **chronic water deficits** and suffer frequently from **droughts**. Strategies for demand reduction include the better land use management, diversification of the water supply sources, rainwater harvesting, artificial recharge of aquifers, exploiting of karstic aquifers and desalinisation. Climate change aggravates the problem. The challenges are manifold; strategic planning, financial assistance for implementation of preventive measures and promotion of innovative technologies are some of the needed measures.

Floods and excess water pose frequently severe problems to the people affected as well as to governments; special problems are encountered in some 'Transformation Countries' due to the breakdown of drainage systems. The need for appropriate preventive measures, e.g. on the sector of land use planning, restoration of wetlands and floodplains, is unquestioned. Water is largely a shared resource and thus a wide variety of actions is involved in its management; their decisions frequently have inter-basin or even trans-regional impacts.

REFERENCES

1. Falkenmark, M. (1998). "Dilemma when Entering 21st Century - Rapid Change but Lack of Sense of Urgency", Water Policy No. 1 (1998), reprint 2, Stockholm International Water Institute, pp.421-436
2. IWMI (2000). "Water Issues for 2025 - A Research Perspective", International Water Management Institute, Colombo, Sri Lanka
3. Loucks, D.P. (1994). Sustainability: What Does This Mean for Us and What Can We Do About It? In: Proceedings for Intl. Symposium Water Resources Planning in a Changing World (Karlsruhe/Germany, June 28-30, 1994). Intl. Hydrological Program of UNESCO, Koblenz, Germany.
4. Prinz, D. (1999a). Policies, Strategies and Planning for Integrated Rural Water Management. In: Hamdy, A., Lacirignola, Tekinel, O., Yazar, A. (eds) Proceedings, Advanced Short Course on “Integrated Rural Water Management: Agricultural Water Demands”. Adana, Turkey, 20 Sept. - 02 Oct. 1999, p. 93 – 126.

KEYNOTE

A STUDY ON HYDRO-GEOLOGICAL ENVIRONMENT AT ANDHRA UNIVERSITY CAMPUS, VISAKHAPATNAM USING VERTICAL ELECTRICAL SOUNDINGS AND PUMPING TESTS

Srinivasa Rao G.V.R[1], Abbulu Y.[1], Sreejani T.P.[2], Balaji B[3] and Siddardha N.[3]

[1]Professor, [2]JRF, DST Project, [3]M.Tech student, Department of Civil Engineering, Andhra University

ABSTRACT

An attempt has been made in this present work to determine the groundwater dynamics at Andhra university campus using Vertical Electrical Sounding (VES) tests in combination with pumping tests. A total of five VES using Schlumberger configuration with a maximum current electrode separation of 100m are carried out in order to investigate the aquifer characteristics using the apparent resistivity values and to evaluate the thickness of geo electric layers. The pumping tests are carried out based on the multiple observation well approach for all the 25 existing bore wells and the drawdowns are measured at specified time intervals. Aquifer Test pro software is used to compute the values of transmissivity from pumping test data. Transmissivity values range from 84.7m^2/day to 153m^2/day. A graphical relationship between aquifer parameter interpreted from VES and pumping test is established in order to estimate the parameterization from surface measurements.

Keywords: VES, Schlumberger configuration, Pumping test, Transmissivity, Aquifer test pro

INTRODUCTION

Groundwater is the water that is present beneath the Earth's surface in soil pore spaces and in the fractures of rock formations. It is stored in and moves slowly through the geologic formations of soil, sand and rocks called aquifers. The aquifer parameters like Transmissivity, Storativity are extremely important for the management and development of groundwater resources. The subsurface lithology controls the occurrence and movement of groundwater (Gulraiz Atcher[5], 2016). Aquifer parameters are necessary for groundwater management especially in description of dynamics of groundwater. The integration of aquifer parameter and subsurface resistivity values can be highly effective since the relationship between hydro-geological and electrical properties of aquifer is possible as both properties are related to pore space structure and heterogeneity (Egbai[4], 2015). However, the conventional methods for determination of hydrogeological parameters are invasive and relatively expensive either to integrate over largest volume of data or to provide information only to the small section of aquifer (Ahmefula[1] 2012). The most important method to estimate aquifer parameters is testing with having a single observation well called Single well test (Dana Khider Mawlood[3] 2016). The analysis of hydro-geological characteristics from the available data ensures the proper documentation of quantifiable results for effective and management of groundwater resources.(Akaha[2], 2008).

STUDY AREA

The study area is the premises of Andhra University campus in Visakhapatnam, Andhra Pradesh. The university has an area extent of 460 acres. It is located in 17^{0}43'45.38" North

Latitude and 83^{0}19’17.61” East Longitude on the uplands of Visakhapatnam. The study area has lot of varied topological features and has good recharge potentials.

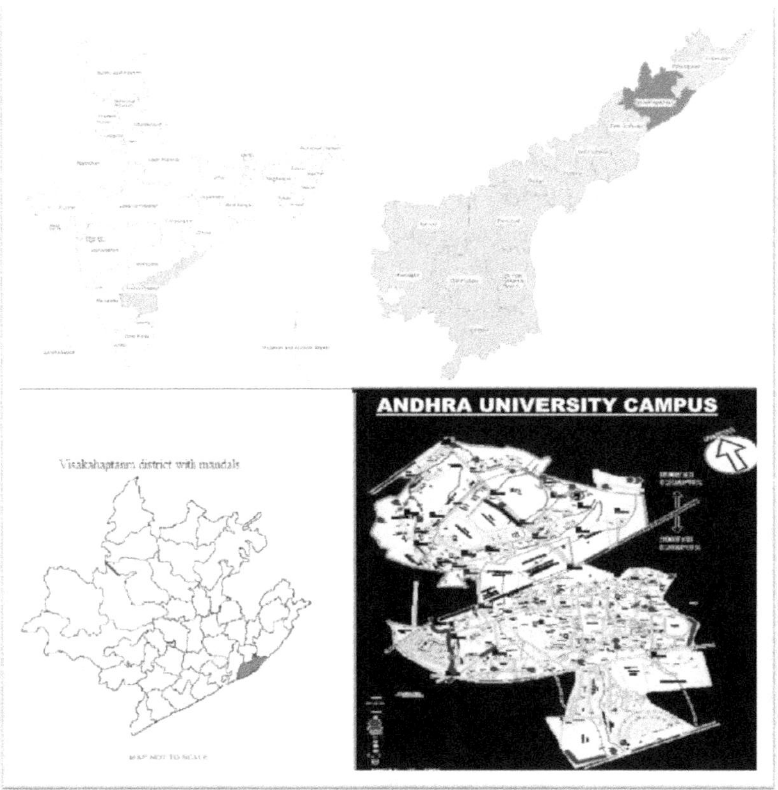

Fig. 1 Location of Study area

METHODOLOGY

Resistivity measurements using Vertical Electrical Sounding (VES), Schlumberger method, are carried out to evaluate the subsurface lithology and to locate the aquifer conditions. The Schlumberger array is used with maximum current electrode separation of 100m. The geophysical equipment employed for the field survey is SSR-MP1 resistivity meter. Resistivity field data is interpreted using software, IPI2Win that gives the output in the form of the number of subsurface layers, their true resistivity values, thickness and depth from the surface. The VES profiles developed using the above cited software are analysed for different zones of resistivity.

The data from the five Piezometric wells established in the study area as a part of DST R&D programme is used to assess the groundwater levels. Pumping tests with duration varying from 6-12 hours are performed to determine the hydro-geological characteristics of the aquifers in the study area. The test is carried out using multiple observation well approach at five locations. The drawdowns are measured based on recorded static water levels in five observation wells at specified time intervals during tests. The rates of discharge from all the 25 bore wells are determined manually. Aquifer Test Pro software is used to compute the values of

Transmissivity from pumping test data in the study area. The drawdown contour maps are developed using the above cited software, for observing the groundwater levels w.r.t static water levels in the observation wells.

RESULTS AND DISCUSSIONS

The results from the geophysical survey are presented in the form of sounding curves vide figure 2, depth tables vide table 1 and pseudo-sections vide figure 3. Field data obtained by resistivity soundings is quantitatively interpreted using IPI2Win software. The field curves generally show three to five layers in the study area. The aquifer is mostly found in the third layer except at VES-2 where it occurred in the second layer. The layer with low resistivity values are recognized as aquifer at the respective VES location. The results showing VES sounding curve and resistivity depth table of a location and pseudo-sections of VES locations are displayed below.

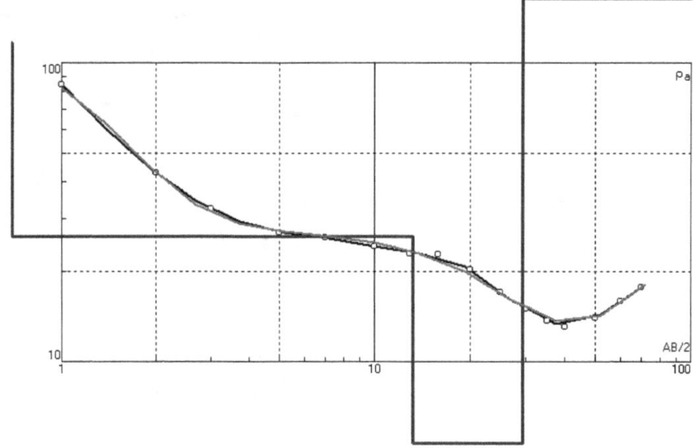

Fig. 2 VES Curve at Dispensary, AU

Table 1 Resistivity, thickness and depth

N	ρ	h	d	Alt
1	117	0.599	0.599	-0.5987
2	26.3	12.7	13.3	-13.26
3	4.63	16.2	29.4	-29.43
4	2399			

The values of Transmissivity estimated from Aquifer test pro are given in the table 2. Minimum and maximum transmissivity values for the study area are 84.7m^2/day to 153m^2/day respectively with an average value of 129.94m^2/day. Transmissivity values are observed to be high in the observation well 5 (153 m^2/day) located in eastern part of the study area due to the presence of sand or gravel indicating a large amount of groundwater in this zone. Transmissivity values are found to be low in the observation well 1 (84.7 m^2/day) located in the southern part of the study area which indicates the presence of groundwater is low.

Table 2 Transmissivity and Resistivity values in the study area

Observation Wells	Location in study area	Transmissivity (m^2/day)	Resistivity (Ohm-m)
OBW-1	Near Assembly Hall, AU	84.7	44.9
OBW – 2	Near Dispensary, AU	128	4.63
OBW – 3	Near CSE Department	141	3.41
OBW – 4	Near NCC Camp Office	143	18.6
OBW - 5	Near Samatha Hostel	153	13.5

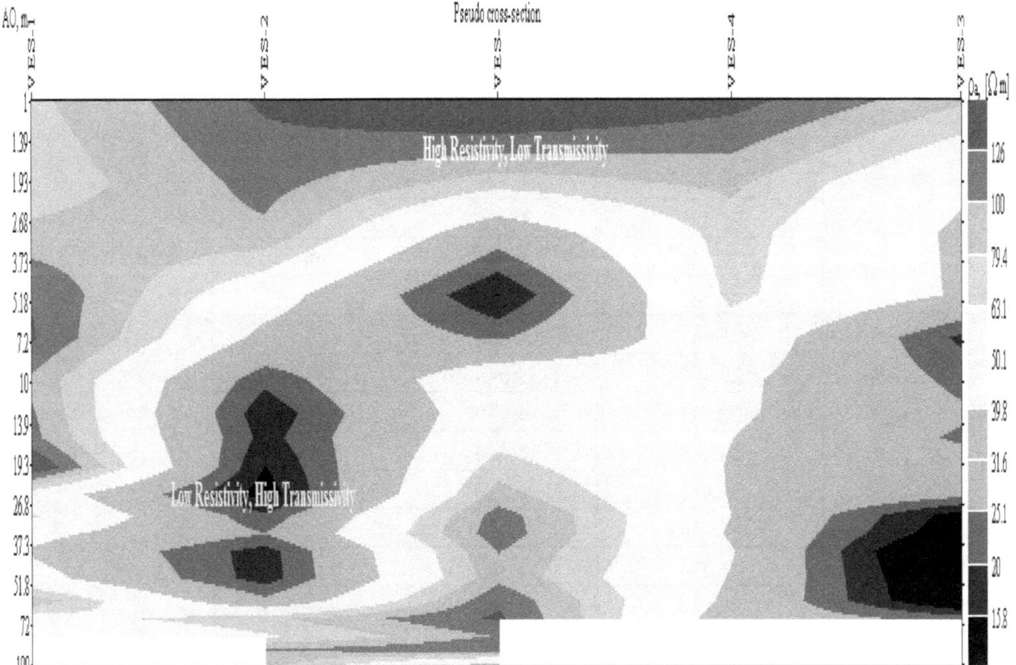

Fig. 3 Pseudo-depth section of VES profile 1 to 5

Table 3 Water levels and drawdowns during pumping test

Static water level-25.75(mbgl)		
Water level (mbgl)	**Drawdown (m)**	**Time (minutes)**
25.747	0.002	1
25.764	0.0190	2
25.785	0.040	3
25.787	0.042	4
25.794	0.049	5
25.825	0.080	6
25.852	0.107	8
25.878	0.133	10
25.905	0.160	12
25.934	0.189	14
25.955	0.210	20
26.026	0.281	25
26.074	0.329	30
26.137	0.392	35
26.201	0.456	40
26.252	0.507	45
26.317	0.572	50
26.409	0.664	60
26.423	0.678	70
26.483	0.738	80
26.546	0.801	90
26.581	0.836	100
26.641	0.896	130
26.790	1.045	160
26.875	1.130	190
26.925	1.180	220
26.971	1.226	250
27.019	1.274	280
27.074	1.329	310
27.139	1.394	370
27.226	1.481	420
27.328	1.583	480
27.436	1.691	540

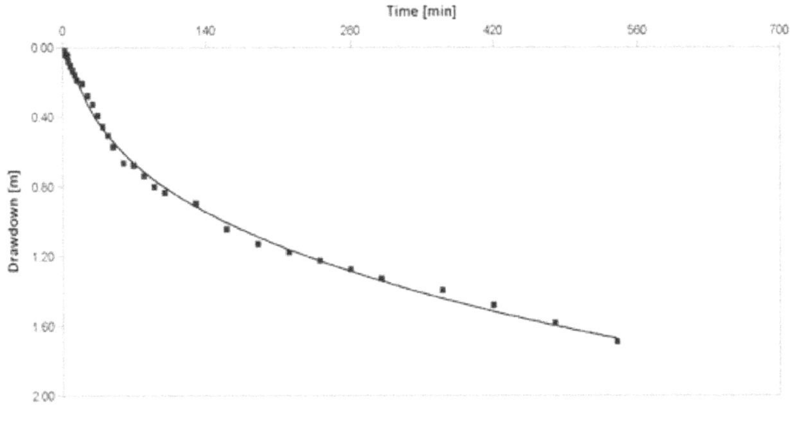

Fig. 4 Drawdown curve at OBW-2 located at Dispensary, Andhra University.

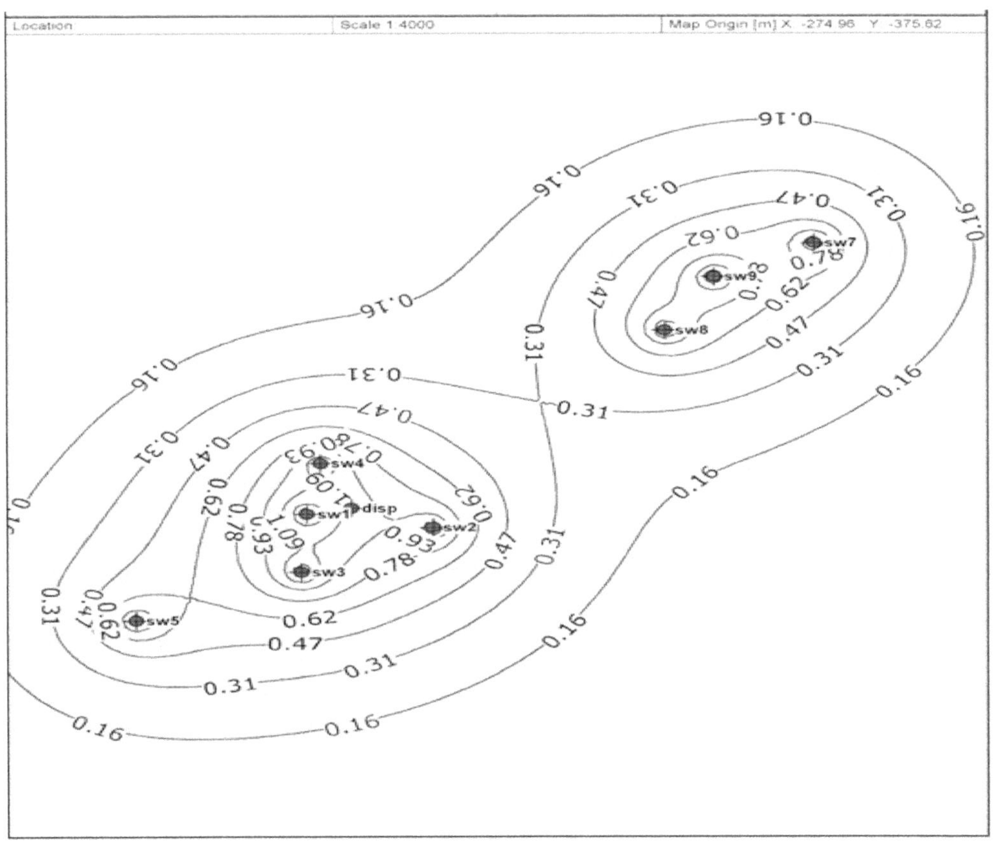

Fig. 5 Drawdown contours at a location in study area

The above figure shows the variations of drawdowns during the pumping test conducted at observation well 2 for 9 hours of period at specified intervals. The groundwater level variations observed during the test w.r.t observation well are found to vary from 0.002 to 1.690 m.

CONCLUSIONS

1. The subsurface lithology at Andhra University Campus consists of Red Sandy Soil, Red Loamy and Gravel at the top layer followed by Weathered Rock, Fractured Roack and Hard Rock formations of Khondalite, Liptinite and Charnockite.
2. The Aquifer zone is found to be occurr in the range of 25 to 40 m in the entire study area.
3. The static water levels observed for all the observation wells are varying from 12.05m to 26.32m. Maximum drawdowns observed from the pumping tests vary from 0.949 m to 1.690 m and the minimum drawdowns vary from 0.002 m to 0.01m.
4. The Minimum and maximum Transmissivity values in the study area are found to be 84.7m^2/day and 153m^2/day respectively with an average value of 129.94m^2/day.

ACKNOWLEDGEMENTS

The authors acknowledge Department of Science and Technology (DST), New Delhi and Andhra University, Visakhapatnam for sponsoring a Research project under which the present work is carried out.

REFERENCES

1. Ahmefula U.Utom, Benard I. Odoh, Anthony U. Okoro, *Estimation of Aquifer Transmissivity Using Dar-Zarrouk Parameters Derived from Surface Resisitivity Measuremnents: A Case History form parts of Enugu Town (Nigeria)*, Journal of Water Resource and Protection, 2012, 4, 993-1000.
2. Akaha. C.TSE, Promise A. Amadi, *Hydraulic Properties from Pumping Test Data of the Aquifers in Azare Area, North Eastern Nigeria*, Journal of Applied Sciences, Environment Management, Vol 12 (4), pp 67-72.
3. Dana Khider Mawlood, Jwan Sabah Mustafa, *Performing Pumping Test Data Analysis Applying Cooper-Javob Method for Estimating of Aquifer Parameters*, Mathematical Modelling in Civil Engineering, Vol 12, No 1, pp 9-20, 2016.
4. Egbai, J.C, Iserhien Emekeme, R.E, *Aquifer Transmissivity Dar Zarrouk Parameters in Groundwater Flow Direction in Abudu, Edo State, Nigeria*, International Journal of Science Environment and Technology, Vol 4, No. 3, 2015, pp 628-640.
5. Gulraiz Atcher, M.Hasan, *Determination of Aquifer parameters using Geoelectrical Sounding and Pumping Test Data in Khanewal District, Pakistan*, Open Geoscience, Vol 8 pp 630-638, 2016.

PRECISION FARMING – A SCIENTIFIC APPROACH FOR SUSTAINABLE WATER AND ENVIRONMENT MANAGEMENT

Abdul Hakkim

V.M., Professor & Head, Department of Land & Water Resources and Conservation Engineering (LWRCE), KCAET (Kerala Agricultural University), Tavanur, Kerala

abduhakkim19@gmail.com

INTRODUCTION

Precision agriculture uses the inputs most efficiently and judiciously to maximize productivity and profitability with minimum impact on soil and environment. Precision in terms of both time and quantity of inputs and agronomic practices, envisages a prospect, which can help in decreasing the cost of production and not having any adverse effect on soil and environmental health. Thus the intent of precision farming is to match agricultural inputs and practices to localized conditions within a field to do the right thing in the right place, at the right time and in the right way. Precision farming basically depends on measurement and understanding of variability.

Spatial Variability of Soil Properties

Management of spatial and temporal variability, through the application of technologies and principles associated with all aspects of agriculture for improving production and environmental quality is known as Precision Agriculture. Accuracy in assessment of the variability, its management and evaluation in the space-time continuum in crop production decides the success of Precision farming. To understand the variability and to give site specific agronomic recommendations management of variability is essential, for which the available technologies such as Remote Sensing (RS), Geographical Information System (GIS), Global Positioning System (GPS), Soil Testing, Yield Monitors and Variable Rate Technology are largely used.

The cause of nutrient depletion is due to the imbalance between input and output of a soil system. Maintenance of proper nutrient status in soil is the key factor for high yield production. Significant variation in crop growth and yield can be achieved if the management practices are recommended based on the information derived from the GIS Database. Recent research in precision agriculture has focused on use of management zones as a method to more efficiently apply crop inputs such as fertilizers across variable agricultural landscapes. Management zones, in the context of precision agriculture, are field areas possessing homogenous attributes in landscape and soil condition, which can potentially serve as a basis for variable fertilizer application. While using management zones to characterize spatial variability in soil and crop properties is important in site-specific studies, it is equally important to consider the temporal effects of climate variability on expression of spatial variation in crop yields. To maximize the yield to a great extent, supplying nutrients as such in the soil will not serve, as the whole quantity of applied nutrients are not supplied to the plants and the requirement varies with different zones in the soil. Therefore it is a must to analyse the soil nutrient status in various zones and the nutrients are to be supplied according to the site specific requirement. Soil fertilizer recommendations in modern crop production rely on laboratory analysis of representative soil samples. The accuracy and precision of the fertilizer recommendation can be improved by considering the factors influencing nutrient variability.The fertilizer and water management varies with different zones and it plays a vital role in determining the yield and quality of farm produce.

The term "fertility" refers to the inherent capacity of a soil to supply nutrients to plants in adequate amounts and in suitable proportions. The term "nutrition" refers to the inter-related steps by which a living organism assimilates food and uses it for growth and replacement of tissue. Previously, plant growth was thought of in terms of soil fertility or how much fertilizer should be added to increase soil levels of mineral elements. Most fertilizers were formulated to account for deficiencies of mineral elements in the soil. The amount of fertilizer, lime, and other amendments recommended for soil improvement should allow optimum growth without undue risk of polluting the natural run-off. It is important not to apply more than is recommended; this will assure greatest plant response with the least chance of plant damage or drainage water pollution. The purpose of a soil test is to supply enough information to make a wise fertilizer and soil amendment choice. When the fertilizer is applied in right dose based on soil test crop response is excellent. Plant nutrition management studies are highly helpful to the farming community, since they provide information on optimum requirement of nutrition to cultivate a crop. It also minimizes the wastage of fertilizer application to a greater extent.

Micro Irrigation and Fertigation

Efficient use of available irrigation water is essential for increasing agricultural productivity for the alarming Indian population. With present potential of 114 million hectare meters (mha.m) of water, only 97 mha is under irrigation in India, for the total cultivated area of 145 mha. Efficient management of water resources is essential to meet the increasing competition for water between agricultural and non-agricultural sectors and the present day share of 80 per cent of water used for agriculture is anticipated to be reduced by 70 per cent in the coming decade. This necessitates scientific management of available water resources, particularly in agricultural sector. Sustainability of any system requires optimal utilization of resources such as water, fertilizer and soil. There is a need to develop agro technologies, which will help in sustaining the precious resources and maximize the crop production, without any detrimental impact on the environment. Fertilizer management is the most important agro-technique, which controls development, yield and quality of a crop. Fertilizer use efficiency is only less than 50 per cent in conventional practice of soil application. Location specific fertilizer management practices are essential for increasing fertilizer use efficiency for optimizing the fertilizer input and maximizing the productivity.

Precision farming or hi-tech agriculture uses the inputs most efficiently and judiciously to maximize productivity and profitability with minimum impact on soil and environment. Precision in terms of both time and quantity of inputs and agronomic practices, envisages a prospect, which can help in decreasing the cost of production without any adverse effect on soil and environmental health. Thus the intent of precision farming is to match agricultural inputs and practices to localized conditions within a field to do the right thing in the right place, at the right time and in the right way. Precision farming basically depends on measurement and understanding of variability. The accuracy in assessment of the variability, its management and evaluation in the space-time continuum for crop production decides the success in Precision farming.

The cause of nutrient depletion is due to the imbalance between input and output of a soil system. Maintenance of proper nutrient status in soil is the key factor for high yield production. Significant variation in crop growth and yield can be achieved if the management practices are recommended based on the information derived from the GIS Database. Recent research in precision agriculture has focused on use of management zones as a method to more efficiently apply crop inputs such as fertilizers across variable agricultural landscapes. Management zones, in the context of precision agriculture are field areas possessing homogenous attributes in landscape and soil condition, which can potentially serve as a basis for variable fertilizer application. While using management zones to characterize spatial variability in soil and crop properties is important in site-specific studies, it is equally important to consider the temporal effects of climate variability on expression of spatial variation in crop yields. To maximize the yield to a great extent, supplying nutrients as such in the soil will not serve, as the whole quantity of applied nutrients are not supplied to the plants and the requirement varies with different zones in the soil. Therefore it is a must to analyse the soil nutrient status in various zones and the nutrients are to be supplied according to the site specific requirement. Soil fertilizer recommendations in modern crop production rely on laboratory analysis of representative soil samples. The accuracy and precision of the fertilizer recommendation can be improved by considering the factors influencing nutrient variability. The fertilizer and water management varies with different zones and it plays a vital role in determining the yield and quality of farm produce.

Although water is a renewable resource, its availability in appropriate quality and quantity is under severe stress due to increasing demand from various sectors. Agriculture is the largest user of water, which consumes more than 80% of the country's exploitable water resources. The overall development of the agriculture sector and the intended growth rate in GDP is largely dependent on the judicious use of the available water resources. While the irrigation projects (major and medium) have contributed to the development of water resources, the conventional methods of water conveyance and irrigation, being highly inefficient, has led not only to wastage of water but also to several ecological problems like water logging, salinization and soil degradation making productive agricultural lands unproductive. It has been recognized that use of modern irrigation methods like drip and sprinkler irrigation is the only alternative for efficient use of surface as well as ground water resources. The scheme on Micro Irrigation (MI) aims at increasing the area under efficient methods of irrigation. Apart from the economic considerations, the adverse effect of injudicious use of water and fertilizers on the environment can have far reached implications. There is a need to develop agro technologies, which will help in sustaining the precious resources and maximize the crop production, without any detrimental impact on the environment. Water availability for irrigation is going to be a major constraint for agriculture in the near future. Efficient management of available water resources is hence necessary for expanding the area under irrigation. Bringing more area under irrigation would depend largely upon efficient use of water. In this context, micro irrigation has the most significant role to achieve not only higher productivity and water use efficiency but also to have sustainability with economic use and productivity.

Need for Efficient Irrigation

The development of irrigation is given top priority in Indian economy as agriculture contributes to about 50% of the Gross National Product. Efficient use of irrigation water is an important means for increasing the productivity. Surface irrigation methods, which is widely practiced in India leads to enormous loss of water due to seepage and evaporation. This is due to poor distribution of water in farm due to inadequate land preparation and lack of farmer's knowledge in the application of water, which leads to excess application and deep percolation. Generally, in case of surface irrigation, only less than one third of the water released reaches the plants. The unscientific use of water results in not only the wastage of water but also has caused soil erosion, salination and water logging which ultimately degrade the quality of the two vital natural resources- soil and water.

Types of Irrigation Systems

There are three general types of irrigation methods; viz. surface irrigation, sprinkler irrigation and micro irrigation. Selection of suitable irrigation system depends upon water availability, topography, soil characteristics, crop requirements, cost and cultural practices. For example, surface irrigation would not be appropriate for fruit crops but would be for rice, whereas drip or micro spray irrigation would not be appropriate for rice but would be for fruit crops. Likewise, if the crop to be irrigated is located adjacent to a stream or similar water source, and water abstractions would not adversely affect downstream users, surface irrigation might be used in preference to other, perhaps more water-efficient and more costly, methods of irrigation.

I. Surface Irrigation

In surface methods of irrigation, water is directly applied to the soil surface from a channel located at the upper reach of the field. Water is either diverted or pumped from a river or stream onto the area to be irrigated. The common surface irrigation methods are flood irrigation, border irrigation, check basin irrigation and furrow irrigation.

Limitations of Surface Irrigation

1. Surface irrigation results in inefficient use of water resources, due to the losses by evaporation, infiltration below the root zone, and runoff.
2. Water stress for plants due to water logging at the time of irrigation and scarcity during the off time of irrigation.
3. Nutrients uptake by plants gets slowed down
4. Loss of nutrients through leaching
5. Surface systems tend to be labour-intensive.
6. Difficulty in applying light, frequent irrigations early and late in the growing season of several crops.
7. Plant is more susceptible to soil borne diseases.
8. Salinization / alkalization of soil
9. Proper soil aeration is not possible.

II. Sprinkler Irrigation

In the sprinkler method of irrigation, water is sprayed into the air and allowed to fall on the ground surface somewhat resembling rainfall. The spray is developed by the flow of water under pressure through small orifices or nozzle and the pressure is usually developed by pumping. Water is conveyed from the source under pressure through a network of pipes called main lines and submains, to one or more pipes with sprinklers called laterals. The sprinklers distribute the water over the land surface. With careful selection of nozzle sizes, operating pressure and sprinkler spacing the amount of irrigation water required to refill the crop root zone can be applied uniformly to suit the infiltration rate of soil. Sprinkler irrigation systems can be fixed in place, portable, semi-portable, or mobile. Sprinkler nozzle types and numbers are selected depending on design application rates and wetting patterns.

Sprinkler irrigation, though provides better irrigation efficiency when compared with surface irrigation, has got some limitations. The major limitations of sprinkler irrigation are:

1. It enhances soil erosion, especially splash and sheet erosion.
2. It enhances the weed growth.
3. Fertigation is not possible with sprinkler irrigation.
4. The high pressure water falling over the canopy of the crop may lead to flower and fruit shedding.
5. Chances of pest and disease infestation are more.

Sprinkler irrigation is generally recommended for fodder cultivation, lawns and coffee or tea plantations.

III. Micro Irrigation

To bring more area under irrigation, it has become necessary to introduce new irrigation techniques for economizing the use of water and to increase productivity per unit of water. Micro irrigation is a method of frequent application of water to the soil near the plants through a low pressure distribution system and special flow control outlets. It can be considered as an efficient irrigation method, which is economically viable, technically feasible and socially acceptable. It is the slow and regular application of water directly to the root zone of the plants through a network of economically designed plastic pipes and low discharge emitter. It enables watering the plants at the rate of its consumptive use thereby minimizing the losses such as deep percolation, runoff and soil evaporation. Micro irrigation includes drip, subsurface drip, bubbler or trickle irrigation, all of which have similar design and management criteria. The major characteristics of micro irrigation are

1. Water is applied at a low rate.
2. Water is applied over a long period of time.
3. Water is applied at frequent intervals.
4. Water is applied directly into the crop root zone.
5. Water is applied under low pressure (upto 2 kg/cm^2) through a network of pipes.

Micro irrigation systems deliver water to individual plants or rows of plants. The outlets are generally placed at short intervals along small tubing, and unlike surface or sprinkler irrigation, only the soil near the plant is watered. The outlets include emitters, orifices, bubblers and sprays or micro sprinklers.

Advantages of Micro Irrigation

(i) Water saving
(ii) Enhanced plant growth and yield
(iii) Uniform and better quality of produce

(iv) Efficient and economic use of fertilizers
(v) Less weed growth
(vi) Possibility of using saline water
(vii) No soil erosion
(viii) Flexibility in operation
(ix) Easy installation
(x) Labour saving
(xi) Suitable to all types of land terrain
(xii) Saves cultivable land as no bunds etc. are required
(xiii) Minimum diseases and pest infestation

Types of Micro Irrigation Systems

The common types micro irrigation systems in use are as follows:

(i) Drip Irrigation

It is the system in which emitters and laterals are laid on the ground surface along the rows of crops. The emitting devices are located in the root zone area of trees. The cost of drip irrigation systems is reasonable on wide-spaced crops such as trees. The closer the crop spacing, the higher the system cost per acre.

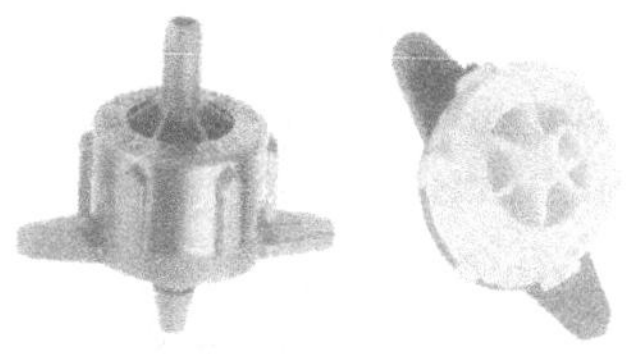

(ii) Bubbler Irrigation

In bubbler irrigation, water is applied to the soil surface as a small stream or fountain. Bubbler systems do not require elaborate filtration systems. These are suitable in situations where large amount of water need to be applied in a short period of time and suitable for irrigating trees with wide root zones and high water requirements. Discharge rates are generally less than 225lph.

(iii) Micro Sprinklers

These are small plastic sprinklers with rotating spinners. The spinners rotate with water pressure and sprinkle the water. These are available in different discharges and diameters of coverage and can operate at low pressure in the range of 1.0 to 2kg/cm^2. Water is given only to the root zone area as in the case of drip irrigation but not to the entire ground surface as done in the case of sprinkler irrigation method.

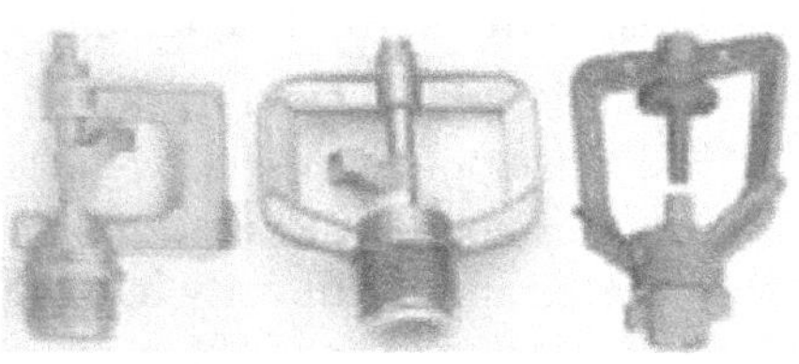

(iv) Spray Jets

The spray pattern of jets is fan type, giving fine droplets and uniform distribution.Jets are mainly used to maintain adequate micro environment in the canopy area. They can be used to irrigate orchards, nurseries, vineyards, green houses and delicate plants such as flowers, vanilla etc. Mature large trunk-trees or trees having wide spread root zone can also be irrigated using jets. The spray pattern is either full circle or half circle.

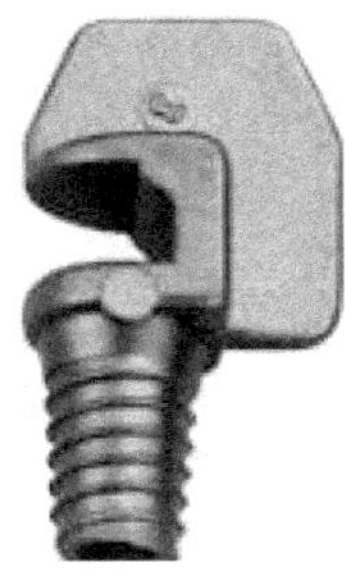

Half circle jet

Full circle jet

(v) Fogger

Foggers are recommended for orchards and green/glass houses requiring a fine mist spray for humidity control. They are suitable for crops which need to maintain micro climate in the canopy area. They are simple in construction and has no moving parts. The spray pattern is misty and the droplets are very fine.

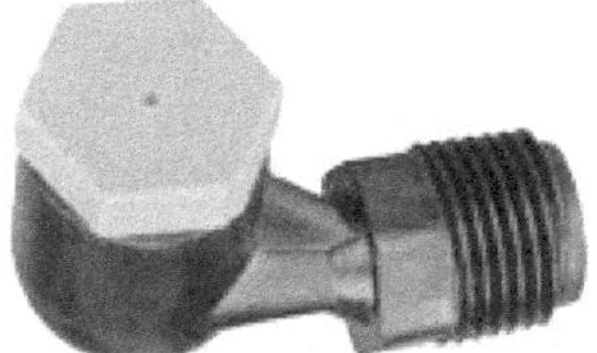

Fogger

Table 1 Comparative Performance of Conventional Irrigation with Micro Irrigation

Performance Indicator	Conventional Irrigation Methods	Micro Irrigation
Water saving	Waste lot of water, Losses occur due to percolation, runoff and evaporation	40 – 70% of water can be saved over conventional irrigation methods. Runoff and deep percolation losses are negligible.
Water use efficiency	30-50%	80-95%
Saving in labour	Labour engaged per irrigation is higher than drip	Labour required only for operation and periodic maintenance of the system.
Weed infestation	High	Less wetting of soil, weed infestation is very less.
Use of saline water	Concentration of salts increases and adversely affects the plant growth. Saline water cannot be used for irrigation	Frequent irrigation keeps the salt concentration within root zone below harmful level
Disease and pest problems	High	Relatively less because of less atmospheric humidity
Suitability in different soil type	Deep percolation is more in light soil and with limited soil depths. Runoff loss is more in heavy soils	Suitable for all soil types as flow rate can be controlled

Contd...

Performance Indicator	Conventional Irrigation Methods	Micro Irrigation
Water control	Inadequate	Very precise and easy
Efficiency of fertilizer use	Efficiency is low because of heavy losses due to leaching and runoff	Very high due to reduced loss of nutrients through leaching and runoff water
Soil erosion	Soil erosion is high because of large stream sizes used for irrigation	Partial wetting of soil surface and slow application rates eliminate any possibility of soil erosion
Increase in crop yield	Non-uniformity in available moisture reducing the crop yield	Frequent watering eliminates moisture stress and yield can be increased up to 15-50% as compared to conventional methods of irrigation

Table 2 Irrigation Efficiencies under Different Methods of Irrigation (Per cent)

Irrigation Efficiencies	Methods of Irrigation		
	Surface	Sprinkler	Drip
Conveyance efficiency	40-50 (canal) 60-70(well)	100	100
Application efficiency	60-70	70-80	95
Surface water moisture evaporation	30-40	30-40	20-25
Overall efficiency	30-35	50-60	90 - 95

Scientific method of cultivation and judicious use of all the inputs, especially water, is called upon to become cost competitive. Keeping in view of acute water scarcity in many basins, efforts were made to introduce most efficient micro irrigation system at farms around 1970. Micro irrigation saved irrigation water by 40%, fertilizer by 25%, enhanced yield up to 50%, improved water – use efficiency by 2.5 times Through the good management of micro irrigation systems, the root zone water content can be maintained near field capacity throughout the season providing a level of water and air balance close to optimum for plant growth. In addition, nutrient levels, which are applied with water through the system (fertigation), can be controlled precisely. Fertigation gives successful results in terms of yield, saving in fertilizer and improvement in quality of the produce. During the dry season in humid areas, micro irrigation can have a significant effect on quantity and quality of yield, pest control and harvest timing.

Fertigation

Efficient crop production requires efficient utilization of soil water and soil fertility. Placement of fertilizers in the zone of moisture availability is important to maximize the fertilizer use efficiency. Fertigation is the method of precise periodic application of water soluble fertilizer along with irrigation water. Fertigation is a pre-requisite for drip irrigation. Since the wetted soil volume is limited, the root system is confined and concentrated. The nutrients from the root zone are depleted quickly and a continuous application of nutrients along with the irrigation water is necessary for adequate plant growth. Fertigation offers precise control on fertilizer application and can be adjusted to the rate of plant nutrient uptake.

Advantages of Fertigation

1. Saving of energy and labour

2. Flexibility of the moment of the application (nutrients can be applied to the soil when crop or soil conditions would otherwise prohibit entry into the field with conventional equipment)
3. Convenient use of compound and ready-mix nutrient solutions containing also small concentrations of micronutrients which are otherwise very difficult to apply accurately to the soil
4. The supply of nutrients can be more carefully regulated and monitored.
5. The nutrients can be distributed more evenly throughout the entire root zone or soil profile.
6. The nutrients can be supplied incrementally throughout the season to meet the actual nutrition requirements of the crop.
7. Soil compaction is avoided, as heavy equipment never enters the field.
8. Crop damage by root pruning, breakage of leaves, or bending over is avoided, as it occurs with conventional chemical field application techniques.
9. Less equipment may be required to apply the chemicals and fertilizers.

The challenge for agriculture over the coming decades will be to use the plant nutrients in a sustainable way. Nutrients mining and imbalanced fertilizer application is adding to the problem of declining soil fertility. This is leading to the growing concerns about the long-term sustainability of agriculture. Poor management of resources has not only damaged the soil health but also the environment. Plants need a fixed quantity and mix of nutrients to flourish and result into economic yield. The higher the yield, the greater is the nutrient requirement. A shortage of one or more nutrients can inhibit or stunt plant growth.

Fertilizer application through the micro irrigation system i.e. fertigation is the most advanced and efficient practice of fertilization. Fertigation combines the two main factors in plant growth and development, water and nutrients. The right combination of water and nutrients is the key for high yield and quality of produce. Fertigation is the most efficient method of fertilizer application, as it ensures application of the fertilizers directly to the plant roots (Rajput & Patel, 2002). In fertigation, fertilizer application is made in small and frequent doses that fit within scheduled irrigation intervals matching the plant water use to avoid leaching.

The efficiency of utilization of chemical fertilizers is very low. It has been reported that nitrogen use efficiency seldom exceeds 40 % under lowland and 60 % under upland conditions. In case of phosphorous and potassium, the efficiency hardly exceeds 20 %. The expensive supply of nutrients in the form of fertilizers was a key factor, along with improved irrigation system and adequate supplies of water with nutrients, in the substantial increase in yields and fertilizer use efficiency.. Drip irrigation enables, the application of water-soluble fertilizers and other chemicals along with irrigation water uniformly and more efficiently in the root zone of crop.

Fertigation Equipments

There are three principal devices of fertilizer injection commonly used for fertigation; namely a) b) fertilizer tank and c) pumps. For successful fertigation the fertilizer injection device must possess high precision, accuracy, and reliability. It is important to select an injection method that best suits the irrigation system and the crop to be grown. Incorrect selection of the equipment can damage parts of the irrigation equipment, affect the efficient operation of the irrigation system and reduce the efficiency of the nutrients.

A. Injector

This is very simple and low cost device. A partial vacuum is created in the system which allows suction of the fertilizers into the irrigation system through venture action The vaccums is created by diverting a percentage of water flow from the main and pass it through a constriction which increases the velocity of flow thus creating a drop in pressure. When the pressure drops the fertilizer solution is sucked into the venturi through a suction pipe from the tank and from there enters into irrigation stream. Although simple and with greater uniformity of dosing than fertilizer tank the venturi cause a high pressure loss in the system which may results in uneven water & fertilizer distribution in the field. The suction rate of venturi is 30 to 120 litre per hour.

B. Fertilizer Tank

In this system part of irrigation water is diverted from the main line to flow through a tank containing the fertilizer in a fluid or soluble solid form, before returning to the main line, the pressure in the tank and the main line is the same but a slight drip in pressure is created between the off take and return pipes for the tank by means of a pressure reducing valve. This causes water from main line to flow through the tank causing dilution and flow of the diluted fertilizer into the irrigation stream. With this system the concentration of the fertilizer entering the irrigation water charges continuously with the time, starting at high concentration. As a result uniformity of fertilizer distribution can be a problem. Fertilizer tanks are available in 90, 120 and 160 litres capacity.

C. Fertilizer Injector Pump

These are piston or diaphragm pumps which are driven by the water pressure of the irrigation system and such as the injection rate is proportional to the flow of water in the system. A high degree of control over the fertilizer injection rate is possible, no serious head losses are incurred and operating costs are low. Another advantage is that if the flow of water stops. Fertilizer injection also automatically stops. This is perfect equipment for accurate fertigation. Suction rate of pumps varies from 40 lit to 160 lit per hour. Various types of fertigation equipments are shown in the figure.

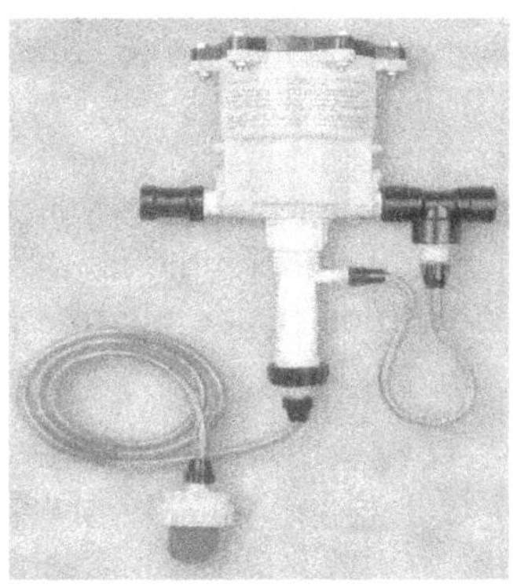

Fertilzer Injector Pump

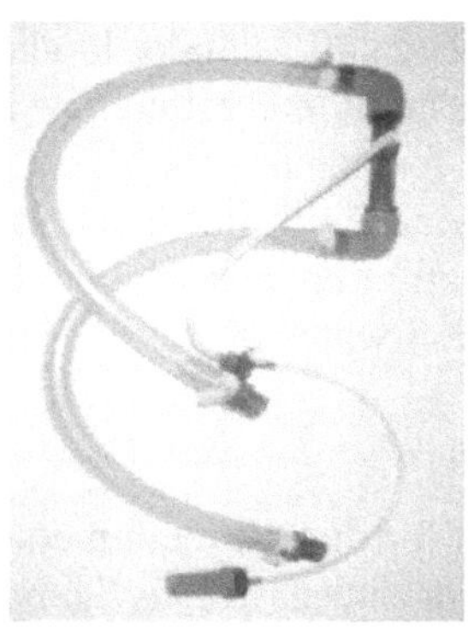

Venturi Injector

Fertilizer Tank

Fertigation Equipments

Present Scenario of Micro Irrigation in India

Micro irrigation is used extensively in many water scare countries like Israel, Arabian States, Mexico, etc. In India, its development is slow in comparison to other developing countries. Experiments and farm trials have been going on India from 1970 onwards. Progressive farmers in Andhra Pradesh, Karnataka, Kerala, Maharashtra and Tamil Nadu adopted this method in the late 70's. The growth of micro irrigation has really gained momentum in recent years. From a mere 1500 ha in 1985, the area under micro irrigation has grown to 0.4 million ha at present (Table 1.3). These developments have taken place mainly in areas acute water scarcity states of Maharashtra, Andra Pradesh, Karnataka, Tamil Nadu and Gujarath. The important high value crops under micro irrigation systems include coconut, grape, banana, mango, sapota, pomegranate, plantation crops, sugar cane, cotton, vegetables and flowers etc.

In 1981, the Government of India established National Committee on use of Plastics in Agriculture (NCPA) under the Ministry of Chemicals and Petrochemicals. Later NCPA was re-named as National Committee on Plasticulture Applications in Horticulture (NCPAH) in 2001 with the Minister of Agriculture as its chairman. NCPAH has dedicated more than two decades in the development and promotion of plasiculture applications in the country. Considering the high cost of plasticulture applications, based NCPAH recommended the Government of India to provide subsidy to eligible beneficiaries for adoption of these applications. Since then, Government of India is providing subsidy to the farmers for the plasticulture applications in water management, greenhouse technology and plasic mulching and other plasticulture applications. The subsidy is channelised through state directorate of horticulture/agriculture. NCPAH is assisting the Government of India in formulating the plan for horticultural development and implementing the subsidy schemes.

CONCLUSION

Precision farming is one of the most scientific and modern approaches for sustainable agriculture that has gained momentum towards the end of 20th century. To make reliable soil interpretations and accurate predictions of soil performance at any particular location, complete knowledge of the variability of soil properties is a must, which forms the basic step for precision agriculture development. Precision farming is the application of technologies and principles to manage spatial and temporal variability associated with all aspects of agricultural production. It is a system for better management of farm resources. Precision farming offers a variety of potential benefits in profitability, productivity, sustainability, crop quality, food safety, environmental protection, on farm quality of life and rural economic development. This scenario forces us to think about efficient irrigation system like micro irrigation combined with fertigaion to have more crop per drop. Precision farming is essential for serving dual purpose of enhancing productivity and reducing ecological degradation.

REFERENCES

1. Babcock, Bruce A. and Gregory R. Pautsch. 1998. Moving from Uniform to Variable Fertilizer Rates on Iowa Corn: Effects on Rates and Returns. Journal of Agricultural and Resource Economics, Vol 23(2):385-400.
2. Bar-Yosef, B. (1991). Fertilization under Micro irrigation. In: Fluid fertilizer science and Technology. (Eds.) Palgrave, Derek A., Marcel Dekker, Inc. New York.

3. Krill, T. 1997. Site Specific Management: Implications to Production Agriculture. Ohio State University Extension Regional Agronomy Meetings. 11 pgs.
4. Swinton, S.M., and J. Lowenberg-DeBoer. 1998. Evaluating the Profitability of Site-Specific Farming. Journal of Production Agriculture, Vol. 11(4):439-446.

COMPUTATION OF SEDIMENT YIELD OVER UN-GAUGED STATIONS USING MUSLE AND FUZZY MODEL

P. Sundara Kumar[1] and M. Anjaneya Prasad[2]

[1]Professor, Department of Civil Engineering, Koneru Lakshmaiah Education Foundation, Vaddeswaram, Guntur, India

[2]Professor, Department of Civil Engineering, University College of Engineering, Osmania University, Hyderabad, Talangana, India

ABSTRACT

Land and water are two most vital natural resources of the world and hence these resources must be conserved carefully to protect environment to maintain ecological balance. Estimation of runoff and sediment yield is one of the prerequisites for conservation and management of water resources and also for many hydrological applications. The present study has been taken up to predict runoff and sediment yield for a reservoir basin viz. Megadrigedda Reservoir situated in Visakhapatnam District, Andhra Pradesh, India which is having a drainage area of 363 Sq.km. Modified Universal Soil Loss Equation (MUSLE) has been used to estimate the sediment yield. The runoff factors of MUSLE were computed by the measured values of runoff and peak rate of runoff at outlet of the reservoir. Topographic factor (LS) and crop management factor(C) were determined using geographic information system (GIS) and field-based survey of land use/land cover. The conservation practice factor (P) is obtained from the literature. Sediment yield at the outlet of the study reservoir has been simulated for fifteen storm events spread over the period of 2014-2017 and is validated with that of measured values. The resulted coefficient of determination value has been obtained as 0.84 for the study area which indicates that MUSLE model is working satisfactory for the selected basin. Fuzzy logic based model has also been developed to estimate sediment yield. The resulted correlation obtained between Fuzzy logic model and MUSLE model is 0.97 and with that of measured value it is 0.937.

Keywords: *Sediment Yield, MUSLE, Topographic factor, GIS, Fuzzy Model.*

1. INTRODUCTION

Soil erosion is an important item for consideration in the planning of watershed development works. It reduces not only the storage capacity of the downstream reservoirs but also deteriorates the productivity of the watershed. Erosion involves the detachment, transport and deposition of soil particles and aggregates. Sediment yield is defined as the total amount of eroded material to be delivered from its source to a downstream control point (Gottschalk, 1964). Thus, sediment yield rates are directly dependent upon both soil loss rates and the transport phenomenon of surface runoff and channel flow. Accurate estimation of sediment-transport rates, in general, depends on an accurate prior estimation of overland flows. Thus, any errors in the estimation of overland flows would be magnified through grossly inaccurate erosion estimations. Sediment yield is a complex phenomenon and the variables involved in erosion modelling makes it difficult to measure and also to predict the sediment yield in a precise manner. Among available soil erosion and sediment yield models, the universal soil loss equation (USLE), the revised version, Revised universal soil loss equation (RUSLE), and its modified version (MUSLE) are widely used in hydrology and environmental engineering for computing the potential soil erosion and sediment yield. The USLE (Wischmeier and Smith 1978) was developed for estimation of the annual soil loss from small plots with an average length of 22 m and its application for individual storm events and large areas leads to large errors. It is reported that its accuracy increases if it is coupled with a hydrological rainfall excess model. In the USLE model, there is no direct consideration of runoff, although erosion depends on sediment that is being discharged with flow and varies with runoff and sediment concentration (Kinnell 2005). It has been observed that delivery ratios to determine sediment yield from soil loss equation can be predicted accurately with considerable variation. The reason for variation may be due to the change in rainfall distribution over time from year to year. Williams and Berndt (1977) proposed MUSLE with the replacement of the rainfall factor with a runoff factor to consider variability of delivery ratio.

The proposed model was intended to estimate the sediment yield on a single storm basis for the outlet of the watershed based on runoff characteristics. It is reported that it is the best indicator for sediment yield prediction (ASCE 1970; Williams 1975a, b; Hrissanthou 2005). MUSLE increases sediment yield prediction accuracy and as well as it eliminates the need for delivery ratios. The MUSLE equation has been used previously by many researchers (Tripathi et al. 2001) and, in some cases, the equation was subjected to different modifications. The sediment yield model like MUSLE is easier to apply because the output data for this model can be determined at the watershed outlet (Pandey et al. 2009). Hikaru et al. (2000) demonstrated by successful application of USLE to mountainous forests in Japan. Tripathi et al. (2001) estimated sediment yield from a small watershed of India using MUSLE and GIS, and the estimated values were very close to the observed values. Based on the reported advantages and applicability of method, the present investigation has been taken up to assess the applicability of the MUSLE for the Meghadrigedda watershed of Visakhapatnam, Andhra Pradesh, India; where there is difficulty in identifying suitable models for estimation of soil erosion and sediment yield at the watershed. The basin also has problems of irregular and discontinuous runoff and sediment data availability.

Fuzzy set theory was originally introduced by Zadeh (1965) and later it became popular in control engineering problems. Many researchers concluded that fuzzy logic is very effective in handling uncertain data and when data also associated with certain vagueness in nature. Despic and Simminivic (2000) presented a general methodology for numerical evaluation of complex qualitative criteria based on the theory of fuzzy set. Mimicking the reservoir operator has been used for modelling the reservoir operation by some researchers. Ross thimothy. (1997) used Fuzzy Model for real time operation. Anjaneya Prasad and Mohan (2006) adopted fuzzy Logic based model for Multi-Purpose Reservoir operation.

2. OBJECTIVES

To develop a model, to predict the sediment yield with greater reliability in watersheds with particularly when deficiency of record in sediment data. To validate sediment yield model by comparing predicted values and observed values. To develop a fuzzy logic based model and to compare the results obtained with MUSLE and also recorded sediment data.

3. STUDY AREA

The geographic location of the Meghadrigedda reservoir catchment is located in the north eastern part of Visakhapatnam district of Andhra Pradesh State and lies between latitudes of 17.43'N-17.57'N and longitudes 83.02'E-83.17'E. The geographical area of the Meghadrigedda reservoir catchment has 363 Sq. Km and the Reservoir has spread to about 6.9 sq.km. Major streams/rivers feeding the reservoir are Meghadrigedda, NarvaGedda and Borramma gedda. Meghdrigedda is flowing from north-west to south-east direction about 17 km, Borrammagedda is flowing from west to east for about 7 km and Narvagedda is flowing from south west to north east for about 6.5 km. Physiographic characteristics of the catchment has varied areas. The catchment area consists of Hilly area, undulating terrain and plains. The hill portion is located in the north west and north east and south-west, undulating areas is located at the foot hill portions and plain areas are located in north and central portions. Major portion of the catchment area is covered by agriculture land which consists of nearly about 51% of the total area, Hill area is covering about 17% and water bodies are consists of 11% in the total area are presented in the Table 1.

Table 1 Land use/Land cover statistics

S. No	LU/LC Type	Area in Sq.KM	Percentage
1	Agriculture Land	186.13	51.28
2	Built-up Land	13.32	3.67
3	Hill	61.65	16.98
4	Plantation	30.28	8.34
5	Waste Land	31.32	8.63
6	Water Bodies	40.30	11.10
	Total	**363.00**	**100.00**

4. MATERIALS AND METHODS

In the present study, MUSLE equation is used to estimate sediment yield for the Meghadrigedda watershed. Runoff factor is a major input into the MUSLE model. It is computed using the runoff and peak runoff rates measured at the outlet of the study area using SCS-CN method. The sediment yields estimated by MUSLE for different events during the years 2014-2017 are compared with the observed sediment yield data collected from the stream ungauged station located at the outlet of the watershed. The model performance is evaluated on the basis of test criteria recommended by the ASCE Task Committee (1993) and graphical performances criteria as suggested by Haan et al. (1982).

4.1 SCS-CN Method of Estimating Runoff Volume

SCS-CN method developed by Soil Conservation Services (SCS) of USA in 1969 is a simple, predictable and stable conceptual method for estimation of direct runoff depth based on storm rainfall depth. It relies on only one parameter, CN. It is a well-established method and widely accepted in USA and many other countries.

Direct Runoff $$Q = \frac{(P - 0.2S)^2}{(P + 0.8S)} \quad .....(1)$$

where, P is the total rainfall, Q is the direct runoff, S the potential maximum retention potential maximum retention S of watershed is related to a CN, which is a function of land use, land treatments, soil type and antecedent moisture condition of watershed. The CN is dimensionless and its value varies from 0 to 100. The Q-value in mm can be obtained from CN by using the relationship of equation (1).

4.2 Estimation of sediment yield using MUSLE

The original USLE is based on soil loss by rainfall energy with slope angle and slope length, which are used as a proxy for the flow detachment process. The MUSLE model (Williams, 1975) improved USLE model (Wischmeier and Smith, 1965) by adding a runoff factor to the driving force. There is no provision for deposition in this model. Therefore, MUSLE is a sediment yield prediction model, for estimating sediment yield from a specified land in a specified cropping and management system. The MUSLE equation (2) is applicable to the point where overland flow enters the streams and all those point are summed up to give the total amount of sediment delivered to the stream network within watershed. It computes the sediment yield for a given site, as a product of seven major variables (William 2005). MUSLE can be adopted for computation of sediment yield from a single storm event.

$$S_y = \frac{11.8}{A}\left(V_Q Q_p\right)^{0.56}.K.LS.C.P \qquad .....(2)$$

Where, Sy is sediment yield in tones :A is Area (ha) : V_Q is Runoff volume (m^3) : Q_P is Peak flow rate (m^3/sec) and K is the soil erodibility factor, which is the erosion rate per unit of erosion index for specified soil in cultivated continuous fallow, having 9% slope and 22.13 m. C is the cropping management factor defined as the ratio of soil loss from a field with a specified cropping and management to that from the fallow condition for which the factor K is evaluated; and finally, P is the erosion control practice factor, which is the ratio of soil loss with contouring, strip cropping, or terracing to that with straight row farming, up and down slope. The data about K, LS, C, and P are adopted from the literature (Arekhi and Kaur 2007). The soil erodibility, crop management, and soil erosion control practice factors, which are more sensitive to temporal variations than other watershed parameters, were assumed to be constant as study period is short as advised by Kinnell (2005). The average weighted values of 0.09 tonnes per hectare, Mj^{-1} mm^{-1}, 3.36, 0.02, and al1 are thus allotted to the watershed factors of K, LS, C, and P, respectively. Subsequently, all the parameters have been substituted in the MUSLE equation in order to derive event-wise sediment yields. Further, the MUSLE model has been validated by comparing the estimated sediment yields with the observed sediment yield for forty storm events occurring from 2014 to 2017. The results of application of the MUSLE model for the storms are shown in Table 2. The Modified Universal Soil Loss Equation (MUSLE) has been widely used to estimate the sediment yield from a single storm event (Williams and Berndt, 1977). MUSLE method has improved accuracy of soil erosion prediction over USLE and RUSLE (Williams 1975a, b; Williams and Berndt 1977).

4.3 Sediment Yield Model using Fuzzy Logic Model

Many researchers have adopted fuzzy logic based model for reservoir management problems and also to address uncertainty and vagueness in variables. In the present study a fuzzy logic based model has been developed to estimate the sediment yield. In the developed model peak rate of flow, rainfall and runoff volume were considered as input to model and sediment yield as output from the model. All these variables which are uncertain and also sometimes vague in nature particularly where measurements are involved were considered as fuzzy variables. The data from 2014 to 2017 events has been considered to develop during calibration of the model. Membership functions have been derived based on the input and output variables from the part of data base for calibration of the model. Various membership forms have been tried for all the input and output variables before arriving to final membership forms for input and output variables. It was found that combination of Triangular, Trapezoidal and gumbell form of memberships were found to perform well for peak flood variable and Triangular and Trapezoidal forms were found to perform well for all other variables. A typical membership form for Peak flow and sediment yield were shown in Fig 1 and 2.

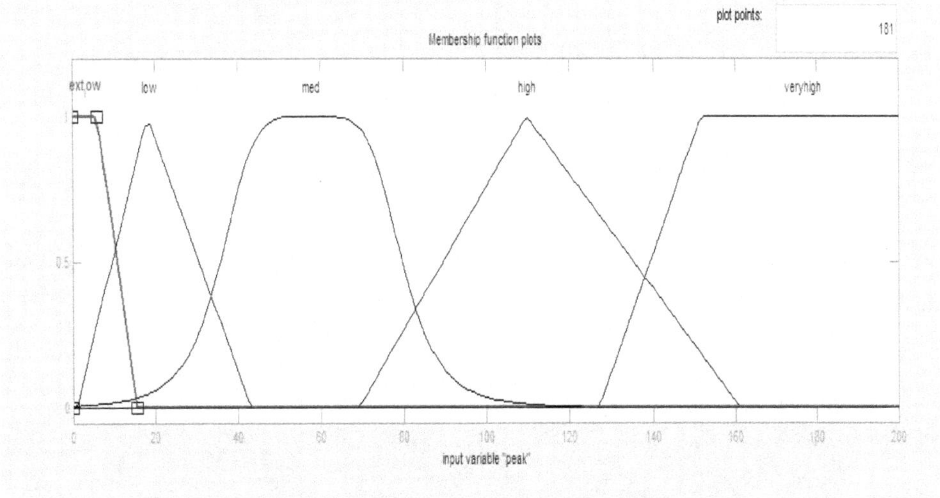

Fig. 1 Membership form of Input Variable "Peak Flow"

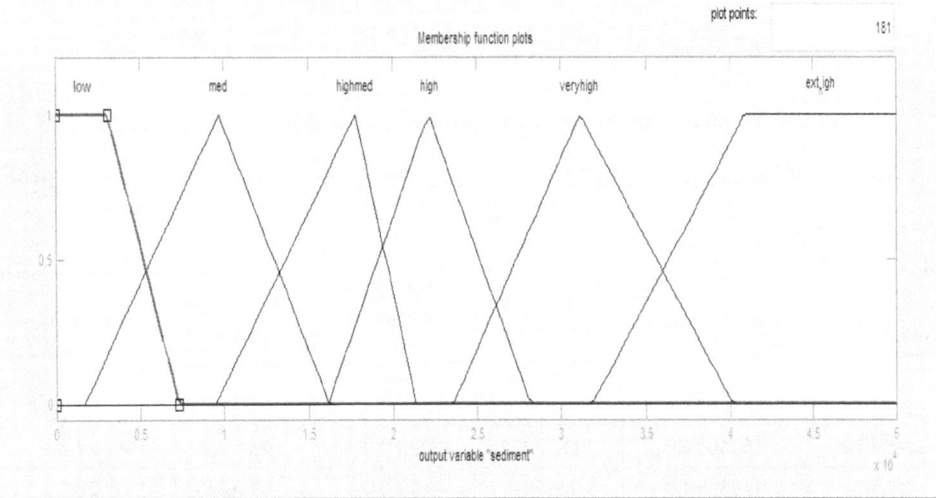

Fig. 2 Membership form of Output Variable "Sediment Yield"

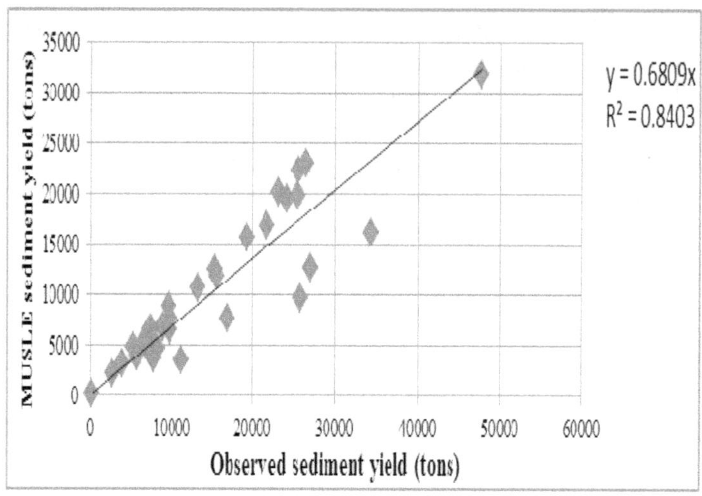

Fig. 3 Observed and computation of Sediment yield using MUSLE method

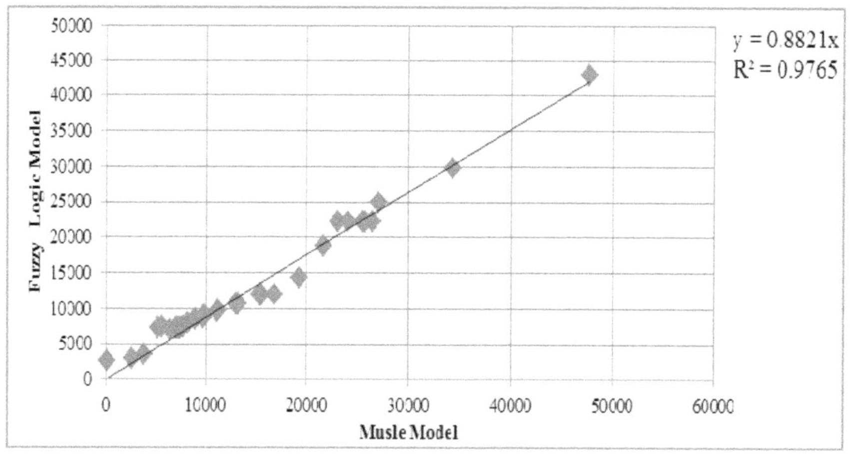

Fig. 4 Validation of computed and Observed Yield by MUSLE model

Table 2. Results of application of MUSLE model at Meghadrigedda watershed (Continued in next page)

Strom date	Runoff Volume (m^3)	Peak discharge (m^3/sec)	Predicted Sediment Yield (t/ha)	Observed sediment yield (t/ha)	Estimation error %
9/3/2014	4849914.6	97.5425	25524.834	19909.37	22.00
9/4/2014	1534044.9	30.853033	7032.0084	5484.9665	22.00
9/5/2014	4193348.9	84.337513	21687.479	16916.233	22.00
6/13/2014	1891392.4	38.04008	8890.7044	6756.9353	24.00

Contd...

Strom date	Runoff Volume (m^3)	Peak discharge (m^3/sec)	Predicted Sediment Yield (t/ha)	Observed sediment yield (t/ha)	Estimation error %
9/19/2014	3116208.1	62.673831	15552.577	11819.959	24.00
9/20/2014	4899416	98.538083	25809.3	9643.095	62.64
9/24/2014	1772801.9	35.654965	8268.7562	4630.5035	44.00
9/29/2014	2309451	46.448163	9809.0964	6572.0946	33.00
9/30/2014	2658379.6	53.465888	16840.26	7731.9	54.09
7/3/2015	2064905.7	41.529817	2655.3429	2150.8278	41.22
7/4/2015	3345580.5	67.287017	6455.0471	5228.5882	44.33
8/5/2015	642978.29	12.931714	145.57307	117.91419	0.01
9/6/2015	1421155.7	28.582582	3887.2226	3148.6503	19.00
10/7/2015	48113.63	0.9676714	34339.395	16139.516	53.00
10/8/2015	903609.85	18.173591	27067.432	12721.693	39.65
9/11/2015	6320633.7	127.12191	7725.1948	3630.8416	-50.84
9/12/2015	5110787	102.78922	24164.684	19602	18.88
8/30/2016	1668374.2	33.554693	9704.1773	8801.6889	9.30
8/31/2016	4618493.5	92.88811	5262.9934	4773.535	9.30
9/27/2016	2045174.4	41.132976	7389.9637	6702.697	9.30
9/28/2016	1184336.4	23.819622	15344.014	12470.08	18.73
9/29/2016	1603580.2	32.251541	13189.309	10718.952	18.73
9/17/2016	3078869.6	61.922871	19346.785	15723.132	18.73
9/20/2016	2689771.9	54.097257	11218.365	3557.4	-46.58
9/21/2016	3786823.3	76.161386	23140.532	20132.263	-14.94
9/22/2016	2327846.4	46.818135	26531.223	23082.164	-14.94
9/16/2016	4443320.8	89.365	232.0428	354.78	34.60
9/17/2016	5020292.7	100.96918	266.04312	304.33	12.58
9/19/2016	1594592.9	32.070788	73.638227	87.99	16.31
2/8/2017	1261261	25.366746	56.628762	32.09	-76.47
5/12/2017	8472291.9	170.39651	478.07554	566.08	15.55
6/24/2017	4870843.3	97.963423	257.18887	156.45	-64.39
7/24/2017	1545904.5	31.091556	71.124642	82.78	14.08

Table 2 Results of application of MUSLE model at Meghadrigedda watershed (Continued from previous page)

Strom date	Runoff Volume (m^3)	Peak discharge (m^3/sec)	Predicted Sediment Yield (t/ha)	Observed sediment yield (t/ha)	Estimation error %
7/14/2017	1448370	29.129921	66.118131	44.45	-48.75
7/28/2017	1049070	21.099116	46.071941	51.03	9.72
8/25/2017	1495560	30.079017	68.535535	45.34	-51.16
9/10/2017	1553640	31.247134	71.523366	87.4	18.17
10/23/2017	1738770	34.970507	81.134712	76.9	-5.51
10/25/2017	1364880	27.450753	61.864472	47.34	-30.68
10/27/2017	1339470	26.939702	60.575981	45.78	-32.32
				Total	7.67

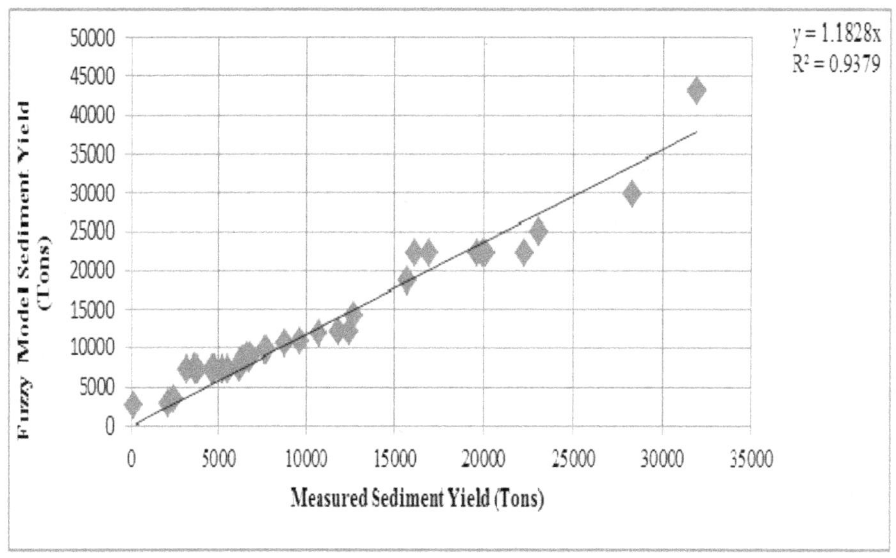

Fig. 5 Validation of Predicted and Observed Yield by Fuzzy model

5. RESULTS AND DISCUSSION

The discharge and sediment yield data from Meghadrigedda watershed were collected for storm events occurring from January 2014 to 2017. All the required information for the application of the MUSLE model such as L=3600m, T_p=13.82 hrs, S=0.0004 and area of the Meghadrigedda basin were extracted with the help of GIS database. Comparison of predicted and estimated values has been carried out and is reported in table 2.0.The good coefficient of determination value (0.75) indicates that good relation exists between observed and estimated values as shown in Fig.3. The percentage deviation of the storm in estimated yield from the observed values and calculated values varied in the range of 0.01% to 53.0% it has presented in table 2.0. The average value of the estimated error for the studied storm was estimated 7.67% for the MUSLE model. The average value of estimated error is within 20%. Hence the value can be considered as the acceptable with level of accuracy for the simulation as per the recommendation of Bingner et al (1989). In other words the model is acceptable for the model process considering the natural phenomena Das (2000). The fuzzy simulation model developed has been calibrated with 30 events of the data base and the balance 10 events were used for validating the model. However the results of both calibration and validation has been given with that of MUSLE model was found to give good correlation between the fuzzy logic based model and MUSLE model. The correlation obtained between the both model sediment yield was found to be 0.97. In order to verify the performance of Fuzzy Logic based model the sediment yield results were also compared with actual measurement of sediment and it has resulted in correlation of 0.937. Fig 4 and Fig 5 shows the correlation obtained from Fuzzy Logic based model and measured sediment separately. Hence it can be concluded that the fuzzy logic based model for sediment yield is performing well and can be adopted as a means to estimate the sediment yield in basin of similar hydrological in nature.

6. CONCLUSIONS

MUSLE model has been successfully used for the estimation of storm-wise sediment yield in the Meghadrigedda Basin with good coefficient of determination of 0.84 which indicates accurate simulation of sediment yield from the MUSLE model. The average error value using MUSLE when compared with measured sediment yield was estimated to be 7.67%. However, the present results can also be used in erosion-based watershed prioritization in the study area. To regionalize the results of the study area, greater numbers of storms events as well as case studies are needed. Hence researchers should consider this aspect. In addition, other simple soil erosion and sediment yield models must be considered with reasonably accurate estimation of system response at the watershed scales, when scarce information exists. The fuzzy Logic based model developed is performing well for sediment computation from the inputs peak flow, rainfall and volume of runoff. The resulted correlation with MUSLE model is 0.97 and with that of measured value it is 0.937.

REFERENCES

1. ASCE (American Society of Civil Engineers)., 1970. Sediment sources and sediment yields. Journal of the Hydraulics Division ASCE 96 (HY6):1283–1329.
2. ASCE Task Committee. 1993. Criteria for evaluation of watershed models. Journal of Irrigation and Drainage Engineering-ASCE 119(3):429–42
3. Bingner RL, Murphee CE, Mutchler CK., 1989. Comparison of sediment yield models on various watersheds in Mississipi. Trans ASAE 32(2):529–534.
4. Das G., 2000. Hydrology and soil conservation engineering. Prentice Hall, India.
5. Durbovin et al. fozzy model real time reservoir operation., 2002. Journal of water resources planning and management, ASCE 128(1), p 66-73.
6. Despic and Simminivic., 2000. Aggregation operator for soft decision making in water resources engineering , fuzzy set and systems Elsevier 115, p11-33
7. FAO/UNEP (1994) Land degradation in south Asia: its severity causes and effects upon the people. FAO, UNEP and UNEP project, Rome

8. Haan CT, Johnson HP, Brakensiek DL. 1982. Hydrological modelling of small watersheds. American Society of Agricultural Engineers, Michigan.
9. Gottschalk, L.C., 1964. Reservoir Sedimentation. Chapter17-1. In: V.T. Chow, éd. *Handbook of Applied Hydrology*, McGraw-Hill Book Co., New York, NY. USA.
10. Hikaru K, Yoichi O, Toshiaki S, Kawanami A., 2000. Application of Universal Soil Loss Equation (USLE) to mountainous forests in Japan. J For Res 5(4):156–162.
11. Hrissanthou V., 2005. Estimate of sediment yield in a basin without sediment data. Catena 64:333–347.
12. Pandey A, Chowdary VM, Mal BC., 2009. Sediment yield modelling of an agricultural watershed using MUSLE, remote sensing and GIS. J Paddy Water Environ (Springer) 7(2):105–113.
13. Kirpich, Z. P., 1940. Time of concentration of small agricultural watersheds. Civil Engineering 10 (6), 362. The original source for the Kirpich equation.
14. Ross Timothy.J., 1997. Fuzzy logic engineering applications McGraw-Hill's international edition
15. Tripathi MP, Panda RK, Pradhan S, Das R K., 2001. Estimation of Sediment yield form a small watershed using MUSLE and GIS. Journal of the Institute of Engineering I 82:40-45.
16. UNEP., 1997. World Atlas of desertification, 2nd edition. Arnold, London, p 77.
17. Walling DE, Webb BW., 1982. Sediment availability and the prediction of storm-period sediment yields. In: Proceedings of the Exeter Symposium, 327–337. IAHS Publ. 137. IAHS Press, Wallingford, UK
18. Williams JR., 1975a. Sediment routing for agricultural watersheds. Water Resources Bulletin 11:965–975.
19. Williams JR., 1975b. Sediment yield prediction with Universal Equation using runoff energy factor. In: Present and prospective Technology for predicting sediment yields and sources, 244–252, Agricultural Research Service, US Department of Agriculture
20. Williams JR, Berndt HD., 1977. Sediment yield prediction based on watershed hydrology. Trans American society of Agriculture Engineers 20(6): 1100–1104.
21. Wischmeier WH, Smith DD., 1978. Predicting rainfall erosion losses. USDA Agricultural Research Services Handbook 537. USDA, Washington, p 57.
22. M. Anjaneya Prasad, S. Mohan.., 2007. Soft Computing Models using Fuzzy Logic for a Multipurpose Reservoir System Proceedings of "International Workshop on Integrated Water Resource Management (IWRM- 2007), Organized by Karnataka Environment Research Foundation, Bangalore, India
23. Zadeh, L. A. (1965). "Fuzzy sets". Information and Control 8 (3): 338.

APPROPRIATE INTRA VILLAGE DRINKING WATER DISTRIBUTION SYSTEMS THROUGH "MISSION BHAGHIRATHA" IN TELANGANA STATE

B. Rajeshwar Rao[1] and Shiva Ram Bandari[2]

[1]Executive Engineer, Rural Water Supply & Sanitation, O/o E-N-C.R.W.S.S. Hyderabad, Telangana State.
[2]M.S. Department of Mechanical Engineering Blekinge Institute of Technology, Karlskrona, Sweden.
rajeshwardesigns@gmail.com; ram.shiva546@gmail.com.

ABSTRACT

The systems established based on the 'Demand Driven Approach' for continuous water supply availability around the clock at consumer level, may be reliable for the socio-economic-environment situations in Developed Countries. The existing systems, establishedin lines ofthe 'Demand Driven Approach' for the domestic drinking water supply systems in Developing Countries are not sustainable and not reliable. The several uncertain parameters are involved based on the socio-economic-environments of underdeveloped countries for reliability in water supply access by round the clock and for sustainability of domestic drinking water supplysystems.

Though there are many factors involved, for 'Equal water distribution' and for 'maintaining the required minimum pressures' in Distribution Zone Areas becomes critical. A few experiments conducted on a pilot basis to examine the causes of problems in the present methodsadopted for the existing pipe line systems and for the investigation of appropriate innovative methods for the drinking water supply to all households simultaneously.

A new method of pipeline system has been developed for maintaining the pressures to reach water supply to all the households at 1st floor level and for overcoming the problems in the existing drinking water distribution systems. The new pipe line methods implemented for the experiments villages,Erravelly,Datarpally, Toopran in Gajwel constituency in Medak district of Telangana State. The required pressure is maintaining in the pipe mains even the situation of all the households in the villages are taking water simultaneously. This method of piping system is an innovative development for India and also suitable to other developing countries. This method is a synchronized pipeline design system for the cumulative discharges with respect to restricted limited with-drawls at household's connections using tampered proof flow control valves (FCV) designed for 5 liters per minute outflow ratesandsuitable for Urban & Rural Areas in India.Now FCVs provision is made mandatory for all the households in all villages of Telangana State and proposed for fixing at the government cost in "Mission Bhaghiratha" program.

INTRODUCTION

Domestic Water Demand as per **Business-as-Usual** Scenario Projections

- The domestic water demand includes the human and livestock water demands.
- The **human water demand** is based on the norms of **150 liters** per capita per day (lpcd) in the **rural areas** and **200 lpcd** in the **urban areas**.
- The livestock water demand is based on the cattle and buffalo population and uses **the norm of 25 liters per day per head water demand.**

Per-capita Water Availability for all requirements ranges: Table 1.

- Water stress condition: from 1000 to 1700 m^3 per year.
- **Water scarcity condition**: less than 1000 m^3 per year.

- As the water available within the country varies widely as a result of rainfall, groundwater reserve and proximity to river basins, most of the Indian States will have reached the **water stress condition by 2020** and water scarcity condition by **2025.**

Table 1 Water Availability in India-Status

Year	Population (Million)	Per-capita Water availability (M^3/year)
1951	361	5177
1955	395	4732
1991	846	2209
2001	1027	1820
2025	1394	1341
2050	1640	1140

Hence need for Water Conservation & Control of water drawls at consumer ends for avoiding wastage /excess usages and for even distribution of water among the populations.

NEED FOR INTERMITTENT DISTRIBUTION SYSTEMS SUITABLE TO INDIAN ENVIRONMENTS

The appropriate methodology for the developing world environments for the design of piped drinking water supply Distribution Systems with limited quantity of water supply in short durationsis essential for the system sustainability for maintaining the required minimum pressure for all consumer ends. The per-capita demands (LPCD) proposed adequately based on system feasibility, practical viability, and economy & water resources availabilities. And daily water delivery proposed in peak hours only, with the provisions of the tampered-proof flow control valve (FCV) of ½ inch size at each end user/house hold connection for restricted outflow rates in short durations. The restricted outflow rates for the provision of FCVs are based on the system feasibility for supply of water in short durations and the end user's minimum acceptable flow rate. The restricted outflow rates are feasible around 5 Liter per minute (LPM) for Rural and Urban Areas in India.

RESTRICTED WITHDRAWLS AT CONSUMER ENDS USING FCVS

The unique design of FCV (FIG-A) is essential for unbiased systems in entire country for the restricted outflows for use of both Rural & Urban populations and accordingly the supplyduration may vary based on the design per capita demands of the towns. In "mission Bhaghiratha" program of TelanganaState, the tampered-proof households FCVs with CNC machine make with solid SS 316 steel bars of ½ inch size are made in economically for service connections and are having simple internal arrangements for free flow of water with floats & debris if any entered in the pipe lines. The outflow rates through the FCVs are pressure dependent even though they are useful for the restricted outflows at small variations and within the limits of residual heads as provisions made in the CPHEEO manual. The FCV of ½ inch size is designed in "Mission Bhaghiratha" program for the restricted discharges of 5 LPM @ 0.5 bar pressure loss, as it is suitable for Rural Areas and it is also suitable for Urban Areas for the discharges of around 7 LPM @ 1.0 bar pressure loss as per the provisions of residual heads in the CPHEEO manual. Accordingly the daily supply duration at peak periods for Rural Areas @ 100 LPCD is 1.5 hours and 2.0 hours for Urban Areas @ 135 LPCD and for the Municipal Corporations/Cities the duration may around 3.0 hours for 150 LPCD.

DISTRIBUTION SYSTEM BASED ON THE CUMULATIVE RESTRICTED WITHDRAWLS AT CONSUMER ENDS

For the pipe sizes in the Distribution Networks, the pipeline design carrying capacities arrived/checked in proportional to the demands envisaged based on number of equivalent household connections and restricted water outflow rate provisions at the end users. The pipe sizes in the Distribution Networks are checked for the minimum terminal heads /residual pressure heads at house hold connections +7m for Rural and +12m for Urban Areas based on the head losses required for the FCVs as proposed at the restricted outflow rates.

METERING PROVISIONS WITH AFFORDABLE CONTROL WATER TARIFF PLANS

The metering provisions with affordable control water tariffs are must for the system reliability, sustainability, accountability, for conservation of water and energy and for even-distribution of water to all households having little variations of average family membersforhouse connections.

ABOUT "MISSION BHAGIRATHA" PROJECT

- The Project outlay is around Rs. 40,000 Cr.
- The project is divided into 26 segments based on the topography, command-ability, proximity and ease of connectivity from various dependable surface sources, i.e., Dams & reservoirs.
- Designed period of the project is up to 2048.
- Water supply will be ensured by 2018.

GOALS OF "MISSION BHAGIRATHA" PROJECT

- It is a unique and most comprehensive project to cover all households in entire State. It envisages treated Surface Drinking Water to every household at their door step at the rate of
- 100 LPCD in rural areas,
- 135 LPCD in Municipalities/ Nagar panchayats
- 150 LPCD in Municipal Corporations.
- 10% to meet the Industrial needs.

Village Water Supply System in "MISSION BHAGIRATHA"PROJECT

- Equitable Distribution of water supply will be ensured by maintaining minimum water pressure of 0.5 bar, duly designing the water distribution system in innovative way for controlled water supply by installing 5 LPM flow control valves in each House Hold connections and thus the water supply for all households in the villages will be at the same time.

INTRA VILLAGE WATER SUPPLY SYSTEM PLANNING

- VILLAGE WATER SUPPLY (VWS) are INTERMITTENT as individual HH TAPS drawls capacity much more Outflows than the Design Peak Demands.
- ½" size 5 LPM FCV fixing in place of ferrule for limiting the Outflows @ all the consumer ends.
- TARIFF-Affordable and control-telescopic Water Tariff Plans required for control of water wastage and to conserve the water in critical situations and for Equal Water Distribution even the HH having distinct family members and for maintaining continuous availability of supply in almost entire day or many hours in a day.

Table 2 Water Tariff for Rural Area May Be Proposed

Basic O&M Charges= Rs.10 per 1KL									
	Water usage			Tariff Rates		Monthly charges			
Block tariffs level	From in LPCD	To in LPCD	Monthly consumption for each HH	Times over basic O&M	Tariff rate In Rs.Per 1KL	Max charges on water meter Reading	Min for each water meter connection O&M	Total water bill in Rs. per Month	Remark s
0	0	55	8kL	Free on water meter reading			30	30	Subsidy up to 55 LPCD
1	56	70	10KL	1	10	20	30	50	Basic @ 70LPCD
2	71	100	15KL	2	20	120	30	150	1ST Control
3	101	135	20KL	4	40	320	30	350	2nd Control
4	136	150	25KL	8	80	720	30	750	3rd control
5	>	150	25KL	8	80	720	30	750	Leads Disconnection

For integrative pipeline design in Rural Areas:

- The water drawls at house-holds reduced to acceptable range of 5 to 6 LPM by fixing tampered proof flow control valves (FCV) at each service connections and
- The main pipeline sizes provided for the peak design discharges revised appropriately w.r.t. the total controlled with-drawls at service points.

Flow control valves suitable for Mission Bhagheeratha as 5 LPM FCV:

- A simple Tampered proof VALVE with Backflow prevention.
- Controls the Excess Water drawls at low lying areas and at areas nearby tanks.
- A single unique design of 5 LPM FCV is suitable designed for both Rural and Urban demands.

- FCV is fixed at place of ferrule to threaded saddle clamps to avoid the bypassing the FCV.
- FCV benefits for also minimizing of UFW/NRW.
- Simultaneous supply can be possible in peak timings to all households as per the synchronized pipe line networks in "Mission Bhaghiratha" Program. Thus the required minimum pressure heads maintained to all households.

Water Conservation:

- Water Usage is almost equals to the Water Wastage?
- What is the effective usage of water?
- How can we control Water Wastage?

How water wastage controls by "MISSION BHAGHIRATHA" Drinking Water Supply Program:

- Water supply evenly for all houses and almost equal quantities if metered.
- Water supply takes place at the same time for all houses in the village.
- Water supply with sufficient pressure to all houses that can be tapped at 1st floor Level.
- Only single valve will be operated for each water tank in the village and all tanks supply will starts and ends supply at same time in the village based on their needs.
- No more valve operations in the middle or any corners of the village. Thus the watermen in the villages can save their time for concentrating on the regular equal supply and for eliminating the leakages and wastages.
- No need of mini water supply systems which were run with single phase motors for coverage of low pressure areas in the village as the water supply can reach all corners with required pressure under the water tanks commands.
- Thus we can save the ground water from excess usage/wastage in addition to the savings in power consumptions of scattered pumping of single phase motors in the villages.
- The roof top water storage tanks for individual houses filled directly without motor pumps for 1st floors.
- Time of collection of water by almost the household's woman hours will be saved thus the utilisation of the saved hours for betterment of their Health & Wealth.
- Thus Power Consumption in peak hours will be saved and indirectly money, water for electricity production will also be saved in addition to the reduction of pollution.

Design of distribution systems (DNS)-comparison-with CPHEEO manual & 24x7 systems			
		24X7 SYSTEM	**INDIAN SYSTEMS-CPHEEO**
Sl.	**Factors involved**	**24x7 method as in developed countries**	**Existing methods, as per CPHEEO , based on 24x7 systems**
1	Design per capita demand in LPCD	May be suitable for >250 LPCD, or designed for more than the actual required demands	Limited nearly 135 LPCD,which is lower than actual demands

Contd...

Design of distribution systems (DNS)-comparison-with CPHEEO manual & 24x7 systems			
		24X7 SYSTEM	**INDIAN SYSTEMS-CPHEEO**
Sl.	**Factors involved**	**24x7 method as in developed countries**	**Existing methods, as per CPHEEO , based on 24x7 systems**
2	Method of supply as per design	Continuous, by 24x7	Continuous, by 24x7
3	Design peak factor for peak hourly demand	Around 2	Around 3
4	Design minimum terminal pressure heads at consumer level	>= 30m	<= 12m
5	Mode of supply as per actuals	Continuous, by 24x7 and water availability for all consumers	Alternate-intermittent supply takes place in short durations
6	Mode of water drawls at consumer ends	Water to be drawls , only when as usage around the clock but not for storage	Water drawls is deferent than the design mode , water drawls in short durations for storage of entire day demands
7	Water storage at consumer ends	No storage as consumers instant uses as demands arisen, as per assumed 100% reliable supply by 24x7	No storage as consumers instant uses as per demands arisen, as per assumed 100% reliable supply by 24x7
8	Consumer ends taps water drawls capacity, for 1/2 bar pressure with 1/2" pipe service	20 -25 LPM , as no control at consumer ends	15-20 LPM , as no control at consumer ends and maintained with very low terminal pressures
9	supply duration as per design	24x7	24x7
10	supply duration as per actual	24x7	With varied short durations
11	Is method of DNS, suitable for under developed countries environments	Not suitable	Not suitable
12	Is water quality deteriorated	No	Yes, required for household level disinfection as intermittent supply takes place
13	Is DNS capable to water with-drawls by the all households at once	Not possible	Not possible
14	Is DNS sustainable after a marginal periods of breakdowns if any in peak demand timings	Not sustainable	Not sustainable
15	Is required for efficient metering system with telescopic water tariff plans	Yes and mandatory	Yes and mandatory
16	Is DNS capable to upgrade as 24x7 system for developing countries environments	Not suitable for underdeveloped countries	Not Sustainable

Design of distribution systems (DNS)-comparison-with INDIAN SYSTEMS-CPHEEO manual &Restricted drawls method using FCV at all consumers			
		INDIAN SYSTEMS-CPHEEO	**RESTRICTED DRAWLS METHOD (RDM FCV SYSTEM)**
Sl	**Factors involved**	**Existing methods, as per CPHEEO , based on 24x7 systems**	**Restricted drawls method using FCV at all consumers**
1	Design per capita demand in LPCD	Limited nearly 135 LPCD,which is lower than actual demands	May be suitable for limited per-capita demands even for 100 LPCD
2	Method of supply as per design	Continuous, by 24x7	Intermittent for limited supply
3	Design peak factor for peak hourly demand	Around 3	12 to 24
4	Design minimum terminal pressure heads at consumer level	<= 12m	<= 12m
5	Mode of supply as per actuals	Alternate-intermittent supply takes place in short durations	Intermittent supply as per designed durations
6	Mode of water drawls at consumer ends	Water drawls is deferent than the design mode , water drawls in short durations for storage of entire day demands	Water drawls as per design with FCV, as when supply arises and water storage takes place
7	Water storage at consumer ends	No storage as consumers instant uses as per demands arisen, as per assumed 100% reliable supply by 24x7	Storage needs for full day requirements
8	Consumer ends, taps water drawls capacity, for 1/2 bar pressure with 1/2" pipe service	15-20 LPM , as no control at consumer ends and maintained with low terminal pressures	Around 5 LPM , restricted by installing flow control valves at each consumer ends
9	supply duration as per design	24x7	1 to 2 hours daily
10	supply duration as per actual	With varied short durations	1 to 2 hours daily and more hours over the design after stabilization
11	Is method of DNS, suitable for under developed countries environments	Not suitable	Suitable
12	Is water quality deteriorated	Yes, required for household level disinfection as intermittent supply takes place	Yes, sometimes required for household level disinfection as intermittent supply takes place
13	Is DNS capable to with-drawls by the all households at once	Not possible	Yes, DNS designed for all HH with-drawls at once
14	Is DNS sustainable after a marginal periods of breakdowns if any in peak demand timings	Not sustainable	Sustainable

Contd...

Design of distribution systems (DNS)-comparison-with INDIAN SYSTEMS-CPHEEO manual &Restricted drawls method using FCV at all consumers			
		INDIAN SYSTEMS-CPHEEO	**RESTRICTED DRAWLS METHOD (RDM FCV SYSTEM)**
Sl	**Factors involved**	**Existing methods, as per CPHEEO , based on 24x7 systems**	**Restricted drawls method using FCV at all consumers**
15	Is required for efficient metering system with telescopic water tariff plans	Yes and mandatory	"Not mandatory" but need for water conservation and for upgrading to 24x7 system
16	Is DNS capable to upgrade as 24x7 system for developing countries environments	Not Sustainable	Yes

CONCLUSIONS

A new method of pipeline system has been developed for intermittent distribution systems for simultaneous supply in short durations to all the households by maintaining the minimum required pressures to reach water supply at 1st floor level and for overcoming the problems in the existing drinking water distribution systems. This method of piping system is an innovative development for India and also suitable to other developing countries. This method is a synchronized pipeline design system with respect to the restricted limited outflows at household's connections using tampered proof flow control valves (FCV) designed for 5 liters per minute outflow rate and is suitable for Urban & Rural Areas in India.

The 1st phase commission of "Mission Bhaghiratha" program of TelanganaState, examined in several villages , the water supply is taking place daily in twice at peak hours in morning and evening with the control provision of the tampered-proof-flow control valve (FCV) of ½ inch size at each house hold connection for restricted outflow rates of 5 LPM.And all the households in the villages are getting water supply evenly and simultaneously in twice in a day. A model village,Erravelly in MedakDistrict, the water supply is taking place at 1st floor level of all houses that are built newly as a specific government program. Now FCVs provision is made mandatory for all the households in all villages and proposed for fixing at the government cost in "Mission Bhaghiratha" program.

This paper is to introduce a new approach suitable for Indian Socio Economic Environments for Design of Distribution Systems, through the integrative approach,

1. By controlling the excess withdrawals by fixing the domestic **tampered proof Flow Control Valves** (FCV) for 5 LPM outflows @ 0.5 bar pressure at all household service connections (HSC),
2. Zoning areas are to be planned with respect to the **distribution mains carrying** capacities, the discharges in main pipelines are the **cumulativecontrolled outflows** at consumer ends based on the number of equivalent HSC, for sustaining the minimum residual heads at all points in entire Distribution Zone Areas. And,

The provision of **Water Meters** with **affordablecontrolWater TARIFF PLANS** implementation for achievements of uninterrupted water supply, for minimizing the wastage/excess usage of water through accountability and for equal water distribution.

Fig. 1a Water scarcity conditions states in india-urban area

Fig. 1b Water scarcity conditions states in india-rural areas

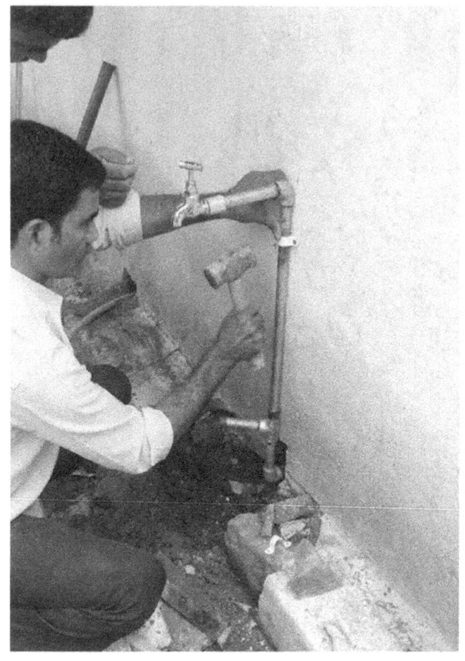

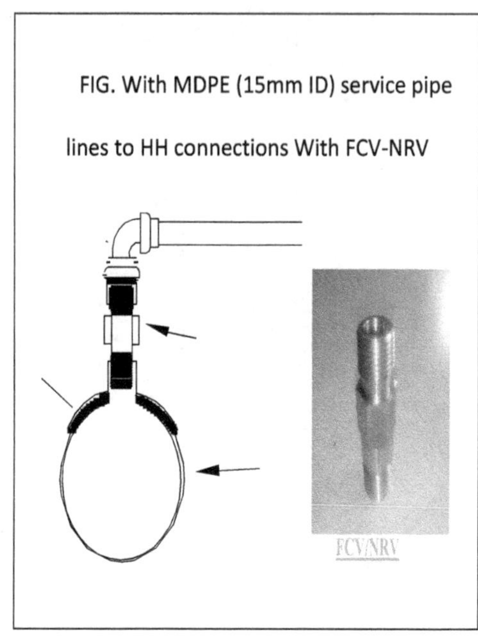

Fig. 2a House connections-at ground level.

Fig. 2b HSC connection with FCV

Fig. 3 Water supply trail run-for phase-i works at 1st floor level

Fig. 4 HH connections-at 1st floor level with sintex tanks

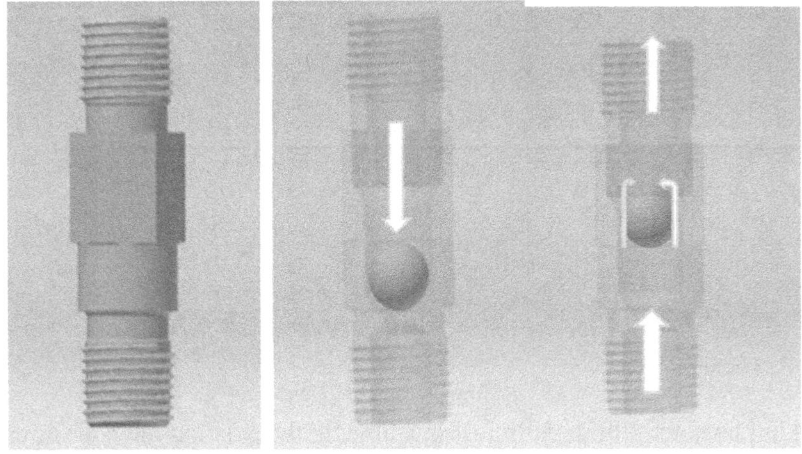

FCV/NRVACT as NRVFLOW Controlling

Fig. 5 Flow Control Valve Parts Arrangements

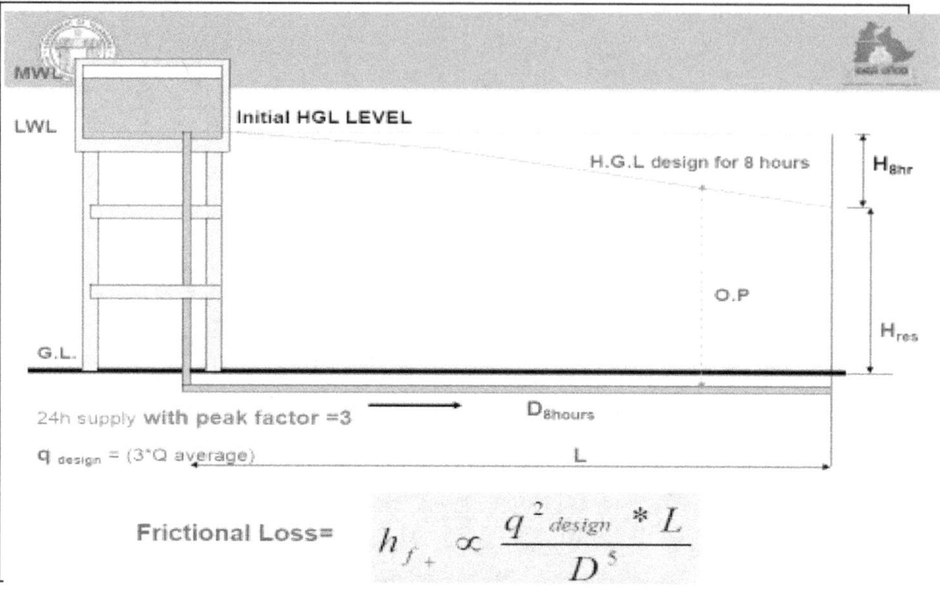

Fig. 6a Designed/assumed hgl lines for the existing methods as per the cpheeo manual for assmed 8 hours supply but fails in operations with unlimitted drawls at hh.

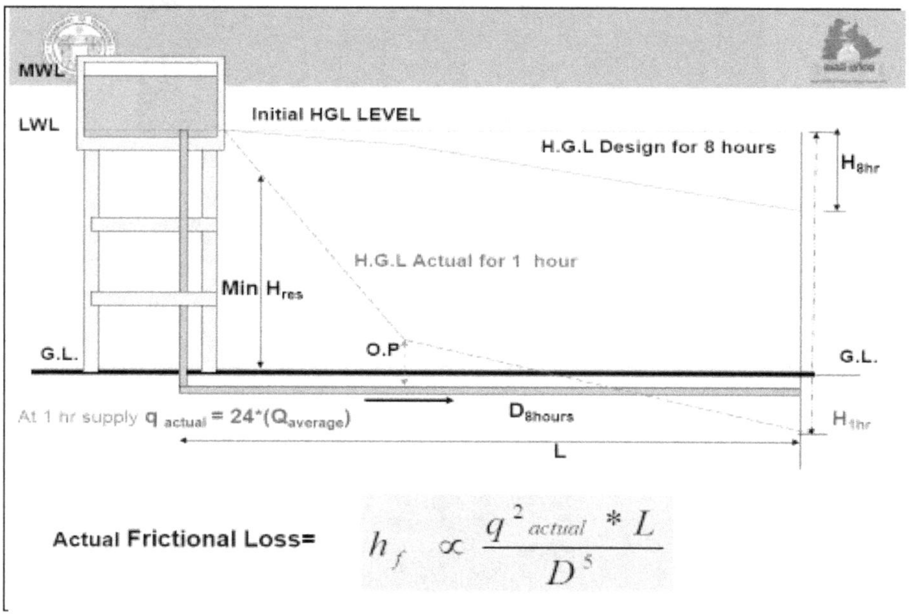

Fig. 6b Actual hgl lines for 1 hour daily for each hh with the existing methods as per the cpheeo manual, minimum pressure will not maintained in entire pipe line network and water will go to down elevated houses in comparision to other adjacent houses.

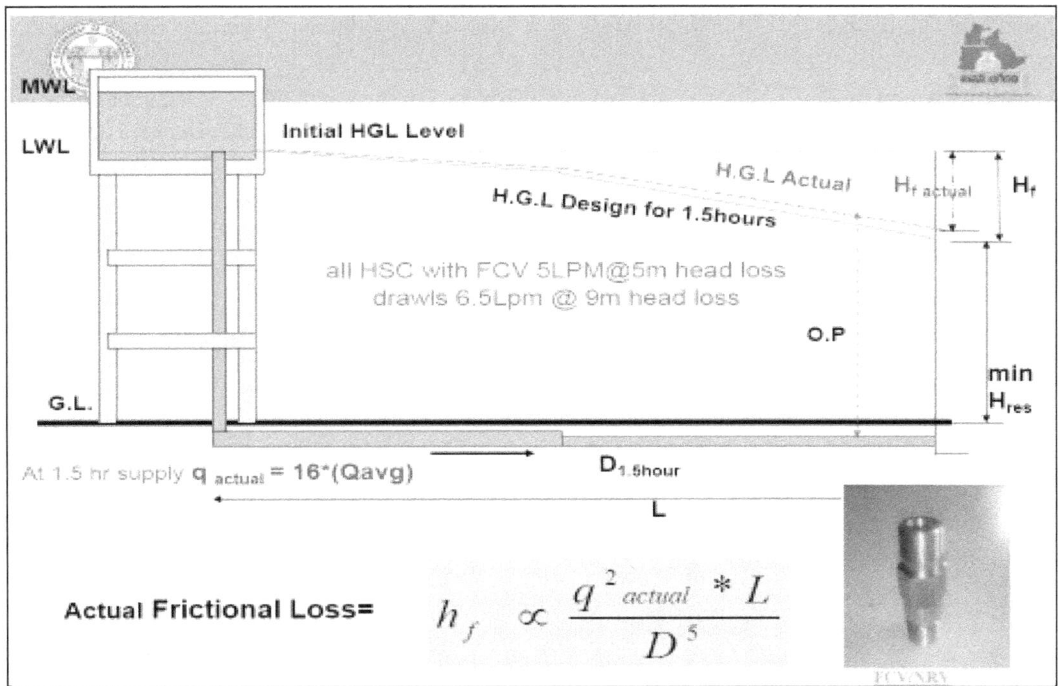

Fig 7 Design &actual hgl lines matches for the new methods for restricted drawls w.r.to pipe sizes designed for 1.5 hours of supply rates for maintainning minimum pressures to all hh simultaneously drawls.

ACKNOWLEDGEMENTS

I am very grateful to Sri Dr.Rajesh Gupta, Professor of V.N.I.T. Nagpur., and, Sri.B.Surender Reddy, the Engineer-In-Chief Rural Water Supply, Hyderabad, Sri.VPrabhakarRao,Retd. Superintending Engineer P.R.RWS, Sri Dr. D.M.Mohan, Retd. Chief Engineer, HMWS&SB Hyderabad., Shri D. Hanumantha Chary,Retd. Dy.Chief Engineer HMWS&SB Hyderabad. And all RWS & PHED Engineers for their motivation & cooperation, useful discussions from time to time enabled me to write this paper.

REFERENCES

1. B.R.RAO & SHIVA RAM BANDARI -An Integrative design approach for domestic water Distribution Systems in Indian cities and villages, 49th Annual Convention of IWWA on "Smart Water Management" January 19-21, 2017.
2. B.R.RAO & SHIVA RAM BANDARI –An Integrative design approach for intermittent supply situations in Indian cities and villages,-innovative technologies for water and waste water management" March 21, 2017-Institute of Engineers (INDIA), Hyderabad.
3. CPHEEO Manual for Water Supply & Treatments from Ministry of Urban Development, Gov't of India.(1999).
4. CPHEEO Manual for O&M from Ministry of Urban Development, Gov't of India.(2005).

5. NRDWP Guide lines -2013 for Rural Areas from Ministry of Drinking Water and Sanitation, Gov't of India
6. Upali A. Amarasinghe, Tushaar Shah, and B.K.Anand , International Water Management Institute, New Delhi, India Consultant, Bangalore, India –"India's Water Supply and Demand from 2025-2050: Business- as- Usual Scenario and Issues" (BAU).

MODEL DEVELOPMENT FOR PREDICTION OF TRAP EFFICIENCY OF SRIRAMSAGAR RESERVOIR USING SOFT COMPUTING TECHNIQUES

Gopal M. Naik[1] and Qamar Sultana[2]

[1]Professor & [2]Research Scholar

Department of Civil Engineering, Osmania University, Hyderabad,

mgnaikc@gmail.com

ABSTRACT

The estimation of sedimentation in the reservoirs has become a significant problem. There are a number of methods for estimation of reservoir sedimentation. However, each model differs greatly in terms of their complexity, inputs and other requirements. In the simplest way, the fraction of sediment deposited in the reservoir can be determined through the knowledge of its trap efficiency. In this study an artificial neural network (ANN) model developed using Matlab software is used to estimate the trap efficiency of the reservoir. Sriramsagar reservoir located at Pochampadu village in Nizamabad district is taken as a case study. The input parameters used are annual rainfall, annual inflow, and age of the reservoir and the output parameter considered is the trap efficiency(T_e) of the reservoir for twenty six years. A conventional regression analysis is conducted, relating the output parameter (T_e) to the input parameters. It is observed from the ANN model results that, the simulated trap efficiency of the reservoir shows the better accuracy and less effort than that of the conventional method. Further since the simulated hydrologic data is very important for the field engineers to manage of the available quantity of water resources of the watershed. The time series analysis of annual reservoir inflows and annual rainfall has been done and subsequently trap efficiency is predicted for the same series data for the next 26 years i.e. from 2013 to 2038 using soft computing technique.

Keywords: Reservoir sedimentation, Trap Efficiency, Artificial Neural Network

1. INTRODUCTION

Reservoirs have well known primary purposes such as water supply, irrigation, flood control, hydropower and navigation. A major portion of the silt that is carried along with the river water settles down in the reservoir, which causes the reduction in its storage capacity and has become a great problem all over the world. In this study, an ANN model has been developed using Matlab tool and the available input parameters for the estimation of the trap efficiency(T_e) in a large reservoir, Sriramsagar reservoir in India. The input parameters such as annual inflow(I_a) , annual rainfall(R_a) and the age of the reservoir (A_g) were considered and the trap efficiency (T_e) was considered as an output parameter.

2. STUDY AREA DESCRIPTION

The Sriramsagar Project (SRSP), formerly known as the Pochampadu irrigation project has been built on Godavari river which is one of the major peninsular rivers in southern India. This project is located at Pochampadu village (18°-58' N latitude and 78°- 20' E longitudes) in Nizamabad district of Telangana State (TS) of southern India at a distance of about 200 km from Hyderabad Metropolitan City. The location and water spread area of Sriramsagar reservoir is shown in Fig.1. The SRSP project has been built to utilize Godavari river water for irrigation and drinking purposes in Telangana state.

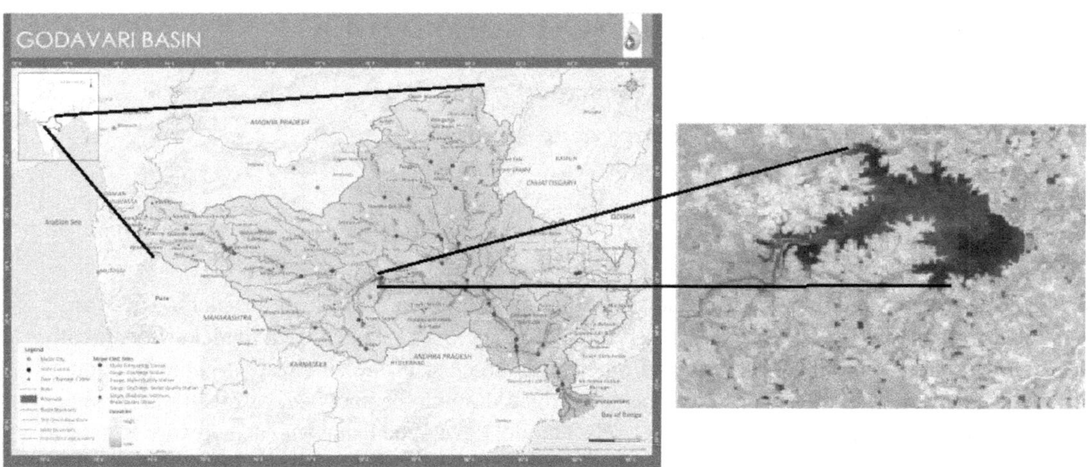

Fig. 1 Location and Water spread area of Sriramsagar Reservoir

3. MODEL DEVELOPMENT

3.1 Artificial Neural Network

Artificial Neural Network (ANN) has an input layer, a hidden layer and an output layer. Each layer consists of several neurons and the layers are interconnected by sets of correlated weights. The neurons receive inputs from the initial inputs or the interconnections and produce outputs by the transformation, using an adequate nonlinear transfer function.

3.1.1 Developing ANN Model:

Designing of ANN model consists of five steps. They are (1) Data collection, (2) Pre-processing of the data, (3) building of the network, (4) training the network and (5) test performance of model. Neural network architecture used for prediction of trap efficiency of the reservoir is shown in Fig.2.

3.1.2 Conventional Regression Method.

A conventional regression analysis is conducted, between the output parameter (T_e) and the input parameters (I_a ,R_a, and A_g), using RegressIt an Excel add-in that performs multiple linear regression analysis.

4. DISCUSSION OF RESULTS

A conventional regression analysis is done relating the output parameter (Trap Efficiency(T_e)) with the input parameters (Annual Rainfall (R_a), Annual Inflows (I_a), Age of the reservoir (A_g)). The regression equation of the three variables with the output parameter which gave the best result is shown in the equation 1.

Estimated Trap Efficiency from regression method =

$$(0.034*A_g) - (0.0000004857*I_a) + (0.004882*R_a) + 80.342 \qquad .....(1)$$

The value of R for three input variables from the Table 4. is 0.207 , which is not satisfactory, and hence an attempt is made by using Artificial Neural Network (ANN) technique for the better results.

A Multilayer Perceptron (MLP) ANN architecture consisting of three layers with 10 neurons in the hidden layer is developed using the MATLAB tool. Since there is a large variation in the data, the data is normalised between 0 and 1 using the equation 1. Twenty six years data is used, out of which 70% is used for testing and the rest 30% is used for training and validation. Based on the values of R generated (0.9216), it is seen that a Feed Forward, Back Propagation (BP ANN) model shown in the Figure 4. with the structure 3-10-1 generated the trend of the trap efficiency values well with TRAINLM - training function, TRAINGDM - learning function and TANSIG - transfer function.

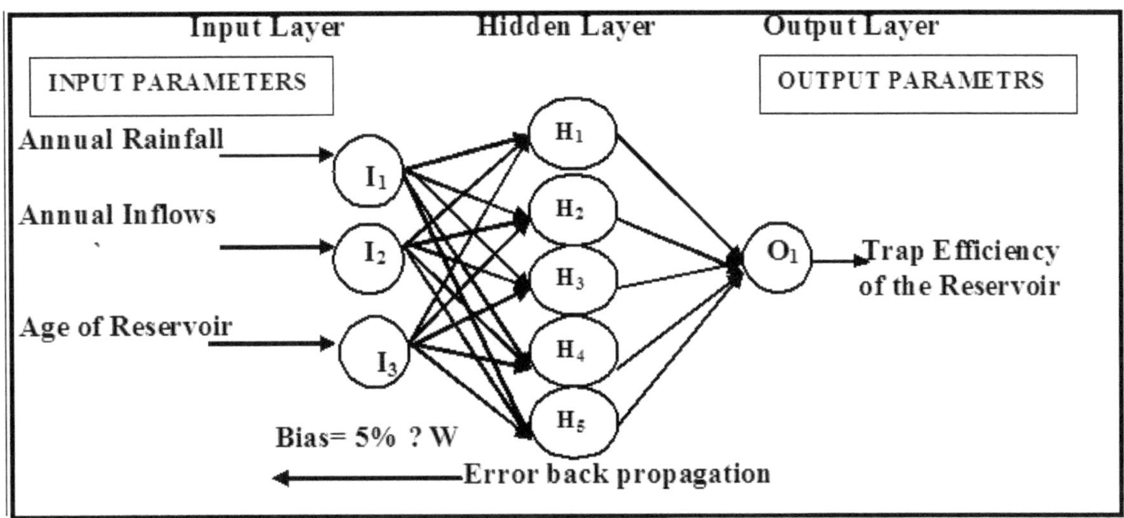

Fig. 2 Neural network architecture used for prediction of trap efficiency of the reservoir.

4.1 Reservoir Annual Inflows, Annual Rainfall and the Trap Efficiency

The ANN model is used to predict the annual inflows , annual rainfall and the trap efficiency for a further period of 26 years i.e. from 2013- 2038.

4.1.1 Reservoir Annual Inflows

Based on the values of minimum MSE value of 0.0052 and the optimum value of R generated (0.666), it is seen that a Feed Forward, Back Propagation (BP ANN) model shown in the Figure 4. with the structure 1-8-1 generated the trend of the trap efficiency values well with TRAINLM - training function, TRAINGDM - learning function and TANSIG - transfer function. Fig. 2 illustrates the time series plot of the annual inflows and annual rainfall for the observed data from (1987-2012) and simulated data (2013-2038).

4.1.2 The Annual Rainfall

Based on the values of minimum MSE value of 0.0072 and the optimum value of R generated (0.7878), it is seen that a Feed Forward, Back Propagation (BP ANN) model shown in the

Figure 4. with the structure 1-10-1 generated the trend of the trap efficiency values well with TRAINLM - training function, TRAINGDM - learning function and TANSIG - transfer function.

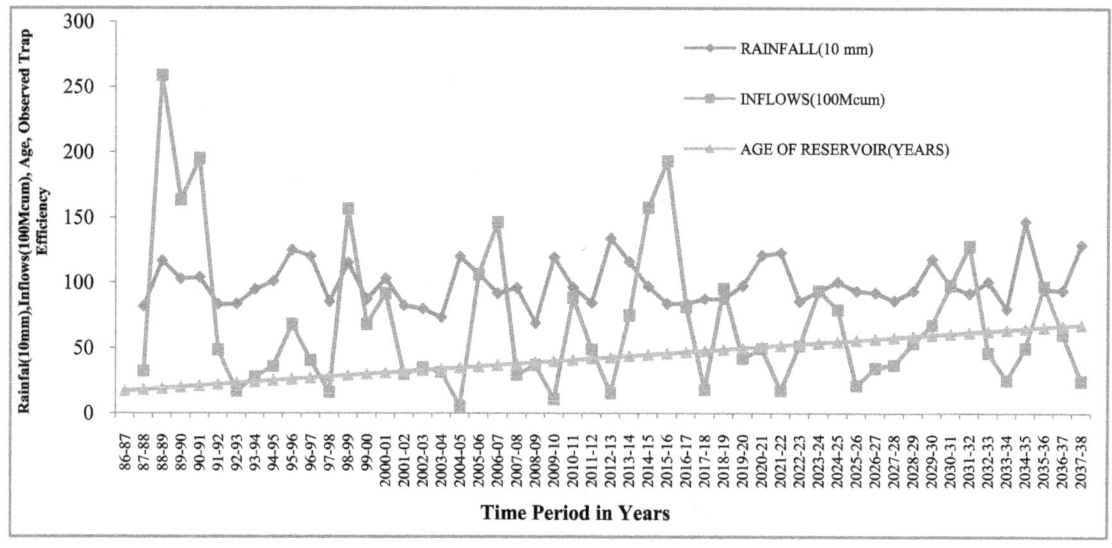

Fig. 3 Time Series Plot of the Annual Inflows and Annual Rainfall for the observed data from (1987-2012) and Simulated data (2013-2038)

4.1.3 Forecast of the Reservoir Trap Efficiency

A Multilayer Perceptron (MLP) ANN architecture consisting of three layers with 12 neurons in the hidden layer is developed. The normalised data of the reservoir annual inflows for a period of 26 years i.e. from 1987-2012 is used, out of which 70% is used for testing and the rest 30% is used for training and validation and the annual inflows data is forecasted for the next 26 years i.e. from 2013-2038. Based on the values of minimum MSE value of 0.0152 and the optimum value of R generated (0.9276), it is seen that a Feed Forward, Back Propagation (BP ANN) model shown in the Figure 4. with the structure 3-12-1 generated the trend of the trap efficiency values well with TRAINLM - training function, TRAINGDM - learning function and TANSIG - transfer function.

A conventional regression method is applied to forecast the trap efficiency values for the next 26 years. The prediction of the trap efficiency of the reservoir from 2013-2038 is shown in Table1. From the Table 1. it can be seen that both the ANN and the Conventional method produces the same average values but from the Fig. 4, it is seen that the trend of the observed values of the trap efficiency is well followed by the ANN model. Thus the ANN model generates better values when compared with the conventional regression method. The performance statistics of this model is shown in the Table 2.

Table 1 Prediction of the Trap Efficiency of the Reservoir from 2013-2038.

Year	Age of reservoir (years)	Annual rainfall (mm)	Annual inflows (10^6m^3)	Observed (Te%)	Predicted by ANN Model (Te%)	Estimated by Conventional Regression Method (Te%)
86-87	17	3828.96	43137.58	97.76	89.75	99.32
87-88	18	818.20	3215.82	82.74	106.48	84.67
88-89	19	1166.99	25900.39	99.99	98.62	86.40
89-90	20	1028.22	16351.96	60.19	60.29	85.76
90-91	21	1040.55	19509.94	75.24	75.49	85.85
91-92	22	833.75	4863.10	92.91	90.43	84.89
92-93	23	835.42	1666.60	90.86	94.58	84.93
93-94	24	947.89	2775.15	89.54	89.63	85.51
94-95	25	1008.65	3586.79	86.33	87.00	85.84
95-96	26	1248.73	6815.24	90.67	92.97	87.05
96-97	27	1201.49	4049.05	96.01	94.38	86.85
97-98	28	853.92	1583.05	93.11	98.92	85.19
98-99	29	1150.78	15672.72	89.47	89.05	86.67
99-00	30	872.89	6815.07	81.72	87.02	85.35
00-01	31	1031.16	9185.72	94.26	92.86	86.15
2001-02	32	823.50	2947.73	86.33	83.67	85.18
2002-03	33	798.66	3505.65	41.80	41.80	85.09
2003-04	34	733.10	3147.25	94.07	93.00	84.80
2004-05	35	1197.80	541.16	80.88	77.52	87.11
2005-06	36	1066.60	10622.35	91.25	94.14	86.49
2006-07	37	917.60	14625.44	98.91	96.87	85.80
2007-08	38	959.30	2920.04	92.84	92.60	86.04
2008-09	39	689.70	3707.05	97.99	101.31	84.76
2009-10	40	1192.70	1061.09	86.07	85.47	87.25
2010-11	41	962.90	8848.57	90.67	91.81	86.16
2011-12	42	845.40	4896.58	63.10	59.45	85.62
2012-13	43	1538.82	1341.70	-	94.33	88.08
2013-14	44	7507.70	1158.88	-	99.99	87.22
2014-15	45	15778.67	970.68	-	50.53	86.33
2015-16	46	19324.39	839.78	-	53.83	85.72
2016-17	47	8147.53	842.17	-	51.69	85.77
2017-18	48	1798.91	876.31	-	92.59	85.98
2018-19	49	9558.85	876.77	-	51.40	86.01

Contd...

Year	Age of reservoir (years)	Annual rainfall (mm)	Annual inflows ($10^6 m^3$)	Observed (Te%)	Predicted by ANN Model (Te%)	Estimated by Conventional Regression Method (Te%)
2019-20	50	4206.63	976.29	-	95.96	86.53
2020-21	51	4960.76	1208.90	-	99.99	87.70
2021-22	52	1721.62	1229.61	-	99.99	87.84
2022-23	53	5185.17	857.09	-	76.32	86.05
2023-24	54	9375.44	935.79	-	61.86	86.47
2024-25	55	7933.48	1005.27	-	86.58	86.84
2025-26	56	2131.20	935.11	-	98.07	86.54
2026-27	57	3459.97	922.95	-	96.87	86.51
2027-28	58	3704.82	863.88	-	94.22	86.25
2028-29	59	5383.52	936.02	-	95.00	86.64
2029-30	60	6759.38	1181.24	-	99.53	87.87
2030-31	61	9813.33	970.41	-	79.49	86.87
2031-32	62	12830.21	917.52	-	57.33	86.65
2032-33	63	4648.42	1007.62	-	98.54	87.13
2033-34	64	2534.38	799.58	-	96.95	86.14
2034-35	65	5001.18	1469.94	-	99.99	89.45
2035-36	66	9725.84	948.22	-	85.66	86.93
2036-37	67	6049.20	940.54	-	97.39	86.93
2037-38	68	2474.56	1292.95	-	99.95	88.69
Average				86.34	86.14	86.61

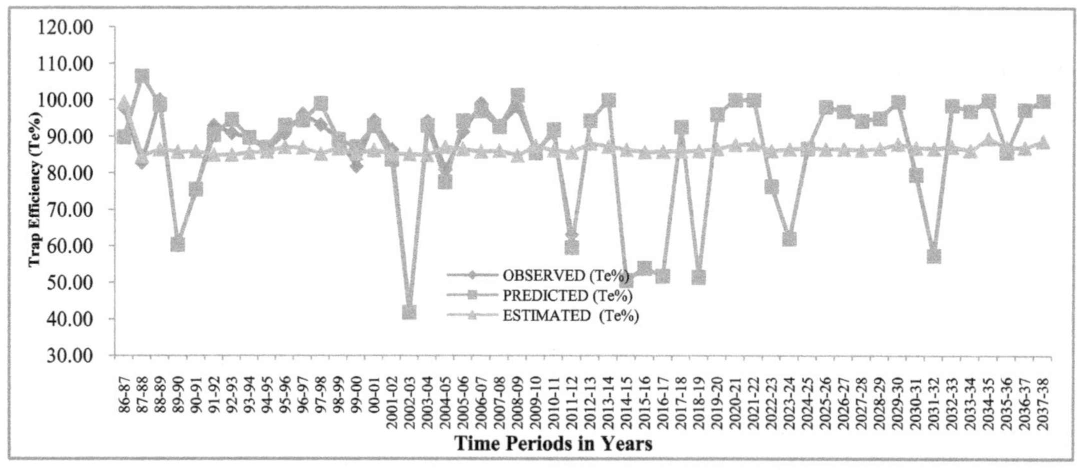

Fig. 4 Plot of the Trap Efficiency for the Observed, Predicted (by ANN Model) and Estimated (by Conventional Regression Method) Data.

Table 2 Performance Statistics of the ANN Model for Prediction of Annual Inflows, Annual Rainfall and Trap Efficiency from 1987-2038

Statistical Parameter	Annual Inflows	Annual Rainfall	Trap Efficiency
R	0.6666	0.7878	0.9276
R^2	0.4443	0.6206	0.8604
MSE	0.0059	0.0072	0.0152
RMSE	0.0768	0.0849	0.1232

5. CONCLUSIONS

In this study, an ANN approach and a conventional regression method are used for the estimation of the trap efficiency in Sriramsagar reservoir on a yearly basis and their results are compared with the observed trap efficiency values for 26 years i.e. from 1987 to 2012 and the same technique is extended for forecasting the reservoir annual inflows, annual rainfall and trap efficiency from 2013 to 2038 . It is seen that the developed ANN model captured the trend of trap efficiency percentages well when compared to the traditional regression approach, which is seen from the trend of the curves in the figures 4. The evaluation criteria Correlation Coefficient (R) of 0.9276 , the Determination Coefficient (R^2) of 0.8604 as shown in the Table 2. confirms the accuracy of the ANN model when compared with the conventional method.

REFERENCES

1. Agarwal A., Rai R.K., Upadhyay A., "Forecasting of Runoff and Sediment Yield Using Artificial Neural Networks" , *J.Water Resources and Protection*, 2009, pp.368-375.
2. Cigizoglu, H. K. “Suspended sediment estimation and forecasting using artificial neural networks.” *Turk. J. Eng. Environ. Sci.*, 26, 2002a, pp. 15–25.
3. Cigizoglu, H. K., and Alp, M., “Generalized regression neural network in modelling river sediment yield.” *Adv. Eng. Software*, 37, 2006, pp. 63–68.
4. Faridah, O., Mahdi, N., " Reservoir inflow forecasting using artificial neural network", *International Journal of the Physical Sciences*, 2011,6(3), pp. 434-440.
5. Garg, V. and Jothiprakash, V., "Trap Efficiency Estimation of a Large Reservoir." ISH *Journal of Hydraulics*, 14(2), 2008, pp.88-101.

SUSTAINABLE IRRIGATION WATER MANAGEMENT PRACTICES IN THE CONTEXT OF NATIONAL WATER POLICY 2012 AND UNITED NATIONS SIXTH SUSTAINABLE DEVELOPMENT GOAL ON "SUSTAINABLE MANAGEMENT OF WATER"

Mohammed Hussain

Professor and Dean(Life Skills and Outreach) , Department of Civil Engineering ,
Gokaraju Rangaraju Institute of Engineering and Technology (GRIET), Hyderabad &
Founder Coordinator for Centre for Water resources Engineering and Management (CREAM) &
Centre for Sustainable Technologies for Eco-Social Resilience to Global Climate Change (CST-ERG),
GRIET, Hyderabad & Former Faculty Member, Water and Land Management Training and Research Institute (WALAMTARI), Hyderabad

ABSTRACT

The present paper describes briefly the practices of sustainable irrigation water management in the world for sustainable agriculture production in the context of National Water Policy 2012 and United Nations sixth sustainable development goal on "Sustainable Management of Water and sanitation for all ". The freely available vast resources(research papers, research reports, User Manuals and softwares, web courses and video lectures in the area of Irrigation Water management) described will be useful for engineering teachers to give guidance to the student projects and research scholars in their research projects. They also will be useful for practicing engineers to update their knowledge and skills for life long excellence in effectiveness and efficiency in their engineering duties.

Technical capacity building and Managerial capacity building of all concerned stakeholders in the area of irrigation management are described. Finally, an action plan for implementation of sustainable irrigation water management in India is discussed. The motto of both JNTUH and AICTE is "Yogaha Karmasu Kaushalam (Yoga is skill in action)". Supramental transformation and evolution of Integral yoga of Sri Aurobindo is needed for sustainable development. Integral yoga is of triple aspects of Aspiration of the heart with the silent mind, Rejection of the movements of the lower nature & Surrender of oneself and all one is and one has to the divine shakti . Choiceless Awareness of Jidddu Krishnamurthy , United Nations Peace Medal Awardee is also needed for deconditioning the mind towards sustainability as conditioned minds have fixed and habitually reinforcing selective perception . This selective perception results in selective listening and selective observation.

1.0 What is sustainable irrigation water management ?

Sustainable development is defined by United Nations World Commission on Environment and Development (Known as Brundtland Commission ,1987 –Our common future report) as development that meets the needs of the present without compromising the ability of the future generations to meet their own needs.

Sustainable Irrigation water management provides the right quantity of water for crops' water requirements in the field at right time with acceptable level of quality and duration ,preventing over-irrigation and erosion in the field . It helps in the sustainable agricultural production.

2.0 National Water Policy 2012 (www.wrmin.nic.in)

Clause 6 deals with demand management and water use efficiency . The Project and the basin water use efficiencies need to be improved through continuous water balance and water accounting studies.Clause 4 deals with the adaptation to climate change.

3.0 Sixth Sustainable development goal of United Nations on " Sustainable Water Management with sanitation" (www.un.org/sustainabledevelopment/sustainable development goals)

According to United Nations,approximately seventy percent of all water abstracted from rivers, lakes and aquifers is used for irrigation.

There are seventeen sustainable development goals.Goal 6 is to ensure access to cleanwater and sanitation for all. There are eight targets to be reached by 2030.

4.0 Freely available vast resources in the area of Irrigation Water Management

The resources in irrigation management available are from Bureau of Indian Standards (BIS), Food and Agricuture Organisation(FAO) of United Nations, International Water Management Institute(IWMI),Utah State University Open courseware in the department of Biological and Irrigation Engineering ,National Institute of Hydrology(NIH) , National Water Academy (NWA),International Crops Research in Semiarid Tropics (ICRISAT),United states Geological Survey (USGS),International Network of Participatory Irrigation Management (INPIM), National Project on Technology Enhanced Learning (NPTEL) courses, MIT Open courseware, sixteen Water and Land Management Institutes (WALMIs)/Irrigation Research Institutes in India , International Commission on Irrigation and Drainage (ICID) ,National Water Mission of Ministry of Water resources , River Development and Ganga Rejuvenation of Government of India & Journal of Irrigation and Drainage Engineering and Journal of Agricultural Water management . The websites of all the above are given in the references.

5.0 Practices of Sustainable Irrigation water management in the world

Professors and students in all colleges may have to be encouraged to do research and degree projects on sustainable development goals. I guided one B.Tech project in the year 2005 on " Civil Engineering Practices to meet Millennium Development Goals of United Nations by 2015".

Two B.Tech projects which the author guided were awarded NIT ,Kozhikode National Award for Best B.Tech Project in Civil Engineering&Architecture during 2003-04 and 2005-06. The names of two projects are as below:

(i) Optimal Surface Irrigation Practices to improve the Water Utilization Efficiency in Irrigation Projects : A case study in Sriramsagar Project (2003-04)

(ii) Synergic Practices of Engineering for Sustainable Development (2005-06)

Two B.Tech projects were guided in collaboration with International Crops Research in Semi-Arid Tropics (ICRISAT)

Hydrological Modelling of Watershed using SWAT (Soil and Water Assessment Tool)(2015-16)

Hydrological Modelling of Bhanur Watershed using SWAT (Soil and Water Assessment Tool) (2016-17).

Field Evaluation of Waste Water Irrigation (2012-13).

Drip Irrigation Evaluation in a farmer's field (2014-15)

Literature survey can be done referring to the above cited organizations.

6.0 Capacity Building of all stakeholders

6.1 Heuristic , Intuitive and Transcendental (HIT) Models developed by the author for Managerial Capacity Building of all stakeholders in effective participatory irrigation management

Heuristic teaching or education encourages a person to learn by discovering things for himself/herself. Intuition is the ability to know something by using feelings rather than considering the facts. Transcendental is going beyond the limits of human knowledge , experience or reason (New Oxford Advanced Learner's Dictionary)

Zero thinking system to develop perceptual skills of Irrigation scheme managers is to be developed (Mohd.Hussain,1993). All the stakeholders of sustainable development may have to be trained in Life skills by developing D.E.E.P. A.C.C.E.S.S. Implementation Quotient (Mohd.Hussain,2013)

Managerial Capacity Building of water resource managers can be done by developing **M.A.A.T.R.U.S.H.R.I.** Enlightenment Quotient (Mohd.Hussain,2006).

Capacity Building can also be imparted on "Universal B.E.S.T. Intuitive Leadership Wisdom model of "Mindfulness attention"(Mohd.Hussain,2010)

Managerial capacity building can be given on " Universal APT self-facing **Environmental Communication and Leadership Wisdom Model"(Mohd.Hussain,2011)** .

Another model for capacity building is on " Selfless-innerself-centred S.P.R.Y. Living Unlearning and Learning Practice Model in the present new minute for sustainable natural resource management"(Mohd.Hussain,2012)

Capacity Building of water resource managers and users can be imparted by developing S.E.L.F. Renewal Quotient in Participatory Irrigation Management (Mohd.Hussain, 2014)

Management capacity building for all stakeholders is to be imparted based on the following MANAGEMENT LECTURES AND TECHNICAL LECTURES delivered by the author at INSTITUTION OF ENGINEERS, A.P. CENTRE, HYDERABAD, INDIA DURING 2001-17

(i) May 4, 2001 - Active Listening Skill for managers

(ii) July 29, 2001 - Improving Communication Skill with Transactional Analysis

(iii) September 20, 2001 - Developing Holistic thinking skill of managers through the re-orientation of self-talk for improving the performance effectiveness and efficiency

(iv) Febraury 6, 2002 - Developing the spontaneous resolution skill of the Psychological conflict

(v) June 6, 2002 – Positive Assertiveness Skill for effective communication and WIN – WIN management .

(vi) November 7, 2002 - Preparation and Implementation of operation plans of all irrigation projects of India : An urgent need (Technical)

(vii) March 5, 2003 - Insights of leadership from Panchatantra

(viii) May 7, 2003 - Annual Practice of Multi-disciplinary Diagnostic Analysis (M.D.A) of irrigation systems : A key task to increase the Crop Productivity and Project efficiency (Technical)

(ix) August 29, 2003 - Anger Management and Release (A.M.A.R) skill for Psycho-biological fitness and socio-organisational fitness.

(x) December 3, 2003 - G.O.O.D. Skill (Genuine and Openminded -Organisational – Deaddicted – Behaviour Skill) to overcome the seven major addictions of 21st Century to improve the organizational productivity

(xi) July 20, 2004 - F.A.C.E. Skill (Fears, Anxieties and Chicken-heartedness Emancipation Skill) to overcome the rational fears and irrartional phobias of 21st century to improve the proactive -decision -making process.

(xii) December 13 , 2004 - Abundance Mentality for the proactive focus of WIN-WIN Management – Scarcity Mentality for the reactive focus of WIN-LOSE Management .

(xiii) January 25, 2006 - A.S.P.I.R.E. Skill to work with Emotional Intelligence for continuous development of individuals and organizations

(xiv) January 18, 2007- Attitude of Self- Enquiry, Self- Respect and Self-Acceptance for even more Effectiveness and Efficiency

(xv) December 10, 2007 – Decision Support Systems for Effective Water Resources Management in the world (Technical)

(xvi) May 15, 2008 – "Preparation and Implementation of Operation plans in Irrigation Projects of India : An urgent need" at Vijayawada Local Centre of Institution of Engineers (Technical)

(xvii) December 7,2010 – " Role of civil Engineers to achieve United Nations Millennium Development Goals by the year 2015" at Institution of Engineers,Hydrabad (Technical)

(xviii) November19,2011-"Engineering Management Practices for the sustainability of quality and quantity of water resources under climate change",Invited lecture in one day workshop on "Recent Advances in Climatic Modeling for water resources planning and management" at BITS-PILANI-HYD.

(xix) March,21 2012-""Engineering Management Practices for the sustainability of quality and quantity of water resources in 21st century",Invited lecture in one day workshop on "Recent Trends in Civil Engineering" at JBIT-HYD.

(xx) September ,4,2014-"Preparation and Implementation of Operation Plans in Irrigation Projects to increase Water Use Efficiency :A case study in India",Invited talk at seminar on Institution of Engineers(India),AP Centre.

(xxi) May , 15 2017 –"Preparation and implementation of operation plans as sustainable water management practice in irrigation projects to increase water use efficiency – A case study in India", Paper presented at seminar on "Sustainable Water Management" at Institution of Engineers , Telengana state Centre , Hyderabad.

(xxii) October , 31, 2017 – "Seventeen Sustainable Development Goals of United Nations to be achieved by 2030 : Action Planning and Implementation for Engineers with a particular reference to Civil Engineers to improve the World Happiness Index", Presented at Institution of Engineers , Telengana state Centre , Hyderabad.

(xxiii) December 21 ,2016 "Seventeen sustainable development goals of united nations to be achieved by 2030, action planning and implementations for civil engineers" Er, R. L. Raju 2nd endowment lecture on 21st Dec 2016 organized by Institute of Engineers (INDIA), Telangana state centre.

(xxiv) 6 January 2017 - "Codal provisions & specifications to mitigate natural disasters" in All India seminar on "Disaster-Mitigation of floods & Urban drainage issues" held at the Institution of Engineers (INDIA), Telangana state Centre on 6 & 7th Jan 2017.

6.2 Technical Capacity Building for all stakeholders

Training modules are available in Technical Capacity Building for Engineers, Agricultural officers, Workinspectors/Laskars and Farmers in Irrigation Management developed by Government of India under National Water Management Project.

Training modules need to be developed separately for Engineers, Agricultural Officers, Workinspectors/Laskars and farmers in Sustainable Irrigation Management under Climatic Change.

7.0 Action plan for implementation of sustainable irrigation water management in all Irrigation Schemes of the world

(i) Flow measurement in canals and rivers are required and necessary infrastructure is to be developed.

(ii) Design operation plans of Irrigation schemes are to be prepared and seasonal operation plans are to be evolved from them as per the principles of National Water Management Project. Reservoir Operation plans are to be developed.

(iii) Management Information systems for effective management of water resources are to be developed.

(iv) Conjunctive use Management of groundwater and surface water is be encouraged.

(v) Four waters Concept of Former United Nations Consultant Er.T.Hanumantha Rao is to be implemented in all watersheds

(vi) Where there is feasibility,interlinking of rivers is to be undertaken.

(vii) (vii)Diagnostic analysis of Irrigation systems is to be annually undertaken

(viii) Capacity building of all concerned with the water resources management is to be undertaken regularly.

(ix) River Basin Planning and management is to be made more scientific for sustainability

(x) Measures are to be taken to improve groundwater recharge

(xi) Participatory management culture is to be nurtured with international coordination.

(xii) Heuristic, Intuitive and Transcendental (HIT) models developed by the author are to be implemented for managerial capacity building of all stakeholders.

(xiii) Awareness to calculate Ecological footprint ,Water footprint and Carbon footprint is to be created (www.footprintnetwork.org) at individual , community, state and nation levels.

8.0 Cultural Models of Attitudes of sustainable management

Human resource development demands knowledge and skills to help people improve their economic performance. Sustainable development requires changes in values and attitudes towards environment and development - indeed, towards society and work at home, on farms, and in factories. **The world's religions could help provide direction and motivation** in forming new values that would stress individual and joint responsibility towards the environment and towards nurturing harmony between humanity and environment (**United Nations Brundtland Commission report ,1987 , known as "Our Common Future – From one Earth to One World "**).

Supramental transformation and evolution of Integral yoga of Sri Aurobindo is needed for sustainable development. Integral yoga is of triple aspects of Aspiration of the heart with the silent mind, Rejection of the movements of the lower nature & Surrender of oneself and all one is and one has to the divine shakti(www.sriaurobindoashram.org). Attitude of Choiceless Awareness of Jidddu Krishnamurthy , United Nations Peace Medal Awardee is also needed for sustainability(www.jkrishnamurti.org).

REFERENCES

1. Mohd. Hussain(2006),"Capacity Building of water resource managers by developing M.A.A.T.R.U.S.H.R.I. Enlightenment Quotient in the successful implementation of Participatory management of water resources towards sustainable development", Paper presented in the "International Conference on Hydrology and Watershed Management", December 5-8,2006 by Jawaharlal Nehru Technological University, Hyderabad.
2. Mohd. Hussain(2010),"Universal B.E.S.T. Intuitive Leadership Wisdom model of "Mindfulness attention"- Paper presented in the "International Conference on Hydrology and Watershed Management",February,3-6,2010 by Jawaharlal Nehru Technological University, Hyderabad
3. Mohd. Hussain(2011),"Preliminary aspects of universal APT self-facing Environmental Communication and Leadership Wisdom Model ",Paper submitted in one day national seminar on "Sustainable Technologies in Civil Engineering: Perspectives and Strategies(STEPS-2011) on 17th December,2011
4. Mohd.Hussain(2012),"Preliminary aspects of selfless-innerself-centred S.P.R.Y. Living Unlearning and Learning Practice Model in the present new minute for sustainable natural resource management", Paper submitted in one day national seminar on "Sustainable Technologies in Civil Engineering:Perspectives and Strategies(STEPS-2012) on 19th December,2012
5. Mohd.Hussain(2014),"Capacity Building of water resource managers and users by developing S.E.L.F. Renewal Quotient in Participatory Irrigation Management",Paper presented in the International conference on "HYDROLOGY AND WATERSHED MANAGEMENT",organized by Centre for Water Resources of Institute of Science and Technology(IST) of JNTUH from 29 October t0 I November 2014.
6. Mohd.Hussain(1993),"Developing perceptual skills of Irrigation Scheme Managers in Conflict Resolution in Social and Psycho-biological Perspective", National Seminar on Human Resource Development in Irrigation Management at WALAMTARI, Hyderabad, September 3-4, 1993.

Bureau of Indian Standards (BIS),www.bis.gov.in

FAO – www.fao.org

www.iwmi.cgiar.org

ocw.usu.edu

nihroorkee.gov.in

nwa.mah.nic.in

www.icrisat.org

www.usgs.gov

www.inpim.org

nptel.ac.in

ocw.mit.edu

Sixteen WALMIs/Irrigation Research Institutes - nwm.gov.in

www.icid.org

National Water Mission- nwm.gov.in

Ascelibrary.org/journal/jidedh

www.journals.elsevier.com/agricultural-water-management

ROLE OF CLIMATE CHANGE ON "TROPICAL STORMS-HURRICANES-TYPHOONS"

S. Jeevananda Reddy

Formerly Chief Technical Advisor – WMO/UN & Expert – FAO/UN
Fellow, Telangana Academy of Sciences [Founder Member]
jeevanandareddy@yahoo.com; jeevananda_reddy@yahoo.com

ABSTRACT

Hurricane Harvey followed by a string of Hurricanes Irma, Jose and Katia in the North Atlantic basin in 2017, has triggered questions on linkage between hurricanes and global warming. The concept is a warmer earth will generate stronger and wetter hurricanes. There is also science, which shows colder world is a stormier world. Classical examples to these are the pre-monsoon summer storms and the post-monsoon winter-storms (Northeast Monsoon season) in India. In the Southern Oscillation generally speaking, droughts are associated with El Nino (warmer) and floods are associated with La Nina (cooler). Before 1997-98 and 2014-16 El Nino events there have been clear cut pauses in global average temperature with no trend. In the satellite era, the global temperature presented clear cut positive and negative peaks with near zero trends in association, respectively, with El Nino and La Nina & Volcanic eruption activity periods. Also, many ignore, misrepresent, or exaggerate the science. There is well-established data about these matters. While doing so, they did not even care to look in to the historical pattern in the occurrence of hurricanes in US. Some counters this by saying global warming is acting through sea level rise in this hurricane belt of US. However, the sea level change in this belt is associated with several localized factors such as "land subsidence" in association with extraction of oil, gas, water, etc. The Greenland ice, so far the lowest in summer was recorded in 2012 and since then it is fluctuating between the mean and 2012 minima. Many a time the word "climate change" is used as de-facto "global warming", which is in reality not true. Climate change is a vast subject. This article looks into these aspects in brief to get the clarity on the role of global warming/climate change on Tropical Storms-Hurricanes-Typhoons.

Keywords: *Climate Change, Global Warming, Hurricanes, Typhoons, Tropical storms, Southern Oscillation/El Nino/La Nina, Land Subsidence*

INTRODUCTION

Natural disasters were once entirely natural. As Hurricane Irma toward Florida, Hurricanes Jose and Katia churn in the Atlantic Ocean and the Gulf of Mexico, and Houston picks up after Hurricane Harvey, the familiar question is being raised: what role does climate change play in all this volatility? Climate scientists say that while a definitive answer is impossible, because the climate is unfathomably complex, however other groups points toward global warming having an impact on the formation and severity of hurricanes without any scientific reasoning.

Most of the stories were not based on data or any kind of quantitative scientific analysis, but a hand-waving argument that a warming earth will put more water vapor into the atmosphere and thus precipitation will increase. A few suggested that a warming atmosphere will cause hurricanes to move more slowly. Model projections of hurricane frequency and intensity are based on climate models. However, none have shown skill at predicting past variations in hurricane activity over years, decades, and longer periods. Much of the propaganda about Harvey and Irma has been directed at Trump. Here, they failed to look in to the historical perspective of Hurricanes or for that matter Typhoons and Tropical Storms. Larry Kummer from Fabius Maximus website: "Millions of words were expended reporting about Hurricanes Harvey and Irma, but too little about the science connecting them to climate change".

Tony Abbott, the former Prime Minister of Australia, delivering a talk at Global Warming Policy Foundation preferred "evidence-based policy rather than policy-based evidence"; and on a 2013 study that showed that 97% of scientists agree humans are driving climate change, he noted that "as if scientific truth is determined by votes rather than facts". He also referred his 2009 assertion that the "so-called settled science of climate change" was "absolute crap". It is pertinent to see two contrasting reports, namely IPCC and NIPCC [Non-IPCC] where in the former the authors involved in writing the report cited their publications more frequently in the compilation; while in the later report such cases were very few. Also, the former give little importance to reviewers' comments and cited journals on the modeling and the later cited journals on paleo-sciences. This lead IPCC to withdrew some of their conclusions.

A new report state that "Billions of dollars of public money was sunk in new fossil fuel projects by the world's major development banks in the year after the Paris climate change deal was agreed, according to campaigners who are calling for the banks to halt their financing of coal, oil and gas". This was the case even before Donald Trump, the US President, has expressed desire to withdraw US from the Paris Agreement. According to a survey by Chapman University [US] the top ten fears of Americans in 2017 are: (1) Corruption of government officials (same top fear as 2015 and 2016); (2) American Healthcare Act/Trumpcare (new fear); (3) Pollution of oceans, rivers and lakes (new in top 10); (4) Pollution of drinking water (new in top 10); (5) Not having enough money in the future; (6) High medical bills; (7) The U.S. will be involved in another r world war (new fear); (8) Global warming and climate change; (9) North Korea using weapons (new fear); (10) Air pollution. We must not forget the fact that high medical bills are associated with pollution. Unfortunately world media and UN agencies are giving hype with billions of dollars of green fund to global warming at the cost of pollution [air, water, soil & food]. Pollution is the major cause for concern in urban India as governments are paying lip-sympathy. I have been advocating this for the last two decades or so.

TROPICAL STORMS-HURRICANES-TYPHOONS

General features

A tropical cyclone is a rapidly rotating "storm system" characterized by a low-pressure center, a closed low-level atmospheric circulation, strong winds, and a spiral arrangement of thunderstorms that produce heavy rain. Tropical refers to the geographical origin of these systems, which form almost exclusively over tropical seas. Thus, cyclones refers to their winds moving in a circle, whirling round their central clear eye, with their winds blowing counterclockwise in the Northern Hemisphere and blowing clockwise in the Southern Hemisphere. The opposite direction of circulation is due to the Coriolis-effect. Depending on its location and strength, a tropical cyclone is referred to by names such as hurricane, typhoon, tropical storm, cyclonic storm, tropical depression and simply cyclone.

A hurricane is a tropical cyclone that occurs in the Atlantic Ocean and northeastern Pacific Ocean; a typhoon occurs in the northwestern Pacific Ocean; and a cyclone occurs in the south Pacific or Indian Ocean – Bay of Bengal & Arabian Sea are part of it. The energy source differs from that of mid-latitude cyclonic storms, such as Nor'easters and European Windstorms, which are fueled primarily by horizontal temperature contrasts/gradients. It is not as simple as this but highly complicated based on the location where they form and season of formation and or general circulation pattern prevailing in that zone and topography [components of "Climate System" as defined by IPCC AR5].

The strong rotating winds of a tropical cyclone are a result of the conservation of angular momentum imparted by the Earth's rotation as air flows inwards toward the axis of rotation. As a result, they rarely form within 5° of the equator. They are far less common south of the Equator, mainly because the African easterly jet. Warmer waters, and areas of atmospheric instability, which gives rise to cyclones in the Atlantic ocean and Americas, occur in the Northern hemisphere. Because of vertical wind shear is much stronger south of the equator, which typically prevents tropical depressions and potential storms from developing into cyclones.

Coastal regions are particularly vulnerable to the impact of a tropical cyclone, compared to inland regions. Therefore these forms are typically strongest when over or near water, and weaken quite rapidly over land. Coastal damage may be caused by strong winds and rain, high waves (due to winds), storm surges (due to severe pressure changes), and the potential of spawning tornadoes. Tropical cyclones also draw in air from a large area—which can be a vast area for the most severe cyclones—and concentrate the precipitation of the water content in that air (made up from atmospheric moisture and moisture evaporated from water) into a much smaller area. This continual replacement of moisture-bearing air by new moisture-bearing air after its moisture has fallen as rain may cause extremely heavy rain and river flooding up to 40 kilometers (25 mi) from the coastline, far beyond the amount of water that the local atmosphere holds at any one time.

Historical facts with reference to US

After major hurricanes Harvey and Irma made landfall in the United States in 2017, there were renewed calls to do something about global warming. The popular perception that land falling hurricanes in the US are becoming more frequent or more severe, however, is shown to be incorrect. History has demonstrated that major hurricanes, sometimes arriving in pairs, have been part of Atlantic and Gulf coastal life for centuries. Even lake-bottom sediments in Texas and Florida reveal more catastrophic hurricane landfalls 1,000 to 2,000 years ago than have happened more recently. Over the last 150 years, the number of major hurricanes hitting Texas has been the same when Gulf of Mexico water temperatures were below normal as when they were above normal. Harvey's record-setting rainfall totals were due to its slow movement, which cannot be traced to global warming (August 2017 was quite cool over most of the US). IPCC state on climate change and hurricanes as "Current data sets indicate no significant observed trends in global tropical cyclone frequency over the past century --- No robust trends in annual numbers of tropical storms, hurricanes and major hurricanes counts have been identified over the past 100 years in North Atlantic basin – In summary, confidence in large scale changes in the intensity of extreme extra-tropical cyclones since 1900 is low ---" [source:

http://rogerpielkejr.blogspot.com.au/2013/10/ coverage-of-extreme-events-in-ipcc-ar5.html]. Number of recorded storms affected US showed a range of 25 in 1880s to 12 in 1970s. Currently in 2010s they are up to now only 8. During 1860s, 1920s, 1960s and 1990s they were 15 in number.

Figure 1 presents the Hurricane Season, showing Atlantic Storms since 1913 between June 1 and November 30. The figure showed a clear normal distribution, peaking at September 10 for combined number of tropical storms and hurricanes and as well hurricanes alone [Source: NOAA/NWS – National Hurricane Centre].

Disaster-damage events

Historically, the central pressure was the predominate factor in determining the strength of a hurricane. Today that policy has changed and now hurricanes are ranked exclusively by the wind. The normal pressure in the U.S. is usually between 1010 and 1030 mb. In the tropics if the pressure drops below 1000 mb, it generally means a Cat 1 hurricane has formed. The pressure in a major Cat 3 hurricane is usually around 950 mb and a Cat 5 occurs when the pressure is below 920 mb. When the pressure drops below 900 mb, you have a super hurricane comparable to the most intense Pacific typhoons.

Table 1 presents all hurricanes with landfall pressures <= 940 mb at time of US land fall. The lowest pressure ever recorded in an Atlantic hurricane was 882 mb while Wilma was in the northwest Caribbean Sea in 2005. The lowest pressure for a land falling hurricane was 892 mb when the 1935 hurricane crossed the Florida Keys. There have been 10 hurricanes with central pressures below 910 mb of which 5 were below 900 mb. Irma did not even make the top 10; therefore, it was not close to being the strongest hurricane ever observed In the Atlantic. Irma was 929 mb and Harvey was 938 mb. With Irma ranked 7th, and Harvey ranked 18th, it's going to be tough for climate alarmists to try connecting these two storms to being driven by CO_2/global warming.

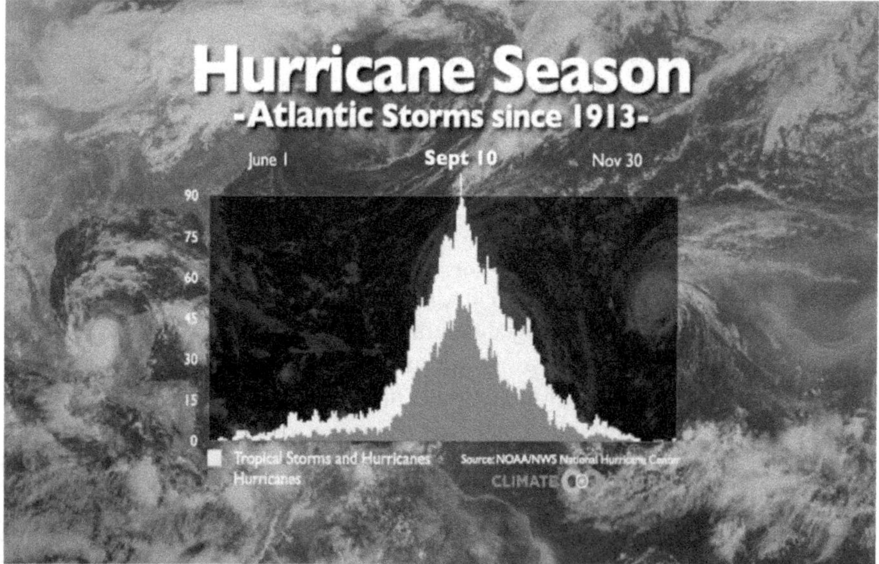

Fig. 1 Hurricane Season – Atlantic Storms since 1913

Figure 2 presents the temporal variation of Atlantic Basin Hurricane Counts [1851-2006] – 5-year running means – for (a) major hurricanes, (b) hurricanes & (c) US land falling hurricanes along with 60-year cycle projection.

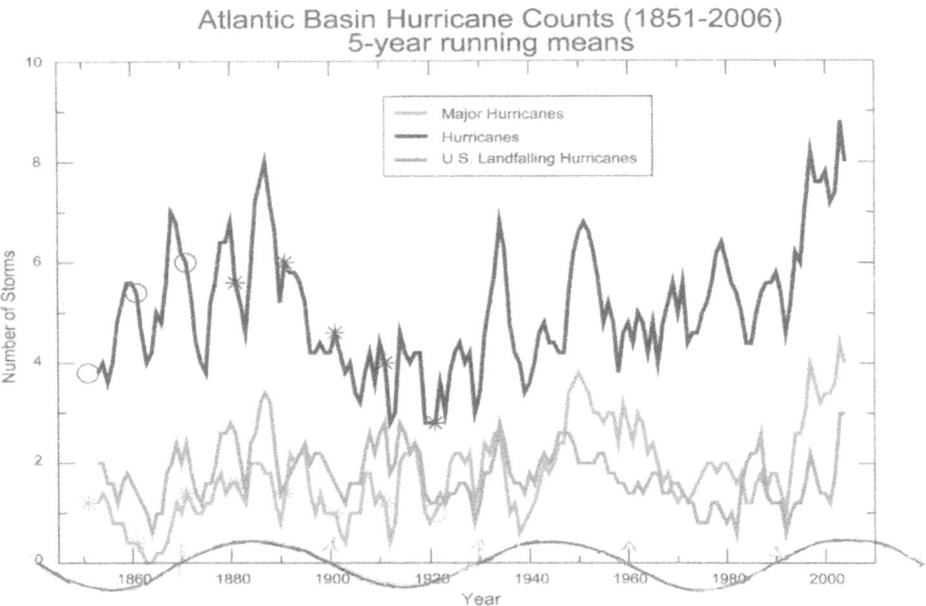

Figure 9b: Five-year Running mean of Atlantic Basin Hurricanes during 1851- 2006

Fig. 2 Atlantic Basin Hurricane Counts [1851-2006] – 5-year running means – for (a) major hurricanes, (b) hurricanes & (c) U.S. land falling hurricanes

But what *has* changed is the number of people and amount of infrastructure at risk along the Atlantic and Gulf of Mexico coastlines. Before 1900, there were virtually no people residing in Florida. Now its population exceeds 20 million. Miami was incorporated in 1896...with only 300 people. Even if there is no long term change in hurricane activity, hurricane damage will increase as coastal development increases. The 1938 New England Hurricane (also referred to as the Great New England Hurricane and Long Island Express) was one of the deadliest and most destructive tropical cyclones to strike Long Island, New York and New England. In addition, it's the fastest tropical cyclone on record worldwide attaining a maximum speed of 70 mph. The storm formed near the coast of Africa on September 9, becoming a Category 5 hurricane on the Saffir-Simpson Hurricane Scale before making landfall as a Category 3 hurricane on Long Island on September 21. It is estimated that the hurricane killed 682 people, damaged or destroyed more than 57,000 homes, and caused property losses estimated at US$306 million ($4.7 billion in 2017).

Table 1 All continental U.S. Hurricanes with <= 940 mb pressure at US landfall

All Continental U.S. Hurricanes with <= 940 mb pressure at US Landfall					
Rank	Year	Storm Name	Landfall Wind (kt)	Landfall Pressure (mb)	SS Category
1	1935	Labor Day	160	892	5
2	1969	Camille	150	900	5
3	2005	Katrina	110	920	3
4	1992	Andrew	145	922	5
5	1886	Indianola	130	925	4
6	1919	Florida Keys	130	927	4
T-7	1928	Lake Okeechobee	125	929	4
T-7	**2017**	**Irma**	**115**	**929**	**4**
T-9	1926	Great Miami	125	930	4
T-9	1960	Donna	125	930	4
11	1961	Carla	125	931	4
12	1916	Texas	115	932	4
T-13	1856	Last Island	130	934	4
T-13	1989	Hugo	120	934	4
15	1932	Freeport	130	935	4
16	1900	Galveston	120	936	4
17	2005	Rita	100	937	3
T-18	1898	Georgia	115	938	4
T-18	1954	Hazel	115	938	4
T-18	2017	Harvey	115	938	4
T-21	1915	Galveston	115	940	4
T-21	1933	Cuba-Brownsville	110	940	3
T-21	1948	Sept. 1948 Florida	115	940	4

Report says that "the two major hurricane strikes endured by the Massachusetts Bay Colony, in 1635 and in 1675, have yet to be rivaled in more modern times. Major hurricane Maria of Dominica and Guadeloupe, is probably no match for the Great Hurricane of 1780 in the Caribbean, which had estimated winds of 200 mph and killed 20,000 people". Regarding Hurricane Irma which terrorized Florida, one may be surprised to learn that it is consistent with a downward trend in both the number and intensity of land falling major Florida hurricanes [**Figure 3**]. Damage from hurricanes has certainly increased over the years. But that is because far more people now live and work in far more expensive communities along America's Atlantic and Gulf coasts. Since 1920, Greater Houston has grown from 138,000 people to 5.7 million; Miami from 43,000 to 6.1 million. Meanwhile, death tolls have declined – at least in countries where fossil fuels, highways and modern technologies enable them to construct stronger buildings, track storms, warn, evacuate and rescue people, and bring in water, food, clothing, and materials to rebuild power lines and buildings in stricken areas. Over 6,000 people perished in the 1900 Category 4 Galveston Hurricane, 2,500 in the 1928 Okeechobee,

Florida Category 4 hurricane and storm surge. More than 1,800 died in Katrina (Category 3), due largely to corrupt and incompetent local and state governments. Thanks to better preparation, warning and evacuation, overall tragic deaths were kept to 82 from Harvey and 93 from Irma. Incredibly, despite the vicious 185-mph winds that reduced most of Anguilla and Barbuda to rubble, Irma killed only one person on those Caribbean islands. Even in recent years, cyclones and hurricanes have brought far more death and destruction to poor nations where modern energy and technology are still limited or nonexistent: 400,000 dead in Bangladesh in 1970, 138,000 in Myanmar in 2008, and 19,000 from Hurricane Mitch in Central America in 1998.

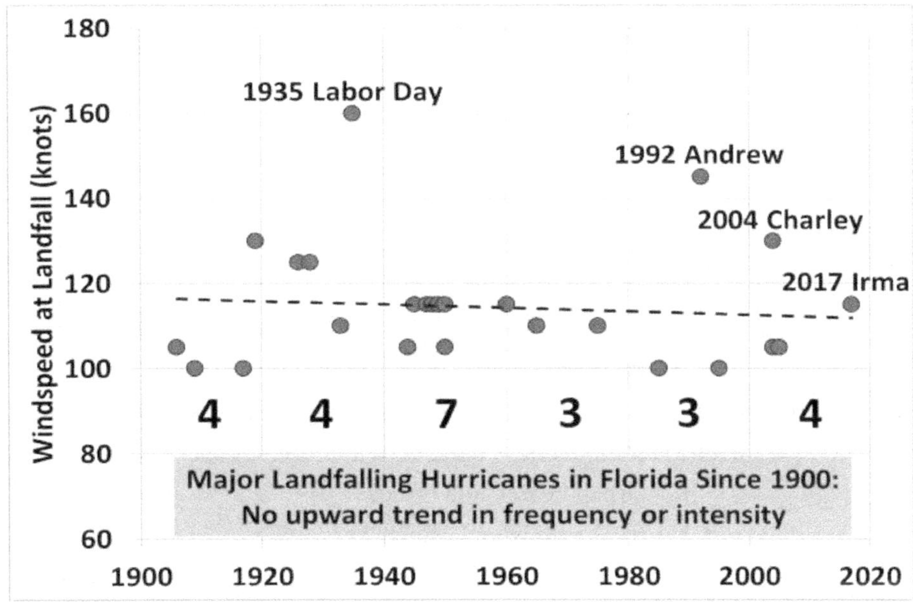

Fig. 3 Major land falling hurricanes in Florida since 1900

It is reported that when the 1921 hurricane hit Tampa, the population was 10,000. Today, it's 3 million. Perhaps the part of "climate" we should be questioning is our penchant for moving gigantic numbers of people onto a peninsula built on limestone and mangrove swamps, where the land is subsiding in the cross-hairs of most every Atlantic cyclone that goes by. Maybe intelligent Conservatives can start framing "climate" as a matter of intelligent building, planning, and insuring instead of futile attempts to propitiate the Angry Gods by a return to Stone-Age living.

CLIMATE CHANGE versus GLOBAL WARMING

What is climate change?

IPCC's AR3 defined climate change as "a statistically significant variation in either the mean state of the climate or in its variability, persisting for an extended period [typically decades or longer]. Climate change may be due to natural internal processes or external forcings or to persistent anthropogenic changes in the composition of the atmosphere or in land use". That means, according to IPCC, climate change can occur naturally or from man-made causes.

UN Framework Convention on climate change [UNFCCC], in its Article 1, defined climate change as "a change of climate which is attributed directly or indirectly to human activity that alters the composition of the global atmosphere and which is in addition to natural climate variability observed over comparable time periods". From these definitions it is clear that climate change consists of several components in addition to global warming component. WMO presented a manual in 1966 to separate natural from manmade. From all these climate change is given as:

1. Natural Variability – beyond human control
    a. Irregular variations
        i. intra-seasonal
        ii. intra-annual variability
    b. Systematic Variations
        i. cyclic variations -- fluctuations
2. Man-induced changes -- represented by trend
    a. Greenhouse Effect [> half of the trend]
        i. Human component -- Global warming – anthropogenic GHG
        ii. Non-human component – volcanic activity, aerosols, etc
    b. Non-Greenhouse Effect – [< half of the trend] -- Ecological Changes
        i. Human component -- Land & Water use & cover changes
            1. Urban heat island effect
            2. rural cold island effect

Here we must keep in mind the fact that the climate is not limited to temperature alone but covers all meteorological parameters including precipitation. See for more details **Reddy (2016).**

What is Global Warming?

From the above it is clear that the increased trend in global average temperature anomaly since 1951 has two components, namely (2a) caused by the greenhouse effect and (2b) caused by non-greenhouse effect [this was there even before 1951]. According to IPCC AR5, that 2a's contribution to the global average temperature trend since 1951 is "extremely likely that more than half" – remember the fact that 50.1% is also more than half; and thus 2b's contribution is less than half. 2a includes human component (2ai) namely global warming caused by anthropogenic greenhouse gases addition to the Atmosphere; and natural/human component related to volcanoes and other aerosols components (2aii). 2b includes human induced components such as urban-heat-island (2bi1) and rural-cold-island (2bi2) effects in association with the ecological changes. From this it is clear that so far ***global warming has not been quantified***. Volumes and volumes were written by thousands of people on radiative forcings that links CO_2 with temperature, known as climate sensitivity factor. IPCC reduced the climate sensitivity factor monotonically from SER to AR5. In AR4 it was 1.95 and in AR5 it is 1.55, though, warmists claimed it as settled science. For doubling of CO_2, the temperature may range between 1.5 to 4.5 °C with a mean of 3.0 °C???

The models actually, inadvertently present one of the strongest disproofs of man-made global warming. **Figure 4a** presents the simplified implication of a figure in IPCC AR5. The colored lines represent the range of results for the models and observations. The trends here represent trends at different levels of the tropical atmosphere from the surface up to 50,000 ft. The gray lines are the bounds for the range of observations, the blue for the range of IPCC model results without extra GHGs and the red for IPCC model results with extra GHGs. The key point displayed is the lack of overlap between the GHG model results (red) and the observations (gray). The non-GHG model runs (blue) overlap the observations almost completely. [Source: https://rclutz.wordpress.com/2017/09/09/warming-from-co2-unlikely/]. It is clear from this that "Climate Sensitivity factor must be far less than 1.55 of AR5 of IPCC and it must not be a constant factor but must be a non-linearly decreasing.

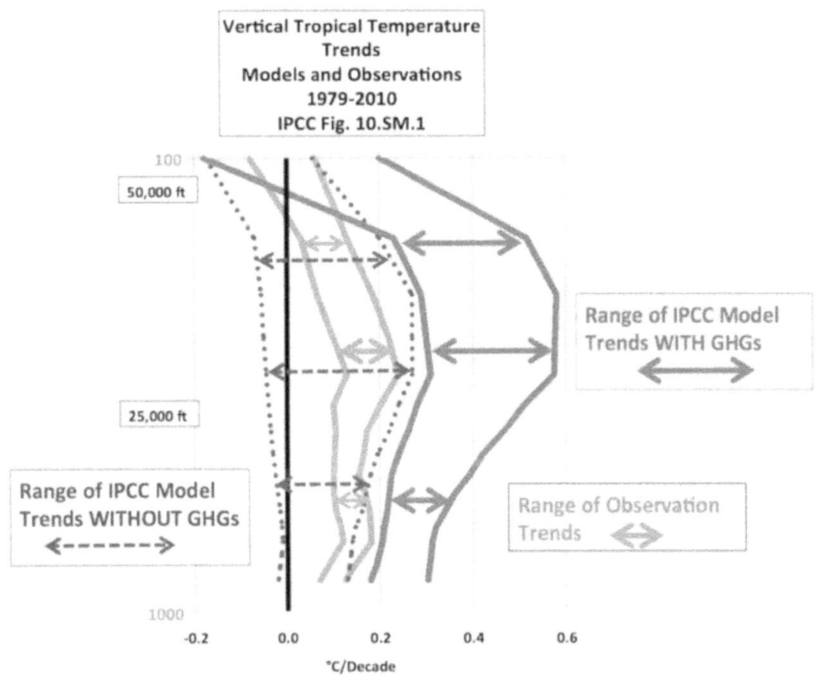

Fig. 4a Vertical Tropical Temperature Trends [Models and Observations, 1979-2010, IPCC Fig. 10.SM.1]

Pause in Global temperature trend

Reports published in September 2017 indicated a pause in global average temperature anomaly trend. They put forth several causes for such a pause. There is no statistical significance in the rate of warming between periods when manmade CO_2 emissions were insignificant and once manmade CO_2 emissions became significant. **Figure 4b** presents the pattern of global average temperature anomaly during the satellite era -- [Source: http://www.drroyspencer.com/latest-global-temperatures/], which presents steep ups and downs. This clearly shows the impact of localized events such as volcanic eruption and Southern Oscillation/El Nino-La Nina on global temperature – referred to irregular variations (1a). Mount Pinatubo volcano in Philippines in June 12, 1991 presents cooler temperatures. Bali's Mount Agung precinct in eastern Indonesia is being threatened by volcanic eruption – last erupted in 1963 that suggested a dip in temperature. Thus, the only real warming in the past 40 years has come from two major El Nino natural events. Before these two events (a) 1980-1997 and (b) 2001-2015 presented zero trends [**Figure 4c** – Figure 1b divided in to two segments after deleting the 1997-98 El Nino and 2014-16 El Nino]. These are part of seasonal & annual variability and other naturally occurring events [Southern Oscillation – El Nino (warm) & La Nina (cold) and volcano eruptions.

An eminent atmospheric scientist says ocean cycles, not humans, may be behind Most Observed Climate Change -- natural cycles may be largely responsible for climate changes seen in recent decades. Natural climatic cycles, includes the well known El Nino cycle [Southern Oscillation] and the less familiar North Atlantic Oscillation [NAO] and Pacific Decadal Oscillation [PDO]. In fact, most of the changes in the global climate over the period of the instrumental record seem to have their origins in the North Atlantic. The report noted that at the start of the 20th century, the NAO pushed the global climate in to a warming phase, and in 1940 it pushed it back cooling mode. The famous pause in global warming at the start of the 21st century seems to have been instigated by the NAO too. The global average temperature data presents not only a trend but on it superposed 60-year cycle varying between -0.3 to and0.3 °C [**Reddy, 2008**].

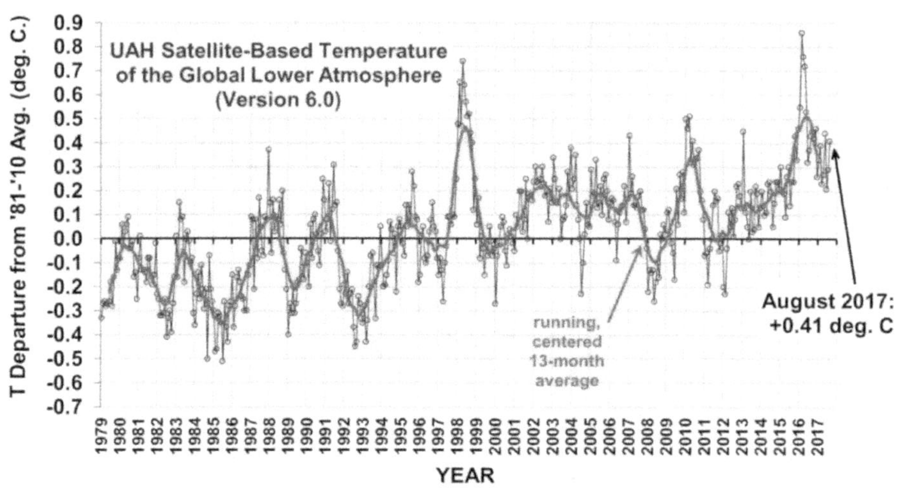

Fig. 4b Satellite-Based Temperature of the Global Lower Atmosphere

Figure 4d presents the North Atlantic Ocean Heat – present cyclic pattern. Over the past 50 years, global surface temperatures and solar output are *negatively* correlated [**Figure 4e**, meaning they're going in opposite directions. While global temperatures have risen rapidly, solar activity has slightly declined. It is clear from the figure that solar irradiance reaching the Earth has no trend but presented a cyclic variation pattern. From this it is clear that net radiation from the Earth follow this only.

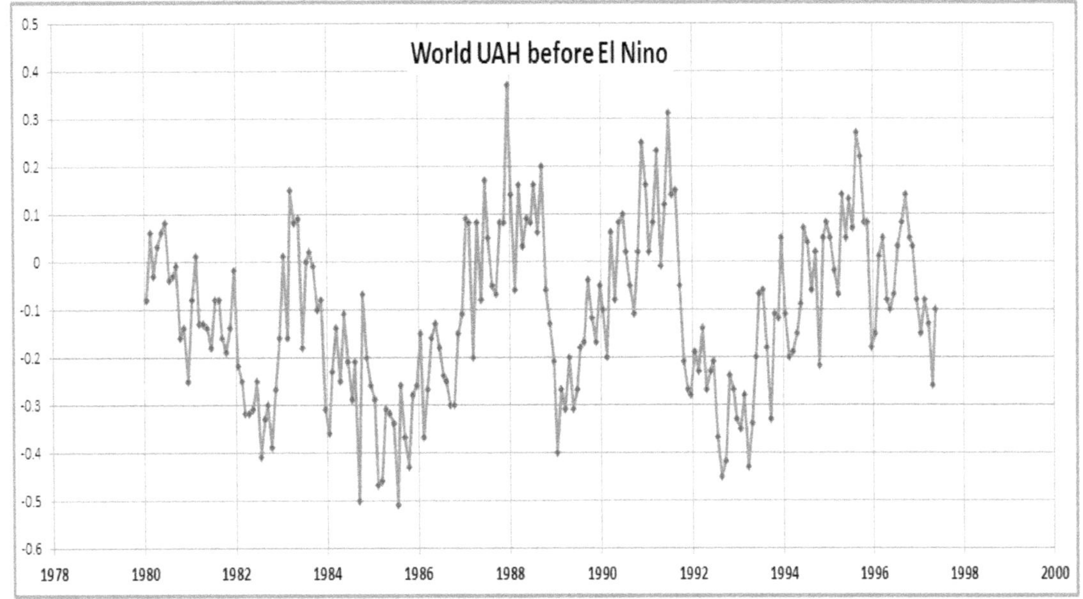

[i] No warming from 1980 – 1997

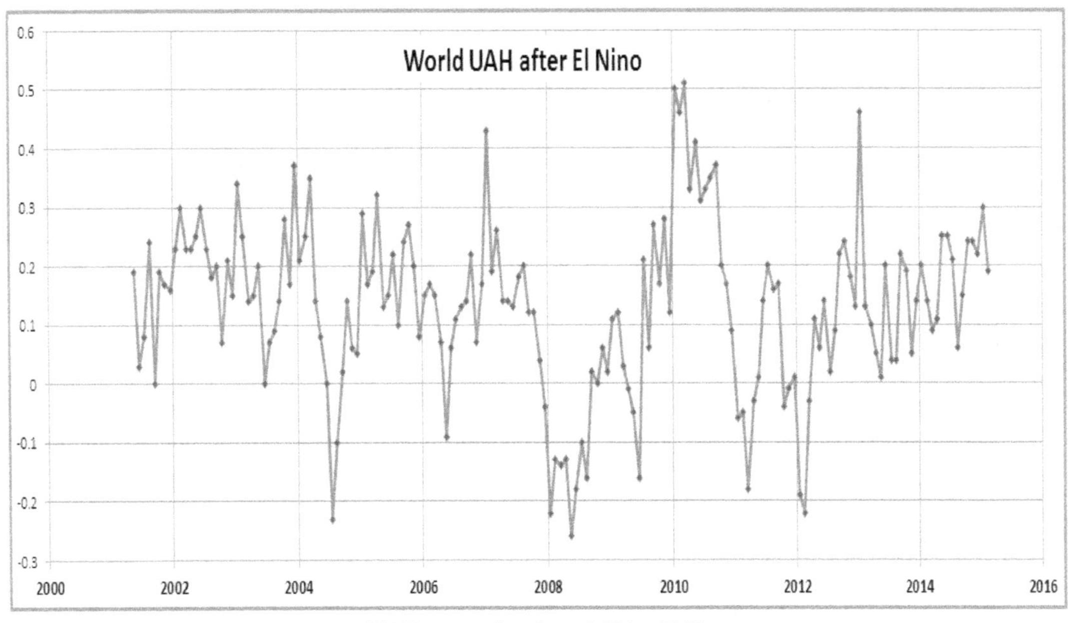

[ii] No warming from 2001 – 2015.

Fig. 4c No warming periods in global average temperature anomaly: (i) 1980-1997& (ii) 2001-2015

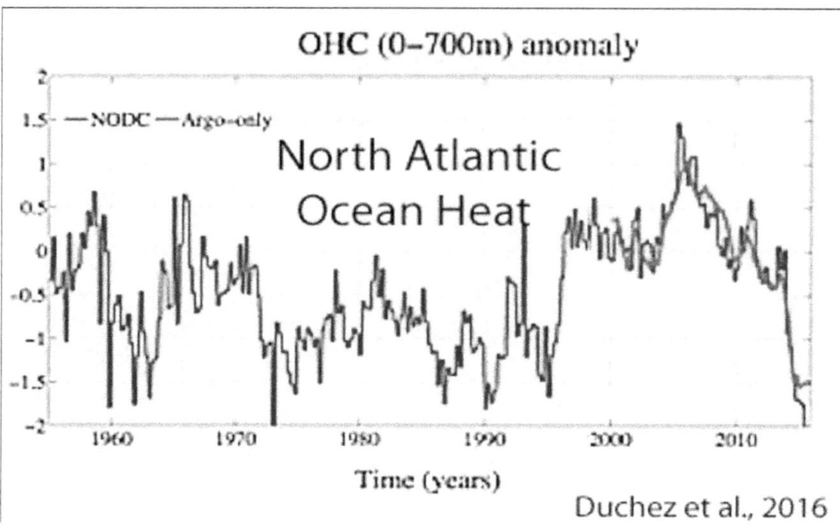

Fig. 4d Northern Atlantic Ocean Heat

Finally it could be inferred that the global warming component is insignificant to influence nature, basically because:

1. The met network, based on which the ground temperature trend built, was unevenly distributed in space in terms of countries and in terms of rural and urban areas. This is not so with the satellite data wherein the trend is insignificant [Figures 4b & c] unlike steep raise in ground data trend [Figure 4e];
2. In reality the land use change part of the trend should be either zero or negative. In ground data, the urban heat island part of this is over emphasized as the major network cover this part of the globe and under emphasized the rural cold island part which cover more than two-thirds of the globe with sparse network [Figure 4e]. This is not so in the satellite data [Figure 4b];
3. All-India annual temperature showed a 0.5 oC raise and this was mainly contributed by the night temperature. That is urban heat island part;
4. Net result is a limit on greenhouse effect which can be seen from Figure 4a wherein model estimates are far away from the satellite observational data series. By increasing anthropogenic greenhouse gases to convert in to temperature has also a limit. Thus as used by modelers, the sensitivity factor cannot be a constant that shows a linear increase in temperature with CO_2. It must be non-linear ending with plateau. The historical ocean temperature and CO_2 cyclic pattern clearly indicate that after raising the ocean temperature, CO_2 released in to the Atmosphere is high and after falling the ocean temperatures CO_2 release in to the atmosphere is low. Here temperature was the driving force for the increase or decrease of CO_2 in the atmosphere before industrialization. Global warming concept is exactly opposite to this.

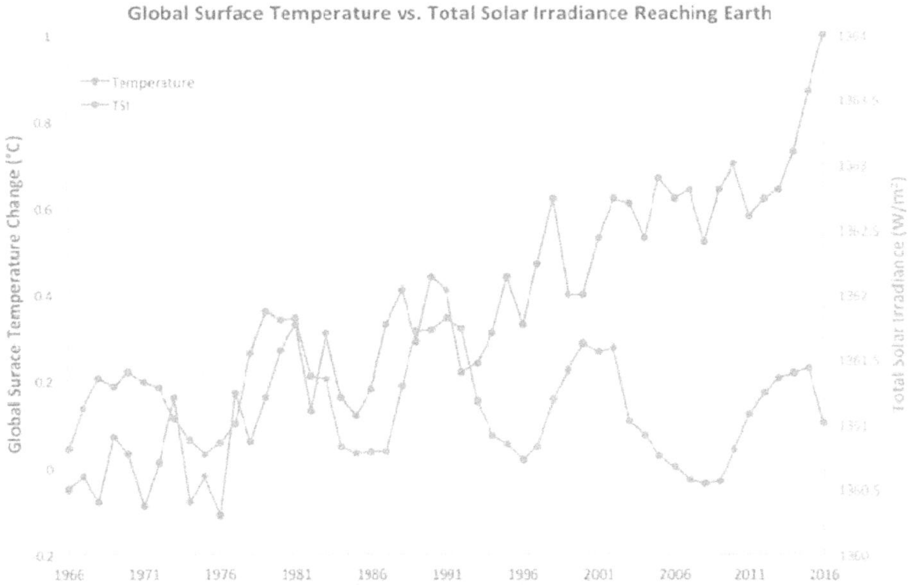

Global average surface temperature (blue - data from Nasa Goddard) versus total solar irradiance reaching Earth (orange - data from the Laboratory for Atmospheric and Space Physics Solar Irradiance Data Center at the University of Colorado). Illustration: Dana Nuccitelli

Fig. 4e Global surface temperature and total solar radiation changes during 1966 to 2016

Thus, the global warming is a science of building castles in the air; and climate change is the science of weather variations with space and time that allows development of sustainable adaptive measures. We must remember the fact that Earth's climate is dynamic and always changing through natural cycles. What we are experiencing now is part of this system.

GLOBAL WARMING versus TROPICAL STORMS

With the August-September, 2017 hurricanes, numerous articles suggest hurricanes Harvey and Irma were the result of global warming. The concept is a warmer earth will generate stronger and wetter hurricanes. There is also science, which shows colder world is a stormier world. Classical examples to these are the pre-monsoon summer storms and the post-monsoon winter-storms (Northeast Monsoon season) in India. A number of people have said Irma was the most intense hurricane in the history of the Atlantic while Harvey was the wettest and both were good examples of what we can expect in the future because of global warming. Harvey has been labeled the wettest hurricane in history; however, the 50 inches recorded in the hurricane is not related to global warming. The reason for the heavy rain is the hurricane stalled for 3 days and unfortunately southeast Texas is where that happened. The amount of rain in a tropical system is not related to the strength of the wind, it depends on the forward speed of motion. If a tropical system is moving 10 mph, expect 10 inches of rain, 20 inches for a system moving 5 mph and if the forward speed is only 2 mph be prepared for 50 inches. That is exactly what happened in Harvey. The hurricane was moving around 2 mph for 3 days and a broad band of 40 to 50 inches of rain covered a large portion of southeast Texas and southwest Louisiana. However, the reality is that Alvin, Texas, was deluged by 43 inches of rain in 24 hours on July 24-25, 1979. That would be more impressive than 52 inches over four days. Commonly with the wind the cloud moves away and cause reduction in rain.

There are numerous examples of stalled tropical systems producing excessive rains. For example, in 1979 tropical storm Claudette stalled for 2 days and generated over 40 inches in a broad area south of downtown Houston. A year earlier, stalled tropical Storm Amelia produced 48 inches in central Texas. In 1967 slow moving Hurricane Beulah moved into in south Texas and generated between 30 and 40 inches inland from Brownsville. If there had been a rain gauge in the area east of the Bahamas where Hurricane Jose stalled for four days, I am sure it would have recorded over 60 inches. They had to come up with a record, and the only record broken by Harvey was the single storm and the four day record. For a real deluge, look at the 1926-27 Mississippi flooding that lasted one year.

The biggest problem for alarmists is there is no upward trend in hurricane frequency or intensity. NOAA doesn't think the alleged impact of anthropogenic CO_2 on storm intensity is detectable. It is premature to conclude that human activities and particularly greenhouse gases that cause global warming have already had a detectable impact on Atlantic hurricane or global tropical cyclone activity. That said, human activities may have already caused changes that are not yet detectable due to the small magnitude of the changes or observational limitations, or are not yet confidently modeled – {source: https://www.gfdl. noaa.gov/ global-warming-and-hurricanes/].

IPCC AR5 WG1, Chapter 14, p.1252: "Although projections under 21st century greenhouse warming indicate that it is likely that the global frequency of tropical cyclones will either decrease or remain essentially unchanged, concurrent with likely increase in both global mean tropical cyclone maximum wind speed and rainfall rates, there is low confidence in region-specific projections of frequency and intensity. Still, based on high-resolution modeling studies, the frequency of the most intensive physical effects, will more likely than not increase substantially in some basins under projected 21st century warming and there is maximum confidence that tropical cyclone rainfall rates will increase in every affected region."

Harvey marked the end of a record 12-year absence of Category 3-5 hurricanes hitting the US mainland. The previous 8-year record was set 1860-1869. NOAA's Hurricane Research Division counts ten Category 4-5 monsters 1920-1969 (50 years) hitting the US, but only three 1970-2016 (46 years). This year has brought two more, and the hurricane season isn't over yet. If Harvey and Irma were caused or intensified by human greenhouse gas emissions, shouldn't those gases be credited for the 12-year lull and half-century decline in Cat 4-5 land falling storms? For Irma's changed intensity and route as it reached Florida and headed north? Certainly not! Indeed, NOAA concludes that neither the frequency of North Atlantic tropical storms and hurricanes, nor their energy level, has displayed any trend since 1950. Despite slightly warmer ocean waters in some regions, global ACE levels in recent years have been at their lowest levels since the late 1970s. When the PDO is in its cyclical positive phase, the tropics, west coast of North America and our Earth overall get warmer; cooling occurs during the PDO's negative phase. The AMO also cycles between warm and cool phases, affecting regional and planetary temperatures, as well as hurricane formation, strength and duration.

If there was a "human factor" in Harvey and Irma, climate alarmists need to explain exactly where it was, how big it was and what role it played. They must present hard evidence to show that fossil fuels and carbon dioxide emissions played a significant role amid, and compared to, the hundreds of natural forces involved in these storms. Their loud rhetoric only highlights their failure and inability to do so. In fact, the Atlantic, Caribbean and Gulf of Mexico are warm enough every summer to produce major hurricanes. But need other conditions, whose origins and mechanisms are still unknown: pre-existing cyclonic circulation off the African coast, upper atmospheric calm, sea surface temperatures that change on a cyclical basis in various regions, to name just a few. The combination of all these factors – plus weather fronts and land masses along the way – determines whether a hurricane arises, how strong it gets, how long it lasts, and what track it follows.

Ophelia becoming Cat1 [11th October 2017] means that 2017 became 10 Atlantic storms in a row reached hurricane strength: Franklin, Gert, Harvey, Irma, Jose, Katia, Lee, Maria, Nate, and Ophelia. However, this is the *fourth time on record* that we have had 10 Atlantic hurricanes. This is nature doing business as usual, with her swings between boom and bust. The last time there were 10 Atlantic hurricanes was in 1893. There were also 10 Atlantic hurricanes reported in 1878 and 1886. Some argued that "for those that want to blame global warming/climate change for what is going on on 2017, they should probably explain why there were three 10 hurricane event years in a short span of time when the planet was noticeably cooler between 1850 and 1900".

Global warming impact on Hurricanes through sea level raise

Some argued that though in the past severe storms occurred, the current severity with hurricanes is contributed by global warming and this is acting through sea level raise. One study [Source: http://www.psmsl.org/data/obtaining/] presented the story of sea level raise along the Gulf Coast [**Figure 5**]. However some others countered that sea level raise might be due to groundwater pumping in the southwestern United States [See... https://geochange.er.usgs.gov/ sw/changes/ anthropogenic/ subside/].

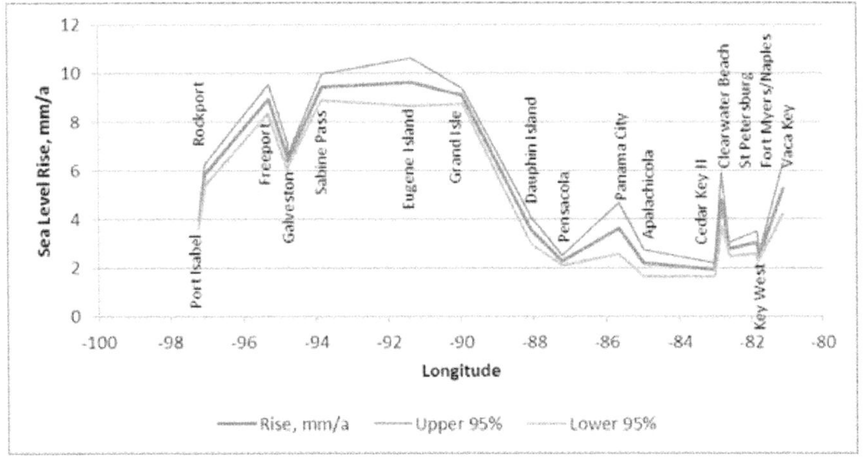

Fig. 5 Sea level rise at stations along the Gulf coast

Land subsidence is lowering land-surface elevation from changes that take place underground. Common causes of land subsidence from human activity are pumping water, oil, and gas from underground reservoirs; dissolution of limestone aquifers (sinkholes); collapse of underground mines; drainage of organic soils; and initial wetting of dry soils (hydro-compaction). Land subsidence occurs in nearly every state of the United States. In areas where climate change [natural variability] results in less precipitation and reduced surface-water supplies, communities will pump more ground water. In the southern part of the United States from states on the Gulf Coast and westward including states of New Mexico, Colorado, Arizona, Utah, Nevada and California, major aquifers include compressible clay and silt that can compact when groundwater is pumped. Also, increased population in the Southwest will increase demands on groundwater supplies, causing more land subsidence in areas already subsiding and new subsidence in areas where subsidence has not yet occurred. In the past, major subsidence areas have been in agricultural settings where groundwater has been pumped for irrigation. In the future, however, increasing population may result in subsidence problems in metropolitan areas where damage from subsidence will be great. Another argument in this direction is that the current level of the sea in the Gulf [see: http://www.nhc.noaa.gov/climo/ images/ AtlanticStorm TotalsTable.pdf] of Mexico will be high due to the circulation: [Source: http://tropic.ssec.wisc.edu /real-time/mtpw2/product.php? color_type=tpw_nrl_colors&prod= conus & timespan =24hrs& anim= html5].

Though the Global Mean Sea Level [**GMSL**] trend of pre- and post- industrialization present the same trend in sea level rise but some areas show rise, some show fall and some no change due to several local conditions. NASA's global climate change website tells **Figure 6** presents a strong correlation between Southern Oscillation and global Mean Sea Level after removing the trend. NASA's Global Climate Change Website tells the story using the satellite data [1993-present] that "For two years, since July 2015, there has been no sustained increase in Global Sea Level, in fact, it appears to have actually fallen a bit".

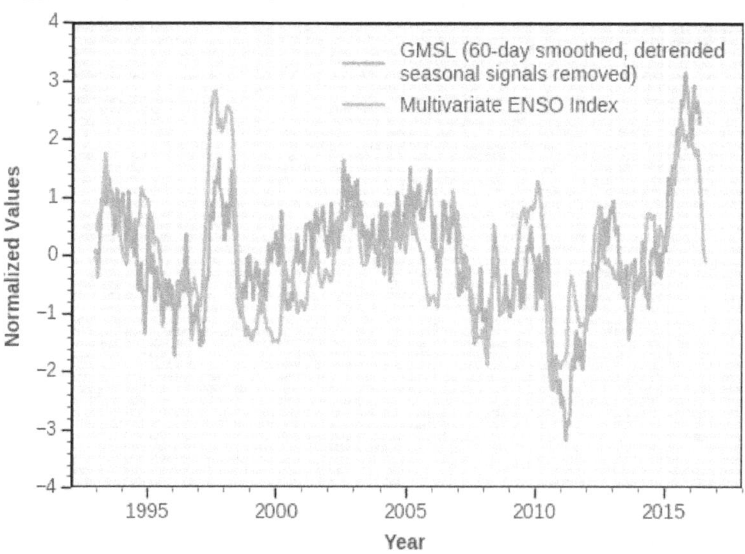

Fig. 6 The correlation between Southern Oscillation [ENSO] and Global Mean Sea Level [GMSL]

Recently, several reports argued that "The common consensus among these studies is a conclusion that future hurricanes will tend to be stronger than those in the present-day climate, assuming that sea surface temperature will continue its current warming trend into the future."

In Figure 7, red dots indicate years of major hurricane strikes in Texas, plotted on average SST departures from normal by year over the western Gulf of Mexico (25-30N, 90-100W). Note: included Hurricane Ike in 2008, which was barely below Cat3, but had a severe impact. As can be seen, major hurricanes don't really care whether the Gulf is above average or below average in temperature. [Source: http://www.drroyspencer.com/2017/08/texas-major-hurricane-intensity-not-related-to-gulf-water-temperatures/].

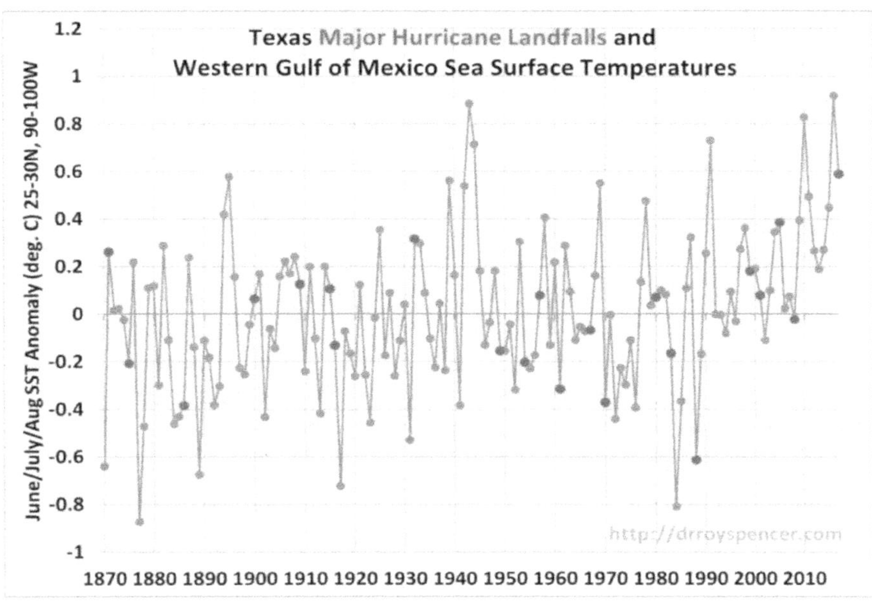

Fig. 7 Texas Major Hurricane Landfalls and Western Gulf of Mexico Sea Surface Temperatures [SST]

Glacial Melt & Arctic Ice Melt

To challenge the so-called accepted science on climate change and the impact of carbon dioxide emissions on the Earth's temperature needs change in the scientific mind set perception on climate and weather. From time to time, doomsday scenarios enter global academic and political discourses. A hallmark of such scenarios is that they rarely come true. The 'Theory of Himalayan Environmental Degradation' predicted an environmental collapse in the world's greatest mountains by the end of last millennium, threatening the life of millions of people. Fortunately, the all-encompassing crisis did not materialize.

Indian scientists [including myself, wrote to ministry of forests & environment; and sent an article to CSE's Down To Earth magazine, just before it is going for printing, announced Noble Prize to IPCC & Al Gore, and thus in its place another article of mine relating to Polavaram irrigation dam project was published] questioned IPCC's conclusion on Himalayan Glaciers melt issue. R. K. Pachauri, the then Chairman of IPCC, dismissed criticism, claim it as "voodoo science". After 2009 December Copenhagen fiasco, IPCC says the Himalayan Glaciers won't melt by 2035 & expressed regret by saying that established standards of evidence not applied properly. In 2014 a study of 2181 Himalayan Glaciers from 2000-2011 showed that 86.6% of the glaciers were not receding [this was also informed to members of parliament in the session by minister concerned after his return from Paris meet in December 2015]. Geological Survey of India monitoring few important glaciers in Himalayan region, Gangotri, is one of them feed the main river Ganga. Due to formation of fault zone the ice started receding; and now it started recovering.

Arctic sea ice, the layer of frozen seawater covering much of the Arctic Ocean and neighboring seas, is often referred to as the planet's air conditioner: its white surface bounces solar energy back to space, cooling the globe. The sea ice cap changes with the season, growing in the autumn and winter and shrinking in the spring and summer. Its minimum summertime extent typically occurs in September. NASA and the NASA-supported National Snow and Ice Data Center (NSIDC) at the University of Colorado Boulder have reported that the Arctic sea ice appeared to have reached its yearly lowest extent on Sept. 13, for 2017. Analysis of satellite data by NSIDC and NASA showed that this year's Arctic sea ice minimum extent is the eighth lowest in the consistent long-term satellite record, which began in 1978 -- below the 1981-2010 average minimum extent. 2013: 7.329 million sq km; 2008: 7.097; 2014: 6.871; 2010: 6.840; 2009: 6.681; 2015: 6.596; 2017: 6.450; 2011: 5.994; 2016: 5.652; 2007: 5.422; 2012: 5.422 -- The two record low years of 2007 and 2012 were tied. Yet 2008 summer low was about average for this period and 2013 high. On the other side of the planet, Antarctica is heading to its maximum yearly sea ice extent, which typically occurs in September or early October -- 2012, 2013 and 2014 all saw consecutive record high maximum extents.

Rainfall extremes

Along with hurricanes rainfalls, Indian media referred Mumbai Rainfall. **Table 2** presents June to September Mumbai rainfall [Normal, extremes]. From this table it is clear that current heavy rains are less than that of 1930. Media simply make statements from the air.

Table 2 Mumbai Rainfall from IMD RED BOOK [1931-60 Normal]:

Month	Normal (mm)	*	Extremes (mm) Month	24-hours
June	520.3	(2)	1103.6/1986	408.9/18th 1886
July	709.5	(2)	1499.8/1907	304.8/03rd 1923
August	439.3	(3)	1265.4/1958	287.0/03rd 1881
September	297.0	(4)	1244.9/1949	**549.1/10th 1930**

*refers to extreme month's rainfall times monthly normal rainfall

If fossil fuels caused Harvey's rainfall, were previous deluges like Hurricane Easy (45 inches in Florida, 1950), Tropical Cyclone Amelia (48 inches in Texas, 1978) and Tropical Storm Claudette (a record 43 inches in 24 hours on Alvin, Texas, July 24-25, 1979) with the result of *lower* fossil fuel use back then? That would be more impressive than 52 inches over four days. They had to come up with a record, and the only record broken by Harvey was the single storm and the four day record. For a real deluge, look at the 1926-27 Mississippi flooding that lasted one year.

SUMMARY & CONCLUSIONS

Hurricane Harvey followed by a string of Hurricanes Irma, Jose and Katia in the North Atlantic basin in 2017, has triggered questions on linkage between hurricanes and global warming. The concept is a warmer earth will generate stronger and wetter hurricanes. There is also science, which shows colder world is a stormier world. Most of the stories were not based on data or any kind of quantitative scientific analysis, but a hand-waving argument that a warming earth will put more water vapor into the atmosphere and thus precipitation will increase. A few suggested that a warming atmosphere will cause hurricanes to move more slowly. Millions of words were expended reporting about Hurricanes Harvey and Irma, but too little about the science connecting them to climate change.

Analysis of trends in Atlantic Hurricane and tropical storm counts over the past 120+ years do not support the notion that greenhouse gas-induced warming leads large increases in either tropical storm or overall hurricane numbers in the Atlantic. Also, it is not new that storms follow one after the other [2 to 5] in a sequence. Disasters associated with the storms are not always follow the intensity of the storm but depends up on several factors, more particularly localized factors.

REFERENCES

1. Reddy, S. J., 2008: "Climate Change: Myths & Realities", www.scribd.com/Google Books, 176p.
2. Reddy, S. J., 2016: "Climate Change and its Impacts: Ground Realities", BS Publications, Hyderabad, Telangana, India, 276p.

Note: ***Some of the information presented in the article was taken from the articles & comments on them presented therein under "Watts Up With That" website. I am herewith acknowledges the same.***

RANKING OF GCMS BASED ON BAYESIAN MODEL FOR INDIAN SUMMER MONSOON RAINFALL

K. Shashikanth[1] and Subimal Ghosh[2]

[1]Associate Professor, Department of Civil Engineering
University College of Engineering, Osmania University, Hyderabad-7 (TS)
[2]Associate Professor, Department of Civil Engineering
IIT Mumbai, Bombay-76 (MS)

ABSTRACT

Impacts of climate change are assessed with General Circulation Models (GCMs), which operate at coarse resolution and hence they are unable to consider local scale features, affecting the precipitation process. GCM simulations must be downscaled to finer resolution, through statistical or dynamic modelling for further use in hydrologic analysis. In this study, we use linear regression based statistical downscaling method for obtaining Indian Summer Monsoon Rainfall (ISMR). We use 20 GCMs of Coupled Model Intercomparison Project Phase 5 (CMIP5) suite and combine them with multi model averaging and Bayesian model averaging. We find spatially non-uniform projected changes at all the resolution for both combinations of projections.

1. INTRODUCTION

India receives more than 80% of its total rainfall in four monsoon months namely June, July, August and September respectively (Webster et al., 1998; Menon et al., 2013). Agriculture, water supply, industrial production and allied activities depend on the Indian summer monsoon rainfall (ISMR hereafter) for sustenance. Impacts of climate change are generally assessed by General Circulation Models (GCMs), which are the reliable tools to replicate the response of the global climate system to increasing the green house gas concentrations (*IPCC*, 2013). Therefore, regional prognosis of ISMR, under changing climate is necessary for adaptation and policy framing purpose. The plausible changes in Indian monsoon rainfall due to global warming is studied by various researchers using climate model outputs with CMIP3 and CMIP5 simulations [*Salvi et al.*, 2013; Shashikanth et al. 2013;]. Although climate model outputs from CMIP5 (Coupled Model Intercomparison Project Phase 5) suite have undergone several improvements in terms of physics and resolutions (IPCC, 2013). Still, outputs from CMIP5 models cannot be directly used for regional scale projections because of the relatively coarse resolutions simulations. Therefore, to obtain the regional projections, they must be downscaled for local scale applications. There are popularly two downscaling approaches are available viz. Dynamical downscaling, Statistical downscaling.

The Statistical downscaling models primarily are the data driven methods. Statistical models are thus predominantly used when sufficient lengths of historical data are available for establishing the requisite stable statistical relationships (Wilby et al., 2004). The statistical downscaling method is based on the development of a relationship between predictors (coarse resolution climate variables/ synoptic scale circulation patterns) and a predictand (e.g., rainfall). The statistical downscaling methodology is relatively easy to apply and has been extensively used for regional applications.

The projections of climate variables are now presented in a probabilistic framework (Raftery et al., 2005). This has led to the development of new methods like Bayesian Multimodel Averaging (BMA) (Tebaldi et al., 2004) which is a better alternative to Multi-Model Averaging (MMA), where equal weights are given to all GCMs for both past and future projections. In the BMA, differential weights are assigned to the GCMs, based on performances during the historic period and convergence (agreement with majority of ensemble members in the projection period).

We apply statistical downscaling (Fig 1), for regional projections of India at $0.25^{0.}$ Here, we use 20 GCMs from the CMIP5 suite, which is a fairly large number of models to project rainfall. The performance of the models is evaluated using PDF based skill scores (Perkins et al., 2007). Further Bayesian methodology is used to know the performance of models (Duan and Philips, 2010) and weights of the models are obtained. The weights indicated the performance of models in the simulation of ISMR under climate change. Both MMA and BMA are computed. The next section presents the details of the data used for this study.

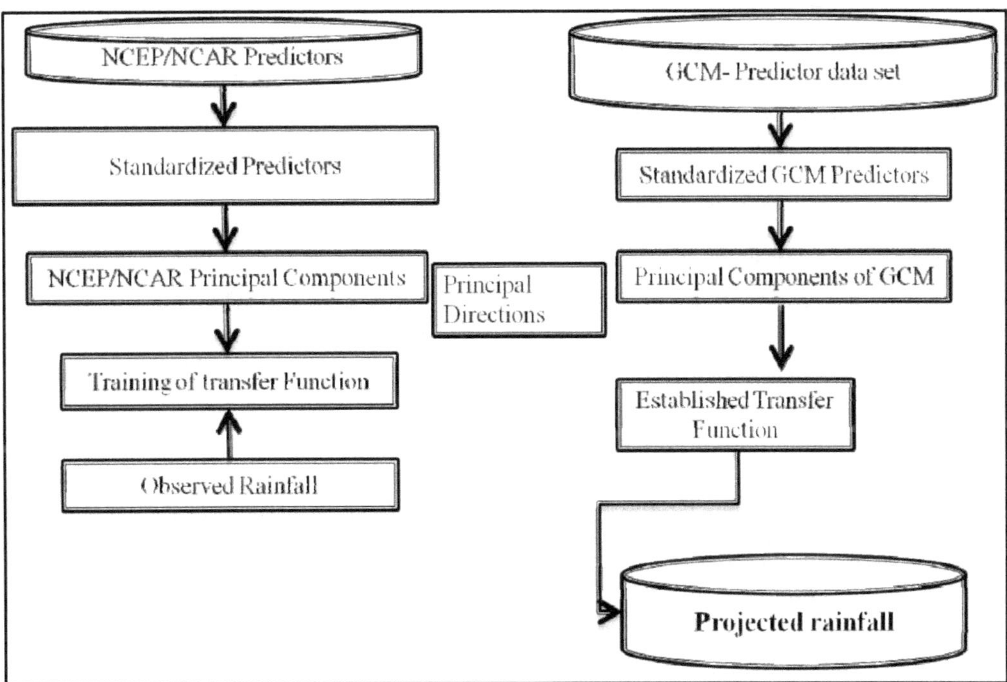

Fig. 1 Flow chart for Statistical Downscaling Algorithm

2.0 DATA

2.0.1 Observed Rainfall Data- APHRODITE

Asian Precipitation Highly Resolved Observational Data Integration towards Evaluation of Water Resources (APHRODITE)), Japan is a daily gridded precipitation data set created from 1961-2004. The rainfall product is based on data collected from dense network of daily rain gauge data from all across Asia including the data from sparse areas like Himalayas and

Mountainous areas of Middle East. The state of art of daily precipitation data is available at 0.5 ° x 0.5 ° and 0.25 ° and 0.25 ° resolution. The Aphrodite body has used an improved interpolation scheme which gives proper weightage to local topographical features to improve the orographic precipitation (Yatagai et al., 2012). Firstly the rainfall product is interpolated to 0.05 degree analysis and regridded to 0.25 degree after ensuring rigorous quality and accurate checking. The website address http://chikyu.ac.jp/precip. The data set is over 40 years so that it can be used for evaluating the long term water resources of Asian region.

2.0.2 Selection of Predictors

Selection of suitable predictors is a crucial step in developing statistical downscaling model. The most basic requirement from a predictor set is that it is informative as well as well simulated by host GCMs. And also should show good correlation with predictand (Wilby et al., 1999). Generally predictors representing the atmospheric circulation, humidity and temperature have been used to downscale precipitation. Predictor selection choice depends on the region and season under consideration (Timbal et al., 2008). Broadly it is assumed that the predictors directly affect rainfall process. Predictors should be so chosen such that in the climate change context the predictors essentially capture effect of global warming. Humidity plays an important role in capturing changes in water holding capacity of atmosphere under global warming (Wilby et al.2004). Temperature , U wind , V wind, Mean sea level pressure (MSLP) add considerable power to predict short and long term changes in precipitation. The predictors selected for this analysis are air temperature, wind velocities (U and V wind) at surface and at 500 hPa, mean sea level pressure, specific humidity at 500 hPa.

2.0.2 NCEP-NCAR reanalysis data

Reanalysis data is surrogate for observed data for any predictor variable. As the numerically solved fundamental equations in case of GCMs result in systematic errors known as bias, it is necessary to correct them. This correction is done based on the observed data.

The NCEP-NCAR Reanalysis data set is a continually updating gridded data set representing the state of the Earth's atmosphere, incorporating observations and numerical weather prediction (NWP) model output dating back to 1948. It is a joint product from the National Centers for Environmental Prediction (NCEP) and the National Center for Atmospheric Research (NCAR), NOAA. For the current projections, the reanalysis data was downloaded. The resolution is 2.5° lat × 2.5° long. The base line period considered for the present study is from 1961-2000 because it is of sufficient duration to establish a reliable climatology. The NCEP/NCAR reanalysis-I data (Kalnay et al., 1996) provide global atmospheric data which is a mixture of physical observations and model forecasts. Kalnay et al., 1996 have used different data assimilated systems such as global rawinsonde data, aircraft data, satellite data, and surface land synoptic data, advanced microwave surface wind speed data etc. to with a T62 resolution and 28 vertical sigma levels to calculate the reanalysis data products for various climate variables. Climatic variables are categorized in to three types (Kalnay et al., 1996). Category A variables are those that are strongly influenced by observations eg. Zonal and meridional wind. Category B variables are influenced by the model and observations, but are not as accurate as category A level variable. Eg. Specific humidity. Category C variables are directly determined by model eg. Precipitation flux. For more information visit the site: http://www.esrl.noaa.gov/psd/data/gridded/data.ncep.reanalysis.html.

For carrying out the work, the data was extracted from lat 5- 40 °N and long 65-100 ° E for the Indian region.

3.0 STATISTICAL DOWNSCALING METHODOLOGY

Statistical downscaling involves development of statistical relationship between large scale climate variables, (which are reliably simulated by GCMs and known as predictors), and fine resolution rainfall (predictands) and then applying the relationship to the GCM simulations.The methodology is presented in Figure 1.The methodology is applied separately to seven meteorologically homogeneous zones (Parthasarathy et al., 1996) in India.

The GCM simulated variables have systematic deviations (bias) with respect to observed NCEP/NCAR reanalysis data. For the current study, standardization method is used for bias correction. The Bias corrected data needs further mathematical treatment because of multidimensionality and multicollinearity (Salvi et al 2013). Principal Component Analysis (PCA) is used to reduce the dimensionality and multicollinearity.

Table 1 List of GCMs used in monthly Rainfall Downscaling

S.No	Name	Institution	Resolution	
			Lat(degrees)	Long(degrees)
1	BCC- CSM1	Beijing Climate Center	2.7906	2.8125
2	BNU-ESM	College of Global Change and Earth System Science, Beijing Normal University	2.79	2.81
3	CCCma-CAN-ESM2	Canadian Centre for Climate Modelling and Analysis	2.8	2.8
4	CMCC-CMS	Centro Euro-Mediterraneo per I CambiamentiClimatici	1.87	1.88
5	CNRM-CM5	Centre National de Recherches Meteorologiques	1.4	1.4
6	CSIR-ACCESS 1-0	CSIRO (Commonwealth Scientific and Industrial Research Organisation, Australia), and BOM (Bureau of Meteorology, Australia)	1.25	1.875
7	CSIRO-Mk3-6-0	Commonwealth Scientific and Industrial Research Organization in collaboration with Queensland Climate Change Centre of Excellence	1.87	1.88
8	FIO-ESM	The First Institute of Oceanography, SOA, China	2.79	2.81
9	GISS-E2-R-CC	NASA Goddard Institute for Space Studies	2.00	2.5
10	Had-GEM2-E2S	Met Office Hadley Centre	1.875	1.25
11	INM-CM4	Inst. for Numerical Mathematics	1.5	2
12	IPSL-CM5A-LR	Institute Pierre-Simon Laplace	1.89	3.75
13	IPSL-CM5A-MR	Institute Pierre-Simon Laplace	1.89	3.75

Contd...

S.No	Name	Institution	Resolution	
			Lat(degrees)	Long(degrees)
14	MIROC5	Atmosphere and Ocean Research Institute (University of Tokyo)and others	1.4	1.41
15	MIROC-ESM	Atmosphere and Ocean Research Institute (University of Tokyo)and others	2.8	2.8
16	MPI-ESM-LR	Max-Planck-Inst. For Meteorology	1.8	1.8
17	MPI-ESM-MR	Max-Planck-Inst. For Meteorology	1.8652	1.875
18	NASA-GISS-E2-R	NASA Goddard Institute for Space Studies	2	2.5
19	NorESM1-ME	Norwegian Climate Centre	1.8947	2.5
20	NorESM1-M	Norwegian Climate Centre	1.5	1.9

Here, we use first few principal components (PCs) which together represent 90% of the variability of original predictors. The principal components (PCs) thus obtained are regressed with the rainfall, at individual grids. The set of predictors and hence the principal components, used as regressors, remain unchanged for all the grids in a homogeneous region. The model is calibrated for the period 1970-2005, for all three spatial resolutions. Applying the principal directions obtained from reanalysis for computing the PCs, to the GCM output may introduce new bias. However, this is a standard procedure (Wilby et al., 2004), which is followed for the GCM outputs, to apply the same relationship, which is obtained between reanalysis predictors and observed predictand. Obtaining the individual principal direction from each GCM will not ensure this and hence is not followed. A possible solution would be merging of reanalysis and GCM predictors for obtaining the PCs, and is a potential area of future research.

The calibrated model is then applied to the GCM simulations after bias correction. GCM simulations are first regridded to the reanalysis grids, and then Wilby et al. (2004) method of standardization is applied. The principal directions obtained from the reanalysis data is applied to the bias corrected GCM predictors for obtaining the principal components (PCs) corresponding to GCMs. These PC are then used in the calibrated linear regression model to obtain gridded rainfall simulations for the historic (here, we consider, 1970-2005) and future (2010-2099) periods (RCP 4.5 scenario).

3.1 Evaluating the GCM Simulations

The downscaled GCM simulations are first evaluated based on the deviations from the long term mean and standard deviations for the historic period. However, such evaluations may not compare the observed and projected simulations in terms of all the statistical characteristics and variability. Hence, here we apply the evaluation metric, suggested by Perkins et al., (2007), where the skill of the simulation is measured with respect to the resemblance of its Probability Density Functions (PDFs) to that of observed.

In this evaluation method, first the entire range of observed/ simulations are divided into multiple bins. This methodology calculates the cumulative minimum value between the model and observed data distribution in each of bin; thereby measuring the common area between two PDFs. If a model simulates the observed data perfectly, the skill score is equal to 1. If it simulates poorly, the skill score is equal to zero. Mathematically, the Skill score (S-score is given by)

$$\text{S-score} = \sum_{i=1}^{n} minimum(Z_m, Z_o) \quad .....(1)$$

where

n = total number of bins.

Z_m = Frequency values in a given bin from model m.

Z_o = Frequency values in a given bin from observed data.

The S-score is computed for each grid, and then the spatial distribution of this score is obtained in terms of the PDF. The PDFs of S-scores, obtained from multiple resolutions are compared.

3.2 Weighted Modeling with Bayesian Approach

Conventional approach for simulating projections with multiple models is to obtain the multi-model average, assuming the positive and negative errors in the models will cancel out each other (Knutti, 2010), resulting in more reliable projections. Following Duan and Philips (2010), here we briefly discuss the steps for assignment of weights to GCMs, with the Bayesian Model Averaging (BMA). The parameters to be determined are $\theta = \{w_k, \sigma_{PK}^2, \sigma_{FK}^2\}$, where w_k are the weights to each GCM (w_k for k^th^ GCM), σ_{PK}^2 are the variance for each of the past/historical downscaled GCMs and σ_{FK}^2 are the variance for each of downscaled future GCMs data.

Step 1: Let us assume the vectors y= (Y_P, Y_F) denote the downscaled data set consisting of past-climate (Y_P) and future simulated climate variable (Y_F) from each of the GCMs. Both data sets are assumed independent and follow Gaussian distributions.

$$g(y|\theta_k) = g(y_P, y_F | \theta_{Pk}, \theta_{Fk}) = g(y_P|\theta_{Pk}).g(y_F|\theta_{Fk}) \quad .....(2)$$

Where θ_{Pk}, θ_{Fk} denote parameters for past and future downscaled simulated GCM data. The above equation is based on Bayes' theorem.

Step 2: The log likelihood function for equation 2, becomes

$$l(\theta) = \Sigma_{(s)} \log(\sum_k w_k.g(Y_{TP}|\theta_{PK}).g(Y_{TF}|\theta_{FK}) \quad .....(3)$$

Where Y_{TP} and Y_{TF} are evidentiary targets (observed) and weighted multimodel consensus estimates for future climate data, respectively.

Since g (Y_P | θ_P) and g (Y_F | θ_F) are Gaussian functions, the maximum likelihood estimates of the means μ_k of the individual model simulations are equal to the mean of the simulations themselves;

i.e. $\mu_{Pk} = f_{Pk}$ and $\mu_{FK} = f_{Fk}$. The f_{Pk} and f_{Fk} are the means of downscaled simulations of the climate variable for the past and future in each GCM.

Step 3: The parameters $\theta = \{w_k, \sigma^2_{PK}, \sigma^2_{FK}\}$ in equation 3 are estimated using Expectation-Maximization (E-M) Algorithm .

3.3 Expectation-Maximization Algorithm

EM algorithm is an iterative method and consists of two steps; the Expectation (E) step and the Maximization (M) step.

It begins with the initial guess of parameter $\theta^{(0)}$ for the parameter vector $\theta = \{w_k, \sigma^2_{PK}, \sigma^2_{FK}\}$. In the E step, a quantity Z_{kst} estimated given present values of parameters; however Z_{kst} takes two values; either 0 or 1. Where $Z_{kst}=1$ if ensemble member k is best forecast for particular (s, t). Otherwise $Z_{kst}=0$;

Thus, the E-step yields,

$$Z_{kst}^{(i)} = (g(Y_{TP}|\theta_{PK}^{(i)}.g(Y_{TF}^{(i)}|\theta_{PK}^{(i)}/\{\Sigma_k g(Y_{TP}^{(i)}|\theta_{PK}^{(i)}).g(Y_{TF}^{(i)}|\theta_{FK}^{(i)})\} \quad ......(4)$$

While the M-step leads to,

$$w_k^{(i)} = \Sigma_s Z_{ks}^{(i)} / S; \quad .....(5)$$

Where (i) and s are iterations and spatial points, respectively, S is the total number of spatial points. For further details refer to Duan and Philips (2010) and Raftery et al. (2005).

3.3.1 BMA consensus Expectation and uncertainty estimation

The BMA expected mean of posterior PDFs for the past and future climates are estimated as a weighted sum of individual model simulations and are calculated as below:

$$E(Y_P \mid f_{p1}, f_{p2}, ...f_{pk}) = \Sigma_k w_k.E(Y_P \mid f_{pk}) = \Sigma w_k.f_{pk} \quad .....(6)$$

$$E(Y_F \mid f_{F1}, f_{F2}, ...f_{Fk}) = \Sigma_k w_k.E(Y_F \mid f_{Fk}) = \Sigma w_k.f_{Fk} \quad .....(7)$$

The total variances are the measures of spreads (uncertainties) of the posterior PDFs and are given by the following expressions:

$$\sigma^2(Y_P \mid f_{p1}, f_{p2}, ...f_{pk}) = \Sigma_k w_k.(\Sigma_k f_{PK} - \Sigma_k w_k.f_{PK})^2 + \Sigma_k w_k.\sigma^2(Y_{TP}| f_{PK}) \quad .....(8)$$

$$\sigma^2(Y_F \mid f_{F1}, f_{F2}, ...f_{FK}) = \Sigma_k w_k.(\Sigma_k f_{FK} - \Sigma_k w_k.f_{FK})^2 + \Sigma_k w_k.\sigma^2(Y_{TF}| f_{FK}) \quad .....(9)$$

4.0 RESULTS AND DISCUSSIONS

A linear regression based statistical downscaling method is adopted for rainfall projections of ISMR at monthly time scale to understand the projections. The results are presented in the following subsections which include a comparison of simulations with observations for the historical period, PDF comparison, comparison between BMA and MMA.

The results of downscaled rainfall product are compared with observed rainfall (Fig 2). The Fig.2 shows the grid wise details of mean and standard deviation and drawn in the order of the observed rainfall, the multi model average from 20 GCMs, difference in mean and difference in standard deviation. The results shows that mean characteristics are matching well with observed data. The most of grids of Indian landmass the percent of errors in mean and standard deviation [Fig 2] are found to be within 10%.. The monthly means are expressed as mm/day. Along with mean, spatial pattern of rainfall is captured well by the statistical model with high degree of accuracy.

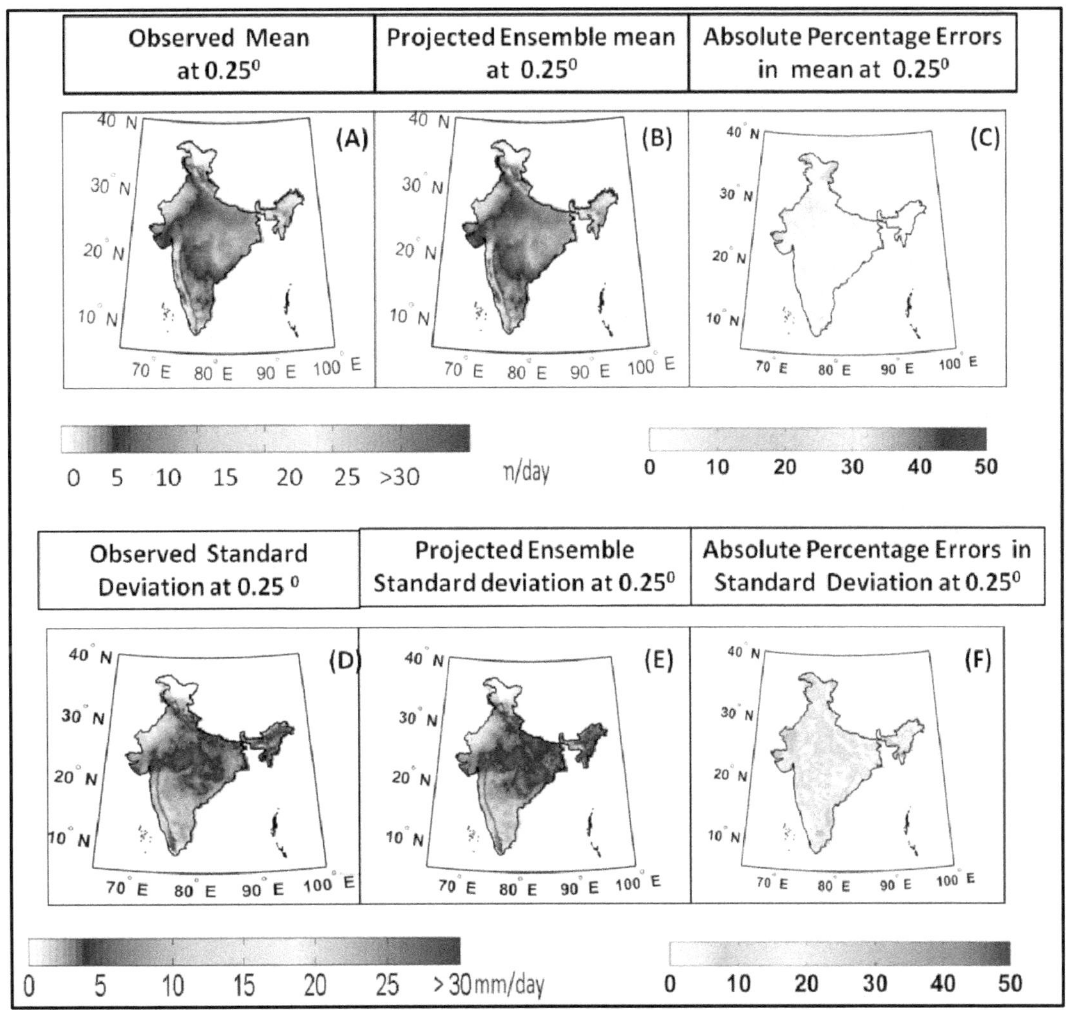

Fig. 2 The mean, standard deviation and percent error plots.

The spatial pattern of monsoon is well captured by the model. The Skill score (S-score) for evaluation, which is based on the similarity of the PDFs computed with both observations and simulations for the historic period (Fig. 3). It is observed from the PDF skill score diagram that the present statistical downscaling method has successfully simulated range of observed values in different regions of Indian landmass.

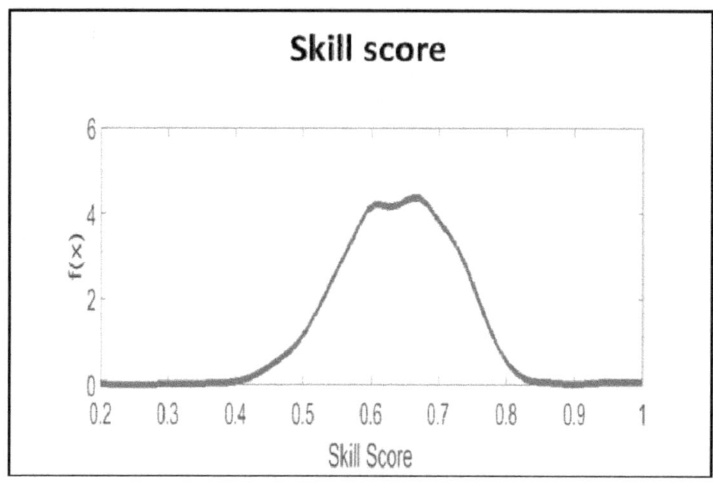

Fig. 3 Plot represents the skill score at 0.25^0 degree resolution. The higher skill score represents better simulations.

To know the plausible changes in future rainfall, we consider downscaled rainfall projections for RCP4.5 scenarios. We consider the projections obtained with both MMA and BMA. The weights obtained by using BMA are presented in Fig. 4 for 2080s. We obtain consistently higher weights for GISS, CMCC, MIROC and MPI. Similarly, consistently lower weights are assigned to BCC, BNU and CCCMA.

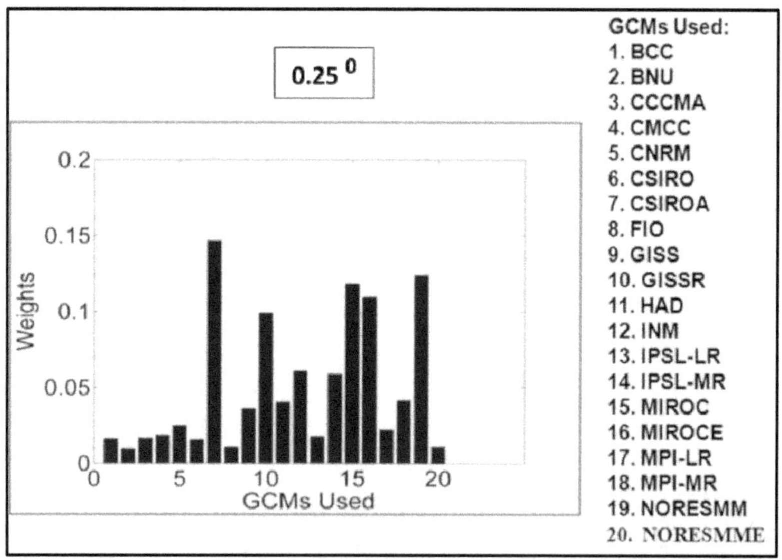

Fig. 4 Weights obtained for different GCMs with Bayesian modeling for 2080s.

The projections obtained with both MMA (Fig. 5a-c) and BMA (Fig. 5d-f) show spatially non-uniform projected changes of ISMR for the 21st century. Increases in precipitation are projected for in the Himalayan foothills, west coast, North regions, while decreasing trends are observed in the Gangetic West Bengal and regions of Central India and mixed results in Southern India. The increase/decrease in rainfall for future time slices is consistent with other studies with fine resolution models (Salvi et al., 2013; Shashikanth et al., 2013; Shashikanth et al.2014). The orographic rainfall is expected to increase in west coast and North-east part of India in future also. We find that the changes projected with MMA are slightly more on the positive side, compared to BMA. This suggests the possibility of the existence of models with low weights projecting high positive changes.

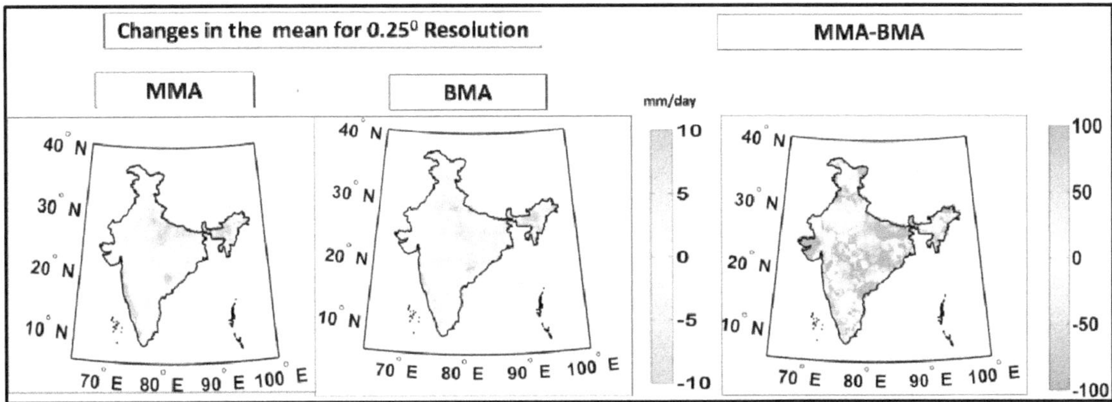

Fig. 5 Projected changes in mean rainfall for a future period (2070-2099) with respect to the base line period (1970-2005), with multi-model average (MMA). The percentage difference between MMA and BMA with respect to MMA are presented with plots

5.0 CONCLUDING REMARKS

In general, the most important meteorological/ climate variable for hydrological impact studies are precipitation and temperature. Therefore, these variables are important from climate change adaptation and policy framing purposes. The major part of this present work contributes towards projections of rainfall for the 21st century, at multiple resolutions on monthly time scales. The statistical downscaling assumes the stationarity relation between predictors and predictand and the same is valid for future. This assumption is prone to risk as this may be due to chosen predictors having failed to produce long term variability.

Other limitations includes requirement of huge sums of data during training of the model and the quality and quantity of observed data set. Of course, these limitations are general in nature and valid for any statistical models.

REFERENCES

1. Duan, Qingyun, Thomas J. Philips, 2010. Bayesian estimation of local signal and noise in multimodel simulations of climate change J.Geophys. Res. 115, D18123, DOI 10.1029/2009/JD013654

2. Fowler, H. J., Blenkinsop, S., Tebaldi, C., 2007. Linking climate change modelling to impacts studies: recent advances in downscaling techniques for hydrological modelling, Int. J. Climatol. **27**: 1547–1578 2007
3. Giorgi, F., Mearns L.O., 2002. Calculation of average, uncertainty range, and reliability of regional climate changes from AOGCM simulations via the reliability ensemble averaging REA method. J Clim 15:1141–1158
4. Hawkins, Ed., Rowan Sutton, 2009. The Potential to Narrow Uncertainty in Regional Climate Predictions. Bull. Amer. Meteor. Soc., 90, 1095–1107. doi: http://dx.doi.org/10.1175/2009BAMS2607.1
5. IPCC (2013) Climate Change 2013: The Physical Science Basis. Contribution of Working Group I to the Fifth Assessment Report of the Intergovernmental Panel on Climate Change [Stocker, T.F., D. Qin, G.-K. Plattner, M. Tignor, S.K. Allen, J. Boschung, A. Nauels, Y. Xia, V. Bex and P.M. Midgley (eds.)]. Cambridge University Press, Cambridge, United Kingdom and New York, NY, USA, 1535 pp, doi:10.1017/CBO9781107415324.
6. Perkins S.E., Pitman, A.J., Holbrook, N.J. Mc. Aneney, J., 2007. Evaluation of the AR4 Climate models simulated Daily maximum temperature, minimum temperature and precipitation over Australia using probability density functions , J.Clim.,Vol.20,4356 DOI10.1175/JCLI4253.1
7. Raftery, A. E., Gneiting, T., Balabdaoui, F., Polakowski, M., 2005. Using Bayesian model averaging to calibrate forecast ensembles, Mon. Weather Rev., 133, 1155–1174, doi:10.1175/MWR2906.1.
8. Shashikanth, K., Kaustubh, S., Ghosh, S., Rajendran, K., 2013. Do CMIP5 simulations of Indian summer monsoon rainfall differ from those of CMIP3? Atmospheric Science Letters, 10.1002/asl2.466, 10.1002.
9. Wilby, R. et al., 2004. Guidelines for use of climate scenarios developed from statistical downscaling methods. http://www.narccap.ucar.edu/doc/tgica-guidance-2004.pdf accessed on 719 10/08/2013
10. Tebaldi, C., Mearns, L. O., Nychkam D., Smith, R. W., 2004. Regional probabilities of precipitation change: A Bayesian analysis of multi-model simulations, Geophys. Res. Lett., 31, L24213, doi: L10.1029/2004GL021276
11. Shashikanth, K., Madhusoodhanan, C.G., Ghosh, S., Eldho, T.I., Rajendran, K., and Murtugudde, R. (2014) Comparing statistically downscaled simulations of Indian monsoon at different spatial simulations J. Hydrol. http://dx.doi.org/10.1016/j.jhydrol.2014.10.042.
12. Yatagai, A. et al. (2012) APHRODITE: Constructing a Long-Term Daily Gridded Precipitation Dataset for Asia Based on a Dense Network of Rain Gauges, Bull. Amer. Meteor. Soc., 939: 1401-1415, 727 10.1175/BAMS-D-11-00122.1.
13. Perkins S.E., Pitman, A.J., Holbrook, N.J. Mc. Aneney, J., (2007) Evaluation of the AR4 Climate models simulated Daily maximum temperature, minimum temperature and precipitation over Australia using probability density functions , J.Clim.,Vol.20,4356 DOI10.1175/JCLI4253.1.
14. Parthasarathy, B., Rupakumar, K., and Munot, A. A. (1996), Homogeneous regional summer monsoon rainfall over India: Inter annual variability and teleconnections, Res. Rep. RR–070, ISSN, 0252–1075.
15. Kalnay, E., M. Kanamitsu, R. Kistler, W. Collins, D. Deavan, L. Gandin, M. Iredell, S. Saha, G. White, J. Wollen, Y. Zhu, M. Chelliah, W. Ebisuzaki, W. Higgins, J. Janowiak, K.C. Mo, C. Ropelewski, J. Wang, A. Leetmaa, R. Reynolds, R. Jenne, D. Joseph (1996), The NCEP/NCAR 40-years reanalysis project, Bull. Amer. Meteor. Soc, 77(3), 437471

AQUASENSE: DEVELOPMENT OF SMART WATER MONITORING SYSTEM

Neelima Satyam

(PI, Aquasense from IIITH), Team Aquasense (IIITH- SCCE-SCIT),

(As a part of Aquasense project funded by ITRA, Govt of India)

ABSTRACT

This paper presents a proposed design, modeling andsimulation of the wireless smart flow meter networks with own power generation. The advantage of this system is that a smart meter node in the network generates power required for its operation within the node, thus eliminating the need for an external power source to be connected or maintained periodically. A power storage unit is also designed which charges during power generation. This unit supplies power to the circuit when generation stops. RF inter-node communication among the smart meters is proposed in this system. the proposed smart meter has a power rectification and regulation system and a power storage system along with a power generator. Super capacitor is used as power storage element. The simulation of the power rectification, regulation, storage and sensor interfacing are done is Proteus 8 software. A laboratory set up has been built and experimentally tested to examine the proposed system. Experimental results are computed and shown to be in good agreement.

Keywords: Smart water Meter, Micro Power Generation, Adhoc Wireless Network, Smart Water Leak Detection System,

1. INTRODUCTION

This project aims to study and develop a suitable model for development of effective and energy efficient wireless smart water meter network. This project aims to identify leakage, and to meter water supply remotely with the help of a wireless sensor network zone. It proposes a novel, scalable, intelligent and incremental architecture which is adaptive to the environmental parameters, quality perception that incorporates intelligence for monitoring the predefined parameters. A vital component of the system shall be a new built-in self power generation with storage, interactive data display and retrieval. This is provided through a web-based control panel. The system is proposed to have an intelligent decision support system for continuous monitoring of the logged data in the wireless sensor zone. Real-time measurements from the sensor network shall be assimilated into hydraulic models. The expected outcome shall be a new model for online, real-time monitoring system that shall be used to improve efficiency of the water supply distribution network. This proposed research project refers to technologies in the move towards next generation quantity monitoring to provide simple, efficient, cost effective and socially acceptable means to detect and analyze distribution regularly and automatically.

2. EXISTING SYSTEM AND ITS LIMITATIONS

There has been lots of research on development of sensors for measuring the different flow parameters & quality parameters. These sensors have been associated with a LCD display device to display the measured value. This can only be used as a standalone device for measuring the parameter on-site. There has been a great progress in the wireless communication research to develop wireless sensor zone using the sensor node comprising of sensor board, processor board and transceiver. To the best of our knowledge, there is no such indigenous technology in the country which integrates the sensors for quantity monitoring using Wireless Sensor Networks. There is still a lack of indigenous developed real-time integrated quantity monitoring system available which can monitor and control the quantity continuously in different places, transmit data to the central server via the wireless communication and generate billing and alarms.

A. Limitations of existing system

The main limitations of existing flow measurement system in public water distribution systems are as follows:

- Flow meters used are stand alone devices
- Real time flow monitoring is not possible
- Leak detection is very difficult
- Readings of each flow meter is to be collected manually

Consumption patterns cannot be predicted

The above limitations increase the amount of manpower to be employed and also decrease the reliability of distribution system.

3. PROPOSED SYSTEM

Taking in to consideration the drawbacks and limitations of existing flow metering in distribution system, we proposed the"WIRELESS SMART METER NETWORKS WITH SELF POWER GENERATION". In this proposed system, we developed a sensor network which will eliminate all the drawbacks of the existing system, which is a standalone system. Our system would integrate the flow sensors with wireless network to make it a remote monitoring network. Since some nodes may be difficult to access to maintain or replace batteries, we have designed a power generation system built in the node, which provides power to all the circuitry. The proposed architecture is adaptive and incremental in nature for intelligent monitoring and to generate data pertaining to utilization, and raise alerts in terms of violation of safety norms.

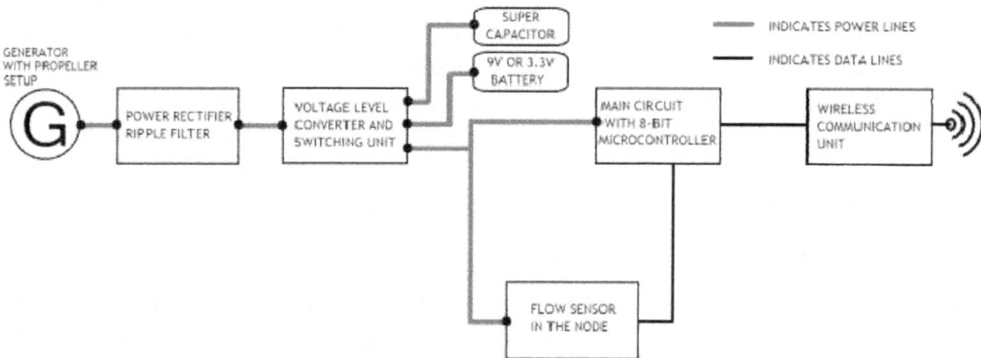

Fig. 1 Block diagram of proposed smart meter

A. Advantages of proposed system

- Real time flow monitoring over an entire network is possible.
- Leak detection in a network is possible
- Built-in power generation in the smart meter avoids replacing batteries periodically.
- Consumption patterns can be designed for future needs.
- Advanced billing system can be implemented.

4. SMART FLOW METER OPERATION

The flow meter consists of a circular sealed chamber with tangential inlet and outlets. A paddle wheel turbine is placed in the chamber such that water flows from inlet to outlet rotating the turbine. The construction diagram of the meter is shown below.

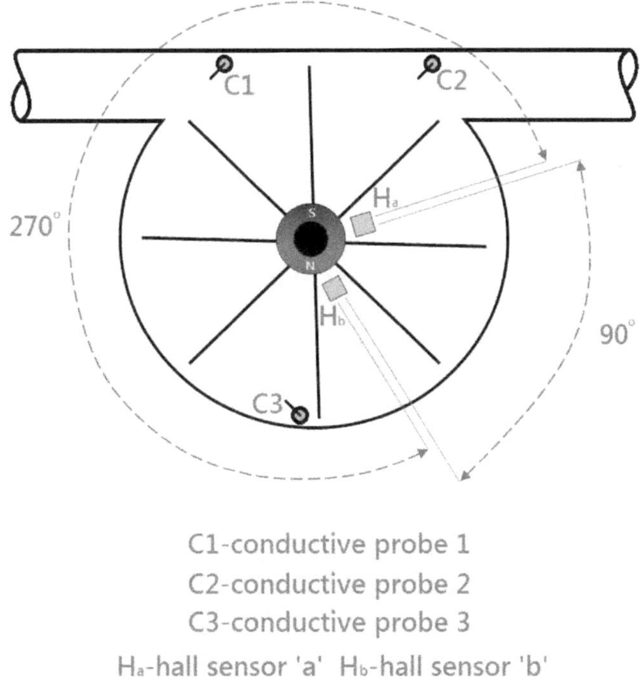

Fig. 2 Flow sensor construction in smart meter.

The paddle wheel turbine axis is positioned to limit contactbetween the paddles and the flowing media to less than 50%of the rotational cycle. This imbalance causes the paddle torotate at a speed proportional to the velocity of the flowing media. A ring magnet is placed at the center of the turbine.When the turbine rotates, the ring magnet also rotates. Hallsensor is placed over the sealed chamber near the center of the turbine. When the turbine rotates, the ring magnet passes under the hall sensor. Hall sensor produces a high pulse when a magnetic north passes near it and a low pulse when a magnetic south passes near it. Thus it produces a square wave when the turbine is rotating. The frequency of this square wave is directly proportional to the fluid velocity.

An 8-bit microcontroller in the smart meter counts this pulses from the hall sensor and stores them in its EEPROM which is a non volatile memory. After every minute, the number of pulses occurred on the sensor output line is added to the memory. The microcontroller has a permanent record of the number of pulses occurred on the sensor output in its memory. This is similar to odometer in automobiles. By counting the number of pulses per minute, the discharge rate or flow rate can be calculated by using the equation Q=f/K, where Q is flow rate, f is the frequency of sensor output and K is kinematic viscosity. By counting all the pulses from the point of installing gives the quantity of water discharged from the meter by using the equation L=p/K where L is the quantity of water in liters, p is the total number of pulses obtained from the sensor from point of installation and K is kinematic viscosity. The microcontroller processes pulse data in the above formulas and displays the flow data information on the LCD display integrated in smart meter.

By using a single hall sensor, there is a drawback. The hall sensor generates pulses irrespective of the direction of rotation of turbine. The direction of flow i.e. downstream or up stream is unknown. To solve this problem a 2 hall sensor meter is proposed. Two hall sensors Ha and Hb placed over the sealed chamber at 90 degrees from each other as shown in fig.2. At regular small time intervals a test is done by the microcontroller to verify the flow direction. During flow measurement, only sensor Ha is used. During flow direction verification, both sensors are used. First, the time taken for one pulse cycle (Ta) on sensor Ha is measured by the controller. Then the controller waits for a low to high transition on sensor Ha. When this happens, the controller starts counting time (Tb) till a low to high transition takes place on other sensor Hb. The time Ta and Tb in two possible cases are shown below.

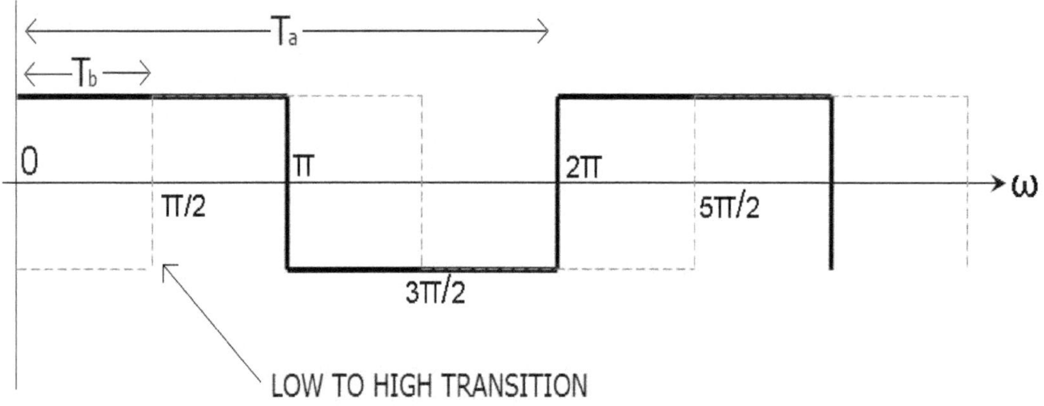

Fig. 3 Output pulses of Ha and Hb during downstream liquid flow

Since the two hall sensors are placed at an angle of 90 degrees in clockwise direction, the time taken by the turbine to move in forward direction from the point Ha to Hb is ¼ thetime taken for one complete rotation. These time periods Ta and Tb are measured by the microcontroller and compared. If Tb~¼Ta as shown in fig. 3, then the smart meter determines that the liquid is flowing down stream.

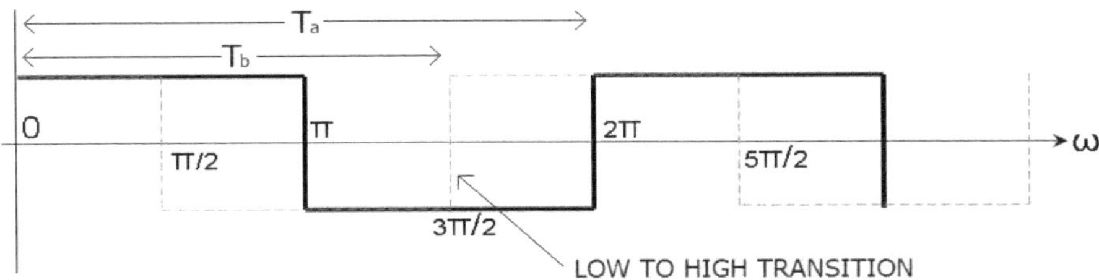

Fig. 4 Output pulses of Ha and Hb during upstream liquid flow

Since the two hall sensors are placed at an angle of 270 degrees in anti clockwise direction, the time taken by the turbine to move in backward direction from the point Ha to Hb is ¾ the time taken for one complete rotation. These time periods Ta and Tb are measured by the microcontroller and compared. If Tb~¾Ta as shown in fig. 4, then the smart meter determines that the liquid is flowing up spring. The smart meter can be configured to keep separate flow data records for up spring and down spring flow.

Sometimes a part of the network is filled with air. When water is pumped in to the network, it pushes the air out of the pipes with high velocity. This flowing air will rotate the turbine in the flow meter. To prevent flow count when air passes through the smart meter, the chamber has conductivity testing probes. The chamber has 3 probes C1,C2 and C3 as shown in fig.2. C1 act as probe on which voltage is given by the microcontroller. C2 and C3 are used as measuring probes on which the voltage due to C1 appears. C2 and C3 are connected to an ADC unit in the microcontroller. At regular time intravels, the probe C1 is toggled high and low many times. This helps in eliminating any stray charges present in the meter. If air is present between C1 and C2 then there will be no voltage appearing on C2 due to the voltage on C1. When the probe C2 has conduction then controller determines that water is present in the meter, only then the turbine rotation is considered for flow measurement. If probe C2 fails the conduction, then even if turbine rotates, it is not considered for flow measurement. Sometimes a bubble might get trapped in the meter even when the water is flowing in the meter. If probe C3 fails conduction and probe C2 has conduction, it indicates that an air bubble might be trapped in the chamber.

5. WIRELESS COMMUNICATION SYSTEM

Every smart meter has a built-in 2.4 GHz RF module. CC2500 is chosen due to its low cost and advanced features. The CC2500 has an internal selection of various modulations like ASK, FSK, G-FSK, OOK etc. the carrier frequency can be selected between 2.4 GHz to 2.5 GHz. The data rate can be up to 512 Kbps. Smart meter transmits data in packet form. The RF module will always be in listening mode. It will be screening the data flow happening in the selected frequency channel. Each smart meter has a unique 4 byte id. The packet data send by a smart meter consists of this unique id along with the flow data parameters and special data parameters. The packet data is encrypted by RSA or other algorithms and is decrypted by the receiver. The circuit of the smart meter with the communication system is shown below.

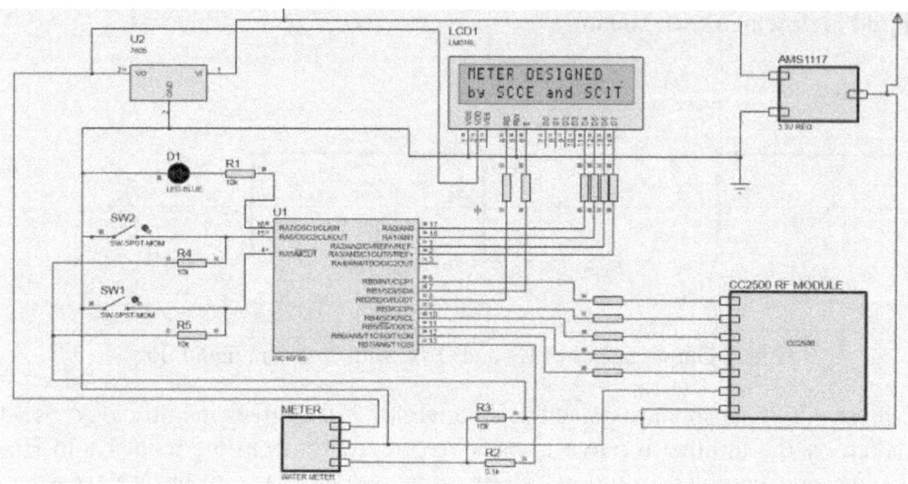

Fig. 5 circuit of the smart meter

Receiving node or the first node in a sub street sends a packet data to a smart meter node. This in-packet data has the unique id of the smart meter along with data flags asking the specific smart meter to send its data. When the smart meter receives, it decrypts this data and verifies the unique id and if it is a match, then a transmit flag is raised by the microcontroller and the flow data along with other data information is encrypted in to an out-packet data and is send by the smart meter. This out-packet data also has the unique id of the receiver, so that other smart meters forward this packet to the receiver through inter-nod communication. The simple flow chart for the operation of the smart meter is shown in fig.6.

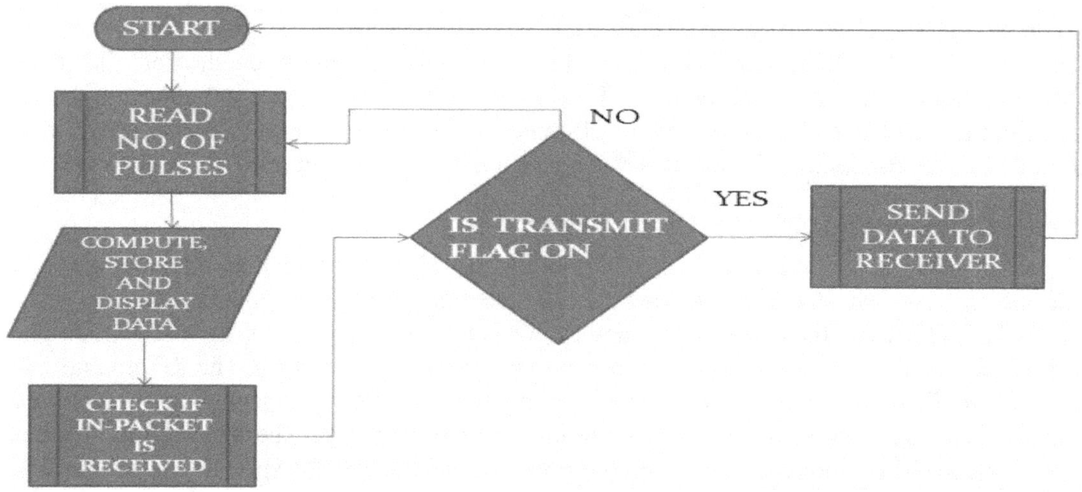

Fig. 6 simple flow chart showing the operation of the smart meter

In a sub street the first smart meter collects the data from all the smart meters in the sub street. The smart meters in the sub street use inter-node communication to transmit data from one smart meter to another in forward direction until it reaches the first smart meter in the sub street as shown in fig. 7. Thisdata from first node of one street is transmitted to first node of another street until it reaches the central receiving station as shown in fig. 7, where data is decrypted and stored in to the online database.

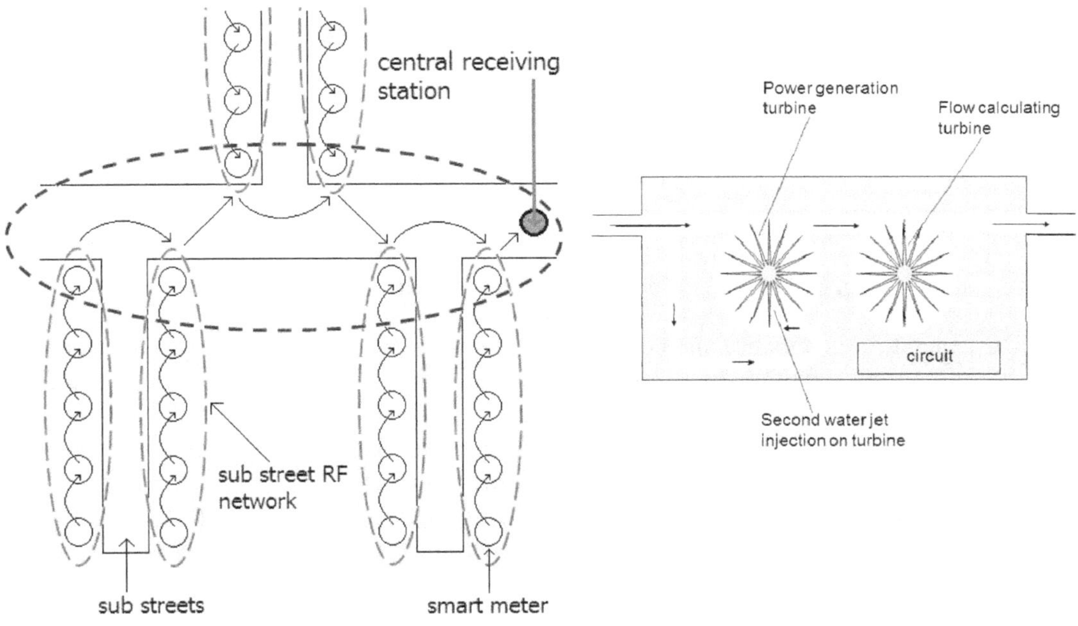

Fig. 7 Inter-node communication among smart meters

Since the first node in the street has to handle most of the communication, it consumes more power. So a dummy first node with solar power source or mains power with node placed on an electric pole can be considered. Since it is a dummy node, it does not have a flow measuring unit but only a communication unit. This dummy node increases the reliability of the smart meter wireless network.

6. POWER GENERATION SYSTEM

The smart meter has a built in power generation and storage system. The smart water flow meter has two separate turbines. One is for flow calculation as shown in fig. 2, another one is for power generation. The turbine generating power offers more resistance to flow of water as torque exerted by the water on the turbine needs to be high to get the turbine rotating. Thus the turbine connected to generator cannot be used for flow calculation as it would not be accurate. So a second turbine which can rotate freely should be used for flow calculation. These two turbines can be stacked side by side in a single package as shown in fig.8.

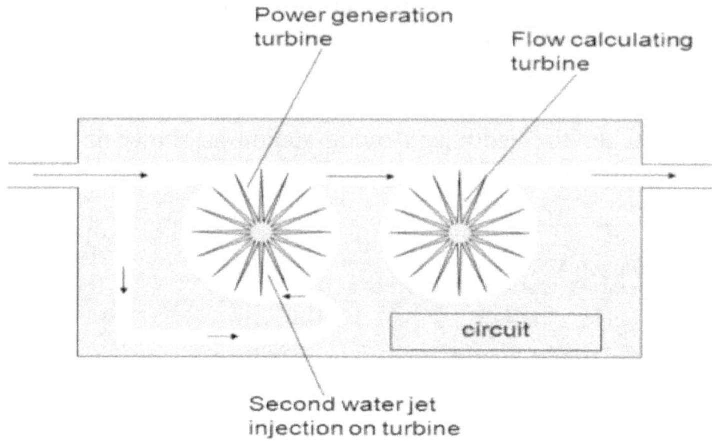

Fig. 8 Power generation turbine and flow calculating turbine

In a DC generator, the brushes and commutator gets worn out after heavy use and render the generator useless until repaired. Thus an AC generator is used. The power generating turbine may have more than one water jet input. The water jet is injected on a turbine with high velocity created by a small nozzle. There may be multiple water jets. A gearing system is necessary to produce sufficient power even at low speed. The turbine is connected to a small gear box. The gear box converts the low speed rotation of the turbine to high speed rotations at its output shaft. The output shaft of the gearbox is connected to the generator as shown in the fig.9.

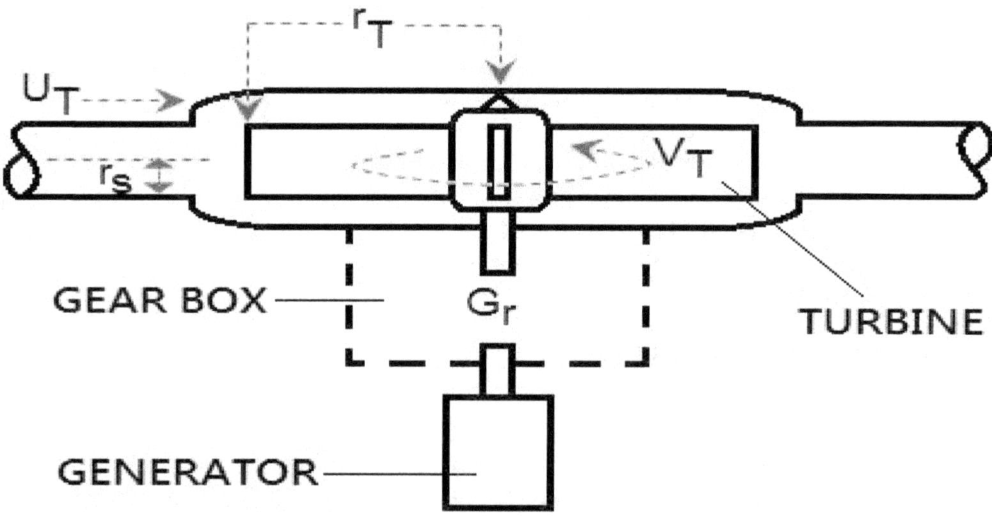

Fig. 9 Power generating turbine with gearbox and generator setup

' r_T ' is the radius of the turbine, ' r_s ' is the radius of the jet,

' V_T ' is the velocity of the turbine,

' U_T ' is the velocity of the water jet,

' G_r ' is the gearing ratio,

Let 'Z' be the number of nozzles, 'Q' be the flow rate and

' H_n ' be the water head.

The power generated by the generator is AC power but the circuit in smart meter needs DC voltage regulated to 5v. ThusAC power generated needs to be rectified and regulated to meet the operational standards of smart meter. The rectification of AC to DC is done with a full bridge diode rectifier. The rectification and regulation circuit of the smart meter is shown in fig. 10.

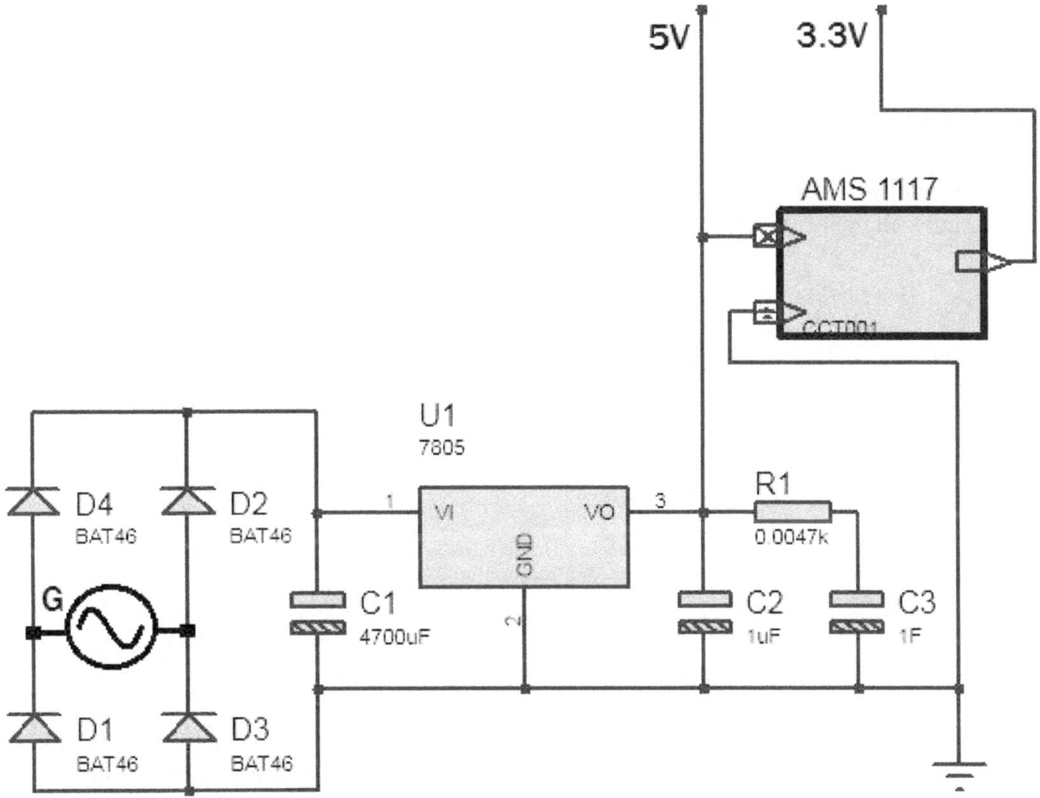

Fig. 10 Power rectification and regulation circuit of smart meter

The rectifier unit consists of a diode bridge rectifier. Normal silicon diodes like 1N4007, 1N4148 etc will have a voltage drop of 0.65V to 0.7V. So for one rectification cycle, there is a drop of nearly 1.4V. To reduce this forward voltage drop, Shottky Barrier Diodes are used in the rectification unit.

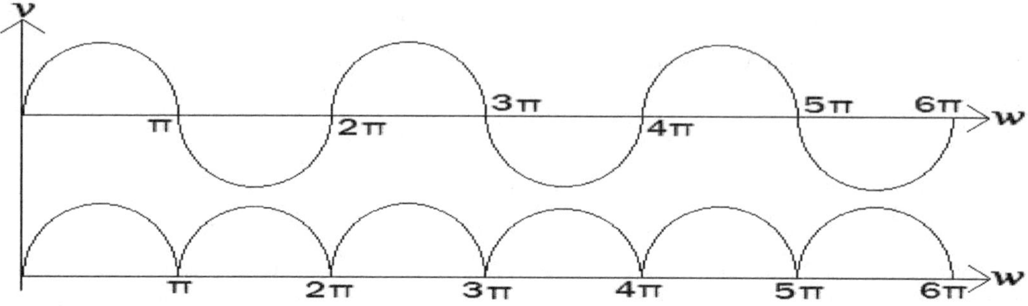

Fig. 11 Pulsating DC output from the rectifier

They have a voltage drop of 0.18V. BAT46 diodes are used in the rectifier unit as they have a forward voltage drop of nearly 0.2V. So for one rectification cycle the maximum drop at the bridge rectifier is only 0.4V. The rectified output from the full bridge rectifier has a continuous series of positive half waves as shown in fig.11. This pulsating output needs to be smoothed in order to be utilized by the circuit. A 4700uF capacitor is used to smooth the pulsating DC. This smoothed waveform still has some ripple content and the generated voltage may be higher than the operating voltage of the circuit.

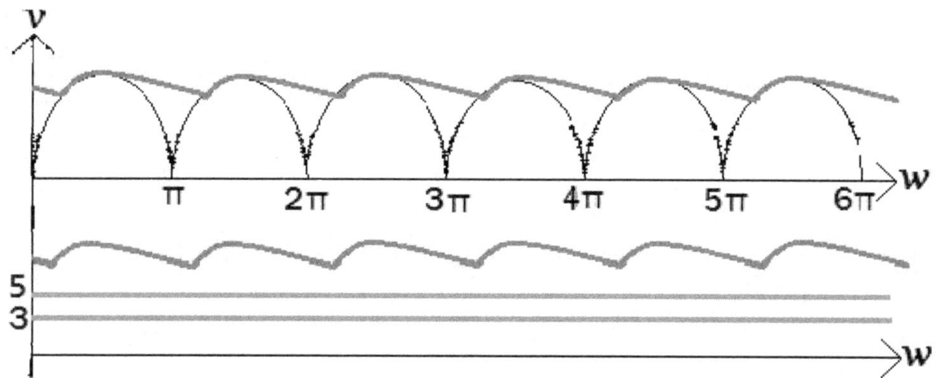

Fig. 12 Smooth waveform and voltage regulated to 5v and 3.3v

The regulation unit has linear voltage regulators connected to the rectified output as shown in fig.10. Some of the components of the node circuit like LCD display, Hall sensor etc need 5V to operate. While components like communication modules need 3V to operate. So two voltage regulators are used. To obtain 5V, LM7805 linear voltage regulator is used. For 3V, AMS1117 linear voltage regulator is used. The microcontroller can operate from 2v to 6v. So it is operated on 3.3v as wireless communication module is operated on this voltage range and is not tolerant to 5v inputs. The smoothed waveform and the regulated voltages are shown in fig.12. In this way power generation and voltage regulation takes place within the the smart meter. This makes the smartmeter ideal for installing at places that are not easilyaccessible like underground pipe systems.

7. LEAK DETECTION IN NETWORK USING SMART METERS

Smart meter network can be employed to detect leaks in the pipe network in real time. Flow parameters like flow rate, velocity and quantity from each smart meter is passed to the receiver and is inserted in to the database in a very short time. These flow rates from all the smart meters in a section of a network can determine if a leak is present in the network and also calculates the quantity of water that has been leaked out. Special software had been designed to read the data of all the smart meters in a network in real time. This program knows whether a smart meter is online of offline, if flow is happening or not and also knows if air is present in a smart meter. The program also has a record of all the pipe diameters and pipe lengths in the network.

The program uses two different algorithms for leak detection in the network. When the pipe network is filled with water, then the program uses the law of continuity to determine leaks. Consider a small section of the network shown in fig.13. M1,M2 and M3 are the three smart meters placed in the network section. Two taps T2 and T3 are located at M2 and M3 respectively. The section is filled with water as shown in the figure. Now when water is pumped in to this section, it rotates the turbines placed in the smart meter and the smart meter starts measuring the flow rate and the quantity of water passed through it. Let Q1,Q2 and Q3 be the flow rates of M1,M2 and M3 respectively. Let L be the leak in the section. Now by law of continuity the flow rate at in M1 should be equal to sum of flow rates in M2, M3 and leak L if present. Thus Q1=Q2+Q3+L. The program analyses the real time values of Q1, Q2, Q3 and verifies if the sum of Q2 and Q3 is equal to Q1. Since there may be slight error in flow readings, the program has a small tolerance to the deviation of Q1 and Q2+Q3.

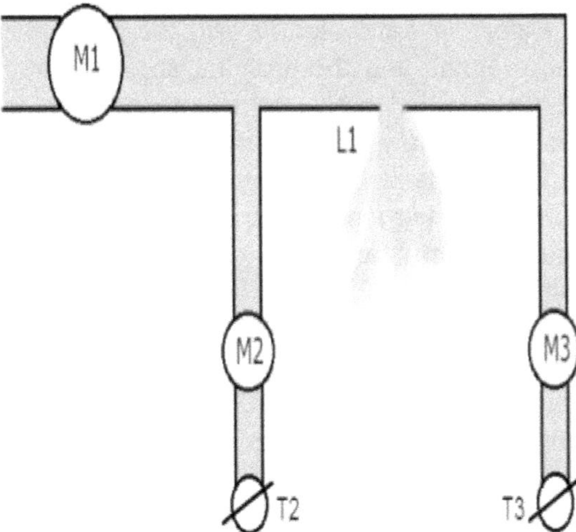

Fig. 13 Section of a network filled with water

If the deviation of Q1 and Q2+Q3 is grater than the tolerance limits, then the programs determines a leak 'L' in the section with a flow rate of Q1-(Q2+Q3). The program then raises an alert that a leak has occurred in the section. The amount of water leaked out and the time of occurrence of leak is stored in the database by the program.

Sometimes a part or section of a network is filled with air. If law of continuity is applied as soon as water is pumped, then it will give false reading that the entire network has leaks even if there are no leaks. This is because the air present in the smart meters fails the conductive test and turbine rotation is not considered for flow counting. So a different procedure is to be applied to detect leaks. A small section of a network with air filled in half of the section is shown in fig.14.

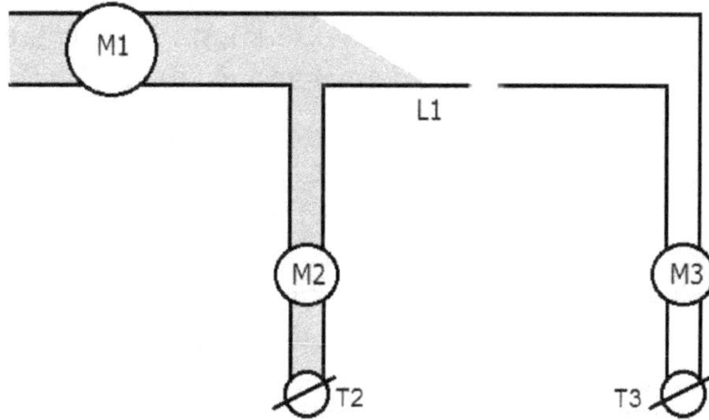

Fig. 14 Section of a network filled with water and air

When water is pumped in to the network, as air is present, it fills the pipes with water and air rushes out of the network. This air rotates the turbine of the smart meter but since it has conductivity probes, turbine rotation is not used for flow count. But this turbine rotation due to air is used for determining leaks. As water starts to fill the pipes, the paths that are filled are immediately detected by the program and the program applies low of continuity only on the smart meters present inthe water filled paths. In smart nodes in which air is flowing, the programs record a time stamp of air flow in that particular smart meter. Since the program has all pipe diameters, lengths in all the paths of the network, the program determines the time that would take by the water to reach the particular smart meter. Since all this calculations are done on real time data, the filling time determined by the program is highly accurate. If water arrives faster than the time predicted by the program, it indicates some water being present in the section when pumping started. Any delay between the predicted time and water flow taking place in the smart meter filled with air indicates a leak. The time delay can be directly converted in to flow rate of the leak. As soon as water flows in a smart meter in which air have flown previously, then the program applies law of continuity in that path up to this smart meter. This method is useful when there are long network paths as leaks in long pipes can be detected in less time than conventional methods even when the pipe is filled with air before pumping.

8. WORKING PROTOTYPE

A prototype model of a smart meter network with 10 smart meters is constructed in the lab and tested. The smart meter is integrated in to existing ordinary turbine flow meters by placing hall sensor near the turbine seal and the circuit along with the lcdmodule is placed on top of the meter. The various components of the smart meter are shown in the figures below.

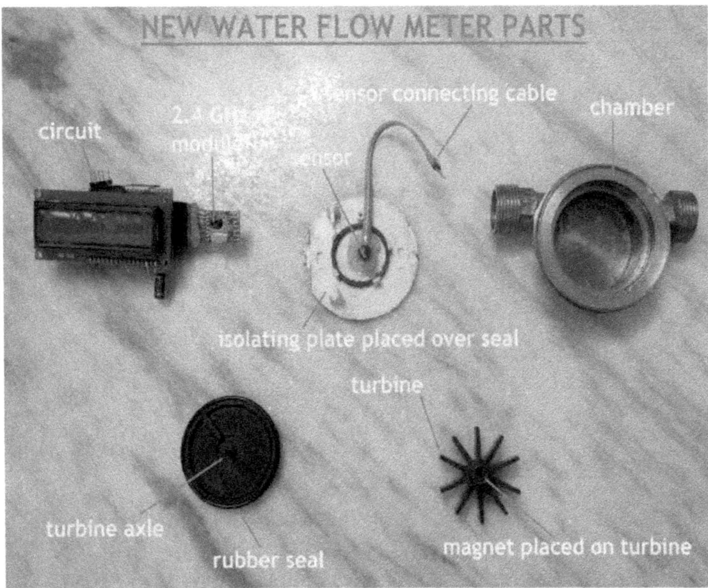

Fig. 15 Various components of the prototype smart flow meter

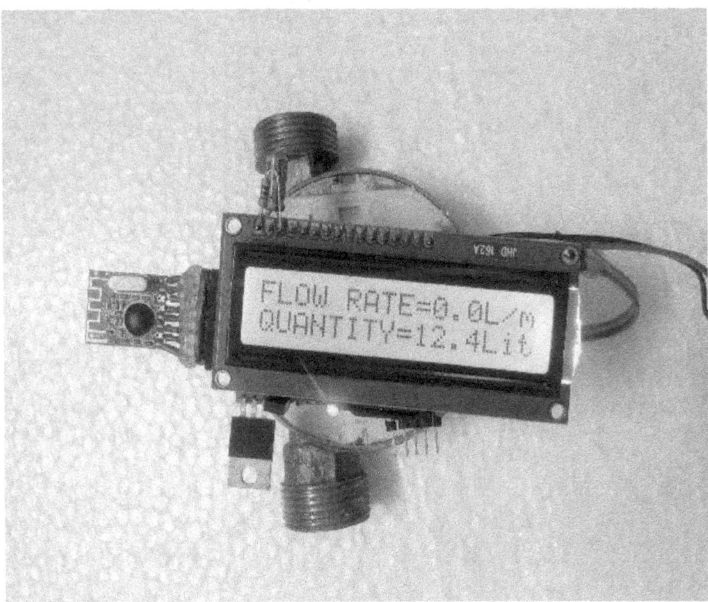

Fig. 16 Prototype smart water meter with 2.4 GHz RF

A smart meter with 433 MHz RF communication system is also constructed for testing purpose. All the components of it including the microcontroller and the RF module are placed in an existing ordinary flow meter as shown in fig. 17.

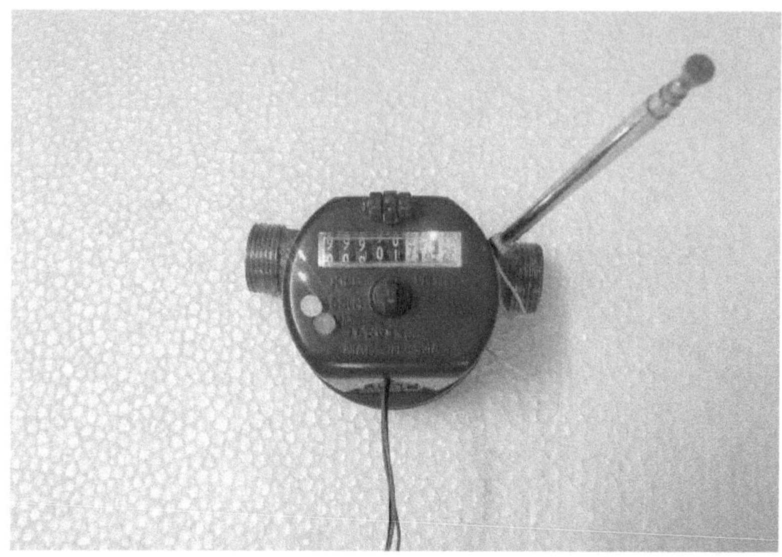

Fig. 17 Prototype smart water meter with 433MHz RF

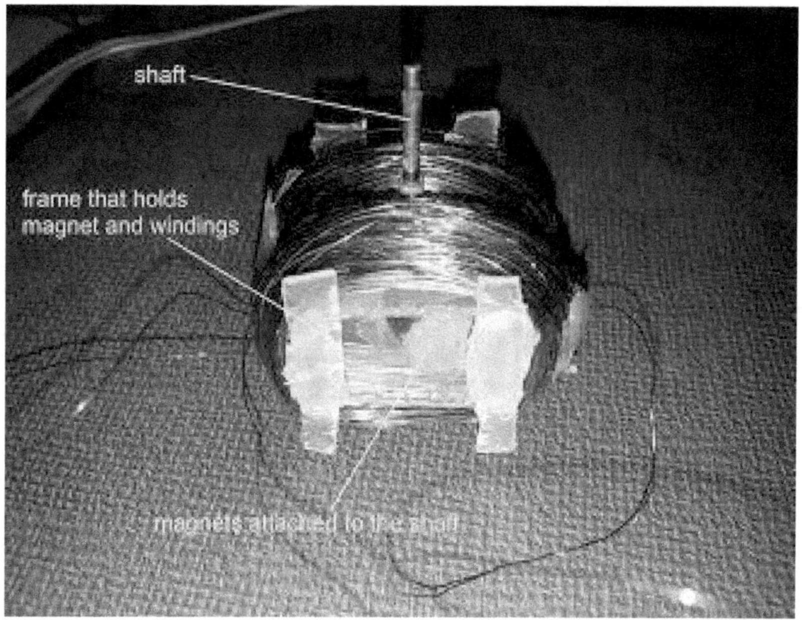

Fig. 18 2-pole AC generator constructed for power generation

The generator shaft is connected to a turbine. Theturbine is placed radial to the water flow direction. Thegenerator and the turbine are sealed completely to avoid any leakage as shown in fig. 19.

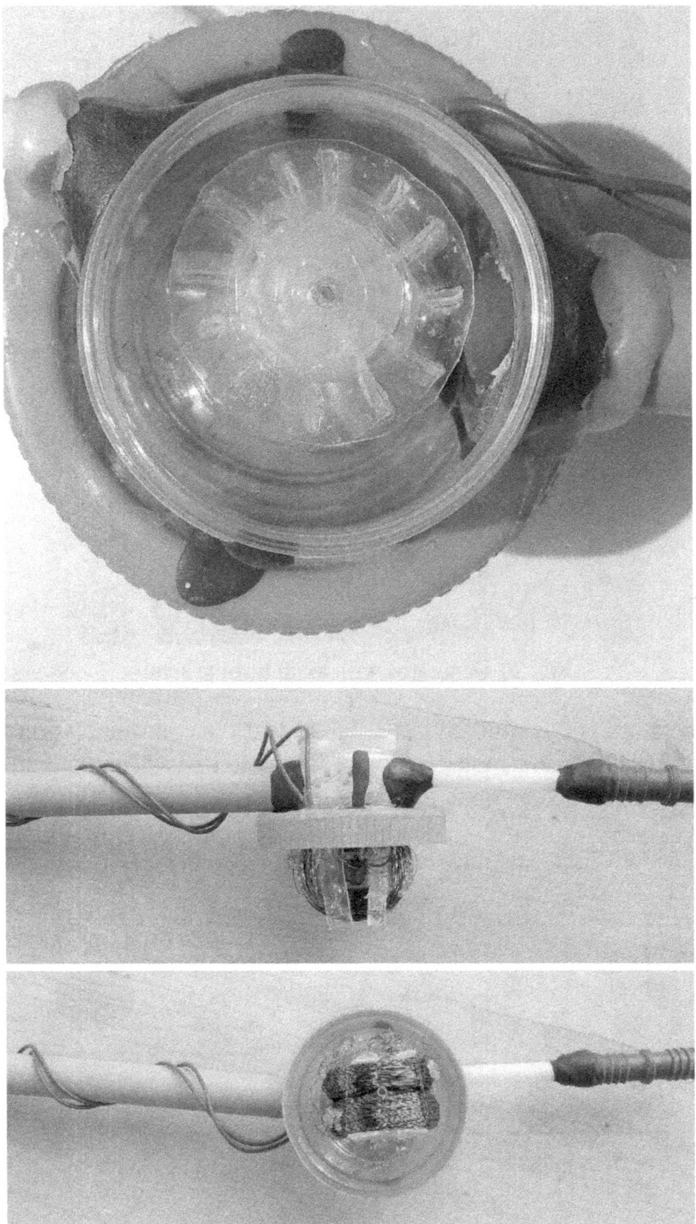

Fig. 19 Generator and radial turbine setup in a closed chamber

A generator with an axial turbine setup is also constructed. A flow obstructing dome with 4 small inlet nozzles is placed over the turbine such that the water jets inject on the turbine vanes. The turbine is connected to a generator. The generator is joined to the walls of the pipe as shown in the fig.20.

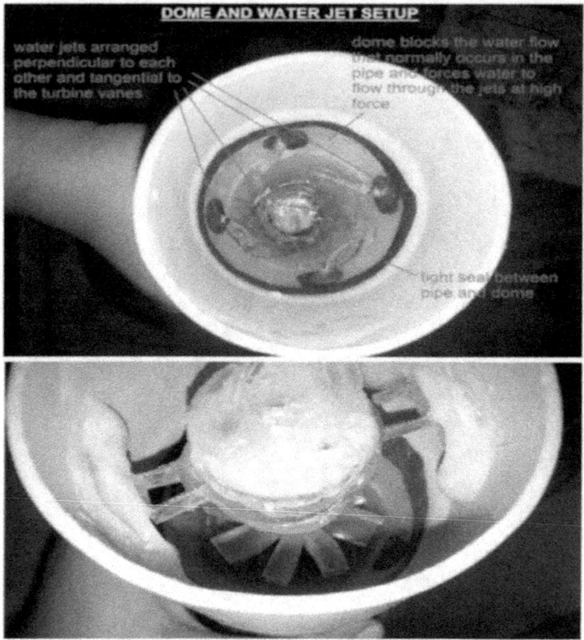

Fig. 20 Generator with axial turbine setup

A smart meter network is constructed in the lab to test the communication and leak detection algorithms. The fig.21 shows the model of the constructed smart meter network. It has 3 sub sections. Taps are placed on the sections to simulate leaks in the network.

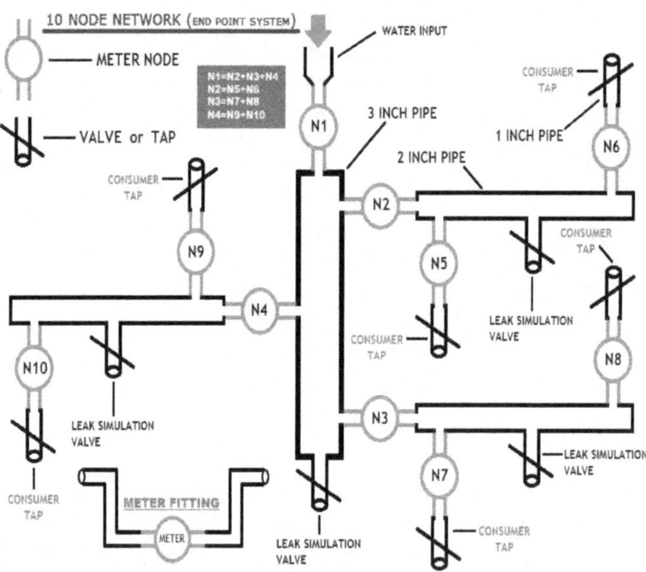

Fig. 21 Prototype smart meter network constructed in the lab

Fig. 22 Smart meter network constructed in the lab with ten smart meters

The smart meter network is run continuously run everyday to test the wireless communication and the accuracy of the meters. Leak detection algorithms are tested rigorously. Drain valves are placed in each section to simulate leaks and are opened at regular intervals to test if leak alert is raised.

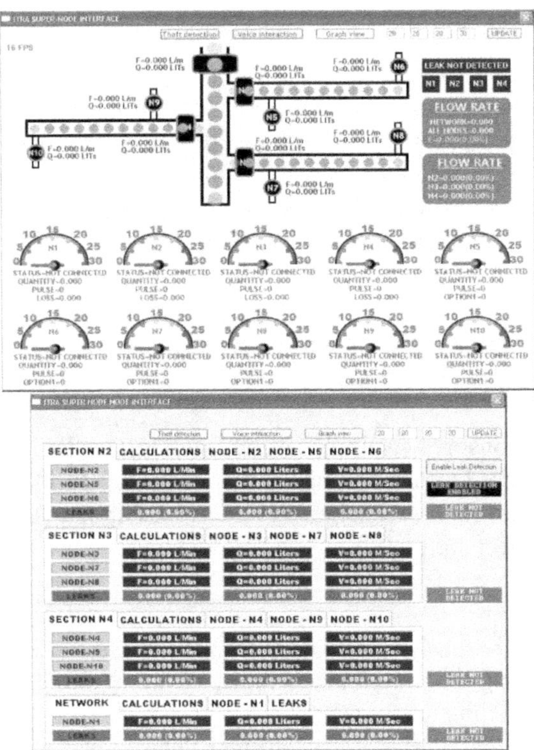

Fig. 23 Software developed to interface the smart meters and detect leaks

Special software shown in fig. 23 has been designed to interface the smart meters. It shows the flow rate, quantity and velocity of water at each smart meter in the network. The network map is also programmed into it so that leak detection algorithms can be effectively computed. The program also stores this data into a database which can be accessed from a mobile device in real time. When leak is detected the program raises an alert to the user about the section in which leak has occurred and also flow rate of leaking water. This system if applied to large scale distribution network can improve the reliability of the network.

9. CONCLUSIONS

Use of Remote and In-situ wireless Sensors and ICT for water quantity monitoring of water supply distribution networks is the need of the hour to reduce Non Revenue Water (NRW). The near real time continuous monitoring of water supply could also identify water theft, pipe burst or water leakage in the entire water supply distribution network.

To this end, we have developed Aquasense – Smart Water Meter as an IOT solution using LPWAN technologies, which operates on a self power driven unit coupled with the circuit.

We have developed this system on a test bed to initially monitor the water quantity and flow rate using a hall sensor, built with RF communication system. The smart meter is supported with a self generating power unit developed using the pipe water flow. Our system architecture enables continuous streaming of sensor data wirelessly, as well as centralized data archival,processing and visualization by web and mobile application.

Our future work extends on several fronts, from AQUASENSE and systems perspective, in the next phase we would extend this project to increase in the number of nodes in the sensor network from 10 to 100, allowing for ad-hoc LPWAN network formation and Routing optimization with respect to bandwidth usage. We intend to develop on-node event detection techniques and localization algorithms for pipe bursts and leaks. We plan to extend the power generation, for a very low flow rates, and conserve the power generated for no flow conditions. From the application domain perspective, we are in the process of adding in-pipe water quality measurements to our system. This will help event detection and localization move from flow rate and quantity related events such as bursts and leaks to contamination events.

ACKNOWLEDGMENT

This research was being sponsored by Information Technology Research Academy – ITRA, a unit of Media Lab Asia, under Department of Electronics and Information Technology, Ministry of Communications and Information Technology, Govt. of India. All the team members from LIN and PIN, students, project staff are highly acknowledged.

REFERENCES

1. Michael Allen, Mudasser Iqbal, Lewis Girod, Ami Preis, Seshan Srirangarajan, Cheng Fu, Kai Juan Wong, Hock Beng Lim, Andrew Whittle, "WaterWiSe@SG: a WSN for Continuous Monitoring of Water Distribution Systems"
2. Thewodros G. Mamo, IlanJuran , " The Potential Future Innovative Application of Municipal Water Supply Database". International Journal of Scientific Engineering and Research (IJSER),www.ijser.in,ISSN (Online): 2347-3878,Volume 2 Issue 3, March 2014.
3. Milind Naphade , IBM Research "Smart Water Pilot Study Report", City of Dubuque

4. Thulo Ram Gurunga, Rodney A. Stewartb, ∗ , Ashok K. Sharmac, Cara D. Beald. “Smart meters for enhanced water supply network modelling and infrastructure planning, 2014 Elsevier.
5. Anshul Bansal, Susheel Kaushik Rompikuntla, Jaganadh Gopinadhan, Amanpreet Kaur, Zahoor Ahamed Kazi “Energy Consumption Forecasting for Smart Meters”

WATER RESOURCES DEVELOPMENT FOR SUSTAINABLE PRODUCTION IN DRYLAND AGRICULTURE THROUGH CATCHMENT MANAGEMENT

R. S. Patode[1], M. B. Nagdeve[2], G. Ravindra Chary[3] and K. Ramamohan Reddy[4]

[1]Associate Professor (Agricultural Engineer),
[2]Chief Scientist, All India Coordinated Research Project for Dryland Agriculture,
Dr. Panjabrao Deshmukh Krishi Vidyapeeth Akola (Maharashtra)
[3]Project Coordinator, AICRPDA, CRIDA, Hyderabad
[4]Professor, Centre for Water Resources, IST, JNTUH, Hyderabad

INTRODUCTION

Agriculture, being the major water user, its share in the total freshwater demand is bound to decrease from the present 83% to 68% due to more pressing and competing demands from other sectors by 2050 AD (GOI, 2013), and the country will face water scarcity if adequate and sustainable water management initiatives are not implemented. The country is already experiencing water shortage and the problem become very acute in the near future unless preventive measures are taken on substantial scale. The recent estimates made by the Central Water Commission indicate that the water resources utilizable through surface structures are about 690 cubic kms only (about 36% of the total). Groundwater is another important source of water. The need for production of food, fodder, fibre, fuel in the crop growing areas have to compete with the growing space require for urbanization. The factors of land degradation, like water logging, salinity, alkalinity and erosion of soils on account of inadequate planning and inefficient management of water resources projects will severely constrain the growth of net sown area in the future. The requirement of 261 BCM by 2025 and 373 BCM by 2050 will be met by utilizing perennial ground water resources as well as from the storages created. As the requirements are dependent on concentration of human habitation and industrial activities, perennial sources of surface and ground waters need to be identified. Also, to ensure conservation, demand should be reduced as far as possible, leakages in the supply network should be minimized and recycling of waste water should be ensured for Agricultural/Horticultural uses. In some areas, desalination may be the only way to provide water security for domestic requirements (Chakraborty, et al., 2012).

Dryland agriculture has been in practice from time immemorial. The area under drylands has shown varied changes in the past decades. Contribution of dryland agriculture is of utmost importance, as 44 per cent of the nation's total food production coming from the drylands. The increase in the irrigated area had an impact on dryland agriculture. The major crops of cotton and oilseeds had shown an increase in the area grown in drylands, where as pulses remain unchanged. These changes are mainly due to the factors of improved irrigation, socio-economic factors etc. and more importantly the impact of green revolution. The area under drylands in India is in a declining rate and is expected to be stabilized by 2050 at 75 million ha (Rao and Ryan, 2004 and Singh et al., 2004). Since 1950 a lot of changes have happened in the dry lands in terms of area and yield. A vast majority of the small scale farmers depend on the dry regions for their livelihood. Dry regions are economically fragile regions which are highly vulnerable to environmental stress and shocks. Degraded soils with low water holding capacities along with multiple nutrient deficiencies and depleting ground water table contributes to low crop yields and further leading to land degradation. In order to ensure long term sustainability for dryland agriculture in India, various components are to be taken into consideration like socio-economic resources, integrated watershed development, improvement of rain water use efficiency, diversification of agriculture through livestock farming alternative land uses and integrated soil–nutrient-water-crop management (Roshni Vijayan, 2016).

CHARACTERISTICS OF DRYLAND AGRICULTURE IN INDIA

Rainfall: The most important contributing character in the drylands is low rainfall, within a range of from 375 mm to 1125 mm which are unevenly distributed, highly erratic and uncertain. The crop production in drylands is mainly dependent on the frequency and intensity of rainfall making it a less productive. **Soil:** The major causes for land degradation include the chemical degradation of soil, loss of soil structure and texture, loss of natural vegetation leading to soil erosion. Sequestration of carbon also turns out to be a major problem, which further degrades the soil making it less productive (ftp://ftp.fao.org). **Occurrence of drought:** The extensive climatic hazards are seen in drylands as the soils are weak and can be subjected to environmental stress to a higher level, leading to further land degradation. Drought is a common scenario in drylands as water availability is less further leading to low productivity. **Extensive agriculture:** Prevalence of monocropping extensively makes farm lands lack of nutrients and result in reduction of yield. **Crops grown:** The crops grown in a particular region will be similar and are not much remunerative compared to the major crops like rice and wheat. When similar crops are grown, all mature at the same time and a large quantity of produce will reach the market leading to glut in the markets. This situation is severely exploited by the traders and the middlemen in the markets. The issue of marketing turns out to be a big problem in dryland agriculture. **Poor economy of farmers:** Economic status and of living of farmers is low in drylands, due to the less choice of the crops that are grown in these areas (Roshni Vijayan, 2016).

Watershed management is a concept which recognizes the judicious management of three basic resources of soil, water and vegetation, on watershed basis, for achieving particular objective for the well being of the people. It includes treatment of land most suitable biological as well as engineering measures. To meet the challenges of sustainability and catchment management requires an approach that assesses resource usage options and environmental impacts integratively. Assessment must be able to integrate several dimensions: the consideration of multiple issues and stakeholders, the key disciplines within and between the human and natural sciences, multiple scales of system behaviour, cascading effects both spatially and temporally, models of the different system components, and multiple databases (Jakeman and Letcher, 2003). Catchment management involves determination of alternative land treatment measures for, which information about problems of land, soil, water and vegetation in the watershed is essential.

METHODOLOGY

1. **For catchment management**
    - Consider overall slope of the micro-catchment
    - Plan across the slope location specific management practices
    - For safe disposal of surface and subsurface runoff prepare common drainage trenches

- The runoff from drainage trenches should be collected in suitable storage structure like farm pond
- The use of harvested water can be possible for protective irrigation to different crops in rabi as well as kharif season based on availability of water in the storage structure.

2. **For water balance estimation**

In hydrology, a water balance equation can be used to describe the flow of water in and out of a system. The method is essentially a bookkeeping procedure, which estimates the balance between the inflow and outflow of water. Water balance shows the conceptual representation of the hydrological cycle. Rainfall is partitioned into various hydrological components as defined by mass balance equation (Walton *et al;* 1970) such as:

Rainfall = Run off from the watershed boundry + Change in ground water storage + Change in reservoir storage + evapotranspiration (evaporation + Transipiration) + Change in soil moisture

The hydrological water balance at Akola location for the duration 2010-11 to 2016-17 was observed as, on an average annually 783.4mm rainfall was received out of which 105mm (13.4%) annual surface runoff was occurred and evapotranspiration losses were 379mm (48.46%), around 230mm (29.41%) recharge takes place annually and other losses were 68.2mm (8.71%).

3. **Application of hydrological model for impact assessment**

Estimation of total water balance is a substantial issue for watershed modelling in order to simulate the major components of the hydrological cycle to determine the stress of different anthropogenic activities on the available water resources within a catchment. Unlike other watershed models, the MIKE SHE model simulates all the processes in the hydrological cycle by fully integrating the surface, subsurface and groundwater flow (Bahaa-eldin E. A. Rahim, et al. 2012). Several studies have been carried out using the MIKE SHE model in different regions, and under diverse soil and climatic conditions. Traditionally, the main aims of hydrological research are to provide an understanding of the water balance operating in catchments or watersheds, the physical processes that control water movement, and the impacts on water quantity and quality for a historical account of pioneering catchment studies (Bachelor *et al.* 1998 quoted in Mungai *et al.* 2004). The MIKE SHE modelling system consists of a water movement module and several water quality modules. The water movement module simulates the hydrological components including ET, soil water movement, OL, streamflow and groundwater flow (Yan & Zhang 2005). Compared with other watershed models, the MIKE SHE model simulates all the processes in the hydrological cycle by fully integrating the surface, subsurface and groundwater flow. Although snowmelt option is also incorporated in the model, however, it is not applicable for tropical climate conditions. It uses the fully dynamic Saint Venant equations to determine surface runoff rather than the Soil Conservation Service curve number (SCS-CN) method. The most recent model enhancement is the development of an integrated surface water and groundwater model by coupling the water movement module of MIKE SHE with the channel flow simulation component of MIKE 11 (Yan & Zhang 2005). It has a modular structure, which enables data exchange between components as well as addition of new components (DHI 2004).

Thus, the integration nature and the ability to account for both surface and subsurface flow systems, and their interaction make MIKE SHE well suited establishing a detailed water balance study for wetland catchments. The MIKE SHE Water Movement module has a modular structure comprising six process-oriented components that describe the major physical processes of the land phase of the hydrological cycle (Fig. 1). One and two-dimensional (2D) diffusive wave approximation Saint Venant equations describe channel and OL, respectively. The methods of Kristensen & Jensen (1975) are used for ET, the 1D equation of Richards (1931) for UZ flow and a 3D equation of Boussinesq (1904) for SZ zone flow. These partial differential equations are solved by finite difference methods, while other methods (interception/ET and snowmelt) in the model are empirical equations obtained from independent experimental research (DHI 2004).

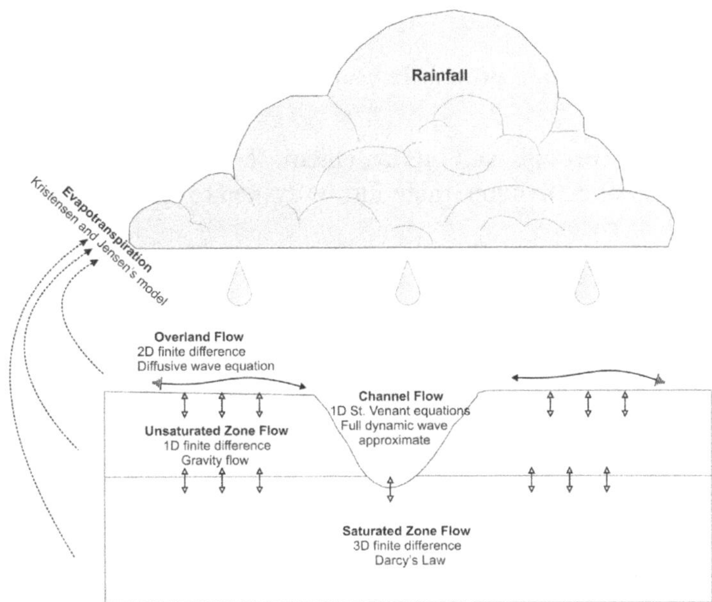

Fig. 1 MIKE SHE components and solutions used in water movement modelling

MANAGEMENT PRACTICES IN DRYLAND AREA

In dryland agriculture, scarcity of water is the main problem. Apart from the low and erratic behavior of rainfall, high evaporative demand and limited water holding capacity of the soil constitute the principle constraint in the crop production in dryland areas. Yield fluctuations are high mainly due to vagaries of weather, often much behind the risk bearing capacity of the farmers. It is surprising to a layman that even humid areas with 2000 mm of annual rainfall not only suffer from moisture stress, but also face drinking water scarcity. Monsoon starts in the month of June and ends in last week of September or sometimes in the first week of October. Most of the rainfall is received during this period. With undulating topography and low moisture retention capacity of the soil, major portion of the rainwater is lost through runoff, causing erosion and adding to the water logging of low lying areas. After the rain stops, very little moisture is left in the profile to support plant growth and grain production. In dryland area deficiency and uncertainty in rainfall of high intensity causes excessive loss of soil through erosion which leaves the soil infertile. Owing to erratic behaviour and improper distribution of rainfall, agriculture is risky, farmers lack resources, tools become inefficient and ultimately productivity is low. The problems can be solved by adopting the following location specific management practices,

- Crop planning
- Planning for aberrant weather
- Crop substitution / Cropping system
- Rainwater management
- Micro-watershed approach
- Resource improvement and utilization
- Alternate land use system

Strategies of Water Conservation in Arable Lands

A) ***Preventive measures***

Water conservation through soil management: Mulch farming, conservation tillage, rough seed bed, contour cultivation, ridge furrow system of planting, formation of tie ridges and soil conservation systems

B) ***Control measures***

(a) ***Slope management through***
(i) Terracing
(ii) Contour bunds

(b) ***Runoff management through***
(i) Surplus water disposal structures
(ii) Grassed Waterways
(iii) Soil Conservation structures
(iv) Rainwater harvesting

RESULTS AND DISCUSSION

At the experimental farm of AICRPDA, Dr. Panjabrao Deshmukh Krishi Vidyapeeth, Akola (Agricultural University) the various research experiments of *in-situ* rainwater conservation and water resource management had been conducted. Based on the results of some of these experiments, following recommendations has been made. These recommendations are useful for water resources development in dryland area.

A. ***Rainwater management***

- For reducing runoff and soil loss and increasing crop productivity, vegetative keyline of vetiver or Leucaena should be developed on contours and cultivation should be done along the vegetative key lines on contour.
- The reduction in runoff was observed to be in the tune of 40-50% whereas, the reduction in soil loss was observed to be 70-75% in case of vegetative barriers as compare to across the slope sowing. Uniform moisture distribution due to contour

cultivation gave higher productivity in case of test crops i.e. sorghum and cotton to the extent of 15 and 20%, respectively.

B. *Alternate land management:* Continuous Contour Trenches for perennial plantations

The results of evaluation of CCTs for perennial plantation for the year 2016-17 are presented here.

Productivity

The green gram was sown as an intercrop at the spacing of 30x10cm in continuous contour trenches (CCT) treated as well as untreated catchments. The computed yield of greengram is given in Table 1. The higher yield of pod and grain of greengram was observed in the treatment of CCT treated catchment, T_2 (785 and 520kg ha^{-1}) as against untreated catchment, T_1 (536 and 360kg ha^{-1}).

Table 1 Yield of greengram

Treatment	Yield (kg ha^{-1})	
	Pods	Grain
Untreated catchment (T_1)	536	360
CCT streated catchment (T_2)	785	520

Runoff

During the season total 6 runoff events was occurred and given in Table 2. The runoff causing rainfall was 296.5mm which had caused surface runoff of 49.6mm (16.72%) in untreated catchment, T_1 and 4mm (1.35%) runoff was observed in the CCT treated catchment, T_2. This indicates that in the CCT treatment catchment the runoff (98.65%) was recharged in the soil and ultimately reached to the groundwater. This will show the usefulness of CCT's on small catchment for *in-situ* conservation of the rainfall.

Table 2 Runoff during the season 2016 through treated and untreated catchment

Date	Rainfall, Mm	Runoff, mm	
		Untreated catchment (T_1)	CCT treated catchment (T_2)
30-06-2016	42.2	7.10	0.0
10-07-2016	91.0	15.70	0.0
11-07-2016	74.1	12.10	4.0
26-07-2016	38.2	7.50	0.0
03-08-2016	21.0	2.90	0.0
17-09-2016	30.0	4.30	0.0
Total	**296.5**	**49.6**	**4.0**

LAI and Interception component of crop and perennial plantation

The leaf area index (LAI) for the crop greengram is calculated by observing the stage wise growth. The canopy interception was calculated based on the LAI during the growth stages. The data of LAI and canopy interception is presented in Table 3. It was observed that the LAI goes on increasing as per the growth of the crop and was highest at the pod initiation (3.92) stages and then decreases as the crop matures. The LAI was more in the greengram crop sown in CCT treated catchment (T_2) compared to untreated catchment (T_1) at every stage of crop growth. As the interception component function directly relates to the LAI, the canopy interception was observed maximum at the pod initiation stage of crop growth and it was maximum in greengram crop sown in CCT treated catchment as compared to untreated catchment. The growth stage wise LAI and canopy interception for green gram is presented in Fig. 2.

The leaf area index (LAI) for the perennial plantation is calculated by observing the stage wise growth. The canopy interception was calculated based on the LAI during the growth stages. The data of LAI and canopy interception is presented in Table 4. It is observed that the LAI goes on increasing as per the growth of the plantation and was highest at the developed fruit stages of Custard apple and Hanuman phal. The LAI was more in the plantation of CCT treated catchment (T_2) compared to untreated catchment (T_1) at every stage of crop growth. As the interception component function directly relates the LAI, the canopy interception was observed maximum at the developed fruit stages and it was observed maximum for plantation in CCT treated catchment as compared to untreated catchment.

Table 3 Growth stage wise LAI and canopy interception for greengram

Crop	Growth stages	Leaf Area Index, LAI		Canopy Interception, I_{max} (mm)	
		CCT Treated catchment	Untreated catchment	CCT Treated catchment	Untreated catchment
Green gram	Initial growth	1.42	1.09	0.07	0.05
	Flowering initiation	2.68	2.14	0.13	0.11
	Flowering	3.74	3.18	0.19	0.16
	Pod initiation	3.92	3.24	0.20	0.16
	Pod development	1.84	1.32	0.09	0.07

Table 4 Growth stage wise LAI and canopy interception of perennial plantation

Growth stages	Leaf Area Index, LAI of Custard apple		Leaf Area Index, LAI of Hanuman phal		Canopy Interception for Custard apple (mm)		Canopy Interception for Hanuman phal (mm)	
	CCT Treated catchment	Untr-eated catchment	CCT Treated catchment	Untr-eated catchment	CCT Treated catchment	Untr-eated catchment	CCT Treated catchment	Untr-eated catchment
New leaves	0.09	0.08	0.58	0.53	0.005	0.004	0.029	0.027
Fruit initiation	2.22	2.18	2.63	2.32	0.111	0.109	0.132	0.116
Fruit development	3.16	3.05	4.52	3.82	0.158	0.153	0.226	0.191
Developed fruits	3.25	3.08	4.72	4.02	0.163	0.154	0.236	0.201
Maturity	2.35	2.12	3.14	2.80	0.118	0.106	0.157	0.140
Leaves shreading	1.32	1.02	1.89	1.52	0.066	0.051	0.095	0.076

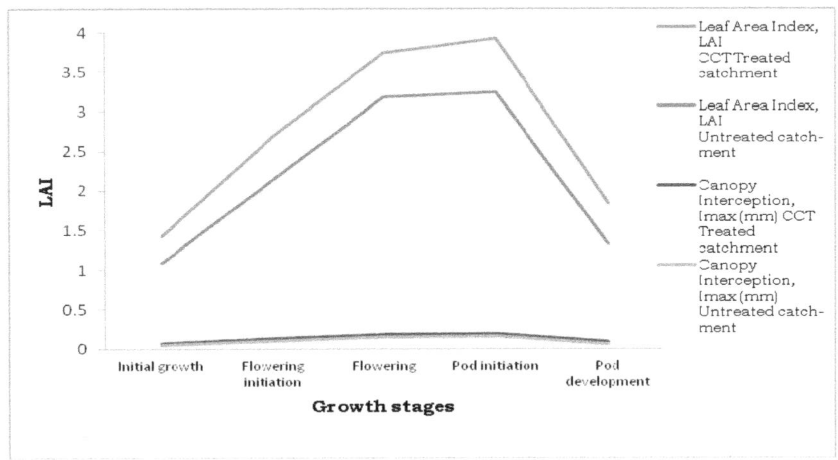

Fig. 2 Growth stagewise LAI and canopy interception of greengram

- The soil moisture status in CCT treated catchment was observed to be better as compared to the untreated catchment at 0-15, 15-30 and 30-45cm depth in every recorded month.
- On an average the ground water recharge in the CCT treated catchment was more by 21.74 % compared to the non treated catchment.
- The CCTs are useful in order to increase infiltration into soil, to control damaging excess runoff and to manage and utilize runoff for groundwater recharge.
- The *in-situ* conservation of rainfall takes place in CCT treated catchment.
- The prolonged moisture in the CCT treated catchment will enhance the growth of perennial plantation.
- The yield obtained during 2012, 2013 and 2015 is depicted in Fig. 3. It was observed that the increase in yield in CCT treated micro-catchment was 47.57, 59.56 and 47.89% over control during the year 2012, 2013 and 2015, respectively.

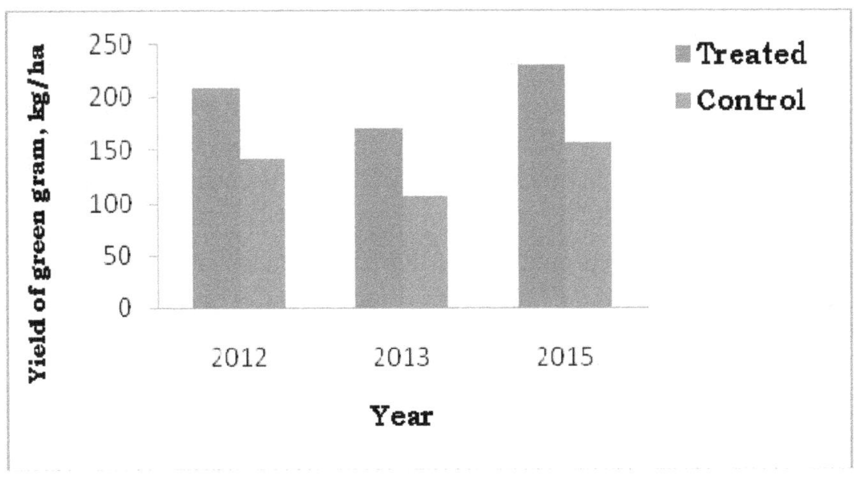

Fig. 3 Comparison of yield of green gram in control and treated micro-catchment

C. *Rainwater Harvesting:* Rainwater harvesting in farm pond and its reuse

Depending upon the amount of rainfall and the intensity of rainfall, the runoff from the catchments can be accumulated in the farm ponds. On the basis of availability of water in the farm ponds and as per need the protective irrigation can be given to the crops in *Kharif*, *Rabi* season as well as to different vegetables. It was observed that one protective irrigation for soybean during *kharif* resulted in yield increase of 23.78% over rainfed crop. Similarly during *rabi* the increase in yield of chickpea over rainfed crop was 45.44%. The farm pond water was used for irrigation to different vegetables by using micro-irrigation systems. It was found that in the vegetable like Okra, Cluster Bean, Brinjal, Sponge Guard, Bitter Guard, Fenugreek, Spinach, Corriander, Carrot, Radish, and Tinda the water use efficiency was in the range of 1.05 - 4.50 kg/m^3. Income generation is possible for small and marginal farmers.

D. *Rainwater Harvesting Structures on Drainage line*

The harvested rainwater was used in *kharif* for soybean crop. One protective irrigation during the dry spell from stored water resulted in significant increase in yield as compared to rainfed condition i.e without irrigation. Overall 31-40% increase in yield was observed due to protective irrigation provided during the critical growth stage of pod filling to soybean crop. Based on the pre and post project implementation data regarding, water storage, reuse of stored water, yield economics, increase in groundwater levels, availability of water in the surrounding wells, cropping pattern it can be recommended to undertake the deepening and widening of existing drainage line network along with construction or repairing of the permanent structure and reuse of harvested water through micro-irrigation in this region.

The well location map was prepared from satellite images with the reference of GPS coordinated system in the Arc GIS 10.00 software. The shape file of wells is attached the groundwater level data for groundwater level maps and impact analysis using IDW techniques. Groundwater level fluctuations data were calculated on the seasonal basis. The groundwater level data was converted in the Arc GIS domain for the impact of rainwater harvesting structure in the groundwater depth and crop production at the surrounding watershed area. The groundwater level fluctuation maps were generated with the help of inverse distance weighted interpolation method (IDW) in the GIS environment. This groundwater maps can be useful for watershed modeling, future prediction of water level and watershed impact changes in the watershed area.

- Observation wells have been monitored, during pre-monsoon period (June 2015) the groundwater depth ranges from 6.7 to 14.2 m. and thereafter in pre monsoon period (June 2016) it was observed to be 3.25 to 9.6m (Fig.4 and 5).

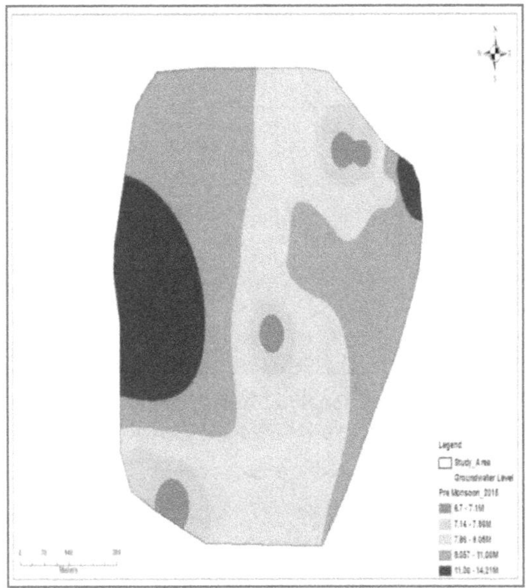

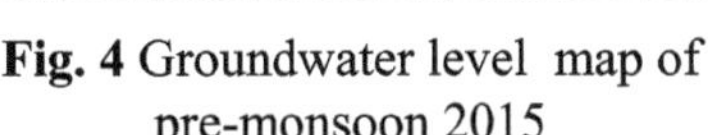

Fig. 4 Groundwater level map of pre-monsoon 2015

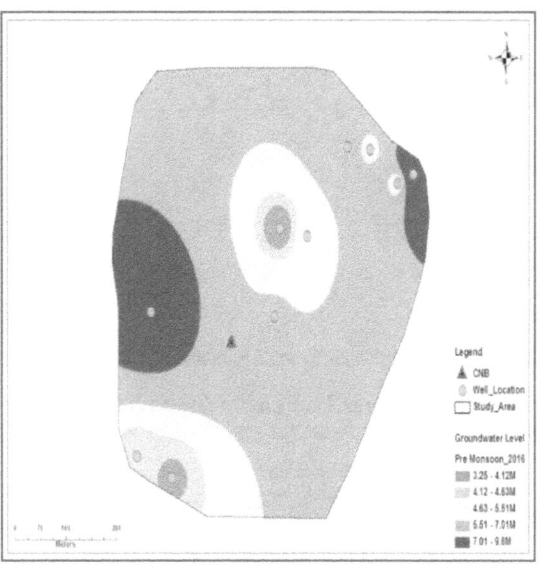

Fig. 5 Groundwater level map of pre-monsoon 2016

E. *Impact Assessment through MIKE SHE model*

Accurate assessment and prediction of effects of alternative land use activities on watersheds through computer modelling techniques is essential. Therefore, the research work on same line was carried out with main objective of demonstrating the advantages of CCT in rainfed area and application of a comprehensive hydrological model in studying the detailed hydrological behavior of the micro-catchments. For this purpose MIKE SHE model was used. The experimental area was divided into two micro-catchments out of which one is treated with CCTs and the adjacent has been kept as untreated. Both micro-catchments have horticultural plantations of Atemoya (*Anona atemoya*) and Custard apple (*Anona squamosa*). The intercrop during *Kharif* has been practiced along the plantation rows. Performance evaluation of the existing CCT shows that its adoption results in about 100, 95.50 and 93.30% reduction in runoff during 2013, 2014 and 2015, respectively. On an average CCT enhances groundwater recharge by about 20.51% in the treated micro-catchment over control. Moreover, it can be an effective measure for utilizing non-arable lands for fruit plantations and for raising agronomical crops between plant rows. The CCTs also helps in increasing the fertility of soils. After the model run, the examples of the results in the form of screen shots are depicted in Fig. 6. These results are presented here for giving the idea of how the MIKE SHE outputs in different forms can be assessed.

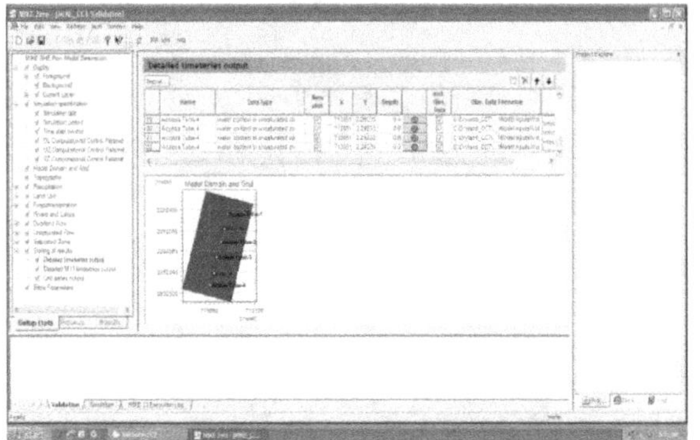

Location of Access Tubes and Observation Wells

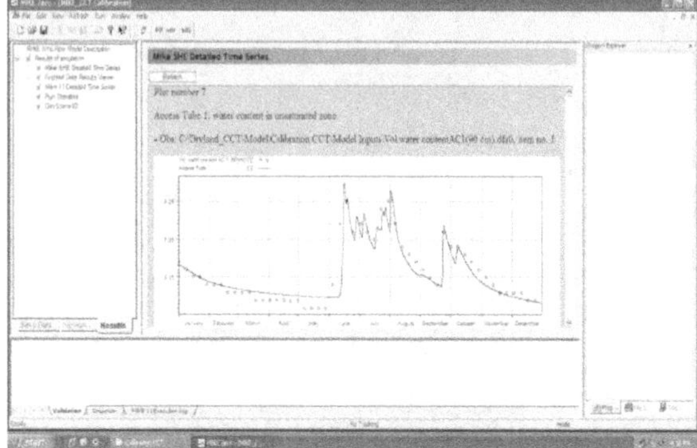

Model Output for Soil Moisture at one Location

Model Output for LAI and Root Depth

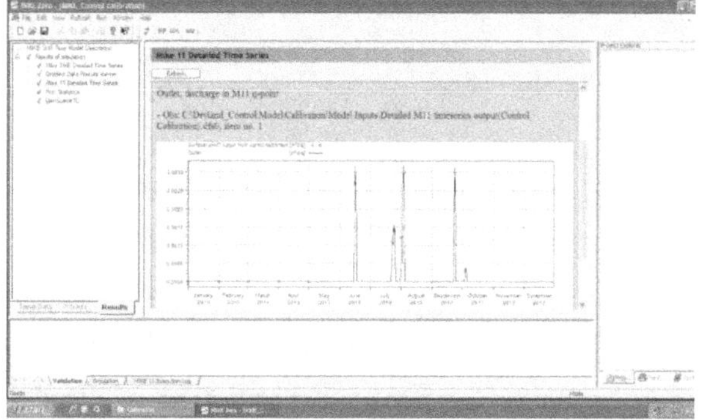

Model Output for Surface Runoff

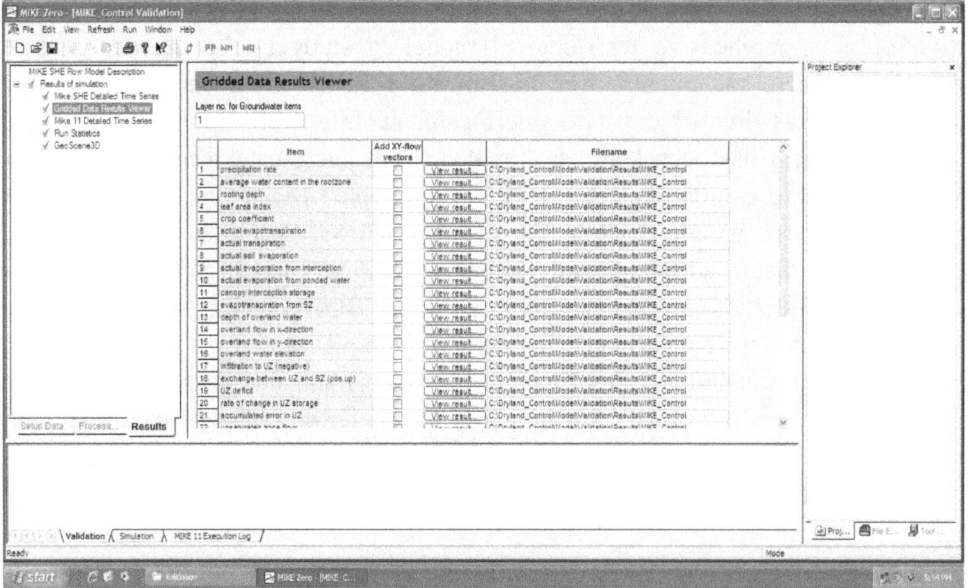

Model Output Results

Fig. 6 Screen Shot Results of MIKE SHE model simulation for the micro-catchment

CONCLUSION

Due to adoption and proper implementation of these techniques the maximum amount of rainwater will be recharged into the ground and ultimately the water resources development and management is possible which ultimately useful for sustainable crop production in dryland area.

ACKNOWLEDGEMENTS

The funds were provided by ICAR- AICRPDA, CRIDA, Santoshnagar, Hyderabad and ICAR-IIWM, Bhubaneswar for the research/projects to All India Coordinated Research Project for Dryland Agriculture, Dr. Panjabrao Deshmukh Krishi Vidyapeeth, Akola, Maharashtra, India. We are thankful to them for the kind help and financial support.

REFERENCES

1. Bachelor, C., Cain, J., Farquharson, F. and Roberts, J. (1998). Improving water utilization from a catchment perspective. SWIM Paper 4, International Water Management Institute, Colombo, Sri Lanka.
2. Bahaa-eldin E. A. Rahim, Ismail Yusoff, Azmi M. Jafri, Zainudin Othman & Azman Abdul Ghani (2012).Application of MIKE SHE modelling system to set up a detailed water balance computation, Water and Environment Journal 26: 490–503.
3. Chakraborty, S.S., Mohan, S., Tyagi, N.K., Sonde, R.R. and Subash Chander (2012). Research Report-Water Resources Management, Indian National Academy of Engineering, New Delhi, India.
4. DHI, (2004). MIKE SHE User Manual. Danish Hydraulic Institute, Hørsholm, Denmark.
5. Gunnell, Y. (2003). Past and present status of runoff harvesting systems in dryland peninsular India: A critical review. Ambio., 32 : 320-324.
6. Jakeman, A. J. and R.A. Letcher (2003). Integrated assessment and modelling: features, principles and examples for catchment management, Environmental Modelling & Software, 18: 491–501.
7. Mungai, D.N., Ong, C.K., Kiteme, B., Elkaduwa, W. and Sakthivadive, R. (2004). Lessons from Two Long-Term Hydrological Studies in Kenya and Sri Lanka. *Agric. Ecosyst. Environ.*, 104 (1), 135–143.
8. Nagdeve, M.B. and R. S. Patode (2012). Protective irrigation through farm pond for enhancing crop productivity. Application Technologies for Harvested Rainwater in Farm Ponds, Proceedings of National consultation meeting held at CRIDA, Hyderabad during 19-20 March, 2012 : 62-67.
9. Patode, R. S., Nagdeve, M. B. and Pande, C. B. (2016).Groundwater Level Monitoring of Kajaleshwar-Warkhed Watershed, Tq. Barshitakli, Dist. Akola, India Through GIS Approach, Advances in Life Sciences 5(24), Print : ISSN 2278-3849, 11207-11210.
10. Peterson, G., Unger, P.W. and Payne, W.A. (2006). American Society of Agronomy, Crop Science Society of America. and Soil Science Society of America. Dryland agriculture. American Society of Agronomy: Crop Science Society of America: Soil Science Society of America, Madison, Wis.
11. Rao, A.S. Water and energy use efficiency in dryland crops. Training Manual (PPT), CRIDA, Hyderabad, 2012.
12. Rao, S.C. and Ryan, J. (2004). Crop Science Society of America. and Symposium on Challenges and Strategies for Dryland Agriculture. Challenges and strategies for dryland agriculture. In : Challenges and strategies for dryland agriculture. Crop Science Society of America, Madison, Wis.
13. Roshni Vijayan, Dryland agriculture in India – problems and solutions, AJESSIAN Journal Of Environmental Science Volume 11, Issue 2, December, 2016: 171-177
14. Shafi, M. and Raza, M. (1987). Dryland agriculture in India. In: Dryland agriculture in India. Rawat Publications, Jaipur (RAJASTHAN) INDIA.

15. Singh, H.P., Sharma, K.P., Reddy, G.S. and Sharma, K.L. (2004). Dryland Agriculture in India. p. 67-92. In: Chalenges and strategies for dryland agriculture. Crop Science Soc of, [S.l.].
16. Sivanappan, R.K. 2006. Rainwater harvesting, conservation and management strategies for Urban and Rural sectors. National Seminar on rainwater harvesting and water management, Nagpur. Pp.1-5.
17. Spartt, E.D. and Chowdhury, S.L. (1978). Improved cropping systems for rainfed agriculture in India. Field Crop Res., 1:103- 126.
18. Subba Rao, I.V. (2002). Land use diversification in rainfed agriculture. CRIDA Foundation day lecturer. Central Research Institute for Dryland Agriculture, Hyderabad (A.P.) INDIA.

REFERENCES

1. Carbon in soils: Available at ftp://ftp.fao.org/agl/agll/docs/ wsrr102.pdf Accessed on 10.29.09.
2. Dryland farming: available at http://www.worldagriculture.com/dry-land-farming/dry-land-farming.php Accessed on 09.30.09.
3. Dryland systems: Available at http://www.millenniumassessment.org/documents/
4. http://oceanworld.tamu.edu/resources/environment-book/aridlanddegradation.html. http://www.indiaenvironmentportal.org.in/content/soilsalinity-threatens-command-areas.
5. Soil Salinity: Available at http://www.informaction.org/cgibin/
6. Yan, J. and Zhang, J. (2005) Evaluation of the MIKE SHE Modelling System. http://s1004.okstate.edu/S1004/Regional-Bulletins/Modeling-Bulletin/MIKESHEfinal

FERTIGATION IN FRUIT CROPS

Surendra R. Patil

Professor, College of Horticulture Dr. Panjabrao Deshmukh Krishi Vidyapeeth Akola (MS).

INTRODUCTION

World agriculture has undergone a major transformation during the past four decades. Production per unit area of land has increased as a result of the adoption of improved farming practices. Per hectare yields of almost all the crops have increased several folds. These remarkable increases have been achieved with a package of practices that includes the use of external inputs and improved crop varieties. This form of energy intensive agriculture uses fertilizers for crop nutrition, agrochemicals for plant protection and mechanisation for farm operation. These practices cater to the needs of short- term economic gains, but lack long -term ecological sustainability.

Sustainability, therefore, is vital to our approach to today's agriculture. Interest in sustainability has been growing over years out of concern for fast depleting natural resources. Of particular concern is agricultural land and irrigation water, the quality and quantity of which have been declining almost unchecked. Unsustainable farming practices have rendered vast areas of land unsuitable for cultivation and plunged groundwater to alarmingly low depths. In such a scenario, the ultimate goal of development is not only to meet the immediate needs of the rural, but also to reserve the trend of natural resource depletion. Achieving this goal requires adoption of sustainable farming practices. Among farming practices, effective usage of water and nutrients play a crucial role in achieving crop productivity. Keeping this in view, an effort has been done to highlight the importance of fertigation technique, which is gaining momentum in the country to produce qualitative yields.

'Fertigation' is a technique for application of fertilizers in the irrigation water. Fertigation can be applied through buried or surface drip lines or through sprinklers. Recent technological developments in the drip and micro irrigation methods have accelerated the adoption of fertigation for a wider range of crops, including fruit trees. The uniform distribution of water by a given injection system is important for maximizing the uniformity of distribution of nutrients delivered through fertigation. Managing irrigation to minimize the leaching of water below the crop rooting depth is critical to minimize their leaching below the root zone. It is generally believed that carefully managed fertigation results in lower nutrient leaching losses than broadcast application of water-soluble granular fertilizers. However, these is dependent on the ability of crop to take up large amount of nutrient immediately following their application, and subsequent redistribute them from the vegetative crop parts into those of economic importance, i.e., fruits, tuber, etc.

Drip fertigation in way can be compared with spoon-feeding to plants. It ensures supply of pant nutrients to the root zone along with micro-irrigation system. Moreover, it offers the viable solution for intensive and economical crop production wherein both water and nutrients are delivered precisely and nutrient use efficiency increases to the larger extent.

Most of the conventional fertilisers are not suitable to apply through drip fertigation as they are not fully soluble and leave precipitation which eventually creates problems of clogging at discharge and reduce system life. In order to avoid the complications, a new class of completely water-soluble specialty fertilisers (WSF) started gaining momentum. This has been already experienced by a large number of progressive farmers of grapes, pomegranate, melons and banana in the states of Maharashtra, Karnataka, Andhra Pradesh, Gujarat and Tamil Nadu.

Irrigation and fertilization are the most important management factors through which we can control plant development, fruit yield and quality. Horticultural crops are long duration so application of fertilizer in split doses results in pronounced increase in the plant nutrient uptake.

Area, Production and Productivity of Fruits

Fruit crop	Area 000' ha	Production 000'MT	Productivity MT/HA
Mango	2309	12750	5.5
Citrus	709	26217	37.0
Grapes	80	1878	23.6
Pomegranate	109	807	7.4
Guava	204	2270	11.1

Growth in fertilizer consumption in India

Fertilizer consumption in India has increased significantly in the last three decades. Total NPK concentration has increased nine folds between (2 million ton to 18million ton) year 1969-1970 to 1999-2000.

After reaching record level in 1999-2000 fertilizer consumption in India has been irregular. It has fluctuated around 17 million ton since 2000-01.

Why use fertigation

1. Your irrigation system does the work, not you.
2. Overfeeding is eliminated.
3. Underfeeding is eliminated.
4. Waste is eliminated
5. Fertilization is continuous, not intermittent.
6. There in no runoff, due to the small amount of fertilizer applied in each drip system cycle.
7. Dry applications rely on subsequent watering to dissolve the granules. Here the fertilizer is liquid and supplied mixed with irrigation water.
8. With desert landscaping it is difficult to manually apply fertilization directly to each plant.
9. With precise application, the overall amount of chemicals applied is reduced vs. dry application.
10. If you have a lawn, there are no stripes where the application of dry fertilizer was uneven.

Advantages

1. Ensures regular flow of both water and nutrients, resulting in increased growth rates and higher yields. Increase in the yield by more than 50% could be achieved with appropriate fertigation schedule.

2. Offers greater versatility in the timing of the nutrient application to meet specific demands at predetermined times according to the crop. It should be noted that application of fertilizer in split doses results in pronounced increase in the plant nutrient uptake.
3. Improves availability of nutrients and their uptake by the plants. The irrigation system is designed to supply both water and nutrients directly to the roots, creating
4. Safer application method as it eliminates the danger of burning the plant root system, since the fertilizer is applied in very low concentration.
5. Improves efficiency as small amounts of fertilizers as applied at frequent intervals and according to different stage of crop growth. This results in substantial savings in quantity of fertilizers (30 % to 50 %).
6. Combining liquid fertilizers with insecticides and herbicides saves labour and machinery for their application separately.
7. Allows crops to be grow on marginal lands, sandy or rocky soils, where accurate control of water and fertilizers in the plant's root environment is critical.
8. Minimizes pollution of soil.
9. Ensure uniform nutrient application.

Disadvantages

1. Both the components (drip and water-soluble fertilizer) are very costly.
2. Maintenance of drip irrigation is difficult. There is possibility of theft and rat infestation.
3. Good quality water is very essential. Clogging of emitters may cause a serious problem.
4. It needs water soluble fertilisers; the availability of these types of fertilisers is limited.
5. Adjustment of fertilizers to suit the need is not easy.
6. Infestation of insect's pest and diseases increases.
7. Area under micro irrigation is now increasing mainly because of subsidy in micro irrigation, if subsidy is withdrawn, the area under micronutrient may also reduce. So also would be the fate of fertigation.
8. Due to fear of yield loss, because of relatively lower dose of fertilisers in fertigation, farmers have the tendency to add additional fertilisers and irrigation water by traditional methods too.

Methods of fertigation system:

I. According to irrigation water applied:

1. Drip fertigation: Fertilizers are applied through drip irrigation system.
2. Sprinkler fertigation: Fertilizers are applied through sprinkler irrigation system.
3. Furrow fertigation: Fertilizers applied through furrow irrigation system to crops.
4. Flood fertigation: Fertilizers applied through flood irrigation system to crops.

II. According to water and fertilization timing:

1. Continuous application: Fertiliser is applied at a constant rate from irrigation start to finish. The total amount is injected regardless of water discharge rate.

2. Three-stage application: Irrigation starts without fertilisers. Injection begins when the ground is wet. Injection cuts out before the irrigation cycle is completed. Remainder of the irrigation cycle allows the fertiliser to be flushed out of the system.
3. Proportional application: The injection rate is proportional to the water discharge rate, e.g. one litre of solution to 1000 litres of irrigation water. This method has the advantage of being extremely simple and allows for increased fertigation during periods of high water demand when most nutrients are required.
4. Quantitative application: Nutrient solution is applied in a calculated amount to each irrigation block, e.g. 20 litres to block A, 40 litres to block B. This method is suited to automation and allows the placement of the nutrients to be accurately controlled.

Choice of fertilizers

1. Choose the fertilizer that is completely soluble in water and does not produce the salt in system.
2. It does not react adversely with salts or other chemicals in irrigation water.
3. It should be safe for field use.
4. The fertilizer applied should avoid corrosion, softening of plastic pipes and tubes or clogging of any component of the system.
5. It should not be leached down below the root zone.
6. It should not change pH of water resulting in precipitation and clogging methods too.

FERTILIZERS IN DRIP FERTIGATION

Straight fertilizers like ammonium sulphate. Other nitrogen source s like urea and ammonium nitrate (NH_4NO_3) do not tend to interact with salts dissolve in irrigation water and their addition does not pose any risk.

Phosphate containing fertilizers used in drip fertigation may react in several ways with salts dissolved in irrigation water. One of the common source of phosphate is phosphoric acid or to be more precise, orthophosphoric acid (H_3PO_4). This is strong acid and by lowering the pH of the irrigation water it causes dissolution of some precipitated salts and thus acts as a cleaning or anti –clogging agent in the system. Mono Ammonium Phosphate ($NH_4H_2PO_4$), a salt of orthophosphoric acid, is used in drip fertigation.

Solubility of potassium salts in water at usual temperature is such that in most cases large concentrations may be injected into the irrigation water. Normally, at 20o C potassium chloride can give a solution of up to 34%, potassium nitrate upto 32%, Mono Potassium Phosphate up to 30%. Past experience apprises that issues like solubility and mobility of the nutrients, mixing compatibility, precipitation, blocking and corrosion occur when straight fertilizers are used in drip fertigation. This reduces the fertiliser use efficiency and also the maintenance of drip irrigation system.

WATER SOLUBLE FERTILISERS IN FERTIGATION

The WSF have more advantages over straight fertilisers. They are completely water-soluble and hence do not create any precipitation problem and further clogging of emitters. Micronutrients are available in most of WSF combinations. There is a possibility of suppyling all the three major nutrient requirements to the crop resulting in increased nutrient use efficiency and finally increased productivity.

Fertilizer for fertigation:

Sr. No	Fertilizers
Water soluble fertilizers	
1	Mono Ammonium Phosphate
2	Poly Feed
3	Mono Potassium Phosphate
4	Potassium Nitrate
5	Sulphate of Potash
6	Ortho Phosphoric Acid
7	Ammonium Nitrate
Conventional Fertilizers	
1	Urea
2	Potassium Chloride
3	Potassium Sulphate
4	Ammonium Sulphate

Trade names of fertilizer:

Ammophos - Ammonium phosphate

Ankur - Aqueous solution of urea and ammonium nitrate

ANL - Mixture of ammonium nitrate and lime

Poly feed - Contain micronutrients

Mahafeed (19:19:19)

Solufeed (19:19:19)

Koromandal (19:19:19)

Mitrophoscu (19:19:19), 0:52:34, 0:0:50

Sea-plus (3-2-2 NPK)

Parker neem

K-Carb 35 liquid fertilizer

K-Blast 36 (0-0-36)

Components

1. **Water Source:** The ideal Drip System is a low-pressure system that uses gravity to increase water pressure. The water source can be an overhead tank placed at a minimum of one meter above ground level for smaller systems up to 400-m2 area. For larger systems, the height of the tank should be increased. If the height of the tank is not increased, the system can be connected to a pump that lifts water from sources such as a well, farm pond, storage tank, or a stream / canal. A manually operated pressure pump also can be used to lift water from a shallow water table (up to 7 meters) and used for the system.

2. **Control Valve:** A valve made of plastic or metal to regulate required pressure and flow of water into the system. There are valves of various sizes depending on the flow rate of water in the system.
3. **Filter:** The filter ensures that clean water enters the system. There are different types of filters. Different sizes of filters are available depending on the flow rate of water in the system.
   - (i) **Hydrocyclone filter:** Hydrocyclone filters are used to remove solid particles of sizes 75 microns (200mesh) or above having density higher than water such as fine sand and silt, etc. entering into the system from the water source. Hydrocyclone filter is essential when the water source is a tube well located in a sandy stratum or a river with high silt load.
   - (ii) **Sand filter:** The surface water sources such as river, pond, lake or canal are exposed to direct sunlight and provide favourable conditions for growth of biological impurities like algae, plankton etc. Such impurities can pass through hydrocyclone or screen filter. Therefore, provision of sand filter is a must when the water from river, pond, lake or canal is used for drip irrigation system.
   - (iii) **Screen filter:** While majority of impurities are filtered by sand filter, very fine sand particles and other small impurities can pass through it. These are filtered out by screen filter. Types: a) super flow screen filter) super clean screen filter, c) brush clean screen filter, d) turbo clean screen filter,
   - (iv) **Disclean filter:** Disc filters are used where lower degrees of organic impurities are to be removed from the water source. It has plastic discs with radial grooves. Discs when stacked together and compressed facilitate three-dimensional filtration.
4. **Pressure gauge:** Maintaing the normal operating pressure in the system is essential to ensure uniformity of irrigation. Pressure gauge is provided at filtration unit to indicate this pressure.
5. **Fertilizer tank:** The fertilizer, after mixing in water, is filled in the fertilizer tank. Due to the pressure difference between the inlet and out let of the tank, the fertilizer gets mixed in the irrigation water and goes to the plant root zone.
6. **Mainline:** Pipe made of poly vinyl chloride (PVC) or polyethylene (PE) to convey water from the source to the sub main line. PE pipe material is normally made from high-density polyethylene (HDPE), low-density polyethylene (LDPE) and linear low-density polyethylene (LLDPE). The size of pipe depends on the flow rate of water in the system.
7. **Sub-main:** Made of PVC / HDPE / LDPE / LLDPE pipe to supply water to the lateral pipes. Lateral pipes are connected to the sub-main pipe at regular intervals. The size of pipe depends on the flow rate of water in the system.
8. **Lateral:** Pipes made of LLDPE or LDPE placed along the rows of the crop on which emitters are connected to provide water to the plants directly. The lateral pipe size is from 12 mm to 16 mm in most ideal Drip Systems.
9. **Drippers:** Drippers is the main component of drip system as it supplies water from laterals to the plant root zone. Type of soil, type of crop, its spacing, age and water requirement are the deciding factors for dripper spacing on the laterals as well as number of drippers to be

provided for each plant. Depending upon topography, drippers of pressure compensating or non-pressure compensating type are used. For hilly, undulating terrain Pressure Compensating (PC) drippers are used. For flate land or land having very mild slope, non-pressure compensating type drippers are used.

10. **Micro jets:** Jets operates at low pressure than sprinklers and apply water at higher rates than emitters. Jets wet a large surface area than strip tubing or emitters as water is sprayed through the air, in either a fan shaped spray or a stream jet pattern. However, because jets possess no moving parts, there is a limit to their distance of throw.

METHODS OF FERTILIZER INJECTION

Ventury injector

This system of injection works on the principle of ventury. Vaccume is created by diverting a percentage of water flow from the main pipeline through a ventury that increases the velocity of flow. The vaccume thus created can be used to initiate suction from a tank through a suction pipe. The ventury injector best suits the fertigation requirements of small farmers as it is very economical and easy to operate. However use of ventury leads to higher-pressure loss in the system which may result in uneven water and fertilizer distribution in the field.

Fertilizer tank

In this system, part of irrigation water is diverted from the mainline into a tank containing fertilizer solution. The water thus diverted, passes through the tank and carries along with it the fertilizer solution and returns to the mainline through the outlet connection. The fertilizer tanks are used of large irrigation systems and/ or crops which need bulk application of fertilizers. However, with this system the concentration of the fertilizer entering the irrigation water changes continuously with the time, with very high concentration at beginning. As a result, uniformity of fertilizer distribution can be a problem.

Fertilizer injector pump

These are piston or diaphragm pumps which are driven by the water pressure of the irrigation system. The injection rate is proportional to the flow of water in the system. A high degree of control over the fertilizer injection rate is possible without much pressure losses. The initial cost of the injector pump will be higher but are suitable for accurate and controlled application of fertilizers. Suction rates of pumps vary from 40 lit. to 160 lit. per hour.

MAINTENANCE OF FERTIGATION SYSTEM

Acid treatment

Precipitation of salts such as calcium carbonate, magnesium carbonate or ferric oxide can cause either partial or complete blockage of the drip systems. Acid treatment is applied to prevent precipitation of such salts in drip system. Acid is also effective in cleaning systems which are already partially blocked with precipitates of salts. Acid may also be used to lower the pH of the water in conjunction with the use of chlorine injection to improve the effectiveness of the chlorine as a biocide.

The most reliable step for deciding on acid treatment is the water analysis. Soil and water samples are collected during the survey and then to recommended acid or chlorine treatment as per the water quality.

The injection of acid into drip irrigation system is carried out to:

1. Presence of large particles as well as suspended silt and clay load in source water
2. Growth of bacteria slime in the system
3. Growth of algae in the water source and in the drip system
4. Bacteria precipitation of iron or sulphur
5. Chemical precipitation of iron
6. Chemical precipitation of dissolved salts

Following acids are used for acid treatment:

1. Hydrochloric acid
2. Sulphuric acid
3. Nitric acid
4. Phosphoric acid

Chlorine treatment

Injecting chlorine into the drip irrigation system can control bacteria and organic growth. Chlorine when dissolved in water acts as a powerfully oxidizing agent and vigorously attacks micro organics such as algae, fungi and bacteria.

The quantity of chlorine required and the frequency of treatment depends on the amount of organic matter present in water. Generally a chlorine concentration of 10-20 ppm is required to control the growth of biological matter in the system. The efficiency of chlorination is more, if the pH of source water is less than 7. The maximum chlorine concentration injected should not exceed 20 ppm; otherwise it may precipitate solids that could clog the drippers.

It is very important to treat the system regularly to prevent blockage. Treatment on a 15-day cycle with chlorine injection at the end of irrigation is a good practice. System should be chlorinated at the end of a crop season and prior to use in the next season to keep laterals/ inlines sterilized.

Prevention of algae at source

Algae can be effectively controlled in surface water by adding copper sulphate from 0.05to 2.0 ppm. The amount of chemical required may be based upon treatment up to 6 feet of water surface. Since algae growth may occur in upper surface of water where sunlight is intense. Copper sulphate can be placed in a bag anchored with a float at various points or can drag on surface after putting it in cotton sacks.

Common source of chlorine

1. Solid granules- Calcium Hypochlorite Ca (OCL) 2

It is also known as "Bleaching Powder". It is available in the form of dry powder or granules and contains 65% freely available chlorine.

2. Liquid –Sodium Hypochloride Na (OCL). It has 15% freely available chlorine.
3. Gas-Chlorine gas (CL_2)

Safety precautions during acid and chlorine treatment:

1. Acids are dangerous. All acids should be handle with care.
2. Never add water to acid. Always add acid to water for dilution.
3. Acid treatment and chlorine treatment should be carried out simultaneously as chlorine gas may be liberated, which is poisonous.
4. Irrigation water mixed with acid or high chlorine concentration is hazardous. Ensure that human beings and animals do not drink the system water during chemical treatments.
5. If acid comes in contact with any part of the body during the treatment, wash the burn using copious water and consult a doctor.
6. Do not inhale acid fumes or chlorine gas.
7. Ensure that equipments used to handle the acid are resistance to acid attack.
8. Back wash sand filter before performing acid or chlorine treatment. This will prevent entry of decomposed/ delayed impurities into the system.
9. After completing chemical treatment rinse/ wash the filtering element of screen filter and fertilizer tank with clean water.

FERTIGATION IN FRUIT CROPS

Fertigation in Mango

Singh and *et al* (2009) found considerable incremental influences on various growth characteristics of Dashehari mango through fertigation. Significant increments were observed due to drip irrigation over conventional method of irrigation in the growth characters, viz. scion height, scion girth, root stock girth.

Effect of fertigation on growth of Mango

Singh and *et al* (2009) concluded that, fertigation with 100% of recommended dose with irrigation at 0.10 levels through drip produced maximum canopy diameter. Fertigation with 50% of recommended dose with irrigation at 0.8 level through drip produced maximum canopy volume. . Fertigation with 100% of recommended dose with irrigation at 0.6 levels through drip produced maximum leaf area.

Effect of fertigation on fruit set and yield of mango

Singh, K.K and *et al* (2006) conclude that, fruit set per plant and yield was maximum in drip fertigation method with 100% recommended dose of fertilizers than conventional fertilizer application.

Fertigation in papaya

Source	Quantity of fertilizer/ Application (g)	No. of Application/ 2 months	Fertilizer (g/plant/2 months)	Nutrient (g/plant/2 months)	Nutr-ient
Urea	13.5	8	108	50	N
Super phosphate	278	1	278	50	P*
Murate of potash	10.5	8	83	50	K

P was applied in soil directly to the plants. 10 liters of water / day + 13.5 g urea and 10.5 g of murate of potash / weak through fertigation and soil application of super phosphate 278g per plant in bimonthly intervals.

Growth attributes of papaya as influenced by fertigation

Jeyakumar *et al.* (2001) concluded that, all growth characters like plant height, plant girth, leaf area, number of leaves, first flowering height and first bearing height was maximum in fertigation treatment than control treatment.

Yield attributes of papaya as influenced by different treatments

Chaudhari *et al.* (2001) revealed that, 50%dose (100:100:100gN.P2O5,K2O per plant) of recommended fertilizer in liquid form through drip was found efficient to achieved compatible yields as that of higher levels of fertilizer dose in liquid form. Number of fruits, yield of fruits per plant and yield per tonnes was also higher in treatment.

FERTIGATION IN BANANA

Fertigation schedule for banana

Total nutrient requirement: N- 150g /plant P- 45g /plant K- 185g /plant

Total quantity of fertilizer require per acre

(Spacing 1.8 x 1.5m, 1452 plants):

Fertilizer/ complex	Quantity/acre
19:19:19	192 kg
0:52:34	59 kg
13:0:46	458 kg
Urea	337 kg

Effect of fertigation and irrigation on bunch weight (kg)

Mahalakshmi *et al.* (2008) found that, fertigation with required N and K with 50L of water per day produced more bunch weight (36.5kg) than rest of treatments.

Effect of fertigation and irrigation on number of fingers per bunch

Mahalakshmi et *al.* (2008) found that, fertigation with required N and K with 50L of water per day produced more number of fingers than rest of treatments.

Effect of fertigation and irrigation on finger weight (cm)

Mahalakshmi *et al.* (2008) found that, fertigation with required N and K with 50L of water per day produced more finger weight (298.6cm) than rest of treatments.

Effect of fertigation and irrigation on finger length (cm)

Mahalakshmi *et al.* (2008) found that, fertigation with required N and K with 50L of water per day produced more finger length (25.5 cm) than rest of treatments.

Effect of fertigation on yield of banana

Anon concluded that, fertigation with water soluble fertilizers- 125%NPK increased yield of banana than surface irrigation with soil application of NF at 100% NPK.

EFFECT OF FERTIGATION ON CITRUS

Effect of fertigation on growth of acid lime

Shirgure *et al.* (1999) from growth observation it was concluded that, there is positive effect of N fertigation on plant height, girth, canopy volume. Fertigation with 60% of N recommended dose increased all growth parameters than band placement of 100% of N recommended dose.

Effect of fertigation on quality of acid lime

Shirgure *et al.* (1999) concluded that, fertigation with 60% of N recommended dose increased fruit weight, diameter and height over band placement of 100% of N recommended dose.

Effect of fertigation on growth of Nagpur mandarin

Shirgure *et al.* (2001) concluded that, incremental plant height (0.46m), girth (19.9 cm), and canopy volume (14.3m3) was more with irrigation scheduled at 20% depletion of available water and 500, 140 and 70 g N,P and K/ plant fertigation.

Effect of fertigation on yield and quality of Nagpur mandarin

Shirgure *et al.* (2001) concluded that, fruit yield, fruit weight, total soluble solids in fruit and juice were higher with irrigation schedule at 20% depletion of available water and 500, 140 and 70 g N, P and K/ plant fertigation.

Fertigation in Sapota

Anon (2010) concluded that, application of 100% recommended dose of N and K (400:450g/tree/year respectively) through drip recorded the superior plant height (3.25m), stem girth (28.85cm) and fruit yield (8.50kg/ tree).

Fertigation in Pomegranate

Rao *et al.* (2009) concluded that, fertigation with 50%recommended dose of N at fortnight interval increased plant height, stem girth, plant spread and fruit yield then rest of the treatment.

CONCLUSION

Fertigation increases water and nutrient use efficiency. Fertigation is another management tool for growers to use in production of fruit crops.

It is an extremely effective method of applying fertilizers and other chemicals via the drip irrigation system.

Fertigation provides essential elements directly to the active root zone, thus minimizing losses of expensive nutrients that ultimately help in improving productivity and quality of farm produce.

Fertigation plays both productive and protective role in crop production.

REFERENCES

1. Anonymous.2009. Indian Horticulture Database,2009, National Horticulture Board, Ministry of Agriculture, Government of India, Gurgaon, India. http.// www. nhb.gov.in
2. Anonymous, 2010, Fertigation studies in sapota, AICRP on Tropical fruit, pp 74.
3. Anonymous, 2006, Effect of fertigation treatments on growth characters of robusta banana, AICRP on Tropical fruit, pp 73.
4. Anonymous, 2005, Fertilizer use by crop in India,www.fao.org.
5. Biswas, B.C. 2010, Fertigation in high tech agriculture. Fertilizer marketing news vol 41 (10): pp 4-8.
6. Chaudhari, S.M., Shinde, S.H., Dahiwalkar, S.D., Danawale, N.J. 2001, Effect of fertigation through drip on productivity of papaya. Journal of Maharashtra Agri Univ, Vol.26 (1-3), 18-20.
7. Jeyakumar, P., Amutha, R. and Balamohan, T.N.2001, Effect of fertigaton on growth of papaya. South Indian Hort, 10(1), 15-20.
8. Mahalakshmi, M., Kumar, N.and Soorianathasundaram.2008, Effect of fertigation and irrigation on the yield of high density plantations of cv Robusta Banana, Info Musa, vol.12 (1).
9. Nanda, R.S.2010, Fertigation to enhance farm productivity. Indian J Fertilizer, vol.6 (2), 13-21.
10. Patel, N and Rajput, T.B.S. 2004, Fertigation- A technique for efficient use of granular fertilizers through drip irrigation. IE (I) Journal- AG (vol-85): 50-54.
11. Rao, K.D and Subramanyam K. 2009, Effect of nitrogen fertigation on growth and yield of pomegranate var mridula under low rainfall zone, Agri Sci Dig, Vol 29 (2), 54-56.
12. Sanjay, K., Singh, C.P.and Rashmi Panwar. 2009, Response of fertigation and plastic mulch on growth characteristics of young dashehari mango. Indian J Hort, 66 (3), 390-392.
13. Shirgure, P.S., Srivastava, A.K., and Shyam Singh. 2001, Growth, yield and quality of Nagpur mandarin in relation to irrigation and fertigation. Indian J Agri Sci, Vol 71 (8), 547-550.
14. Shirgure, P.S., Lallan Ram, Marathe, R.A., and Yadav, R.P. 1999, Effect of nitrogen on vegetative growth and leaf nutrient content of acid lime, Indian J Soil Cons, Vol 27(1), 45-49.

WATER RESOURCE MANAGEMENT FOR SUSTAINABLE DEVELOPMENT

Siddayya

Associate Professor, CNRM, NIRD & PR, Hyderabad.

INTRODUCTION

Water is critical for sustainable development and indispensable for human health and well-being. It is vital for reducing the global burden of disease and improving the health, welfare and productivity of populations. Water is also at the heart of adaptation to climate change, serving as the crucial link between the climate system, human society and the environment. The sustainable development of country depends on adequacy of water needed for irrigation, power generation, navigations, industries, domestic requirements and wild life. The availability of an adequate quantity of good quality of water has numerous implications for economic, social and environmental viability of member regions and is therefore essential to their sustainable development. Management of water resources is among the critical policy issues across the continents. The need for action in this direction is growing day by day, as countries and communities across the globe are increasingly experiencing water stress in various contexts. Water stress often leads to civil strife and conflict. The conflicts can be traced from micro level to global level. At the global level, while water conflicts are well known in the Middle East, sharing of river waters between India and Bangladesh poses a potential conflict situation. There is a global race for economic development across the countries.

Consequently, the depletion of natural resources, particularly, water posed a serious challenge to sustainable development but managed efficiently and equitably, water can play a key enabling role in strengthening the resilience of social, economic and environmental systems in the light of rapid and unpredictable changes. Developing nations are still heavily depending on agriculture which requires enough water resources. Simultaneously, the explosive growth of population also amounts of huge water for drinking purpose. The growth of power sector is also depending on water resources in coming future more than what it is today. Further, the sustainable development of a country certainly depends upon the availability of water resources. It depends on the pattern of usages, other alternative water resource. People participation and more important is that of precaution in the use of resources.

SUSTAINABLE DEVELOPMENT AND WATER

Sustainable development is defined as "development that meets the needs of the present without compromising the ability of future generations to meet their own needs. The basic concept endorses putting in place strong measures to spur economic and social development, particularly for people in developing countries, while ensuring that environmental integrity is sustained for future generations. Sustainable development of water resources refers to reducing the usage of water and recycling of waste water for different purposes such as cleaning, manufacturing, and agricultural irrigation in such a way that water demands of future generations are not hampered.

Agriculture is by far the thirstiest consumer of water globally, accounting for 70% of water withdrawals worldwide, although this figure varies considerably across countries. Rainfed agriculture is the predominant agricultural production system around the world, and its current productivity is, on average, little more than half the potential obtainable under optimal agricultural management. By 2050, world agriculture will need to produce 60% more food globally, and 100% more in developing countries.

Industry and energy together account for 20% of water demand. More-developed countries have a much larger proportion of freshwater withdrawals for industry than less-developed countries, where agriculture dominates. Balancing the requirements of sustainability against the conventional view of industrial mass production creates a number of conundrums for industry. One of the biggest is globalization and how to spread the benefits of industrialization worldwide and without unsustainable impacts on water and other natural resources.

Domestic sector accounts for 10% of total water use. And yet, worldwide, an estimated 748 million people remain without access to an improved source of water and 2.5 billion remain without access to improved sanitation.

Ecosystems. Perhaps the most important challenge to sustainable development to have arisen in the last decades is the unfolding global ecological crisis that is becoming a barrier to further human development. From an ecological perspective, the sustainable development efforts have not been successful. Global environmental degradation has reached a critical level with major ecosystems approaching thresholds that could trigger massive collapse. The growing understanding of global planetary boundaries, which must be respected to protect Earth's life support systems, needs to be the very basis of the future sustainable development framework.

AREAS OF WATER RESOURCE MANAGEMENT

1. **Water Security**: Water security means that people and communities have reliable and adequate access to water to meet their different needs, present as well as in future, are able to take advantage of the different opportunities that water resources present, are protected from water related hazards and have fair recourse where conflicts over water arise. Such water security ensures equity and sustainability. In the context of scarcity allocation of water should be governed by optimality rather than productivity.
2. **Food Security**: In the fragile resource regions environmental degradation is seen as a cause of household food insecurity as a consequence of water insecurity. In other words, food security is linked to water security through environmental degradation in these regions.
3. **Quality and Quantity**: Irrigation development and management assumes paramount importance in the agrarian economies mostly in developing countries.
4. **Maintenance**: Despite ever increasing budget allocations towards major and medium irrigation, funds available for operation and maintenance (O & M) are inadequate resulting in poor maintenance of the systems, unsatisfactory service and ecological problems across globe.

WATER RESOURCE MANAGEMENT FOR SUSTAINABLE DEVELOPMENT: AREAS OF INTERVENTIONS

Set a clear spatial framework

Water is an essential regional resource, so good management of water resources based on surface and groundwater reserves should be a high priority and it should be integrated with the protection of the environment, as well as economic and social progress. Inappropriate development of land can damage or reduce water supplies, increase risk of flooding and contamination of water and make it difficult to supply adequate sanitation. Excessive domestic, industrial and agricultural use puts the water cycle at grave risk. Water is a naturally renewable resource, but its availability may be limited by degradation of the quality or depletion of the quantity. For this reason all sectors of society must be responsible for the sustainable management of water resources to ensure availability for future generations. This is the first thing that regions must acknowledge in order to contribute to a responsible management of water.

Set a management framework and principles

A framework should set out an integrated and sustainable approach to the management of water that can guarantee appropriate abstraction, wisely balanced use, and finally, purification of wastewater that ensures that it does not degrade receiving waters and which may enable it to be re-used. The proper management of the water resources within any catchments ensures that water is provided and used efficiently, that domestic, industrial and agricultural water needs are met whilst, at the same time, safeguarding ecological processes and natural ecosystems and bio-diversity. The management framework should be founded on the principle of equitable access to water resources.

Protect groundwater supplies

Groundwater supplies are increasingly under pressure around the world. Member region's policies and regulations should seek to protect groundwater from contamination or unsustainable abstraction. Inappropriate development of land can contaminate natural water sources. One way of dealing with this threat is by means of groundwater protection zones, within which certain polluting activities are prohibited or controlled. Unregulated or excessive abstraction can deplete supplies. Continued abstraction of groundwater can also in some case cause the spread of pollution and contamination through an aquifer e.g. saline intrusion. The aim is to establish sustainable use, founded on long-term protection of natural water resources and the continued reduction of pollution of subterranean water.

Improve the Quality of Surface Waters

Surface water quality is important to public health, quality of life and biodiversity. The quality of a member region's rivers and lakes will provide a clear indication of how well it is managing the aquatic environment because any waterborne polluting substances are naturally collected by river systems. The values and uses for local waterways need to be determined through a process involving local communities. Water quality objectives then need to be set in support of those values and uses can be determined and plans developed on how the water quality objectives will be attained. These plans would include regulating pollution sources and appropriate catchments management practices. Through this process, requirements should be placed on point source dischargers to ensure that pollutant discharges are consistent with meeting water quality objectives. Requirements should also be placed on those people undertaking activities likely to cause significant diffuse source pollution.

Prevent further deterioration; protect and improve the state of water quality and aquatic ecosystems

Specific measures should be identified aimed at the (sometimes a rapid reduction may be warranted) reduction of polluting discharges and the elimination, or steady reduction, of all releases of dangerous substances to reflect the carrying capacity of the receiving body. This may require all dischargers to find alternative raw materials to those that contain substances that are known to cause intractable environmental problems, improve production processes or to improve treatment prior to discharge.

Reduce the risk of flooding and drought

Plans should be put in place to prevent unsuitable development on flood prone land and to manage flood risk to protect human life and property. There should also be drought plans that address the management of supplies and demands in times of reduced resource availability. These should include a balance between societal needs and the environment. In both cases these plans will need to take into account predictions about how the climate of member regions is changing.

Principles to be adopted

Member Regions should adopt the following principles in their policy and regulations:

(a) Protecting the water cycle

The water cycle is a key element of life and of the ecological equilibrium of the planet. Indeed, the excessive human use of water impacts on nearly all other living creatures. The water we use is a renewable resource with a natural cycle: rain feeds springs, rivers, lakes and subterranean waters and this should be respected in policies. Ensuring sufficient water is retained in water bodies to sustain ecological processes is a fundamental requirement. There should also be recognition that actions and management will affect adjacent marine and coastal environments.

(b) Protecting the rights of future generations

Use of all water resources should be environmentally and economically sustainable, providing the right amount of water for people, agriculture, commerce and industry and improved water-related environment now and in the future.

(c) Meeting human basic needs

Drinking water use comes before any other human use of the same water resource be it surface or subterranean, but other uses may be allowed when the resource is sufficient, conditional on the safeguarding of the resource's water quality for human consumption and on protection of natural ecosystems.

(d) Equality of access

Water is a public asset. Everyone has equal rights to water regardless of financial capability, social status, gender, race and political or religious beliefs; that access to water should never become an element of social, political, or religious pressure or discrimination.

No one has the right to infringe on the ecosystem's ability to function. There also needs to be a fair efficient means of allocating water for irrigation, industrial and commercial purposes consistent with ensuring appropriate returns to the community on the use of its resources.

(e) Respect for other's basic needs

Whilst water catchments areas often do not correspond to administrative boundaries, individual administrations should not use water in a way that prejudices the ability of others to the reasonable use of that water and should act in solidarity with neighbors. Water is distributed in an uneven way in terms of geographical area and time, so management should take into account the principles of respect for the basic needs of other territories, both nationally and locally.

(f) Water conservation

Every effort should be made to maximize conservation of water before additional sources of supply are sought. This should be tackled through a variety of techniques including demand management and leakage control and uptake of these measures should be strongly promoted.

(g) Managing water as a resource

Water is a critical and valuable resource. Care is needed in its management and allocation to protect water dependent ecosystems, provide basic human needs and maximize returns to the community from commercial and non-essential use. Through production, raw materials and ultimately products, we consume invisibly transported water. This hidden consumption has direct impacts upon ecosystems and the ability of people to access that right to water. Recognizing and managing this invisible consumption is an urgent priority. If water for commercial and non basic purposes is not valued, appropriately, it will be subject to wasteful practices. However, care is needed to ensure that large public or private corporations do not monopolize the market and extract unreasonable profits from the use of water. Within a market economy, operational, service and management costs of water supply, including the cost of environmental restoration, under the principles of "the polluter pays and user pays" should always be recovered.

WATER AS A RESOURCE FOR SUSTAINABLE DEVELOPMENT

While the word sustainability can mean different things to different people, it always includes a consideration of the future. We simply cannot know for certain just how sustainability can be achieved. An improvement in welfare over time cannot occur without sustainable water resources management policies and practices, those that can meet society's demands for water and the multiple purposes it serves, now and on into the future and to the fullest extent possible. Sustainable water resources management is a concept that emphasizes the need to consider the long-term future as well as the present. Water resource systems that are managed to satisfy the changing demands placed on them, now and on into the future, without system degradation, can be called .sustainable. Sustainable water resource systems are those designed and managed to fully contribute to the objectives of society, now and in the future, while maintaining their ecological, environmental, and hydrological integrity (ASCE, 1998; UNESCO, 1999). But those involved in managing the resources can still work toward increasingly sustainable levels of development and management. This includes learning how to get more from our resources and how to produce less waste that degrades these resources and systems. We need to develop improved ways of achieving more economically efficient and effective recycling and use of recycled materials. We need to identify new management approaches that are more non-structural and compatible with environmental and ecological life-support systems. In short, we need to constantly improve our processes and procedures of planning, developing, upgrading, maintaining, and paying for a changing infrastructure that we and future generations need in order to obtain the maximum benefits from the resources we manage and use.

Important guidelines for the planning and management of sustainable water resource systems for sustainable development include:

- Developing a shared vision of desired social, economic, and environmental goals benefiting present as well as future generations, and identifying ways in which all parties can contribute to achieving that shared vision.
- Developing coordinated approaches among all concerned and interested agencies to accomplish these goals, collaborating with all stakeholders in recognition of mutual concerns.
- Using approaches that restore or maintain economic vitality, environmental quality, and natural ecosystem biodiversity and health.
- Supporting actions that incorporate sustained economic, socio-cultural, and community goals.
- Respecting and ensuring private property rights while meeting community goals, and working cooperatively with private stakeholders to accomplish these common and shared goals.
- Recognizing that economies, ecosystems, and institutions are complex, dynamic (changing), and typically heterogeneous over space and time, and developing management approaches that take into account and adapt to these characteristics.
- Integrating the best science available into the decision- making process, while continuing scientific research to improve knowledge and understanding.
- Establishing baseline conditions for system functioning and sustainability against which change can be measured.
- Monitoring and evaluating actions to determine if goals and objectives are being achieved.

CONCLUSION

Water is a key driver of economic and social development while it has the basic function in maintaining the integrity of the natural environment. Intergenerational-equity is one of the issues in water management for the sustainable development. The sustainable development of a country certainly depends upon the availability of water resources, its use pattern and the alternative other available resource, people participation and more important is that of precaution in the use of resources. Sustainability is an integrating process. It encompasses technology, ecology, and the social and political infrastructure of society. It is probably not a state that may ever be reached completely, but it is one for which we should continually strive. And while it may never be possible to identify with certainty what is sustainable and what is not, it is possible to develop some measures that permit one to compare the performances of alternative systems with respect to sustainability.

Finally, we must create better ways of identifying and quantifying the amounts and distribution of benefits and costs (however many ways they might be measured) when considering tradeoffs in resource use and consumption among current and future generations as well as among different populations within a given generation.

REFERENCES

1. ASCE Task Committee on Sustainability Criteria. 1998. Sustainability Criteria for Water Resource Systems.. ASCE, Reston, Virginia, USA. 253 pages.
2. Barrow, C.J. 1998. .River Basin Development Planning and Management: A Critical Review.. World-Development 26, No. 1: 171-186. Oxford, United Kingdom.
3. Bender, M.J., G.V. Johnson, and S.P. Simonovic. 1994. Sustainable Management of Renewable Resources: A Comparison of Alternative Decision Approaches.. International Journal of Sustainable Development and World Ecology 1, No. 2: 77.88.
4. Cooper, A.B., A.B. Bottcher. 1993. .Basin-Scale Modeling as a Tool for Water-Resource Planning.. Journal of Water Resources Planning and Management, ASCE, 119, No. 3: 306.323.
5. Engelman, R. and P. LeRoy. 1993 Sustaining Water: Population and Future of Renewable Water Supplies. Population and Environment Program, Population Action International, Washington, DC, USA. 57 pages.
6. Falkenmark, M. 1988. .Sustainable Development as Seen from a Water Perspective.. In Perspectives of Sustainable Development. Stockholm Studies in Natural Resources Management, No. 1: 71.84.
7. Flyvbjerg, B. 1996. .Practical Philosophy for Sustainable Development: The Phronetic Imperative.. In M. Rolen, ed. Culture, Perceptions, and Environmental Problems: Interscientific Communication on Environmental Issues: 89.109. Swedish Council for Planning and Coordination of Research, Box 7101, S-10387, Stockholm, Sweden.
8. Frederick, K.D., D.C. Major, E.Z. Stakhiv. 1997. .Water Resources Planning Principles and Evaluation Criteria for Climate Change: Summary and Conclusions.. Climate Change 37, No.1: 29.313.
9. Gleick, P.H., P. Loh, S. Gomez, and J. Morrison. 1995. California Water 2020: A Sustainable Vision. Pacific Institute for Studies in Development, Environment, and Security, Oakland, California, USA. 113 pages.
10. Haimes, Y.Y. 1992. .Sustainable Development: A Holistic Approach to Natural Resources Management.. IEEE Trans actions on Systems, Man, and Cybernetics, Vol. SMC, No.3: 413.417.
11. Holling, C.S., ed. 1978. Adaptive Environmental Assessment and Management. New York, New York, USA: John Wiley & Sons.
12. Hufschmidt, M.M. and K.G. Tejwani. 1993. Integrated Water Resources Management: Meeting the Sustainability Challenge. UNESCO IHP Humid Tropics Programme Series No. 5, UNESCO, Paris, France.
13. Institution of Engineers, Australia. 1989. Policy on Sustainable Development. Barton, ACT, Australia. July.

14. Jordaan, J., E.J. Plate, E. Prins, and J. Veltrop. 1993. Water in Our Common Future: A Research Agenda for Sustainable Development of Water Resources. Committee on Water Research (COWAR), IHP, UNESCO, Paris, France.
15. Pearce, D.W. and J.J. Warford. 1993. World Without End. Economics, Environment and Sustainable Development. Oxford, UK: Oxford University Press.
16. Pezzey, J. 1992. Sustainable Development Concepts, An Economic Analysis. World Bank Environment Paper Number 2, World Bank, Washington, DC, USA. 71 pages.
17. Plate, E.J., ed. 1994. Proceedings of the Conference on Water Management in a Changing World, Karlsruhe, Germany, June 28-30.
18. Serageldin, I., S. Mink, M. Cernea, C. Rees, M. Munasinghe, A. Steer, and E. Lutz. 1993. Sustainable Development.. A series of articles on sustainable development in Finance & Development 30, No. 4. World Bank, Washington, DC, USA.
19. Simonovic, S.P. 1996. .Decision Support Systems for Sustainable Management of Water Resources.. Water International 21, No. 4: 223.244.
20. Svedin, U. 1988. .The Concept of Sustainability.. In Perspectives of Sustainable Development. Stockholm Studies in Natural Resources Management, No. 1: 1.18.
21. Toman, M.A. and O. Crosson. 1991. Economics and .Sustainability,. Balancing Trade-offs and Imperatives. ENR91-05.Washington, DC, USA: Resources for the Future. Mimeograph. 37 pages.
22. UNESCO Working Group M.IV. 1999. Sustainability Criteria for Water Resource Systems. Cambridge, UK: Cambridge University Press.
23. Yan den Bergh, J.C.J.M. and J. van der Straiten, eds. 1994. Toward Sustainable Development. Washington, DC, USA: Island Press. 287 pages.
24. World Bank. 1994. Making Development Sustainable. Washington, DC, USA: International Bank for Reconstruction and Development. 270 pages.

DECISION SUPPORT SYSTEM FOR INTEGRATED WATER RESOURCES MANAGEMENT

G.K. Viswanadh

Professor of Civil Engineering & OSD to Vice-Chancellor JNTUH

gkviswanadh@jntuh.ac.in

ABSTRACT

Integrated Water Resources Management (IWRM) is the process of formulating and implementing shared vision planning and management strategies for sustainable water resources utilization with due consideration of all spatial and temporal interdependencies among natural processes and water uses. Maximum development of water resources from a basin based on the quantitative information for planned beneficial use is the main scope of the integrated water resources management. It involves present status of development, socioeconomic considerations and policy formulation. Management issues like globalization, free trade, energy security, technological developments, information and communication evolution, changing population, urbanization dynamics and immigration will have major impacts on water in the future.

Policy makers develop consensus and decide on shared vision strategies based on information generated and communicated by Decision Support Systems (DSS) and associated processes. Thus, the role of DSS is to leverage current scientific and technological advances in developing and evaluating specific policy options for possible adoption by the IWRM process. DSSs are developed and used by research institutions, government agencies, consultants, and the information technology industry. This paper describes the concepts of IWRM , various processes of designing, developing, and implementing effective decision support systems and to bring together the necessary disciplines for achieving effective results in integrated water resources management.

1.0 INTRODUCTION

By its nature, IWRM is a process where information, technology, natural processes, water uses, societal preferences, institutions, and policy makers are subject to gradual or rapid change. To keep current, IWRM should include a self-assessment and improvement mechanism. This mechanism starts with monitoring and evaluating the impacts of decisions made. These evaluations identify the need for improvements pertaining to the effectiveness of the institutional set-up, the quality and completeness of the information generated by decision support systems and processes, and the validity and sufficiency of the current scientific knowledge base.

2.0 PRINCIPLES OF IWRM

Water resources management has to be addressed with an integrated context and on a basin-wide level. Integrated watershed management of the available water resource, hydropower, water quality and flood protection constitutes major challenges. It focuses on understanding the conditions of the water resources in order to meet requirements of various usages in the watershed. Numerous aspects have to be considered in order to respond to changes in natural phenomena and provision of adequate and good quality water at a reasonable cost to the public. In- creased focus on watershed protection and sound use of water requires that scientists, planners, managers and decision makers are able to quickly produce reliable estimates, assess impacts and efficiency of potential strategies.

Integrated water resources management deals with the management of water under normal conditions as well as under flood and drought conditions. It is accepted fact that it is not possible under all circumstances to prevent floods from occurring and a shift has been noticed from flood prevention to flood preparedness and flood management. One important aspect to account for in such management is the trade-off between different interests. Flood prevention initiatives may run counter to irrigation or other agricultural or ecological water use priorities in a river basin and as such IWRM may not favour a maximum possible prevention of floods. IWRM can be characterized as a process, not a product scale independent - applies at all levels of development. It is a tool for self-assessment and program evaluation; a tool for policy, planning, and management. It is a mechanism for evaluating competing demands, resource allocation and tradeoffs.

The knowledge to support planning and management decisions resides in various disciplines including climatology, meteorology, hydrology, ecology, environmental science, agro-science, water resources engineering, systems analysis, remote sensing, socio-economics, law, and public policy. Policy makers (such as politicians, judges, government agencies, financial institutions, Non-Governmental Organizations, citizen groups, industries, and the general public) are often in a position to make critical decisions that reflect society's shared vision for water resources utilization.

Use and integration of sophisticated web based and GIS enabled graphical user interfaces with relational databases, visualization techniques, analysis tools and decision logic greatly enhances and promotes the decision process. It enables users and decision makers focus on transparency and accessibility of results to a broad range of the public including governmental institutions, private and public stakeholders and communities involved/interested in environmental and water resources issues.

3.0 CONCEPT OF IWRM

IWRM is an empirical concept which is built up from the on-the ground experience of practitioners, a flexible approach to water management that can adapt to diverse national and local contexts, thus it is not a scientific theory that needs to be proved or disproved by scholars. and it requires policy-makers to make judgments about which reforms and measures, management tools and institutional arrangements are most appropriate in a particular cultural, social, political, economic and environmental context.

IWRM is a process which promotes the coordinated development and management of water, land and related resources, in order to maximize the resultant economic and social welfare in an equitable manner without compromising the sustainability of vital ecosystems. What does it really mean? More coordinated development and management of Land and water - Surface water and ground water Upstream and downstream interests

The main water resources management functions are Water allocation, Flood and drought management, Basin planning, Information management, Stakeholder participation, Pollution control Monitoring and Financial management.

4.0 WHY IWRM?

Globally accepted and makes good sense. It is a key element in national water policy and incorporates social and environmental considerations directly into policy and decision making. Also, it directly involves the stakeholders. Hence it is a tool for optimizing investments under tight financing climate.

5.0 CHALLENGES

IWRM processes can lead to great successes just as they can cause costly failures. In a world where water disputes are on the rise and the delay between science and technology advances and their consideration by management practices widens, IWRM suffers the following challenges:

- Lack of integrative tools to support planning and management decisions;
- Segmentation of institutions responsible for water resources planning and management;
- Limited participation of stakeholders in decision making processes;
- Lack of disinterested self-assessment and improvement mechanisms;
- Continuing specialization of science and engineering education at the expense of interdisciplinary training.

In order to achieve a more efficient management and sustainable development of water resources for multiple uses according to the principles of IWRM, a decision support system will be an effective tool. The idea of an action plan, as opposed to a traditional master plan, is to create a flexible planning instrument that takes the resource, not the projected use of one or more specific sectors, as the starting point.

6.0 DECISION SUPPORT SYSTEM

Decision is a reasoned choice among alternatives. Computer based model together with their interactive interfaces are typically called decision support systems (DSSs) or Computer-based systems integrating tools and databases that assist a decision-maker in making informed decisions and analyse consequences.

Why do we need a DSS?

- Structured approach to problem solving
- Large volume of information
- Integrate many information sources
- Models are difficult to use
- Deal with trade-offs: social, economic, biophysical, legislation
- Identify preferred options for further follow up

DSS does not take decisions, it provides timely information. It is easy to have comprehension of abstract information and communicates the result to a larger audience with open and unbiased working. It is a tool that helps with the age-old problem of how to choose between alternatives when there are conflicts and trade-offs with respect to time, costs, direct consequences and indirect consequences.

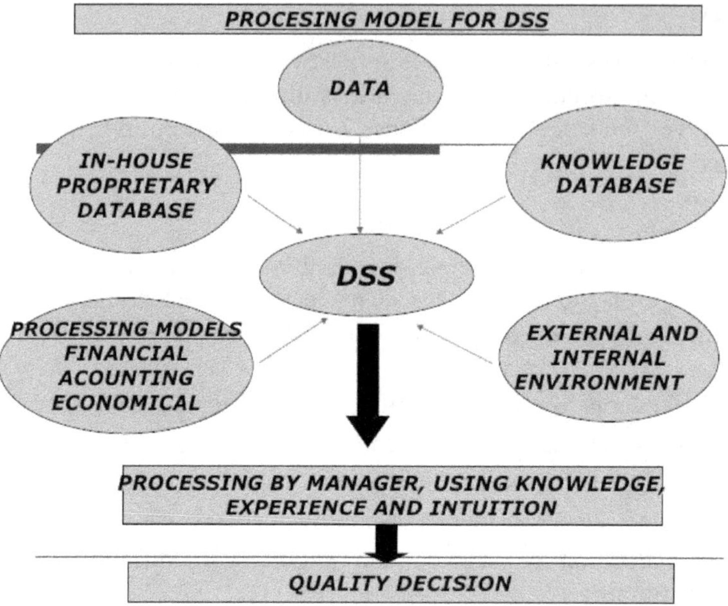

7.0 COMPONENTS OF A DSS

It is a paradigm that helps user to make decisions, when there are many and conflicting objectives. It contains databases, GIS, simulation models, economic analyses, decision models and GUI, Optimization, decision theory, expert systems, discussion groups. DSS is a different concept to different people

Components of a DSS

- Databases – Temporal, spatial
- GIS for spatial data
- Mathematical models
  - Empirical
  - Calibration
  - Optimization, simulation
- Expert systems
- Statistical, graphical software, spreadsheets
- User interface

8.0 KEY PRINCIPLES

- Make best use of available tools and information
  - Don't wait for data to become available
- Transparent working
- Choose appropriate time frame
- Employ the best technology
- Improves capacity building

The system reviews and discusses different scenarios and adjusts ranking of criteria, adjust constraints including additional criteria or options

9.0 DESIGNING A DSS

- Clearly understand the problem, objective
- Issues, needs, preferences of users
- Associate users in development
- It is a team work
- Should be user friendly

Process and Components

Process: Data Analysis, Stakeholder Participation , Discussion of Outcomes

Components: Statement of Issue, Identify Stakeholders, Define Resource Use Options, Identify Criteria, Ranking Criteria, Carry Analysis, View and Discuss Results

10. DECISION SUPPORT TOOLS

The use of integrated decision support tools is a prerequisite in proper IWRM. DSS provides a custom, flexible and dedicated management system, to assist managers, decision makers and policy makers in:

- provide timely, transparent, well informed and reproducible answers to important questions
- quickly and effectively streamline workflow, reduce time and cost requirements
- transform data and information into knowledge and produce understand- able results and decisions

Development and implementation of a DSS is typically structured around a number of phases with well-defined functionality and scope. The DSS focuses on specific needs supporting and enabling development and production of timely, well informed and reproducible answers to essential questions. Initial phases focus on data and information management. WEB and GIS technologies are used to link and integrate databases. The water information management system empowers users to transform data and information into knowledge and provide the basis for analysis and decision support. Subsequent phases may involve adding new data, incorporating and linking to other databases, applying more advanced analysis and modelling tools and expanding the decision support functionality. This evolving and dynamic business process mirrors the ever changing requirements of our society and environment.

Typical DSS interactive and integrated components are:

- Data and information management. The data and information component is key and central in developing a DSS. The focus is integrating database and connecting data islands into a dynamic framework with advanced display, mapping, query and presentation capabilities.
- Analysis and modeling. The data framework provides the basis for further analysis and interpretation of data and information. Depending on stage and scope of the DSS the analysis can range from simple to complex including statistical and numerical models, economic and cost/benefit as well as User Defined and Custom tools

- Scenario management and alternative formulation. The DSS framework is capable of supporting and providing information (costing and prioritization) for project feasibility and planning projects as well as design and implementation. Upon implementation the project may have an operations component that requires real time and online decision making.
- Decision making. Customizable GIS and Web based interfaces are tailored to meet specific needs and requirements. Advanced graphics, on-line access, custom rules and interpretations can be embedded into the DSS to support and provide the basis for decision makers to make timely, reproducible and well informed decisions

11.0 CONCLUSION

Decision support systems are integral parts of IWRM processes facilitating the use of science and technology advances in public policy. Although generic DSS development principles exist, much is system specific and a thorough understanding of the interdependence of natural processes, water uses, institutional setting, and decision maker objectives is necessary for a successful DSS design. The DSS can, depending on specific needs, remain specific in scope to support a very focused and dedicated decision process. The DSS may, on the other hand, also evolve into an enterprise DSS to support a wide range of users and a broad management scope. Decision support systems are built within the framework of a Geo- graphical Information Systems (GIS), which provide a convenient platform for handling, compiling and presenting large amounts of spatial data essential to integrated water resource management. Since GIS technology is often linked to information and knowledge management systems and is readily available to most governmental entities, a high degree of transparency in decision-making for stakeholders can be achieved.

REFERENCES

1. Brumbelow, K., and A. Georgakakos (2001) An Assessment of Irrigation Needs and Crop Yield for the United States under Potential Climate Changes. Journal of Geophysical Research – Atmospheres 106(D21), 27383-27406.
2. Tsuji, G.Y., G. Uehara, and S. Balas (eds.) (1994) Decision Support System for Agrotechnology Transfer, Version 3. University of Hawaii, Honolulu.
3. Aris P. Georgakakos (2004) Decision Support Systems For Integrated Water Resources Management with an Application to The Nile basin.
4. Sharad K Jain (2006), Introduction to Decision Support Systems.
5. Carlo Giupponi and Alessandra Sgobbi (2013), Decision Support Systems for Water Resources Management in Developing Countries: Learning from Experiences in Africa.
6. Joaquim Comas (2009) Decision Support Systems for Integrated Water Resources Management Under Water Scarcity.
7. G.Anzaldi et.al. (2014) Towards an Enhanced Knowledge-based Decision Support System (DSS) for Integrated Water Resource Management (IWRM)

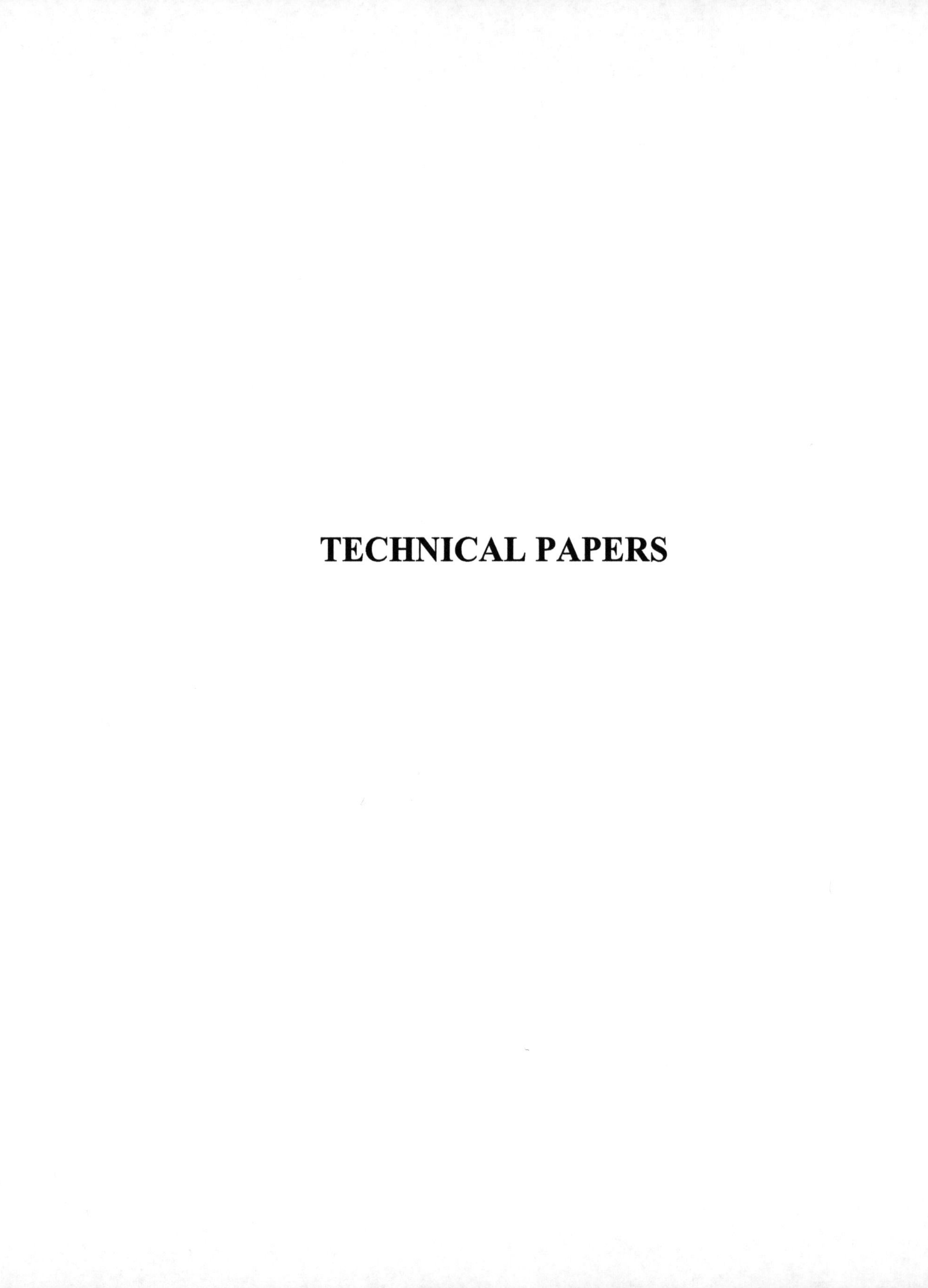

TECHNICAL PAPERS

APPLICATION OF LIFE CYCLE ASSESSMENT IN LANDFILLING

Shilpa Mishra[1] and Mohua Salilkanti Banerjee[2]

[1]Assistant Professor, Dept. of Civil Engineering, MVSR Engineering College, Hyderabad, India
[2]Assistant Executive Engineer, Maharashtra Jeevan Pradhikaran
shilapatiwarimishra0605@gmail.com, mohuab85@gmail.com

ABSTRACT

Life cycle assessment is a tool used to finds the impact of products on environment at different stages of their life cycle *i.e.* from resources extraction, through the production of materials, production of product and product parts and the use of the product after it is discarded. The total system of unit processes involved in the life cycle of a product is called as "product system". This paper discussed the use of LCA to analyze the potential environmental burden of landfill process.

Keywords: Landfilling, Leachate, Treatment.

1. APPLICATION OF LCA IN LANDFILLING

According to the ISO standards, a Life Cycle Assessment is carried out in four distinct phases: goal and scope, life cycle inventory, life cycle impact assessment and interpretation (Lehtinen *et al.* 2011; Azapagic *et al.*, 2009; and Guiyuan Han *et al*).

1.1. Goal and Scope

It defines the purpose and method of including life cycle environmental impacts into the decision-making process. The goal for landfilling process may be evaluating the environmental performance of the process and identifying critical factors relating to the environmental performance of the system.

The scope would be the methodology which should adopted for achieving the goal. It comprises of choices, assumptions and limitations. It includes fixing the system boundaries. While analyzing a process through life cycle perspective, it is not always clear how far one should go in including processes that are inter-related to the process concerned in complex way. In case of municipal solid waste (MSW) landfilling process, the MSW is first to be collected from the individual houses to the community bins then by using smaller trucks, they are taken to the transfer stations. In the transfer stations, the MSW is transferred to bigger trucks which then are transported to the landfill site. At the landfill site, the waste is compacted with the help of diesel compactors. Each of these steps which are related to the landfill process includes input of energy and emissions of gases. Also the truck, compactors and other equipments used in landfilling also have a life cycle. To produce a truck steel is needed, to produce steel, coal is needed, to produce coal, trucks are needed etc. It is difficult to trace all inputs and outputs to product systems. Therefore, one has to frame boundaries around the system. Hence, for practical reasons a line must be drawn for accurate and optimum results. This is what the system boundary is about. For fixing up the system boundaries in landfilling process, a flow chart of the process is required to be chalked out. It is helpful to draw a diagram of the system and to identify the boundaries in this diagram as shown in Figure 1.

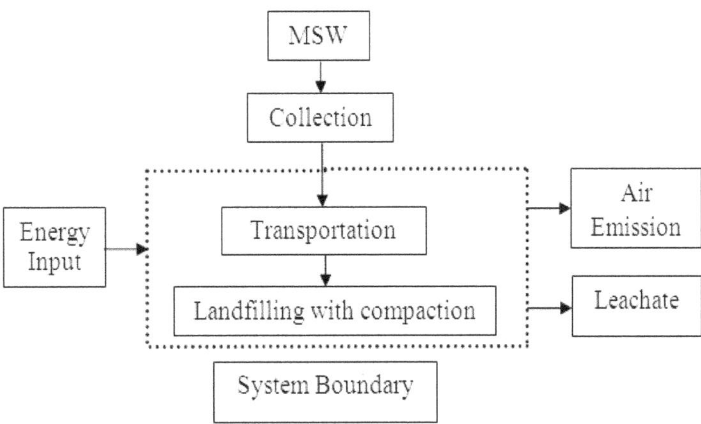

Fig. 1 Flowchart of landfill process showing system boundary

1.2. Life Cycle Inventory (LCI)

LCI involves the modeling of the system, data collection and the description of data. This implies that data for inputs and outputs for all affected unit processes that compose the process system are available. The inputs and outputs include inputs of materials, energy and outputs in the form of air emissions, water emissions or solid waste. In case of landfilling process, the energy input and output for transportation and compaction stages is calculated. The gaseous emissions in the landfill process are also evaluated. The leachate quantity and quality is also determined in this stage (Pradeep Jain *et al.*, 2014 and Abeliotis 2011).

1.2.1. Transportation Stage

The waste collected is transferred to community bins and then they are transported to the transfer stations. Here the waste is transferred from short range transport vehicles to large long range vehicles. The latter carries the waste to the disposal site. The energy consumed in transportation of municipal solid waste to the disposal site depends upon the total distance traveled by the vehicle during transportation.

$$T_r = (MSW_t \times MSW_f / H_{au}) \times X \quad .....(1)$$

In equation 1, T_r is total run of vehicle during transportation in *km/yr*, MSW_t is Total MSW generated in *tonnes/yr*, MSW_f is the fraction of MSW disposed at the disposal site since some part of it is reduced due to recycling at source, waste not reaching landfill site due to insufficient solid waste management, H_{au} is Actual Hauling

Capacity of vehicle in tonnes/trip and X is the distance travelled by the vehicle during transportation in *km/trip*. The total energy consumption during transportation is given by equation 2 where E_i is energy input during transportation in KJ / yr, AC is average consumption of fuel in km / lit, ρ is density of the fuel used in the vehicle in kg / lit and Dcv is Calorific Value of the fuel in MJ / kg.

$$E_i = (T_r / AC) \times \rho \times D_{cv} / 1000 \quad .....(2)$$

Complete oxidation of hydrocarbon fuel yields only CO_2 and H_2O as the product of chemical combination. When air is used as a source of the oxygen required for combustion, some of the oxygen and nitrogen combine to form nitric oxide. Under this condition other products are also formed. These include carbon monoxide (CO), hydrocarbons (HC), etc. Also, it includes particulate matter (PM). Thus for the output analysis the emission of the gaseous pollutants during transportation is given by

$$E_{tr} = T_r \times \text{Emission } i / 10^6 \quad .....(3)$$

Where, E_{tr} is emission of the gaseous pollutant during transportation and ***Emission i*** is the emissions of gas *i* for a diesel vehicle in *gm /km.*

1.2.2. Compaction Stage

Sanitary landfilling includes compaction process in which after dumping the MSW on the land it is compacted by using compactors. In this paper, it is assumed that the heavy-duty 165 hp compactor is used. The energy consumption of this compactor is one litre of diesel per ton of MSW land filled. Therefore, the energy input can be calculated using equation 4 based on the fuel consumption during compaction process

$$E_{ci} = MSW_t \times MSW_f \times D \times \rho \times (D_{cv}/1000) \quad .....(4)$$

Where, E_{ci} is energy input during compaction in sanitary landfilling in *GJ / yr* and ***D*** is fuel consumption rate of MSW compacted in *lit/ton.*

Energy Output: In this stage, there are emissions of gaseous pollutants such as CO_2, N_2O, SO_x, NMVOC, CO, PM, NO_x. If *A* is the amount of gas emitted by a 165hp compactor in *kg/ lit*, then the total gaseous emission in *tones / yr* (T_{ge}) due to compaction of MSW is given in equation 5.

$$T_{ge} = A \times MSW_t \times MSW_f \times D / 1000 \quad .....(5)$$

1.2.3. Landfilling Stage

Biological degradation of the waste in landfill in aerobic condition, anaerobic condition, or in facultative anaerobic condition generates landfill gases. Considerable heat is generated by these reactions with CH_4, CO_2, and other gases as the byproducts. CH_4 and CO_2 are the principal gases produced. Solid waste landfills produce CH_4 as bacteria decompose organic wastes under anaerobic conditions. CH_4 accounts for approximately 45 % to 50 % of landfill gas, while CO_2 and small quantities of other gases comprise the remaining 50 % to 55 %. CH_4 production may last for decades, depending on disposal site conditions, waste characteristics and the amount of waste in the landfill. CH_4 migrates out of landfills and through zones of low pressure in soil, eventually reaching the atmosphere. During this process, the soil oxidizes approximately 10 % of the CH_4 generated by a landfill and the remaining 90 % is emitted as CH_4 unless captured by a gas recovery system and then used or flared. The production and release of CH4 from landfill site to the atmosphere lead to global warming and the emissions need to be check and estimated. The amount and rate of CH_4 production over time at a landfill depends on five key characteristics of the landfill material and surrounding environment (Sunil Kumar *et al.,* 2004). These characteristics are

(i) ***Quantity of organic material*:** The most significant factor driving landfill CH_4 generation is the quantity of organic material, such as paper, food and yard wastes, available to sustain methane producing microorganisms. The amount of CH_4 produced and release from landfill is directly proportional to quantity of organic waste present in the landfill. Amount of CH_4 generation as the quantity of waste increase in waste disposal site and it gradually start decreasing when the landfill site stop receiving waste. However, landfills may continue to generate methane for decades after closing.

(ii) ***Nutrients:*** Methane generating bacteria need nitrogen, phosphorus, sulfur, potassium, sodium, and calcium for cell growth. These nutrients are derived primarily from the waste placed in the landfill.

(iii) ***Moisture content*:** The bacteria also need water for cell growth and metabolic reactions. The water present in landfills is from surface water infiltration, water from incoming waste, water produced by decomposition of material such as sludge and ground water infiltration. Another source of water is precipitation. In general, CH_4 generation occurs at faster rates in non-arid climates than in arid climates.

(iv) ***Temperature*:** Warm temperatures in a landfill speed the growth of methane producing bacteria. The temperature of waste in the landfill depends on landfill depth, the number of layers covering the landfill, and climate.

(v) ***pH:*** Methane is produced in a neutral environment *i.e.* close to pH = 7. The pH of most landfills is between 6.8 and 7.2. For pH = 8.0 and above, methane production is negligible.

According to the IPCC Guidelines there are two methods for estimation methane emissions from solid waste disposal. The first method is based on mass balance equation *i.e.* on the theoretical gas yield, while theoretical first order kinetic methodologies is the second method which is used to introduces the "First order decay model" (FOD). The main difference between the two methods is that the first method does not reflect the time variation in SW disposal and the degradation process as it assumes that all potential methane is released the year the SW is disposed. The timing of the actual emissions is reflected in second method. The first method is preferable used to estimate emission when the yearly amounts and composition of waste disposed and disposal practices have been approximately constant for long period. Increasing amounts of waste disposed will lead to an overestimation, and decreasing amounts correspondingly to underestimation, of yearly emissions. Second method gives a more accurate estimate of the yearly emissions. In my study the second method i.e. First order decay method is used for estimation of methane gas emitted from the landfill taken for the study. The data required for this method is annual amount of MSW to be landfill, methane generation potential ***Lo***, and the methane generation rate constant ***k*** which determine the rate of gas production are functions of site-specific conditions. The methane emission rate in m^3/year from the landfill in case of FOD method is given by the equation 6

$$Q_{CH4} = Lo \times k \times MSW_t \times MSW_f \times e^{-kt} \quad .....(6)$$

In case of ***Lo***, a range between less than 100 m^3 / Mg SW and more than 200 m^3 / Mg SW is presented in the IPCC Guidelines. In this paper, the value of Lo is taken as 100 m^3 / Mg SW. In case of ***k***, the value is dependent on waste composition and climatic conditions. It describes the rate of degradation process. A very wide range of values between 0.005 and 0.4 is given for ***k*** in the IPCC Guidelines. The factor ***k*** is primarily a function of waste moisture content. There are two default values for k, for non-arid area, k = 0.04 / yr and for arid region, k = 0.02 / yr. We consider semi-arid region, therefore, the value of k is taken as 0.03 / yr. The major quantity of the gaseous emission is CH_4 and CO_2, in which some part of it gets dissolved with the water present in the MSW while some part of CH_4 gets oxidized. The quantity of gases other than CH_4 is calculated considering the percentage of that gas present in the emission. The quantity is calculated taking percentage of the quantity of methane gas.

1.2.4. Leachate Generations

The quantity of leachate generated in a landfill is strongly dependent on the quantity of infiltrating water. This, in turn, is dependent on weather and operational practices. The amount of rain falling on a landfill to a large extent controls the leachate quality generated. Precipitation depends on geographical location (Mishra *et al.,* 2017). The leachate generation during the operational phase from an active area of a landfill may be estimated by a water balance approach in a simplified manner as follows:

Leachate volume = (volume of precipitation) + (volume of pore squeeze liquid)
– (volume lost through evaporation)
– (volume of water absorbed by the waste) .....(7)

The important factors which influence leachate quality include waste composition, elapsed time, temperature, moisture and available oxygen. In general, leachate quality of the same waste type may be different in landfills located in different climatic regions. Landfill operational practices also influence leachate quality. To assess the leachate quality of a waste, the normal practice is to perform laboratory leachate tests.

1.3. Life Cycle Impact Assessment (LCIA)

The community is exposed to the environment through location, occupation and behavior. The environment is changed by the project. New health hazards may be introduced and old health hazards may disappear. The changes may take place immediately or over a timescale of ten or more years. Hot, humid and moist environment is congenial for most of the disease causing organisms. Various components such as air, water, land, noise and socioeconomic are considered during environmental and health impact studies.

The third phase LCIA is aimed at evaluating the contribution to impact categories such as global warming, acidification, etc. Potential environmental impacts are calculated on the basis of the data from the inventory analysis. For impact assessment, the inventory data are classified into the environmental impact categories according to effects they contribute and characterized.

The impact assessment methods themselves are described in ISO 14042. In this standard a distinction is made between:

- **Obligatory elements**, such as classification and characterization
- **Optional elements,** such as normalization, ranking, grouping and weighting

This means that according to ISO, every LCA must at least include classification and characterization. If such procedures are not applied, one may only refer to the study as a life cycle inventory (LCI).

1.3.1. Selection of methods and impact categories

An important step is the selection of the appropriate impact categories. The choice is guided by the goal of the study. An important help in the process of selecting impact categories is the definition of endpoints. Endpoints are major factor of environmental concern such as extinction of species, human health and resources availability for future generation etc. Before selection of impact categories, careful selection and definition of endpoints is required, however endpoints are not recommended by the ISO.

1.3.2. Classification

LCA inventory result consists of hundreds of various emissions and resource extraction parameters. Once the relevant impact categories are determined, these LCI results must be assigned to these impact categories. For example CO_2 and CH_4 emissions in case of landfilling are both assigned to the impact category "Global Warming", while SO_x and NH_3 are both assigned to an impact category "Acidification". We can assign emissions to more than one impact category at the same time. For example, SO_x can also be assigned to an impact category such as "Respiratory diseases or Human health".

1.3.3. Characterization

It is necessary to define characterization factors after the impact categories are defined and the LCI results are assigned to these impact categories. These factors should reflect the relative contribution of an LCI result to the impact category indicator result. For example, if we consider a time scale of 100 years, then the contribution of 1 kg CH_4 towards global warming is 42 times more than the emission of 1 kg CO_2. This means that if the characterization factor of CO_2 is 1, the characterization factor of CH_4 is 42. Thus, the impact category indicator result for global warming can be calculated by multiplying the LCI result with the characterization factor.

Normalization, grouping and ranking, weighing are used to simplify interpretation of the result. These steps are regarded as optional steps in ISO 14042. The relations between the Inventory Table, Classification and Characterization and Weighting are illustrated in Figure. In this paper, the classification and characterization is done in the LCIA stage for landfilling. The impact potential depends upon the mid-point score which is calculated by the equation 8

$$\text{Mid-point score (category)} = \Sigma \text{ (Emission (gas)} \times \text{Characterization Factor)} \quad 8$$

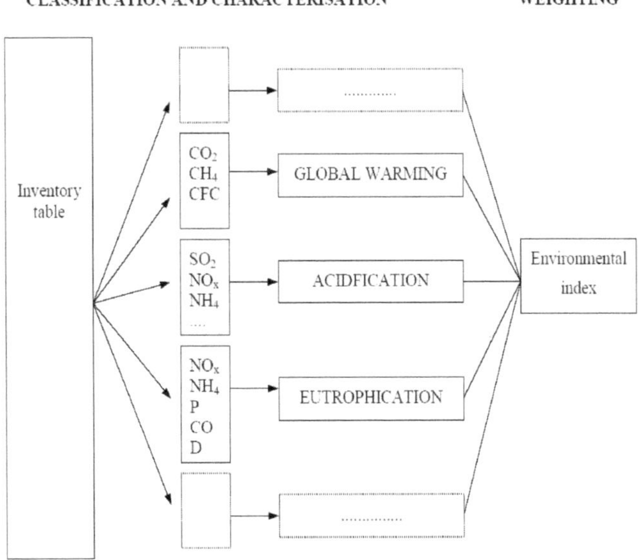

Fig. 2 Steps involve in LCIA

1.4. Life Cycle Interpretation

This is the most important stage in LCA process. The results from inventory analysis and impact assessment are discussed in accordance with the goal. All conclusions and recommendations are drafted during this phase. Results of life cycle inventory analysis and life cycle impact assessments are analyzed with respect to various aspects such as completeness, sensitivity, and consistency. In addition, key issues that contribute significantly to the environmental impact of the product system are also identified. Key issues in this context can mean key processes, materials, activities, and components or even a life cycle stage. From these analyses, conclusions are drawn and recommendations made as to the environmental aspects of the product, possible areas for improvement or key environmental information that could be communicated to the consumer, all depending on the goal of the LCA study.

There are three key elements in life cycle interpretation as defined by ISO 14043. First is the identification of key issues, second is the evaluation, and third is development of conclusions together with recommendations.

1.4.1. Identification of key issues

Key issues are activities, processes, materials, components, or life cycle stages which have a significant impact on the total impact of a product system. One of the objectives of performing LCA is to identify weak points of a product system or service. The identification of key issues is a must in any LCA aimed at improving environmental aspects of the product or service.

A method called "contribution analysis" is used for the identification of key issues or weak points of a product system or service for which characterized impact data is required.

1.4.2. Evaluation

Basic premises for performing an LCA study such as data quality, goal, major assumptions, system boundary setting, etc is determined during the goal and scope definition phase. Data are then collected during the inventory analysis phase, and their impact on the environment assessed during the life cycle impact assessment phase. During the first part of the life cycle interpretation phase, key issues were also identified. However, all these results are based on basic premises defined earlier, such as assumptions, data quality, and methodologies employed. It is necessary to perform a systematic evaluation of all these results in order to check completeness, sensitivity and consistency.

1.4.3. Conclusions and recommendations:

The objective of this section is to draw conclusions from the LCA study and then make recommendations based on those conclusions for the intended audience.

2. VARIOUS CHALLENGES IN DEALING WITH THE LIFE CYCLE OF MSW MANAGEMENT

The MSW management using LCA is a very difficult and challenging task due to the following reasons (K. Vamsi Krishna *et al.*, 2015):

(i) Purpose of every single waste management facility is to make environment clean and green. However, it requires huge amount of land, non renewable natural resources such as fuels and electricity for their operation, harmful gaseous emission and leachates generation. Therefore, proper planning and action is required for management of MSW facilities so that their effect on natural environment can be minimize. The LCA involve study of trade-offs between environmental gains and burdens.

(ii) The MSW facilities may produce some useful products such as different sorts of cardboard and paper, plastics and glass etc. A mechanical biological treatment facility of MSW generates RDF which is used as a fuel in cement kilns and compost can be used as a fertilizer in agricultural activities. The energy produce from the thermal treatment of MSW can be used for electricity and heat.

(iii) MSW treatment process involves various degrees of uncertainties such as lack of quality data with respect to MSW management practices. The various uncertainties related to landfilling are related to the time frame of the impacts. The data available from collection, recycling and treatment of MSW are more reliable than data from landfills which partially have to be modelled and where estimations are necessary.

3. CONCLUSION

The LCA results are useful in determining where the improvements could be made in the process to obtain a sustainable municipal solid waste management system. For decision makers, LCA could serve as an invaluable tool for such an analysis. The results obtained from LCA can be used for policy decisions as well as strategic decisions on waste management systems

REFERENCES

1. Hannele Lehtinen, Anna Saarentaus, Juulia Rouhiainen, Michael Pitts and Adisa Azapagic (2011), “A Review of LCA Methods and Tools and their Suitabiltiy for SMEs”, BIOCHEM project under work package 3, May 2011.
2. Azapagic, A. and H. Stichnothe (2009), “A Life Cycle Approach to Measuring Sustainability”, Chemistry Today, Vol-27 (1) 44-46.
3. Guiyuan Han and Jelena Srebric, “Life –Cycle Assessment Tool for Building Analysis”, Research brief: RB0511, The Pennsylvania Housing Research Center.
4. Pradeep Jain, Jon T. Powell, Justin L. Smith, Timothy G. Townsend, and Thabet Tolaymat (2014), “Life-Cycle Inventory and Impact Evaluation of Mining Municipal Solid Waste Landfills”, Environ. Sci. Technol., 48 (5), pp 2920–2927.
5. Konstadinos Abeliotis (2011), “Life Cycle Assessment in Municipal Solid Waste Management”, Volume I, Mr. Sunil Kumar (Ed.), InTech, DOI: 10.5772/20421.
6. Sunil Kumar, S.A. Gaikwad, A.V. Shekdar, P.S. Kshirsagar and R.N. Singh (2004), “Estimation method for national methane emission from solid waste landfills”, Journal of Atmospheric Environment Vol. 38 (2004), pp. 3481–3487.
7. Shilpa Mishra, Santennagari Praveen, Shwetha Kaushik (2017), “A Review on Various Landfill Leachate Treatment Methods”, 1st National conference on Advances of Construction Engineering for Sustainability, Oct 2017, Hyderabad.
8. K. Vamsi Krishna, Dr Venkateshwar Reddy and Dr P Rammohan Rao (2015), “Municipal solid waste management using landfills in Hyderabad city”, *I. J.* of Engineering & Technology, Vol. 4 issue 02, pp. 1047-1054.

SUSTAINABLE CONSERVATION PRACTICES FOR DRYLAND FARMING IN VIDARBHA REGION OF MAHARASHTRA

M.B. Nagdeve, R.S. Patode, V.V. Gabhane and M.M. Ganvir

All India Coordinated Research Project for Dryland Agriculture,
Dr. Panjabrao Deshmukh Krishi Vidyapeeth, Akola (Maharashtra)

ABSTRACT

The field experiment was conducted at the field of AICRP for Dryland Agriculture, Dr. Panjabrao Deshmukh Krishi Vidyapeeth, Akola. The experimental layout had been arranged in split plot design with five replications. Main plots and sub plots include different tillage and nutrient management treatments. The long term impact of low tillage treatments is being observed in terms of productivity, energy management and resource conservation. Here in this paper results for the year 2015-16 regarding resource conservation are presented. The conventional tillage treatment (T_1) has recorded 32.81% more runoff compared to low tillage with hand weeding (T_2). The runoff in the low tillage treatments was less as compared to conventional tillage treatment (T_1). Also the low tillage treatments, T_3 and T_2 has less soil loss (0.64 and 0.92tons ha^{-1}) as compared to conventional tillage (1.44tons ha^{-1}) treatment (T_1). Rainwater use efficiency was observed to be highest in the in low tillage with hand weeding treatment, T_2 (6.08) followed by low tillage with herbicides, T_3 (5.53) and conventional tillage, T_1 (4.73). It is concluded that treatment combination of low tillage with hand weeding along with 50% recommended dose of fertilizer through organic and inorganic each was superior over other treatment combination.

Keywords: Conservation, low tillage, runoff.

INTRODUCTION

The success of dryland farming mainly depends on the evenly distributed rainfall during crop growing period. The root zone soil moisture is utilized for transpiration, when the rainfall becomes insufficient to meet the potential needs to transpiration. This causes depletion in soil moisture storage and a situation which may be designated as agricultural drought. In terms of crop groups, 77% of pulses, 66% of oilseeds and 45% of cereals are grown under rainfed conditions. Therefore, a breakthrough in dryland agriculture is an imperative for poverty alleviation, livelihood promotion and food security in India (Abrol, 2011). In order to meet the food requirements of the growing population, it is essential to develop strategies for crop, land and water productivity improvement through resource conservation. In this scenario the improved crop productivity in less intensively cropped and land degraded dryland areas may play vital role. Low till conservation farming strategy is aimed to reduce the tillage input for better resource conservation including energy. If low till planting is practiced for long period of 4-5 years and the crop residues are recycled, the soil ecology will be build up to such an extent to minimize the adverse effect of low till. The crop stubble and organics like gliricidia besides improving the soil organic matter status should also help in moisture conservation and water intake. Conservation tillage (CT) reduces soil loss on cropland by retaining crop residue on the soil surface and minimizing deep tillage. Reduced tillage has resulted in a greater dependence on herbicides for weed control, which increases costs and the risk of developing herbicide resistant weeds. In order to know the effect of different tillage methods on runoff, soil loss, yield and to quantify the energy estimates the experiment was undertaken.

MATERIALS AND METHODS

The site is situated at the latitude of 20^0 42' North and Longitude of 77^0 02' East. The altitude of the place is 307.4m above MSL. The climate of the place is subtropical and characterized by hot dry summer and cool winter. Rains are mostly received from South-West monsoon during June to October. Mean annual rainfall is 824.8mm, which is generally received in 41 days. The soil belongs to Vertic Inceptisols. Inceptisols are developed from basalt and they are very shallow to shallow. Inceptisols show vertic characteristics, whereas, the Vertisols are developed in basaltic alluvium brought out by rivers. These soils are medium to heavy in texture, high in lime content with high base saturation.

Treatments

The experimental layout had been arranged in split plot design with five replications. Main plots and sub plots includes following treatments.

Tillage treatments

T_1 : Conventional Tillage, (CT) Ploughing (after every three years) + Two intercultural operations + Two hand weeding.

T_2 : 50% of CT (one intercultural operation + one hand weeding).

T_3 : 50% of CT (one intercultural operation + pre emergence herbicides).

Nutrient Management Treatments

N_1 : Recommended dose of sorghum through inorganic fertilizers,

N_2 : 50% recommended dose through inorganic and 50% through organic (FYM/Glyricidia) and

N_3 : Recommended dose through organic sources (FYM/Glyricidia).

RESULTS AND DISCUSSIONS

This is the long term experiment however here the results for the year 2015-16 are presented to know the effect of low tillage treatments on resource conservation.

Runoff and soil loss

The runoff and soil loss observed in different treatments is given in Table 1. During the season total 4 runoff events were occurred out of which only two events were major. The highest total runoff of 52.58mm was observed in conventional tillage treatment (T_1) and lowest total runoff of 32.13mm was observed in low tillage with herbicides treatment (T_3). The total runoff of 39.59mm was observed in low tillage with hand weeding treatment (T_2). Runoff in conventional tillage treatment (T_1) was 63.64% more than that of low tillage with herbicides treatment (T_3). The conventional tillage treatment (T_1) has recorded 32.81% more runoff compared to low tillage with hand weeding (T_2). The runoff in the low tillage treatments was less as compared to conventional tillage treatment (T_1) and therefore it can be concluded that the low tillage treatments has conserved more water as compared to conventional treatment. Also the low tillage treatments, T_3 and T_2 has less soil loss (0.64 and 0.92tons ha^{-1}) as compared to conventional tillage (1.44tons ha^{-1}) treatment (T_1).

Table 1 Runoff and soil loss during the season 2015 for sorghum crop

Date	Rainfall, mm	Runoff, mm			Soil loss, tons ha^{-1}		
		T_1	T_2	T_3	T_1	T_2	T_3
04-08-2015	194.0	40.69	30.74	26.86	1.22	0.79	0.57
12-08-2015	28.0	3.04	1.95	0.00	0.00	0.00	0.00
15-09-2015	57.0	4.08	3.41	2.27	0.10	0.06	0.03
17-09-2015	78.5	4.77	3.49	3.0	0.12	0.07	0.04
Total	**357.5**	**52.58**	**39.59**	**32.13**	**1.44**	**0.92**	**0.64**

Rainwater use efficiency (Table 2) was observed to be highest in the in low tillage with hand weeding treatment, T_2 (6.08) followed by low tillage with herbicides, T_3 (5.53) and conventional tillage, T_1 (4.73). In case of nutrient management, the treatment half recommended dose through organic and inorganic treatment, N_2 (5.79) was observed to be highest followed by recommended dose through organic, N_3 (5.40) and recommended dose through inorganic, N_1 (5.14).

Table 2 Rain water use efficiency of sorghum as influenced by different treatments.

Treatment	Rain water use efficiency (Kg/ha/mm)
(A) Tillage	
T_1	4.73
T_2	6.08
T_3	5.53
(B) Nutrient Management	
N_1	5.14
N_2	5.79
N_3	5.40

Soil Moisture

The soil moisture at the depths 0-15 and 15-30cm is given in Table 3 and presented in Fig. 1. The soil moisture was observed to be low at the initial stages of crop growth; it was better at middle stages of crop growth and goes on decreasing at later stages. It was found to be better in all treatment combinations during flowering stages of crop growth at the depths 0-15 and 15-30cm.

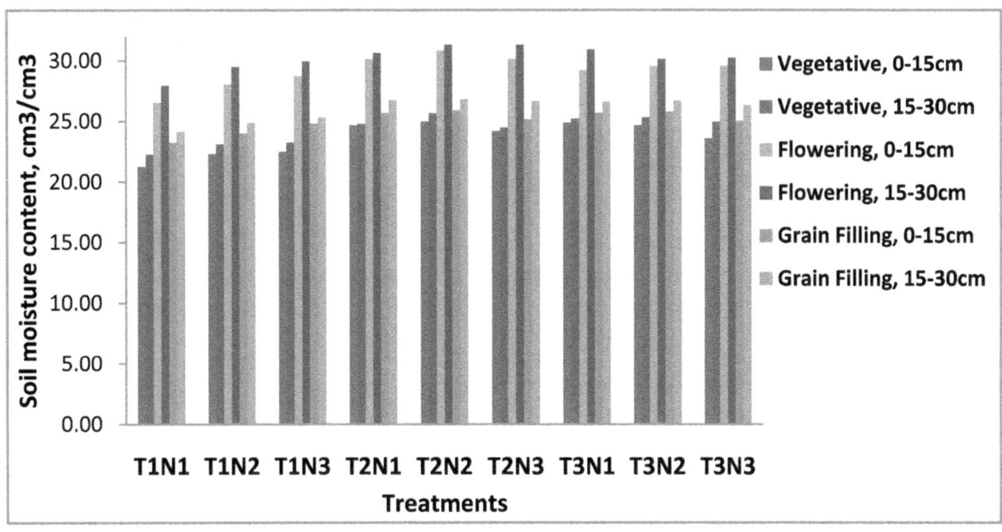

Fig. 1 The soil moisture at 0-15 and 15-30cm depth in different treatment combinations

Table 3 Soil moisture content (cm^3/cm^3) at different crop growth stages

Treatments	Depth (cm)	Soil moisture content (cm^3/cm^3) at stage		
		Vegetative	Flowering	Grain Filling
T_1N_1	0-15	21.24	26.52	23.26
	15-30	22.24	27.92	24.16
T_1N_2	0-15	22.31	28.03	24.02
	15-30	23.10	29.51	24.86
T_1N_3	0-15	22.50	28.74	24.81
	15-30	23.26	29.94	25.32
T_2N_1	0-15	24.66	30.13	25.67
	15-30	24.78	30.64	26.73
T_2N_2	0-15	24.96	30.85	25.90
	15-30	25.66	31.31	26.80
T_2N_3	0-15	24.17	30.16	25.14
	15-30	24.47	31.32	26.65
T_3N_1	0-15	24.86	29.21	25.68
	15-30	25.20	30.92	26.61
T_3N_2	0-15	24.65	29.56	25.79
	15-30	25.32	30.14	26.68
T_3N_3	0-15	23.56	29.59	25.02
	15-30	24.95	30.24	26.30

Inferences

- In tillage treatments, the treatment 50% of conventional tillage and hand weeding, T_2 has recorded significantly highest grain yield (3843kg ha^{-1}) over conventional tillage, T_1 (2987kg ha^{-1}) and it was at par with 50% of conventional tillage and herbicides, T_3 (3501kg ha^{-1}). Highest fodder yield (10069kg ha^{-1}) was observed in treatment, T_2.
- In nutrient management, the treatment 50 % recommended dose through organic and inorganic, N_2 recorded statistically significant grain yield (3664kg h^{a-1}) over the treatment recommended dose through inorganic, N_1 and at par with treatment recommended dose through organic, $_{N3}$ (FYM/Glyricedia). Significantly highest fodder yield (10155kg h^{a-1}) was recorded in N_2.
- Treatment combination of low tillage with hand weeding along with 50% recommended dose of fertilizer through organic and inorganic, $_{T2N2}$ was superior over other treatment combination.

CONCLUSION

In tillage management, low tillage treatments (50% of conventional tillage + hand weeding) have highest rainwater use efficiency and; highest grain and fodder yield. As well as, in nutrient management the treatment 50% recommended dose through organic and 50% through inorganic (N_2) have highest rainwater use efficiency and gave highest grain and fodder yield. Hence, it is concluded that treatment combination of low tillage with hand weeding along with 50% recommended dose of fertilizer through organic and inorganic each was superior over other treatment combination.

REFERENCES

1. Abrol, I. P. 2011. Natural resource management and rainfed farming. Report of the XII Plan Working Committee, New Delhi.
2. De, Dipankar. 2000. Concepts of energy use in Agriculture, Research Report, CIAE Bhopal.
3. Sharma K.L., K. Srinivas, U.K. Mandal, K.P.R. Vittal, K.J. Grace. and G.R. Maruthi Sankar. 2004. Integrated nutrient management strategies for sorghum and greengram in Semi-Arid Tropical Alfisol. Indian J. Dryland Agric. Res. & Dev. 19(1): 13-23.
4. Wang, X.B., D.X. Cai, W. B. Hoogmoed , O. Oenema, U.D. Perdok. 2007. Development in conservation tillage in rainfed regions of north china. Soil & Tillage Research, (93), pp. 239-250.
5. Zhang, X.C. 1998. Effects of surface treatments on surface sealing, runoff and inter rill erosion. American Society of Agricultural Engineers 4 (41): 989-994.

SOIL MOISTURE MONITORING: A USEFUL INDICATOR FOR IMPACT ANALYSIS OF CONSERVATION MEASURES ADOPTED CATCHMENT

R.M. Wankhade, R.S. Patode, M.B. Nagdeve and K. Ramamohan Reddy[1]

All India Coordinated Research Project for Dryland Agriculture
Dr. Panjabrao Deshmukh Krishi Vidyapeeth, Akola (M.S.)
[1]Centre for Water Resources, IST, Jawaharlal Nehru Technological University, Hyderabad, (Telangana)

ABSTRACT

Field experiment has been carried out at All India Co-ordinated Research Project for Dryland Agriculture, Dr. Panjabrao Deshmukh Krishi Vidyapeeth, Akola during 2014-15. Area of experimental plot was divided into two parts (50x100 m^2) each. The catchments A and C are treated with continuous contour trenches (CCTs) and B and D are non-treated. The catchment A and B are having custard apple (*Annona squamosa*) plantation and catchment C and D are having atemoya (*Annona cherimola*) plantation. In this paper we examine the variation in distribution of soil moisture at different depths of soil profiling. The soil moisture content of custard apple and atemoya plantation in the CCT treated micro-catchment (T_1) was observed to be more by 25.22% and 35.78% respectively over untreated (T_2) micro-catchment at all depths in every recorded month. However, the soil moisture was observed highest at the depth of 70-80 cm and lowest at the depth of 0-10 cm in all recorded months in the micro- catchments. The enhanced soil moisture in CCT treated (T_1) micro-catchments was observed to be useful for better growth of the perennial plantations.

Keywords: Conservation, depth, moisture, plantation.

INTRODUCTION

The understanding of the soil water regime of rainfed regions is important for efficient rainwater conservation and for its optimum uses for practical soil water management. Soil moisture depletion studies are needed by many researchers in order to describe the availability of soil water to plants and to model the movement of water and salts in unsaturated soils. Soil texture and the properties it influences, such as porosity, directly affects water and air movement in the soil with subsequent effects on plant water use and growth. The proportion of pores filled with air or water varies, and changes as the soil wets and dries. When all pores are filled with water, the soil is 'saturated' and water within macropores will drain freely from the soil via gravity. Whenever rainfall-runoff event occurs, runoff begins and flows down from the slopes causing erosion giving not much chance for water to infiltrate down the soil. In such situations CCTs are adopted for reducing runoff and enabling the water to infiltrate down to the ground. In the top portion of catchment area, contour trenches can be excavated all along a uniform level across the slope of the land. Bunds can be formed downstream along the trenches with material taken out of them to create more favorable moisture conditions and thus accelerate the growth of vegetation. Contour trenches breaks the velocity of runoff and for small catchments the infiltrated water can be helpful for increasing the soil moisture regimes.

Study area

The present field experiment was conducted at All India Co-ordinated Research Project for Dryland Agriculture, Dr. Panjabrao Deshmukh Krishi Vidyapeeth, Akola during 2014-15. The site is situated at the latitude of 20^0 43' North and Longitude of 77^0 02' East. The altitude of this place is 307.41m above MSL. The climate of the place is semi-arid and characterized by hot dry summer and cool winter.

MATERIALS AND METHODOLOGY

Area of experimental plot is 1 hectare and it is divided into two parts (50x100 m^2) each. The micro catchments A and C are treated (T_1) with continuous contour trenches (CCTs) and B and D are untreated (T_2). The micro-catchment A and B are having Custard apple (*Annona squamosa*) plantation and micro-catchment C and D are having Atemoya (*Annona cherimola*) plantation. There are four observation wells viz., well A, well B, well C and well D prepared in each micro-catchment A, B, C and D respectively. Eight moisture tubes viz., M_1, M_2, M_3, M_4, M_5, M_6, M_7 and M_8 are also installed in these micro-catchments; Two moisture tubes in each micro-catchment i.e. M_1 and M_2 in A, M_3 and M_4 in C, M_5 and M_6 in B and M_7 and M_8 in D. Green gram has been taken as an inter crop in the perennial plantation in both the micro-catchment. The details of treatments in the micro-catchments are given below.

Treatments:

T_1- CCT Treated micro-catchment (0.5 ha)

T_2 - Untreated micro-catchment (0.5 ha)

Determination of soil moisture with GOPHER soil moisture profiler

The Gopher Soil Moisture Profiler (Fig. 1) uses the proven and sensitive technique of measurement of the dielectric constant of the soil plus water to determine the moisture content

Fig. 1 Gopher Soil Moisture Profiler.

of the soil. As the water content of the soil increases, the resultant measured dielectric constant increases. The Gopher Soil Moisture Profiler is a microprocessor controlled measurement system with an LCD dot matrix display, for display of graphs and information, and a 16-key keypad for operator interface. The digital soil moisture profiler directly indicates the soil moisture at different depths when the probe was inserted in the access tubes. For this purpose the access tubes needs to be inserted in the ground at appropriate depths so that the moisture assessment can be done with the probe.

RESULTS AND DISCUSSION

Weekly monitoring of soil moisture was done by Gopher soil moisture profiler at different soil depths for the months of August, September, October, November, December, January and February. Monthly soil moisture was obtained by averaging weekly soil moisture of that month at given depth. The soil moisture was recorded at different depths *viz.*0-10,10-20, 20-30, 30-40, 40-50, 50-60, 60-70 and 70-80 cm for CCT treated (T_1) as well as untreated (T_2) micro-catchment. The soil moisture content on volumetric basis (SMC) at different depths recorded in different months is presented in Fig 2.

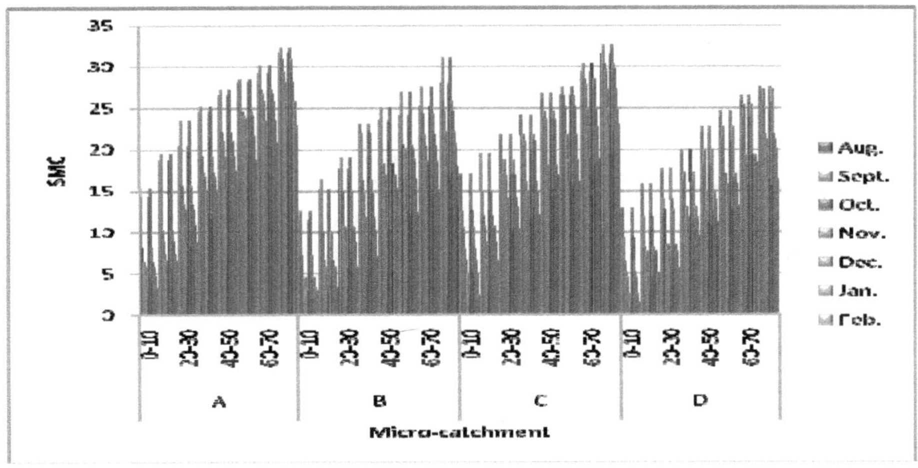

Fig. 2 Soil moisture content in soil profile in different months.

The result shows that soil moisture was observed more in treated (T_1) micro-catchment as compared to untreated (T_2) micro-catchment in both Custard apple and Atemoya plantation. The enhanced soil moisture in CCT treated (T_1) micro-catchments was observed to be useful for better growth of the plantations. Average percent increase in SMC in Treated (T_1) micro-catchment over untreated (T_2) micro-catchment is given in Table 1 and presented in Fig. 3.

Table 1 Average percent increase in SMC in treated (T_1) micro-catchment over untreated (T_2) micro-catchment

Months	Aug.	Sep.	Oct.	Nov.	Dec.	Jan.	Feb.
Increase in SMC in treated (T_1) micro-catchment (%)	17.38	21.02	22.67	41.74	29.83	38.70	42.16

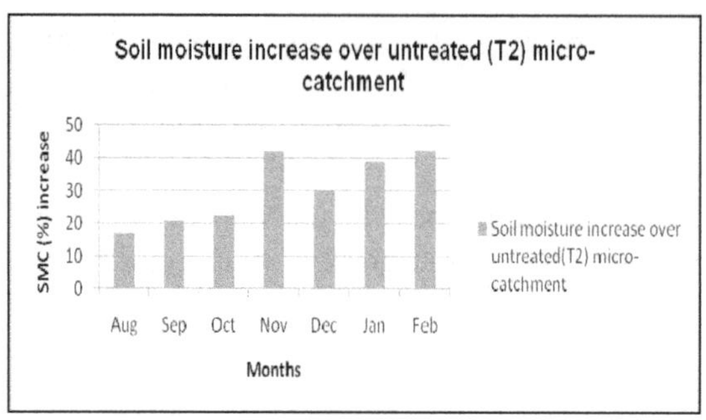

Fig. 3 Average percentage increase in soil moisture content in treated (T_1) over untreated (T_2) micro- Catchment

In micro-catchment A and C percentage increase in soil moisture was observed to be 30.50 % over untreated (T_2) micro-catchment. Soil moisture increase over untreated micro-catchment is highest in month of November and February in treated (T_1) micro-catchment over untreated (T_2) micro-catchment.

CONCLUSION

The soil moisture content of custard apple and atemoya plantation in the CCT treated micro-catchment (T_1) is observed to be more by 25.22% and 35.78% respectively over untreated (T_2) micro-catchment at all depths in every recorded month. However, the soil moisture is observed highest at the depth of 70-80 cm and lowest at the depth of 0-10 cm in all recorded months in the micro-catchments. From these results it can be concluded that response of CCTs on the soil moisture is observed to be better as compared to untreated micro-catchment.

REFERENCES

1. Agrawal,O. P.and Indrapati Singh, 1970. Effect of contour cultivation on soil conservation and yield of sugarcane. J. soil and water conservation in india.18 (3&4):29-33.
2. Barai,U. N. and B. M. Patil. 1991. Effect of conservation practices on soil moisture and crop yield. Indian J. Soil Cons., 17(2):55-57.
3. Bhanawase, D. B. 2007. Effect of in-situ moisture conservation measures on yield of pigeon pea in southern zone of Maharashtra. Indian J. Soil Cons., 35(3):230-233.
4. Chavan, B. N. 1990. Evaluation of low cost in situ water conservation techniques under different cropping systems in vertisols. Phd. (Agronomy) Thesis. Marathwada Agri. Univ., Parbhani. India. 170.
5. Dalvi, V. B., M. B. Nagdeve and L. N. Sethi. 2009. Rainwater harvesting to develop non-arable lands using continuous contour trench (CCT). Assam University Journal of Science and Technology; Physical Sciences and Technology, 4(II): 54-57.
6. Sadgir P. A, G. K. Patil, V. G. Takalkar, 2006. Sustainable Watershed Development by Refilled Continuous Contour Trenching Technology. National Seminar on Rainwater Harvesting and Water Management, Nagpur.

RAINWATER MANAGEMENT THROUGH *IN-SITU* SOIL AND WATER CONSERVATION TECHNIQUES AND UTILIZATION OF HARVESTED WATER THROUGH FARM POND

R.S. Patode, M.B. Nagdeve, N.R. Palaspagar and G. Ravindra Chary[1]

All India Coordinated Research Project for Dryland Agriculture,
Dr. Panjabrao Deshmukh Krishi Vidyapeeth Akola (Maharashtra)
[1]Project Coordinator, AICRPDA, CRIDA, Hyderabad

ABSTRACT

Rainwater harvesting is essential in view of the fact that rainfall, which is a source of fresh water, occurs in very short spells and runs off as a waste unless arrangements are made for its storing. In this paper the importance of location specific farm pond with proper design considerations and results of enhancement of productivity of *Kharif* and vegetables by using stored rainwater are presented. The experiment was conducted at the field of AICRP for Dryland Agriculture, Dr. Panjabrao Deshmukh Krishi Vidyapeeth, Akola. From the results of the experimentation it was observed that, during the *Kharif* season treatment T_2 (Two protective irrigations) have shown better yields as compared to treatments T_1 (One protective irrigation) and T_3 (No irrigation). The water use efficiency and B:C ratio was also higher in T_2 over treatments T_1 and T_3. During *rabi* season, for chickpea, highest yield and B:C ratio was recorded in the treatment of two protective irrigations of 50mm depth each with sprinkler set from stored pond water (T_2). For vegetable crops the water use efficiency was found in the range of 2.50 – 5.60kg/m^3. The total income from small vegetables plots during rabi season was Rs. 9375. Computed total income from these vegetables is around Rs.101241ha^{-1}. Thus from these results it can be concluded that if rainwater is harvested in the farm pond and if utilized judiciously then the sustainability in production can be achieved.

***Keywords*:** Farm pond, Harvesting, Income, Rainwater, Use.

INTRODUCTION

The demand for water is increasing day by day not only for Agriculture, but also for household and Industrial purposes. Water need for drinking and other municipal uses is estimated to be increased from 3.3 Mha-m to 7.00 Mha-m in 2020/25. Similarly the demand of water for industries will be increased by 4 fold *i.e.* from 3.0 Mha-m to 12.00 Mha-m during this period. At the same time more area should be brought under irrigation to feed the escalating population of the country, which also needs more water. The perennial rivers are becoming dry and groundwater table is depleting in most of the areas. Country is facing floods and drought in the same year in many states. This is because, concrete action to conserve, harvest and manage the rainwater efficiently is lacking. The theme paper on Water vision 2050 of India, prepared by Indian Water Resources Society (IWRS) has indicated that storage of 60 Mha-m is necessary to meet the demand of water for irrigation, drinking and other purposes. But the present live storage of all reservoirs put together is equivalent of about 17.5 Mha-m which is less than 10% of the annual flow of the rivers in the country. The projects under construction (7.5 Mha-m) and those contemplated (13 Mha-m) are added, it comes only 37.50 Mha-m and hence we have to go a long way in water harvesting to build up storage structures in order to store about 60 Mha-

m. Therefore there is an urgent need to take up the artificial recharge of the rainwater for which water harvesting and water conservation structures are to be build up in large scale. In most of the rainfed areas, rainwater conservation measures cannot conserve all the rainwater and a certain amount of runoff is bound to occur. This runoff can be collected and stored in tanks for a life saving irrigation to rainfed crops. Therefore the farm ponds should be constructed at specific location depending upon the catchment and thereafter size may be decided after estimation of runoff from the particular catchment.

Consideration for making a farm pond

The detailed features of the water contributing area, possible utilization of the stored water, suitable site for the pond and economics in terms of benefits have to be studied before making a farm pond. Catchment selection is one of the important aspects in designing the farm pond. The runoff from the catchment will depends on several factors like, rainfall, topography, soil types and land use. The construction of runoff harvesting ponds involves consideration of yield of watershed, storage available at site, water requirements for different needs and ground water conditions. The design considerations shall include relationship between watershed area and capacity of pond, choice and design of outlets, size of embankment and cost benefit ratio.

Rainfall analysis

Rainfall is one of the most important and critical hydrological input parameter for the design of rainfall harvesting structure. Its distribution varies both spatially and temporally in the semi arid regions. The quantity of surface runoff depends mainly on the rainfall characteristics like intensity, frequency and duration of its occurrence. The high intense rainfall exceeding infiltration capacity of soil can produce more runoff than the event with low intensity for longer duration. Apart from the physical characteristics of the catchment area, the rainfall analysis is very critical for optimal economic design. The variability of rainfall in arid and semi-arid areas is considerable. An analysis of only 5 or 6 years of observations is inadequate as these 5 or 6 values may belong to a particularly dry or wet period and hence may not be representative for the long term rainfall pattern.

Runoff

Three different catchments were considered at the experimental field of All India Coordinated Research Project for Dryland Agriculture, Dr. Panjabrao Deshmukh Krishi Vidyapeeth, Akola. The site is situated at the latitude of $20^0$42' North and Longitude of $77^0$02' East. The altitude of this place is 307.41m above MSL. The amount of runoff for these catchments were computed by using the following formula and given in Table 1;

$$R = \frac{kPA}{10} * A$$

where, R = Amount of runoff in cu meter

k = Expected runoff in per cent of the total rainfall

P = Average annual rainfall in mm

A = Catchment area in hectares

Table 1 Runoff from different catchments.

Catchment No.	Expected runoff in per cent of the total rainfall, k	Catchment area in hectares, A	Amount of runoff from the catchment area (m^3)
1	15	3.5	4154
2	12	5.0	4748
3	10	2.0	1582

Average annual rainfall = 791.28mm

Dimensions of farm ponds for different catchments

Based on the runoff from three catchments, the capacity of the farm pond has been decided. Accordingly the location for construction of the farm pond had been chosen and the dimensions were decided and construction/reshaping of three farm ponds for three different catchments was done and the details are given in Table 2.

Table 2 Dimensions of farm pond

Catchment no. /Farm pond no.	Capacity (cum)	Top dimensions (m x m)	Bottom dimensions (m x m)	Depth (m)	Side slopes
1	2753	45 x 27	36 x 18	3.0	1.5:1
2	4265	60 x 30	51 x 21	3.0	1.5:1
3	370	18 x 11	12 x 5	3.0	1:1

Farm pond no. 1 and 2 are having embankments on all sides however farm pond no. 3 have embankments on two sides only and other two sides are without embankments which will acts as inlets for sheet flow in the pond. The sheet flow is to be passed from the vegetative key lines surrounding the sides of the farm pond. All the three farm ponds are having pipe outlets located at suitable places for safe disposal of the excess water when the farm ponds will be filled to its fullest capacity. The berms are provided on all sides of the farm ponds for easy access and for stability of the embankments.

RESULTS AND DISCUSSION

At the experimental farm of AICRPDA, Dr. Panjabrao Deshmukh Krishi Vidyapeeth, Akola (Agricultural University) the various research experiments of *in-situ* rainwater conservation and water resource management had been conducted. Based on the results of some of these experiments, following recommendations has been made. These recommendations are useful for water resources development in rainfed area.

A. Results of rainwater management through in-situ techniques

- On sloppy fields (up to 3% slope), cotton (Rajat) in the lower toposequence covering about 30% area is replaced by soybean (PKV-1) - chickpea (ICCV-2) sequence cropping instead of cotton. For demarcation of such area, Vetiver keyline is established in the beginning of the system which also helps in moisture conservation.
- The yields of soybean (915kg ha^{-1}) and chickpea (410kg ha^{-1}) are significantly higher under lower toposequence compared to sole cotton (498kg seed cotton ha^{-1}). Monetary

returns from this toposequence based cropping are 40% higher over the farmers practice of sole cotton in entire field. This improved system, reduced the runoff and soil loss to an extent of 24 and 20%, respectively. This improved system besides providing higher monetary returns also reduces soil loss and runoff due to canopy architecture of soybean.

- For reducing runoff and soil loss and increasing crop productivity, vegetative keyline of vetiver or Leucaena should be developed on contours and cultivation should be done along the vegetative key lines on contour.
- The reduction in runoff was observed to be in the tune of 40-50% whereas, the reduction in soil loss was observed to be 70-75% in case of vegetative barriers as compare to across the slope sowing. Uniform moisture distribution due to contour cultivation gave higher productivity in case of test crops i.e. sorghum and cotton to the extent of 15 and 20%, respectively.

B. *Results of adoption of Continuous Contour Trenching system*

- Contour trenches are usually made in the top portion of the catchment by excavating trenches along the contours and forming bunds downside of the trenches. The *in-situ* conservation of rainwater is possible due to these trenches. In Vidarbha region of Maharashtra the continuous contour trench of 60 cm width and 30 cm depth is commonly adopted. For *in-situ* percolation of rainwater and to restrict the erosion of soil, the work of Continuous Contour Trenches have been done under various schemes *viz.*, EGS, SGRY, RLEGP, etc. by various departments. In order to have actual quantification of water conservation, recharge, soil moisture fluctuations due to adoption of CCTs the research work has been undertaken on micro-catchment basis so that the evaluation and monitoring of CCTs is possible and with the help of modelling the quantification of different parameters is possible which will be useful for researchers, field officers, farmers and NGOs.

1. **Impact of CCT on seasonal crop productivity**

   The increase in yield in treated micro-catchment was 47.57, 59.56 and 47.89% over control during the year 2012, 2013 and 2015 respectively.

2. **Impact of CCT on fruit production**

   On an average the fruit production was more in CCT treated micro-catchment as compared to control in Custard apple (92.48%) as well as Atemoya plantation.

3. **Impact of CCT on nutrient status of the soil**

   It was found that the available N, P, K (nitrogen, phosphorus, potassium) and organic carbon content in the soils under custard apple plantation in treated micro-catchment was more by 33.38, 14.14, 11.19 and 0.03% over soils of the control micro-catchment respectively. Similarly for soils in Atemoya plantation, the available nitrogen, phosphorus, potassium (N,P,K) and organic carbon content in treated micro-catchment was more by 26.08, 12.65, 3.12 and 0.04% over soils of the control micro-catchment respectively.

C. *Results of the utilization of water from farm pond*

During the *Kharif* season the treatment T_2 (Two protective irrigations) have shown better yields as compared to treatments T_1 (One protective irrigation) and T_3 (No irrigation). The water use efficiency and B:C ratio was also higher in T_2 over treatments T_1 and T_3. During the *rabi* season, for chickpea, highest yield and B:C ratio was recorded in the treatment two protective irrigations of 50mm depth each with sprinkler set from stored pond water (T_2). For vegetable crops the water use efficiency was in the range of 2.50 – 5.60kg/m^3. The total income from small vegetables plots during the season is Rs. 9375. Computed total income from these vegetables is around Rs.101241ha^{-1}.

d. Other technologies of rainwater management and sustainable production

- Opening of furrows in each row at 30-35 DAS in soybean and cotton for in-situ moisture conservation resulted higher yields of soybean and cotton as compared to farmers practice (without opening of furrows).
- Adoption of intercropping systems *viz*; soybean + pigeonpea (4:2) resulted in higher soybean equivalent yield and cotton + greengram (1:1) resulted in higher cotton equivalent yield as compared to farmers practice (sole soybean and sole cotton).
- Supplemental irrigation to rainfed crops from harvested rainwater through farm ponds in soybean and chickpea resulted in higher yields as compared to farmers practice.

CONCLUSION

Harvesting of rainwater is of utmost important. Judicious mix of ancient knowledge, modern technology, public and private investment and above all, people's participation will go a long way in reviving and strengthening water harvesting practices throughout the country. Availability and storage of water in reservoirs and lakes depends ultimately on yearly rainfall. Natural conservation of water and efficient use of this natural storage and at the same time making arrangements for additional recharge of groundwater aquifer by one way or other, to replenish the used groundwater becomes our responsibility.

REFERENCES

1. Bangar, A. R., A. N. Deshpande, V. A. Sthool and D. B. Bhanavase. 2003. Farm pond – A boon to agriculture, ZARS, Solapur (MPKV) : 32-37.
2. Nagdeve, M.B. and R. S. Patode. 2012. Protective irrigation through farm pond for enhancing crop productivity. Application Technologies for Harvested Rainwater in Farm Ponds, Proceedings of National consultation meeting held at CRIDA, Hyderabad during 19-20 March, 2012 : 62-67.
3. Patode, R. S., M. B. Nagdeve and K. Ramamohan Reddy. 2014. Impact assessment of continuous contour trenches on soil moisture regime. Watershed Development: ISBN 978-81-926207-1-8, Proceedings of All India Seminar on Recent Advances in Watershed Development Programme, Published by Ahmednagar Local Centre, The Institution of Engineers (India), pp. 116-120.
4. Singh, R. P. 1986. Farm Ponds. CRIDA, Hyderabad Project Bulletin No. 6.
5. Sivanappan, R. K. 2006. Rainwater harvesting, conservation and management strategies for urban and rural sectors. National Seminar on Rainwater Harvesting and Water Management, Nagpur. Pp.1-5.
6. Sthool, V. A., S. K. Upadhye, J. D. Jadhav, R. V. Sanglikar, V.U.M. Rao. 2013. Farm pond- A boost for sustainability in Dryland under climate change situation. MPKV/Res. Pub. No. 80/2013.
7. Taley, S. M., R. S. Patode, M. G. Dikkar and V. D. Hedau, 2009. Rainwater management in deep black soils under rainfed agroecosystem. Green Farming Vol.2 (12): 816-820.
8. Vora, M.D., H. B. Solanki and K.L. Bhoi , 2008. Farm pond technology for enhancing crop productivity in Bhal area of Gujrat. Journal of Agricultural Engineeering Vol.45(1): 40-46.

STORM WATER MANAGEMENT USING SWMM: A CASE STUDY ON INDRAPRASTHA AREA, BELAGAVI

B. Komalakshi[1] and Chandrashekarayya. G. Hiremath[2]

[1]PG Student, VTU-PG studies, Belagavi, Karnataka, India

[2]Assistant Professor. Department of water and land management, VTU, Belagavi, Karnataka, India

ABSTRACT

Evaluation of surface runoff in urban catchments and reduction of overflow, to protect the storm water and water quality Storm Water Management Model (SWMM) has been used in urban catchment Indraprastha area, Belagavi, India. The runoff simulation may be on the basis of continous or event wise rainfall data. Green-Ampt model has been used for infiltration modeling. The present paper focuses on a systematic methodology for runoff analysis, validation. For validating the performance of the programs, few objectives were utilized which include rational method to find-out the discharge, the coefficient of determination R^2, Nash-sutcliffe, Root Mean Square Difference (RMSD) and Standard Error Estimation (SEE). The outcomes represented that the observed and simulated discharges are comparable.

Keywords: SWMM, LID Control Structures, Rational Method.

I. INTRODUCTION

Urbanization is increasing at its own pace, as people from rural regions are highly attracted to urban areas. According to the estimation, above more than half of the population dwells in the urban areas. Nearly 500 cities make it possible for 1 million people to get shelter. Increased rate in flood is caused due to high intensity rain events, global warming, climate change (Deepak Singh Bisht et al., 2016). The urban areas are of high importance for the study as the risk of occurrence of flood is more in such areas is due to combined result of urbanization and global climate change. The worst outcome of urbanization is that it leads decrease in the pervious surface which in turn alters the land cover, this consequences in a combination alters the water cycle. Often experienced hydrological effects are the rise in overall and direct runoff capacities and peak flows related with quicker reaction time, and therefore it decreases the infiltration rate and the ground base flow. (Mingfu Guan1 et al., 2015). More often the problem faced in the urbanized area during rainy seasons is flooding. This problem is mainly caused due the more appreciation in storm water drains and even due to non-provision of such drains, as this drains help in conveying the excess storm water safely to the outfalls. The consequences faced due to the urban flood is that it paves way for the deterioration of roads, buildings and for times the excess stagnation of water on the moving area becomes an un-avoidable problem during the rainy season in relation to routine lifestyle.

Hence to combat this problem, it becomes necessary to have a proper drainage system especially in low-lying areas. A manual storm water drainage system shall never reach to the expectations and shall not early decrease the problem of inundation. A systematized and accurate designed storm water drainage system, it is necessary to know on what basis the urban drain works. Urban drain comprises of two type of fluids i.e. firstly the waste water and the other being storm water. The water is used in for the life support e.g. for all domestic purposes, the water that need at the industries for the processing to get end product and these are many such sectors from where water let out is not used until recycled . Such water should be collected and transported safely without causing any nuisance any hazardous issues. So to design a storm water drainage it is necessary to keep this kind of fluid, its sources, it's outlets in consideration. The other type of fluid taken into consideration is storm water. This water is the runoff during precipitation which does not need any prior processing for conveying into drainage system.

In developing country like India, is highly necessary to have a well-designed besides economical, for the management of storm water (Needhidasan .S et al., 2013).

In the current study we are utilized Storm Water Management Model (SWMM). The input information needed for SWMM tool are area of the sub-catchment, width, percentage of slope which is prepared by DEM in ArcGIS tool and percentage of pervious/ imperviousness these parameters plays very vital role in estimation of runoff and rainfall data are required. The main objective of this examination using SWMM demonstrate surface overflow volume and some different approaches like Low-Impact Development (LID) control structures to reduce the surface runoff, water quality protection and it also helps in groundwater recharging.

II. STUDY AREA

The Indraprastha Nagar is a part of Belgaum City, situated in north part of Karnataka was selected for the study. Belgaum is 4th largest city in Karnataka having a total population of 4, 88,157 (As per census2011). Study area having pleasant atmosphere throughout the year. It is at coldest in winter season from November – February temperature dropping 9 °c, and this encounters continuous storm amid July- September. The yearly normal precipitation of the city is more than 2000 mm. The proposed study area Indraprastha Nagar is lies between 15.84° N latitude and 74.50° E longitude. It has an elevation of 751 meters. The study area is situated near the railway station of Belgaum city. The Indraprastha Nagar spread in an area of 70.08 hectares. The land use / land cover of Indraprastha Nagar distributed as residential, commercial, open space and transportation, parks. On other hand small portion of that area comprises of agricultural fields. Development activities are taking in the area.

DATA USED

Distinctive sorts of information required for modeling have been gathered from different associated department. The hydro-meteorological data, digital elevation model (DEM), physical data (such as satellite imagery) and the existing building footprints, road network were digitiized from the base map, daily rainfall data has been collected and related information of the study area. (1) Digitization is process in which first base map is added in Arc-GIS software then shape files are created using polygon features for boundaries and line feature for roads. (2) Slope Map the digital elevation model (DEM) data Cartosat-1: DEM- Version 2-R1 (2014) of 32m resolution was downloaded by Indian Space Research Organization (ISRO) then DEM data was brought into the Arc GIS 10.1 platform for preparation of slope map of the study area. (3)Pervious/ imperviousness of study area is estimated using land use/land cover distribution of the study area. (4) Rainfall data**:** The hydro-meterological data i.e. daily rainfall data of 14 years starting from 2001- 2014 was collected from Statistical Department, Government of India office in Belgaum district, Karnataka**.** For the current examination we utilized one day annual maximum rainfall data that is 2007 (120mm).

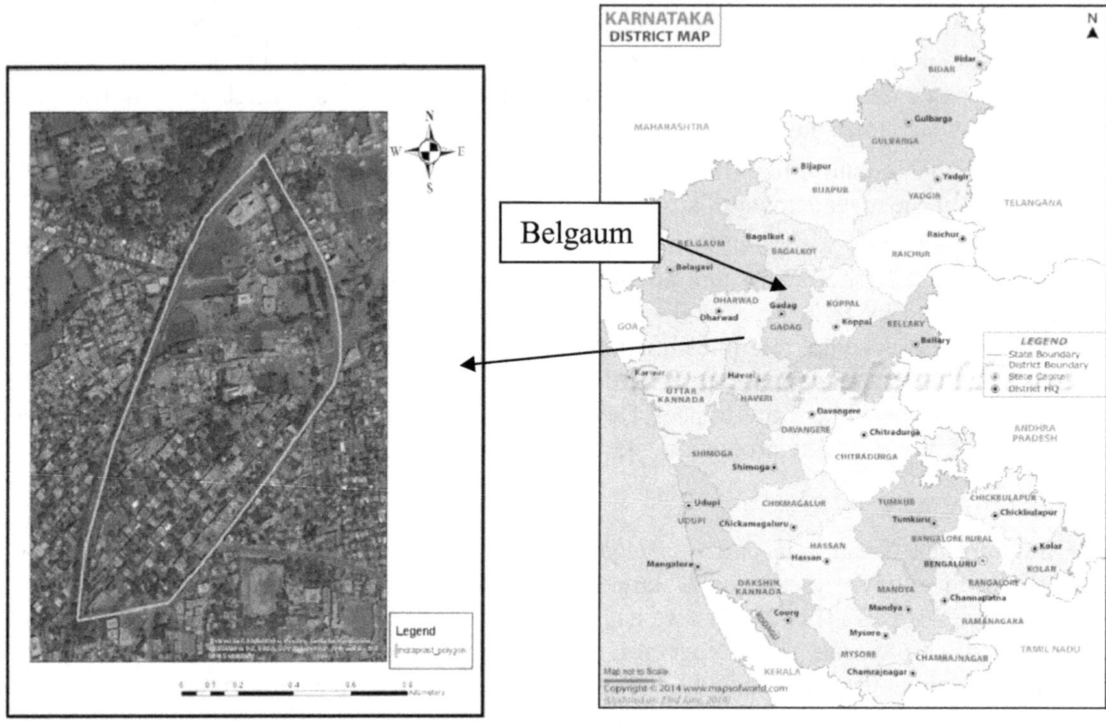

Fig. 1 Location map of Study area in Belgaum District in Karnataka

METHODOLOGY ADOPTED

In the present study SWMM is utilized as hydrological model because of its success in modeling the urban catchments. SWMM model was first created in 1971 by US Environmental Protection Agency (EPA, USA). The SWMM is a dynamic precipitation-overflow simulation model utilized for single event or continuous simulation of runoff quantity from fundamentally urban regions. The SWMM demonstrates the quantity and quality of run-off which is created within each sub-catchment, and flow rate, flow depth, and quality of water in each channels amid a simulation time included various time setups (Source : SWMM user's manual version 5.1). It consist of different block to be simulated separately (Sanat Nalini Sahoo and P. Sreeja 2013). The runoff parameters describe the physical characteristics such as area, slope, width and hydrologic characteristics such as percentage of imperviousness, depression storage, Manning's roughness coefficient of the catchment.

The study area is divided into 20 sub-catchments to best capture the effect of topography variation, landuse variation, soil variation on the runoff. Each sub-catchments is then modeled as a nonlinear reservoir with rainfall data as input. Both the pervious and impervious areas of the catchments are characterized by a value of maximum depression storage, which represents the capacity of the non-linear reservoirs (ASCE mannual 1992). Surface runoff occurs only when the depth of water in the reservoir exceeds the maximum depression storage. In pervious areas, infiltration has been modeled by Green-Ampt method whose parameters reviewed from literature (Rawls etal.1983).

Low-Impact Development which portray a land arranging, and engineering plan to deal storm water runoff as component of green infrastructure. SWMM model comprise of various sorts LID controls such as rain gardens, bio-retention cells, infiltration trenches, green roof, rain barrel, vegetative swales, preferring the rain gardens and bio-retention cells in open areas of the study area.

RESULTS AND DISCUSSION

The delineated area has been sub-divided into 20 different zones called as sub-catchments (S) with uneven perimeters. Each sub-catchment is designed with sewer lines by providing proper slope at intermediate junctions by connecting conduits. The overall runoff which was delivered from all the sub-catchments was discharged to outlet point through conduits with proper slope and depth. From the present simulation model S1 to S20 indicates sub-catchments, J indicates junctions between the nodes and C represents conduits which connects flow between junctions. Surface runoff, storm sewer design parameters, water elevation graphs were obtained by simulating the SWMM model.

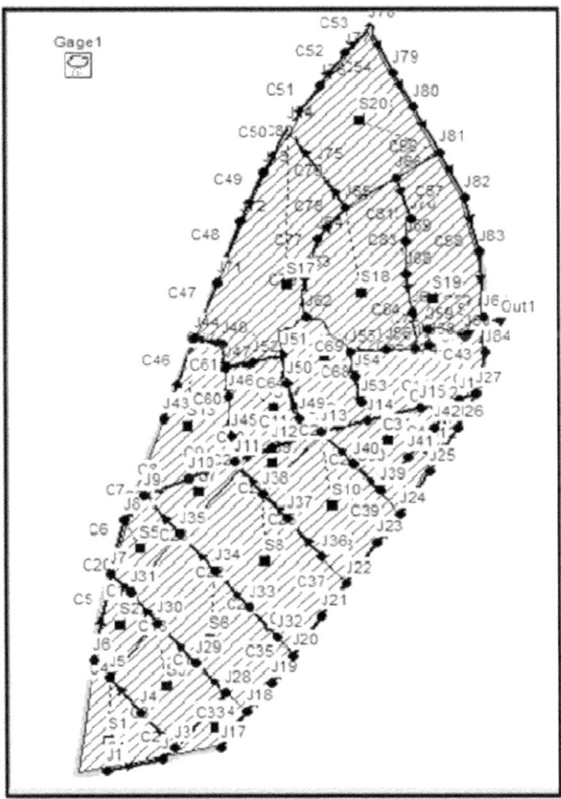

Fig. 2 Discretization Map of the study area

Based on the data obtained from the ArcGIS (% Slope, % Perviousnesss/imperviousness area) the rainfall runoff simulation is conducted using SWMM model.

Model Set-up:

- Input data: DEM, % Slope of the study area, Area of sub-catchments, Perviousness /Imperviousness of the study area.
- Precipitation data: From prepared data one day annual maximum rainfall 9th July 2007 (120mm) was selected from the 14 years rainfall data.
- No. of sub-catchments: 20
- Number of nodes:85
- No. of junctions: 84
- Manning Roughness co-efficient: 0.013 to 0.08 depending on land use/land cover
- Method for loss of runoff: SCS Curve Number method.
- Simulation time: 24 hours

Table 1 below shows the runoff results derived SWMM model and by rational method. The results from both methods are comparable.

Table 1 Surface runoff values from SWMM model and Rational Method

Sub-Catchments	Peak runoff by model (CMS)	Using Rational method (CMS)
S1	0.59	0.49
S2	0.33	0.28
S3	0.83	0.71
S4	0.2	0.19
S5	0.28	0.2
S6	1.36	1.03
S7	0.32	0.28
S8	1.63	1.28
S9	0.19	0.15
S10	1.11	1.09
S11	0.57	0.49
S12	0.32	0.27
S13	0.65	0.52
S14	0.83	0.73
S15	0.84	0.71
S16	0.7	0.61
S17	0.56	0.48
S18	1.3	1.02
S19	1.47	1.3
S20	1.32	1.12

The R^2 value of 0.92 was obtained indicated a best fit of results between simulated versus calculated values.

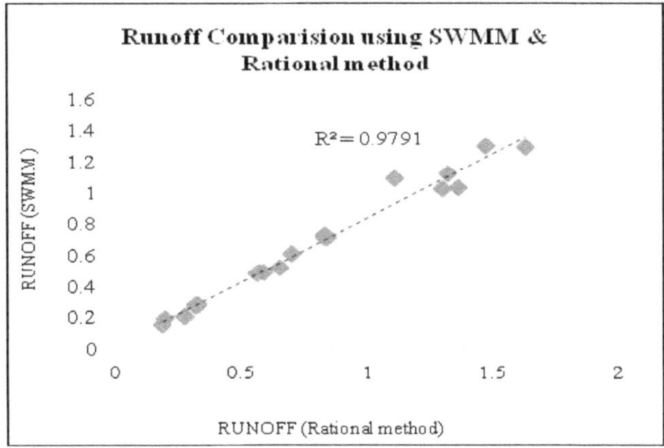

Fig. 3 Scatter plot of runoff obtained from SWMM and Rational method

SWMM model performance was determined through simulated outcomes and validated outcomes. Several methods utilized to check the model performance, Viz. Coefficient of determination (R^2) and Nash-Sutcliffeefficiency (E), Maximum absolute difference (MXE), Mean absolute difference (MAE), and Root mean square difference (RMSD), Standard error estimation (SEE) (Krebs, 2016).The R^2 value is an relationship between the observed and simulated values. The values of the Nash-Sutcliffe can ranges from negative infinity to 1. R^2 coefficient greater than 0.75 are "good", whereas values between 0.75- 0.5 as "satisfactory" (Rahman,2013). The Ttable:2 showing the model performace results values by different methods.

Table 2 Model performance by different methods

Sl. No	Different error estimation methods	Values
1	Nush-Sutcliffe efficiency (E)	0.97
2	Maximum absolute difference (MXE)	0.34
3	Mean absolute difference (MAE)	0.122
4	Root mean square difference (RMSD)	0.5
5	Standard error estimation (SEE)	0.0016

IMPLEMENTING LOW IMPACT DEVELOPMENT (LID) CONTROL

Out of many LID controls- rain gardens and bio-retention were selected as they were suitable for the study area. In SWMM, the selected LID controls were installed in appropriate locations as shown in Figure 4.

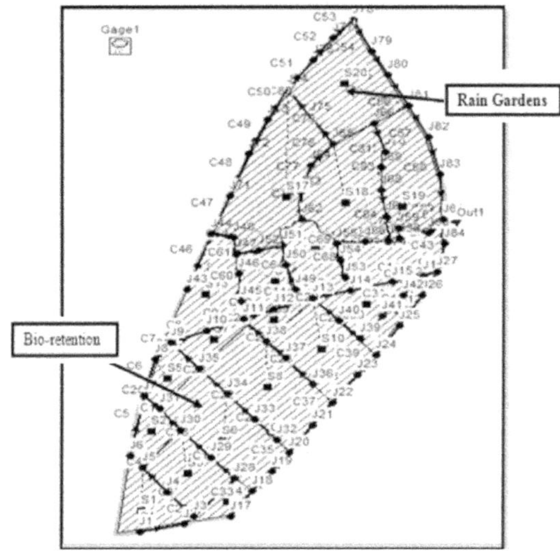

Fig. 4 Location of LID control structures implemented in the study area

The results indicate that the runoff from the study area has decreased after the implementation of LID controls. The results are appreciable in sub-catchment 3, 6, 8, 13, 14, 15, 19 and 20. The amount of runoff decreased in different sub-catchments is presumed to be lost majorly through groundwater recharge and to some extent by evapotranspiration process. Hence implementing such kind of LID will not only handle the runoff at the source but will also help to augment the groundwater resources, which can be reused during dry periods.

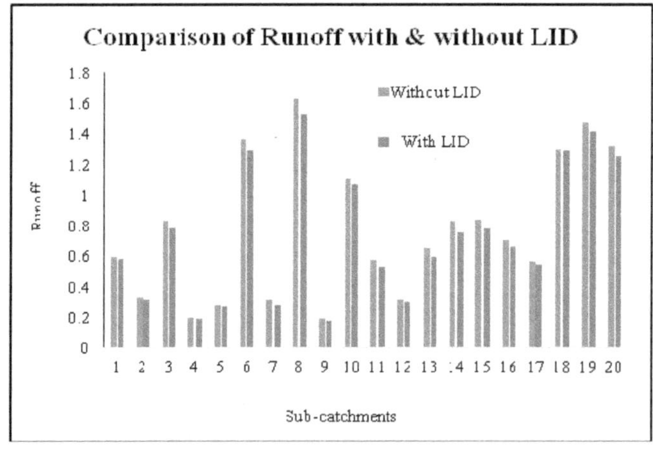

Fig. 5 Comparison graph of runoff without LID and with LID.

6. CONCLUSION

In the present study, runoff has been estimated for a Indraprastha Nagar Belgaum urban catchment based on event wise rainfall data. A comparision has been made for simulated and validated runoffs correspoing to rainfall event. The peak runoff by SWMM is 6.67 m^3/s and by rational method is 6.18 m^3/s. SWMM model is dynamic model and is equipped for demonstrating the amount of water flow through the catchment. Regularly inundating during stormy season in urban catchments regularly causes water stagnation problems are sorted.

Sustainable storm water management using LID techniques like bio-retention and rain gardens applied in open area in each sub-catchments was attempted successfully.The discharge values were reduced after applying LID. Hence simulated models information is satisfactory.

REFERENCES

1. ASCE(1992). Design and construction of urban strom water management system. ASCE – Manuals and reports on Engineering Practice No.77, New Yark.
2. CPHEEO. Govt. of India, Manual on Rainfall Analysis for Storm Design System (2013)
3. Deepak Singh Bisht., Chandranath Chatterjee., Shivani Kalakoti., Pawan Upadhay., Manaswinee Sahool., Ambarnil Pandal (2016). "Modeling Urban Floods and Drainage Using SWMM and MIKE URBAN". Journal of Springer Secience.
4. Elliott A.H., Trowsdale S.A (2007). "A Review of Models for Low Impact Urban Storm water Drainage". Model softw 22: 394-405.
5. Fluid Mechanics and Hydraulic Machines Handbook for a Guide to Understand Most Economical Section of Channels. Dr. R. K Bansal.
6. Gaurav V., Jain. , Ritesh Agrawal., R.J. Bhanderi., P. Jayaprasad., J.N. Patel., P.G Agnihotri., B.M.Samtani (2015). "Estimation of Sub-catchments Area Parameters for Storm water Management Model (SWMM) using Geo-Informatics. Journal of Geocarto International, 2016 Vol.31, No 4, 462-476.
7. G. Krebs. , T. Kokkonen. , M. Valtanen. , Koivusalo., H. Setala (2013). "A High Resolution Application of Storm water Management Model (SWMM) Using Genetic Parameter Optimization". Journal of Urban Water. , Vol.10, No. 6, 394-410.
8. Guan. , M. Sillanpaa. , N. Koivusalo (2015). "Modeling and Assessment of Hydrological Changes in a Developing Urban catchment." Journal of Hydrological Processes, 29 (13). PP. 2880-2894.
9. Harshal Pathak., Pravin Chaudhari (2015), "Simulation of Best Management Practices Using SWMM". Vol. No. 02.
10. Hydrology Handbook for frequency analysis A Guide to Understand flood data. Dr. P. Jaya Rami Reddy (2012).
11. Krebs Gerald., (2016)." Spatial Resolution and Parameterization of an Urban Hydrological Model: Requirements for the Evaluation of Low Impact Development Strategies at the City Scale". Department of Built Environment, Aalto University publication series DOCTORAL DISSERTATIONS,pages-79, ISBN: 978-952-60-6780-3
12. Needhidasan S. Manoj Nallanathe (2013). "Design of Storm water Drains by Rational Method an Approach to Storm water Management for Environmental Protection". Vol. No 4.
13. P. Sundara Kumar., T. Santhi., P. Manoj Srivastav., S.V. Sreekanth Reddy., M. Anjaneya Prasad., T.V. Praveen (2015). "Storm Water Drainage Design". Journal of Earth science and Engineering Vol.08. No.02.
14. Rawls, W.J., Brakensiek, D.L., and Miller, N. (1983). "Green Ampt infitration parameters from soils data". J.Hydraul. Eng.,109(1),62 -70.

15. Reddy, P. J. (2014). A textbook of hydrology: (in M.K.S and S.I Units). New Delhi: University of Science Press.
16. Sanat Nalini Sahoo., P. Sreeja (2013). "Role of Rainfall Events and Imperviousness Parameters on Urban Runoff Modelling". Journal of Hydraulic Engineering, 19:3, 329-334.
17. Wenting Zhang et al., (2013). "Analysis and Simulation of Drainage Capacity of Urban Pipe Network". Research Journal of Applied Sciences, Engineering and Technology 6(3): 387-392.

PRE-HISTORIC CULTURAL REMAINS IN WATER SHED

Kshirsagar S.D.

Dept. of A. I. H. C and Archaeology, Deccan College P.G. and R.I. (Deemed University) Pune.

shivajigeoarchnew@gmail.com

ABSTRACT

Pre-Historic Culture period to till date water is important aspect of every one life. Animal, Plant, Human etc are depending upon the water. Rain water is main source of the water. Natural water storage is important form of natural reservoir i.e. lakes, ponds etc. Pre-Historic man was used these natural recourses for their survival. Today so many small and big reservoirs, water sheds, dams and bund are constricted for artificial store and ground water recharge. In the case study of the natural ponds and lakes of middle reaches of Bhima basin and its tributary Sina, Man, Bori and Bor river Basin of Western Maharashtra. Pingali Man Dahiwadi, Talsangi village of Mangalwedha tehsil of Satara and Solapur District etc sites are recovered important evidence of Prehistoric Cultural material of Mesolithic period Microlithic stone tools. Agate, Chart and Chalcedony etc locally available raw material are used for the Microlithic tool making. Blade, Flakes, Core and debitages are recovered from above mention natural lakes or ponds but now it is construct for water conservation reservoir. These tools are very important evidence of natural water storage activity and settlement of hunting gathering society settlement. So, now many places are used for modern water shed constriction.

Keyword: Pre-Historic-Water Shed.

INTRODUCTION

Rain water and ground water is main resource of life especially on rain shadow zone of Western Ghat, Maharashtra. For example Satara districts of Man rivers source area in Man Dahiwadi tehsil. (Figures 1A, 1, 2).

Study Area

Water shed of Pingali Bk/Kh is situated in between of the this two village, 2 km SW of Dahiwadi town on the right bank of the Man river small tributary and district Satara, Maharashtra. Survey of India (SOI) topo-sheet no. 47K/10 on 720 AMSL, scale 1:50000 year 1977.

Geology of study area is form of Upper Cretaceous to Lower Eocene age. The pahoehoe flows comprise several units. Each unit flows a basal section with pipe amygdales, a middle section of dense basalt and a top section, which is highly vesicular or amygdule. While the vesicles in the aa flows are elongated or twisted, the vesicles in pahoehoe flows are characteristically spherical. The vesicles contain zeolites, heulandites and stilbite, secondary silica like chert, banded chalcedony or quartz, altered glass and in rare cases chloritic material. Rainfall is very low below 500 mm to low 500-1000 mm. Soil is shallow black.

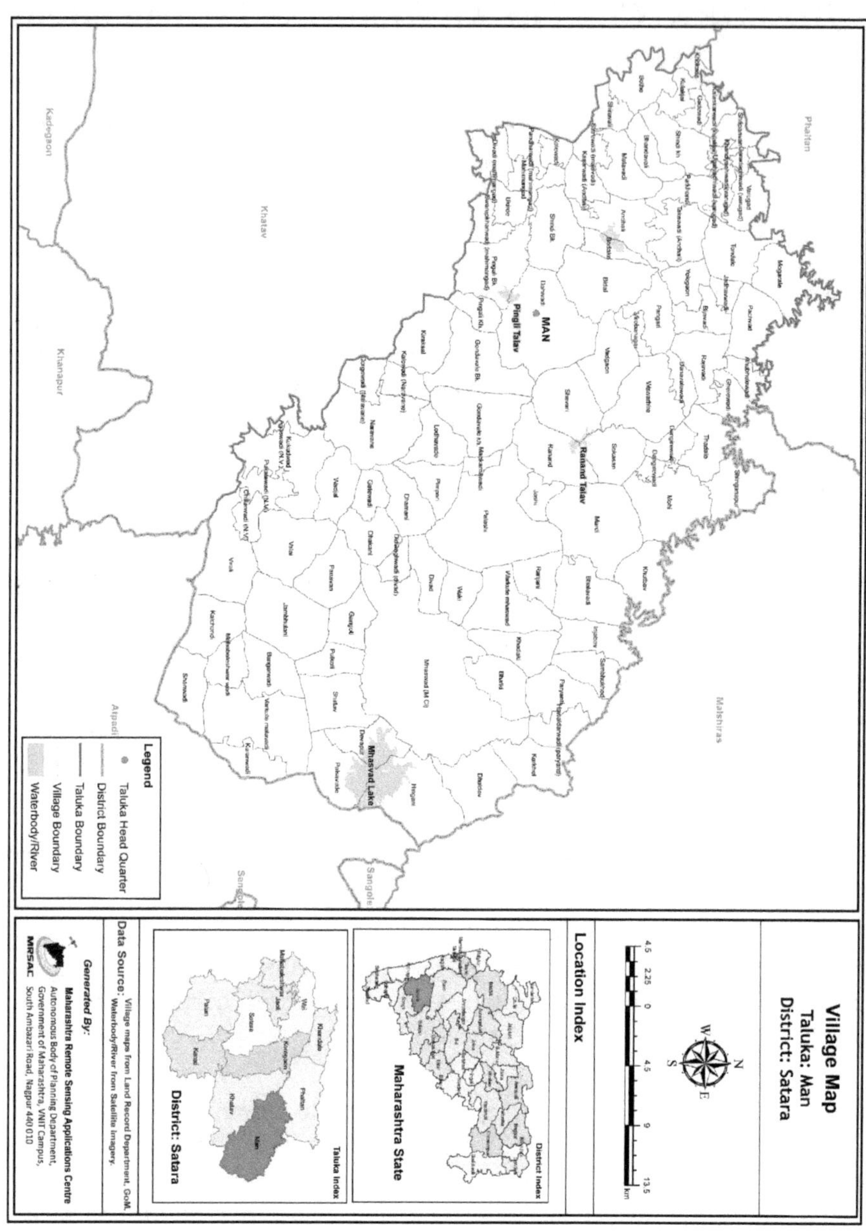

Fig. 1A Pingalii (Dhahiwadi) Water Shed google earth view

Fig. 1 Pingalii (Dhahiwadi) Water Shed Catchment narea

Man River

The Man River rise on 900 MSL in Hira hills of Kulakjai SW of at Tathavade village tahsil Man, Statara district. Man River total length is160 Km, 64 Km in Man tahsil, 80 Km in Solapur district and 16 Km in Sangli district. The Man River has only about 16 km of course within or on the border of the Sangli district but along with its tributaries is responsible for draining the north-eastern parts of Khanapur and tehsil and northern part of the Jath tehsil into the Bhima River. To the west of the river are a number of tributaries draining the slopes of the Khanapur plateau eastwards into it, from north to south they are the union of the Bhandora Sikir and Dabucha streams, while two other in upper courses in the Sangli district.

Ground Water

The alluvial patches are good aquifer. The trap rocks are generally poor as far as there capacity to store and yield groundwater is concerned. Wherever the traps are weathered, the joints and fractures become open and help in groundwater movement. The horizontal joints found at the basal section of flows help in lateral migration of water. Wells which pass through such jointed or fractured or weathered parts of flows, wherever they occur under confined conditions, yield fairly good amount of water. Contacts of flows, often show seepages of water or springs, as the breccias tops of aa flows are impervious when they are fresh. Figures 9.

Previous Work

In previous work by the authors (Deo et al., IAR, 2005; Kshirsagar, 2016) highlight on the Mesolithic cultural material of microlithc stone tools of Blades, Flakes, Scrapers, Cores made of up Chalcedony, Agate, Chert etc raw material in this study area with the sandy silt gravel.

Aims and objects

To high light the water resource which is using by ancient man to till date

To understand the knowledge of about the ancient water management

To understand the natural resource use of Man

To understand an idea about the water storage system

Methodology

Field survey of natural lakes and water shed with help of Survey of India toposheet and district resource map (DRM) of Geological Survey of India. Google Earth, Global maper DEM and Bhuvan 2 D and 3 D view of NRSC Hyderabad.

Field work data

Field work of Pingali and Talsangi village water shed was carried out by author (Kshirsagar 2016, 2017) for his Post Doctoral research and Deo et al., IAR. 2005, Carried out geo-archaeological exploration in Man River. New light on Mesolithic cultural material of microlithic stone tools of chalcedony, agate and chert etc are used for raw to making the tools. Blades, flakes, cores and debitages etc.... tool types are recovered from these sites. The water is most important factor of from begging of our life. Early human society depends on the hunting and gathering of animal. Three months are rainy session of this area. Remaining day's water source is natural lakes, springs, ponds etc. This natural features are using till. Geo-morphological futures are use for natural water storage.

Pingali water shed catchment area is 59.310 Km^2 (Figures 2). Terrain and Slope map of the study area shows in (Figures 4 and 7). Pingali (Dhahiwadi) Water Shed DEM model (Figures 5 and 6.). Survey of India topo-sheet no. 47 K/10 part of NW corner shows in Figures 8.

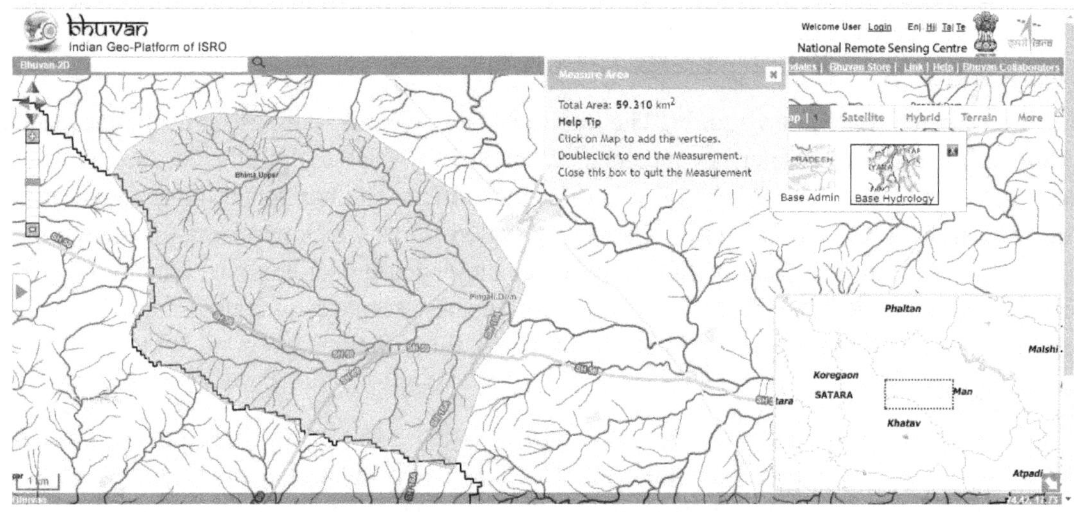

Fig. 2 Pingalii (Dhahiwadi) Water Shed Catchment narea

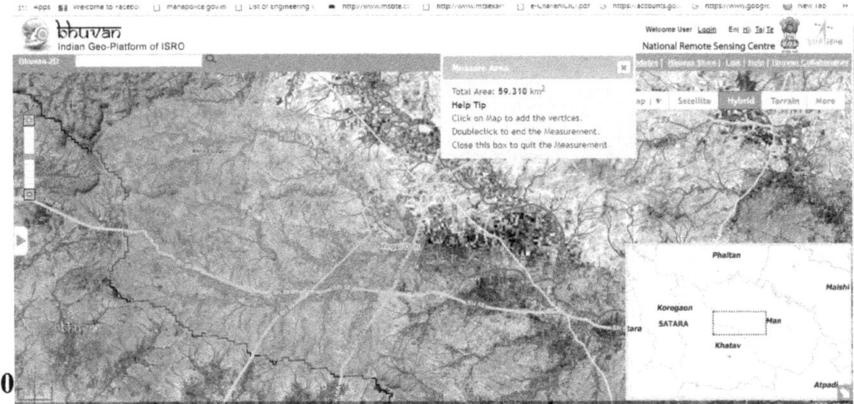

Fig. 3 Pingalii (Dhahiwadi) Water Shed Catchmentn area

Fig. 4 Pingalii (Dhahiwadi) Water Shed Slope map

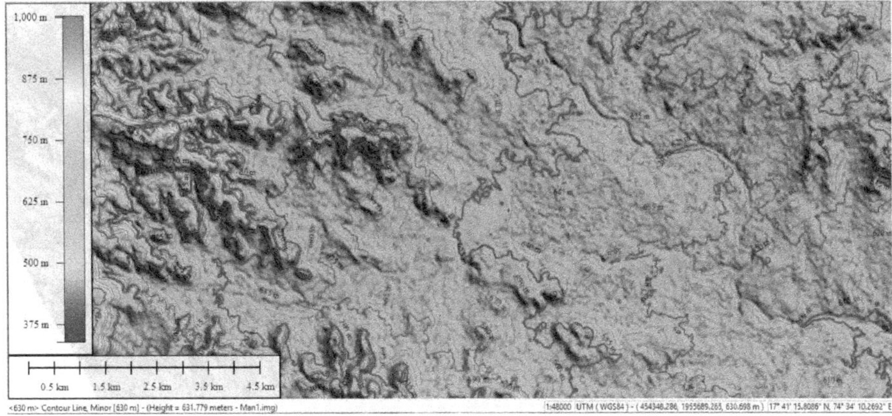

Fig. 5 Pingalii (Dhahiwadi) Water Shed countur map

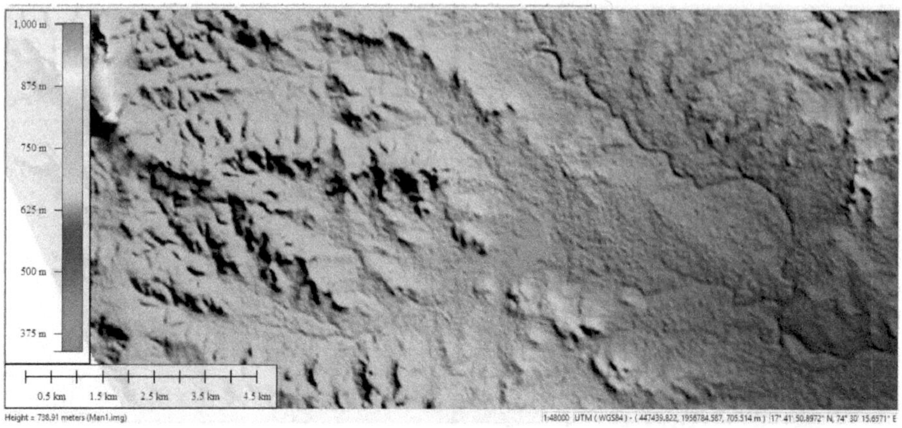

Fig. 6 Pingali (Dhahiwadi) Water Shed DEM model

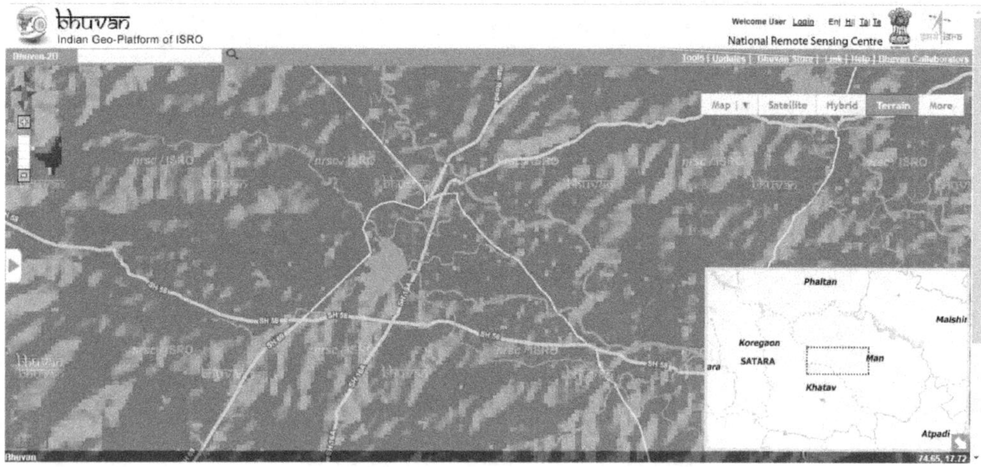

Fig. 7 Pingalii (Dhahiwadi) Water Shed Catchmentn area

Sediments

Sandy Silty Clay mixed with calcrit noodal and microlithic stone tools are recoverd form the foot of water shed bund eastern area.

Other sites

Ranand Talav and Bodake-Andhali lake are other two site are recovered from this area as well as Malavadi surfae site on right bank of Man river.

Summary

Geographical, Geological and Climatic setting of these stone industries and a clear description of the raw materials utilized brings together essential information of study area.

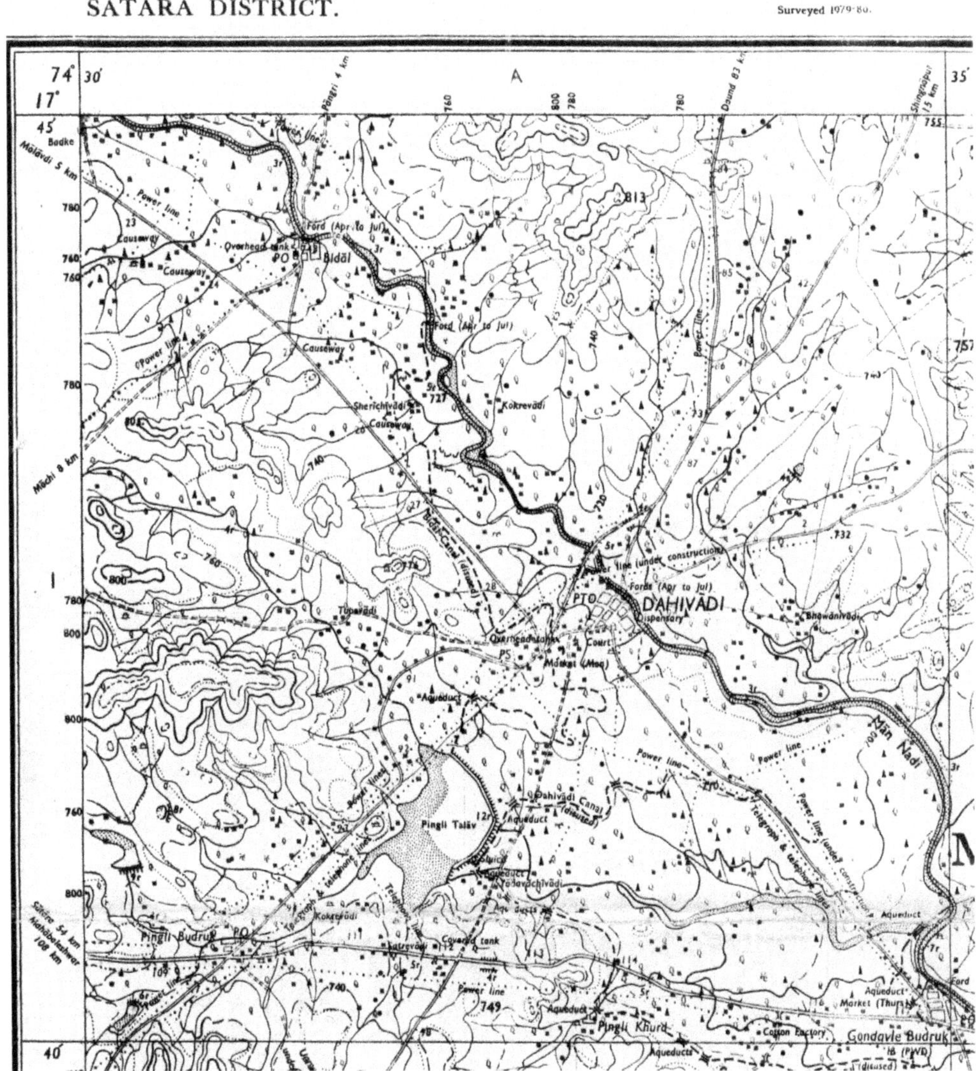

Fig. 8 Pingalii (Dhahiwadi) Water Shed Survey Of India Toposheet (1:50000)

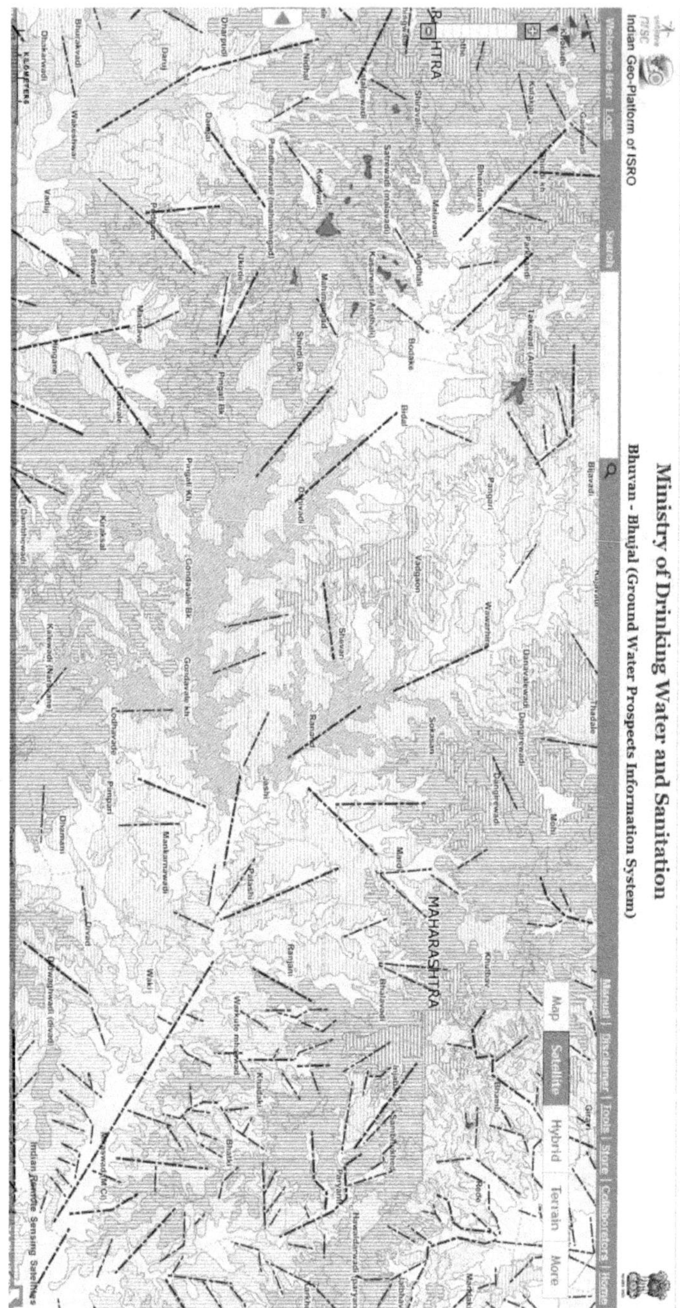

Fig. 9 Pingalii (Dhahiwadi) Water Shed Graundwater Map

REFERENCES

1. Malik S. C. 1959. Stone Age Industries of the Bombay and Satara Districts. M.S. University Archaeological Series No. 4 Baroda.
2. Kshirsagar S. D. 2016. Further Geo-archaeological Investigation in Bori and Bor river of Western Maharashtra. Unpublished Post-Doctoral Research Report Submitted to Deccan College P.G. and R.I. Pune.
3. Indian Archaeology: A Review 2004-05. Archaeological Survey of India, New Delhi.
4. Deo Sushama G, Shivaji Kshirsagar, S. N. Rajaguru, Santosh Hampe and P. P. Jogalekar 2005. Geoarchaeological Observation of Man Basin in Solapur District, Maharashtra. Puratattva Vol. 35, Pp. 111-117 (Plate- 1-4), Archaeological Society of India, New Delhi.
5. Gazetteer of Satara District. 1972. Government of Maharashtra, Mumbai.

ESTIMATION OF GROUND WATER STORAGE CHANGES IN USING SATELLITE DATA

Ballu Harish[1], K. Manjulavani[2], L. Ravi[3], V. Madhava Rao[4]

[1]Lecturer, [3]Ph.D scholar [4]Student, Department of Centre for Spatial Information Technology, JNTUH-IST
[2]HOD & Professor, Department of Civil Engineering, JNTU Hyderabad

ABSTRACT

Hyderabad region part of Telangana state, India Study focus: Impact of Global warming has an effect on water resource management. Hencemonitoring of changes in terrestrial water storage has become significantly important. From the combination of GRACE and MAIRS data the changes of Terrestrial water storage in Hyderabad region has been found from January 2009 to December 2014. In this paper we use GRACE monthly gravity data to track water storage change from January 2009 to December 2014. The main theme of the work is to use GRACE data set to see ground water storage levels and its changes in Hyderabad region and to do water balance analysis using GIS and RS techniques which is obtained by

(i) Time series comparisons of satellite data over field based data in same spatial scale at various temporal cycles.

(ii) Generation of GIS maps with ground water levels between 2009 to 2014.

In Hyderabad region it is found that

Average ground water depletion is 0.61 per year from 2009-14.

Average recharge rate of ground water from pre to post monsoon is 17.3203 kg/cm3.But in 2014 it is 9.623 kg/cm3 i.e. recharge rate is decreased. The water potential is less than 1 kg/cm3, below ground water levels occupied 73.96 sqkm2 i.e. 66.66% of study area. Hence it is predicted that study area is falling under low ground water potential.

Keywords: GRACE, MAIRS, Ground water storage (GWS), Total water storage, Terrestrial water storage, GIS.

INTRODUCTION

Ground water is largest accessible source of fresh water and it plays a critical role in maintaining adequate water supplies for human needs, agriculture usage. Some of the factors affecting water resources are increase in population growth, climate variability. Linking the spatial temporal changes in climate with ground water will minimise the risk of stakeholders about water, energy, food. Hydrological cycle is being affected by factors like precipaitation, snowmelt, streamflow which causes occurrence of drought [2].Precipitation is main atmospheric factor forcing of hydrological catchment response. Studing of water balance & modelling become meaningless without estimation of precipitation. Small error on precipitation estimates may leads to higher error on estimation of stream flow [4].

Studies have been started about the linking of spatial temporal patterns with climate oscillations in some areas like west coast of United states which has shown good result. i.e Teleconnection pattern present in surface hydrologic process is affecting ground water of united states west coast [3].

Ground water will be sustained only if quality and quantity is accessed properly[7]. GIS techniques can be used for providing ground water quality zones for different usages such as irrigation, domestic needs. GIS can also be used to prepare layers of maps based on water quality &availability [6]. Some of the reasons for getting poor quality of water is mainly i)Anthropogenic Activities: Industralisation, Mining which is quickly effecting the quality of water compared to natural process(Geology. Ground water movement).ii) Increase in population, growing urbanisation, irrigated agriculture. Distinguishing of natural & anthropogenic factors from chemical composition of ground water is difficult[1].

Better separation of GWS changes from GRACE derived TWS changes can be done if improvements are made in study area [5]. Grace data has been used across worldwide for finding the changes in TWS. It is important to obtain spatial and temporal variability in water storage on continental scale [9,8].

In the next section, the methodology is discussed, followed by a description and discussion of the results in sections III. Finally, the discussions and conclusions are summarized in section IV.

II. METHODOLOGY

A. Water budget Concept

Total Water Storage (TWS) tends to be dominated by snow and ice in polar and alpine regions, by soil moisture in mid-latitudes, and by surface water in wet, tropical regions . A basic equation is formulated by explaining the components of total water storage explained in Eq. 1

TWS=G+S1+S2+W+SI---- (1)

TWS= Total Water Storage

G=Groundwater

S1=Soil moisture

S2=Surface water

W= Wet biomass

SI=Snow &ice

For a particular environment one or more of these parameters can have impact on TWS. For the present study area it is assumed that soil moisture and groundwater has more impact over TWS or in turn over GWS.

This is because the region is primarily based on agriculture.

Therefore for the present study area Eq. No. 1 is modified suitably to exclude other parameters. Hence, the modified equation for the change of TWS is as follows:

TWSC= SMSC+ G WSC---- (2)

TWSC= Change in Total Water Storage

SMSC= Change in Soil Moisture Storage

GWSC= Change in Groundwater Storage

Here, the storage terms in Equation (2) are spatial averages over the study area..

The change in GWS is the difference between storage anomalies of any two successive time stamps.

MAIRS data is used to compute global estimates of soil moisture (SM), from which soil moisture storage change (SMSC) is computed.

B. Methodology Flow Chart

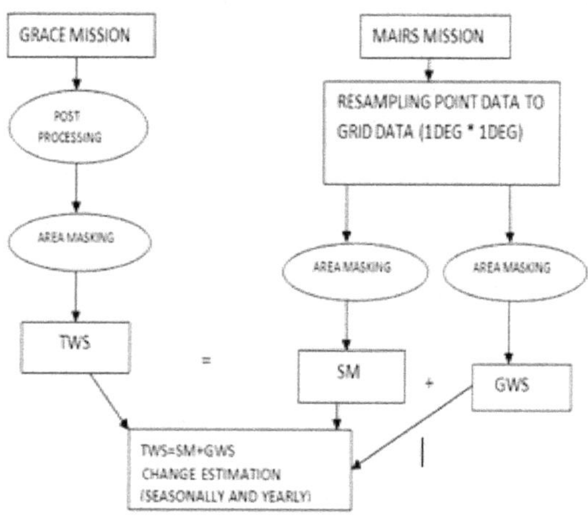

Fig. 1 Methodology flow chart.

TWS (Total water storage) include soil moisture, groundwater, surface water, snow and ice, and biomass. In this selected study area soil moisture and groundwater are the primary contributors to TWS and the water contribution from snow, ice, biomass, and surface water are relatively minor.

Therefore, in this study TWS is assumed to be controlled primarily by soil moisture, and groundwater.

$$\Delta TWS = \Delta\text{Ground water storage (GWS)} + \Delta\text{Soil moisture (SM)} \quad .....(3)$$

where Δ is change (e.g., monthly, seasonal, or annual changes),SM in this context refers to volumetric soil moisture content, which is the volume of water stored within the soil column. If TWS and SM are known. Above equation (i.e. equation 3) can be reorganized as follows so that GWS is estimated

$$\Delta GWS = \Delta TWS - \Delta SM$$

Here, GRACE month-to-month gravity fields are inverted for total water storage anomaly (TWSA), from which ΔGWS is computed .GWS anomaly is the residual storage content at a given time. Then change in GWS is the difference between storage Anomalies of any two successive time stamps.

MAIRS data is used to compute global estimates of soil moisture (SM), from which soil moisture storage change (ΔSM) is computed.

III. Results and analysis of GWS changes in Hyderabad region

A. Ground Water Storage spatial distribution

The spatial distributions in Figure 2 suggest overall groundwater storage levels in the study area for the year 2014.

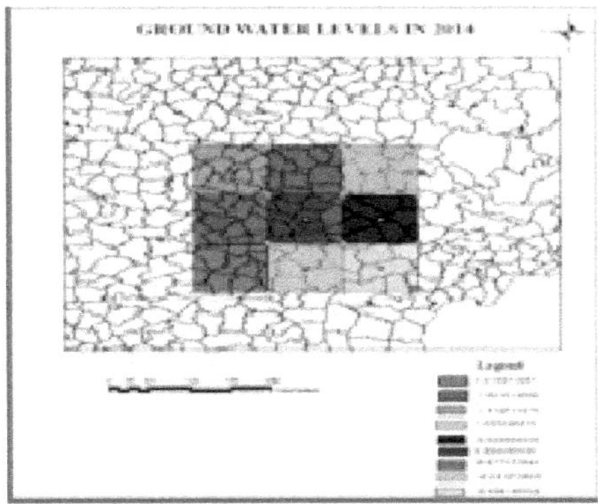

Fig. 2 Spatial distribution of ground water levels in the year 2014.

Groundwater table map has been prepared based on the attribute data of groundwater depth in the study area. The above GWS map has been prepared for study area using one degree regular grids data from

GRACE total water levels and MAIRS soil moisture. For each grid the mean value of groundwater has been calculated, that may varies on three parameters used in the study. Parameters used in study are total water storage, groundwater storage and soil moisture.

The water tables i.e. <1 kg/cm^3 below ground levels (figure 2) occupied about 66.66% of the study area. Thus, majority of the study area can be predicted as low groundwater potentials. The spatial distribution of groundwater levels are shown in below table.

Table 1 Groundwater levels in 2014

Latitude	Longitude	GWS (kg/cm^3)	Variation Of GWS with standard GWS(106 kg/cm^3)	Greater/lesser than standard GWS value
16.5	77.5	104.7833	1.21667	Less than standard GWS
16.5	78.5	105.1138	0.88619	Less than standard GWS
16.5	79.5	105.7581	0.24188	Less than standard GWS
17.5	77.5	106.4775	0.47753	Greater than standard GWS
17.5	78.5	106.55001	0.55001	Greater than standard GWS
17.5	79.5	106.9498	0.9498	Greater than standard GWS
18.5	77.5	107.6165	1.6165	Greater than standard GWS
18.5	78.5	107.6519	1.6519	Greater than standard GWS

B. Generation of GWS maps

Monthly GWS maps for GRACE data and MAIRS soil moisture are created using 1 degree regular grids. These maps give the storage variations over different seasons within the basin and to see the shift of surface mass with each month. The multi-temporal cycle analyses are used only for the purpose of comparisons, multi-level decision-making and policy/strategy measures.

It may be noticed that the monthly cycles of three parameters used in the study which includes Total water storage, Ground water storage and Soil Moisture constitute for a time span of 2009–2014 averaged over the study area. Then the average monthly cycles constitute a two level averaging – both monthly and yearly for 2009–2014 over the entire study area. The seasonal and average seasonal cycles are similarly processed.

Table 2 Pre monsoon GWS levels in 2014

Latitude	Longitude	GWS (kg/cm³)	Variation Of GWS with standard GWS(106 kg/cm³)	Greater/Lesser than standard GWS value
16.5	77.5	97.50423	8.49577	Less than standard GWS
16.5	78.5	97.77448	8.22552	Less than standard GWS
16.5	79.5	98.22036	7.77964	Less than standard GWS
17.5	77.5	98.0945	7.9055	Less than standard GWS
17.5	78.5	97.80152	8.198418	Less than standard GWS
17.5	79.5	97.74256	8.25744	Less than standard GWS
18.5	77.5	97.7667	8.23373	Less than standard GWS
18.5	78.5	97.17669	8.82331	Less than standard GWS
18.5	79.5	9652519	9.47481	Less than standard GWS

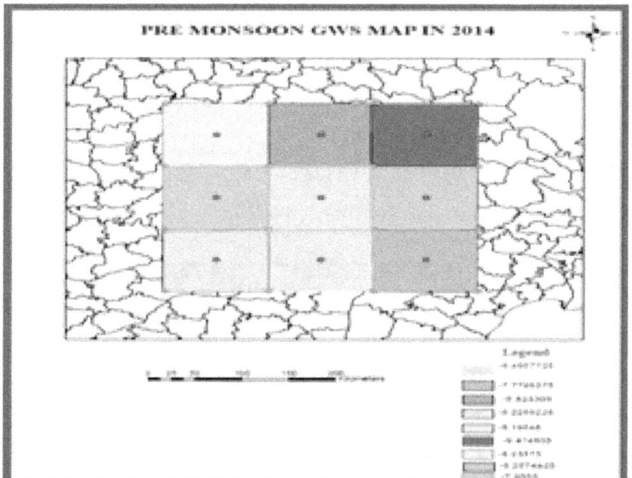

Fig. 3 Spatial distributions of pre monsoon GWS changes in the year 2014.

Table 3 Post monsoon GWS levels in 2014

Latitude	Longitude	GWS (kg/cm3)	Variation Of GWS with standard GWS(106 kg/cm3)	Greater/Lesser than standard GWS value
16.5	77.5	105.63104	0.36896	Less than standard GWS
16.5	78.5	105.66729	0.33271	Less than standard GWS
16.5	79.5	106.25486	0.254865	Greater than standard GWS
17.5	77.5	107.2759	1.2759	Greater than standard GWS
17.5	78.5	107.21478	1.214783	Greater than standard GWS
17.5	79.5	107.68801	1.688015	Greater than standard GWS
18.5	77.5	108.6544	2.654458	Greater than standard GWS
18.5	78.5	108.70864	2.708645	Greater than standard GWS
18.5	79.5	108.67005	2.67005	Greater than standard GWS

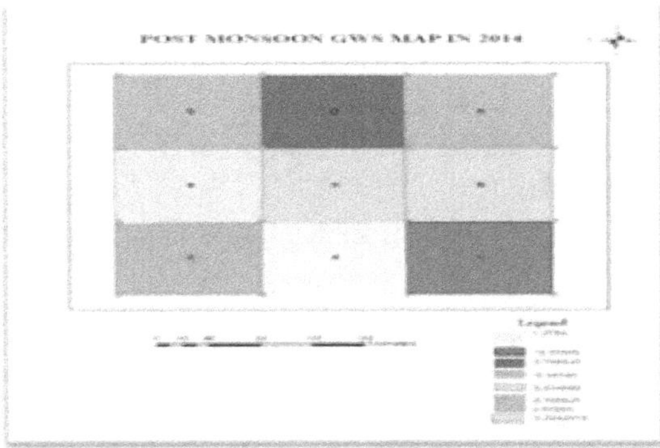

Fig. 4 Spatial distributions of post monsoon.

GWS levels in the year 2014The spatial distributions in Figure 3 and 4 suggest overall groundwater storage change in the study area over pre and post monsoon season of 2014.Pre monsoon groundwater levels are very low because of dry summer season, in monsoon season it starts slowly increasing due to rains, at the end of monsoon season it reached peak. It changes are clearly observed in above figure 3 and 4, in pre monsoon all groundwater values are lesser compared to standard GWS value observed in table 2 that ranges from 7.7796 cm to 9.4748 cm but after monsoon season almost all these value are increased compared to standard GWS in table 3 which ranges from 0.33271 to 2.7065. The water tables i.e. >1 kg/cm^3 above ground levels about 55.5% of the study area increased after monsoon season.

Depending on the seasonal rainfall GWS recharge rate is changed from one spatial location to other and also over usage of ground water also make changes in recharge of ground water. The unbalanced recharge and usage of ground water cause depletion of groundwater from year to year.

C. GWS variation

Time-series analysis of groundwater level and total water storage in the study area during 2009-2014 at average seasonal and yearly cycles are shown in figure 5.

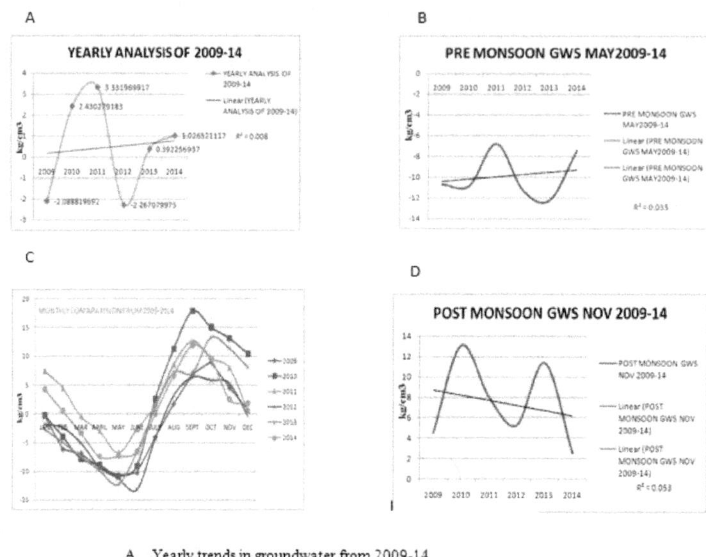

A. Yearly trends in groundwater from 2009-14
B. Pre monsoon trends in groundwater changes from 2009-14
C. Monthly trends in groundwater changes from 2009-14
D. Post monsoon trends in groundwater changes from 2009-14

Fig. 5 Plots of annual trends in groundwater.

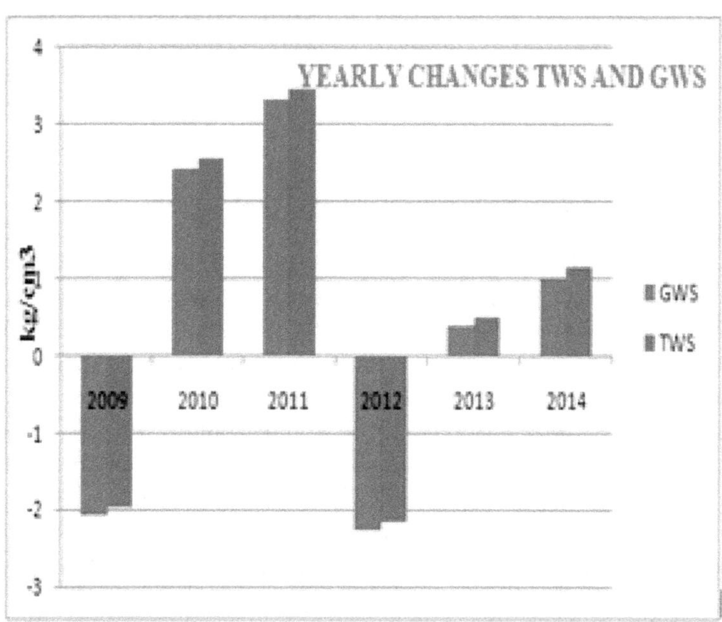

Fig. 6 GWS and TWS from 2009.

The seasonal yearly trends in ground water change (figure 5 A) have highest peak in 2010 and also very low in 2012. In 2014 groundwater levels 1.026 kg/cm^3, it was just above the average GWS i.e. 0.6123 kg/cm^3.

Table 4 GWS and TWS from 2009-14

YEAR	Variation of GWS with respect to standard GWS(106 kg/cm³)	Variation of TWS with respect to standard TWS(106 kg/cm³)	Variation Of GWS & TWS
2009	3.22828	1.96221	Lesser than standard value
2010	1.231334	2.563495	Greater Than standard value
2011	2.249037	3.452296	Greater Than standard value
2012	3.21462	2.1618	Lesser than standard value
2013	0.68111	0.51152	Lesser than standard value
2014	0.03335	1.144285	Greater Than standard value

The seasonal trends of pre monsoon and post monsoon is shown in figure 5.The entire pre monsoon trends are negative, trends in year 2013 has most negative value which is 12.22 kg/cm^3.

In post monsoon trends are positive trends in that 2014 has lowest value which is 2.524 kg/cm^3 it suggest that storage was peak in the post monsoon months.

D. Comparison of variation between TWS and GWS

The GRACE-derived TWSC is compared with groundwater storage average seasonal and yearly cycles. Amplitude ranges generally decrease at higher temporal scales. The comparisons are generally favourable with stronger agreements at higher temporal and averaging levels. The agreements are strongest at the average seasonal and weakest at the monthly cycles.

IV. DISCUSSION AND CONCLUSIONS

The trends in the monthly, seasonal and yearly curves in figure 5 confirm the depletion of groundwater. The trends suggest both narrowing soil moisture storage & also increasing total water storage loss. All the seasonal plots depict a clear seasonality of storage in the region. While the trends in SMSC and TWSC are negative at the monthly and seasonal cycles, those in GWSC are positive. All the trends are negative at the yearly cycle. The favourable comparisons of total water storage in Figure 3 and 4 at two different seasonal scales shows 55.5% of study area's groundwater levels are increased more than one kg/cm^3, it suggest that GRACE sufficiently detects storage signal in the study area. Seasonal storage dynamics is equally critical for sustainable water resources management strategies in the study area. The yearly negative trends in the storage change are therefore more reliable reflections of storage dynamics in the study area. These trends largely reflect the prevailing agro-climatic conditions. Storage loss in the region could have negative implications for the agriculture and people's livelihood. The spatial distributions in Figure 5 A suggest the total water storage depletion in the region. By Further analysis, the average storage depletion in the study area is 0.61 cm/year, following the concepts of equation (3). This finding could negatively impact the study area that depends on groundwater discharge system and groundwater recharge system. The spatial distributions in figure 2 suggest 66.6% of study area groundwater levels are below one kg/cm^3 i.e. the majority of the study area has been predicted as low groundwater potentials.

REFERENCES

1. Deepesh Machiwala, MadanK.Jha. (2015) "Identifying sources of groundwater contamination in a hard-rock aquifer system using multivariate statistical analyses and GIS-based geostatistical modeling techniques".
2. Elzie M. Velascoa, Jason J. Gurdaka,, Jesse E. Dickinsonb, T.P.A. Ferréc, ClaudiaR. Corona.(2015) "Interannual to multidecadal climate forcings on groundwaterresources of the U.S. West Coast".
3. Frank Herrmann, Luise Keller, Ralf Kunkel, Harry Vereecken, Frank Wendland. (2015) "Determination of spatially differentiated water balancecomponents including groundwater recharge on the FederalState level – A case study using the mGROWA model in NorthRhine-Westphalia (Germany)".
4. Gilles Drogue, Wiem Ben Khediri.(2016) "Catchment model regionalization approach based on spatialproximity: Does a neighbor catchment-based rainfall inputstrengthen the method".
5. LongweiXianga,b, HanshengWanga, HolgerSteffenc, PatrickWud, LuluJiae, LimingJianga, QiangShena (2016) "Groundwater storage changes in the Tibetan Plateau and adjacent areas revealed from GRACE satellite gravity data" .
6. Nadia Babiker Ibrahim.Shakak (2015) "integration of remote sensing and gis in ground water qualityassessment and management".
7. N. Tangdamrongsub , P.G. Ditmar , S.C. Steele-Dunne , B.C. Gunter , E.H. Sutanudjaja (2016) "Assessing total water storage and identifying flood events over Tonlé Sap basin in Cambodia using GRACE and MODIS satellite observations combined with hydrological models".
8. Wang Hanshenga, Xiang Longweia,b, Jia Luluc, Wu Patrickd,Steffen Holgere, Jiang Liminga, Shen Qianga (2015) "WaterStorage changes in North America .retrieved from GRACE gravity and GPS data".
9. Zhao Qiana, Wu Weiweia,b, Wu Yunlongc(2015). "Variations in China's terrestrial water storage over the past decade using GRACE data".

RADON MEASUREMENTS IN KOZHIKODE COAST, SW INDIA AND ITS IMPLICATIONS

Mintu Elezebath George, Akhil T, Rafeeque M K and Suresh Babu DS

Coastal Processes Group, National Centre for Earth Science Studies, Akkulam, Thiruvananthapuram
mint.george@gmail.com, akhiltaurus2@gmail.com

ABSTRACT

To distinguish as well as detect the presence of fresh and re-circulated Submarine Groundwater Discharge the best tracers that can be used are radon and salinity. Due to its conservative nature, short half-life, high abundance in groundwater compared to surface water, radon (^{222}Rn) acts as a good indicator of groundwater characteristics. Salinity (>35 PSU) differentiates sea water and groundwater with values <2 PSU reflects the presence of freshwater. Kozhikode coast of SW India is very dynamic due to its tidal influence. There are regions, where saline water intrudes much towards inland (upto 500m) and there are locations, where groundwater discharges to sea and often this mechanism varies seasonally. We have investigated such variations and fluctuations using spatial and temporal measurements of radon in air and water. Tidal fluctuations may affect aquifer recharging and thus the discharge of water to sea. Characteristics of 35 km coastal zone in Kozhikode, Kerala state with respect to the radon reflections have been brought out.

KEYWORDS Radon, RAD7, Submarine Groundwater Discharge, coastal aquifer, Kozhikode coast, Kerala

INTRODUCTION

Radon-222 is a naturally occurring radioactive element with a half-life of 3.8 days. Sediments and rocks, containing uranium-bearing materials such as limestone and phosphatic material, acts as a constant source of ^{222}Rn(Noble et al., 2009, Cable et al., 1996; Burnett et al., 2008; Dulaiova et al., 2008; Burnett et al., 2006). Groundwater constantly being in contact with these materials gets enriched with radon and is often about two to three orders of magnitude higher than most surface waters. Radon dissolves in groundwater irrespective of its composition (fresh water or seawater) and thus acts as a tracer of fresh as well as recirculated submarine groundwater discharge (Dulaiova et al., 2008, Noble et al., 2009). The chances of the presence of groundwater in a coastal environment is likely to be only because of radon input of significant magnitude to surface water, thus making this tracer very useful in identifying areas of groundwater discharge into lakes, rivers and the coastal ocean (Cable et al., 1996; Burnett et al., 2008; Burnett et al., 2006).

The concentration of radon in the water column will depend on several factors like (1) In-situ production by in growth from ^{226}Ra, radon's radioactive parent dissolved in water/biogeochemical reactions; (2) inputs by diffusion, sediment resuspension, bioturbation, or gas ebullition from sediments; (3) input by groundwater discharge; (4) removal by exchange with open ocean water (i.e., dilution with low radon/methane offshore water); (5) removal by evasion from water to the atmosphere; (6) losses by radioactive decay/biogeochemical reactions (Dulaiova et al., 2010). Being a gas, ^{222}Rn does not build up in the surface water but escapes directly to the atmosphere (Burnett et al., 2003, Dulaiova and Burnett 2006; Burnett et al., 2006). To have a detailed study on SGD, its related processes and understand the total fraction of SGD, it requires the use of different environmental tracers, i.e., naturally occurring hydrochemical indicators or dissolved tracers that show substantial gradients at the groundwater/seawater interface, as these species are inert in water and have comparable half-lives to groundwater residence time (Moore et al. 1996). Spatial data of radon activity can be used to visualize and demarcate the zones with highest potential SGD(Tamborski 2013). Catlin et al. (2015) strongly suggested a coupling between shallow pore water nutrient concentrations with surface water ^{222}Rn to effectively identify the hotspots of SGD nutrient loading. They also developed a multivariate regression model that describes 79% of variability in shoreline nutrient discharge. As a non-reactive noble gas radon's only losses from the water column are due to radioactive decay and evasion to the atmosphere (Noble et al., 2009). The purpose of this paper is to identify the amount of radon in different coastal groundwater and link with the dynamism of fresh and salt water interface.

1. METHODS

The area chosen for present study is a very dynamic coastal region between Koyilandy and Beypore (~35km) in Kozhikode district, Kerala, India (Fig.1). This is a very dynamic coast with three zones of identified submarine groundwater discharges. It has different coastal features like tidal inlets at Beypore, Kallayi and Elathur, have harbours at Puthyappa, Koyilandi and Beypore, promontories etc. The water sampling of 20 open wells of identified submarine groundwater discharge zones was performed in March 2017. The water level of these wells varied from 5-10 m bgl.

Water samples for radon analysis were collected in special glass bottles (250 mL capacity) designed for radon in-water activity measurement ensuring minimum radon loss by degassing based on the procedures explained by Dulaiova et al. The water sampled must be representative of the water being tested without any contact with air. ^{222}Rn measurement of ground water samples was carried out using a radon-in-air monitor RAD-7 (Durridge Co.Ltd) using RAD H_2O technique with closed loop aeration concept (Lee and Kim, 2006). RAD H_2O accessory consists of the RAD7 or radon monitor, the water vial with aerator and the tube of desiccant, supported by the retort stand. Wat-250 protocol of RAD7 was selected for the vial of 250 mL capacity and the same was used for water sampling. ^{222}Rn activities are expressed in Bq m^{-3} disintegration per hour per m^3 with 2 r-uncertainties. At the end of the run (30 min after the start), the RAD7 prints out a summary, showing the average radon reading from the four cycles counted a bar chart of the four readings, and a cumulative spectrum. The radon level is calculated automatically by the RAD7 (Noble et al., 2015). In situ water salinity measurement along these wells in coast of Kozhikode was done simultaneously using a multi-parameter water quality probe, (Aquaread AP-2000 – Advanced portable multi-parameter Aquaprobe).

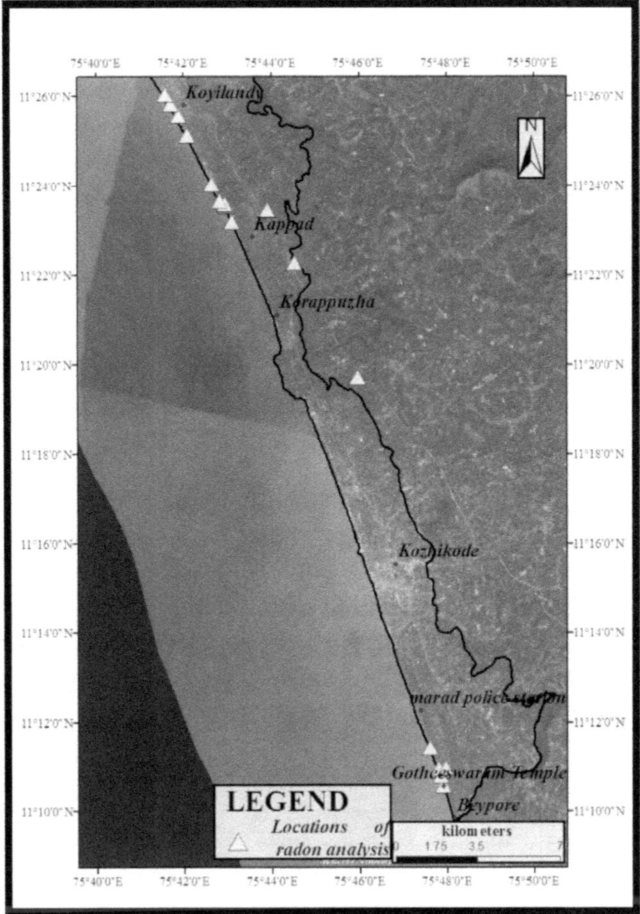

Fig. 1 Locations of radon sample analysis along Kozhikode coast, Kerala, SW India.

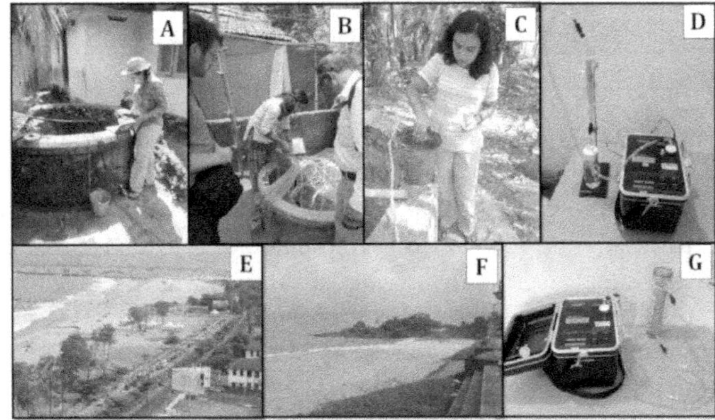

Fig. 2 (A) Checking groundwater salinity (B) and (C) Sample collection for radon analysis (D) Measuring radon using RAD7 (E) Overview of Kozhikode (F) Overview of Thoovappara (a promontory at Kappad Beach) (G) Purging of RAD7 after sample analysis

2. RESULTS AND DISCUSSION

Out of the 120 km^2, 1.92 km^2 area falls under residential and 0.33km^2 under commercial purposes. 4.9km^2 form beaches and the rest are different crops and marshes. This shows there is no much urbanization in the coastal sector of Kozhikode coast 90% of coastal community is fishermen whose source of water is shallow open wells. Hence the urbanization impact does not significantly affect the recharge or discharge of that coastal aquifer. The nearest tidal observation station is located in Cochin which is 180km towards south of study area. The tidal variation is ~ 0.7m. Coastal land elevation rises up to 20-30m from msl, where the highest points are on Nandi Hills and Thoovappara in North of study area. The vertical fluctuation of water table in the coastal aquifer under consideration varies between 5m to 12m bgl based on the data collected during the period from June'2016 to March'2017 in 118 observation wells.

During SW Monsoon (June 2016), water table was 1m below the water table of NE Monsoon (September 2016). Water table stood highest in NE Monsoon due to the monsoonal recharge. It decreased up to 1m in Post Monsoon (January 2017) and to 0.5m in Pre Monsoon (March 2017) due to the draught condition of 2016-2017. Laterite, sand layers and weathered/fractured rock with a maximum depth of ~20m bgl acts as coastal aquifers in Kozhikode coast. Gotheeswaram, Kappad and Koyilandi were identified as potential zones of groundwater discharge in the Kozhikode coastal aquifer using different field observations. Sandy aquifer at Marad, Vengalam and Beypore showed presence of seawater intrusion. Based on radon values in groundwater and surface water, the presence of groundwater discharges could be identified. The results from radon analysis in few wells of zones of groundwater discharge are broadly classified into five groups as given below

Table 1 Variations of radon and salinity

Type	Reason
Higher radon and low salinity	Fresh groundwater
Low radon and low salinity	Groundwater in abandoned or non-pumping wells
Medium radon and high salinity	Wells influenced by tide/ wave but is also flushed with groundwater and is thus recharged periodically
Medium radon and low salinity	Recirculated saline ground water
Low radon and high salinity	Influence of sea water

From the radon survey in 20 open wells at Kozhikode coast the results are grouped under five categories. First group represent locations inland to the coast, where lateritic aquifer feeds the wells and these samples show high radon content (>1000 B) and lower salinity (0-0.5 PSU) indicating fresh groundwater. The next set of samples belong to coastal belt with low salinity (0-1 PSU) shows the presence of groundwater as salinity of >2PSU gives brackish water. But low radon value (<100B) do not support the presence of groundwater in those wells. The possible reason for low value of radon in the wells is that, these wells are abandoned or non-pumping wells and is not at all influenced by tide/wave (low salinity). The groundwater in those wells were depleted in radon since the half-life of radon is only three days and this might be decayed, as these wells are non-pumping or is not being recharged on a frequent basis. Radon value of (200-500B) in coastal stretch indicates presence of groundwater mixed with brackish water. Higher salinity supports the idea that these wells are of seawater mixed with groundwater. Here tidal mixing dominates. The next type of radon value within 300-500B, and with low salinity shows wells that are tidally/wave influenced and fresh groundwater mixed in a 1:1 proportion. Low salinity value shows groundwater influence is higher that seawater. This can be explained as this well is flushed in with high tide/ wave and is recharged with groundwater during low tide. Low radon and high salinity indicates influence of seawater, flushed and mixed thoroughly with the seawater.

CONCLUSIONS

Based on the differences in values of radon and salinity, fresh groundwater and recirculated seawater zones have been distinguished. Presence of higher value of radon near coast and low salinity values shows the presence of groundwater closest to the ocean, supporting freshwater discharge to sea through aquifer medium. Similarly low radon value and high salinity value marks the re-circulated saline SGD. Stagnant water seen in wetland zones of the area as well as in selected coastal wells of the site, with no signature of radon release could be due to complete escape of radon at the time of measurement. It is inferred that the tidal fluctuations may affect the coast to a distance of 100-200m from the shoreline in summer.

ACKNOWLEDGMENTS

We thank Director, NCESS for extending research facilities and funds. This project was made possible by a student research grant from Kerala State Council for Science Technology and Environment (KSCSTE). Thanks are also due to the Coastal Processes Group team members for their assistance and support all through the work and especially in field.

REFERENCES

1. Burnett, W.C., Peterson, R., Moore, W.S., de Oliveira, J., 2008. Radon and radium isotopes as tracers of submarine groundwater discharge - Results from the Ubatuba Brazil SGD assessment inter-comparison. *Estuarine Coastal and Shelf Science*, 76(3): 501-511.
2. Burnett, W.C., Aggarwal, P.K., Aureli, A., Bokuniewicz, H., Cable, J.E., Charette, M.A., Kontar, E., Krupa, S., Kulkarni, K.M., Loveless, A., Moore, W.S., Oberdorfer, J.A., Oliveira, J., Ozyurt, N., Povinec, P., Privitera, A.M.G., Rajar, R., Ramessur, R.T., Scholten, J., Stieglitz, T., Taniguchi, M., Turner, J.V., 2006. Quantifying submarine groundwater discharge in the coastal zone via multiple methods. *Science of the Total Environment.* 367, 498–543.
3. Burnett, W. C. and Dulaiova, H., 2003. Estimating the dynamics of groundwater input into the coastal zone via continuous radon-222 measurements. *J. Environ. Radioact.* 69, 21–35.
4. Caitlin Young, Joseph Tamborski and Henry Bokuniewicz , 2015. Embayment scale assessment of submarine groundwater discharge nutrient loading and associated land use. *Estuarine, Coastal and Shelf Science*, 158, 20-30
5. Cable, J. E., Burnett, W. C., Chanton, J. P. and Weatherly, G. L., Estimating groundwater discharge into northeastern Gulf of Mexico using radon-222. Earth Planet. Sci. Lett., 1996, 144, 591– 604.
6. Dulaiova, H., M. E. Gonneea, P. B. Henderson, and M. A. Charette, 2008. Geochemical and physical sources of radon variation in a subterranean estuary-Implications for groundwater radon activities in submarinegroundwater discharge studies.*Mar. Chem.*110(1–2), 120–127.
7. Dulaiova, H., and Burnett, W.C., 2006. Radon loss across the water-air interface (Gulf of Thailand) estimated experimentally from ^{222}Rn and ^{224}Ra. *Geophysical.Research. Letters,* 33, L05606.

8. Dulaiova, H., Richard Camilli, Paul B. Henderson and Matthew A. Charette, 2010.Coupled radon, methane and nitrate sensors for large-scale assessment of groundwater discharge and non-point source pollution to coastal waters. *Journal of Environmental Radioactivity*, 101, 553–563.
9. Gonneea, M.E., Morris, P.J., Dulaiova, H. and Charette, M.A., 2008.New perspectives on radium behavior within a subterranean estuary. *Marine Chemistry*, 109(3-4), 250-267.
10. Gopal Krishan, SomeshwarRao, M., Kumar C.P., Sudhir Kumar and Ravi AnandRao M. (2015) Study on identification of submarine groundwater discharge in northern east coast of India. *Aquatic Procedia*. 4, 3 – 10
11. Lee, J.M., and Kim, G., 2006. A simple and rapid method for analysing radon in coastal and ground waters using a radon-in-air monitor. *J. Environ. Radioact.* 89, 219-228
12. .Kim, G., and D. W. Hwang, 2002. Tidal pumping of groundwater into the coastal ocean revealed from submarine ^{222}Rn and CH_4 monitoring. *Geophysical Research Letters,* 29(14), 23–27.
13. Moore, W.S. 1996. Large groundwater inputs to coastal waters revealed by ^{226}Ra enrichments. *Nature,* 380 (6576), 612–614.
14. Noble Jacob, D. S. Suresh Babu and K. Shivanna. 2009. Radon as an indicator of submarine groundwater discharge in coastal regions. *Current Science,* 97 (9), 1313-1320.
15. Noble Jacob, Md. Arzoo Ansari and C. Revichandran. 2015. Environmental isotopes to test hypotheses for fluid mud (mud bank) generation mechanisms along the southwest coast of India. *Estuarine, Coastal and Shelf Science*, 164 , 115-123
16. Tamborski, J. 2013. Submarine Groundwater Discharge: Understanding Flux Signatures in Thermal Infrared Data in Port Jefferson and Stony Brook Harbors, Long Island, New York. in Hanson, G. N., Chm., 20th Annual Conference on Geology of Long Island and Metropolitan New York, 13 April 2013, State University of New York at Stony Brook, NY, Long Island Geologists Program with abstracts, 7 p.
17. Webster I.T., Gary J. Hancock and Andrew S. Murray, 1995.Modelling the effect of salinity on radium desorption from sediments. *Geochimica et Cosmochimica Acta* , 59(12), 2469-2476

GIS & RS APPLICATION FOR HYDROLOGICAL & ENVIRONMENTAL PLANNING - A CASE STUDY

Bipin Chand Pandey
IISM, Survey of India.
bcppandey@gmail.com

ABSTRACT

Water is one of our most important natural resources. Without it, there would be no life on earth. The supply of water available for our use is limited by nature. Although there is plenty of water on earth, it is not always in the right place, at the right time and of the right quality. Adding to the problem is the increasing evidence that chemical wastes improperly discarded yesterday are showing up in our water supplies today. Hydrology has evolved as a science in response to the need to understand the complex water systems of the Earth and help solve water problem. Polluted ground water is less visible, but more insidious and difficult to clean up, than pollution in rivers and lakes. Ground water pollution most often results from improper disposal of wastes on land. Major sources include industrial and household chemicals and garbage landfills, industrial waste lagoons, tailings and process wastewater from mines, oil field brine pits, leaking underground oil storage tanks and pipelines, sewage sludge and septic systems. Human settlement with their propensity to create hard, impermeable surfaces for building houses and roads, and the need of water intake and overflow in a variety of forms, are not in harmony in a natural hydrological cycle. The adverse effect of creating impervious surface cover in urbanized watershed, reducing the groundwater recharge and consequent reduction in the base flow of stream and river flow the area. Sewage and water supply system having dense settlement can further interfere with ground water and surface hydrology. Urbanization in India and other developing countries is taking place at a faster rate than in the rest of the world. Urban water supply, storm water and watershed management is at a critical juncture all over the world. Methods must evolve in response to urban development, population growth and diminishing natural resources.

Keywords: Watershed, Hydrology, groundwater, natural resources, pollution.

INTRODUCTION

This paper is to develop a powerful, innovative and emerging concept about GIS Mapping for dissolving the forthcoming challenges in hydrological and environmental sectors. However using a GIS tool, authority can integrate the issues related with hydrological feature and also give a best suitable option to make a strategy for effective implementation on root level. The purpose of this paper to find out the goal to achieve GIS application for improving the water level area in the city, those are facing the issues related with water conservation.

Surveying technology is enhancing the process of old method of collecting geospatial data, through coming new hardware and software surveying technology. Previously the data collection and handling both are very tedious work, but now a day's both are going to be easier and faster for a trained person. If data quality is good as per user demand & makes its availability of data is easy to some extent. So it will be a new era for real analysis of ground in any field. So the functioning & accepting the challenges can also be manageable.

The objective of GIS operation is to assemble a database that contains all the parameters to manipulate data through models and other decision making procedures to yield a series a output to solve a problem. GI uses information from multiple sources in different way that can help in different types of analysis. GIS data represents real world objects in digital form.

The study of Uppal, Hyderabad area for this paper is basically have done to keep in mind about the problem facing mostly in the unprecedented developing urban areas , where population is growing blindly in unsystematic manner due to socio-economical issues. Which can't be rearranged properly in future due to improper management and the current reality ground is not accessed quickly and ultimately the society will be sufferer.

AREA OF INTEREST

The study and analysis area of this paper is lying between latitude of 17 : 19 : 0.5 to 17 : 30 : 07 (approx.) and longitude of 78 : 27 : 30 to 78 : 38 : 29 (approx). It covers an area of 400 Sq. Km. The area falls in Survey of India OSM Sheet No. E44M7 & E44M11 and Polyconic Sheet No. 56 K / 7 & 56 K / 11.

DOCUMENTS USED FOR THE STUDY

❖ Hardcopy / Softcopy Used

- Survey of India Hard Copy of Hyderabad Guide Map of edition 2014.
- Soft Copy of Hyderabad Guide Map and OSM Map of new edition.
- Survey of India Topo Sheet No. 56K/7,11 and E44M7,11
  OSM Sheet of Scale 1:50,000.
- Survey of India Topo Sheet No. 56K/7,11 (NE,NW,SE,SW)
  of Scale 1:25,000.
- Imageries (having good resolution) downloaded from Google Earth Pro for digitization / updation.
- DEM downloaded using USGS website.

❖ Hardware / Software Used

- A high configuration computer system to carry out the project.
  (Processor- i3, Hard Disk- 500 GB, RAM- 2 GB, Latest Monitor Screen with keyboard & mouse).
- HP Design jet 500PS Plotter to take hard copy of imageries/ Map/ necessary plots.
- Arc GIS software available / new version, Erdas Imagine 10.1, Microstation V8.
- Internet connection for downloading imageries (if latest imagery not available for updation).
- All software tool of Microsoft office available / new version.

RESEARCH METHODOLOGY

The Preparation study & Analysis (using Microstation V8 and Arc GIS) of specified area in terms of spatial and non- spatial data and preparation of stream feature using ASTER-DEM is a challenging task. The heavy data volume cannot be handled and analysis by traditional system of record keeping. Since the map are easily destroyed and displaced and analysis with map is a very tedious. The whole geospatial data is prepared in such a manner that it should be used for further any type of GIS analysis. Here we used to discuss some basic tools, which are using during hydrological analysis in GIS.

DEM

Digital elevation model (DEM) is a digital representation of ground surface topography or terrain. It is also widely known as a digital terrain model (DTM). A DEM can be represented as a raster (a grid of squares) or as a triangular irregular network. DEMs are commonly built using remote sensing techniques, but they may also be built from land surveying. DEMs are used often in geographic information systems, and are the most common basis for digitally-produced relief maps.

FILL

The Fill tool uses to locate and fill sinks. The tool iterates until all sinks within the specified z limit are filled. As sinks are filled, others can be created at the boundaries of the filled areas, which are removed in the next iteration.

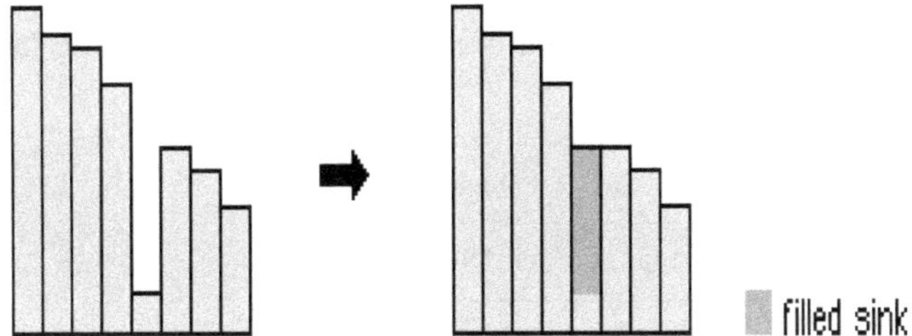

Profile view of a sink before and after running Fill

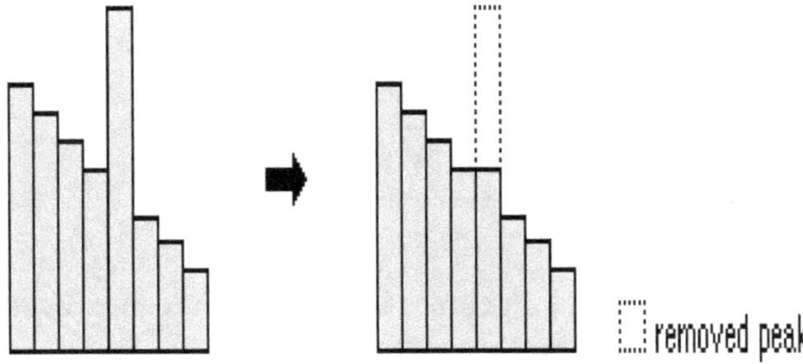

Profile view of a peak before and after running Fill

Flow Direction

One of the keys to deriving hydrologic characteristics of a surface is the ability to determine the direction of flow from every cell in the raster. This is done with the Flow Direction Tool.

1014	1011	1004
1019	1015	1007
1025	1021	1012

⇩

+1	+4	+11
-4		+8
-10	-6	+3

⇩

	⇨	

Flow Accumulation

The Flow Accumulation tool calculates accumulated flow as the accumulated weight of all cells flowing into each down slope cell in the output raster. If no weight raster is provided, a weight of 1 is applied to each cell, and the value of cells in the output raster is the number of cells that flow into each cell.

1014	1011	1004
1019	1015	1007
1025	1021	1012

⇩

↖	→	→
↖	→	→
↖	↗	↗

Map Algebra

Map Algebra is a simple and powerful algebra with which you can execute all Spatial Analyst tools, operators, and functions to perform geographic analysis.

Stream Order

The output of Stream Order will be of higher quality if the input stream raster and input flow direction raster are derived from the same surface. The results of the Flow Accumulation tool can be used to create a raster stream network by applying a threshold value to select cells with a high accumulated flow.

Pour Point

Pour Point placement is very important in process of watershed delineation. The point used for deriving contributing watershed.

Watershed

An Area that drains water or other substances to a common outlet. A drainage basin or catchment area is any area of land where precipitation collects and drains off intia common outlet, such as into a river, bay, or other body of water.

OBSERVATION

The people's representatives give all encouragement and support and stand by the people in introduction of new technology and in new ways of doing things, irrespective of their political affiliations. The Central and State planning department plays an important role to consider advancement of new mapping technology and make some solid steps about revaluation of mapping concept.

The study and analysis of hydrology tool for stream feature generation & some Geo database for comparison and analysis of output with ground reality , it is found that-

Final stream feature obtained from hydrological analysis is maximum percentile related with the Survey of India Maps.

SCOPE OF PAPER

The study of this paper presented a very good scope to compare and study the different type of available DEM to compare with the Topo sheets or other map available in hard copy / soft copy format and also watch out and do needful activity for changing or diversion of watersheds, so that the various problem like drought, flood, irrigation, canal, drainage etc. can be sort out with an effective planning and implementations.

OUTPUT

- From figure 1 to figure 7 - Stream Network Analysis.
- From figure 8 to figure 9 - Watershed Analysis.
- From figure 10 to figure 11 – Comparison with GIS data & Topo Sheets.

Fig. 1 ASTER DEM

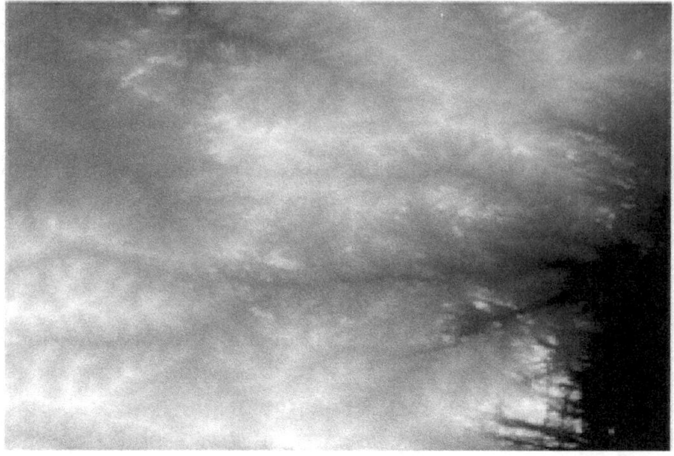

Fig. 2 Fill Raster.

Fig. 3 Flow direction Raster.

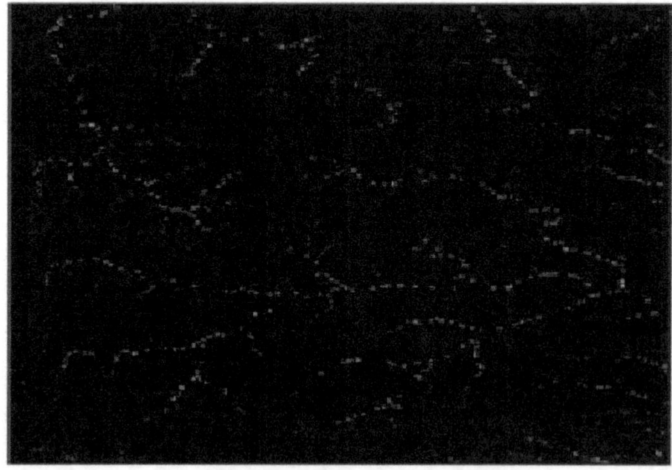

Fig. 4 Flow accumulation Raster.

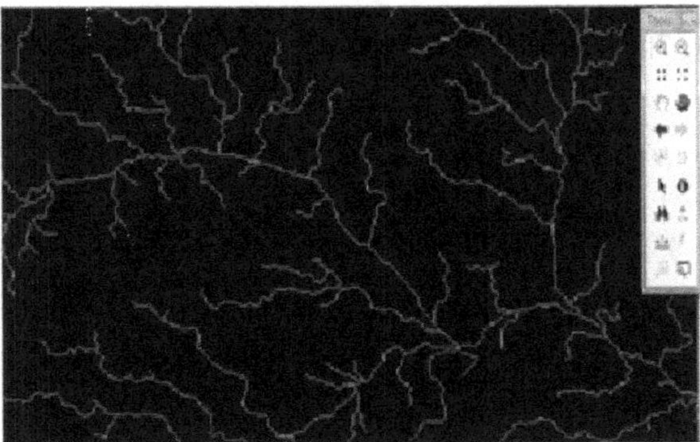

Fig. 5 Single Output of Map algebra.

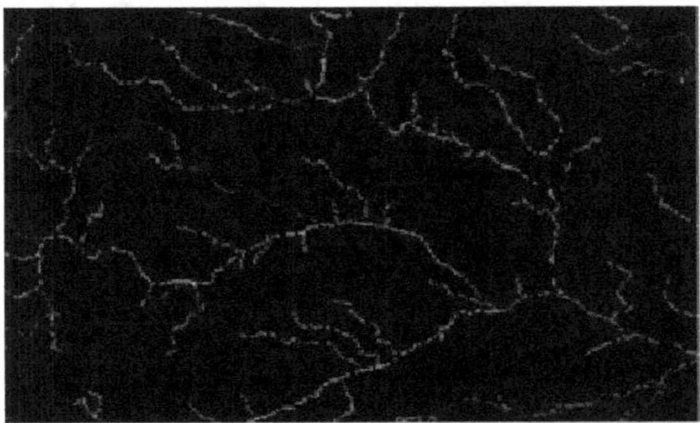

Fig. 6 Stream Order.

Fig. 7 Stream Link.

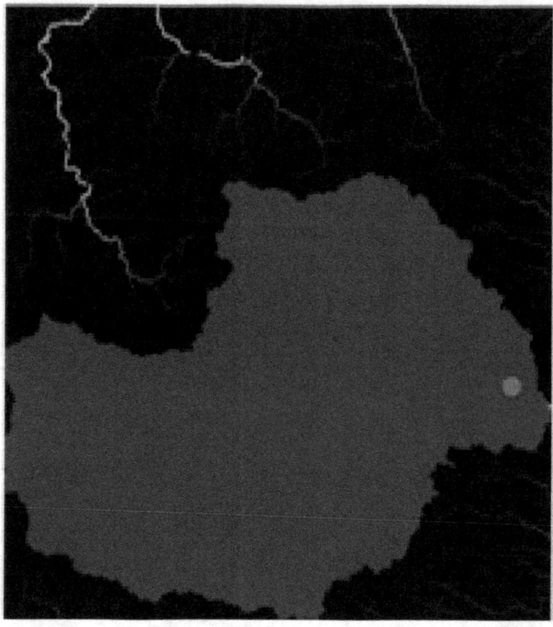

Fig. 8 Watershed Area.

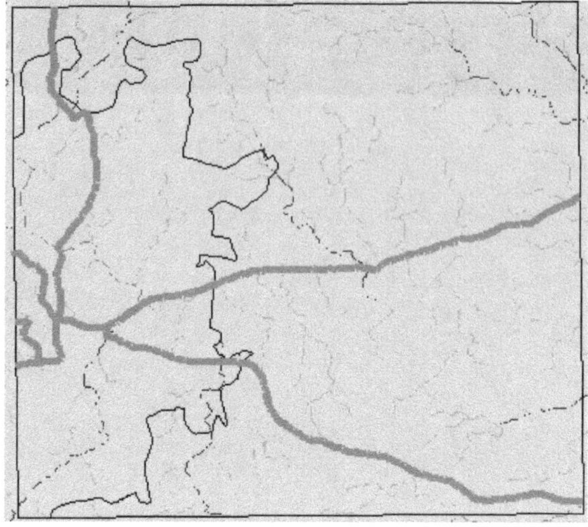

Fig. 9 Watershed Area, Strem Order in study area.

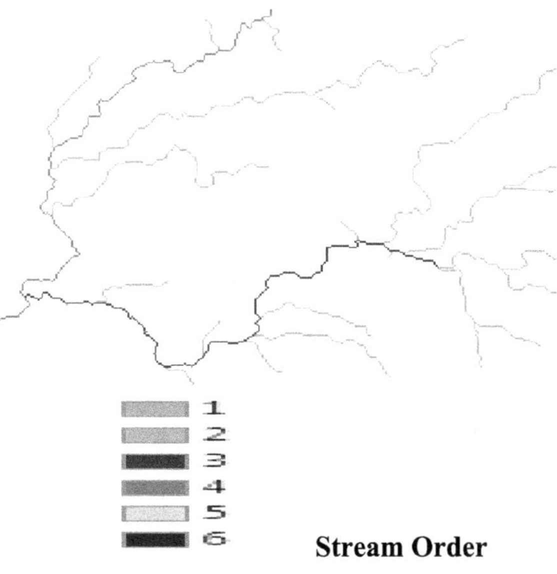

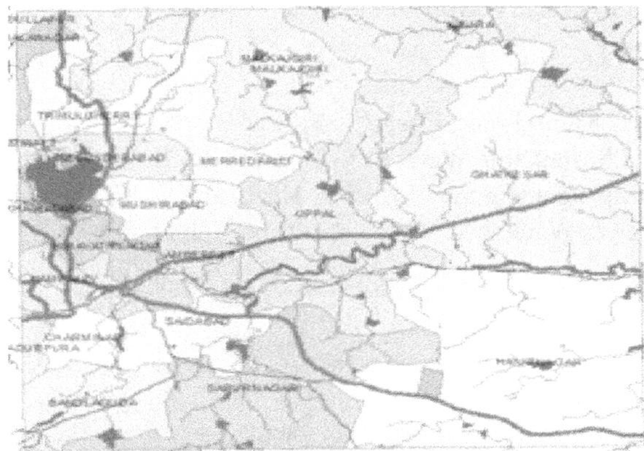

Fig. 10 –Stream in Study area

Fig. 11 Comparison of stream with SOI Map

CONCLUSION

- Quality of watershed can be managed through their stream order quality to manage the health of ground water quality and environment also.
- If quality of DEM (may be prepared using point cloud gathering information UAV technology ie; modern drone technology) is good, then generated streams can be used on any scale for strategic planning.
- Streams can also be compare with different types of Hard copy Map/soft copy Map for different analysis purposes.
- The further city development layout plan can be done in strategic way to control the encroachment in hydrology due to human participation.
- The output streams can be exported in digital form and used as spatial information to locate & identify the problems.
- Sustainable behavior of environment can be managed by knowing the originality of water level existence in concerned area.
- The flood management can also be organized using this GIS technology.
- The encroachment of outlet of water feature can also be indentify.
- The study will help in construction of canal & improve the irrigation technique.

REFERENCES

1. Introduction to GIS – by Kang-Tsung-Chang
2. Modern Hydrology and Sustainable Water Development- By S. K. Gupta
3. Integrated Watershed Management – by Isobel W Heathcote
4. GIS for Water Resource & Watershed Management- by John Grimson Lyon
   http://water.usgs.gov
   http://pro.arcgis.com

GREEN TREATMENT OF ACID MINE DRAINAGE (AMD) EFFLUENT FOR THE REMOVAL OF CU(II) USING *CHILLI STALKS*

M.Padmaja[1] and R.Pamila[2]

[1]Department of Civil Engineering, Mahatma Gandhi Institute of Technology, Hyderabad.
[2]Department of Civil Engineering, Sri Sairam Engineering College, Chennai.
padmajamegham@gmail.com; pamilaramesh@gmail.com

ABSTRACT

Acid mine drainage (AMD) is found to be a key environmental issue in the mining industry in the current scenario. Among a variety of metals found in AMD discharges, copper is observed to be in high concentrations i.e. greater than 3 mg/l (MoEF stds.) which are quite alarming to the mining industry, as the danger it causes to the human health, animals and ecological systems. A part from various conventional methods, adsorption has taken its place in heavy metal removal because of its advantages over other methods. The intention of this work is to assess the ability of *chilli stalks* in the removal of Cu(II) from AMD effluent. Kinetics and equilibrium models have been developed to describe adsorption isotherm relationships, the two main isotherm models used in this work are the Langmuir and Freundlich models.

Keywords: AMD, Adsorption, Chilli stalks, Kinetics, Langmuir and Freundlich.

INTRODUCTION

Effluents released from AMD usually contain metal ion concentration much higher than the permissible limits (1). These metals include cadmium, lead, mercury, chromium, nickel, copper, zinc and cobalt which are harmful to human health even in very low concentrations (2). Heavy metals are bio-accumulative, toxic at high concentrations, have neurological impacts, and some are carcinogenic. Copper(II) is one of the heavy metals most toxic to the living organisms and it is one of more widespread heavy metal contaminants of the environment. The increase levels of copper in environment are posing a serious threat to mankind. It can cause harmful biochemical effects, toxicity and hazardous disease in human beings. Prescribed limit for copper in drinking water is 0.05mg/L as per WHO norms and also 0.05 mg/L as per ISI prescribed limit. Some of the polluting agents have deleterious effect on human health.

Researchers have focused upon a sustainable way of development using techniques that are environment friendly as well as cost-effective and practical. This has directed attention to the use of biological wastes. Several biosorbents such as rice husks Jaman *et al.* 2009), wheat straws (Dang *et al.* 2009), tea factory waste (Wasewar *et al.*2008), orange peel (Annadurai *et al.* 2002), fish scale (Espinosa *et al.* 2001), castor seed hull (Sen *et al.* 2010), wheat shell (Basci *et al.* 2009), coffee husk (Oliveira *et al.* 2008), maple sawdust (Rahaman & Islam 2009) has been investigated for adsorbing the ion from aqueous solutions. Bhatnagar and Sillanpää (2010) reviewed the use of various agro-based waste materials as low-cost adsorbents.

In this study, the effectiveness of onion skin and garlic skin for removal of Cu2+ from aqueous solutions was investigated.*chilli stalks* (CS) are commonly generated from both households and food-processing industries. Although the amount of these organic wastes generated from households is negligible, that generated from food-processing industries is large, as they are a major by-product in these industries. These wastes can be potentially used as low-cost adsorbent materials. In addition, using these wastes in various other potential applications will eliminate them from the environment and reduce solid-waste handling, which will add some value to these wastes [2,3].

2.0 MATERIALS AND METHODS

2.1 Adsorbent Materials

Raw Chilli Stalk Powder (RCSP)

The Chilli Stalks were obtained from a vegetable market in Hyderabad (Telangana). They were air- dried and powdered in a grinder. The raw dry biomass was crushed into granules, sieved to different particle sizes, and then preserved in desiccators for use. (i.e., RCSP).

2.2 Activated Chilli Stalk Powder (ACSP)

Air-dried and powdered chilli stalks were soaked in concentrated H_2SO_4 for 12 hours and washed thoroughly with distilled water till they attained neutral pH and soaked in 2% $NaHCO_3$ overnight in order to remove any excess acid present. Then the material was washed with distilled water several times and dried at 110±5° C. The dry biomass was roughly grinded and was activated in the muffle furnace at 400 ° C for 12 hours, then cooled in dessicator , sieved to different particle sizes, and then preserved in air tight bags for use.(i,e.,ACSP).

2.3 PREPARATION OF ADSORBATE SOLUTIONS

Metal solutions

Stock solution of 10 mg/l Cu (II) ion is prepared dissolving copper sulphate pentahydrate ($CuSO_4.5H_2O$). To do this 39.28 mg $CuSO_4.5H_2O$ is added in distilled water contained in 1000 ml volumetric flask. Hydrochloric acid and Sodium hydroxide were used to adjust the solution pH. Distilled water was used throughout the experimental studies.

2.4 BATCH MODE ADSORPTION STUDIES

Batch mode adsorption studies for individual metal compounds were carried out to investigate the effect of different parameters such as adsorbate concentration, adsorbent dose, agitation time and pH. Solution containing adsorbate and adsorbents were taken in 250 mL capacity beakers and agitated at 170 rpm in a mechanical shaker at predetermined time intervals. The adsorbate was filtered and separated from the adsorbent using a filter paper.

2.5 EFFECT OF PH ON CU(II) ADSORPTION

The effect of solution pH on adsorption of Cu(II) was studied by mixing 0.5 g of individual adsorbent with 100 ml of Cu(II) solution having concentration of 1.15 mg/L at different pH value (2 – 12) at room temperature. The pH was adjusted with 1 N NaOH or 1 N HCl solutions. Agitation was made at a constant stirring speed of 170 rpm for 60 minutes. The remaining concentration of Cu(II) after adsorption was measured using AAS. The percentage uptake of Cu(II) was calculated according to the following equation:

$$\text{Percentage uptake (\%)} = \frac{C_o - C_t \times 100}{C_o}$$

Where, C_o is the initial concentration and C_t is the concentration at time *t*.

2.6 EFFECT OF CONTACT TIME ON CU(II) ADSORPTION

The effect of solution Contact Time on adsorption of Cu(II) was studied by mixing 1.5 g of both adsorbents with 50 ml of Cu(II) solution having concentration of 20 mg/L of Copper concentration at pH value of 6 at room temperature. Agitation was made at a constant stirring speed of 170 rpm. The remaining concentration of Cu(II) after adsorption was measured at different time intervals of 30, 60, 90,120 and 180 minutes using AAS. The percentage uptake of Cu(II) was calculated according to the following equation:

$$\text{Percentage uptake (\%)} = \frac{C_o - C_t \times 100}{C_o}$$

Where C_o is the initial concentration and C_t is the concentration at time *t*.

2.7 EFFECT OF ADSORPTION DOSE ON CU(II) ADSORPTION

The effect of adsorption dose on Cu(II) adsorption was investigated by different amount of adsorbents 0.5 gm, 1.0 gm, 1.5 gm, 2 gm and 2.5 gm in 50 ml of Cu(II) solution having initial concentration of 20 mg/l of Copper. Agitation was made at a constant stirring speed of 170 rpm for 60 minutes. The remaining concentration of Cu(II) after adsorption was measured using atomic absorption spectrometer (AAS).

3.0RESULTS AND DISCUSSION

3.1 Effect of pH on Cu(II) Adsorption

The pH value of aqueous solution is an important parameter in adsorption process because it affects the surface charge of the adsorbent, the degree of ionization and specification of the adsorbate. The batch equilibrium studied for mixed metal solutions having concentration of 20 mg/L of Copper concentration at different pH value ranging from 2 to 9 were carried at room temperature. Fig.1 shows that maximum percentage of Cu(II) adsorption on ACSP were observed at pH 6.

Table 1 Effect of pH on the adsorption of Cu(II)) by RCSP and ACSP

S.No	pH	Quantity of RCSP (gm)	Adsorption Efficiency (%)	Quantity of ACSP (gm)	Adsorption Efficiency (%)
1	2	2.5	55.9	2.5	62.1
2	4	2.5	65.9	2.5	72.3
3	5	2.5	92.6	2.5	95
4	6	2.5	98.4	2.5	98.7
5	9	2.5	81.8	2.5	82.5

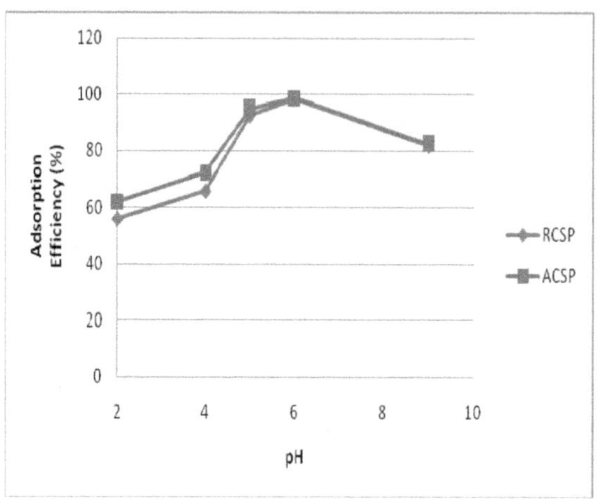

Fig. 1 Effect of pH on the adsorption of Cu(II) by RCSP and ACSP.

3.2 Effect of Contact Time on Cu(II) Adsorption

Contact time plays an important role in adsorption process and the effect of contact time on adsorption capacity has been studied by varying the contact time from 30 to 180 minutes. The Copper adsorption percentage at different contact time by ACSP is shown in Fig 3.Results indicated that the Cu(II) adsorption by ACSP reached at most 99% at 120 minutes contact time.

Table 2 Effect of contact time on the adsorption of Cu(II) RCSP and ACSP

S.No	Contact Time (min)	Quantity of RCSP (gm)	Adsorption Efficiency(%)	Quantity of ACSP (gm)	Adsorption Efficiency(%)
1	30	2.5	63.98	2.5	66.5
2	60	2.5	79.33	2.5	74.3
3	90	2.5	89.56	2.5	88.2
4	120	2.5	98.56	2.5	99
5	180	2.5	98.56	2.5	99

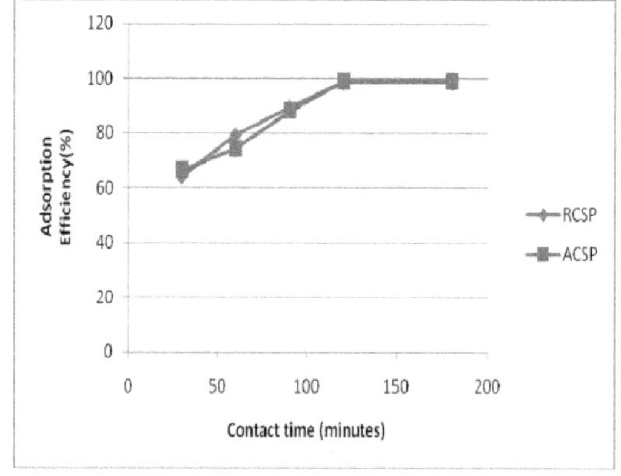

Fig. 2 Effect of contact time on the adsorption of Cu(II) by RCSP and ACSP.

3.3. Effect of Adsorbent Dose on Cu(II) Adsorption

The effect of adsorbent dosage was studied by varying the amount of adsorbent from 0.1 gm to 0.5 gm in 100 ml of Cu(II) solution for a contact time of 60 minutes. After equilibrium the solution was analyzed for the amount of Cu(II), the results indicate that adsorption increased with increase in adsorption dosage.

Table 3 Effect of adsorbent dose on the adsorption of Cu(II) by RCSP and ACSP

S.No	Quantity of RCSP (gm)	Adsorption Efficiency(%)	Quantity of ACSP (gm)	Adsorption Efficiency(%)
1	0.5	67.78	0.5	70
2	1.0	74.96	1.0	79.6
3	1.5	89.2	1.5	90.6
4	2.0	93.2	2.0	96.6
5	2.5	98	2.5	99

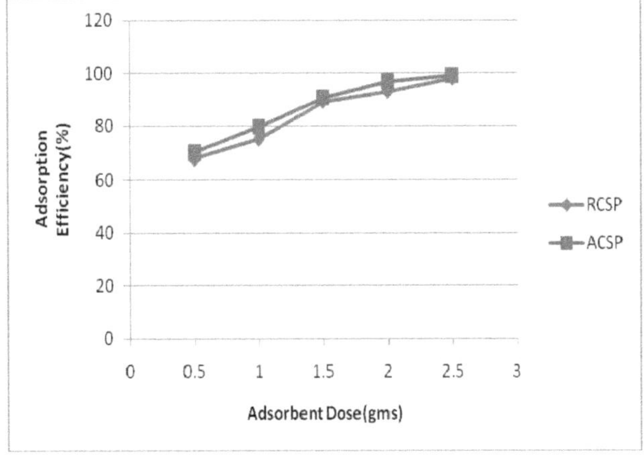

Fig. 3 Effect of adsorption dose on the adsorption of Cu(II) by RCSP and ACSP

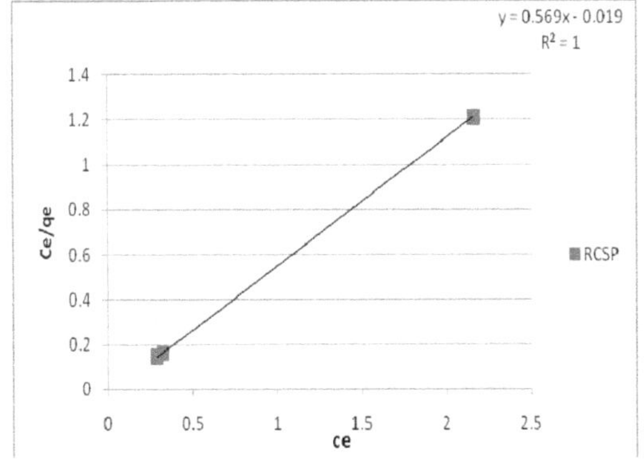

Fig. 4 Langmuir Isotherm for equilibrium conc. of Cu(II) adsorbed on RCSP

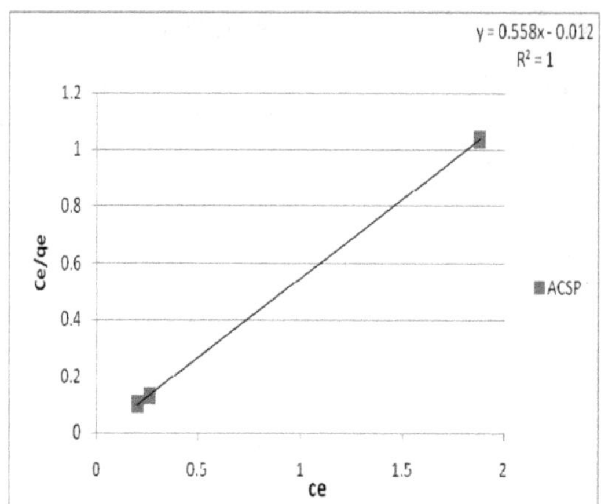

Fig. 5 Langmuir Isotherm for equilibrium conc. of Cu(II) adsorbed on ACSP

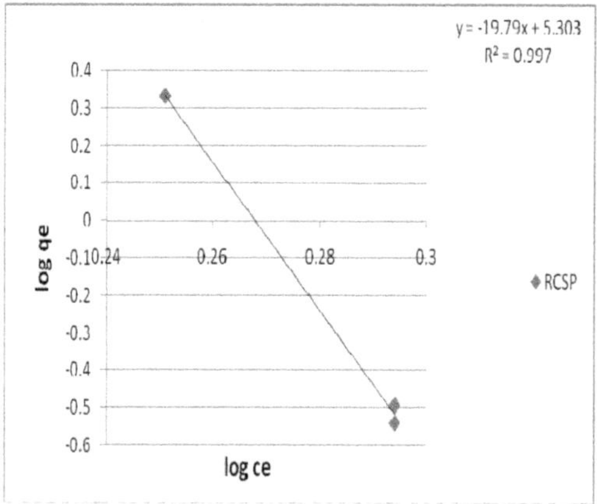

Fig. 6 Freudlich Isotherm for equilibrium conc. of Cu(II) adsorbed on RCSP

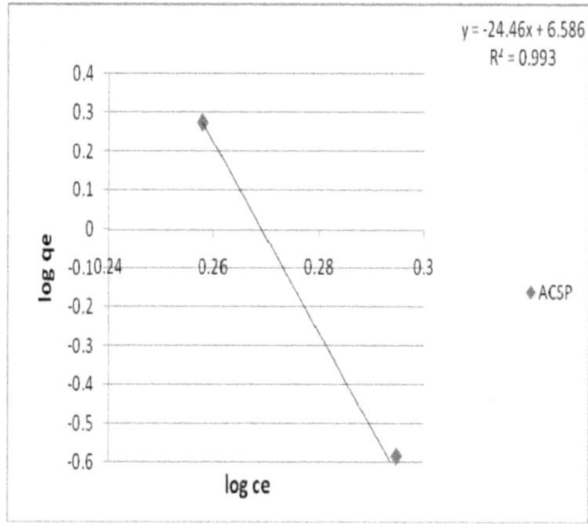

Fig. 7 Freudlich Isotherm for equilibrium conc. of Cu(II) adsorbed on ACSP

4.0 CONCLUSION

The removal of Cu(II) from AMD waste water using RCSP and ACSP has been experimented under several conditions such as at different pH, contact time and adsorption dose. The optimum pH for copper adsorption was found at pH 6. The optimum contact time was found to be 120 minutes at an agitation speed of 170 rpm. The adsorption data were fitted to different isotherm models and the Langmuir model was found to be the best model for both adsorbents with R^2= 1 respectively. Increase in adsorption dose increased the adsorption of Cu(II). The results showed that RCSP and ACSP are very efficient in the elimination of Copper from AMD in the order ACSP > RCSP.

REFERENCES

1. Moreno, Natalia, Querol, Xavier, Ayora, Carles, Utilization of zeolites synthesized from coal fly ash for the purification of Acid MineWaters. Environ. Sci. Technol. 35, (2001),3526–3534.
2. Abu Al-Rub FA, El-Naas MH, Ashour I, Al-MarzouqiM. (2006) Biosorption of copper on *Chlorella ulgaris* from single, binary and ternary metal aqueous solutions. *Proc Biochem* 41: 457–464.
3. Aksu Z, İşoğlu İA. (2005) Removal of copper(II) ions from aqueous solution by biosorption onto agricultural waste sugar beet pulp. *Proc Biochem* 40:3031–3044.
4. Aksu Z, Kutsal T. (1998) Determination of kinetic parameters in the biosorption of copper(II) on
5. *Cladophora sp.*, in a packed bed column reactor.*Proc Biochem* 33: 7-13.
6. Annadurai JFA, Juang RS, Lee DJ. (2002) Adsorption of heavy metals from water using banana and
7. orange peels. *Wat Sci Technol* 47: 185–190.
8. Basci N, Kocadagistan E, Kocadagistan B. (2004) Biosorption of copper (II) from aqueous solutions by wheat shell. *Desalination* 164: 135-140.
9. Bhatnagar A, Sillpanää M. (2010) Utilization of agro-industrial and municipal waste materials as potential adsorbents for water treatment—A review.*Chem Eng J* 157: 277-296.
10. Chowdhury S, Saha P. (2010) Biosorption of Methylene Blue onto Tamarind Fruit Shell: Comparison of Linear and Nonlinear Methods. *Bioremediation J*14: 196–207.
11. Dang VBH, Doan HD, Dang-Vu T, Lohi A. (2009) Equilibrium and kinetics of biosorption of cadmium(II) and copper(II) ions by wheat straw. *Bioresour Technol* 100: 211-219.
12. Espinosa V, Esparza MH, Ruiz-Treviño FA. (2001) Adsorptive properties of fish scales of *Oreochromis.*
13. Goertzen SL, Thériault KD, Oickle AM, Tarasuk AC, Andreas HA. (2010) Standardization of the Boehm titration. Part I. CO2 expulsion and endpoint determination. *Carbon* 48: 1252 –1261.
14. Hameed BH, Ahmad AA. (2009) Batch adsorption of methylene blue from aqueous solution by garlic peel, an agricultural waste biomass. *J Hazard Mater*164: 870–875.
15. Iftikhara AR, Bhattia HN, Hanifa MA, Nadeema R. (2009) Kinetic and thermodynamic aspects of Cu(II) and Cr(III) removal from aqueous solutions using rose waste biomass. *J Hazard Mater* 161: 941–947.
16. Jaman H, Chakraborty D, Saha P. (2009) A Study of the Thermodynamics and Kinetics of Copper Adsorption Using Chemically Modified Rice Husk.*Clean* 37: 704-711.
17. Kumar P, Dara SS. (1981) Binding metal ions with polymerized onion skin. *J Poly Sci* 19: 397-402.
18. Kumar YP, King P, Prasad VSRK. (2006) Equilibrium and kinetic studies for the biosorption system of copper(II) ion from aqueous solution using *Tectona grandis* L.f. leaves powder. *J Hazard Mater* B137:1211–1217.
19. Rahman MS, Islam MR. (2009) Effects of pH on isotherms modeling for Cu(II) ions adsorption using maple wood sawdust. *J Chem Eng* 149: 273–280.

20. Saeed A, Akhter MW, Iqbal M. (2005) Removal and recovery of heavy metals from aqueous solution using papaya wood as a new biosorbent. *Sep PurifTechnol* 45: 25–31.
21. Sun GG, Shi W. (1998) Sunflower stalks as adsorbents for the removal of metal ions from wastewater.
22. Varshney R, Bhadauria S, Gaur MS. (2011) Biosorption of Copper (II) from electroplating wastewaters by *Aspergillus terreus* and its kinetics studies. *WATERJournal* 2: 142-151.
23. Wang X, Li ZZ, Sun C. (2009) A comparative study of removal of Cu(II) from aqueous solutions by locally low-cost materials: marine macroalgae and agricultural by product. *Desalination* 25: 146–159.
24. Wasewar KL, Atif M, Prasad B, Mishra IM. (2008) Adsorption of Zinc using Tea Factory Waste: Kinetics, Equilibrium and Thermodynamics. *Clean* 36:320 – 329.
25. Zhu C-S, Wang Li-P, Chen W-B. (2009) Removal of Cu(II) from aqueous solution by agricultural byproduct:Peanut hull. *J Hazard Mater* 168: 739–746.

TREND ANALYSIS OF REFERENCE EVAPOTRANSPIRATION IN A HOT AND HUMID COASTAL LOCATION IN TAMILNADU STATE

Murugappan, A[1]©, Manikumari, N[2] and Mohan, S[3]

[1©, 2]Department of Civil Engineering, Annamalai University, Annamalaingar

[3]Environment and Water Resources Division, IIT Madras, Chennai

profam@gmail.com; aumani67@yahoo.in; smohan@iitm.ac.in

ABSTRACT

Evapotranspiration is a key component of the hydrologic cycle. In the context of global warming and climate change, it becomes imperative to study the nature of variations on a short-term and long-term basis on evapotranspiration on regional and local levels. The general belief is that due to global warming, there will be an increase in evaporation or evapotranspiration. However, some studies reported in the literature reveal that despite the increase in ambient air temperature due to global warming, evaporation and/or evapotranspiration tends to decrease in certain parts of the world. In the present study, the trend in reference evapotranspiration (ET0) computed using the FAO-56 Penman-Monteith method was investigated for a hot and humid coastal location in Tamilnadu State namely, Annamalainagar, Chidambaram. Non-parametric and parametric methods have been employed for detecting trends in reference evapotranspiration at the study location.

Keywords: Reference evapotranspiration, P – M method, hot and humid location, trend analysis

INTRODUCTION

The investigations on climate change due to increasing emissions and buildup of greenhouse gases in the atmosphere have risen much in the recent years. Climate-related concerns have started to dominate and influence policy decisions at various levels of governance in both developed and developing countries. Global warming characterized by increasing temperature has come out as one of the serious environmental concerns of the present century. The global average temperature has increased by 0.6°C over the last 100 years, with 1998 being the warmest year (IPCC, 2007). Increasing global surface temperatures are expected to enhance the water holding capacity and water vapour transport in the atmosphere, which in turn results in alterations in atmospheric circulation (Bates et al., 2008).

Evapotranspiration (ET), the sum of evaporation and plant transpiration, is one of the most dynamic and complex hydrological mechanisms that connect water balance and land surface energy balance in an ecosystem (Xu and Singh, 2005). The process of evapotranspiration governs the transport of moisture between soil and the atmosphere in a catchment and is greatly influenced by surface land-use changes and climate variations. Reference evapotranspiration (ET_O) is often used to estimate actual ET in water balance studies (Xu and Chen, 2005).

The common belief is that global warming will direct to a rise in evaporation or evapotranspiration which is a main constituent of the hydrologic cycle. However, some studies reported in the literature show that although with the rise in air temperature, evaporation and/or evapotranspiration decreased in some regions across the globe. This explains that apart from air temperature, there are other climatic parameters such as wind speed, relative humidity and radiation which can counteract the influence of rise in temperature on evaporation and/or evapotranspiration thereby effecting decrease in observed evaporation and/or evapotranspiration. As per Donahue *et al.* (2010), even though increases in temperature produced the largest increase in evaporation, each of the other climate variables acted to reduce evaporation, thereby resulting in an overall reduction in evaporation.

Chattopadhayay and Hulme (1997) reported decreasing trends in both pan evaporation and potential evapotranspiration over India. Lawrimore and Peterson (2000) observed simultaneous occurrences of reductions in pan evaporation and increases in rainfall during the warm-season months in parts of the United States.

Xu *et al.* (2006) observed decreasing trends in both ET_O and E_{pan} in the catchment of Changjiang, China. Zhang *et al.* (2007) observed decreases in E_{pan} and ET_O at 47% and 38% of the respective stations over the Tibetan plateau. Jhajharia *et al.* (2009) showed decreasing trends in E_{pan} in humid NE India. Bandyopadhyay *et al.* (2009) also found decreasing trends in ET_O all over India, which was chiefly caused by a considerable increase in the relative humidity and a consistent and considerable decrease in the wind speed throughout the country. Wang *et al.* (2010) reported that the wind speed and relative humidity were commonly accepted as the chief driving forces for the decreasing trends in $\mathrm{ET_O}$ in the plain and the mountain areas of the Haihe River basin (China).

These researches highlighted about the knowledge of evapotranspiration and the significance of its trend in long-term water resources planning and proper water management. These studies also demonstrated the importance of Mann-Kendall test for trend analysis. The objectives of this study were (1) to estimate ET_O using the Penman-Monteith (PM) method at annual and seasonal time scales over the hot and humid costal location namely, Annamalainagar, in Tamilnadu State, (2) to investigate trends in ET_O using the Mann-Kendall's trend test and (3) to obtain the magnitude of trends in ET_O using linear regression.

STUDY AREA AND DATA BASE

Annamalainagar is a Special Grade Panchayat town in Cuddalore District of Tamil Nadu State. The latitude and longitude of Annamalainagar are 11.4°N and 79.7°E. The elevation of Annamalainagar is 5.79 m above mean sea level. The Koppen-Geiger climate classification is Aw. The average annual rainfall is 1248 mm. The maximum, minimum and average temperatures in the last decade was found to vary in the ranges 39°C to 28°C, 29°C to 22°C and 35°C to 26°C respectively. The average wind speed was found to vary between 13.4 mph to 6.9 mph in the last decade. Daily data on weather parameters namely, maximum and minimum temperatures in°C , maximum and minimum relative humidity in %, average wind speed at 3 m height above ground level in km/h, and actual hours of bright sunshine were collected from the India Meteorological Observatory at Annamalainagar for a period of 21 years from 1995 to 2015, for the present study.

METHODOLOGY

Penman–Monteith method

The Penman–Monteith method has been regarded as a global standard method for computation of ET_O by Food and Agriculture Organization of the United Nations (FAO)

(Allen etal.1998). The method is physically based and explicitly incorporates both physiological and aerodynamic parameters. In this method, the calculation of ET_O is given as

$$ET_o = \frac{0.409\Delta(R_n - G) + \gamma \frac{900}{T_a + 273} u_2 (e_s - e_a)}{\Delta + \gamma(1 + 0.34) u_2} \quad \text{.....(1)}$$

where ET_O is reference evapotranspiration (mm day^{-1}), R_n is net radiation at the crop surface (MJ m^{-2} day^{-1}), G is the soil heat flux density (MJ m^{-2} day^{-1}), T_a is mean daily air temperature at 2 m height (°C), u_2 is wind speed at 2 m height (ms^{-1}), e_s is saturation vapour pressure (kPa), e_a is actual vapour pressure (kPa), $(e_s - e_a)$ the saturation vapour pressure deficit (kPa), Δ is slope of the vapour pressure (kPa $°C^{-1}$) and γ is psychrometric constant (kPa $°C^{-1}$). The computation procedure given in Chapter 3 of the FAO paper 56 (Allen et al. 1998) had been used in the present study.

Trend Analysis

The significance of the trends in the annual and seasonal ET_O computed using the P-M method were evaluated by the Mann – Kendall test technique (MK test). The MK test is a rank-based nonparametric method, which has been extensively applied for trend spotting in hydro-climatic time series because of its sturdiness against the effect of abnormal data and specifically its consistency for biased variables. The implementation of MK trend test is started from the calculation of the statistic:

$$S = \sum_{i=1}^{n-1} \sum_{j-i+1}^{n} \operatorname{sgn}(x_j - x_i) \quad \text{....(2)}$$

where

$$\operatorname{sgn}(x_j - x_i) = \begin{cases} +1(x_j > x_i) \\ 0(x_j = x_i) \\ -1(x_j < x_i) \end{cases} \quad \text{.....(3)}$$

where x_i and x_j are the sequential data values and n is the length of the data set. The statistics S is approximately normally distributed when $n \geq 8$, with the mean and the variance as follows:

$$E(S) = 0 \quad \text{.....(4)}$$

$$Var(S) = [n(n-1)(2n+5) - \textstyle\sum_{i=1}^{n} ti\ i(i-1)(2i+5)\ 1/18 \quad \text{.....(5)}$$

where t is the extent of any given time.

The standardized statistics (Z) for one-tailed test is formulated as:

$$Z = \begin{cases} \dfrac{(S-1)}{\sqrt{Var(S)}} & S>0 \\ 0 & S=0 \\ \dfrac{(S+1)}{\sqrt{Var(S)}} & S<0 \end{cases} \quad .....(6)$$

The null hypothesis of no trend is rejected if $|Z|>1.96$ at the 0.05 significance level and rejected if $|Z| > 2.32$ at the 0.01 significance level. A positive value of Z represents an increasing trend, and a negative value matches to a decreasing trend.

A normal linear regression model in the form of $y = \alpha t + \beta$ is used to estimate the rate of change α, with t as the time (year), y being the annual or seasonal ET_O.

RESULTS AND DISCUSSION

The average annual ET_O at Annamalainagar during the 21-year study period (1995 – 2015) was 1668.9 mm. The maximum and minimum annual ET_O were 1524.1 mm and 1803.9 mm recorded respectively in the years 2009 and 2002. The standard deviation of annual ET_O was 98.7 mm. Table 1 shows the linear equations fitted to annual and seasonal ET_O.

Table 1 Linear equations fitted to annual and seasonal ET_O.

Period	Fitted linear equation	Annual rate of change	R-squared value
Annual	y = - 13.859t + 1821.3	-13.859	0.759
Monsoon (Jun – Dec)	y = - 7.423t + 1022.5	-7.423	0.544
NE monsoon (Oct – Dec)	y = - 1.421t + 325.2	-1.421	0.181
SW monsoon (Jun – Sep)	y = - 6.002t + 697.4	-6.002	0.535
Summer (Mar – May)	y = - 4.477t + 547.3	-4.477	0.524
Winter (Jan – Feb)	y = - 0.798t + 244.8	-0.798	0.135

The negative sign for the rate of change α indicates the decreasing trend in ET_O. Annual and all seasonal ET_O show decreasing trends. As per the fitted linear equations, it is found that the annual ET_O decreases at the rate of 13.859 mm/year. The rate of decrease in ET_O was found to be maximum at 6.002 mm/year for SW monsoon season followed by 4.477 mm/year for summer season. The decrease in ET_O for NE monsoon and winter seasons were less pronounced at 1.421 mm/year and 0.798 mm/year respectively. The low R^2 values obtained for the fitted linear equations for NE monsoon and winter seasons indicate the varied fluctuations in ET_O which could be attributed to high variability in rainfall during NE monsoon and high variability in mean air temperature and/or rainfall during winter season. In the context of global warming with the air temperatures on the rise, the trends detected for ET_O are decreasing for all seasons instead of the expected increasing trends in the present study location. The significant decreasing trends in ET_O observed for summer and south-west monsoon seasons might be attributed to reduction in wind speeds and vapour pressure deficits.

Table 2 shows the Z – value of Mann-Kendall trend test for the calculated reference evapotranspiration on seasonal and annual basis.

Table 2 Z – values of Mann-Kendall trend test for the calculated reference evapotranspiration on seasonal and annual basis

Period	Annual	Monsoon	NE monsoon	SW monsoon	Summer	Winter
Z - value	-3.775**	-3.473**	-1.842	-3.412**	-3.412**	-3.775**

** delineates significance at 0.01 level

* delineates significance at 0.05 level

The negative Z – values for annual and all seasons indicate negative trends based on the MK test. As $|Z| > 1.96$ and $|Z| > 2.32$ for annual and all seasonal ET_O except for NE monsoon ET_O, the null hypothesis of no trend is rejected at both 0.05 and 0.01 significance levels. As Z values are negative in all cases, they depict decreasing trends in annual and seasonal ET_O.

CONCLUSION

Both the non-parameteric (MK trend test) and the parameteric (linear regression) tests clearly demonstrate decreasing trends in annual and seasonal ET_O in the hot and humid coastal location of Annamalainagar.

REFERENCES

1. Allen RG, Pereira LS, Raes D, Smith M. 1998. Crop evapotranspiration – guidelines for computing crop water requirements. *FAO Irrigation and Drainage Paper* **56**. Food and Agriculture Organization, Rome, Italy.
2. Bandyopadhyay A, Bhadra A, Raghuwanshi NS, Singh R. 2009. Temporal trends in estimates of reference evapotranspiration over India. *Journal of Hydrologic Engineering* **14**(5): 508–515.
3. Bates BC, Kundzewicz ZW, Wu S, Palutikof JP. 2008. Climate change and water. Technical Paper of the Intergovernmental Panel on Climate Change, IPCC Secretariat, Geneva, p 210.
4. Chattopadhyay N, Hulme M. 1997. Evaporation and potential evapotranspiration in India under conditions of recent and future climate change. *Agricultural Forest Meteorology* **87**: 55–73.
5. Donohue RJ, McVicar TR, Roderick ML. 2010. Assessing the ability of potential evaporation formulations to capture the dynamics in evaporative demand within a changing climate. *Journal of Hydrology* **386**: 186–197.
6. IPCC. 2007. Summary for policymakers. In *Climate Change 2007: The Physical Science Basis*, Solomon S, Qin D, Manning M, Chen Z, Marquis M, Averyt KB, Tignor M, Miller HL (eds). Cambridge University Press: New York, USA.
7. Jhajharia D, Shrivastava SK, Sarkar D, Sarkar S. 2009. Temporal characteristics of pan evaporation trends under the humid conditions of northeast India. *Agricultural Forest Meteorology* **149**: 763-770.
8. Lawrimore J, Peterson T. 2000. Pan evaporation in dry and humid regions of the United States. *Journal of Hydrometeorology* **1**: 543–546.
9. Wang W, Peng S, Yang T, Shao Q, Xu J, Xing W. 2010. Spatial and temporal characteristics of reference evapotranspiration trends in the Haihe River basin, China. *Journal of Hydrologic Engineering*. DOI:10.1061/(ASCE)HE.1943–5584.0000320.
10. Xu CY, Chen D. 2005. Comparison of seven models for estimation of evapotranspiration and groundwater recharge using lysimeter measurement data in Germany. *Hydrological Processes* **19**: 3717–3734.

11. Xu CY, Singh VP. 2005. Evaluation of three complementary relationship evapotranspiration models by water balance approach to estimate actual regional evapotranspiration in different climatic regions. *Journal of Hydrology* **308**: 105–121.
12. Xu CY, Gong LB, Jiang T, Chen DL, Singh VP. 2006. Analysis of spatial distribution and temporal trend of reference evapotranspiration and pan evaporation in Changjiang (Yangtze River) catchment. *Journal of Hydrology* **327**: 81–93.
13. Zhang Y, Liu C, Tang Y, Yang Y. 2007. Trends in pan evaporation and reference and actual evapotranspiration across the Tibetan Plateau. *Journal of Geophysical Research* **112**: D12110. DOI:10.1029/2006JD008161.

CLIMATE CHANGE: EFFECTS ON SUSTAINABLE SCENARIO IN INDIA

Srisailam Gogula[1] Krishna Kishore Jyothi[2] and Sunder Kumar Kolli*

[1]Department of Chemistry, Govt.City College, Hyderabad, Telangana

[2]St.Ann's College of Engineering and Technology, Chirala, Andhra Pradesh

*Department of Chemistry, Annamacharya Institute of Technology & Sciences, Hyderabad

ABSTRACT

Climate change is one of the major challenges of our time and adds considerable stress to our societies and to the environment. The impacts of climate change are global in scope and unprecedented scale. It is a consequence of Global warming with increased temperatures and seasonal variations. Many issues like that Climate Change, Environmental degradation and Displacement have been the major challenges to the entire human populations. A range of human activities which mainly include the burning of fossil fuels, industrial wastes, deforestation, population rise are substantially increasing the concentrations of greenhouse gases in the atmosphere resulting in adverse change of climate. Climate change is a serious threat to human security and national economy. Scientists predict that climate change would increasingly impact the humans by severe heat waves, floods, storms and forest fires causing as many as 5,00,000 deaths a year by 2030 and thus it has become the greatest humanitarian issue engaging the attention of all people in the third world (Liu Jie and IPCC Report 2007). Forced displacements leading to loss of livelihoods, drought, famine, protracted conflicts and unresolved disputes over land and property led them to marginality. The lukewarm response of the Government's and the private entrepreneurs towards the displaced people is often extemporized and largely insufficient and they frequently find themselves in extremely vulnerable situations. The recent instances of displacement across the country met with vehement resistance bears testimony to the people's struggle and keenness to protect environment and their livelihoods. These protest movements illustrate the people's concerns and priorities for their well-being and sustainable livelihoods rather than for sheer economic considerations.

Keywords: Environmental degradation, fossil fuels, human population and national economy.

INTRODUCTION

Many of India's 1.3 billion people a fifth of the world's population face pollution that is cutting short lives, stunting children's cognitive development and putting public health under terrific stress. Air pollution is the leading risk factor for most deaths and disabilities in India, a country that's home to 13 of 20 of the world's most polluted cities. To lift millions from poverty, it will require ever more energy. But most of India's electricity is generated by coal-burning power plants. Millions of new cars choke the roads each year. Add to the mix the burning of garbage and crops, and it's a toxic cocktail that makes India the third-largest contributor of greenhouse gas emissions in the world, after China and the United States. A study on the impact of climate change by the International Monetary Fund (IMF) shows countries in the tropics will be the worst affected as a result of global warming. "For the median emerging market economy, a 1°C increase from a temperature of 22°C lowers growth in the same year by 0.9 percentage point," says the report. India is one of the worst affected, with its per capita output expected to fall by 1.33 percentage points, these predictions underline the importance of policies to combat the impact of climate change in countries like India.

Climate change may have contributed to the suicides of nearly 60,000 Indian farmers and farm workers over the past three decades, according to new research that examines the toll rising temperatures are already taking on vulnerable societies. An increase of 5°C on any one day was associated with an additional 335 deaths, the study published in the journal PNAS. In total, it estimates that 59,300 agricultural sector suicides over the past 30 years could be attributed to warming. Also supporting the theory was that rainfall increases of as little as 1cm each year were associated with an average 7% drop in the suicide rate. So beneficial was the strong rainfall that suicide rates were lower for the two years that followed, researcher Tamma Carleton found. One drought-hit state, Maharashtra, reported 852 farmer suicides in the first four months of this year 2017, while in 2015, one of the worst years on record, about 12,602 farmers killed themselves across India. Overall, more than 300,000 farmers and farm workers have killed themselves in the country since 1995.

Relatively likely and early effects of small to moderate warming

Rise in sea level due to melting glaciers and the thermal expansion of the oceans as global temperature increases. Massive release of greenhouse gases from melting permafrost and dying forests, a high risk of more extreme weather events such as heat waves, droughts and floods global incidence of drought has already doubled over the past 30 years. Severe regional impacts. Example: In Europe River flooding will increase and in coastal areas the risk of flooding, erosion and wetland loss will increase substantially. Natural systems, including glaciers, coral reefs, mangroves, Arctic ecosystems, alpine ecosystems, Boreal forests, tropical forests, prairie wetlands and native grasslands, will be severely threatened. The existing risks of species extinction and biodiversity loss will increase. The greatest impacts will be on the poorer countries least able to protect themselves from rising sea levels. There will be spread of disease and declines in agricultural production in the developing countries of Africa, Asia and the Pacific. At all scales of climate change, developing countries will suffer the most.

Longer term catastrophic effects if warming continues

Greenland and Antarctic ice sheets are melting. Unless checked, warming from emissions may trigger the irreversible meltdown of the Greenland ice sheet in the coming decades, which would add up to seven meters rise in sea-level over some centuries. New evidence showing the rate of ice discharge from parts of the Antarctic means that it is also facing a risk of meltdown. The slowing, shifting or shutting down of the Atlantic Gulf stream current is having dramatic effects in Europe, disrupting the global ocean circulation system. Catastrophic releases of methane from the oceans are leading to rapid increases in methane in the atmosphere and the consequent warming. Never before has humanity been forced to grapple with such an immense environmental crisis. If we do not take urgent and immediate action to stop global warming, the damage could become irreversible. Extreme weather events are costing India $9-10 billion annually and climate change is projected to impact agricultural productivity with increasing severity from 2020 to the end of the century. In a recent submission to a parliamentary committee, the agriculture ministry said productivity decrease of major crops would be marginal in the next few years but could rise to as much as 10-40% by 2100 unless farming adapts.

Wheat, rice, oilseeds, pulses, fruits and vegetables will see reduced yields over the years, forcing farmers to either adapt to challenges of climate change or face the risk of getting poorer. Adaptation will need different cropping patterns and suitable inputs to compensate yield fluctuations. The challenge is particularly urgent for Indian agriculture where productivity for crops like rice does not compare even with neighbors' like China. The possibility of a further dip due to climate change will be particularly worrying as it could turn India into a major importer of oilseeds, pulses and even milk. By 2030, it may need 70 million tons more of food grains than the expected production in 2016-17. The economic survey, in its latest mid-year report, says "estimates indicate that currently India incurs losses of about $ 9-10 billion annually due to extreme weather events. Of these, nearly 80% losses remain uninsured". It pointed out those 2014 floods in Kashmir cost more than $ 15 billion and Cyclone Hudhud the same year cost $ 11 billion.

Impacts of global warming on climate of India

The effect of global warming on the climate of India has led to climate disasters as per some experts. India is a disaster-prone area, with the statistics of 28 out of 36 states being disaster prone, with foods being the most frequent disasters. The process of global warming has led to an increase in the frequency and intensity of these climatic disasters. According to surveys, in the year 2007-2008, India ranked the third highest in the world regarding the number of significant disasters, with 18 such events in one year, resulting in the death of 1103 people due to these catastrophes. The anticipated increase in precipitation, the melting of glaciers and expanding seas have the power to influence the Indian climate negatively, with an increase in incidence of floods, hurricanes, and storms. Global warming may also pose a significant threat to the food security situation in India. According to the Indira Gandhi Institute of Development Research, if the process of global warming continues to increase, resulting climatic disasters would cause a decrease in India's GDP to decline by about 9%, with a decrease by 40% of the production of the major crops. A temperature increase of 2° C in India is projected to displace seven million people, with a submersion of the major cities of India like Mumbai and Chennai.

Rehabilitation and Resettlement problems

On October 2 in 2017, India became a signatory to the Paris Agreement on climate change, hence moving a step closer to achieving its goal of reducing carbon emissions. However, neither the climate pact nor the recently concluded United Nations Summit for Refugees and Migrants addressed the direct human cost of climate change: the displacement of millions by natural disasters and slow-onset environmental changes. It is well established that climate change often forces the affected populations to move from their habitual place of residence. According to the Internal Displacement Monitoring Centre, 19.3 million people were displaced worldwide in 2014 due to climate change, with studies indicating that the number could be anywhere between 250 million and one billion by 2050.The geographically diverse Indian subcontinent is particularly vulnerable to a wide variety of natural disasters, and India, as the largest country in the region, is the destination to move to for those displaced by these disasters. Floods, storm surges, saltwater intrusions and cyclones have pushed millions of people from rural Bangladesh into India. Earthquakes and water-induced disasters in Nepal, droughts in Pakistan and Afghanistan, and the rise in sea levels around the Maldives are also likely to cause large-scale migration into India in the future. Climate change will significantly affect migration in three distinct ways: The effects of warming and drying in some regions will reduce agriculture potentials and undermine ecosystem services such as clean water and fertile soil, the increase in extreme weather events – in particular, heavy rainfall or snowfall and resulting floods, Sea level rise will permanently destroy extensive and highly productive low-lying coastal areas that are home to millions of people.

Reasons for displacement of people

People are forced to move out of their land due to both natural and man made disasters. Natural disasters like earthquake, cyclones, tsunamis, volcanic eruptions, prolonged droughts conditions, floods, hurricanes etc. Manmade disasters like industrial accidents (e.g. Bhopal gas tragedy), nuclear accidents (Current disaster in Japan), oil spills (Exxon Valdez oil spill), toxic contamination of sites etc. In search of better employment opportunities. Developmental projects like: construction of dams, irrigation canals, reservoirs etc. Infrastructural projects like flyovers, bridges, roads etc. transportation activities like roads, highway, canal etc. Energy related project like power plants, oil exploration, mining activities, pipelines like HBJ pipeline etc. agricultural Projects, national parks, sanctuaries and biosphere reserves displace people from their home. Thus, resettlement refers to the process of settling again in a new area.

Resettlement issues

As per the World Bank estimates, nearly 10 lack people are displaced worldwide for a variety of reasons: Little or no support: Displacement mainly hits tribal and rural people who usually do not figure in the priority list of any political authorities or parties. Meager compensation: The compensation for the land lost is often not paid, it is delayed or even if paid, is too small both in monetary terms and social changes forced on them by these mega developmental projects. Loss of livelihood: Displacement is not a simple incident in the lives of the displaced people. They have to leave their ancestral land and forests on which they depend for their livelihood. Many of them have no skills to take up another activity or pick up any other occupation. Usually, the new land that is offered to them is of poor quality and the refugees are unable to make a living. Lack of facilities: When people are resettled in a new area, basic infrastructure and amenities are not provided in that area. Very often, temporary camps become permanent settlements. It is also a major problem of displacement or resettlement that people have to face. Increase in stress: Resettlement disrupts the entire life of the people. They are unable to bear the shocks of emptiness and purposelessness created in their life. Payment of compensation to the head of the family often lead to bitter quarrels over sharing of compensation amount within the family, leading to stress and even withering of family life. Moreover, land ownership has a certain prestige attached to it which cannot be compensated for even after providing the new land. With the loss of property and prestige, marriages of young people also become difficult as people from outside villages are not willing to marry their daughters to the refugees. Increase in health problems: Lack of nutrition due to the loss of agriculture and forest based livelihood, lead to the general decline in the health of the people. People are used to traditional home remedies. But the herbal remedies and plants gets submerged due to the developmental projects. Secondary displacement: Occupational groups residing outside the submergence area but depending on the area for the livelihood also experience unemployment. Village artisans, petty traders, laborers etc. lose their living. Loss of identity: Tribal life is community based. The tribal are simple people who have a lifestyle of their own. Displacement has a negative impact on their livelihood, culture and spiritual existence in the following ways: Break up of families and communities are the important social issues of displacement. The women suffer the most as they are deprived even a little compensation. Inter-community marriages, cultural functions, folk songs and dances do not take place among the displaced people. When they are resettled, it is generally individual based resettlement, which ignores communal character. Resettlement increases the poverty of the tribal due to the loss of land, livelihood, food insecurity, jobs, skills etc.Loss of identity of individuals and the loss of connection between the people and the environment is the greatest loss in the process. The indigenous knowledge that they have regarding the wildlife and the herbal plants are lost. The land acquisition laws do not pay attention to the idea of communal ownership of property which increases stress within the family. The tribal people are not familiar with the market trends, prices of commodities and policies. As such, they are exploited and get alienated in the modern era.

Objectives of rehabilitation

The following objectives of rehabilitation should be kept in mind before the people are given an alternative site for living: Tribal people should be allowed to live along the lives of their own patterns and others should avoid imposing anything on them. They should be provided means to develop their own traditional art and culture in every way. Villagers should be given the option of shifting out with others to enable them to live a community based life. Removal of poverty should be one of the objectives of rehabilitation. The people displaced should get an appropriate share in the fruits of the development. I should say that it is really a good move by ISC to share its profits among the active contributors. The displaced people should be given employment opportunities. Resettlement should be in the neighborhood of their own environment. If resettlement is not possible in the neighbor area, priority should be given to the development of the irrigation facilities and supply of basic inputs for agriculture, drinking water, wells, grazing ground for the cattle, schools for the children, primary healthcare units and other amenities. Villagers should be taken into confidence at every stage of implementation of the displacement and they should be educated, through public meetings, discussion about the legalities of the Land Acquisition act and other rehabilitation provisions. The elderly people of the village should be involved in the decision making.

Climate changes due to Fossil fuels

Consumption of fossil fuel resources leads to global warming and climate change. In most parts of the world little change is being made to slow these changes. If the peak oil theory proves true, and more explorations of viable alternative energy sources are made, our impact could be less hostile to our environment. The scientific consensus on global warming and climate change is that it is caused by anthropogenic greenhouse gas emissions, the majority of which comes from burning fossil fuels with deforestation and some agricultural practices being also major contributors. A 2013 study showed that two thirds of the industrial greenhouse gas emissions are due to the fossil-fuel (and cement) production of just ninety companies around the world (between 1751 and 2010, with half emitted since 1986).Although there is a highly publicized denial of climate change, the vast majority of scientists working in climatology accept that it is due to human activity. The IPCC report *Climate Change 2007:* Of all the most polluting nations – US, China, Russia, Japan and the EU bloc – only India's carbon emissions are rising: they rose almost 5% in 2016. No one questions India's right to develop, or the fact that its current emissions per person are tiny. But when building the new India for its 1.3 billion people, whether it relies on coal and oil or clean and green energy will be a major factor in whether global warming can be tamed.

Human Population Growth and Climate Change

The largest single threat to the ecology and biodiversity of the planet in the decades to come will be global climate disruption due to the buildup of human-generated greenhouse gases in the atmosphere. People around the world are beginning to address the problem by reducing their carbon footprint through less consumption and better technology. But unsustainable human population growth can overwhelm those efforts, leading us to conclude that we not only need smaller footprints, but fewer feet.2009 study of the relationship between population growth and global warming determined that the “carbon legacy” of just one child can produce 20 times more greenhouse gas than a person will save by driving a high-mileage car, recycling, using energy-efficient appliances and light bulbs, etc. With more people in the state, it has become very difficult for many Indians to obtain a well-paying job. Therefore, many of India’s people have begun farming for themselves in order to provide sustenance for their families (Ninkovic et al., 2013). In fact, it is estimated that farming is the main livelihood of 75% of India’s total population (Shukla 196). Since such a high proportion of Indians are farmers, it has begun to take a noticeable effect on India’s environment. The process of environmental degradation in India has also only accelerated with the growing effects of climate change. The primary people in India affected by environmental degradation are the farmers because they rely on the land to provide them with the resources they need in order to survive. The majority of India’s people who engage in farming are only able to utilize the available natural resources around them because they live in rural areas, but their reliance on the environment has often had negative consequences in their lives (Nagdeve 5). One reason many farmers cannot sustain themselves without the help of the environment is because they do not have the money to do so. Originally, many Indian farmers utilized natural underground reservoirs to water their crops (Nagdeve 7). However, as more individuals have begun farming, these natural sources of water are used more frequently. Because of constant usage, Indian farmers have overexploited India’s main water reservoirs, so they no longer serve as a viable source of water (Ninkovic et al., 2013). Indian farmers mainly rely on the annual monsoons to water their crops, but because of climate change the yearly monsoon patterns have changed (Ninkovic et al., 2013). Lack of water often results in insufficient crop growth and farmers not being able to produce the expected number of crops. For those who farm commercially, they fail to meet their sales obligations and are usually forced to take a loan to account for their loss of sales (Ninkovic et al., 2013). However, if the monsoons do not arrive at the correct time in the subsequent year and the same process occurs once more, commercial farmers are then unable to pay off their loan from the previous year (Ninkovic et al., 2013). In this situation, it is common for them to have to sell their land. For the farmers who rely on their crops as their main source of food, they often starve if the monsoons do not occur at the right time or do not produce enough rainfall. Unfortunately, one result of the difficulty to grow crops is that approximately 17,500 of the farmers who rely on their harvest as a food source throughout India commit suicide annually (Shukla 196). The rate of suicide is staggeringly high because of India's overpopulation problem, which forces many poor families to solely depend on their agriculture.

Economic impacts of climate change

In a literature assessment, Smith *et al.* (2001:957-958) concluded, with medium confidence, that: Climate change would increase income inequalities between and within countries. A small increase in global mean temperature would result in net negative market sector impacts in many developing countries and net positive market sector impacts in many developed countries. With high confidence, it was predicted that with a medium (2-3 °C) to high level of warming (greater than 3 °C), negative impacts would be exacerbated, and net positive impacts would start to decline and eventually turn negative. The farm sector in India is in distress and several state governments have responded with loan waivers, which could affect their fiscal math and the ability to push capital expenditure at a time when the Indian economy has slowed significantly. This comes after India faced deficient rainfall for two consecutive years in 2014 and 2015. According to estimates, production of kharif crops in the current year is expected to decline by 2.8% because of an uneven monsoon. The possibility of such weather events is likely to increase in the future. And that means a serious challenge for a country like India where about 50% of the population directly or indirectly depends on agriculture for a livelihood.

Weather does not affect the agriculture sector alone, it affects productivity in general. Research shows that productivity starts declining strongly after peaking at an average annual temperature of about 13 degrees Celsius. Therefore, countries located in areas with higher temperature will face a disproportionate impact of global warming. Loss of output and lower productivity also affects capital formation, which has a bearing on medium- to long-term growth prospects. The IMF, for example, notes: "The results suggest that having the right policies and institutions in place may help attenuate the effects of temperature shocks, to some extent. The instantaneous effect of a temperature shock is slightly smaller in countries with lower public debt, higher inflows of foreign aid, and greater exchange rate flexibility." India is relatively better off in this context, but it needs to preserve and further strengthen macroeconomic stability to be able to deal with such shocks. Over the years, India has done well to reduce its dependence on the monsoon, which is evident from the fact that two successive years of drought did not result in runaway inflation. However, more needs to be done to enhance productivity in the agriculture sector. Financial losses can be reduced by higher penetration of insurance products.

CONCLUSIONS

Climate change is a big issue in India as it has more population and most polluted cities. Coal burning power plants, deforestation and air pollution are causing climate changes, thus monsoon changes and it is giving low rain fall, impacting the crops and it is the cause of suicides of the farmers. India is disaster prone area, so need to take many preventive activities, Government and NGOs should educate people about the disasters and climate changes around them. Rehabilitation or resettlement has to take care by the government, importance should be given to their own lifestyle and environment at new area, many opportunities need to provide for the settlement and their identity should not be lost, and authorities should not repeat previous resettlement problems where some people are struggling even today. Over-population is the real cause of climate change – it's killing us all off, despite all the warnings of global warming and imminent disaster, it is unlikely that we will change our ways until a real catastrophe actually occurs. In the past six decades, India's total population has doubled twice. However; the rapid rate of population growth has many negative consequences on India's people and its environment. The necessary steps to minimize the impact of climate change will have to be taken at both the individual country level and the global level. In order to reduce the impact of changing weather patterns, emerging market and low-income economies will have to build significant macroeconomic resilience. So, can India's leaders bring light to its poorest people, build clean, green cities for its billion-strong population and end the plague of air pollution. And it is individual responsibility to sustain the environment for the generations.

REFERENCES

1. Mint Press News Deck. "India's Government Increasingly Embraces GMOs Despite High Rate of Farmer Suicide." Mint PressNews, Mint Press, 14 Aug. 2015,
2. www.mintpressnews.com/indias-government-increasingly-embraces-gmos-despite-high-rate- of-farmer-suicides/208663/. Accessed 30 Nov. 2016.
3. Nagdeve, Dewaram A. "Population Growth and Environmental Degradation in India." Princeton University, paa2007.princeton.edu/papers/7192.
4. Ninkovic, Nina, and Jean-Pierre Lehmann. "India's Food Crises: A Close-Up." The Globalist, 1 Sept. 2013, www.theglobalist.com/indias-food-crises-close-up/.Accessed 30 Nov. 2016.
5. "Overpopulation Causes Environmental Degradation in India and Pakistan." South Asia Investor Review, 29 July 2010, southasiainvestor.blogspot.com/2010/07/overpopulation-causes-environmental.html. Accessed 30 Nov. 2016.
6. Shukla, Priyadarshi R. Climate Change and India: Vulnerability Assessment and Adaptation. Universities Press, 2003.

GRAPHENE FOR SUSTAINABLE WATER

Shivarajappa[1] and Mohd. Hussain[2]

[1]Assistant Professor in Civil Engg. Dept., AVN Institute of Engineering & Technology, Ibrahimpatnam. RR Dist.

[2]Professor in Civil Engg. Dept., Gokaraju Rangaraju Institute of Engineering & Technology, Hyderabad

ABSTRACT

However, the use of graphenic materials for large-scale and down to earth application like water purification is limited. This is mainly because of the difficulty in large-scale synthesis. The ability to make GO through chemical methods and its subsequent reeducation to reduce grapheme oxide (RGO) opened up the possibility for the mass production of grapheme in solution phase. The properties of GO or RGO can be easily enhanced through chemical modifications. Several attempts have been made to produce GO and RGO-composites. Recent literature suggests that RGO, GO and their composites are getting into environmental remediation. RGO- magnetite and GO-ferric hydroxide composites were used for the removal of arsenic from water. Iron based oxides and hydroxides are known to remove arsenic from drinking water. The report shows that RGO and GO supported materials have higher binding capacity compared to free nanoparticles. A study Hy Hu et al. also showed that RGO is antibacterial and this property may help in preventing the development of biofilm on the filter surface due to bacterial growth, which can cause unwanted tastes and odors or prematurely clogging of filters.

Important aspects to be considered for the large-scale production of RGO-composites are ease of synthesis and post synthesis purification. In most of the existing methods of composite preparation, constituents were separately prepared and mixed or external aids were employed for the production of composites, which has many limitations in large-scale synthesis. Another aspect for using such composites for application such as water purification is the ease of solid-liquid separation and post treatment-handling. Laborious processes like high speed centrifugation, membrane filtration, or magnetic separation are not practical for many end-users. A practical adsorbent material has been under study here which shows that RGO-based materials are also field adaptable.

In this report, we propose a simple strategy to synthesize monodispersed and uncapped nanoparticles of silver, gold, platinum, palladium and manganese oxide on the surfaces of RGO. An in situ homogenous reduction strategy utilizing the inherent reducing properties of RGO to produce composite materials was explored, at room temperature without any external aids. The simple methodology adopted here permits to make large-scale composites with good control over the particle size. The process uses the inherent reduction ability of RGO, simplifying the post-synthesis treatment and thereby increasing the liability in commercial applications. Among the materials prepared one metal (Ag) and one metal oxide (Mno_2) based composites were selected, considering their possible utility in mitigating range of contaminants from water. The applicability of RGO-Ag and RGO-Mno2 were demonstrated for removing heavy metals from water. Hg (II), one of the most toxic metals found in the environment, was used as the model pollutant. The effects of mercury on humans and the environment have been documented. Considering the practical difficulty in using RGO and its composites as such in water purification, a simple methodology was developed to immobilize the composites on a cheap and inert support like river sand (RS). Chitosan (Ch), an abundantly available and environment- friendly biomaterial was used as a binder for this process. The supported RGO-composites were also demonstrated for Hg (II) uptake and their applicability in the field of water purification. Various microscopic and spectroscopic techniques were used to probe the composite formation and attachment of Hg (II) onto the composites.

Scientists have reported that membranes made from grapheme oxide appear to be highly permeable to water while being impermeable to all other liquids and gases. The membranes consist of millions of small flakes of grapheme oxide with nanometer-sized empty channels (or capillaries between the flakes that favour the passage of monolayer of water and resist other substances. Grapheme oxide is similar to ordinary grapheme but is covered with molecules, such as hydroxyl groups (OH).

Copper oxides and its salts are now widely used as pesticides to control fungal and bacterial diseases of field crops. Copper toxicity is often a major contributor of human health problems caused through accumulation of excess copper ions in various organs via drinking water, fruits and vegetables. So, detection and estimation of cupric ions in biological organs, drinking water, fruits and vegetables are extremely important. Recently, a fluorescence based sensor using coumarin dye (high quantum yield) has been proposed to detect micro-molar cu++ ion in biological organs. But major problem with coumarin dye is that it is insoluble in water and undergoes dye-dye aggregation in organic solvents. We proposed here a synthetic scheme of preparation of grapheme oxide conjugated coumarin dye derivative which would be water dispersible and expected to be an ideal candidate for Cu^2+ ion estimation in biological organs and drinking water.

In this paper, we would propose to design synthetic scheme of coumarin conjugated Graphene Oxide (GO) nanomaterials. This design will take advantage of the high selectivity of the coumarin derivative towards Cu2+ ions as reported in the litterateur and minimize the self-quenching problem. This is a feasible strategy as the coumarin dye will be covalently attached to GO surface and this attachment will hinder the formation of J-aggregation of coumarin dye molecules. The use of GO to prevent such J- aggregation has been reported in the recent time Graphene, GO and nonmaterial research has gained a tremendous momentum because of their potential applications in material science. Grapheme has two-dimensional (one atom thick) crystals of SP2 bonded carbon atoms densely packed in a honey comb crystalline lattice. The C-C bond length of grapheme is nm and Grapheme sheets are stuck to form graphite with inter planner spacing of nm. It possesses some unique properties such as high surface area high electronic conductivity (electron mobility 20,000 cm/second under unit potential gradient) low resistivity (specific resistance) high mechanical and chemical stability.

Calculation based on an initio shows that Graphene is thermodynamically unstable if it contains less than 6000 atoms (by Shenderovaab et al.) Graphene is a hydrophobic material and it has no binding sites available for ions. Graphene is chemically converted to GO (via oxidation)/GO derivatives for increasing its hydrophillicity by introducing suitable functional groups. GO can be dispersed in aqueous solution and they carry hydroxyls, acids and epoxy groups on the surface. Functional groups on the GO surface can be further modified to attach suitable ligands. In general, GO is considered as a biocompatible materials. However, there are reports showing cytotoxic effect of GO to human fibroblast cells above 50 μg/ml concentration.

EXPERIMENTAL SCHEME

SYNTHETIC STRATEGY

The basic idea is to attach GO, a high surface area substrate material to the Cu ion selective fluorescent dye. For such attachment, Cu ion selective coumarin dye derivative will be further modified to obtain a terminal primary amine group.

The amine functionalized coumarin dye will be covalently attached to GO surface through epoxy ring opening reaction, forming the GOCD. Since the coumarin dye molecules are covalently attached to the GO substrate, it is anticipated that dye-dye interaction (J-aggregation) will be minimized. This is important for boosting the sensitivity for the Cu2+ ion detection in the nanomolar range. Moreover, the proposed GOCD is a new hybrid material.

SCHEMATIC REPRESENTATION OF OVERALL SYNTHETIC STRATEGY

SYNTHESIS OF GRAPHENE-OXIDE COUMARIN CONJUGATE (GOCD)

The proposed GOCDE will involve a multi-step synthesis process.

Step 1 is the preparation of the GO from graphite flakes following Hummar method. This method uses chemical and sonication techniques to exfoliate GO from purified natural graphite. The resulting exfoliated GO from colloidal suspension of individual GO sheet in water. Then following the procedure of Wang et al, pH assisted selective sedimentation will be done to separate GO of different size range.

Step 2 is the synthesis of primary amine functionalized Cu2+ ion selective coumarin dye (compound #4,) an amine containing precursor, methyl 4-amino-2hydroxybenzoate (compound 1, Sigma-Aldrich; catalogue #PH001949) will be reacted with the protected az-ir-idine to make compound #2. This addition reaction involves nucleophilic ring opening of az-ir-idine by the NH_2 group of the compound #1 under mild acidic reaction condition to control reaction rate. Compound #3 (a cyclic ester) will be then prepared by treating the compound #2 with acetonitricle in presence of SmI_2 catalyst.Using $LiA1H_4$, selective reduction reaction will be carried out to convert the compound #3 to the corresponding aldehyde (compound#4). Similar synthetic route has been reported in the literature where Diethylamino-2-hydroxy-benzaldehyde was used as the precursor starting material instead of compound #1 and therefore this proposed synthesis step is feasible.

Step 3 is the GO conjugation with the compound#4 forming compound #5. The GO which is produced via Hammer method is intrinsically acidic and it has surface epoxy groups. It is expected that the compound #4 will readily react with the epoxy containing GO when combined together. The terminal amine group of the compound #4 will readily react with the epoxy containing GO when combined together. The terminal amine group of the compound #4 will take part in the epoxy ring opening reaction. Similar epoxy ring opening reactions involving primary amine containing molecules and the GO have been reported in the literature.

Step 4 is a simple acidcatalayzed Michael type addition reaction between the dimethyl malonate and the compound #5 forming the compound #6. Then the compound #6 will be converted to the compound #7 in acidic condition where GO is expected to serve as a catalyst. This proposed synthesis step is based on a recent research reported by the Bielawski group. Their findings confirmed that GO served as a catalyst in auto tandem oxi-condensation and hydride elimination step.

Step 5 is the final step where the compound #7 will be converted to the compound #8 (the proposed GOCD). In this step, 2-aminomethyl pyridine addition and condensation reaction catalyzed by GO will be carried out in acidic reaction condition.

CONCLUSION

Synthetic steps proposed for synthesis of GOCD are conventional and uses of GO as catalyst are also well reported. Fluorescence of coumarin in GO-coumarin derivative (GOCD) will quench upon binding with Cu ions similar to its quenching of coumarin derivative as studied previously. Presence of GO in GOCD will not affect the strong fluorescence emission of coumarin derivative as GO is reported to exhibit a weak fluorescence property and it is covalently boded to CD. GOCD should exhibit two distinct absorption bands characteristic to GO (200-450 nm absorption) and the coumarin dye (430 nm). Based on the literature reports, it is evident that the absorption bandposition of GO largely depends a number of factors including particle size, pH and the degree of oxidation. GO usually emits in the range 400-700nm. Depending on the pH, GO particle size and extent of GO oxidation, emission of GO could vary anywhere between the 370-650nm. Since GO is water soluble, GOCD will also be water dispersible and minimize dye-dye aggregation. Thus, GOCD will be an ideal fluorescence sensor for estimation of nanomolar concentration of cupric ion in aqueous solutions and biological organs.

REFERENCES

1. Bowen, H.J.M. and Hutzinger, D. (1985) The Natural Environment and the Biogeochemical
2. Cycles. The Handbook of Environmental Chemistry. Springer-Verlag, New York, 1-96
3. Donnelly, P.S., Xiao, Z.G. and Wedd, A.G. (2007) Copper and Alzheimer's Disease. Current Opinion in Chemical Biology, 11, 128-133.
4. Dong, Y., Koken, B., Ma, X, Wang L., Cheng Y. and Zhu, C. (2011) Polymer-Based Fluorescent Sensor Incorporating 2,2'- Bipyridy1 and Benzo[2,1,3] Thiadiazole Moieties for Cu2+ Detection Inorganic Chemistry Communications, 14, 1719-1722.
5. Eisler, R. (1998) Copper Hazards to Fish, Wildlife, and Invertebrates: A Synoptic Review Geological Survey, Washington DC.
6. Forstner, U and Witmann, G.T.W. (1979) Metalpollution in the Aquatic Environment Springer-Verlag, Berlin, 1-486.
7. Frigoli, M., Ouadahi, K. and Larpent, C. (2009) A Cascade FRET-Mediated Ratiometric Sensor for Cu2+ Ions Based on Dual Fluorescent Ligand-Coated Polymer Nanoparticles. Chemistry-A European Journal, 15, 8319-8330.
8. Harris, Z.L. and Gitlin, J.D. (1996) Genetic and Molecular Basis for Copper Toxicity. American Journal of Clinical Nutrition, 63, 836-841.
9. Helal, A., Rashid, M.H.O., Choi, C.H. and Kim, H.S. (2011) Chromogenic and Fluorogenic Sensing of Cu2 Based on Coumarin Tetrahedron, 67, 2794-2802.
10. Kalis, E.J.J., Weng, L., Dousma, F., Temminghoff, E.J.M. and Van Riemsdijk, W.H. (2006) Measuring Free Metal Ion Concentrations in Situ in Natural Waters Using the Donnan Membrane Technique. Environmental Science Technology, 40, 955-961.

11. Minami, T., Sohrin, Y. and Ueda, J. (2005) Determination of Chromium, Copper and Lead in River Water by Graphite Furnace Atomic Absorption Spectrometry after Coprecipitation with Terbium Hydroxide. Analytical Sciences, 21 1519-1521.
12. Mulazimoglu, I.W. (2012) Electrochemical Determination of Copper (II) Ions at Narigenim-Modified Glassy Carbon Electrode: Application in Lake Water Sample. Desalination and Water Treatment, 44,161-167.
13. Multhaup, G.(1997) Amyloid Precursor Protein, Copper and Alzeimer's Disease. Biomedicine Pharmacotherapy, 51,105-111.
14. Richardson, H.W. (1997) Handbook of Copper Compounds and Applications. Marcel Dekker, Inc., New York, 1-432.
15. Shao, N., Zhag, Y., Cheung, S.M., LYang, R.H., Chan W.H., Mo, T., et al. (2005) Copper Ion-Selective Fluorescent Sensor Based on the Inner Filter Effect Using a Spiropyran Derivative Analytical Chemistry, 77, 7294-7303.
16. Sirilaksanapong, S., Sukwattanasinitt, M. and Rashatasakhon, P. (2012) 1,3,5-Triphenylbenzene Fluorophore as a Selective Cu2+ Sensor in Aqueous Media. Chemical Communication. 48, 293-295.
17. Wang, W.D., Fu, A., You., J.S., Gao, G., Lan, J.B. and Chen, L.J. (2010) Squaraine-Based Colorimetric and Fluorescent Sensors for Cu2+ Soecufuc Detection and Fluorescence Imaging in Living Cells. Tetrahedron, 66, 3695-3701.
18. Yin, S., Leen, V., Van, S.S., Boens, N. and Dehaen, Wo (2010) A Highly Sensitive, Selective, Colorimetric and Near-Infrared Fluorescent Turn-On Chemosensor for Cu2+ Based on BODIPY. Chemical Communications, 46, 6329-6331.
19. Zhao, X.H, Ma, QJ. Zhang, XB., Huang B., Q. and Zhang, J. (2010) A Highly Selective Fluorescent Sensor for Cu2+ Based on a Covalently Immobilized Naphthalimide Derivative. Analytical Sciences, 26, 585-590.
20. Zhou, Y. Wang, F., Kim Y., Kim, S.J. and Yoon, J. (2009) Cu2+ - Selective Ratiometric and "Off-On" Sensor Based on the Rhodamine Derivative Bearing Pyrene Group. Organic Letter, 11, 4442-4445.

TO PURIFYING OF GROUND WATER AT SUB GROUND LEVEL BY NATURAL METHODS

M. Kavitha Yadav

Assistant Professor, Mahaveer Institute of Science and Technology, Hyderabad.
kavi.dhora@gmail.com

ABSTRACT

Groundwater typically becomes polluted when rainfall soaks into the ground, comes in contact with buried waste or other sources of contamination, picks up chemicals, and carries them into groundwater. Sometimes the volume of a spill or leak is large enough that the chemical itself can reach groundwater without the help of infiltrating water. Heavy metals occur in the earth geological structures, and therefore entire water resources through natural process. For example, heavy rains or flowing water can leach heavy metals out of geological formations. Arsenic occurs in many minerals usually in combination with sulphur and metals, and zinc is chemically similar to magnesium, zinc deficiency such as retardation of growth in children, mail reproduction, according to this two chemicals lead, cadmium plays a major role in groundwater due this pollution. Humans affected with several health disorders like reduce in blood cell production, break up red blood cells in circulation and brain damage. in this circumstances by using natural methods we can clean the ground water and sub ground level by using natural methods with natural products, like corn, coal powder neem bark, wood activated carbon, alum rice husk, and gravel by this method we can purify ground water and control the entering of chemicals into the food chain then we can control the food born diseases.

Keywords: Groundwater, contamination, Arsenic, zinc, corn, coal powder and gravel.

1. INTRODUCTION

Groundwater typically becomes polluted when rainfall soaks into the ground, comes in contact with buried waste or other sources of contamination, picks up chemicals, and carries them into groundwater. Sometimes the volume of a spill or leak is large enough that the chemical itself can reach groundwater without the help of infiltrating water.

Groundwater tends to move very slowly and with little turbulence, dilution, or mixing. Therefore, once contaminants reach groundwater, they tend to form a concentrated plume that flows along with groundwater. Despite the slow movement of contamination through an aquifer, groundwater pollution often goes undetected for years, and as a result can spread over a large area. One chlorinated solvent plume in Arizona, for instance, is 0.8 kilometers (0.5 miles) wide and several kilometers long

The growing population and an increase of industrialization and agricultural production in numerous countries require more and more water of adequate quality. In many regions there is a lack of surface water and severe water contamination is to be found. Shallow groundwater resources are often of insufficient quality and over-exploited. Therefore, it is of high priority to take into consideration all the proved water techniques that could help to reduce the existing disaster.

Artificial groundwater recharge is an approved method that has been improved during the last decades. It has been found that also the new kinds of polluting agents, especially organic compounds, can be minimized or even removed by natural purification processes in the subsurface.

Until recently, this view may have been mostly true. But now groundwater investigators have found contaminants in groundwater supplies, such as industrial and municipal wastes; leaking sewer or septic tank effluent; animal feedlot runoff; Ground water get polluted in many ways in that due to the globalization industries well developed the industries are factories are discharged directly are in directly in to the near water body's that particular contaminants affects the plants and animals living in these in most all cases the effect is damaging not only to individual and populations but also the nature

2. OBJECTIVES

1. To clean the ground water by natural methods.
2. Presently the ground water get polluted by heavy metals from industries and agriculture sector because of pesticides and chemicals.
3. Another hand because of landfill sites the ground water is heavily polluted. Due to lizchade.it is nothing but combination of chemicals and plastic.
4. The lizchade contains heavy metals like arsince, zink, cadmium.

3. IDENTIFICATION OF POLLUTED GROUND WATER

We can identify easily by with naked eye most probably, in that first one is

1. By Oder of the water.
2. By color.
3. By taste

3.1 SOURCE OF POLLUTION

Source of pollution mainly two types

1. Point source of pollution.
2. Non point source of pollution.

3.2 POINT SOURCE OF POLLUTION

In point source of pollution the pollutants discharge into a particular water body from a single source.

3.3 NON POINT SOURCE OF POLLUTION

In non point source of pollution the pollutants discharge into particular water through many ways that means not from the single way.

And ground water pollution due to heavy metals from municipal and E- electronic waste.

The municipal and E-electronic waste release the some heavy metals into the ground this are mainly

3.4 IMPORTANTHEAVYMETALS CONTAMINATED

1. Arsenic
2. Zinc
3. Lead
4. Cadmium

3.5 MEASUREMENTES OF GROUND WATER POLLUTION:

1. Low BOD
2. High DOB
3. High alkanity
4. Low pH

4 HEAVY METALS ENTER IN THE GROUND WATER

Heavy metals occur in the earth's geological structures, and can therefore enter water resources through natural processes. For example, heavy rains or flowing water can leach heavy metals out of geological formations. Such processes are exacerbated when this geology is disturbed by economic activities such as mining.

These processes expose the mined-out area to water and air, and can lead to consequences such as acid mine drainage (AMD). The low pH conditions associated with AMD mobilize heavy metals; including radionuclide's where these are present. Mineral. Processing, operations can also generate significant heavy metal pollution, both from direct extraction processes (which typically entail size reduction - greatly increasing the surface area for mass transfer - and generate effluents) as well as through leaching from ore and tailings stockpiles.

4.1 HEALTH EFFECTES CAUSED BY HEAVY METALS

Soluble inorganic arsenic can have immediate toxic effects. Ingestion of large amounts can lead to gastrointestinal symptoms such as severe vomiting, disturbances of the blood and circulation, damage to the nervous system, and eventually death. When not deadly, such large doses may reduce blood cell production, break up red blood cells in the circulation, enlarge the liver, Color the skin, produce tingling and loss of sensation in the limbs, and cause brain damage Deficiency in zinc interferes with the cell division in sperm. Young boys going through puberty need higher levels of zinc to develop health reproductive organs.

5. PLAN OF THE WORK

Plan of my work is identify the good irrigated land and check that land is suitable for the ground water cleaning by some methods. If that particular land is may be a block soil land it is not completely good for the ground water cleaning process, and if that particular land is may be sand soil. Also not good for the ground water cleaning process because this particular two types of land not having the good capacity to hold the water. another types of soils are good for ground water purification in this project main objective is purify the ground water in ground by natural methods. Plan for this project is we want suitable irrigated land and 10 cement ring And sand and small gravels, cast iron turnings and wood activated carbon, alum oxidation zone,. Neem bark, rice husk ash. In this first step is select the particular land and dig the well with 10 fits depth and in the second step arrange the cement rings one by one without gapes. In the first layer we can spread the sand and gravel, along with the alum .In the next layer we can spred the sand, charcoal, and wood activated carbon. In the 3rd layer add the neem bark and rice husk along with the sand ,in the next level we can treat the water with in oxidation zone. In the fourth step give connections to another empty well. In the fifth step collect ground water from ground by bore well and give the connection to newly form well which one is filled with sand and gravel. Natural purification effects within filter layers and in the subsurface are caused mainly by filtration, sedimentation, precipitation, oxidation-reduction, sorption-desorption, ion-exchange and biodegradation. The oxygen content of the water is decisive for oxidation processes and activities of microorganisms.

The presence of reducing substances such as humic matter, causing a lack of oxygen, is responsible for chemical reductions. pH-value and redox-potential influence these reactions, too. Dissolved compounds, among them also contaminants, can be adsorbed especially by clay minerals, iron-hydroxides, amorphous silicic acid, and organic substances. If the chemical composition of the water changes, desorption may happen Ion exchange processes take place mainly in the presence of organic matter and clay minerals. One kind of ion is exchanged against another in stoichiometric relation.

In this way, contaminating ions can also be fixed at underground.

The forming of ionic and molecular complexes changes the solubility, precipitation and sorption of substances such as heavy metals and organic compounds. Within the layer of filter sand and the aquifer, a great variety of natural microorganisms exist, which are highly involved in rehabilitation processes (Balke and Griebler, Biodegradation, the decay of organic compounds by microorganisms, reduces the amount of organics, no matter they are of natural origin or stemming from contaminations.

6. METHODOLOGY

In this process we totally follow the natural methods the polluted water enter into the newly formed well by pipe lines in this weal the polluted ground water filtered through four layers. The first layer is combination of sand and gravel along with the alum. when the polluted water flow through sand layer the major contaminants are blocked in the sand layer and below the sand layer maize layer is their the maize's are mainly formed with cellulose compounds light in weight .In the second layer we add the sand, charcoal and wood activated carbon aling with maize, and this layer is called as the maize layer. when the water flow through the maize layer some micro organisms and contaminants which are having related compounds of cellulose are blocked here From bottom to second layear spred the sand and neem bark along with the rice husk after the complication of this three layer filtering the can enter in to the oxidation zone. In the fourth layer of gravel qualify the water from micro organisms. In the last step the purified ground water entered in to the newly formed well by using the pipe line give the direct connection to irrigated lands. increases, as well as the removal efficiency.

7. RESULTS

Table1 Sample unpurified water

Sample(un purified water)	P^H	Turbidity NTU	Arsenic (μg/l)	Zinc (μg/l)	Lead (μg/l)	Cadmium (μg/l)
Tap water	9	10	12	15	11	9
Ground water	10	13	13	12	19	12
Surface water	9	11	11	16	16	2

Table 2 Sample purified water

Sample (purified water)	P^H	Turbidity NTU	Arsenic (μg/l)	Zinc (μg/l)	Lead (μg/l)	Cadmium (μg/l)
Tap water	6	4	7	5	9	5
Ground water	7	4	5	1	13	8
Surface water	6	3	5	1	16	4

8. CONCLUSION

For tape water: pH value for un purified water is 9 by using layers method the change in P^H value is 6 and the heavy metals in the water con be decrease for un purified water the arsenic is 12 and after the purification the value of arsenic is 7 and finally the remaining heavy metal concentration is reduces in results zinc for unpurified water is 15 after purification is 5 along with this the cadmium and lead are decreases.

REFERENCE

1. Gokcen, N. A (1989). "The As (arsenic) system". Bull. Alloy Phase Diagrams.
2. Ellis, Bobby D.; MacDonald, Charles L. B. (2004). "Stabilized Arsenic Iodide: A Ready Source of Arsenic Iodide Fragments and a Useful Reagent for the Generation of Clusters". Inorganic Chemistry.
3. Cverna, Fran (2002). ASM Ready Reference: Thermal properties of metals. ASM International.
4. Lide, David R., ed. (2000). "Magnetic susceptibility of the elements and inorganic compounds". Handbook of Chemistry and Physics (PDF) (81 ed.). CRC press.
5. Standard Atomic. Thornton, Papers and Lectures Online. Archetype Publications.
6. Wolf, L., Nick, A., Cronin, A. (2015). How to keep your groundwater drinkable: Safer siting of sanitation systems – Working Group 11 Publication. Sustainable Sanitation Alliance
7. wolf, jennyfer; prüss-ustün, annette; cumming, oliver; bartram, jamie; bonjour, sophie; cairncross, sandy; clasen, thomas; colford, john m.; curtis, valerie; de france, jennifer; fewtrell,

ASSESSING THE CONTRIBUTION OF CLIMATE CHANGE IN AN INTENSE WEATHER EVENT: A CRITICAL RISK MANAGEMENT

Sumaiyah Tazyeen[1], B L Shivakumar [2], Shivakumar J Nyamathi [3]

[1]Research Scholar, [2] Professor, Department of Civil Engineering, R V College of Engineering, Bengaluru, India

[3]Associate Professor, Department of Civil Engineering, UVCE, Bangalore University, Bengaluru, India

sumaiyah_fz@yahoo.co.in

ABSTRACT

Pertaining to the occurrence of extreme weather events in certain locations, it has become immensely significant to perceive the influence of the burgeoning climatic changes on such events to establish a scientific frontier on the deliberate human-engendered activities. While the observational records provide inaccurate and deficient amount of data concerning climate change, there exists a snowballing curiosity from the scientific commonality to facilitate the grappling of the perception that the anthropogenic actions have infinitely aggravated and modified the natural climate resulting in detrimental calamities. It is therefore substantial to apprehend and ascertain the extent of this climate change on the magnitude of the extreme events. An instinctive primary phase in event attribution, perhaps, is to investigate the observations for the purpose of establishing the uncommonness of the event in the past, otherwise to probe into the distribution and relevant characteristics of the condition of the climate which had predominated at the time of the event. In view of the fact that there are strong verifications present, portraying the human influence mounting the probability of several extremely warm seasonal temperatures as well as diminishing the probability of extremely cold seasonal temperatures in various regions across the world. However, the data for human influence on the probability of extreme rainfall events, droughts, and storms seems to be varied. Despite the fact that the study of event attribution has expounded swiftly in the present day, geographical analysis of events continues to be inconsistent as it was grounded on the interests and competences of the distinct research groups. In order to assess the event attribution in a precise manner, the outstanding scientific ambiguities must be strongly weighed and the outcomes could be further interconnected. Reviewing the sequences of past data, trend analysis, and models from previous studies, we will endeavor to clarify the impact of the human activities on the climate that instigated in the extreme event in this paper.

INTRODUCTION

Climate change has unequivocally been proclaimed as the biggest hazard to human health in this era (Mitchell et al., 2016). Ascribing to the ceaseless development of extreme weather events, it has become highly obligatory to corroborate the contribution of the climate change in these events. On every occasion of an extreme weather event, there is habitually an insatiable necessity for a concrete investigation as well as attribution. It further becomes noteworthy to ascertain if the weather event was a result of anthropogenic climate change. Event attribution is substantial in assessing the relative contributions of manifold underlying influences to a climate change with an obligation of statistical confidence (Hegerl et al., 2010). Figure 1 depicts the extreme events that took place across the world in 2015.

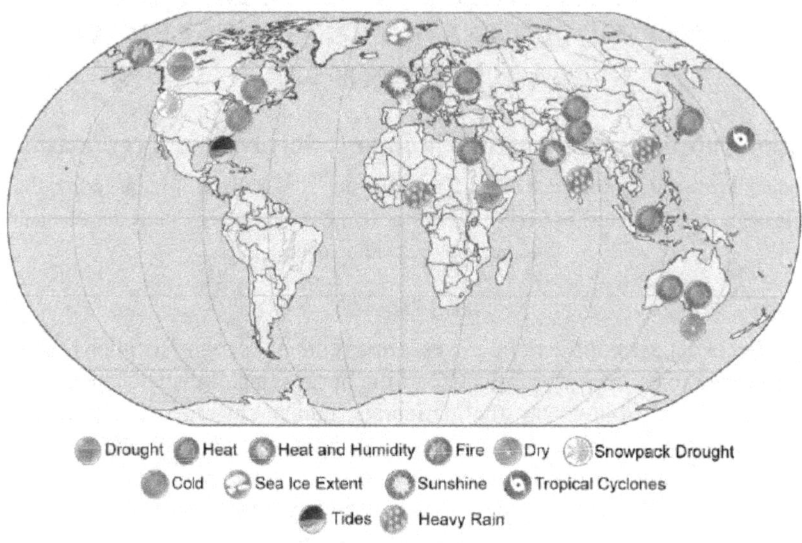

Fig. 1 Extreme events analyzed in 2015 (Herring et al., 2015).

The constraints for risk management mostly are diverse for an extraordinary inimitable event with the exception of a little trend. Event attribution propelled by the climatic changes perhaps has explicit real-world aftermaths i.e. the sooner the contribution of climate change is detected to be implicated in a natural calamity, the more effective the reason for the respective organizations to instigate appropriate measures for acclimatization.

Event attribution to anthropogenic climate change is perhaps arduous operationally, wherein the relevant experts who are required to understand the justifications of an extreme event are not oriented with determining the extent of the event. It is contemplated to be extremely demanding to attribute one extreme weather event to its impelling force subsequent to the incidence. Consider the rise in mean temperature across the world which has become apparent since several years. By means of an array of global climate models, the idea of attributing the warming of the climate to the anthropogenic climate change through simulations is conceivable. An association is plausible if the warming is simulated in case several other aspects such as the examined changes in greenhouse gas and aerosol concentrations, and land-use changes besides the natural drivers are employed in the climate models. This is because, when the natural drivers solely are used, there is no increase detected in these models.

El Niño and La Niña

Extreme events that are habitually accompanying with considerable amount of damage, might differ from year to year as a result of preponderant climatic conditions for instance *El Niño* and *La Niña*. *El Niño* and *La Niña* are two forms of correlated major climatic events that arise from the tropical Pacific, and emerge again in some years as a function of a naturally-occurring cycle. Both these events have significant effects outspreading the entire world, and the related changes in local weather could encompass critical aftermaths (Davey et al. 2011).

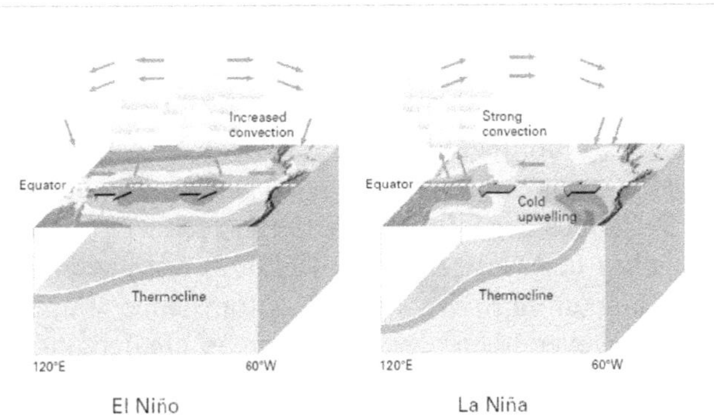

Fig. 2 Circulation patterns during *El Niño* and *La Niña* (WMO, 2014)

El Niño events usually commence in the mid-year with extensive warming of surface water in the central and eastern equatorial Pacific Ocean accompanied by changes in the tropical atmospheric circulation, in other words, winds, pressure and rainfall. The *El Niño*/Southern Oscillation (ENSO), a naturally transpiring event that encompasses oscillating oceanic temperatures in the same region, possesses a huge effect on the climate patterns in several parts of the world. On the whole, *El Niño* attains its highest point between November and January, subsequently deteriorating over the beginning half of the next year. This event befalls generally every two to seven years and survives up to 18 months. Intense and mild *El Niño* events pose a warming result on average global surface temperatures. However, *La Niña*, the counterpart of *El Niño,* meaning "little girl", denotes extensive cooling of the ocean surface temperatures in the same region in the equatorial Pacific, in addition to a turnaround of the superimposing atmospheric situations. In several places, particularly in the tropics, *La Niña* generates the contrary climate disparities to *El Niño* (WMO, 2014). Figure 2 depicts the circulation patterns of both the events.

In this paper, we will analyze different scenarios of extreme events, deviously budging from the causes of the events to the prediction of the events. Discrepancies and tendencies in extreme climatic events have not long ago obtained copious attention mainly due to the ascending economic losses, besides rising number of fatalities as a consequence of such events occurring persistently. A concerning issue in investigating the climate record for changes in extreme events is the insufficiency in finding accurate, long term data.

Literature Review

Outlining the significance of the relationship between the extreme events and climate change, several researchers have studied various attributes and have developed models in order to highlight the risks and pertaining risk management techniques. In this section, we will review some of these researches comprehensively to aid in the interpretation of the contribution of anthropogenic climate change in the extreme weather events.

It has been apparent that climatic changes contribute to extreme temperature and rainfall events that are liable for significant consequences. The extreme hot events result in wildfires additional to upsetting the ecosystem carbon storage (Mills et al., 2013a). Reductions in the number of extreme cold events may as well give rise to springs beforehand, lengthier cultivating seasons along with better crop productivity that has been discerned in the last few decades (Hicke et al., 2002). All at once, it can also cause reductions in the number of chilly days while intensifying of the frost-free zones that could surge the sustenance of several insects and pests in addition to spreading of crop diseases (Bale et al., 2002). Temperature changes in extremes perhaps could even effect human morbidity as well as transience (Luber and McGeehin, 2008; O'Neill and Ebi, 2009). However, there can probably be challenging consequences with an adverse effect from mounting extreme hot temperatures and a positive impact from decreasing extreme cold temperatures (Mills et al. 2013b). Correspondingly, changes in extreme temperature can impinge on energy demand, with rising extreme hot events instigating more running of air conditioners (Miller et al., 2008; McFarland et al., 2013). Nevertheless, declining extreme cold temperatures can lessen the consumption of fuels for heating purposes (Mansur et al., 2008). Further, it was noted that an intensification in extreme rainfall could afflict crop harvests (Rosenzweig et al., 2002) and cause detriment to infrastructures (Wright et al., 2012).

According to the research carried out by Feng and Houser (2015), the longstanding changes in the water cycle unpredictability were analyzed by developing a group of spatially and temporally scalable Water Cycle Indicators (WCI) which were deliberated to compute the changes of water cycle in the circumstances of the warming climate. The study investigated the period between 1979 and 2013 in the contiguous US (CONUS) wherein the usage of the WCI alongside Modern Era Retrospective-Analysis for Research and Applications (MERRA) reanalysis product was established. MERRA postulates the assessments counting over a duration of time of atmospheric conditions and land surface fields. The indicators comprise of six water balance variables that observe the mean conditions and intense facets of the fluctuating water cycle including precipitation, evaporation, runoff, terrestrial water storage, moisture convergence flux, and atmospheric moisture content. Firstly, the mean values are ascertained as the daily total value, during which extremes incorporate wet and dry extremes. These extremes are delineated as the upper and lower 10th percentile of daily distribution. Further, the drifts are measured for yearly as well as seasonal indicators at numerous but diverse spatial degrees. The authors found that there were substantial changes ensuing in a majority of the WCI, likewise the recorded changes seemed to be contingent on geographical and seasonal aspects.

Fischer and Knutti (2015) appraised the portion of the entire world having heavy precipitation and heat intensities that were attributable to warming. The research employs a robust technique on account of the global outlook. The models used were impervious to model prejudices and were explanatory about the mitigation policy, and in that way harmonizing to single-event attribution. The research indicated that even though the current warming was recorded to be around 0.85°C, approximately 18 per cent of the average day-to-day rainfall extremes all across the land were attributable to the perceived rise in temperature from the duration of pre-industrial times that consecutively and largely are the outcomes of human influence. It was further noticed that for every 2°C of warming, a proportion of precipitation extremes was attributable to human influence escalating to nearly 40 per cent. Similarly, in the present day, it was found that around 75 per cent of the average everyday heat extremes across the land were attributable to warming. The researchers concluded that the human-induced activities were for the most part accountable for the extreme events and this ratio intensified exponentially with the warming.

In line with the above, Mitchell et al. (2016), unambiguously enumerated the part of human influence on climate change and human mortality in an extreme heat event using an event attribution framework. The researchers have thoroughly examined the temperature response in 2003 in the entire Europe, as well as the local responses from London and Paris. A number of climatic simulations were executed of a high-resolution regional climate model which let a collection of a wide-ranging statistical report of the 2003 events and the contribution of human activities on them. The study employed the outcomes as an input to a health impact assessment model of human deaths. It was learnt that an extensive dynamical mode of atmospheric variability would continue to be for the most part unaffected under anthropogenic climate change, thus indicating that the direct thermodynamic response was principally answerable for the risen mortality. The anthropogenic climate change was accountable for an upsurge of the heat-related mortality in Central Paris by ~70 per cent and by ~20 per cent in London that had faced lower extreme heat, in the summer of 2003. The total number of deaths related to the heatwave recorded in these two cities were 735 and 315 respectively, of which about 506 in the former and 64 in the latter cities verified were due to the anthropogenic climate change. This implied that the anthropogenic activities driving climate change were largely liable for human catastrophe.

Oldenborgh et al. studied the 2015 Chennai floods that totaled the amount of losses approximately to $3 Billion and the city was affirmed a disaster area following flooding of several areas in just a day's precipitation. The observational analysis carried by the researchers observed no signal for a positive tendency in extreme one-day rainfall at the southeastern coast of India in either between the years 1900 and 1970, nor between 1970 and 2014. Combined models indicated higher amounts of extreme one-day rainfall events from 1970 to 2015, although a huge amalgamation of SST-forced models recurrently depicted no rise in the prospect of extreme one-day rainfall as a result of anthropogenic emissions. One credible cause could be the deficiency of rise in the Sea Surface Temperature (SST) in the western Bay of Bengal in the last four decades that could not be simulated suitably by the combined models, however was found to be highly accurate in the SST-forced model. Contrary to the aforementioned studies, this research disqualifies the attribution of the floods to anthropogenic causes, perhaps to a great amount, because of two major pollutants, greenhouse gases and aerosols, producing antagonistic consequences. While there exists a minor but strong surge in probability of extremes in the SST-forced regional model that is allied with *El Niño* and other SST incongruities, in the annotations, the ENSO signal was as well existing although not statistically noteworthy.

FINDINGS AND CONCLUSION

In this brief review, the present significance of research into the event attribution of intense weather events based on anthropogenic climate change, besides diverse methodologies are appraised. While no single methodology ensures an accurate outcome, the findings were found to be enriched and coherent due to a concrete basis and understanding of the researchers. However, the usage of multiple approaches has facilitated the studies with the evaluations producing the results, majorly indicating the liability of the anthropogenic influence on climate change, portraying evident data for human-induced activities on extreme temperature events relative to extreme precipitation events, droughts and storms. More certainty in attribution of extreme temperature events arise owing to a strong observational basis, the effectiveness of climate models to epitomize the pertinent processes and confirmatory studies reproducing the outcomes.

The multifaceted interface involving the several changes in extreme events and the many effects on economic sectors makes a broad investigation of the climate change effects challenging. This accentuates the necessity for multi-impact model projects to accomplish reliable assessment of climate change effects in India. Further, a robust scientific progress on the understanding and modelling of ENSO can improve the prediction skills approximately months in advance, helping societies to prepare well in advance for the accompanying dangers such as floods and droughts. Also, the improvement of operational event attribution can grant a well-timed and systematic production of attribution evaluations instead of formerly attained on an impromptu basis. For event attribution appraisals to be utmost effective, outstanding scientific ambiguities must be vigorously measured and the outcomes distinctly imparted. Hence, the relentless development of methodologies to measure the consistency of event attribution results is noteworthy along with understanding the conceivable efficacy of event attribution for stakeholders and decision makers.

While event attribution science be that as it may, is rather fledgling, several matters still remain as to existing abilities to strongly ascribe the contribution of anthropogenic climate change to the risk of numerous extreme climate events and it is imperative that climate models endure to be evaluated and enhanced. As this science develops, event attribution must be comprehended as an essential element of climate services to apprise adaptation and mitigation programs across the world and to facilitate climate risk management.

REFERENCES

1. Mitchell D., Heaviside C., Vardoulakis S., Huntingford C., Masato G., Guillod B.P., Frumhoff P., Bowery A., Wallom D., Allen M., 2016, Attributing Human Mortality During Extreme Heat Waves to Anthropogenic Climate Change, Environmental Research Letters 11/074006.
2. Hegerl G.C., Hoegh-Guldberg O., Casassa G., Hoerling M., Kovats S., Parmesan C., Pierce D., Stott P., 2010, Good Practice Guidance Paper on Detection and Attribution Related to Anthropogenic Climate Change, In: Stocker TF, Field C, Dahe Q, Barros V, Plattner G-K, Tignor M, Pauline Midgley P, Ebi K, eds. Meeting Report of the Intergovernmental Panel on Climate Change Expert Meeting on Detection and Attribution of Anthropogenic Climate Change. IPCC Working Group I Technical Support Unit. Bern: University of Bern; 8 pp.
3. Herring S.C., Hoell A., Hoerling M.P., Kossin J.P., Schreck III C.J., Stott P.A., 2016, Explaining Extreme Events of 2015 from a Climate Perspective, Bulletin of the American Meteorological Society, Vol. 97, No. 12, S1–S145.

4. Davey M., Huddleston M., Brookshaw A., 2011, Global impact of El Niño and La Niña, The Lighthill Risk Network, London, United Kingdom.
5. World Meteorological Organization (WMO), 2014, El Niño/ Southern Oscillation, ISBN 978-63-11145-6, WMO, Geneva, Switzerland.
6. Mills D., Jones R., Carney K., St Juliana A., Ready R., Crimmins A., Martinich J., Shouse K., DeAngelo B., Monier E., 2013a, Quantifying and Monetizing Potential Climate Change Policy Impacts on Terrestrial Ecosystem Carbon Storage and Wildfires in the United States. Clim Chang.
7. Hicke J., Asner G., Randerson J., Tucker C., Los S., Birdsey R., Jenkins J., Field C., 2002, Trends in North American Net Primary Productivity Derived from Satellite Observations, 1982–1998. Glob Biogeochem Cycles 16(2):1018.
8. Bale J., Masters G., Hodkinson I., Awmack C., Bezemer T., Brown V., Butterfield J., Buse A., Coulson J., Farrar J., Good J., Harrington R., Hartley S., Jones T., Lindroth R., Press M., Symrnioudis I., Watt A., Whittaker J., 2002, Herbivory in Global Climate Change Research: Direct Effects of Rising Temperature on Insect Herbivores, Glob Change Biol 8(1):1–16.
9. Luber G., McGeehin M., 2008, Climate Change and Extreme Heat Events., Am J Prev Med 35(5):429–435.
10. O'Neill M.S., Ebi K.L., 2009, Temperature Extremes and Health: Impacts of Climate Variability and Change in the United States., J Occup Environ Med 51(1):13–25.
11. Mills D., Schwartz J., Lee M., Sarofim M., Jones R., Lawson M., Deck L., 2013b, Climate Change Impacts on Extreme Temperature Mortality in Select Metropolitan Areas of the United States. Clim Chang.
12. Miller N.L., Hayhoe K., Jin J., Auffhammer M., 2008, Climate, Extreme Heat, and Electricity Demand in California. J Appl Meteor Climatol 47(6):1834–1844.
13. McFarland J., Zhou Y., Clarke L., Schultz P., Sullivan P., Colman J., Patel P., Eom J., Kim S., Kyle G., Jaglom W., Venkatesh B., Haydel J., Miller R., Creason J., Perkins B., 2013, Modeling Climate Change Impacts on Energy Supply and Demand in the United States. Clim Chang.
14. Mansur E.T., Mendelsohn R., Morrison W., 2008, Climate Change Adaptation: A Study of Fuel Choice and Consumption in the US Energy Sector. J Environ Econ Manage 55(2):175–193.
15. Rosenzweig C., Tubiello F.N., Goldberg R., Mills E., Bloomfield J., 2002, Increased Crop Damage in the US from Excess Precipitation Under Climate Change. Glob Environ Change 12(3):197–202.
16. Wright L., Chinowsky P., Strzepek K., Jones R., Streeter R., Smith J., Mayotte J., Powell A., Jantarasami L., Perkins W., 2012, Estimated Effects of Climate Change on Flood Vulnerability of U.S. Bridges., Mitig Adapt Strat Glob Change 17(8):939–955.
17. Feng X., Houser P., 2016, Developing and Demonstrating Climate Indicators for Monitoring the Changing Water Cycle, Hindawi Publishing Corporation, Advances in Meteorology, Volume 2016.
18. Fischer E.M., Knutti R., 2015, Anthropogenic Contribution to Global Occurrence of Heavy-Precipitation and High-Temperature Extremes, Nature Climate Change, Institute for Atmospheric and Climate Science, Zurich, Switzerland.
19. Oldenborgh G.J.V., Friederike E.L.O., Haustein K., Achutarao K., 2016, The Heavy Precipitation Event of December 2015 in Chennai, India, American Meteorological Society, BAMS S87-S91.

QUANTITATIVE AND MORPHOMETRIC ANALYSIS OF SELECTED WATERSHEDS USING GIS

S.P.Nikam[1], P. K. Singh and Pravin Dahiphale

Department of Agricultural Engineering Engineering, College of Agriculture, MPKV, Dhule (MH), India.

ABSTRACT

Development of morphometric techniques was a major advance in the quantitative description of the geometry of the drainage basins and its network which helps in characterizing the drainage network, comparing the characteristics of several drainage networks and examining the effect of variables such as lithology, rock structure, rainfall etc. Development of a geomorphic response model requires some of the important geomorphological characteristics which are to be evaluated for the watersheds. The geomorphologic parameters of a river basin play an important role in modeling various hydrological processes for the determination of soil loss and runoff. Morphometric analysis and their their relative parameters have been quantitatively carried out for the twelve selected watersheds of Tapi basin, Maharashtra, India. The quantitative analysis of the morphometric characteristics of the basin include average slope of the watershed (Sa), elongation ratio (Re), circulatory ratio (Rc), basin shape factor (Sb), relief ratio(Rf), relative relief (Rr), ruggedness number (R_N), main stream channel slope (Sc), drainage factor (Df), stream length ratio (Rl), bifurcation ratio (Rb), and length width ratio (Lbw). These parameters then can be used according to their importance in the basin for development of geomorphological models to study various hydrological processes. The study would help the local functionaries to utilize the resources for sustainable of the basin area.

Keywords: *Geomorphological parameters, Geomorphological Information system, Modelling.*

INTRODUCTION

Geomorphological characteristics of a watershed are commonly used for development of regional hydrological models to resolve various hydrological problems such as prediction of runoff and sediment yield of unguaged watershed or inadequate data situation (Pandey *et al.*, 2004, Sarangi *et al.*, 2004 and Sharma *et al.*, 2010). For predicting runoff and estimating accurate sediment production rate from the known causative factors, it is important to include topographic or geomorphic characteristics which reflect directly or indirectly on climate, geology and transportation processes from the watershed. The rainfall and watershed characteristics in the form of geomorphic parameters can be utilized in the development of reliable response model for predicting runoff and sediment yields from watersheds which are not gauged (Kumar, 1991 and Singh *et al.*, 2009). Leopold and Miller (1956) obtained a geometric progression between discharge and Horton order, by combining the Horton's law of basin area with an empirical relationship between mean annual stream discharge and basin area.

It is extremely difficult and monotonous for the users to derive the geomorphological parameters from toposheets. Thus, it discourages the users from adopting the various regional approaches. To overcome this difficulty, now a day, Geomorphological Information System (GIS) software like ILWIS, ERDAS, Arc INFO and ArcGIS etc are available for derivation of these characteristics in a less time consuming and simplified manner. The use of GIS is increasing in various hydrological applications (Olivera and Maidment, 2004; Jain and

Kothyari, 2000 and Pandey *et al*, 2004). In the recent past, Geographical Information System (GIS) has emerged as comprehensive tool for description of hydrological processes at basin scale, and facilitates easy and accurate determination of the hydrologic and geomorphic characteristics of a watershed using Digitized Elevation Models (DEM) and/or other hydrological data. Binjolkar (2007) and Sharma (2010) have used the GIS software for quantification of various geomorphological parameters of the watersheds.

In recent past, Geographical Information System (GIS) has emerged as comprehensive tool for description of hydrological processes at basin scale and facilitates easy determination of the hydrologic and geomorphic characteristics of a watershed.

MATERIALS AND METHODS

The drainage maps of all the ten watersheds were delineated in the ArcMap 9.3 software. Survey of India Toposheets 46K/4, 46K/10, 46L/10, 46/14, 46P/13, 46P/14, NF43/11 and NF43/12 of Tapi basin were used as references. Digitization and analysis of drainage has been carried out using GIS software (ArcGIS 9.3). Raster data is commonly obtained by scanning toposheets or collecting aerial photographs and satellite images. Scanned map datasets don't normally contain spatial information (either embedded in the file or as a separate file).Thus, in order to use some raster data sets in conjunction with other spatial data, it often needs to align it, or georeferenced it to a map. A map coordinate system is defined by using a map projection (a method by which the curved surface of the earth is portrayed on a flat surface).

A raster dataset, that is, various toposheets of the study area are added to the GIS ArcMap for georeferencing using georeferencing toolbar. After rectification a new dataset will form in GRID, TIFF or ERDAS IMAGINE format. These rectified maps are then further used for creating new digitized layers of watershed boundary, drainage lines and contour lines of selected watersheds.

The attributes were assigned to create the digital database for drainage layer and contour layer of all the watersheds. Selected geomorphological parameters were computed using Arc GIS 9.3 software following the formula suggested by Horton (1945) and Strahler (1957) and well known relationships as presented in Table 1.

Table 1 Selected geomorphological parameters

S. No.	Geomorphological parameter	Formula
S_a	Avg. slope of the watershed	$S_a = \frac{H \sum_{i=1}^{n} L_{ci}}{10 n A}$
R_e	Elongation ratio	$R_e = \frac{2\sqrt{A}}{L_b \sqrt{\pi}} = 1.12838 \sqrt{A / L_b}$
R_c	Circulatory ratio	$R_c = \frac{2\sqrt{\pi A}}{L_p} = 3.544 \frac{\sqrt{A}}{L_p}$
S_b	Basin shape Factor	$S_b = L_b^2 / A$

Contd...

S. No.	Geomorphological parameter	Formula
R_f	Relief ratio	$Rf = \frac{H}{L_b}$
R_r	Relative ratio	$R_r = \frac{H}{L_p}$
R_n	Ruggedness number	$R_n = \frac{H \ D_d}{1000}$
S_c	Main stream channel slope	$S_c = \frac{H_e}{1000 \ L_{ms}} \times 100 = \frac{H_e}{10 \ L_{ms}}$
D_f	Drainage factor	$D_f = F_s / D_d^2$
R_l	Stream length ratio	$\log_{10} \bar{L}_u = a + b\,u;\ R_l = anti\log b$
R_b	Bifurcation ration	$\log_{10} N_u = a - b\,u;\ R_b = Anti\log b$
L_{bw}	Length width ratio	L_b/L_w

RESULTS AND DISCUSSIONS

The morphometric analysis requires measurement of the linear features, gradient of channel network, and contributing ground slopes of the drainage basin. The measurement of various geomorphic parameters of the selected watersheds has been carried out in the GIS environment.

It is observed that the order of main stream channel ranges from 1 to 5. It is revealed from Table 1 that number of streams of a particular order decreases with the increase in stream order. It means that the number of streams of any given order is less than that of immediate lower order but more than the next higher order. However, in general, the average length of the stream of a particular order increases as the order of the stream increases, which means that the mean length of a stream of a given order is greater than that of immediate lower order but lesser than that of the next higher order. Large variations are observed in case of length of perimeter from 21.82 km (W5) to 59.18 km (W6), maximum length of watershed from 7.14 km (W10) to 22.24 km (W8), width of watershed from 3.28 km (W5) to 10.55 km (W6) and the length of main stream channel from 9.72 km (W5) to 26.67 km (W6). The drainage area of the selected watersheds varies from 19.92 sq. km (W7) to 185.81 km2 (W6). The stream frequency and drainage density of the watersheds vary from 0.34 (W2) to 4.16 (W10) and from 0.93 (W2) to 2.69 (W8) respectively. The watersheds under study are observed to varying from very high relief 328.00 m in case of W6 watershed to relief as low as 70.00 m in case of W7 watershed showing a large variation in the elevation of the catchments. The average length of the contour varies from a minimum of 2.59 km in W7 watershed to a maximum of 25.87 km in W6 watershed.

The parameters circulatory ratio and elongation ratio are the indicative parameters of shape of the watershed. They are observed to range from 0.613 (W8) to 0.922 (W2) and 0.482 (W7) to 0.853 (W2) respectively. The elongation ratio for most of the watersheds is greater than 0.5 indicating that the shape of the watersheds is not close to circles. The length-width ratio of the watershed varies greatly between 1.685 (W1) to 4.399 (W8). The basin shape factor varies from

1.751 (W2) to 5.475 (W7). All these factors are indicative of the various shapes of the watershed.

Table 2 (a) Linear and Areal Geomorphic Parameters

S. No.	Watershed	Number of the streams in the order specified					Total no. of streams ∑Nu	Stream length in the order specified, Km					Total stream length, Km ∑Lu
		1st	2nd	3rd	4th	5th		1st	2nd	3rd	4th	5th	
1	**W1**	36	10	3	1	–	50	62.45	28.12	15.88	0.96	–	107.42
2	**W2**	31	8	4	2	1	6	63.33	38.01	13.04	8.89	1.50	124.76
3	**W3**	185	44	13	2	1	245	109.48	51.89	22.40	18.67	4.06	206.50
4	**W4**	94	24	5	1	–	124	70.61	25.02	19.41	5.72	–	120.76
5	**W5**	72	16	4	2	1	95	36.73	18.30	9.82	3.96	1.15	69.96
6	**W6**	48	13	4	2	1	68	126.22	50.76	12.49	18.05	0.85	208.36
7	**W7**	36	7	2	1	–	46	23.94	5.24	3.89	8.28	–	41.34
8	**W8**	216	42	12	4	1	275	148.78	54.03	28.08	22.05	10.66	263.60
9	**W9**	115	27	9	3	1	155	78.65	30.12	19.77	13.97	2.75	145.26
10	**W10**	77	21	3	1	–	102	38.16	13.36	9.21	4.07	–	64.79

Table 2 (b)

S. No.	Water shed	Length of Perimeter, Lp (km)	Maximum Length of Watershed , Lb (km)	Width of Watersh ed, Lw	Main Stream Length, Lms	Drain age Area, A (Sq	Stream Frequen cy, Fs (km^{-2})	Drainage Density, Dd (km^{-1})	Maxi mum Wate rshed	Aver age Cont our
1	**W1**	43.97	14.36	8.52	14.47	99.90	0.50	1.08	284.0	12.8
2	**W2**	44.53	15.34	8.57	16.69	134.32	0.34	0.93	180.0	8.81
3	**W3**	39.41	14.91	7.16	17.96	84.72	2.89	2.44	136.0	14.5
4	**W4**	33.49	13.07	4.62	15.26	51.96	2.39	2.32	274.0	6.58
5	**W5**	21.82	8.58	3.28	9.72	27.15	3.50	2.58	90.00	8.67
6	**W6**	59.18	20.83	10.55	26.67	185.81	0.37	1.12	328.0	25.8
7	**W7**	24.76	10.44	2.55	11.88	19.92	2.31	2.08	70.00	2.59
8	**W8**	57.15	22.24	5.05	25.50	97.85	2.81	2.69	177.0	12.7
9	**W9**	37.58	11.39	5.46	13.36	64.85	2.39	2.24	237.0	3.47
10	**W10**	22.82	7.14	3.99	9.96	24.49	4.16	2.65	157.0	3.62

Table 3 Selected Dimensionless Geomorphic Parameters

	S_a	R_e	R_c	S_b	R_F	Rr	R_N	S_c	D_f	R_l	R_b	L_{bw}
W1	3.652	0.785	0.806	2.064	0.020	0.0065	0.305	1.223	0.433	0.811	3.303	1.685
W2	1.180	0.853	0.922	1.751	0.012	0.0040	0.167	0.217	0.397	0.849	2.280	1.791
W3	2.332	0.697	0.828	2.624	0.009	0.0035	0.332	0.499	0.487	1.127	3.863	2.082
W4	3.472	0.622	0.763	3.286	0.021	0.0082	0.637	0.476	0.442	0.879	4.570	2.826
W5	2.875	0.685	0.846	2.712	0.010	0.0041	0.232	0.518	0.527	1.117	2.890	2.613
W6	4.566	0.738	0.816	2.335	0.016	0.0055	0.368	0.374	0.291	0.800	2.917	1.975
W7	0.909	0.482	0.639	5.475	0.007	0.0028	0.145	0.464	0.536	1.042	3.319	4.101
W8	2.297	0.502	0.613	5.053	0.008	0.0031	0.477	0.448	0.387	1.148	3.707	4.399
W9	1.269	0.798	0.760	2.001	0.021	0.0063	0.531	0.527	0.476	0.968	3.213	2.084
W10	2.317	0.782	0.769	2.080	0.022	0.0069	0.415	0.103	0.595	0.955	4.467	1.788

Bifurcation ratio, stream length ratio and drainage factor are all indicative of the drainage profile of a watershed and range from 2.280 (W2) to 4.570 (W4), 0.799 (W6) to 1.148 (W8) and 0.203 (W5) to 0.546 (W2) respectively (Table 2). The values of bifurcation ratio show large variation within all the watersheds.

The parameters which involve elevation difference are relief ratio, relative relief, ruggedness number, average slope and main stream channel slope. It is seen from computed values (Table 2) that average slope of the watersheds varies from 0.909 per cent (W7) to 4.565 per cent (W5). The value of relief ratio, relative relief, ruggedness number and main stream channel slope are observed to range from 0.0067(W7) to 0.022 (W10), 0.0028 (W7) to 0.0082w4) and 0.145(7) to 0.636 (W4) respectively.

CONCLUSIONS

The GIS based approach is more appropriate than conventional methods for the evaluation of geomorphological parameters. The conventional morphometric analysis is error prone, time consuming and tiresome. It is found that the number of streams of any particular order were more than the next higher order but less than the immediate lower order. However, the mean length of stream of a particular order is greater than that of immediate lower order but less than the next higher order. The similar trends were observed and reported by several other research workers for other watersheds. Large variations were observed in basin shape factor, main stream channel slope and length width ratio. Out of ten watersheds, except the W1, the main stream channel slope has been found less than one per cent. It is observed that watershed relief varies from very high in case of W6 (328 m) to low in case of W7 (70 m). The average length of contour varies from a minimum of 2.59 km for W7 watershed to a maximum of 25.87 km for W6 watershed. The shape of W2 watershed is observed approaching to circle with their elongation ratios close to unity. The shape of the watersheds W1, W3, W5 and W6 is observed to be an oval as elongation rations are between 0.8 to 0.9 while the watersheds W4, W9 and W10 are found to be less elongation. Watersheds W7 and W8 are elongated in shape as elongation ration ranges between 0.5 to 0.7.

These parameters according to their importance can be used for development of geomorphic response models for the basin.

REFERENCES

1. Binjolkar, P. and A. K. Keshari. 2007. Estimating geomorphological parameters using GIS for Tilaiya reservoir catchment. *IE(I) J.* – CV. 88: 21-26.
2. Horton, R.E. 1945. Erosional development of streams and their drainage basin, Hydrological approach to quantitative morphology. Geol. Soc. Am. Bull. 56: 275-370.
3. Jain, M.K. and U. C. Kothyari. 2000. Estimation of soil erosion and sediment yield using GIS. J. Hydrol. Sci. 45(5): 771-786.
4. Kumar, V. 1991. Deterministic modeling of annual runoff and sediment production rate for small watersheds of Damodar Valley Catchment, Indian J. of Soil Cons. 19(1 & 2): 66-74.
5. Leopold, L.B. and J.P. Miller. 1956. Ephemeral streams – Hydrualic factors and their relation to drainage net. *U.S. Geol. Surv. Prof.*: 282-A.
6. Olivera, F. and D. Maidment. 2004. Geographical Information System (GIS) - based spatially distributed model for runoff routine. *Water Resour. Res.* 18(6):1155-1164.
7. Pandey, A., Chowdhary, V. M. and Mal, B. C. 2004. Morphological analysis and watershed management using GIS. J. *of Hydrology*. 27(3), 71-84.
8. Sarangi, A., C.A. Madramootoo and D.K.Singh. 2004. Development of ArcGIS assisted user interfaces for estimation of watershed morphologic parameters. *J. of Soil and Water Cons.* 3(3&4): 139–149.
9. Sharma, S. K., Rajput, G. S., Tignath, S. K. and Pandey, R. P. 2010. Morphometric analysis and prioritization of a watershed using GIS. *J. Indian Water Resource Society,* 30(2), 33-39.
10. Singh, P.K., V. Kumar, R.C., Purohit, M. Kothari and P.K. Dashora. 2009. Application of principle component analysis in grouping geomorphic parameters for hydrological modelling. *Water Res. Manage.* 23: 325-339.
11. Strahler, A.N. 1957. Quantitative analysis of watershed geomorphology. Trans. Am. Geophys. Union 38: 913-920.

PREDICTION MODELING OF COMBINED THERMOPHILLIC COMPOSTING (IN-VESSEL) AND VERMICOMPOSTING IN THE BIOCONVERSION OF VEGETABLE MARKET WASTE

C. C. Monson[1], A. Murugappan[2], K. Rajan[3] and S. Gnanakumar[4]

[1]Former Research Scholar [2] Professor, [4]Associate Professor,

Department of Civil Engineering, Annamalai University, Annamalainagar,

[3]HOD, Department of Computer Engineering, Muthiah Polytechnic College, Annamalainagar, Tamilnadu.

profam@gmail.com

ABSTRACT

Pre-composting (thermophillic) was carried out to sanitize the vegetable market waste for 14 days in a rotary In-vessel and later introduced to earthworms for 16 days of vermicomposting. The integrated approach was carried out to combine the pertinent approaches of both these composting techniques of pre-composting in an In-vessel followed by vermicomposting, to enhance the overall process and product qualities. The substrate taken was vegetable market waste and was composted in seven trials of each inoculums of cow dung, sheep manure, pig manure, sewage sludge, Horse manure, poultry manure, Pancha kavya, along with bulking agents, sawdust, and dry leaves. The outcome of the results with combined In-vessel composting and vermicomposting were taken for modeling using artificial neural network (ANN) approach to quantify the volatile solids with respect to time. The observed physical and chemical parameters were given as input for the model prediction and the Volatile solids and C/N ratio has been derived as output. The results indicate that the system will be of much help in predicting the outcome in the case of large scale composting operations where the input is often heterogeneous and the recipe has to be modified to derive the right outcome. One model was developed with 50% of total data for training and remaining 50% for validation and another model was developed with 75% of total data for training and remaining 25% for validation. Thereby the predicted values of modeling assures to provide a model efficiency of 84.22% for the first and 90.16 for second model, ensuring the combined integrated approach of composting helps in shortening stabilization time of compost and improving its quality.

***Keywords*:** Aerobic composting, Bioconversion, Bulking agents, In-vessel composting, Thermophillic composting, Vermicomposting. Artificial neural network

1.INTRODUCTION

The Predominant problem the world over, that now persists with the society is the management of heterogeneous solid waste, which has reached its peak due to Industrialization, together with the upsurge inhuman population in towns and cities. The processing of organic solid wastes alone into organic manure via traditional thermophillic composting is a technique usually been used to address the issues of environmental pollution, non-reliance on chemical fertilizers, and to sustain natural soil fertility (**Ismail, 1993).** Though, there is continuing search for alternate disposal, composting still becomes the preferable alternative for organic municipal solid waste, that can be adopted for conversion and for its safe disposal (Kaviraj and Satyawati Sharma, 2003; Kim et al,2008; De Luca T.H and De Luca D.K, 1997).

The main problem in such traditional composting is its long process duration, leading to a cause of concern environmentally, but the technology of In-vessel composting when adopted maintaining a controlled atmosphere, provides the high temperature required for the pathogen kill, so that the organic matter get composted in a much shorter duration, in a better oxygenated environment compared to other traditional methods (Haug, 1993). However, due to the heterogeneity of solid wastes, the required quality of major nutrients and minor nutrients may not be achieved by thermophillic composting alone. Therefore, the material that is composted via In-vessel has to be further processed viably through vermicomposting to attain a more refined quality with macro and micro nutrients as well as to obtain a low C/N ratio which is acceptable as plant growth media. Though the rate of In-vessel composting affect the cost effectiveness to a certain extent, it would take care of odors both at the time of processing and process residue levels as well as pass through USEPA-40 part 503 rules for pathogen reduction, besides providing a more stable end product rapidly adaptable for feeding the earthworms.

Pre-composting of wastes before vermicomposting is essential as the initial pH of the substrate is very low due to the initial acidification of the wastes and by the excessive leachate release, when fed to the earthworms which makes them intolerable, but by feeding the stabilized form of substrate, they are able to consume a wide range of organic residues rapidly converting them to humus like, soil building substance in a shorter time period (Tognetti et al, 2005; Singh et al, 2005; Thomas Herlihy, 2000; Ismail, 1997; Ndegwa et al, 2001).

2. Materials and Methods

In this study, composting of vegetable wastes generated from the market in Chidambaram, located at the centre of the town is taken for this study; the central market contributes a major share of nearly 12 tonnes out of the total generation of 39 tonnes daily collected from the entire town. The organic fraction available in the market waste is about 75% and is taken for the study and investigated by pre-composting through In-vessel and further stabilized by vermicomposting. The substrate taken was vegetable market waste along with sawdust and dry leaves as bulking agents.

3. IN-VESSEL COMPOSTING FOLLOWED BY VERMICOMPOSTING

In-vessel composting

The cylindrical vessel used for the bio conversion of the wastes was designed and fabricated for a capacity of 1000 litres as shown in Figure 1. It was made of fibre reinforced glass material to take care of the load, as well as the temperature and chemical changes taking place at the time of the conversion process.

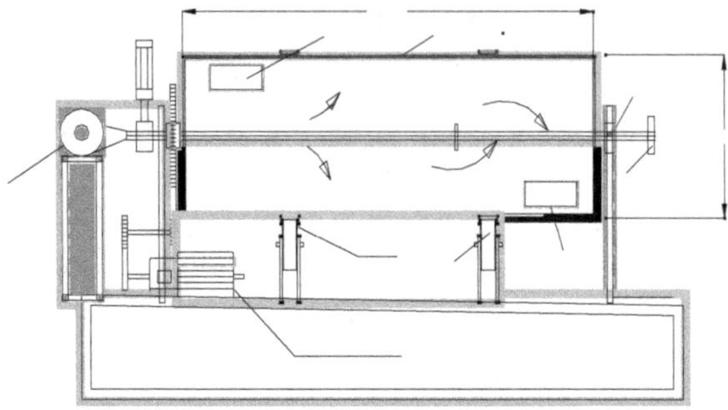

Fig. 1 Rotary In-vessel Composter.

The airflow was given by a blower through a perforated central pipe regulated in such a way that the requirement of 1 gram of air to 1 gram of substrate is met in an aerobic composting, so that the optimal quantity of air was supplied to degrade the substrate (**Haug, 1993; Diaz et al, 1993).** The vessel was rotated by a large gear chain driven by a 1 H.P. motor at a normal speed of about 3 rpm, facilitated by a reduction gear arrangement and was rotated intermittently for a total period of 12 hours in a day, run in four spells with 3 hours rotation and 3 hours idle condition.

Waste material collected from the vegetable market was sorted manually, to ensure that it contains no oversized materials and undesirable materials. The material was manually shredded to a size of 20-30 mm, to give a better exposure for microbial treatment and to expedite the ensuing metabolic process.

In this method, the market vegetable waste was composted in the In-vessel for 14 days in seven trials each with inoculums of cow dung, sheep manure, pig manure, sewage sludge, Horse manure, poultry manure, Pancha kavya, added along with bulking agents, sawdust, and dry leaves cow dung along with bulking agents, namely dry leaves and saw dust. The proportions of feed components are adjusted to satisfy the energy balance to avoid rate limitations caused by lack of moisture, lack offree air space or low nutrient levels **(Haug, 1993)**. The feed stock composition of vegetable market waste, manure and bulking agents were arrived with the constituents of Carbon to Nitrogen well below 35 as per **Tom Richard, 2000.** In-vessel composting was carried out for 14 days and the parameters of the substrate were periodically analysed once in every two days.

4.VERMICOMPOSTING

The release of ammonia and the sudden lowering of pH due to the acidification in the initial stages necessitated the pre-composting before the administering of earthworms (**Nair et al, 2006**; **Singh et al, 2005**). The vermicomposting was carried out in concrete tubs of 50 litre capacity covered with a lid of wire mesh to provide shade as well as to prevent birds and insects to feed on the worms as shown in figure 2.

The problems due to ants were well catered by greasing the outer edge of the concrete tub and the lid. To collect the vermiwash, a tap was provided at the bottom and natural aeration is facilitated through the openings beneath the covered lid for the circulation of air. Periodical watering was done to make up the moisture content nearly to 60-5%. Excess water was prevented by collecting the vermiwash periodically, as it will endanger the survival of the species. (**Thomas E.Hearlihy, 2000).**

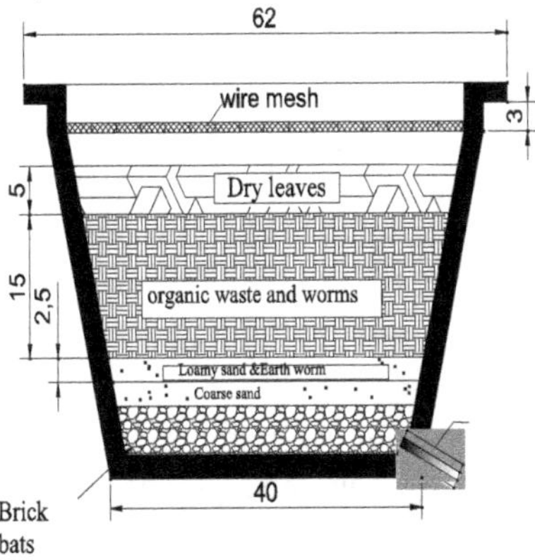

Fig. 2 Vermicomposting Bin.

The pre-composted waste from the in-vessel was transferred to the vermibins when the substrate has cooled down. The earthworms were fed into a 10 kg of In- vessel pre- composted waste laid over a bed of brick bats and fresh river sand. ***Eudrilus eugeniae*** is the species of earthworms suggested to survive in our environment and consume large amount of waste As per the recommendations given by **Ndegwa et al (2001)** that 1.6 kg worms /m^2 being the optimal worm stocking density, nearly half a kg of worms were fed to the waste in the bin. Nearly 20 to 30 earthworms of adult variety per kg of waste were provided in each bin.

5.MODELING APPROACH

The modeling of combined composting and vermicomposting process is done to quantify the volatile solids degradation with respect to time. The modeling approach uses Artificial Neural Network (ANN) technique to predict and determine the volatile solids remaining in the system at any time. The data are obtained from the trials conducted to determine the optimum inoculate of different animal manures, where combined composting in an In-vessel for 14 days and vermicomposting with earthworms for 16 days, were carried out. The outcome of this combined composting was taken for this modeling. The observed physical and chemical parameters were given as input for the model prediction. The measured data given as inputs for modeling were inoculum, days, temperature (° C), pH, and moisture content (%) and C/N ratio taken during the composting and vermicomposting process. The volatile solids content was the output derived from modeling.

An ANN model was developed to illustrate the process of composting and vermicomposting combined together. Two trials were carried out, one with 50 % of data collected for training the ANN and the remaining data for the validation of the model and in the second trial, 75% data collected was taken for training and the remaining 25% for validation. The model performances of two trials were determined using statistical indices like Root Mean Square of Error (RMSE) and model efficiency.

6. ARTIFICIAL NEURAL NETWORK MODEL

Artificial Neural Network technique was used to predict the volatile solids present in the system at any time. Artificial Neural Network models were developed using the MATLAB R2007b software. Literatures suggest the usage of greater than 50% to 80%of the total data for training (Patterson D. 1996). The greater the percentage of data for training, the better would be the predictions in validation part. In this work, one ANN model was developed with 50% of the total data for training and the remaining 50% for validation and in another model 75% of training data was taken for training and the remaining 25% for validation. Among all parameters used in the model, C/N ratio varies over a wide range, with moisture content, temperature and pH. The sensitivity of the model to C/N ratio was verified by the two trials. The normalization of data was done prior to segregation into training as well as validation part in order to avoid the possibility of error due to extrapolation.

Single hidden layer is used. the training parameters used for transfer function, number of epochs, maximum gradient, learning rate coefficient and minimum gradient assigned were 'Transig', 6000,1.0e^{-}3,0.05,and1.000e-0.010 respectively. These were assigned based on trial and error. The numbers of neurons used in the hidden layers are 12.

Once the training of ANN model was over, the remaining dataset was used for the validation of the model. The output was compared with the measured data to find the efficiency of ANN model. Simple statistical indices namely Root Mean Square of Error (RMSE) and model efficiency were calculated to determine model performance. The following formulae were used to calculate the statistical indices.

$$\text{RMSE} = \sqrt{\frac{\sum (y-\hat{y})^2}{n}},$$

$$\text{Model\_efficiency} = 1 - \frac{\sum (y-\hat{y})^2}{\sum (y-\bar{y})^2}$$

Where $y, \hat{y}, \bar{y}$ are measured, predicted and mean values of the volatile solids of validation dataset and n is the number of observations. The Performance indices of ANN model attempted are given in Table 1.

Table 1 Performance of ANN models

Model	Description of data set	Correlation Coefficient		RMSE	Model Efficiency (%)
		Train data	Test Data		
Trial 1	Trial 1 -50% Data for training and 50% Data for Validation	0.9762	0.9462	2.4316	84.22
Trial 2	Trial 2 -75% data for Training and 25% Data for validation	0.9767	0.9504	2.062	90.16

6.1 PERFORMANCE OF ANN MODELING

The Figure 3 and 4 shows the trial 1 plot of measured VS% vs. predicted VS% for the trained and validated data. The statistical indices for the trained data points taken for trial 1 are slightly higher than that of trial 2 shown in figure 5 and 6 and their values of correlation coefficient are 0.9767 and 0.9762. Similarly Figure 4 shows the trial 1 plot of measured VS% versus predicted VS% of test validated data which are found to be 0.9462 slightly lesser than 0.9504.The Root Mean Square Error are calculated as shown in equation 1 and the model efficiency is arrived from the equation in 2. Both the models showed that the lesser the number of data chosen for training, the performance of the model decreases and if higher percentage of data are chosen, the model efficiency achieves a value of 90.16%. In the second trial the model efficiency reaches to 84.22% which has taken only 50% data for training and 50% for validation.

6.2 CONCLUSION

The Figure 8 shows the sample input details of inoculate as 1(cow dung), day as 6, temperature as 56 °C, pH as 6.5, moisture content as 60% and C/N ratio as 23. The predicted Volatile solids reduction was 53.32 for the input sample given. This is shown in the output message box shown in Figure 8. The volatile solids prediction through ANN modeling has the advantage particularly in large scale composting facilities where the incoming waste is heterogeneous as the expected time duration has to be known before commencing the operation. Similarly The C/N prediction through ANN modeling shown in figure 7, has also the advantage particularly in large scale composting facilities where the incoming waste is heterogeneous as the expected time duration has to be known before commencing the operation.

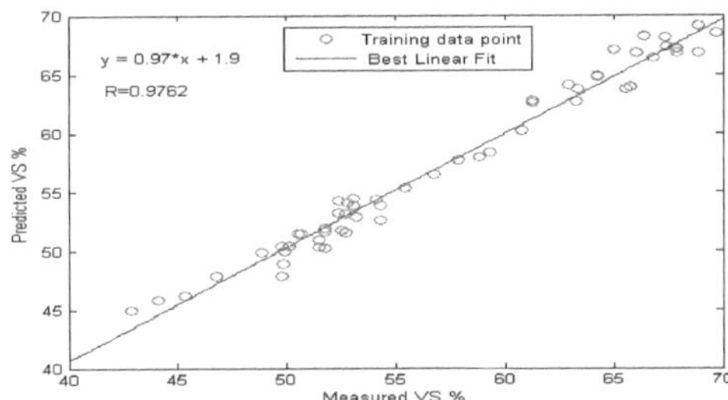

Fig. 3 Trial 1- Plot of Measured VS% versus Predicted VS% (50% Data points selected for training)

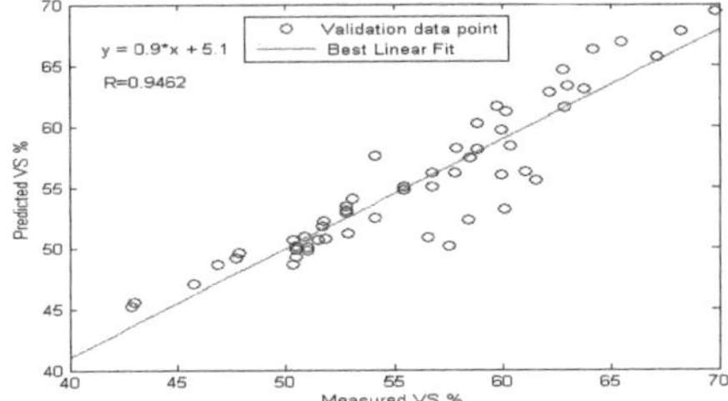

Fig. 4 Trial 1- Plot of Measured VS% versus Predicted VS% (50% Data points selected for validation)

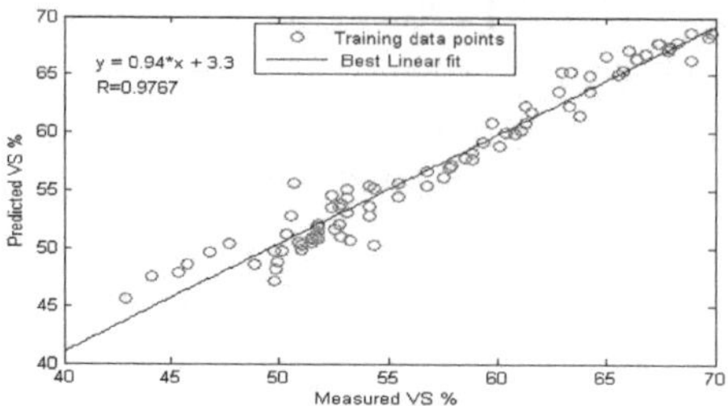

Fig. 5 Trial 2- Plot of Measured VS% versus Predicted VS%

(75% Data points selected for training)

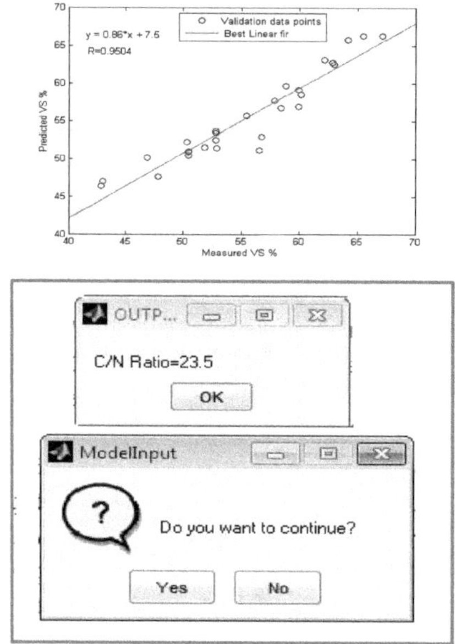

Fig. 6 Trial 2 Plot of Measured VS% versus Predicted VS%(25% Data points selected for Validation)

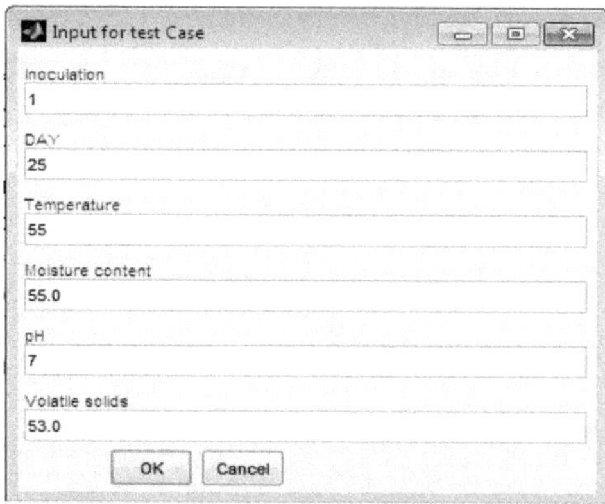

Fig. 7 Input and output of prediction modeling using Volatile solids as input and C/N ratio as output

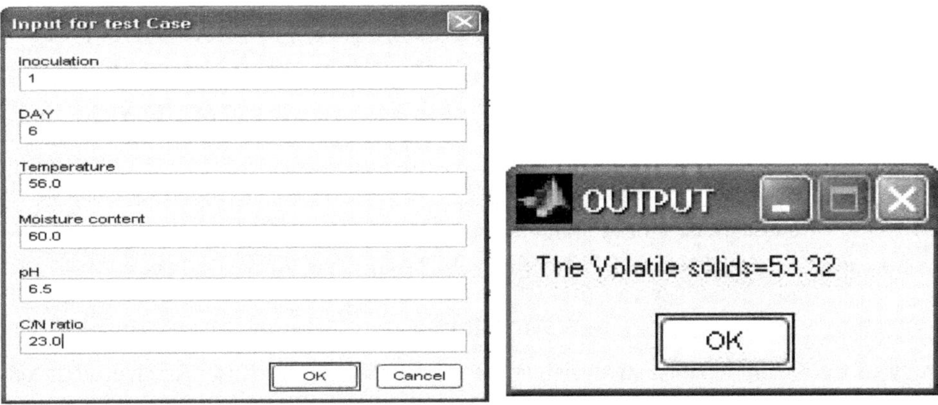

Fig. 8 Input and output of prediction modeling using C/N ratio as input and Volatile solids as output

REFERENCES

1. APHA-AWWA., (2005). Standard Methods for the Examination of Water and Waste water.20th edn. APHA /AWWA/WPCF, Washington. DC.
2. Deluca, T.H. and Deluca, D.K. (1997). "Composting for feedlot manure management and soil quality". *Journal of production in Agriculture*, 102: 235-241.
3. Haug, R.T. (1993). The practical handbook of compost engineering, Lewis Publishers, Florida, USA.
4. Ismail, S.A. (1997). "Vermicology - The Biology of Earthworms", Orient Longman, Chennai, India. ISBN 81-250-10106. p 90.
5. Kaviraj Pradhan and Satyawati Sharma, (2003), "Municipal solid waste management through vermicomposting employing exotic and local species of earthworms", Bioresource Technology 90: 169-173.
6. Kim Joung-Dae, Joon-Seok Park, Byung-Hoon In, Daekeun Kim Wan Namkoon (2008). "Evaluation of pilot-scale in-vessel composting for food waste treatment", Elsevier publications, *Journal of Hazardous Materials*, 154: 272–277
7. Nair J., Seikiozoic V. and Anda M. (2006), "Effect of pre-composting on vermicomposting of kitchen waste", *Bioresource Technology*, 97, 2091-2095.
8. Ndegwa P.M. and S.A.Thompson. (2001). "Integrating composting and vermicomposting in the treatment and bio-conversion of biosolids", *Bioresource Technology*, 76, 107-112
9. Thomas E. Herlihy (2000). "Vermicomposting of organic wastes", Joyce Engineering 2301.w Meadow view Rd, Suite 203, Greensbors K C 27407.
10. Tognetti, C., Mazzarino, M.J. and Laos, F. (2007). "Improving the quality of municipal organic waste compost", *Bioresource Technology*, 98 (5), 1067-1076.
11. Tom Richard and Nancy Trumann (2000). Dept of Agricultural & Biological Engineering, Cornell composting science and Engineering, Cornell University
12. USEPA - 40 CFR Part 503 USEPA (1994). Land Application of Biosolids [online]. Availablefrom:http://www.epa.gov/owm/mtb/biosolids/503pe

GEOSPATIAL APPROACH FOR ANALYSIS OF GEO-MORPHOMETRIC PARAMETERS IN THE PURNA WATERSHED IN AKOLA DISTRICT OF MAHARASHTRA

Kanak Moharir[1], C. B. Pande, R. S. Patode, M. B. Nagdeve and Ranee Wankhade

[1]Sant Gadge Baba Amravati University Amravati, (MS), India

All India Coordinated Research Project for Dryland Agriculture, Akola, Dr. PDKV, Akola

ABSTRACT

The study of Geo-Morphometric analysis can be used to take information for hydrology related project such as watershed planning, agriculture development and artificial recharge conservation structures in the saline zone area. In this study various landforms like younger and older alluvial, mesa and butte has been identified. During morphometric analysis parameters were conceded out from digital elevation model using geospatial technology. The drainage characteristics have been classified based on linear, aerial and relief aspects using survey of India toposheet and satellite data. The watershed boundary and drainage network lines have been delineated from SRTM with 30 m resolution through Arc hydro tools in the extension of Arc GIS 10 software. The drainages pattern has been observed such as dendritic and sub-dendritic drainage types in the Purna watershed area. The stream order map was prepared from drainage network with reference of LISS-III satellite images. The drainage data were checked by ground data using GPS instruments. This study based on geospatial approach is useful for analysis of different morphometric characterization for development of groundwater and thereby the agricultural development. In this watershed stream orders ranges has been observed such as first to fourth orders using Arc GIS 10.3 software. These results should be applicable for planning of artificial recharge structures, groundwater recharge structures and thereby sustainable agriculture development in the Purna watershed area.

Keywords: *Geo-morphometric, GIS, Recharge, Watershed.*

INTRODUCTION

The groundwater and surface water issues are now being faced by the society is growing in world's population and its impact on the availability of fresh water and water planning and management. Presently high inhabitant's expansion, fast urbanization and climate change along with the irregular frequency and intensity of rainfall make difficulty for appropriate water management and storage plans. Therefore, there is an urgent need for the assessment of water resources planning and management because they play an important role in the sustainability of living being and regional economics throughout the world (Singh et al. 2011, 2013). In the present day, Geographical Information System (GIS) and remote sensing technology has been used in examining various topography area and morphometric parameters were used of the watersheds development and drainage basins analysis, as they provide a flexible environment and a powerful tool for stored information from of spatial and non-spatial data (Nag and

Chakraborty 2003). Morphometric analysis is considered to be the most appropriate technique because it enables (i) an understanding of the association of different aspects within a drainage basin characteristics (ii) a relative estimation to be made of different drainage basins parameters has been developed for the different geo-morphological and groundwater regimes (iii) the description of certain useful variables of drainage basins analysis in arithmetical terms (Krishnamurthy et al. 1996). The watershed development and management was related to uplands, low lands, land use, geomorphology, slope and soil parameters have been most favorable for soil and water conservation structures. Soil and water conservation are the main themes in the watershed management while delineating watersheds from digital elevation model with the help of spatial analysis tools. Drainage basin analysis is totally depend on drainage network is very significant role in the micro-watershed planning. Morphometry analysis is the measurement and arithmetical analysis of the configuration of the Earth's surface, shape and dimension of its landforms (Clarke 1966). The drainage basin analysis has based on linear, areal and relief aspects of watershed characteristics (Biswas et al. 1999). The results of this study can be utilized for control of soil erosion, groundwater quality improvement, groundwater regime and suggested rainwater harvesting activity in the Purna region.

STUDY AREA

The Purna watershed is located in the Akola District of Maharashtra and is between 20°32′30"N latitude and 77°19'0''E longitude. The total area is 35.65 Sq. Km (Fig.1). The study area has been occupied by younger and older alluvium and Deccan basalt rocks region which are horizontally disposed and it is traversed by well-developed sets of joints. The study area consists of black and red soils.

METHODOLOGY

The Purna watershed map boundary was prepared from SRTM data with the reference of Survey of India Topographic Maps on a 1:50,000 scale (55D/7 and 55D/9) (Table 1). During drainage network lines were delineated from digital elevation model with the reference of satellite data and Survey of India (SOI) topographical maps. The results have been used by Strahler's system for classification labels a stream segment with no tributaries as a first-order stream using Arc GIS 10.3 software. Where two first-order streams segments have been joined together then created a second-order stream segment and so on. The fourth stream order as the highest stream has measured using geospatial technology. The drainage network map was updated from satellite data and field verification. During the geo-morphometric parameters analysis based upon standard mathematical formulas has been used to assess the different drainage characteristics likes stream number, stream order, stream length, stream length ratio, bifurcation ratio and basin length have been analyzed.

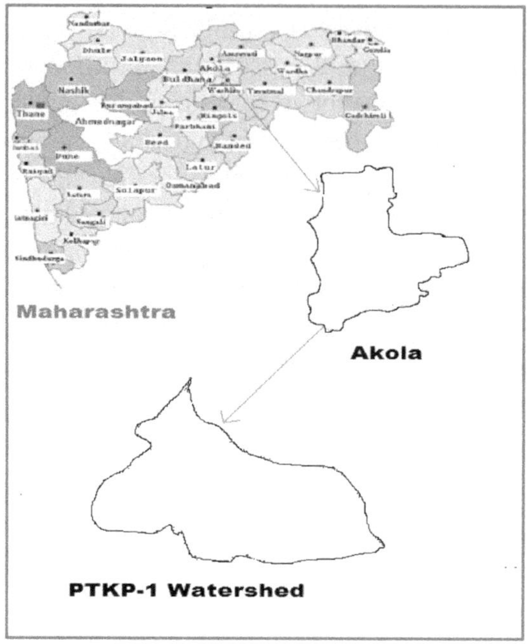

Fig.1 Location map of Purna watershed

Table 1 Details of data product and source

S. No.	Data Product	Source
1	Satellite Data National Remote Sensing, LISS- III of IRS IC and IRS - P6 (Raw data) Agency Government of India.	Bhuvan Portal freely available satellite data in National Remote Sensing IRS ID-PAN, LISS-III of IRS1D Agency Government of India site.
2	Details of Topo-sheets data	Survey of India. Government of India, Topo-sheets no. 55 D/7, 55 D/9
3	SRTM DEM	30 m Resolution
4	Field Data	GPS through intensive field work feature verification.

RESULT AND DISCUSSION

Geo-morphometric parameter analysis

The present study of morphometric parameters such as stream order, stream length, mean stream length (Lsm), bifurcation ratio (Rb), stream number and weighted mean bifurcation ratio etc. has been measured using geospatial technology (Fig. 2). During analysis of bifurcation ratio (Rb) may be defined as ratio of the number of stream segments of a given order to the number of segments of the next higher. Bifurcation ratio values ranging between 3.96 and 9.66 are considered to be characteristics of the watershed, which have experience minimum structural disturbances (Strahler, 1964). The, mean bifurcation ratio is observed as 5.54 from stream numbers and orders using spatial technology. This indicates that the drainage pattern of the basin has not been affected by structural disturbances and the observed Rb is not the same from one order to its next order. These irregularities depend upon the geological and lithological development of the watershed.

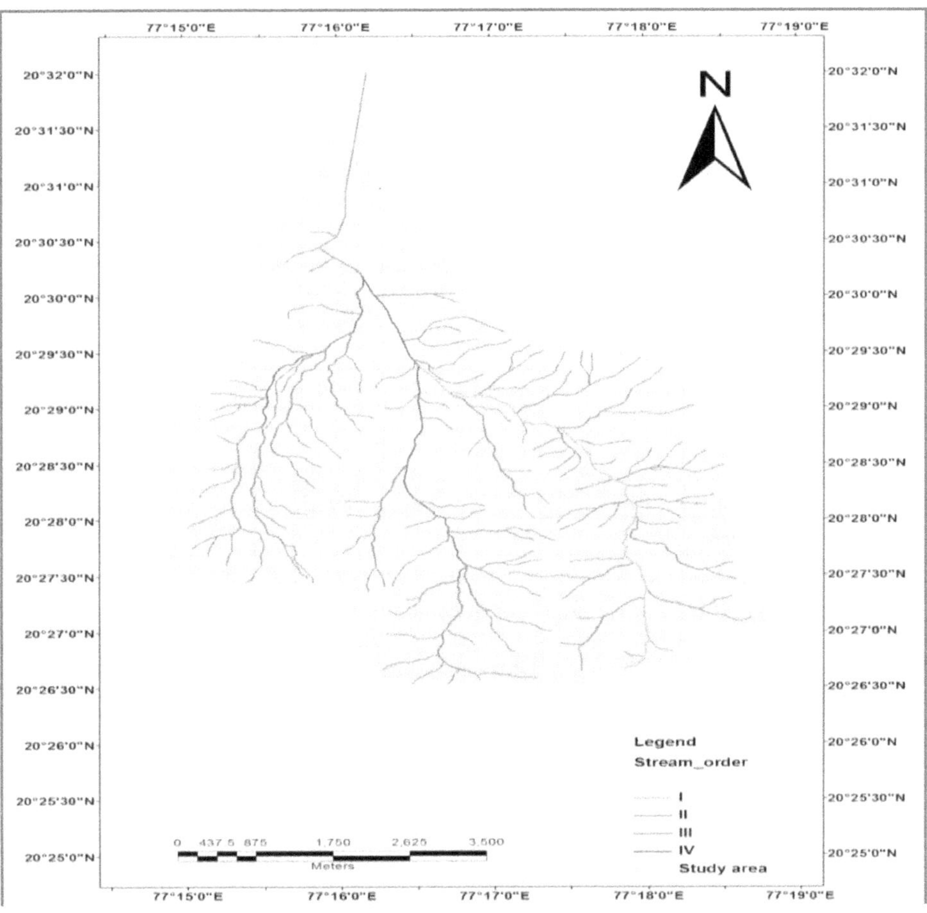

Fig. 2 Stream order map of Purna watershed

The first step for drainage-basin analysis is designation of stream orders in Purna watershed. The stream number first introduced by Horton and modified by Strahler method has been adopted during stream ordering (Schumms 1956). Each segment of the stream was numbered starting from the first order to the maximum order present in each of the sub-basins. After numbering, the drainage-network elements are assigned their order and numbers, the segments of each order are counted to yield the number Nu of segments of the given order. The ratio of stream number of segments with a given order Nu to the number of segments with the higher order Nu+1 has been termed the bifurcation ratio Rb, thus Rb = Nu/Nu+1. According to Horton's law of stream numbers, a plot of stream order against stream numbers has been plotted on a semi-log sheet reveals the slope of the fitted regression of stream order vs numbers of stream segments. The bifurcation ratios normally shows range between 3.0 and 5.0 was observed from drainage network and watershed characteristics.

The mean length Lu of stream segment of order u is a dimensional property, which reveals the characteristic size of components of a drainage network and its contributing basin surfaces. Each of the channel lengths was measured using a digital curve meter. According to Horton's law of stream lengths a plot of logarithm of stream length (ordinate) as a function of order (abscissa) will yield a set of points lying along a straight line. This indicates that the basin evolution follows the erosion laws acting on geologic material with homogenous weathering-erosion characteristics. Any deviation in the points may be due to structural control of the streams. A graph of stream order against stream length (ordinate) plotted on a semi-log sheet reveals a linear relationship (Table 2 and 3).

Table 2 Details of Drainage basin characteristics

S_u	N_u	R_b	N_{u-r}	R_b*N_{u-r}	R_{bwm}
I	115	---	---	---	
II	29	3.96	144	570.24	
III	3	9.66	32	309.12	4.95
IV	1	3	4	12	
Total	148	16.62	180	891.36	
Mean		5.54			

(S_u: Stream order, N_u: Number of streams, R_b: Bifurcation ratios, R_{bm}: Mean bifurcation ratio*, N_{u-r}: Number of stream used in the ratio, R_{bwm}: Weighted mean bifurcation ratios)

Table 3 Details of Drainage length parameters ratio

S_u	L_u	L_u/S_u	L_{ur}	L_{ur-r}	L_{ur}*L_{ur-r}	L_{uwm}
I	61.84	1.85				
II	11.36	2.55	2.24	73.2	163.96	6.39
III	19.80	0.15	3.10	31.16	96.59	
IV	60.43	0.016	3.5	80.23	280.80	
Total	153.43	4.56	8.84	84.59	541.34	
Mean			2.21			

(S_u: Stream order, L_u: Stream length, L_{ur}: Stream length ratio, L_{urm}: Mean stream length ratio*, L_{ur-r}: Stream length used in the ratio, L_{uwm}: Weighted mean stream length ratio)

CONCLUSION

Geo-morphometric analysis of drainage basin parameters has been requirement to any hydrological project. Thus, purpose of drainage networks' behaviour and their interrelation with each other is of great significance in any sustainable water resources development studies. Geospatial satellite data and GIS techniques have been proved to be a useful tool in watershed boundary, drainage parameters and drainage delineation. The drainage pattern shows dendritic and sub-dendritic drainage pattern was observed using satellite data and field verification. The difference of stream length ratio might be due to variations of elevation and slope changes. The results of morphometric analysis provide information about catchment growth on priority basis and areas susceptible for land degradation in the Purna watershed.

REFERENCES

1. Biswas S., Sudhakar S. and Desai V. R. (1999), Prioritization of sub-watersheds based on morphometric analysis of drainage basin, district Midnapore, West Bengal, Journal of Indian Society of Remote sensing, Vol. 27(3) pp. 155-166.
2. Burrough P. A. (1986), Principles of geographical information systems for land resources assessment, Oxford University Press, New York, pp. 50.
3. Clarke J. I. (1966), Morphometry essays in geomorphology. Elsevier Publ. Co., New York, pp. 235–274.
4. Horton R. E. (1932), Drainage basin characteristics. Trans Am Geophysics Union vol. 13, pp. 350–361
5. Kanak N. Moharir, Chaitanya B. Pande (2014), Analysis of morphometric parameters using remote-sensing and GIS techniques in the Lonar nala in Akola district, Maharashtra, India. International Journal for Technological Research in Engineering, vol. 1, Issue 10.
6. Krishnamurthy, J., Srinivas, G., Jayaraman, V. and Chandrashekar, M. G., (1996), Influence of rock types and structures in the development of drainage networks in typical hard rock terrain. ITC Journal, vol. 3-4, pp. 252-259.
7. Miller, V. C. (1953), A quantitative geomorphic study of drainage basin characteristics in the clinch mountain area, Technical Report 3, Department of Geology, Columbia University.
8. Nag S. K, Chakraborty S. (2003) Influence of rock types and structures in the development of drainage network in hard rock area, Journal of Indian Society Remote Sensing, vol. 31(1), pp. 25-35.
9. Schumms, S. A. (1956), Evaluation of drainage systems and slopes in Badlands of perth Amboy, New Jersey. Geological Society of America, Bulletin, vol.67, pp. 597-646.
10. Singh P, Kumar S, Singh U (2011), Groundwater resource evaluation in the Gwalior area, India using satellite data: an integrated geo-morphological and geophysical approach, Hydro-geological Journal, vol. 19, pp. 1421–1429.
11. Singh P, Thakur J. K., Singh U. C. (2013), Morphometric analysis of Morar river basin, Madhya Pradesh, India, using remote sensing and GIS techniques. Environ Earth Science Springer vol. 68, pp. 1967–1977
12. Strahler A. N. (1964,) Quantitative geomorphology of drainage basins and channel networks. In: Chow VT (ed) Handbook of applied hydrology. McGraw Hill Book Company, New York, section pp. 4–11.

OCCURRENCE AND DISTRIBUTION OF GROUNDWATER IN WARDHA RIVER SUB-BASIN, CHANDRAPUR DISTRICT, MAHARASHTRA STATE, INDIA: A GIS APPROACH

Manish S. Deshmukh[1] and Nalanda G.Taksande[1]*

[1]Post Geaduate Department of Geology, Rashtrasant Tukadoji Maharaj Nagpur University, Nagpur, Maharashtra State, India

taksande_nalanda@Rediffmail.com

ABSTRACT

During this study WR-16 watershed of wardha river basin flowing through Chandrapur district, Maharashtra state, India has been studied with respect to pre-monsoon and post-monsoon groundwater levels, seasonal groundwater fluctuation, type of aquifers and lithological control over the occurrence and distribution of groundwater in the area. The groundwater assessment reveals that the watershed falls in safe category with 18.66% stage of development, indicating further scope for groundwater development which is attributed to the assured rainfall of the area (GSDA,2014).Well inventory of 34 wells has been carried out with collection of hydrogeological information regarding depth, diameter of the wells, pre-monsoon and post-monsoon water levels, lithology, type of aquifer, water level fluctuation, pumping capacity, canal command area if any, area irrigated by the well etc. After analysing hydrogeological data, it is observed that pre-monsoon water levels ranges between 2.5 to 17.4 meter (bgl), post-monsoon water levels ranges between 0.5 to 9.4 meter and seasonal water level fluctuation ranges between 1.8 to 10 meter.Accordingly pre-monoon and post-monsoon water level maps has been prepaired to study variation in ground water levels within the watershed. It is observed that the occurrence and distribution of groundwater in this area is dominantly controlled by the lithological variations and not by the topography, slope and relief.

Keywords: Ground Water, Wardha River, GIS, Aquifer.

INTRODUCTION

The demand for drinking water has increasing day by day and will increase with rapid growth of population, agriculture and industry.As a result, fresh water resources depleting day by day.The requirement of drinking water per person is about 2.7 liture/day,thus the global requirement is about 16.5 billion liture/day.The agriculture is one of the major consumer of freshwater resources(Santra,2005).Nearly 97.2% of water on the Earth is saline,2.15% in the form of ice, 0.6% occurs as groundwater, and only 0.01% of fresh water in streams and lakes. But groundwater levels are depleting due to excessive withdrawal specially for irrigating cash crops and for industries. During this study WR-16 watershed of wardha river basin has been studied with respect to pre-monsoon and post-monsoon groundwater levels, seasonal fluctuation in groundwater levels, type of aquifer, lithological control over the occurrence and distribution of groundwater to study the groundwater availability and behaviour in the area.

Study Area:

The WR-16 watershed selected for study is a part of wardha river basin, flowing through Chandrapur district of Maharashtra state, India. The geographical area of the watershed lies between the Latitude 19°38’54" N to 19°53’45" N and Longitude 79°09′30"E to 79°20′45"E (Figure 1). The watershed covers an area of about 298 Sq.Km.

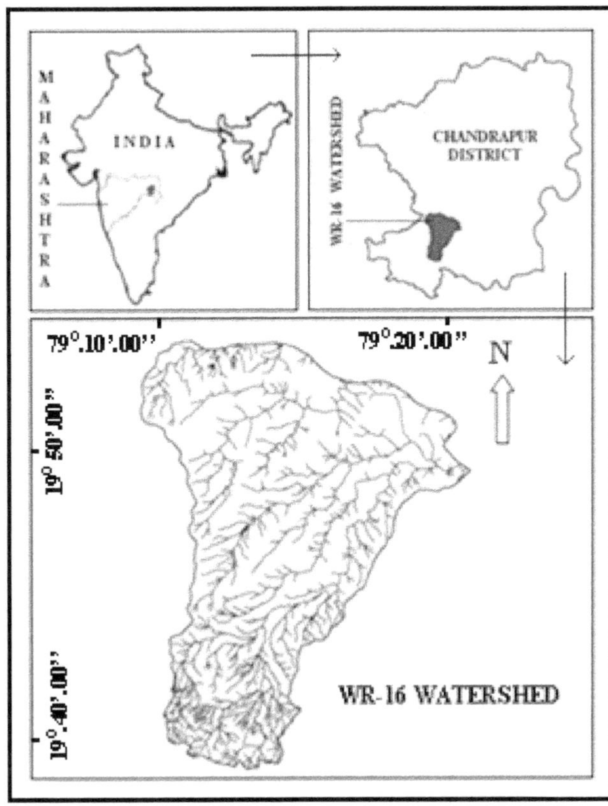

Fig. 1 Location map of WR-16 watershed.

METHODOLOGY

Hydrogeological survey of the area has been carried out with the help of detailed well inventory of 34 open dug wells and bore wells, representing complete area of the WR-16 watershed. During this study information regarding depth and diameter of the wells, pre and post monsoon groundwater waterlevels, type of aquifer, water level fluctuation, pumping capacity, canal command area if any, area irrigated by the well etc. has been collected. The GPS is used to identify exact location of the wells and also to measure altitude of the well with respect to mean sea level. These well locations are transferred on the base map by using Arc-Map 10.2 software. Accordingly different hydrogeological maps have been prepared to study groundwater occurrence and distribution in the watershed.

GEOLOGY

Geologically the watershed area is very complex comprising rocks belonging to the Vindhyan Formation, Gondwana Formation, Deccan Traps and Quaternary sediments. The Vindhyan formation comprising limestone, shale and shaley limestone; Lower Gondwana formation comprising sandstone, shale, carbonaceous shale and coal; Deccan Trap comprising hard, compact, massive, fine grained basalt lava flows with vesicular top. An exposure of plagioclase phyric to giant phenocryst basalt (GPB) is also observed in the South part of the watershed. The alluvium composed of sand, silt and clay, occurring towards North part of the watershed along Wardha river bank (DRM,GSI,2001). These lithological variations dominantly control occurrence and distribution of groundwater in the watershed.These lithological variations in the watershed consist of valuable mineral deposits. Lithological variations of Chandrapur district comprises rich economic mineral deposits and contributing about 29% of the value of total mineral production of the Maharashtra state, India (CGWB, 2013).

Physiography and Drainage:

The WR-16 watershed comprises maximum up to VIth order drainages and exhibit dendritic drainage pattern in the southern part, indicating presence of hard rock and steep slope, whereas drainage pattern in the middle and northern part is parallel to sub-parallel indicating presence of soft rocks. The highest elevation is 478 meters towards South and lowest elevation is 177 meters amsl, towards northern part of the watershed.General slope of the ground is towards North with presence of hilly area towards South.The physiographic set up, lithology and rainfall controls the occurrence and distribution of groundwater in the area.

Hydrogeology:

The water level of the aquifer is one of the important parameter in groundwater hydrology and detailed analysis of its spatio-temporal variation reveals useful information on the aquifer system.The occurrence and distribution of groundwater depends mostly upon lithology, landforms, structures and rainfall.The major water bearing formations in the watershed are Gondwana formation, Vindhyan formation, Deccan basalts and Quaternary formations.The results of groundwater assessment of WR-16 watershed indicates that the watershed falls in safe category with 18.66% stage of development, indicating further scope for groundwater development which is attributed to the assured rainfall of the area. The annual net groundwater recharge is 2931.57 hectare meter, while annual draft from the existing wells is 480.36 hactare meter (GSDA, 2014).

Table 1 Groundwater assessment of WR-16 watershed,Wardha river sub-basin.

Sr.No.	Groundwater Assessment of WR-16 watershed	
1	Annual net groundwater recharge	2931.57 ham
2	Balance available for future usage	2784.99 ham
3	Annual groundwater net withdrawal	146.58 ham
4	stage of Development of the watershed	18.66%
5	Status of watershed	Safe

Complete well inventory of 34 representative dug wells has been carried out during hydrogeological survey with reveals, depth of the dug wells in the area ranges between 5.3 to18.8 meters below ground level (bgl).The diameter of dug wells ranging between 1.2 to 9 meters. The static water level of the study area varies between 0.5 to 9.4 meter bgl during post-monsoon and 2.5 to 17.4 meter bgl during pre-monsoon season. The yield of dugwells in Deccan basalt are less,but yield of dugwells in alluvium are comparatively high, about 110 m^3/hr (CGWB,2014).Pre-monsoon water level map shows that NE part of the watershed has 10 to 12 meter water level bgl (Figure 2,zone 5).This zone is characterized by alluvium consisting of unconsolidated material.This zone represents storage of the watershed.Towards NE and western part of the watershed groundwater occurres 8 to 10 meter bgl(Figure 2,zone 4), comprising mostly alluvium, limestone and shale.Towards NW and SE part of the watershed groundwater occurres 6 to 8 meter bgl (Figure 2,zone 3). This zone dominantly composed of limestone and shale.Towards NW, SW and eastern part of the watershed, groundwater occurres at 4 to 6 meter bgl (Figure 2,zone 2).Towards South shallow groundwater level occurres i.e. 2 to 4 meter bgl with very less recuperation rate (Figure 2,zone 1) mainly due to the presence of basaltic lava flows.

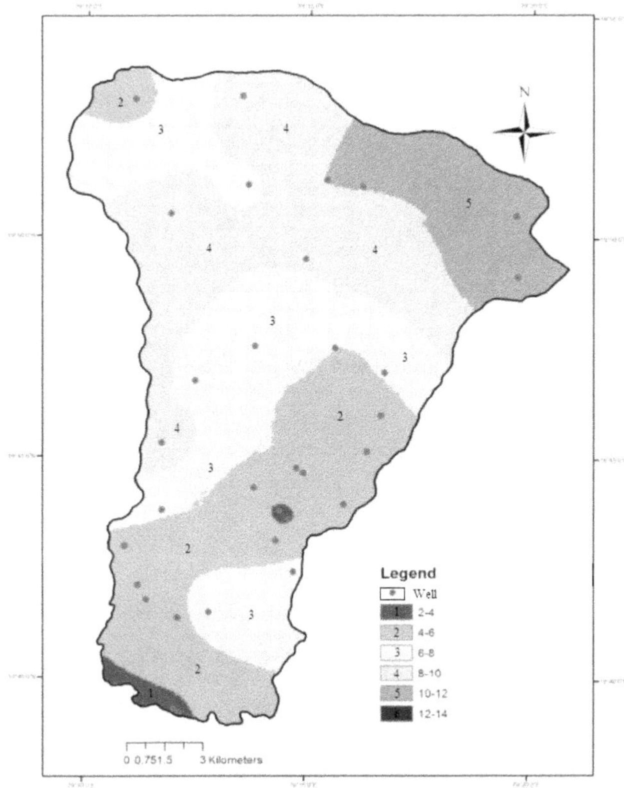

Fig. 2 Pre-monsoon water level of WR-16 Watershed.

The post monsoon water levels of WR-16 watershed represents the picture of natural recharge which takes place after the monsoon rainfall. In post-monsoon period, groundwater occurs at 8 to 10 meter bgl (Figure 3, zone 5).Towards North part of the watershed, groundwater occurres 6 to 8 meter bgl(Figure 3,zone 4) attributed to the presence of soft rocks. The area towards East, North and NW part of watershed comprises alluvium, sandstone, shale, pebble, tillite and shale with limestone bands (Figure 3,zone 3)where groundwater occurres at 4 to 6 meter bgl.

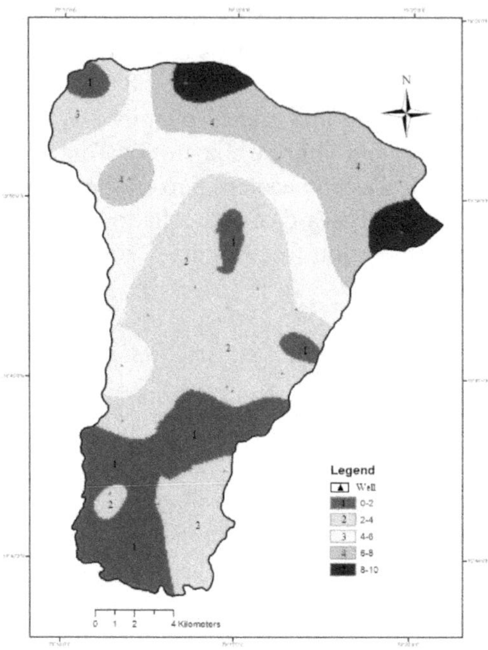

Fig. 3 Post-monsoon water level of WR-16 Watershed.

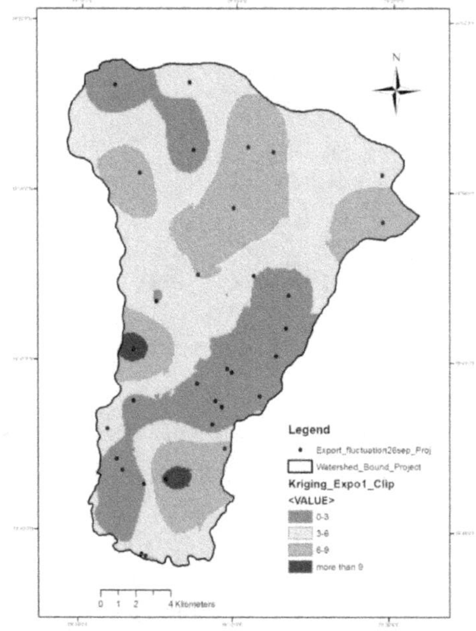

Fig. 4 Water level fluctuation in WR-16 Watershed.

Seasonal ground water level fluctuation is mainly depends on the difference in pre-monsoon and post-monsoon water levels, which can be clearly related to recharge and discharge of groundwater. The magnitude of the water table fluctuation also depends on climatic factor, drainage, topography, and geological conditions (Karanth,1997).The pre and post monsoon water level fluctuations are calculated on the basis of 34 representative dug wells in the study area. The water level fluctuation map has been prepared using spatial analyst tool of Arc-Map 10.2 software. The seasonal water level fluctuation ranges between 1.8 to 10 meter in different part of the watershed (Figure 4) which is dominantly controlled by the lithological variations

Table 2 Well inventory data of representative dug wells in WR-16 watershed,Wardha river basin.

Sr. No.	Locality/ Village	Well No.	Latitude	Long-itude	Elevation in meters (amsl)	Depth of well in meter (bgl)	Pre-monsoon water level in meter (bgl)	Post-monsoon water level in meter (bgl)	Water level fluctuation in meters	Lithology
1	koalgaon	KW/3	N19°51.13"	E79°20.09"	224.2	18.8	17.4	9.4	8	alluvium
2	Babapur	BW/8	N19°50.51"	E79°20.05"	195.7	13.7	10.4	6.3	4.1	Alluvium
3	Antergaon	ANW/16	N19°47.35′	E79°17.11"	187.1	8.9	5.4	3.8	1.6	Alluvium
4	Kurli	KW/66	N19°53.11"	E79°13.29"	198.8	13.5	12	8.7	3.3	Alluvium
5	Nandgaon	NAW/74	N19°50.31"	E79°11.53"	214.4	14.9	12.9	7	5.9	Alluvium
6	Charli	CHW/64	N19°51.09"	E79°16.11"	197.5	11.7	11.7	6.1	5.6	Sandstone
7	Nirli	NIW/65	N19°51.18"	E79°15.23"	196.6	13.2	13.2	5.2	8	Sandstone
8	Sakhari	SAW/11	N19°49'51"	E79°15'32"	197.2	9.8	9.6	1.5	8.1	Shale, pebble
9	Sakhari	SAW/11	N19°49'51"	E79°15'32"	197.2	9.8	9.6	1.5	8.1	Shale, pebble
10	Lakhmapur	LAW/22	N19°45.20"	E79°11'44"	225.6	15.4	15.4	5.4	10	Shale, limestone
11	Thutra	TUW/24	N19°43'48"	E79°11'46"	231.2	7.8	5.1	2.9	2.2	Sandstone
12	Pandharpauni	PW/26	N19°45'10"	E79°16'20"	197.7	5.3	5.3	3.1	2.2	Shale, limestone
13	Mangi	MW/42	N19°42'25"	E79°14'04"	222.7	12.3	9.7	3.3	6.4	Limestone
14	Lingaguda	LiW/45	N19°43'58"	E79°15'05"	209.6	9.4	2.9	0.5	2.4	Shale,limestone
15	Gopalpur	GW/48	N19°42'59"	E79°10'54"	204.4	8.6	3.3	1	2.3	Limestone
16	Isapur	IW/52	N19°41'46"	E79°11'25"	253.9	10.1	5	3.2	1.8	Shale,limestone
17	Salegud	SAW/55	N19°41'50"	E79°13'23"	248.9	13.1	11	2	9	Shale,limestone
18	Chikali Burg	CBW	N19°39'14"	E79°12'11"	426.3	6.7	2.5	1.5	5.1	Basalt

CONCLUSION

Watershed WR-16 has its own unique characteristics regarding physiography,drainage and lithology.The hydrogeological survey of the WR-16 watershed,wardha river basin has been carried out with the help of 34 representative dug wells by collecting information in well inventory proforma.Accordingly pre-monsoon and post-monsoon groundwater levels have been studied. The observed depth of the dug wells in the area ranges between 5.3 to18.8 meters below ground level (bgl).The diameter of dug wells ranging between 1.2 to 9 meters. The static water level of the study area varies between 0.5 to 9.4 meter (bgl) during post- monsoon and 2.5

to 17.4 meter (bgl) during pre-monsoon season.The seasonal water level fluctuation ranges between 1.8 to 10 meter. The deeper water levels are observed in the soft rocks of WR-16 watershed. The yield of dug wells in basalt are less, but yield of dug wells in sandstone and alluvium are comparatively high. Lithologically the watershed area is very complex comprising rocks belonging to the Vindhyan formation, Gondwana formation, Deccan Traps and Quaternary sediments. Also the assessment data revels that annual net groundwater recharge is 2931.57 ham, annual draft from the existing wells is 480.36 ham, stage of development is 18.66% and watershed included in safe category indicating further scope for groundwater development.Apart from topography, slope and relief, the occurrence and distribution of groundwater within the WR-16 watershed is dominantly controlled by the lithological variations.

REFERENCES

1. CGWB (2013) Ground Water Information Report of Chandrpur district Maharashtra State, India., Central Ground Water Board, Ministry of Water Resources Government of India,24p.
2. Chen, Z.S, Osadetz MK (2002). Prediction of average annual groundwater levels from climate ariables: An empirical model. J. Hydrol., 260: 102-117.
3. Gleick, P.H. (1989). Climate change, hydrology, and water resources. Rev. Geophys., 27(3): 329344.
4. Karanth,K.R.(1997).Ground water Assessment Development and Management. McGraw-Hill, New Delhi.,720p.
5. Lewis, J.E. (1989).Climate change and its effects on water resources for Canada: A review Can. Water Resour. Jour., 14: 35-55.
6. Maathuis H, Thorleeifson L.H. (2000). Potential impact of climate change on prairie ground water supplies: Review of Current Knowledge. SackatchewanResearch Council (SRC), Publication No. 11304-200.
7. DRM,GSI(2001).District Resource Map of Chandrapur district,Maharashtra state,India,Geological Survey of India Publ.,Central Region,Nagpur(M.S.).
8. G.S.D.A. and C.G.W.B.(2014) Report on the dynamic groundwater Resources of Maharashtra (2011-12),Groundwater Surveys and Development Agency,Govt.of Maharashtra and Central Ground Water Board,Govt. of India Publ., 907p.
9. Santra,S.C.(2005) Environmental Science,New central book agency,London,1244p.

EFFECT OF DIFFERENT ENVIRONMENTAL CONDITIONS ON PERFORMANCE OF SAPOTA SOFTWOOD GRAFTS WORKED ON INVIGORATED KHIRNI ROOTSTOCK

S. R. Patil*, A. M. Sonkamble*, Ashutosh A.*, M.M. Ganvir and R.S. Patode

College of Horticulture, Dr. PDKV, Akola (MS)

AICRP on Dryland Agriculture, Dr. PDKV, Akola (MS)

ABSTRACT

An experiment entitled "Effect of different environmental conditions on performance of sapota softwood grafts worked on invigorated khirni rootstock." was carried out at Commercial Fruit Nursery, Nagarjun Garden, Dr. Panjabrao Deshmukh Krishi Vidyapeeth, Akola, during the year 2016-2017 with the objectives to study the effect of different environmental conditions on performance of sapota softwood grafting on invigorated Khirni rootstocks and to find out the suitable environmental condition for higher success and better growth of sapota grafts on invigorated Khirni rootstocks. Experiment was laid out in Factorial Randomized Block Design with eight treatment combinations comprising factor A four different environmental conditions C_1 (Open condition), C_2 (Partial shade condition), C_3 (Partial shade (tree shade) condition), C_4 (Poly house condition) and factor B comprised of two decaping height of invigorated khirni rootstock viz., 10 cm and 15 cm from ground level and these were replicated five times.

The treatment combination of poly tunnel and decaping height at 15 cm from ground level took minimum days for bud sprouting (15.80) days, initial graft success (83.00 %), length of scion shoot (20.20 cm), sprout length (19.76 cm), leaves per graft (24.40), leaf area (51.40 cm^2), Average growth rate (0.40), fresh and dry weight of grafts (9.46 and 6.20 g respectively), final survival of grafts (81.00%) and per cent saleable grafts (79.00 %) which was at par with treatment combination green shade net tunnel 50% and decaping height at 15 cm from ground level. Whereas, open condition and decaping height at 10 cm from ground level exhibited minimum values regarding above parameters. For utilization of the invigorated khirni rootstock seedlings which were failed during last year, can be reused by grafting in poly tunnel with decaping height at 15 cm from ground level, for getting maximum grafts success.

INTRODUCTION

Sapota is a hardy rainfed fruit crop.The area under this crop is increased at fast rate in recent year. The cost of graftable rayan stock plant is very high (Rs. 10 per plant). Normally percentage of graft take ranges between 65 to 80 and the rest 20 to 35 per cent of stock plants from unsuccessful grafts can not be reused successfully for grafting. It is felt necessary to find out the possibilities of utilization of these plants for re-grafting. Madalageri *et al.*, (1990) reported the possibility of utilization of stock plants from unsuccessful grafts. Hence, the present investigation was aimed to find out the suitability of rayan stock plants from unsuccessful grafts for grafting by invigoration technique under different environmental conditions.

Success of grafting in sapota may be possible. To overcome such problem of climate, the technique for establishment of sapota propagation under greenhouse condition (Poly house, Green Shade net house) may be adopted in climate of Vidarbha region of Maharashtra because due to hot and dry climate especially in summer season. The enhancement and growth under control climatic condition has been reported by many researchers (Takakura *et al.*, 1971 and Dorland and Went 1947). Maintenance of micro climate in open condition is a major problem lies in subtropical belt and thus sapota grafts mortality observed after grafting due to hot and dry climate situation. There is vital role of temperature, relative humidity, light intensity and carbon dioxide content for success and survival of sapota grafts. Therefore, the present study was undertaken to ascertain the comparison to extent of success and behaviour of grafts under different environmental conditions.

MATERIALS AND METHODS

The experiment was conducted at Commercial Fruit Nursery, Nagarjuna Garden, Dr. Panjabrao Deshmukh Krishi Vidyapeeth, Akola during the year 2016-2017. Two-year-old khirni rootstock seedlings which were used previously for grafting and failed were beheaded leaving 4-5 cm stump to get new shoots. Such rootstock seedlings with 5-6 month old invigourated shoots were used as invigorated rootstock seedlings. For grafting, the scion sticks were defoliated before 10 days of grafting operation. Current season, about 5-6 months old, healthy, 6-8 mm thick, greenish brown coloured, round shaped and unsprouted scion sticks with well developed buds were detached from the selected mother tree in the morning on the day of grafting

Softwood grafting was carried out on 6th of September to 13th September 2016. The sapota grafts were kept under four environmental conditions (*viz.,* Open condition, Partial shade (Tree shade), Green shade net tunnel 50% and Poly tunnel) with two decaping height *viz.,* 10 cm and 15 cm from ground level in invigorated khirni rootstock and arranged in Factorial Randomized Block Design. All observations were noted at 30 days interval from the date grafting.

RESULTS AND DISCUSSION

Days required for bud sprouting

The interaction effect of different environmental conditions and decaping height of invigorated khirni rootstock was found to be significant in days required for bud sprouting. However, minimum days required for sprouting (15.80) days was recorded in poly tunnel and decaping height at 15 cm from ground level and maximum (22.46) days required for bud sprouting was recorded in the treatment combination open condition and decaping height at 10 cm from ground level.

This might be due to the higher cambial activity at 15 cm height for sprouting inside the polyhouse could be attributed to congenial environment condition owing to rapid callusing and early contact to cambium layers, thus enabling the graft to heal quickly and make a strong union. Ultimately results into more vegetative growth of grafts which might be helpful in early sprouting in sapota grafts. These results are in close agreement with the finding of Patel *et al.* (2007) in mandarin, Tandel and Patel (2009) in sapota, Islam *et al.* (2003) in Jackfruit, Panchabhai *et al.* (2005) and Roshan *et al.* (2008) in anola.

Length of scion

At 180 days after grafting, the maximum length of scion (20.20cm) was exhibited by the poly tunnel and decaping height at 15 cm from ground level, which was at par with (11.68 cm) green shade net tunnel – 50% and decaping height at 15 cm from ground level. Whereas, minimum length of scion (12.68 cm) was recorded in open condition and decaping height at 10 cm from ground level.

This might be due to the fact that, it promotes the rate of transpiration, keeps the guard cells turgid and the stomata open which may have resulted in earlier production and accumulation of carbohydrate, protein and earlier completion of other physiological process involved in development of rapid growth between the stock and scion. Dhungana (1989) in mango, The difference in days to sprouting, graft success, plant survival and scion growth seems to be due to a built in mechanism and inherent potential or physiological condition of the rootstock for initial success and early sprouting and storage of more metabolites for survival and growth of the sprout as reported by Dubey *et al.* (2002). Ultimately results into more vegetative growth of plants which might be helpful in the length of scion in sapota graft. This line is corroborated with the finding of the scientist Singh (1977), Pampanna and Sulikheri (2000) in sapota, Nair *et al.* (2002) in mango. Hadli and Raijadhav (2010) in lime.

Sprout length

At 180 days after grafting maximum sprout length (19.76 cm) obtained by the poly tunnel and decaping height at 15 cm from ground level, which was at par with treatment combination green shade tunnel – 50% and decaping height at 15 cm from ground level. However, minimum length (11.04 cm) was obtained by open condition and decaping height at 10 cm from ground level.

This might be due to the fact that, the better growth of grafts and weather condition like temperature and humidity, which played important role in growth of grafts. Yelleshkumar *et al.* (2008). This could be attributed to the vigorous growth of stock, which increased the growth and loads to maximum accumulation of stored metabolites at the time of grafting. (Devechandra, 2006). Ultimately results into more vegetative growth of plants which might be helpful in length of sprout in sapota grafts. This results are line with the finding of Singh (1977), Dewangan and Raut (2014), Patel *et al.* (2007).

Branches per grafts

The maximum branches per graft produced (5.92) in poly tunnel and decaping height at 15 cm from ground level, which is at par with treatment combination green shade tunnel-50% and decaping height at 15 cm from ground level. However, minimum branches per graft (1.84) obtained from treatment combination open condition and decaping height at 10 cm from ground level.

Table 1 Effect of different environmental conditions and decaping height of invigorated khirni rootstock on days required for bud sprouting, length of scion shoot and branches per graft at 180 days after grafting.

Treatments	Days required for bud sprouting	Length of scion shoot (cm) 180 DAG	Sprout length (cm) 180 DAG
Environmental conditions			
C_1 (Open condition)	21.70	13.23	11.60
C_2 (Partial shade (Tree shade)	20.09	14.81	13.13
C_3 (Green shade net tunnel 50%)	17.28	17.84	16.57
C_4 (poly tunnel)	16.32	18.77	17.59
F-Test	Sig	Sig	Sig
SE(m) ±	0.16	0.17	0.18
CD at 5%	0.48	0.51	0.52

Contd...

Decaping height			
H_1 (10 cm from ground level)	19.60	15.02	13.41
H_2 (15 cm from ground level)	18.09	17.30	16.03
F-Test	Sig	Sig	Sig
SE(m) ±	0.11	0.12	0.12
CD at 5%	0.34	0.36	0.36
Treatment combinations			
C_1 x H_1	22.46	12.68	11.04
C_1 X H_2	20.94	13.78	12.16
C_2 X H_1	20.66	13.70	12.26
C_2 X H_2	19.52	15.92	14.00
C_3 X H_1	18.44	16.36	14.92
C_3 X H_2	16.12	19.32	18.22
C_4 X H_1	16.84	17.34	15.42
C_4 X H_2	15.80	20.20	19.76
F-Test	Sig	Sig	Sig
SE(m) ±	0.23	0.25	0.25
CD at 5%	0.69	0.73	0.73

Table 2 Effect of different environmental conditions and decaping height of invigorated khirni rootstock on branches, leaves per grafts and leaf area at 180 days after grafting

Treatments	Branches per graft at 180 DAG	Leaves per graft at 180 DAG	Leaf area (cm^2) at 180 DAG
Environmental conditions			
C_1 (Open condition)	2.30	20.19	38.72
C_2 (Partial shade (Tree shade)	2.98	21.24	37.59
C_3 (Green shade net tunnel 50%)	4.54	21.92	47.19
C_4 (poly tunnel)	5.08	23.03	48.46
F-Test	Sig	Sig	Sig
SE(m) ±	0.11	0.34	0.40
CD at 5%	0.31	1.00	1.18
Decaping height			
H_1 (10 cm from ground level)	2.88	21.03	40.80
H_2 (15 cm from ground level)	4.57	22.15	45.18
F-Test	Sig	Sig	Sig
SE(m) ±	0.07	0.24	0.28
CD at 5%	0.22	0.71	0.83
Treatment combinations			
C_1 x H_1	1.84	20.16	37.28
C_1 X H_2	2.76	20.22	40.16
C_2 X H_1	2.16	21.04	36.14
C_2 X H_2	3.80	21.44	39.04
C_3 X H_1	3.28	21.28	44.26

Contd...

C_3 X H_2	5.80	22.56	50.12
C_4 X H_1	4.24	21.66	45.52
C_4 X H_2	5.92	24.40	51.40
F-Test	Sig	Sig	Sig
SE(m) ±	0.15	0.49	0.57
CD at 5%	0.45	1.42	1.67

Branches per graft increased significantly with when kept under the poly tunnel with 15 cm from ground level. This might be due to the fact that wide variation in the temperature and humidity under the different conditions. These factors influence the branches per plant, sprouting, grafts success and plant growth. Patel *et al.* (2007). This result are in line with the finding of Singh (1977), Anon. 1993, Reddy and Kohli (1985), Chattopadhyay (1994).

Leaves per grafts

At stage 180 days after grafting the maximum leaves per graft were observed (24.40) on poly tunnel and decaping height at 15 cm from ground level, followed by (22.56) green shade tunnel - 50% and decaping height at 15 cm from ground level. However, minimum leaves per graft (2016) obtained from open condition and decaping height at 10 cm from ground level.

The leaves per graft increased significantly when kept under the poly tunnel with 15 cm from ground level. This might be due to the fact that, the vigorous growth of grafts includes by stimulative organs and also influenced by maximum number of sprout leading to maximum number of leaves. Yelleshkumar *et al.* (2008) due to the better environmental condition in side polyhouse ultimately results into more vegetative growth of plants which might be helpful in leaves per grafts in sapota. This result are in agreement with Singh (1977), Reddy and Kohli (1985), Chattopadhyay (1994), Pampanna *et al.* (1995), Pampanna and Sulikheri (2000)

Leaf area

At the stage of 180 days after grafting, maximum leaf area was recorded in poly tunnel and decaping height at 15 cm from ground level (51.40 cm^2), which was at par with green shade tunnel – 50% and decaping height at 15 cm from ground level with (5012 cm^2). However, the treatment combination partial tree shade and decaping height at 10 cm from ground level, showed minimum leaf area (37.28 cm^2).

Leaf area increased significantly when kept under the poly tunnel with decaping height at 15 cm from ground level. This might be due to the fact that, the production and accumulation of more food material in the grafts as well as increased hormonal activities in polyhouse, which resulted in quick cell division and enlargement of already existing cells. These results are line with the findings of Chattopadhyay (1994) and Kamachuk and Golovatkaya (1998) due to the better environmental condition under polyhouse ultimately results into more vegetative growth of grafts which might be helpful in increasing the leaf area of sapota grafts. These observations are in conformity with those of Mir and Kumar (2011), who reported that maximum Leaf area (cm^2) was recorded in wedge grafting of walnut under poly house condition. Angadi (2012) in Jamun, when treated with poly mist house condition and Nair *et al.* (2002) in mango under red polyhouse condition.

Grafts success (%)

Significantly maximum graft success was statistically recorded by poly tunnel and decaping height at 15 cm from ground level (83.00%), followed by green shade net tunnel and decaping height at 15 cm from ground level (74.00 %), whereas, minimum graft success was recorded in partial shade (tree shade) condition and decaping height at 10 cm from ground level (61.00 %).

Higher graft success per cent in mango was due to the favorable environmental condition which could have resulted in maximum cambial activity in both stock and scion. Besides, the scion seemed to be in a physiologically active condition for better sap flow at that time (Sivudu *et al* 2014). Similar results were obtained by Hartmann and Kester (1979) in mango, Islam *et al.* (2003) in Jackfruit and Maheshwari and Nivetha (2015) in softwood grafting of jackfruit under Gujarat conditions.

Stock: scion ratio

Numerically maximum Stock : Scion ratio obtained by poly tunnel and decaping height at 15 cm from ground level (1.90), followed by green shade net tunnel – 50% and decaping height at 15 cm from ground level (1.47). However, optimum stock: scion ratio obtained from treatment combination partial shade condition and decaping height at 15 cm from ground level (1.04).

Salable grafts (%)

Maximum percent of saleable grafts was obtained in poly tunnel and decaping height at 15 cm from ground level (79.00 %), followed by green shade net tunnel – 50% and decaping height at 15 cm from ground level (70.00 %). However, minimum percent of saleable grafts obtained in partial shade (tree shade) and decaping height at 10 cm from ground level (58.00 %).

Table 3 Effect of different environmental conditions and decaping height of invigorated khirni rootstock on grafts success, stock: scion ratio and saleable grafts percent

Treatments	Graft success (%)	Stock : Scion ratio at	Saleable grafts (%)
Environmental conditions	**150 DAG**	**180 DAG**	
C_1 (Open condition)	64.50 (53.43)	0.86	61.50 (51.65)
C_2 (Partial shade (Tree shade)	62.50 (52.24)	0.98	59.00 (50.18)
C_3 (Green shade net tunnel 50%)	72.00 (58.05)	1.47	68.00 (55.55)
C_4 (poly tunnel)	77.50 (61.68)	1.72	74.00 (59.34)
F-Test	Sig	Sig	Sig
SE(m) ±	1.09	0.08	0.36
CD at 5%	3.16	0.25	1.05
Decaping height			
H_1 (10 cm from ground level)	66.50 (54.63)	1.16	63.25 (52.59)
H_2 (15 cm from ground level)	71.75 (57.86)	1.35	68.00 (55.55)
F-Test	Sig	Sig	Sig
SE(m) ±	0.77	0.06	0.25
CD at 5%	2.24	0.18	0.74

Contd...

Treatment combinations			
C_1 x H_1	63.00 (52.53)	0.84	60.00 (50.77)
C_1 X H_2	66.00 (54.33)	0.88	63.00 (52.53)
C_2 X H_1	61.00 (51.35)	0.92	58.00 (49.60)
C_2 X H_2	64.00 (57.86)	1.04	60.00 (50.77)
C_3 X H_1	70.00 (56.79)	1.36	66.00 (54.33)
C_3 X H_2	74.00 (59.34)	1.58	70.00 (56.79)
C_4 X H_1	72.00 (58.05)	1.54	69.00 (56.17)
C_4 X H_2	83.00 (58.05)	1.90	79.00 (62.72)
F-Test	Sig	NS	Sig
SE(m) $\pm$	1.54	0.12	0.51
CD at 5%	4.48	-	1.49

* figures in parentheses indicated are arc sin transformed values

REFERENCES

1. Angadi, S.G. and Rajeshwari Karadi, 2012. Standardization of Softwood Grafting Technique in Jamun under poly mist house conditions. Mysore Journal of Agricultural Sciences; 46(2): 429-432
2. Anonymous, 1993, Annu.Rep. All India Co-ordinated Research Project on Cashew (1992-93), National Research Centre for Cashew, Puttur, pp. 74-91.
3. Awasthi, O.P., P.L. Saroj and D.G. Dhandar, 2005. Standardization of time on success of patch budding in tamarind (*Tamarindus indica*) under arid conditions. Progressive Hort. 37(2): 294-297.
4. Chattopdhyay, T.K., 1994. A text book on pomology Vol. 1, Fundamentals of fruit growing, Kalyani publication, Calcatta, West Bengal, India pp 44-45.
5. Devechandra, D., 2006. Synergetic effect of AM fungi in combination with bioformulations on germination; Graft-take, growth and yield of Jamun, M.Sc. (Hort.) Thesis, Univ. Agric. Sci., Dharwad (India).
6. Dewangan, R.K. And U.A. Raut, 2014. Performance of mango grafts under different environmental conditions. 3rd International Conference on Agri. & Hort.2;4.
7. Dhungana, D.B., M. Avavindakshan and K. Gopikumar, 1989. Standardization of stone grafting in mango. Acta Horticulture, 231: 170-174.
8. Dorland, R.F. and F.W. Went, 1947. Plant growth under controlled conditions. Am. J. Bot. 34: 393 -401.
9. Dubey, A.K., M. Mishra and D.S. Yadav, 2002b. Studies on new technique for Khasi mandarin (*Citrus reticulata* Blanco). Indian J. Citriculture. 1(1) : 103-107.
10. Hadli, J., and S. B. Raijadhav, 2010. Root Stock Influence on the Growth of Scion Cv. Mosambi of Rangpur Lime Strains. Karnataka Journal of Agricultural Sciences, 18 (4).
11. Hartmann , H.T. and D.S. Kester, 1979. Plant propogation ,Principles and practices. 4th edn.Prientice- Hall of India pvt Ltd.New Delhi. Pp: 320-321.
12. Islam. M.M., A. Haqque and M.M Hossain, 2003. Effect of age of rootstock and time of grafting on the success of epicotyls grafting in jackfruit (*Artocarpus Heterophyllus* L.). Asian Journal of Plant Sciences 2(14): 1047-1051.

13. Kamachuk, R.A. and I.F. Golovastskaya, 1998. Effect of spectral composition on hormonal balance growth and photosynthesis in plant seedling. Russian J. Plant Physiology. 45(6): 805-813.
14. Madalageri M.B., G.S. Sulikeri, N.C. Hulamani and V.S. Patil. 1990. Studies on greenwood wedge grafting on sapota. Paper Presented at the XXII International congress, 27th August- 1st September. Firenze Italy.
15. Maheshwari, Uma. T and K. Nivetha, 2015. Effect of age of the rootstock on the success of softwood grafting in jackfruit (*Arto Carpus Heterophyllus* L). Plant Archives Vol.15 No.2.PP 823-825.
16. Mir Muzaffar, and A. Kumar, 2011. Effect of different methods, time and environmental Conditions on grafting in walnut. International Journal of Farm Sciences 1(2):17-22.
17. Mulla, B.R., S.G. Angadi, J.C. Mathad, V.S. Patil and U.V. Mummigatti, 2011. Studies on softwood grafting in jamun (*Syzgium cumini skeels*). Karnataka J. Agirc. Science, 24 (3) : 366-368.
18. Nair, H., B.S. Baghel, R. Tiwari and B.K. Nema, 2002. Influence of coloured poly house/light and methods of epicotyl grafting on vigour of mango grafts. JNKVV Research Journal; 36 (1/2): 51-54.
19. Pampanna, Y. and G.S. Sulikeri, 2000. Effect of season on the success and growth of softwood grafts in sapota on invigorated Rayan rootstock. Karnataka Journal of Agricultural Sciences. 13: 779-782.
20. Panchbhai, D.M, R.K. Roshan , V.K.Mahorkar and S.M. Ghawde, 2005. Effect of rootstock age and time of grafting on grafting success in Aonla.Current Sci.47:468-469.
21. Patel, R.K., K.D. Babu, and A. S. Yadav, 2007. Softwood grafting in mandarin - A novel vegetative propagation technique. Int. J. of Fruit Sci., 10 (1): 54-64.
22. Pathak, R.K., 1991. Propagation of fruit plants (In Hindi), ICAR, New Delhi.
23. Raghavendra, V.N., S.G. Angadi, T.B. Allolli, C.K. Venugopal and U.V. Mummigatti, 2011. Studies on soft wood grafting in wood apple (*Feronia limonia* L.). Karnataka Journal of Agricultural Sciences; 24 (3): 371-374.
24. Reddy, Y.T.N. and R.R. Kohli, 1985. Rapid multiplication of mango by epicotyl grafting. Second International Symposium on mango. Bangalore.
25. Singh, A., 1977. Practical plant physiology, Kalyani Publishes, New Delhi, Ludhiana.
26. Sivudu, B.V., M.L.N. Reddy, P. Baburatan and A.V.D. Dorajeerao, 2014. Effect of structural conditions on veneer grafting success and survival of mango grafts (*Mangifera Indica* cv. Banganpalli*)*. Plant Archives. Vol. 14 No 1: 71-75.
27. Syamal, M.M, R. Kumar and Mamata Joshi, 2012. Performance of wedge grafting in guava under different growing conditions. Indian J. Horti.69 (3):424-427.
28. Takakura, T., K.A. Jorden and L.L. Boyd, 1971. Dynamic simulation of plant growth and environment in the green house. Trans. ASAE, 14 : 964-971.
29. Tandel, Y.N and C.B. Patel, 2009. Effect of scion stick storage on growth and success softwood grafts of sapota cv. Kalipati. The Asian Journal of Horticulture, Vol .4 No (1) : 198-201.
30. Yelleshkumar, H.S., G.S.K. Swamy, C.P. Patil, V.C. Kanamadi and Prasad Kumar, 2008. Effect of Pre-Soaking Treatments on the Success of Softwood Grafting and Growth of Mango Grafts. Karnataka J. Agric. Sci., 21(3): 471-472.

IMPACT OF MUNICIPAL SOLID WASTE DUMPSITE ON GROUND WATER QUALITY

K. Syamala Devi[1], K. Venakateswara Rao[2] and A.V.V.S.Swamy[3]

[1]G.Narayanamma Institute of Technology & Science (for Women), Hyderabad,
[2]Andhra Pradesh Pollution Control Board, [3]Acharya Nagarjuna University, Guntur

ABSTARCT

Land filling is the preferred method of Municipal Solid Waste (MSW) disposal. However, poorly designed land fill leads to contamination of ground water, soil and air. As water percolates through the landfill, contaminants are leached from the solid waste. Leachate is produced when moisture enters the refuse in a landfill, extracts contaminants into the liquid phase. Leachate is generated in a landfill as a consequence of the contact of water with solid waste. Leachate contains dissolved or suspended material associated with wastes disposed off in the land fill, as well as many by-products of chemical and biological reactions. Leachate tend to migrate in surrounding soil may result in contamination of underlying ground water and soil. The rate at which it percolates depends on the soil, texture and depth of the aquifer. It is established that the ground water was contaminated by the leachate in many cities not only in India and but in many other countries. The peri-urban areas face severe problem of ground water contamination from MSW dumpsites. Owing to the importance of the topic, the present work has been carried near Kapuluppada dumpsite in Visakhapatnam, Andhra Pradesh. Visakhapatnam is the largest city in Andhra Pradesh with a population of 20 lakhs. An area of 100 acres at Kapuluppada village was selected as the dumpsite in 2004 for Landfilling located 15 Km away from the city. Visakhapatnam city generates 980 tonnes of garbage every day. The landfill site is located at a distance of 15 km to the west of the city, spread across 100 acres of land having an elevation of 45 m. Four stations have selected as sampling stations namely Paradesipalem, Kamala Nagar colony, Kothapalem Village, and main dumpsite i.e., Kapullapada Dumpsite during the study period. The physico-chemical analysis was done for four years at all the six stations selected for the present study. Six samples were collected in alternate months in a year. A total of 144 samples were collected during the study period from these six stations. The parameters studied were: pH, total dissolved solids, total hardness, chlorides, nitrates, sulphates, phosphates, phenols, cyanide, iron, zinc, nickel, copper, chromium and cadmium. The results of the present study revealed that the total dissolved solids, total hardness and chlorides were in very high concentrations. These parameters, though not directly affect the human health, indicate the increase of ions in the ground water through leachate contamination worsen the quality of water, to prevent future contamination of heavy metals and organic and inorganic materials, few recommendations are suggested.

Keywords: Leachate, Heavy metals, Landfills.

INTRODUCTION

The generation of solid waste is not a new phenomenon. It is directly related to human civilization. The first organized dump of solid waste is reported to have been set up outside the Athens in 500 B.C. (Krishnamoorthy, 2001). The problems associated with solid waste management in all major cities of developing countries are more acute than in the major cities of developed countries. The pressure on ground water has been increasing as the urbanization is gaining momentums. There is a general notion that ground water is safer than surface water. Several studies have proven this is fabe, as the earlier studies in many places reported the contamination of ground water by various source. The contaminants from the surface soils also reach the ground water by infiltration and percolation. A relatively recently identified source of contamination of ground water is the leachate from the municipal solid waste dumpsites. Many studies (Chavan and Zambare, 2014; Omofonmwan and Eseigbe, 2009; Dharmarathne and Gunatilake, 2013, Raman and Sathiyanarayanan. 2011; Boobalan *et al.*, 2015, Kamboj and Choudhary, 2013; Arafath *et al.*, 2014, Jamnami and Singh, 2009, Bundala *et al.*, 2012, Odunlami, 2012, Nandwana and Chhipa, 2014. Rajkumar *et al.,* 2010) have reported that the leachate generated in MSW dumpsite entered into the ground water through infiltration and percolation. The leachates bring about abnormal changes in TDS, TH, Chlorides, Sulphates, Nitrates and many heavy metals inding the contamination, that affect the potability and palatability of ground water.

The present study has been carried out in Visakhapatnam from March 2011 to January, 2015. Visakhapatnam is designated as one of the fast growing cities in Asia, and is the largest city in Andhra Pradesh, India with a population of 20, 35, 922 as per the census of 2011. Visakhapatnam city is spread in 540 Sq. Km with a solid waste generation of 980 Tonnes Per Day (TDP) municipal waste is sent to Kapuluppada Dump Site located at 15 Km to the west of Visakhapatnam.

Four stations have selected as sampling stations namely Paradesipalem (Station - I), Kamala Nagar colony (Staation - II), Kothapalem Village (Station - III), and main dumpsite i.e., Kapullapada Dumpsite (station - IV) during the study period.

The physico-chemical analysis was done for four years at all the six stations selected for the present study. Six samples were collected in alternate months in a year. A total of 144 samples were collected during the study period from these six stations. The parameters studied were: pH, total dissolved solids, total hardness, chlorides, nitrates, sulphates, phosphates, phenols, cyanide, iron, zinc, nickel, copper, chromium and cadmium.

RESULTS & DISCUSSION

STATION - I: Annual Means of the Ground Water Quality Parameters at the outskirts of Paradesipalem:

The ground water samples were collected near a church at the outskirts of Paradesipalem (Station-I). pH, TDS, Chlorides, TH, Nitrate, Sulphates, phenols, cyanide, Lead, Iron, Zinc, Nickel, Copper, Chromium and cadmium were estimated for 4 years in alternate months. The values of different parameters were recorded and compared with BIS standards (2012). The P^H of groundwater was within the permissible limit of 6.5 – 8.5. At this level of pH, the potability of ground water is not affected.

The Total Dissolved Solids (TDS) were estimated and expressed as mg/l. The TDS concentration was higher than the limit of 500 mg/l prescribed by BIS (2012). Beyond 500 mg/l the palatability decrease and may cause gastrointestinal irritation.

The chloride concentrations were less than the acceptable limit of 250 mg/l, during the period of study from 2011-12 to 2014-15. The variation in the concentration of chlorides was very narrow (179 to 192 mg/l) with a difference of only 13 mg/l between minimum and maximum values. The mean concentration of chlorides was 186.25 mg/l for the total study period.

The total hardness (mg/1) as $CaCO_3$ was higher during the study period in the ground water. The total hardness has exceeded the acceptable limit of 200 mg/l in all the years of study. Beyond this limit encrustation in water supply structure and adverse impact on domestic use of water sets in. The concentration of total hardness ranged between 330 mg/l and 356 mg/l with a combined mean for the study period of 344 mg/l.

The nitrates as NO_3-N in the ground water at station I were far below the acceptable limit of 45 mg/l. The present study has recorded values between 6.4 and 6.93 mg/l in 2011-12 and 2013-14 respectively, with a combined mean of 6.73 mg/l. The continued exposure beyond 45 mg/l cause methaemoglobinamia. But the lesser values in the present study pose no threat of health to the users.

Sulphates in the ground water at Station I were very less (46.7 mg/l) compared to the prescribed acceptable limit of 200 mg/l (BIS, 2012). The sulphates concentrations ranged between a minimum of 45.9 mg/l in 2012-13 and a maximum of 48.5 mg/l in 2011-12 with a combined mean of 46.7 mg/l. The sulphates beyond 200 mg/l concentrations, cause gastrointestinal irritation when sodium or magnesium is present. But at present levels there is no threat to human health by sulphates in ground water.

The Iron (Fe) concentrations were higher than the acceptable limit of 0.3 mg/l prescribed by BIS (2012). The Iron concentrations ranged between a minimum of 0.42 and a maximum of 0.48 mg/l during the 4 years of study. The combined mean for the study period was 0.46 mg/l. There was no relaxation of acceptable limit of 0.3 mg/l because beyond this limit taste and appearance of water are affected and has adverse effect on domestic uses and water supply structures and promotes iron bacteria. As there was marginal difference between acceptable limit and the actual values of iron recorded, the quality of water does not affect the potability.

The Zinc concentration in ground water at station I was far below the acceptable limit of 5 mg/lt. The concentrations recorded were between 0.296 mg/l and 0.321 mg/l during the 4 years of study with a combined mean of 0.307 mg/l. There was no source of zinc inputs into a ground water as there were no agricultural farms. The concentration above 5 mg/l of zinc cause astringent taste and are opalescence in water, however, in the absence of alternate source the zinc concentrations were relaxed up to 15 mg/ l. At the levels recorded in the present study, Zinc does not pose any threat to the health of human beings and do not to attribute any taste of water. These concentration reported in the present study does not affect the potability or palatability.

The phenols and cyanide were found in below detectable limit (BDL) throughout the present 4 year study. Though phenols and cyanide were permitted upto 0.001 mg/l and 0.05 mg/l , respectively, the present study showed values at below detectable levels. The Lead, Nickel, Copper, Chromium and Cadmium also were estimated throughout the study but not detected (ND), in any of the samples during the 4 year study. So the ground water at the outskirts of Paradesipalem safe for drinking.

The parameters like pH, Chlorides, Nitrates, Sulphates, Phenols, Cyanide, Lead, Zinc, Nickel, Copper, Chromium, Cadmium were within acceptable limit for drinking water by BIS

(2012). But so far the acceptability in terms of taste and lather production, the ground water at station - I is fit for all domestic uses including drinking.

Table 1 Annual Mean Concentrations of different water quality parameters compared with BIS Standards at Station - I (outskirts of Paradesipalem) during the study period.

S. No.	Parameter`	2011-12	2012-13	2013-14	2014-15	Mean	BIS STANDARD
1	pH	6.88	6.98	6.85	6.93	6.91	6.5-8.5
2	Total Dissolved Solids	833	818	838	824	828.25	500 mg/l
3	Chlorides	192	179	185	189	186.25	250 mg/l
4	Total hardness as $CaCO_3$	330	356	351	339	344	200 mg/l
5	Nitrates Nitrogen (as NO_3.N)	6.4	6.7	6.93	6.89	6.73	45 mg/l
6	Sulphates	48.5	45.9	47.6	46.2	46.7	200 mg/l
7	Phenols	BDL	BDL	BDL	BDL	BDL	0.001 mg/l
8	Cyanide	BDL	BDL	BDL	BDL	BDL	0.05 mg/l
9	Lead as Pb	ND	ND	ND	ND	ND	0.05 mg/l
10	Iron as Fe	0.48	0.42	0.47	0.46	0.46	0.3 mg/l
11	Zinc	0.321	0.308	0.296	0.302	0.307	5 mg/l

Station - II: Annual Means of the Ground Water Quality Parameters at Kamalanagar Colony:

The pH at Station - II (Kamalanagar Colony) was within the range of acceptable limit prescribed by BIS (2012). All the samples during the four year study period have shown a pH nearer to neutral pH. The lowest annual mean was 6.86 and the highest was 7.16 with a combined mean of 7.0.

The Total Dissolved Solids (TDS) were less than the acceptable limit of 500 mg/l. A lowest annual mean of 390 mg/l and a highest of annual mean of 402 mg/l were recorded in the years 2011-12 & 2012-13 and 2013-14, respectively. There was a relaxation of value for TDS upto 2000 mg/l in the absence of alternate source. However, the present study revealed the TDS was within the permissible limit and hence the TDS will not be responsible for any change of quality of water in terms of palatability.

The chloride concentrations were within the permissible limits. The acceptable limit was fixed by BIS (2012) as 250 mg/l where as the present study recorded chloride concentrations between 96 and 106 mg/l in 2013-14 and 2011-12, respectively, with a combined mean of 100 mg/l. The chlorides at the concentrations in the present study cannot attribute any taste; affect palatability and corrosive effects in the drinking water.

The total hardness was slightly higher in the ground water during all the four years than the acceptable limit of 200 mg/l (BIS 2012). The lowest total hardness recorded was 258 mg/l in 2011-12 and a highest was 274 in 2014-15 with a combined mean of 268 mg/l. All the total hardness values were higher than the acceptable limits but less than the maximum permissible limit of 600 mg/l. under conditions of no alternate source.

The nitrate concentrations were far below the acceptable limit of 45 mg/I (BIS, 2012). A nitrates exhibited a range of 3.54 mg/l to 3.66 mg/l during the four year study period with a combined mean of 3.63 mg/l.

The sulphate concentrations were also very less compared to the acceptable limit of 200 mg/l. The lowest annual mean of 22.9 mg/l and highest 24.2 mg/lt were recorded in 2012-13 and 2011-12 respectively, with a combined mean of 23.5 mg/l.

The Iron (Fe) concentrations were much higher than the acceptable limit of 0.3 mg/l (BIS, 2012). The water samples at this station have shown concentrations < 0.9. The lowest iron concentration recorded was 0.962 mg/l in 2011-12 and 0.982 mg/l. in 2014-15, with a combined mean of 0.974 mg/l.

The Zinc concentrations in the ground water at Station - II were meagre when compared to the acceptable limit of 5 mg/l (BIS, 2012). A lowest concentration of 0.082 mg/l and a highest of 0.093 mg/l were recorded in the present study at Station - II. The zinc concentration at the levels reported in the present study will not have impact on taste of water. The combined mean of the study period was 0.089 mg/l. The phenols and cyanides were present in below detectable limits where as lead, Nickel, Copper, Chromium and Cadmium were not detected in the water samples at Kamalanagar Colony (Station – II). Only total hardness (268 mg/l) and iron (0.974 mg/l) have exceeded the acceptable limit in the samples of present study and all other parameters studied were within the acceptable limits.

Table 2 Annual Mean Concentrations of different water quality parameters compared with BIS Standards at Station - II (Kamalanagar Colony) during the study period.

Sl. No.	Parameter`	2011-12	2012-13	2013-14	2014-15	Mean	BIS STANDARD
1	pH	6.86	6.92	7.1	7.16	7.0	6.5-8.5
2	Total Dissolved Solids	390	390	402	401	398	500 mg/l
3	Chlorides	106	98	96	104	100	250 mg/l
4	Total hardness as $CaCO_3$	258	264	268	274	268	200 mg/l
5	Nitrates Nitrogen (as NO_3.N)	3.54	3.59	3.66	3.6	3.63	45 mg/l
6	Sulphates	24.2	22.9	23.8	23.4	23.5	200 mg/l
7	Phenols	BDL	BDL	BDL	BDL	BDL	0.001 mg/l
8	Cyanide	BDL	BDL	BDL	BDL	BDL	0.05 mg/l
9	Lead as Pb	ND	ND	ND	ND	ND	0.05 mg/l
10	Iron as Fe	0.962	0.973	0.969	0.982	0.974	0.3 mg/l
11	Zinc	0.092	0.093	0.086	0.082	0.089	5 mg/l

STATION - III: Annual Means of the Ground Water Quality Parameters at Kothapalem Village:

The Borewell samples at Kothapalem village, have exhibited neutral pH. All the samples collected during 2011-12 to 2014-15 have shown pH values near to neutral. The lowest pH recorded was 6.8 (2014-15) and 7.2 (2012-13) with a combined mean of the study period 7.0. At this level of pH the ground water at Kothapalem Village was fit for drinking, domestic and industrial purposes.

The ground water at Kothapalem village was having high Total Dissolved Solids (TDS) compared to acceptable limit of 500 mg/l. The annual mean concentrations ranged between a minimum of 636 mg/l (2011-12) and a maximum of 652 mg/l (2013-14) with a combined mean of 648 mg/l. The TDS beyond 500 mg/l decreases the palatability and may cause gastrointestinal irritation.

The chloride concentrations at station - III were within the acceptable limit of BIS (250 mg/l). The chlorides ranged between a minimum of 101 mg/l and a maximum of 113 mg/l with a combined mean of 103 mg/l during the four year study period.

The total hardness in ground water at Kothapalem village was higher than the permissible limit of 200 mg/l (BIS, 2012). The hardness recorded in the bore well of Kothapalem Village ranged between a minimum annual mean of 328 mg/l and a maximum of 346 mg/l in 2011-12 and 2012-13, respectively, with a combined mean of 340 mg/l. Though the hardness is higher than the acceptable limit, it was lesser than the permissible limit in the absence of alternate source (600 mg/l).

Nitrates were present in very less quantities compared to the acceptable limit of 45 mg/l (BIS, 2012). The nitrate, concentrations at station - III ranged between a minimum annual mean of 7.6 mg/l to 8.2 mg/l in 2012-13 and 2014-15, respectively, with a combined mean of 7.9 mg/l.

Sulphates in ground water at Kothapalem village were present within the permissible limit of 200 mg/l. But the actual sulphate concentrations recorded in the Kothapalem village ranged between a minimum annual mean of 36.5 mg/l and a maximum of 42.0 mg/l with a combined mean of 39.5 mg/l.

The iron concentration in the ground water of Kothapalem village was within the permissible limit of 0.3 mg/l (BIS 2012). The concentrations ranged between a minimum annual mean of 0.048 mg/l and a maximum of 0.056 mg/l in 2011-2012 & 2013-14, respectively with a combined mean of 0.053 mg/l.

The concentrations of Zinc were far below the acceptable limit of 5 mg/l. The annual means of Zinc concentration ranged between 0.029 mg/l and 0.038 mg/l in 2013-14 and 2012-13, respectively, with a combined mean of 0.032 mg/l.

Phenols and Cyanide were below detectable limits during the 4 year sampling period and Lead, Nickel, Copper, Chromium and Cadmium were not detected in any of the samples during the entire study period.

Table 3 Annual Mean Concentrations of different water quality parameters compared with BIS Standards at Station -III (Kothapalem Village) during the study period.

Sl. No.	Parameter`	2011-12	2012-13	2013-14	2014-15	Mean	BIS STANDARD
1	pH	7.1	7.2	6.9	6.8	7.0	6.5-8.5
2	Total Dissolved Solids	636	646	652	650	648	500 mg/l
3	Chlorides	106	102	113	101	103	250 mg/l

Contd...

Sl. No.	Parameter`	2011-12	2012-13	2013-14	2014-15	Mean	BIS STANDARD
4	Total hardness as $CaCO_3$	328	346	332	338	340	200 mg/l
5	Nitrates Nitrogen (as NO_3.N)	7.8	7.6	7.9	8.2	7.9	45 mg/l
6	Sulphates	36.5	42.0	40.6	38.9	39.5	200 mg/l
7	Phenols	BDL	BDL	BDL	BDL	BDL	0.001 mg/l
8	Cyanide	BDL	BDL	BDL	BDL	BDL	0.05 mg/l
9	Lead as Pb	ND	ND	ND	ND	ND	0.05 mg/l
10	Iron as Fe	0.048	0.052	0.056	0.054	0.053	0.3 mg/l
11	Zinc	0.036	0.038	0.029	0.032	0.032	5 mg/l

STATION - IV: Annual Means of the Ground Water Quality Parameters at Kapuluppada MSW dumpsite:

Station - IV was a bore well located within the dumpsite at Kapuluppada. Dumping was started in 2004 at this site. In order to establish the formation and dispersion of leachate, and its impact on ground water, this station was selected. The bore well water is not in use by public. The quality of ground water is assessed. The pH of ground water ranged between 7.1 and 7.4 with a combined mean of 7.4. The water at this station was slightly alkaline and the pH was within the permissible range of 6.5 – 8.5 (BIS-2012) specified for drinking water.

Total Dissolved Solids (TDS) have exceeded the acceptable limit of 500 mg/l but within the permissible limit of 2000 mg/l. The lowest concentration of TDS were recorded during 2012-13 (1490 mg/l) and highest was during 2013-14 (1550 mg/l) with a combined mean of 1532 mg/l during the 4 year study period. The water at this site at the reported levels of TDS is not fit for drinking.

The concentrations of Chlorides have exceeded the acceptable limit of 250 mg/l. The lowest annual mean recorded in the study was 495 mg/l and highest was 520 mg/l with a combined mean for the study period of 509 mg/l. The chlorides beyond 250 mg/l will interfere with taste, palatability and cause corrosion.

The total hardness was very high at Station - IV when compared with the acceptable limit of 200 mg/l. The hardness recorded in this station varied between a minimum annual mean of 785 mg/l and a maximum of 814 mg/l with a combined mean of 810 mg/l. The nitrate concentration were within the acceptable limit of 45 mg/l and fairly low. The annual means during the study period ranged from a minimum of 10.86 mg/l and a maximum of 12.22 mg/l with a combined mean of 11.65 mg/l. At these levels, the nitrates in ground water at this station does not interfere with potability of water. However, this bore well has not been in use for a long time.

The concentrations of sulphates were within the permissible limit of 200 mg/l (BIS, 2012). The lowest annual mean recorded during the 4 year study period was 135 mg/l (2011-12) and the highest was 152 (2013-14) with a combined mean of 146 mg/l.

The iron concentrations were far below the acceptable limit of 0.3 mg/l (BIS, 2012). The annual means of the study period varied between 0.058 mg/l and 0.066 mg/l in 2014-15 and 2011-12 respectively, with a combined mean of 0.062 mg/l.

The concentrations of zinc were fairly low in the bore well at dumpsite. The Zinc concentration was always below 0.05. The lowest annual mean recorded was 0.024 mg/l and 0.032 mg/l in 2013-14 and 2012-13, respectively, with a combined mean of 0.027. The phenols and cyanides were present in below detectable limits while lead, copper, Nickel, Chromium and cadmium were not detected in any of the samples. The Total Dissolved Solids, Chlorides and total hardness were higher than the acceptable limits.

Table 4 Annual Mean Concentrations of different water quality parameters compared with BIS Standards at Station - IV (at Kapuluppada MSW dumpsite) during the study period.

Sl. No.	Parameter`	2011-12	2012-13	2013-14	2014-15	Mean	BIS STANDARD
1	pH	7.1	7.4	7.3	7.2	7.4	6.5-8.5
2	Total Dissolved Solids	1506	1490	1550	1538	1532	500 mg/l
3	Chlorides	495	520	506	512	509	250 mg/l
4	Total hardness as $CaCO_3$	785	814	802	806	810	200 mg/l
5	Nitrates Nitrogen (as NO_3-N)	10.94	11.62	12.22	10.86	11.65	45 mg/l
6	Sulphates	135	144	152	146	146	200 mg/l
7	Phenols	BDL	BDL	BDL	BDL	BDL	0.001 mg/l
8	Cyanide	BDL	BDL	BDL	BDL	BDL	0.05 mg/l
9	Lead as Pb	ND	ND	ND	ND	ND	0.05 mg/l
10	Iron as Fe	0.066	0.061	0.064	0.058	0.062	0.3 mg/l
11	Zinc	0.028	0.032	0.024	0.030	0.027	5 mg/l

SUMMARY & CONCLUSION

The composition of MSW and leachate change from place to place and a seasonal change of composition was also reported. The dumped solid wastes gradually release its initial interstitial water and some of its decomposition by products gets into water moving through the waste deposit. This complex liquid is known as leachate. This leachate accumulates at the bottom of the landfill and percolates through the Soil. Areas near to landfill have a greater possibility of ground water contamination. Such contaminations of ground water contamination pose a substantial risk to local resource user to the environment (Zhu *et al.,* 2008).

Recently India is becoming a dumping ground for electronic waste especially used computers from the developed countries which contains hazardous metals such as lead, cadmium and mercury. About seventy percent of the heavy metals found in landfills come from electronic wastes (Toxics Link, 2003). The composition of landfill leachate varies from time to time and site due to the differences in waste composition, amount of precipitation,moisture content, climate changes, waste compaction, incineration of leachate with the environment etc (Kulikowska and Klimiuk, 2008; Umar *et al.,* 2010). The waste management policies and strategies are still struggling with the conflicts arising between developmental and environmental goals. The draft National Environmental Policy of 2005, which incorporates the concept of the 3R's is currently under consideration (MOEF, India). The ever increasing demand for larger space for the disposal of domestic and industrial wastes generated from urban areas makes landfill site a necessary component of the urban life cycle. These low lying disposal sites, being devoid of a leachate collection system, landfill gas monitoring and collection equipment, can hardly be called sanitary landfills and are the potential threat for water resources, especially ground water.

The results of the present study revealed that the total dissolved solids, total hardness and chlorides were in very concentrations. These parameters, though not directly affect the human health, indicate the increase of ions in the ground water through leachate contamination. The Kapuluppada dumpsite is a landfill and not scientifically designed. Hence, to prevent future contamination of heavy metals and organic and inorganic materials, the following recommendations are made:

A sanitary landfill site shall be planned, designed and developed with proper documentation of construction plan as well as closure plan in a phased manner.

1. Waste processing units should be permitted in the vicinity.
2. A state of the art incineration plant shall be established to minimize solid waste problem.
3. Steps must be taken to segregate e-waste from MSW.
4. A proper leachate collection system should be installed with immediate effect.
5. The Landfilling must be compacted using heavy compactors
6. Proper drainage system shall be provided to divert run-off away from the active cell of the landfill.
7. Preferably a land fill gas recovery system be established.
8. A Permeable Reactive Barriers (PRB) Technology for in situ ground water remediation shall also be considered, that passively capture a plume of contaminants and removes or breakdowns the contaminants, finally releasing uncontaminated water.

REFERENCES

1. Arafath, Y., Praveen Kumar., Vigneshwaran and Banupriya, 2014. Analysis of Pollutants present in the ground water due to Leachates at Thuraipakkam Dumpyard, Chennai. Civil and Environmental Research. 6 (6): 55-60.
2. Boobalan, C., Gurugnanam, B. and Suresh, M. 2015. Assessment of Groundwater Quality in Sarabanga Sub-Basin, Cauvery River, Tamilnadu, India. International Journal of Current Advanced Research. 4(11): 504-508.
3. Bundela, P.S., Anjana Sharma, Akhilesh Kumar Pandey, Priyanka Pandey and Abhishek Kumar Awasthi. 2012. Physicochemical Analysis of Ground water Near Municipal Solid Waste Dumping Sites Jabalpur. International Journal of Plant, Animal and Environmental Sciences. 2 (1): 217-222.
4. Chavan, B.L. and Zambare, N.S., 2014. Ground Water Quality Assessment near Municipal Solid Waste Dumping Site, Solapur, Maharashtra, India. International Journal of Research in Applied,
5. Cheboterev. II., 1955. Metamorphism of natural waters in the crust of weathering-I. Geochimica et Cosmochima Acta.8(1-2): 22-48.
6. Dharmarathne, N. and Gunatilake, J., 2013. Leachate Characterization and Surface Groundwater Pollution at Municipal Solid Waste Landfill of Gohagoda, Sri Lanka. International Journal of Scientific and Research Publications. 3(11): 1-7.
7. Jhamnani, B. and Singh, S. K., 2009. Groundwater Contamination due to Bhalaswa Landfill Site in New Delhi. International Journal of Civil and Environmental Engineering. 1(3): 121-125.

8. Kamboj, N. and Choudhary, M. 2013. Impact of Solid waste disposal on ground water quality near Gazipur dumping site, Delhi, India. Journal of Applied and Natural Science 5(2): 306 – 312.
9. Krushnamoorthy, B. 2001. Environmental Management. Abhinav kumar Publications, Mumbai. 80-81.
10. Kulikowska, D. and Klimiuk, E. 2008. "The effect of landfill age on municipal leachate composition". Bioresour. Technol. 99(13): 5981-5985.
11. Nandwana R. and Chhipa, R C. 2014. "Impact of Solid Waste Disposal on Ground Water Quality in Different Disposal Site at Jaipur, India", International Journal of Engineering Sciences and Research Technology. 3 (8): 93-101.
12. Odunlami,M.O., Member, IACSIT 2012. Investigation of Groundwater Quality near a Municipal Landfill Site (IGQMLS). International Journal of Chemical Engineering and Applications, 3(6): 366-369.
13. Omofonmwan, S.I. and Eseigbe, J.O. 2009. Effects of Solid Waste on the Quality of Underground Water in Benin Metropolis. Nigeria. J. Hum. Ecol. 26(2): 99-105.
14. Rajkumar, N., Subramani, T. and Elango, L. 2010. Groundwater contamination due to municipal solid waste disposal – A GIS based study in erode city. International Journal of Environmental Sciences. 1: 39-55.
15. Raman, N. and Sathiyanarayana, D. 2011 Quality Assessment of Ground Water in Pallavaram Municipal Solid Waste Dumpsite Area Nearer to Pallavaram in Chennai, Tamilnadu. Rasayan J. Chem. 4(2): 481-487.
16. Toxic link. 2003. Scrapping the Hitech myth: computer waste in India. http://www.toxicslink.org/ovrvw.
17. Umar, M., Aziz, H. A. and Yusoff, M. S. 2010. Variability of Parameters Involved in Leachate Pollution Index and Determination of LPI from Four Landfills in Malaysia. International Journal of Chemical Engineering. Hindawi Publishing Corporation. 10(6): 1-6.
18. Zhu, D., Asnani, P.U., Zurbrugg , C., Anapolsky , S., Mani,S. 2008. Improving municipal solid waste management in India. A source book for policymakers and Practitioners. The World Bank.

EFFECT OF *IN-SITU* RAINWATER CONSERVATION PRACTICES ON MOISTURE USE AND YIELD OF PIGEON PEA

S.R. Weladi[1] and S.D. Payal[2]

Vasantrao Naik Marathvada agricultural University, Parbhani (M.S)

snehaweladi82@gmail.com

ABSTRACT

The field investigation was conducted at All India Co-ordinated Research Project (AICRP), for Dry Land Agricultural Farm, Vasantrao Naik Marathwada Agricultural University Parbhani, for Pigeon Pea crop during kharif season of 2013. Performance of opening of furrow across the slope after two rows in pigeon pea imparted significant effect on moisture conservation in the root zone of the pigeon pea growth, plant height, number of branches, and no. of pods are found significantly superior. Among the *in-situ* rainwater conservation practices, significantly higher mean consumptive use(724.45mm) and rain water use efficiency (4.95 kg/ha-mm) was recorded in opening of furrow after two rows (T_3). Also the Highest yield was recorded i.e., T_3 (3583.3kg/ha) over rest of treatments with higher net return (138930Rs./ha) and B: C ratio (7.4:1).

Keywords: Opening of furrow after different rows, Water use efficiency, Consumptive use, Yield, Cost benefit.

INTRODUCTION

Soil is the earth's fragile skin that anchors all life on Earth. It is comprised of countless species that creates a dynamic and complex ecosystem and is among the most precious resources to humans. India is the largest producer of pulses in the world with 25% share in the global production and has the largest area (3.6 M ha) under pigeon pea. Pigeon peas are very drought resistant. Pigeon peas are a food crop, and a forage/ cover crop. In this crop there is large variation for maturity that offers its cultivation under different environments and cropping systems. Hence the greater attention is now being given to managing crop because it is in high demand at remunerative prices. Conservation practices are generally those that reduce wind speed, reduce rate and amount of water movement, and/or increase soil organic matter levels.

MATERIAL AND METHODS

A field experiment entitled "Effect of *in-situ* rainwater conservation practice on moisture use and yield of pigeon pea" was conducted during 2013-2014. The experiment was conducted in medium black soil. Field experiment was planned comprising of opening of furrow as the main factor in RBD with seven treatments each having three replications. The pigeon pea was sown on 27th June 2013. Soil moisture was calculated during sowing. After sowing, treatment wise different numbers of furrows were open i.e. T1- Sowing of pigeon pea along the slope, T2 - across the slope ,T3- Opening of furrow after two rows across the slope, T4 - three rows across the slope, T5- four rows across the slope, T6 - five rows across the slope,T7 - six rows across the slope to achieve moisture conservation. The observations were recorded on different growth stages i.e 0-45 days : Branching stage,45-110 days : Flowering stage, 110- 120 days : 50% Flowering stage, 120-140 days : Pod formation, 140- 160 days : Grain development ,160-180 days : Maturity stage. Soil water content (%) in the profile was worked out by water content in the soil profile layers, i.e. 0-15, 15-30 and 30-45cm by gravimetric method at the time of sowing and harvesting of crop. The soil water content was then converted on volume basis (mm) as

$$d = \sum (Mi \div 100)\, DiAi + ER$$

Where,

Mi = is soil water content (%) of ith layer.

Di = is soil depth of ith layer.

Ai = Apparent specific gravity of ith layer.

The difference in soil water content at sowing minus soil water content at harvest was computed for consumptive use (CU) by the crop.

Rain water use efficiency (RWUE) (kg $ha^{-1}mm^{-1}$) was worked out by dividing the grain yield (kg ha^{-1}) by crop consumptive use (mm).

$$\text{Rain water use efficiency (RWUE)} = \frac{\text{Grain yield (kg ha}^{-1})}{\text{Consumptive use (mm)}}$$

The data for individual treatments were recorded and subjected to statistical analysis in RBD.

RESULT AND DISCUSSION

During the study important findings on soil moisture content, grain yield, strove yield and economics of treatments (gross monetary return, net monetary return and B:C ratio) of pigeon pea as influenced by various rain water conservation practices. Replication wise soil samples were collected at crop growth stages at 0-15 cm, 15- 30 cm and 30-45 cm depths.

Table 1 Mean moisture content during Kharif 2013-2014 as influenced by rain water conservation practices.

Treatments	Per-cent moisture
T_1	13.92
T_2	17.01
T_3	21.86
T_4	19.12
T_5	18.24
T_6	17.77
T_7	16.24
Mean	17.74
SE±	0.34
CD at 5% level	1.06

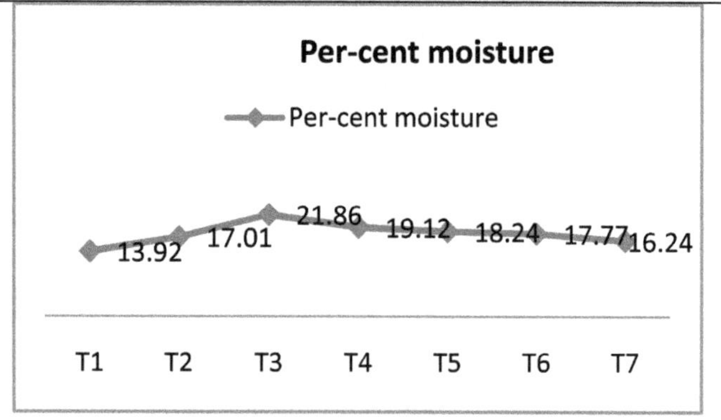

Fig 1 Mean soil moisture during the kharif 2013-2014

As per table no.1 data on soil moisture conserved during the kharif season 2013 overall up to 45cm depth indicated that the moisture contents in opening of furrow across the slope in pigeon pea crop were significantly superior over control (T_1).

Yield and yield attributing parameters

The data indicated that pigeon pea yield significantly higher (3583.3 kg/ha) with opening of furrow across the slope after two rows. On the contrary the yield obtained in sowing along the slope (control) was reduced to near 11.7 per cent shown in table no.2.

Consumptive use and rain water use efficiency

Mean consumptive use and rain water use efficiency are represented in table no. 2

Table 2 Mean effect of opening of furrows across the slope on consumptive use, grain yield, water use efficiency, GMR, NMR and B: C ratio in pigeon pea crop Kharif- 2013.

Treatments	Consumptive use (mm)	Grain yield (kg/ha)	Water use efficiency(kg/ha-mm)	Cost of cultivation	GMR	NMR	B:C Ratio
T_1	697.67	2406.0	3.44	17009	105860	88855	5.20
T_2	723.64	2545.0	3.51	17009	111980	94971	5.50
T_3	724.45	3583.3	4.95	18634	157564	138930	7.40
T_4	719.41	2625.0	3.57	18309	115500	97191	5.30
T_5	716.75	2572.0	3.61	17984	113170	95184	5.30
T_6	716.68	2517.0	3.42	17659	110750	93089	5.26
T_7	716.47	2453.0	3.50	17334	75099	90598	5.20
Mean	716.44	2671.1	3.71	17705	112850	99831	5.59
SE±	1.54	20.60	0.02	158.06	126005	255.6	0.05
CD at 5% level	4.7	63.40	0.07	486.30	38784	786.5	0.16

Biometric parameters

During harvesting biometric observations of fully matured pigeon pea were recorded from five plants randomly selected from each treatment each replication and control field. The mean biometric observations of pigeon pea influenced by different rain water conservation practices are presented in table no. 3.

Table 3 Mean biometric observation as influenced by opening of furrows and cropping systems kharif -2013.

Treatments	No. of Branches	Height of Plant(cm)	No. of Pods
T_1	8.46	160.93	86.26
T_2	9.53	166.87	96.40
T_3	12.00	183.73	153.93
T_4	11.66	178.73	151.47
T_5	11.53	178.33	150.93
T_6	11.46	178.07	139.27
T_7	10.00	173.07	112.87
Mean	10.66	174.25	127.30
SE±	1.03	5.55	17.27
CD at 5% level	3.17	1.70	53.13

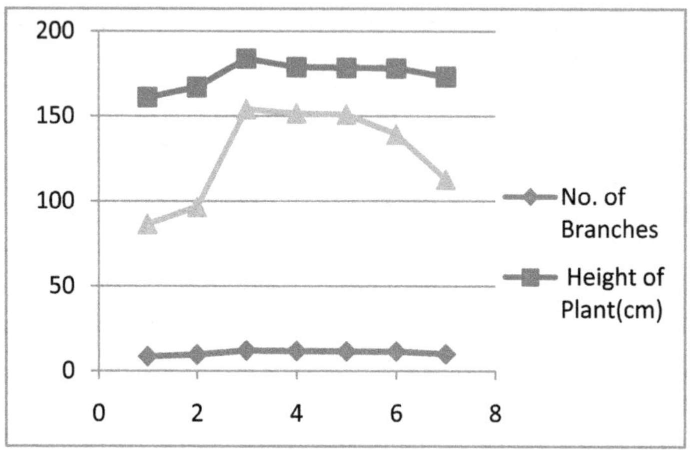

It was noted that the per cent increase in height of plant (14.17) and number of pods (78.44) in opening of furrow after two rows (T_3) over the control treatment T_1.

Economics

Analysis of different moisture conservation practices in pigeon pea obviously reflects the superiority. It was also found that per-cent increase in net monitory return in opening of furrow across the slope after two rows in pigeon pea (T_3) over the control treatment (T_1) was recorded 56.35%. All the *in situ* rain water conservation practices caused higher net return and B: C ratio over the sowing of pigeon pea crop along the slope (T_1).

REFERENCES

1. Akhtar J., S. M. Mehdi, Obaid-Ur-Rehman, K. Mahmood and M. Sarfraz (2005). Effect of deep tillage practices on moisture preservation and yield of groundnut under rainfed condition. J. Agri. Soc. Sci., Vol. 1, No. 2.
2. Allolli T. B, Hulihalli U. K and Athani S. I. (2007). Influence of in situ moisture conservation practices on the performance of dry land chili. Department of Horticulture University of Agricultural sciences, Dharwad - 580 005, Karnataka, India.
3. Devakant, Tiwari P. N. (2000). Effect of *in-situ* moisture conservation system on yield and soil moisture conservation. Bhartiya Krishi Anusandhan Patrika. 15 (1/2): 65-68.
4. Gupta Dilip kumar and Suraj Bhan (1993). Effect of *in-situ* moisture conservation practices and fertilizer on root characteristics and yield of rain fed maize. Indian J. Soil Cons. Vol. 21 pp. 22-24.

WATER SUSTAINABILITY AND ENVIRONMENTAL MANAGEMENT

Sk. S. Alisha and P. Rohith

Vishnu Engineering College, Bhimavaram,
subhansk3@gmail.com and rohithnani134@gmail.com

ABSTRACT

Sustainability development is a hot issue facing corporations. Studies showed that financial accounting could not fully support sustainability development since the highly regulated financial accounting had specific accounting rules that resulted in incomplete capturing and presentation of environmental costs. In the relatively less regulated accounting application, the management accounting, studies found that environmental costs were usually absorbed in overheads. The communication between accountants and environmental experts were usually limited and this lead to misallocation or incorrect calculation of environment costs. As a result, managers did not have the correct environmental information for managing environmental costs for sustainability development. To address the limitations of management accounting, environmental management accounting (EMA) was developed. EMA could address both monetary and physical aspects of environmental accounting. Physical EMA included the flow of water, energy, while monetary EMA measured the costs of the firm's consumption of natural resources and the costs for controlling or preventing environmental damages.

INTRODUCTION

Throughout this document, the term "Environmental Management System," or EMS, is used to describe the organizational structure, responsibilities, practices, processes and resources for implementing and maintaining environmental management. To be considered for this alternative, Ecology has determined that pollution prevention, as defined herein, must be explicitly considered in such a system.

EMA was developed to recognize some limitations of conventional management accounting approaches to environmental costs, consequences, and impacts. For example, overhead accounts were the destination of many environmental costs in the past. Cost allocations were inaccurate and could not be traced back to processes, products, or process lines. Wasted raw materials were also inaccurately accounted for during production.

Water sustainability could be defined as supplying or being supplied with water for life or, perhaps more precisely, as the continual supply of clean water for human uses and for other living things. It is does not specify exactly how much water we have, nor does it imply the unrestrained, infinite availability of water. Rather, it refers to the sufficient availability of water into the foreseeable future. Water is, after all, a renewable resource, so sustaining its uses should be possible, shouldn't it? But it turns out that we can have too much water or too little water to meet our needs.

Basic concepts

- Water can be saved through the combination of three important measures:

- Firstly, overall water use can be reduced through the installation of highly efficient fixtures, appliances and systems, throughout your property.
- Secondly, rainwater and / or greywater should be used in preference to drinking water for purposes such as toilet flushing, laundry and irrigation where appropriate.
- Thirdly, minimise the volume of external water features and pools. Reuse water in water features and utilise pool covers.

Literature review

- Within the European Community, last decades were marked by a continue increasing of preoccupations concerning generalization of the sustainable growth. The identification of instruments, methods and techniques for decoupling the economic activities by their negative impact on the environment and society constitutes permanent objectives of decisional and legislative factors of the European institutions. Renewed EU Sustainable Strategy restated the necessity to achieve a growth of a sustainable type in which the environment protection beside the social cohesion and economic prosperity are essential objectives of a continuous action for improve the life's quality for the present and future generations (European Council, 2006). Nowadays, the European Commission proposes that the action directions to accelerate the sustainable development to follow up the transition to a resource-efficient Europe (SEC 1067, 2011). The efficient use of resources will lead to increase the grade of sustainability of the productive activities on the whole national economy, on branches and companies. In the literature there are numerous studies which present the relation between economic performance and environment performance as an inverted U shape function (Lankoski, 2000).
- These researches show that an increasing of the environment performance over one optimum point could diminish the company's profit (Salzmann et al., 2005). Other case studies demonstrate that business sustainability is not a utopia and cannot generate the costs increasing; contrary it contributes to their reduction, obtaining important gains on the long period of time (Hogevold and Svensson, 2012). Also, documents issued by the European institutions consider that actions which aim diminishing the ambient environment's pollution can produce favourable economic and social effects.

CASE STUDY

Water is life. Growing pressure on water resources – from population and economic growth, climate change, pollution, and other challenges – has major impacts on our social, economic, and environmental well-being. Many of our most important aquifers are being over-pumped, causing widespread declines in groundwater levels. Major rivers – including the Colorado River in the western United States and the Yellow River in China – no longer reach the sea in most years. The California drought is exacerbating the large and growing gap between the state's water use and the available water supply. Half of the world's wetlands have been lost to development. The world's water is increasingly becoming degraded in quality, threatening the health of people and ecosystems and increasing the cost of treatment. Some 780 million people around the globe still lack access to clean water and thousands perish daily for lack of it. Most countries have water resource management plans that address both supply and demand.

Water pricing is one of the measures used to reduce water demand. The Water Framework Directive requires EU Member States to ensure that by 2010 the proportion of the cost of water services – such as pumping, weirs, dams, channels, supply systems – with a negative impact on the environment – must be paid by the users (e.g. agriculture, hydropower, households, navigation). Member States are required to split the costs according to the 'polluter-pays' principle in order to reduce the impact on the environment and promote economic instruments to tackle the decline of natural resources. If Member States fail to include other infrastructures than drinking water supply and wastewater treatment in their economic analyses, there is a major risk that such infrastructures already identified as creating major environmental problems will be exempt. Consequently, the economic burden of water bodies reaching 'good status' by 2015 will remain with citizens, who already pay high prices for water services.

EMA systems can help environmental managers justify these cleaner production projects, and identify new ways of saving money and improving environmental performance at the same time.

The results of improved costing by EMA may include: Different pricing of products as a result of re-calculated costs; Re-evaluation of the profit margins of products; Phasing-out certain products when the change is dramatic; Re-designing processes or products in order to reduce environmental costs; Improved housekeeping and monitoring of environmental performance.

Environmental Management Systems (EMS) according to the ISO standard The ISO14001 standard requires the evaluation of environmental aspects during the planning phase of the environmental management system. In ISO 14001 environmental aspects are "elements of an organization's activities, products and services that can interact with the environment."

The company shall: Identify the aspects which have an impact on the environment and Assign a level of significance to each environmental aspect "When establishing and reviewing its objectives, an organization shall consider the legal and other requirements, its significant environmental aspects, its technological options and its financial, operational and business requirements, and the views of interested parties".

An EMA system can separate end-of-pipe costs from prevention costs. It also helps in calculating the savings gained through the reduced use of raw materials and energy. Without these data from environmental programmes, companies will continue to think of environmental management as a strictly non-profit-generating part of business that always costs money. Cleaner production can save money and thereby increase profits. With an EMA these savings can be captured and reported. EMA generated data improves the bargaining power of environmental managers with a company's top managers and shareholders, to create or obtain funding for environmental programmers, CP projects and EST investments. It will also provide precise numbers on environmental costs, when required by external stakeholders. While shareholders are concerned about their liabilities, external stakeholders (authorities, civil societies, NGOs, etc.) are interested in seeing the company's efforts toward environmental management supported by substantial environmental expenditures. Data generated by an EMA will help demonstrate these efforts.

Endeavour to maintain stock quality water quality and where use restrictions are required, ensure appropriate notification and management is in place Objective

Establish final land use objectives which are consistent with land capacity and the surrounding social context Objective 6. Engage in Community/Stakeholder consultation to a degree that there is general acknowledgement that the process has been satisfactory.. The criteria used to assess impacts are described below. Each activity of the project will identify potential impacts, and propose mitigation methods for the identified impacts. and the proposed mitigation measures.. To ensure delivery of this policy Arafura Resources strives to achieve the following objectives: Objective

- Ensure the site is physically safe and does not pose a human health risk Objective
- Ensure that land is left in a stable condition that minimises long-term environmental impacts Objective
- Rehabilitate disturbed land such that it promotes sustainable ecosystems Objective
- Measuring water use is a prerequisite for water prices reducing consumption. Households with water meters installed generally use less water than households without meters. In Europe, household and industrial water metering continues to increase. Many of the NWE countries already meter the majority of water uses. However, in many countries and in relation to agriculture water use metering is still limited.
- When addressing water charges, focus should also be placed on households and agriculture that have difficulties with paying for water for essential purposes (since it is generally recognised that no one should have to compromise personal hygiene and health). The Water Framework Directive requires an affordable price to guarantee a basic level of domestic water supply

Over the past 10 years there has been a marked increase in the amount of information provided to consumers (e.g. water-efficiency labels for households' appliances, information on efficient lawn watering and gardening practices, etc.) and agriculture. Many countries, NGOs, large municipalities, water companies and international organisations have dedicated home pages to water conservation and water use behaviour.

Standards such as ISO 14001 take a comprehensive view of all of the processes of an organization - hence they are system dependent, and not person-dependent. An EMS creates a structured management system, from which a cycle of continual improvement can be established. It brings the many environmental issues of concern expressed by stakeholders into day-to-day operations and development of long term work plans and programmes. It also improves the understanding amongst an organisation's personnel of where operations interact with the natural environment and the role that various groups play. An EMS can result in both business and environmental benefits, eg helping to: • Improve environmental performance; • Enhance compliance; • Prevent pollution and conserve resources; • Reduce/mitigate risks; • Attract new customers and markets (or at least retain access to customers and markets with EMS requirements); • Increase efficiency/reduce costs; • Enhance employee morale (including the possibility of enhanced recruitment of new employees); • Enhance image with public, regulators, lenders, investors; • Achieve/improve employee awareness of environmental issues and responsibilities.

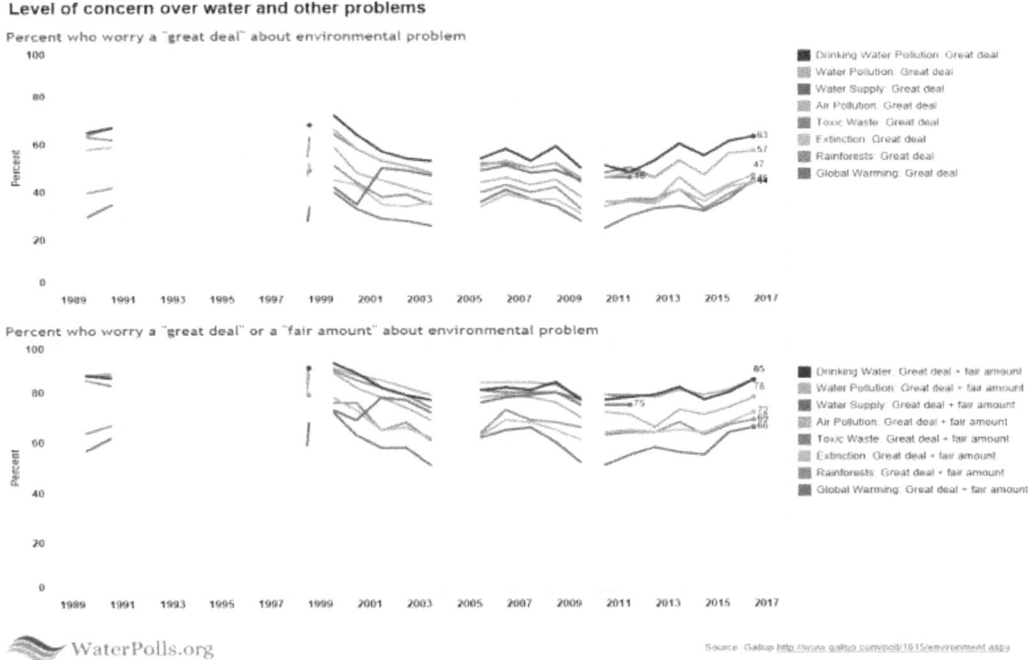

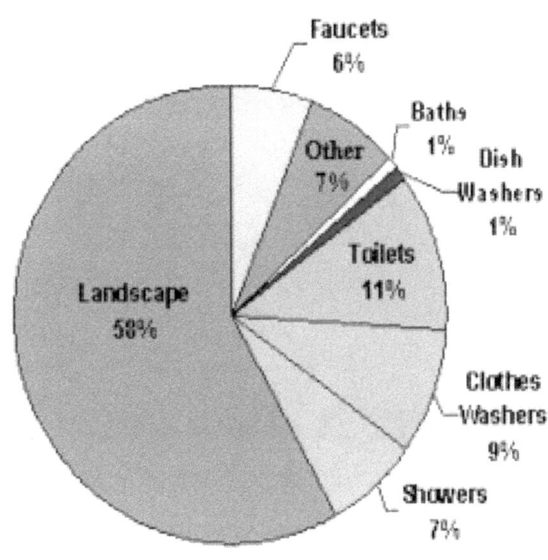

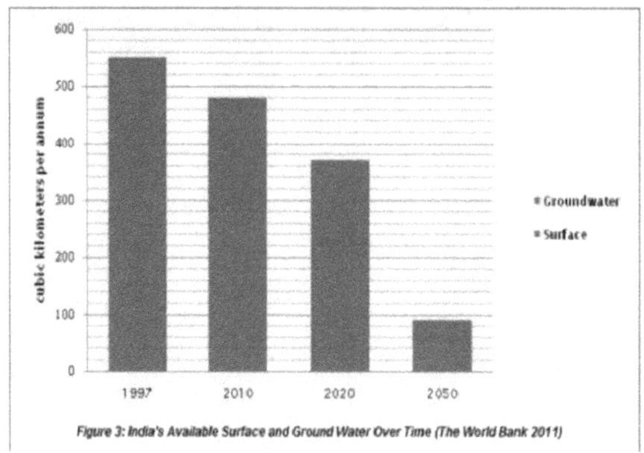

Figure 3: India's Available Surface and Ground Water Over Time (The World Bank 2011)

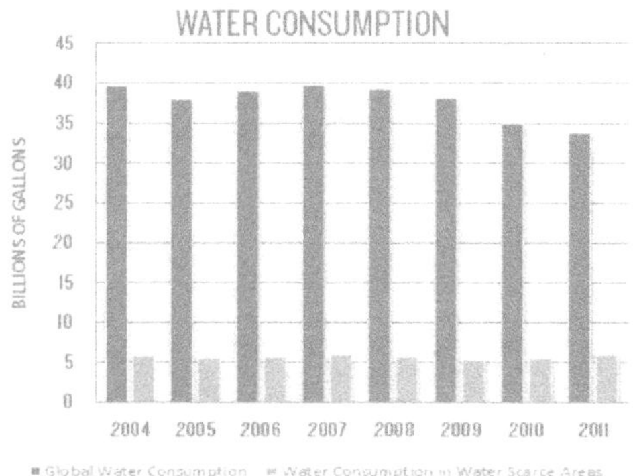

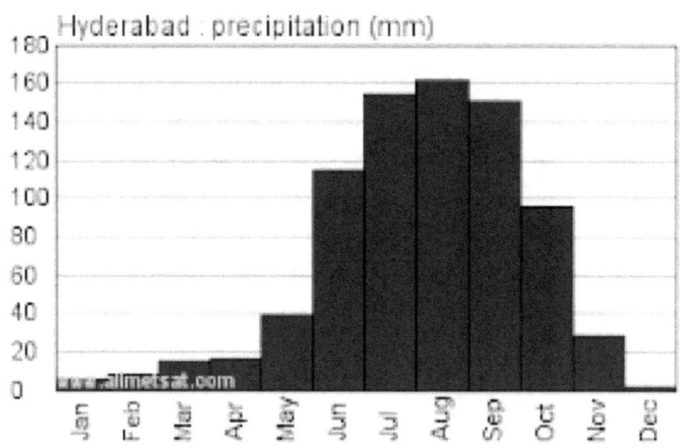

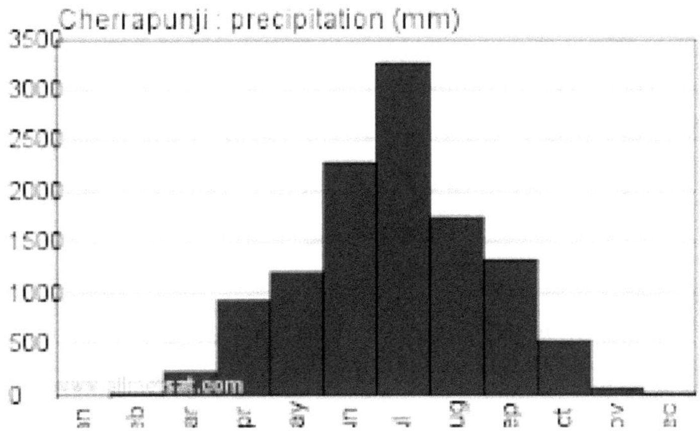

Purpose	Litres/ person/day
Drinking	03
Cooking	04
Bathing	20
Flushing	40
Washing clothes	25
Washing utensils	20
Gardening	23
Total	**135**

Table 1: Per Capita Water Requirement

Water use

Water withdrawal			
Total water withdrawal	2008	338	million m^3/yr
- irrigation + livestock	2008	318	million m^3/yr
- municipalities	2008	17	million m^3/yr
- industry	2008	3	million m^3/yr
• per inhabitant	2008	482	m^3/yr
Surface water and groundwater withdrawal	2008	338	million m^3/yr
• as % of total actual renewable water resources	2008	0.43	%
Non conventional sources of water			
Produced wastewater		-	million m^3/yr
Treated wastewater		-	million m^3/yr
Reused treated wastewater		-	million m^3/yr
Desalinated water produced		-	million m^3/yr
Reused agricultural drainage water		-	million m^3/yr

CONCLUSIONS

Environmental management systems are going to emerge as prominent management systems in the coming days. Their administrative structure and management framework introduce the concrete mechanism that can be used to execute an environmental agenda in a robust way. However, these tools fall short of incorporating any definition of environmental sustainability

OPERATION STUDY FOR RELIABLE WATER SUPPLY OF HYDERABAD SYSTEM (A CASE STUDY ON SINGUR RESERVOIR SYSTEM)

M. Anjaneya Prasad[1], M. Satyanarayana[2] and S. Santosh Kumar[3]

[1]Research Scholar and [2]Professor of Civil Engineering, College of Engineering, Osmania University.
[3]Executive Director in Hyderabad Water Supply & Sewerage Board (HMWSSB).
[3]Dy.GM(E) in HMWSSB.

ABSTRACT

The study on Reservoir Operation Models is a challenging task and researchers have adopted various optimization and simulation techniques to evolve the best release policies. Singur reservoir is one of the key storage reservoirs for Hyderabad Drinking water supply system. This paper presents the simulation model study on Singur reservoir operation based on simulation basis using fuzzy rules as basis and the results were compared with that of crisp model releases and the developed fuzzy rule based model is found to be superior over the crisp simulation mode. The deficits from the developed model was found to minimized the deficits from the reservoir.

Keywords: Reservoir Operation, Simulation Models, Design performance, Fuzzy Logic.

INTRODUCTION

The reservoir system management for operation for releases of drinking water, Irrigation water, power releases, mandatory releases etc is considered in the literature for the study for maximizing the efficiency and improved the performance of the reservoir. Researchers have evolved various reservoir operation policies with several conventional and mathematical techniques including optimization and simulation using various modeling techniques.

The modelling techniques and their applications some of the real time reservoirs operations are studied by various research scholars. Durbovin et al (2002)[1], Real time reservoir operation model based on total fuzzy similarity and compared with fuzzy inference method. Vedula & Mohan (1991)[5], Real-time Multipurpose Reservoir Operation for Irrigation and Hydropower generation for case studies of Bhadra reservoir system in Karnataka using Stochastic Dynamic Programming SDP. Panigrahi and Mujumdar (2000)[2], Studied fuzzy rule based model for a single reservoir operation using (SDP) for framing rule base. Shrestha (1996)[6], developed fuzzy relations for input and output of reservoir operating principles and defuzzified to get the crisp outputs. Mousavi et al (2004)[7], Reservoir operation using a Dynamic Programming Fuzzy rule based approach to establish the general operating policies.

The present study is exclusively on performance evaluation reservoir on drinking water supply after meeting it's mandatory releases if any and to evaluate its performance in meeting its demands. In this paper, the Singur reservoir system has been taken for the reservoir operation analysis. The Singur reservoir is operated for Drinking Water Supply and also for mandatory releases to the downstream reservoir for bulk release which acts as a master storage reservoir. The inflow data was obtained from the department of central water commission office from the Saigaon flow measuring gauge station. The Singur reservoir built across the River Manjeera is in the Godavari basin being operated for drinking water supply of Hyderabad city and provides bulk mandatory releases to the downstream reservoir of Nizamsagar for irrigation purpose and acts as storage reservoir. The reservoir has a active storage capacity of 847.5 Mm^3.

The objective of the study is to apply the proposed methodology to evolve the reservoir operating policies which indicates substantial increase of model releases compared to the historical releases. The operating policy considered thus provided improved releases from Singur reservoir system which helps in minimizing the supply deficits in the demand system. The proposed operation methodology consists of three phases of modeling. In the first phase, the release policy for the given initial storage and known inflows with defined operating policy for a given simulation period is determined using sequential process simulation. In the second phase, the release policy was determined using fuzzy logic rule based simulation. The released results were compared for the historic data, pure simulation and fuzzy rule based simulation. The model releases obtained from fuzzy simulation in the second phase model and the demand deficit strategies were studied to develop strategies for minimizing deficits in the third phase. Comparison of the monthly operation releases with that of historical operation, crisp model operation and fuzzy model operation for the strategies adopted for the system demonstrated that the proposed methodology has improved performance in meeting its demands.

SYSTEM DATA

Monthly inflow data of Singur reservoir system for 23 water years (1994-2016, water year beginning 1st January and ending 31st December) and monthly withdrawals data for 23 years (1994-2016) were used in the present study. A fixed downstream river releases was allowed from Singur reservoir for the purpose of filling the Nizamsagar reservoir. In the Table-1 the annual inflows in Mm3 for Singur reservoir for the period 1994-2016 are presented.

Evaporation loss data for the period from 1994-2016 were used in deriving the relationship between the evaporation and average storage in each month by least squares fitting and evolved the evaporation curves & used in the model. The initial Storage for Singur Reservoir was considered for the month of January 1994.

Table 1 Annual Inflows in Mm^3

Year	Singur	Year	Singur
1994	4250	2006	33733
1995	33285	2007	16485
1996	100140	2008	53714
1997	8301	2009	8247
1998	165623	2010	107392
1999	40140	2011	34830
2000	88966	2012	6035
2001	25549	2013	27207
2002	5745	2014	2749
2003	11714	2015	688
2004	3616	2016	112994
2005	36962		

Source:- HMWSSB Reservoir Log Records– (2016) [3]

WATER DEMAND

Monthly demands in the command area of entire Singur reservoir and for the total Hyderabad Water Supply System were computed for all nodal demand centers based on the LPCD guidelines given by the CPHEEO, GoI. In the computation of the demands, basis was considered as people's access through service connection to the system, category wise consumption pattern such as domestic slum consumption pattern and quantities, domestic general, commercial, industrial, mobile supply etc in conformity with the approved norms of the applied system. The demands for various categories of domestic slums, domestic general, commercial, industrial, mobile supplies etc were show in Table-II for each month.

In the present study, the reservoir operation with node wise demands and releases were taken for monthly simulation run for Singur reservoir.

Table 2 Demands in Mm^3

Month	Domestic slum	Domestic general	Commercial	Industrial	Mobile supply	Total
January	4.000	40.837	4.925	4.116	0.631	54.510
February	3.621	36.931	4.454	3.731	0.570	49.307
March	4.022	40.956	4.932	4.131	0.631	54.672
April	3.901	40.011	4.772	3.996	0.617	53.297
May	4.041	41.501	4.932	4.128	0.637	55.240
June	3.922	40.221	4.779	4.043	0.617	53.581
July	4.067	41.625	4.944	4.178	0.637	55.451
August	4.075	41.670	4.946	4.178	0.641	55.509
tember	3.953	40.376	4.787	4.105	0.623	53.844
October	4.094	41.785	4.959	4.243	0.644	55.724
November	3.972	40.502	4.800	4.106	0.623	54.003
December	4.119	41.929	4.962	4.246	0.644	55.900
Total	47.787	488.344	58.192	49.201	7.515	651.038

Source:- HMWSSB Revenue Data – (2016) [4]

Reservoir Operation Model

Storage Continuity Equation:-

$$S_{t+1} = (S_t + I_t - R_t - E_t - O_t) \quad (1)$$

Capacity Constraints:-

$$S_{\square+1} \leq S_{Cap} \quad (2)$$

Spill Constraints:-

$$Spill_t = S_{\square+1} - S_{cap} \quad (3)$$

$$Spill_t \geq O_t \quad (4)$$

Release Constraints: -

$$R_t \ \square \ D_t \quad (5)$$

$$Deficit = D_t - R_t$$

$$Min\ Def = \sum (D_t - R_t)^2 \quad (6)$$

Generally: $R_t = f(S_t, I_t)$ (7)

The releases from reservoir is a function of S_t, I_t. But it can also be operated as R_t is independent of I_t.

RESERVOIR OPERATING POLICY

The Reservoir Operating Policy for each month for known values of initial storage and inflow sequences for the Singur reservoir was obtained with crisp values of class interval of active storages / Levels based on the general formulation rules from time to time were observed historically and based on the operator's experience on the performance of the system. The class intervals selected for this purpose are limited to five classes to study the system. The rules were then generated based on "if then" principles using sequential processing simulation and is referred as "crisp model".

The inflows were considered as deterministic as obtained from the measured records of the respective reservoir. The operating policy derived from the crisp model is a set out of rules specifying the storage at the beginning of the next period for each combination of initial storage and inflow for the current period thus specifying the release for the current period. The objective of the model is to obtain the release values as per the defined release policy based on the crisp concepts. The Model Release Curves are as below:

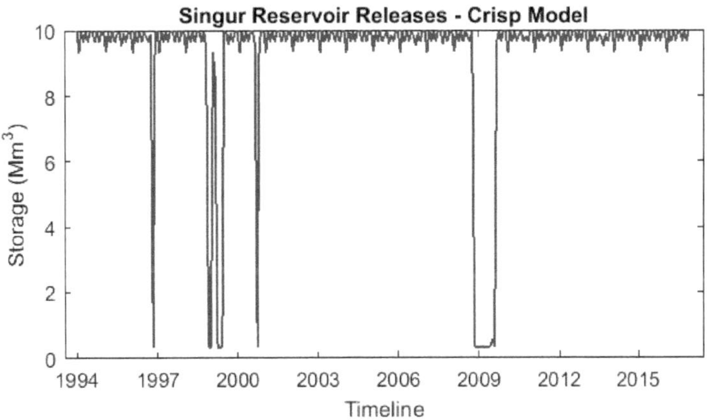

Fig. 1 Crisp Model Releases.

In figure-I the model releases based on the simulation are shown through graphical plots for Singur reservoir are found to be satisfactory as per the historical operation.

In the second phase, the fuzzy model simulation was performed to derive the monthly operating policy using the storages, inflows, release values into fuzzy sets, fuzzy intervals with associated membership functions, fuzzy rules, fuzzy inferences and defuzzification for obtaining the final storages and release values. In the both model the evaporation equations were used in the storage continuity equation with the storage constraints, release constraints, demand constraints etc. The typical fuzzy rule base is shown in table III. The screenshot of fuzzy based model is shown in figure 2.

Table 3 Typical Fuzzy Rules

S. No.	Typical Rule	Strength of Rule
1	If (Storage is very low) and (Inflow is low) then (Releases is very low)	1
2	If (Storage is low) and (Inflow is low) then (Releases is low)	1
3	If (Storage is medium) and (Inflow is low) then (Releases is medium)	1
4	If (Storage is high) and (Inflow is low) then (Releases is very high)	1
5	If (Storage is medium) and (Inflow is high) then (Releases is very high)	1
6	If (Storage is high) and (Inflow is high) then (Releases is very high)	1

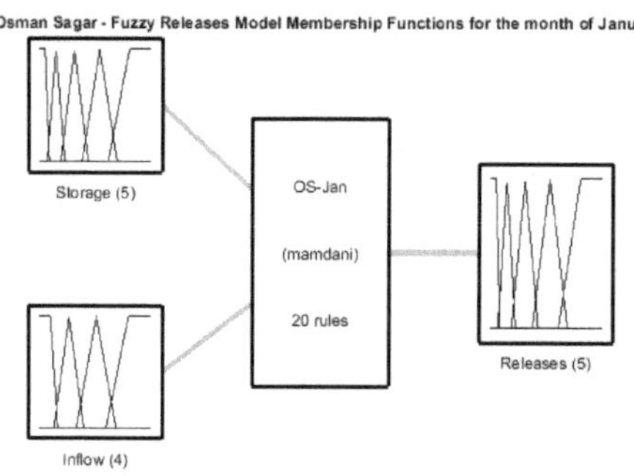

Fig. 2 Fuzzy Model Screenshot.

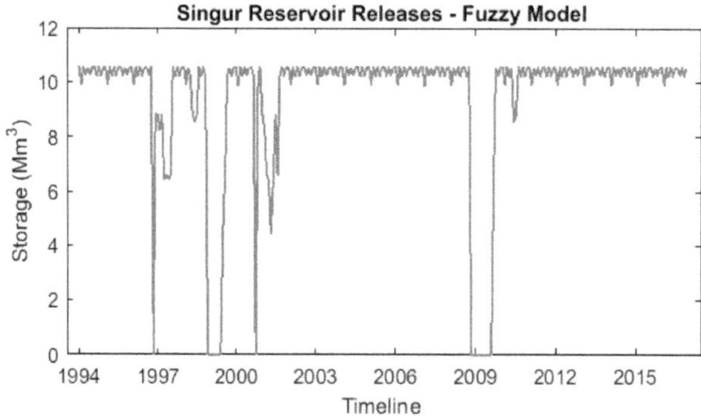

Fig. 3 Fuzzy Model Releases.

In figure-3 shows graphical plots are based on the fuzzy model releases for Singur reservoir which are found to be superior over the crisp model releases.

Model Results

In Table-IV the model release results of crisp and fuzzy simulation for typical two years are presented which are found to be superior over the historical releases.

Table 4 The comparison of monthly release for the consistently performed simulation period.

In Mm^3

Month	Singur	
Year	**Crisp**	**Fuzzy**
Jan-12	9.96	10.56
Feb-12	9.1	9.66
Mar-12	9.95	10.55
Apr-12	9.63	10.2
May-12	9.97	10.56
Jun-12	9.65	10.22
Jul-12	9.98	10.57
Aug-12	9.98	10.57
Sep-12	9.65	10.22
Oct-12	9.96	10.56
Nov-12	9.66	10.23
Dec-12	9.95	10.54

Month	Singur	
Year	**Crisp**	**Fuzzy**
Jan-11	9.97	10.56
Feb-11	9.33	9.89
Mar-11	9.95	10.54
Apr-11	9.49	10.06
May-11	9.97	10.56
Jun-11	9.66	10.23
Jul-11	9.78	10.56
Aug-11	9.76	10.4
Sep-11	9.52	10.12
Oct-11	9.95	10.55
Nov-11	9.66	10.23
Dec-11	9.98	10.57

Table 5 Model Release Mean & Variance in Mm^3

Reservoir	Model Releases	Mean Value	Variance Value
Singur	Crisp	9.7028	1.22017
	Fuzzy	9.9089	1.64397

Table 5 shows the model releases statistical performance was evaluated through mean and variance values for Singur reservoir which are found to be in order. The evaluation, the monthly releases are recommended from the Fuzzy Model study to the supply nodes through integrated system along with the fixed releases.

Table 6 Monthwise Model Releases - Mean Variance in Mm^3

Month-Year	Fuzzy Releases	Mean	SD	Variation	Upside-down	Threshold	Lower Threshold	Upper Threshold	Average Threshold
Jan-94	15.17	13.95	1.48	4.01	12.74	9.71	11.95	15.4958555	13.72
Feb-94	12.55	13.95	1.98	4.01	15.36	9.71	11.95	15.4958555	13.72

Contd...

Month-Year	Fuzzy Releases	Mean	SD	Variation	Upside-down	Threshold	Lower Threshold	Upper Threshold	Average Threshold
Mar-94	13.10	13.95	0.73	4.01	14.81	9.71	11.95	15.4958555	13.72
Apr-94	11.33	13.95	6.90	4.01	16.58	9.71	11.95	15.4958555	13.72
May-94	11.34	13.95	6.83	4.01	16.57	9.71	11.95	15.4958555	13.72
Jun-94	10.97	13.95	8.92	4.01	16.94	9.71	11.95	15.4958555	13.72
Jul-94	11.05	13.95	8.43	4.01	16.86	9.71	11.95	15.4958555	13.72
Aug-94	11.06	13.95	8.39	4.01	16.85	9.71	11.95	15.4958555	13.72
Sep-94	10.69	13.95	10.63	4.01	17.21	9.71	11.95	15.4958555	13.72
Oct-94	10.99	13.95	8.80	4.01	16.92	9.71	11.95	15.4958555	13.72
Nov-94	13.31	13.95	0.41	4.01	14.60	9.71	11.95	15.4958555	13.72
Dec-94	13.39	13.95	0.32	4.01	14.52	9.71	11.95	15.4958555	13.72

In table-V the statistical analysis of mean, standard deviation and upper and lower threshold values are presented which indicates the results are satisfactory limits.

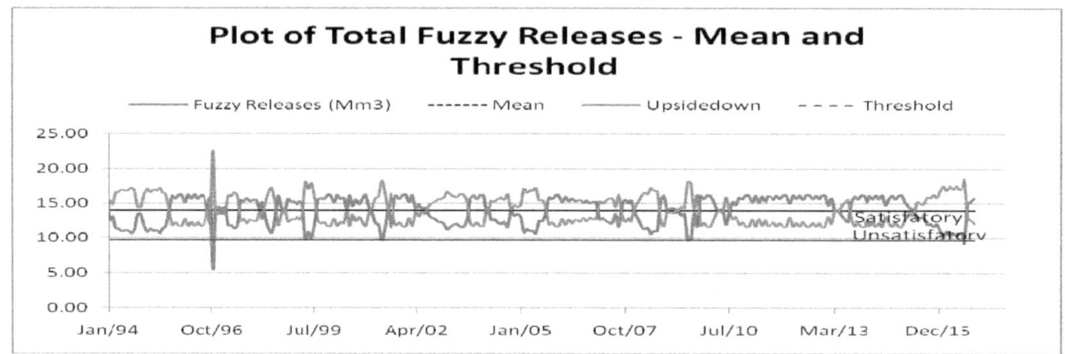

Fig. 4 Model Release Mean & Variance.

The Figure-4 shows the fuzzy releases with mean and threshold limit and found to be most of the results are within the threshold limits except two exceeding in full simulation periods.

The Table-VI the monthly releases derived from the fuzzy simulation model recommended for reservoir operation for ensuring the improved water supply in the system.

In the third phase with the model releases and fixed releases into integrated transmission system accessible to the nodal centers are allocated as per the preference to the nearby source connectivity, departmental approved norms, using rational methods, the releases are allocated to the nodes. From the already estimated demands for each nodal centers with the given integrated releases, the demand deficits are calculated and the strategies for reduction of commercial and industrial and tradeoff to domestic slum and domestic general categories are made and evaluated the original percentage of deficit nodes, reduced percentage of deficit nodes on account of tradeoff strategies are studied and recommended the suitable strategies.

Table 7 Monthly Releases Recommended from fuzzy Model Study in Mm^3

Month / Year	Singur
January	10.56
February	9.89
March	10.54
April	10.06
May	10.56
June	10.23
July	10.56
August	10.4
September	10.12
October	10.55
November	10.23
December	10.57

CONCLUSIONS

A mathematical simulation model has been successfully developed for performance evaluation of a reservoir Singur in meeting the drinking water needs of parts of Hyderabad city successfully. The developed simulation model is performing satisfactorily for reservoir operation in meeting its demands and also reducing deficits by 2.12% over conventional operation of reservoir system.

ACKNOWLEDGEMENT

The authors gratefully acknowledged the help rendered by Hyderabad Metropolitan Water Supply & Sewerage System in providing the necessary field data for the study.

REFERENCES

1. Durbovin et al (2002), Real time reservoir operation model based on total fuzzy similarity and compared with fuzzy inference method.
2. Panigrahi and Mujumdar (2000), Studied fuzzy rule based model for a single reservoir operation using Stochastic Dynamic Programming (SDP) for framing rule base.
3. Hyderabad Metropolitan Water Supply & Sewerage Board, Reservoir Log Records– (2016).
4. Hyderabad Metropolitan Water Supply & Sewerage Board, Revenue Data – (2016)
5. S. Vedula & S. Mohan (1991), Real-time Multipurpose Reservoir Operation for Irrigation and Hydropower generation for case studies of Bhadra reservoir system in Karnataka using SDP.
6. Shrestha (1996), developed fuzzy relations for input and output of reservoir operating principles and defuzzified to get the crisp outputs.
7. S.J. Mousavi, K. Ponnambalam & F. Karray (2004), Reservoir Operation using a Dynamic Programming Fuzzy Rule Based Approach to establish the general operating policies.
8. M. Satyanarayana, Ph D scholar under the guidance of Prof. M. A. Prasad, OUCE – Research Design and Progress Report 1 & Report 2 Submitted to Civil Engineering Department, Osmania University.

Table 7 Monthly Releases Recommended from fuzzy Model Study in Mm^3

Month / Year	Singur
January	10.56
February	9.89
March	10.54
April	10.06
May	10.56
June	10.23
July	10.56
August	10.4
September	10.12
October	10.55
November	10.23
December	10.57

CONCLUSIONS

A mathematical simulation model has been successfully developed for performance evaluation of a reservoir Singur in meeting the drinking water needs of parts of Hyderabad city successfully. The developed simulation model is performing satisfactorily for reservoir operation in meeting its demands and also reducing deficits by 2.12% over conventional operation of reservoir system.

ACKNOWLEDGEMENT

The authors gratefully acknowledged the help rendered by Hyderabad Metropolitan Water Supply & Sewerage System in providing the necessary field data for the study.

REFERENCES

1. Durbovin et al (2002), Real time reservoir operation model based on total fuzzy similarity and compared with fuzzy inference method.
2. Panigrahi and Mujumdar (2000), Studied fuzzy rule based model for a single reservoir operation using Stochastic Dynamic Programming (SDP) for framing rule base.
3. Hyderabad Metropolitan Water Supply & Sewerage Board, Reservoir Log Records– (2016).
4. Hyderabad Metropolitan Water Supply & Sewerage Board, Revenue Data – (2016)
5. S. Vedula & S. Mohan (1991), Real-time Multipurpose Reservoir Operation for Irrigation and Hydropower generation for case studies of Bhadra reservoir system in Karnataka using SDP.
6. Shrestha (1996), developed fuzzy relations for input and output of reservoir operating principles and defuzzified to get the crisp outputs.
7. S.J. Mousavi, K. Ponnambalam & F. Karray (2004), Reservoir Operation using a Dynamic Programming Fuzzy Rule Based Approach to establish the general operating policies.
8. M. Satyanarayana, Ph D scholar under the guidance of Prof. M. A. Prasad, OUCE – Research Design and Progress Report 1 & Report 2 Submitted to Civil Engineering Department, Osmania University.

COMPUTATION OF RUNOFF BY SCS-CN METHOD USING REMOTE SENSING AND GIS

M. Anil Kumar[1] and M.V.S.S. Giridhar[2]

[1]M.Tech (HWRE) Student, Department of Civil Engineering, JNTU Hyderabad

[2]Associate Professor & Head, Centre for Water Resources, Institute of Science & Technology, JNTU Hyderabad.

ABSTRACT

Watershed management plays vital role in water resources engineering. It is necessary to plan and conserve the available resources. Remote Sensing (RS) and Geographic Information System (GIS) techniques can be used effectively to manage spatial and non-spatial data base that represent the hydrologic characteristics of the watershed. A study was conducted to estimate the runoff by SCS-CN method using RS & GIS technique. The study area considered is Dhulapally watershed in Malkajgiri district of Telangana State with an area of 80.86 sq.km. The spatial and non-spatial data were collected from various departments and thematic layers of land use, hydrologic soil group were prepared and overlaid with one another, the overlaid output results were assigned curve numbers with respect to soil and land use categories. The most prominent land use classes were cultivated/open land, forestland, built up area and hydrologic soil group for the project area is identified as C. Finally, runoff is calculated by based on the past 31 years' rainfall data i.e., from 1978 to 2008. The result obtained by SCS-CN method show that the average annual runoff depth of watershed is 294.8 mm & total runoff volume is 23.87 Mm^3. Regression equations for Exponential, Linear, Logarithmic, Polynomial and power assuming the rainfall as an independent variable and runoff as the dependent variable is plotted, reliability and performance of the relation obtained by graph was checked by computing the correlation coefficient and found that Polynomial rainfall runoff empirical relation gives the highest coefficient of correlation in all the time periods.

Keywords: *Rainfall, Runoff, SCS-CN Method, RS, GIS.*

INTRODUCTION

Of all planets, Earth deserves to be called as the water planet. The potential source of all fresh water is precipitation, which is a natural process results from the earth's un ending Hydrologic Cycle. Hydrology plays a vital role in protection and management of water and other environment resources associated with the occurrence and distribution of water above and below the land surface. Rainfall and runoff are the important components contributing significantly to hydrological cycle, design of hydrological structures and morphology of drainage system. Rainstorms generate runoff, and its occurrence and quantity are dependent on the characteristics of the rainfall event i.e., intensity, duration and its distribution. The rain falling on the catchment undergoes number of transformation and abstractions through various component processes such as interception, detention, evapotranspiration, overland flow, interflow, percolation, sub-base flow etc., which depends on the various watershed characteristics and emerges as runoff at catchment outlet. Estimation of the same is required in order to determine, forecast its effects and also for better management of water resources.

There are number of empirical formulae, curves, tables developed based on rainfall runoff relation but SCS-CN is widely used for computing direct runoff for a given rainfall event. This method was originally developed by US Department of Agriculture, Soil Conservation Service and documented in detail in the National Engineering Handbook, Section 4. The main reasons for its wide acceptance is that it accounts for many of the factors affecting runoff generation including soil type, land use and treatment, surface condition and antecedent moisture condition, incorporating them in a single parameter called curve number (CN).

NEED FOR STUDY

Runoff is one of the important hydrologic variables used in most of the water resources applications during its planning, designing and management. Urban Centers are facing a typical situation with regard to water now days on one had there is acute scarcity of water during the dry season resulting in over exploitation of ground water and on the other hand streets are often flooded during the monsoon, requires managerial efficiency to use the surplus water during the rainy season to overcome the deficiency in other season. The main cause for such emerging problems in the urban area is the changing land surface from pervious to impervious which results in reduction in infiltration rate and decrease travel time which significantly increases peak discharge and runoff. Hence Statistics and relation between rainfall and runoff are mandatory for effective Management of Watershed. Remote sensing and GIS is increasingly used as sophisticated database management system for efficient storage, retrieval, manipulation and analysis of spatially referenced data used for computation of runoff. Hence in this study and attempt is made to compute the Runoff generated from a watershed using the SCS-CN Method and GIS.

OBJECTIVE OF THE STUDY

The objective of the study is to

(i) To Prepare Geo spatial data base of land use and land cover using RS data and GIS.
(ii) To find the Hydrologic parameters of catchment by using GIS and Remote sensing
(iii) To estimate the surface runoff generated from the watershed by using SCS- CN Method.

STUDY AREA

The study area (Fig. 1) considered is Dhulapally watershed which falls in Survey of India topo sheets No. 56 K/6/SE, 56 K7, 56 K/10 & 56 K/11 with geographical extents 17^0 35' 30" to 17^0 28' 00" latitude 78^0 25' 30" to $78^0$31 00" longitude and lies in Malkajgiri District of Telangana State in India. The extent of watershed is 80.96 km^2 and the elevation ranges from +620 m to +527m above mean sea level.

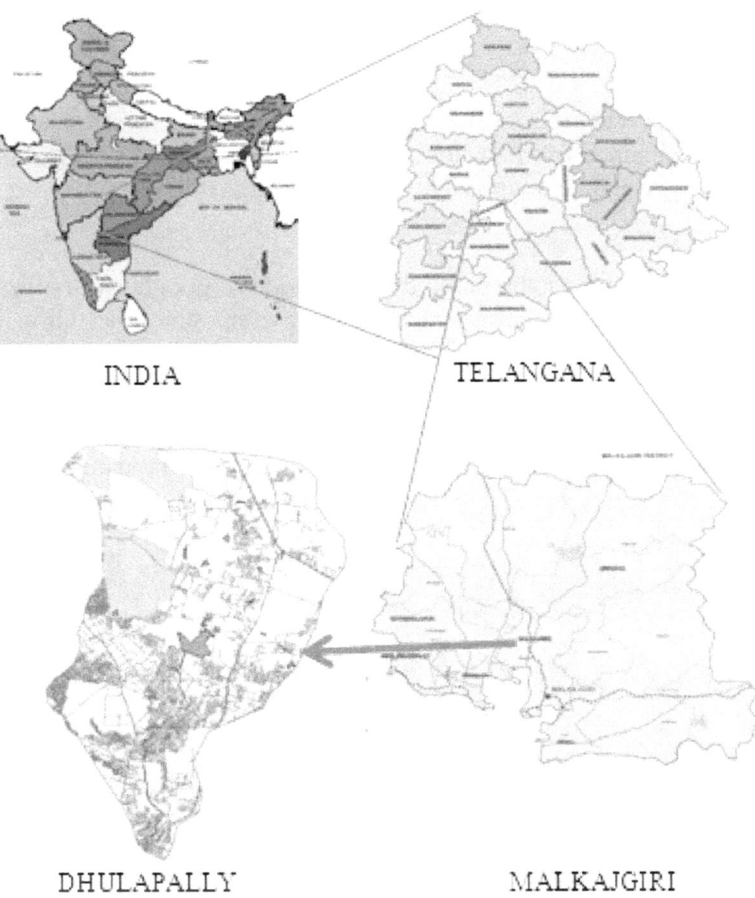

Fig. 1 Location of study area

METHODOLOGY

The methodology adopted for conducting this study is divided into the following three stages:

(i) Collection of spatial and non-spatial data from different sources,
(ii) Preparation of thematic layers of land use, hydrologic soil group and overlaid with one another and creation of CN map,
(iii) Computation of runoff depth based on rainfall in the study area using SCS-CN method

DATA COLLECTION

Rainfall Data

The historical daily rainfall data of nearest Rain gauge station i.e., Begumpet rain gauge station located in the premises of Begumpet Airport is collected for the years between 1978 and 2008 (31 years from IMD. The collected data is rearranged to find monthly, Annual rainfall and seasonal distribution. It is found that the annual average rainfall in the watershed is 841mm, which ranges from nil to minimum in December and January, maximum in August. August is the wettest month of the year.

Soil Map and Topo sheets

The Topo sheets were procured from Survey of India and used to delineate the watershed boundary. Soil map is obtained from Geological survey of India and found that the project area contains only one soil type and belongs to hydrologic soil group C.

Land use and Land Cover

The land use and land cover map is prepared by digitization of google earth image downloaded from google earth. Considering the requirement of the present study the following major land use types as listed in table 1 below were digitized and areas are calculated for each type of land use from the attribute of the respective shape files is presented in the table 1 shown below:

The shape files of different land use created by digitizing i.e., buildings, roads, forests, water bodies were combined and land cover map of the project area is prepared which is shown in Fig. 2

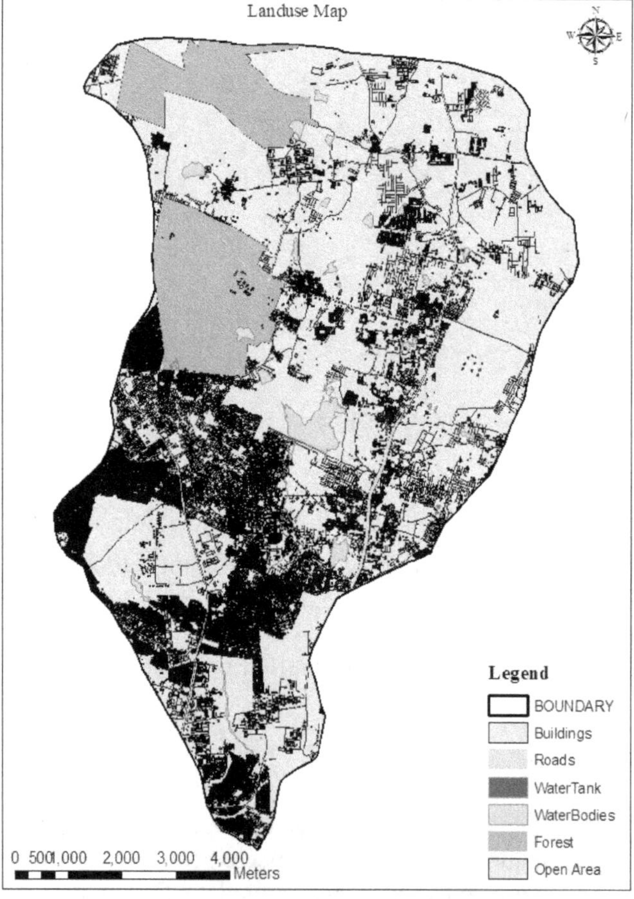

Fig. 2 Land use and land cover Map

Table 1 Details of Land use and land cover

S. No.	Land use	Area in Sq.km	Percentage
1	Built up Area	7.42	9.16
2	Roads	3.44	4.25
3	Open Land (cultivated & others)	59.57	73.57
4	Forest Area	9.14	11.29
5	Water Bodies	1.39	1.72
	Total	**80.96**	**100.00**

Software used

The Software used for this project is ArcGIS 9.3, MS Office 2016, Auto cad 2007.

SCS-CN Method

SCS-CN method is a simple predictable and stable conceptual method for estimation of direct runoff based on storm rainfall depth. The SCS Curve number method relates a calculated curve Number (CN) to runoff, accounting for the initial abstraction losses and infiltration rates of soils. It is a dimension less number and ranges from 0 to 100, is determined from a table based on land cover, Hydrologic Soil Group (HSG) and Antecedent Soil Moisture (AMC). HSG is expressed in terms of four groups (A,B, C and D) based on the rate of infiltration and transmission of water through soil after prolonged wetting, soil with high rate of infiltration and transmission is classified as Group A while a soil with very low rate of infiltration and transmission is classified as Group D. AMC is expressed in three levels (I, II and III) according to the rainfall limits dormant and growing seasons. A low CN indicate a dry antecedent soil moisture condition (AMC I) and high CN indicate a wet condition (AMC III) and average curve number indicate normal condition (AMC II). The rainfall-runoff equation of SCS method which is based on the water balance equation is

$$Q = \frac{(P - I_a)^2}{(P - I_a) + S}$$

Where

Q = runoff in mm
P = precipitation in mm
S = potential maximum retention of watershed in mm
I_a = Initial abstraction in mm

Initial abstraction (Ia) is a factor that determines all losses before runoff begins. It includes water retained in surface depressions, water intercepted by vegetation, evaporation and infiltration. Ia is highly variable but generally is correlated with soil and soil cover parameters. Through the studies of many small watersheds, Ia is approximated by the following empirical equation

$$I_a = 0.2S$$

Hence Rainfall Runoff Relationship can also be given as

$$Q = \frac{(P - 0.2S)^2}{(P + 0.8\ S)}$$

The potential maximum soil retention S depends on the soil vegetation, land use antecedent soil moisture condition of watershed prior to commencement of rainfall event. For conveyance in practical application CN is related to S as

$$S = \frac{25400}{CN} - 254$$

COMPUTATION OF RUN OFF

The calculated curve numbers for normal, dry and wet conditions are 81.685, 65.62 and 91.70 respectively. The daily rainfall database for the Dhulapally watershed from 1978 to 2008 (31 Years) and the curve numbers were inputs to the SCS formula and the results are obtained from the daily runoff values are added up to get monthly, seasonal and yearly runoff. The yearly rainfall and runoff values are presented in table 2 and figure 3 and average monthly and seasonal values is shown in figure 4 and 5 below.

Table 2 Annual rainfall and runoff of watershed

Year	Annual	Runoff (mm)	Year	Annual	Runoff (mm)	Year	Annual	Runoff (mm)
1978	1087.1	338.5	1989	1007.6	506.8	2000	1045.5	528.3
1979	703.8	228.2	1990	920.0	229.8	2001	825.6	319.1
1980	509.5	83.3	1991	770.4	358.8	2002	663.8	186.2
1981	1003.7	397.7	1992	764.6	254.0	2003	747.7	257.2
1982	775.3	230.8	1993	721.1	214.2	2004	684.1	168.8
1983	920.3	331.2	1994	809.9	235.8	2005	1131.9	435.6
1984	780.1	335.5	1995	1223.8	419.9	2006	988.6	362.4
1985	357.6	78.9	1996	971.7	348.2	2007	760.9	232.9
1986	621.3	162.2	1997	765.0	243.8	2008	1151.6	464.0
1987	963.0	375.4	1998	942.2	371.1			
1988	917.5	326.2	1999	564.1	113.3			

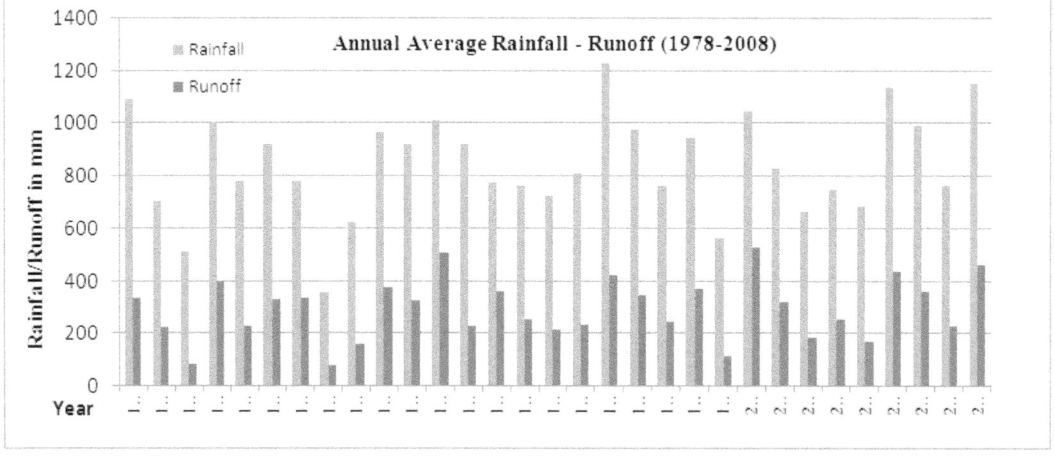

Fig. 3 Average Annual Rainfall Runoff (1978-2008)

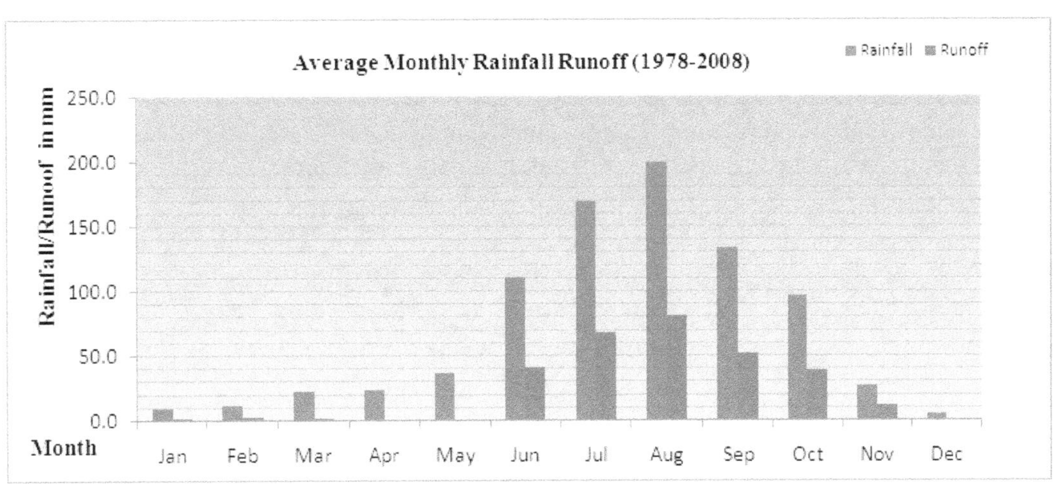

Fig. 4 Average Monthly Rainfall Runoff (1978-2008)

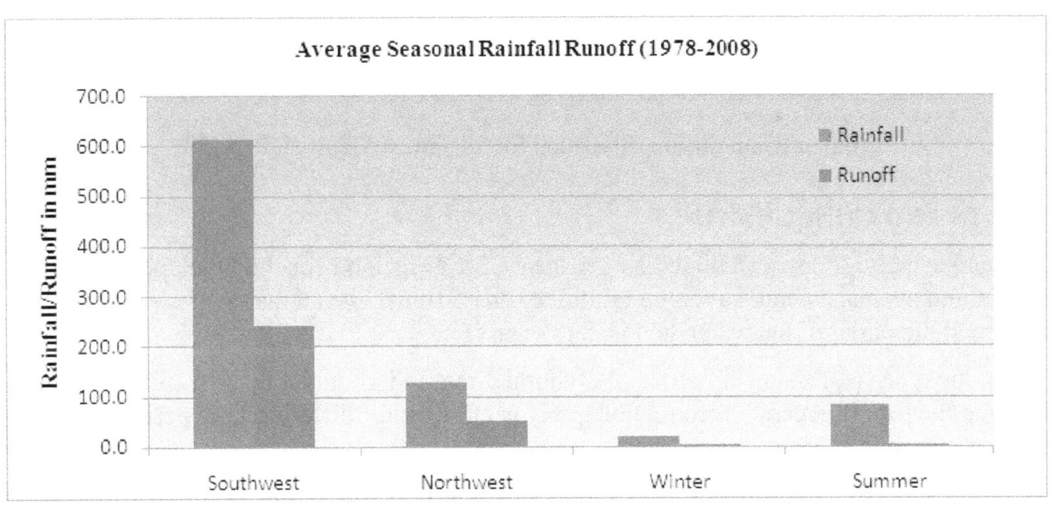

Fig. 5 Average Seasonal Rainfall Runoff (1978-2008)

Since the Southwest monsoon contributes to 73% of Annual Rainfall and 84% of the Annual Runoff in the Watershed. Graphs for southwest monsoon period is plotted considering rainfall in mm on X axis and the corresponding runoff in mm on Y-axis. Regression equations for Exponential, Linear, Logarithmic, Polynomial and power considering the rainfall as an independent variable and runoff as the dependent variable is plotted and are shown in Figures 6 below. The reliability and performance of the relation obtained by graph was checked by computing the correlation coefficient (R). Correlation coefficient represents the accuracy between the relationship in the present case between the observed rainfall and runoff. It is found that the polynomial rainfall runoff empirical relation gives the highest coefficient of correlation in all the time periods for the present watershed.

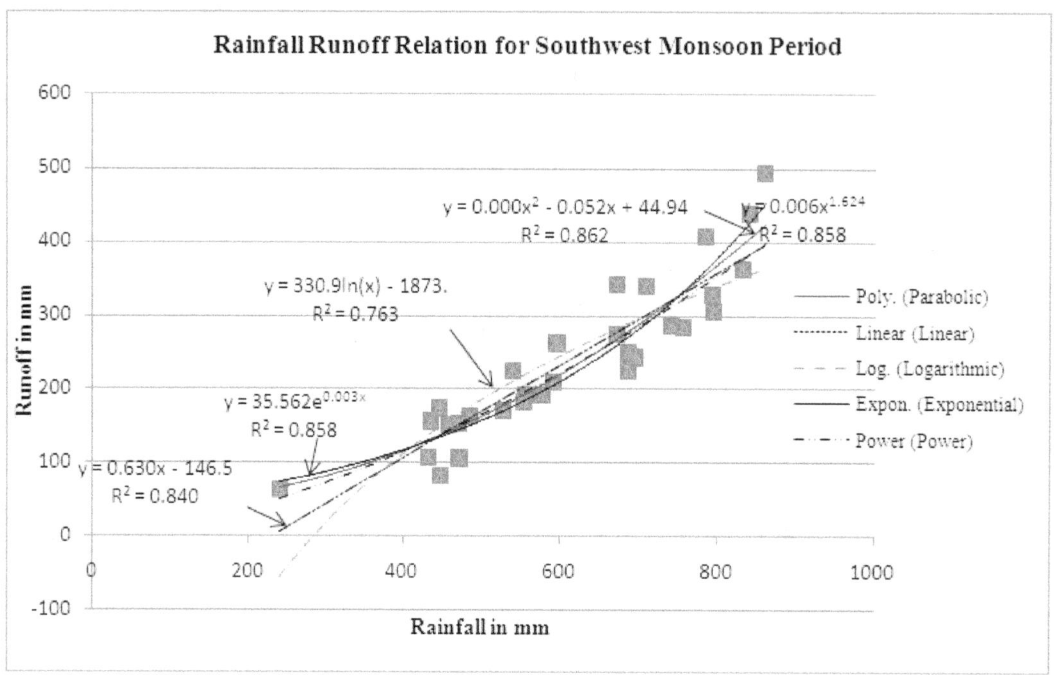

Fig. 6 Rainfall Runoff Relation for Southwest Monsoon Period

RESULTS AND CONCLUSION

The annual average runoff calculated as per the SCS CN method for the Dhulapally water shed is 294.78 mm against a annual average rainfall of 841.91mm. The annual average runoff volume generated at the outlet of the watershed is 23.87 Mm^3.

Of all the types regression equations, Polynomial regression model is most suitable for computing the runoff generated from Dhulapally catchment for different time periods of southwest monsoon season.

By integrating Remote sensing data and application of SCS CN method in GIS environment provides a powerful tool for computation of runoff. This tool can be used for finding the hydrologic characteristics of the catchment, for assessment of runoff such that the results can be used for planning and management of watershed efficiently.

Based on this study the following conclusions can be made:

GIS based SCS CN model can be used effectively to compute runoff from ungauged watersheds.

GIS based SCS CN model can also be used to study the impact of urbanization on runoff.

The estimated runoff data can be used to plan for proper water and land management in the study area

REFERENCES

1. National Engineering Handbook (Part 630), Natural Resources Conservation Service, United States Department of Agriculture.
2. Remote Sensing and Geographical Information System, Third Edition by M. Anji Reddy, Published by BS Publications, Hyderabad
3. Irrigation water Resources and Water Power Engineering, Seventh Edition, by Dr. P.N.Modi, Published by Standard Book House, Delhi.
4. Dr. M.V.S.S. Giridhar and Dr. G.K.Viswanadh, "Runoff Estimation in an un gauged watershed using RS and GIS" by Centre for Water Resources, Institute of Science and Technology, J N T U H, Hyderabad.
5. Vidyapriya.V, Nagalakshmi.R, Ramalingam. M "Assessment of Urban Runoff using Remote and GIS", National journal on Advances in Building Sciences and Mechanics, Vol.1, No.1, March 2010.
6. P. Sundar Kumar et.al. "Analysis of the Runoff for watershed using SCS-CN Method and Geographic Information System" International Journal of Engineering Science and Technology Vol.2(8),2010,3947-3654.
7. A.C.Lalitha Muthu and M. Helen Santhi, "Estimation of Surface Runoff Potential using SCS-CN Method integrated with GIS", Indian Journal of Science and Technology, vol 8(28), DOI:10.17485/ijst/2015/v8i28/83324, October 2015.
8. Vaishali S.Bhuktar and Dr.D.G.Regulwar, "Computation of Runoff by SCS-CN Method and GIS, International Journal of Engineering Studies and Technical Approach.
9. Centre for studies in Resource Engineering, Indian Institute of Technology, Bombay, Pune and Institute of Computer Technology, Pune, "Analysis of surface runoff from Yerla River basin using SCS-CN and GIS", International Journal of Geomatics and Geosciences, Volume4, No.3,2014.
10. College of Engineering and technology, Palestine Polytechnic University, P.O.Box 198, Hebron-West Bank, Palestine and Institute of Environmental and water Studies, Birzeit University P.O. Box 14, Birzeit-west, Palestine, " Estimation of Runoff for Agricultural watershed using SCS Curve Number and GIS", Thirteen International Water Technology Conference, IWTC 13 2009, Hurghada, Egypt.
11. K.P.C.Rao, T.S.Steenhuis, A.L.Colge, S.T. Srinivasan, D.F.Yule, G.D.Smith "Rainfall infiltration and runoff from an Alfisol in semi-arid tropical India.II. Tilled Systems" .
12. Sarva Mangala, Pratibha Toppo and Shradda Ghoshal " Study of Infiltration Capacity of Different Soils", International Journal of Trend in Research and Development, Volume 3 (2), ISSN:2394-9333.
13. Technical Release -55, June, 1986 Urban Hydrology for Small Watersheds, Natural Resources Conservation Service, United States Department of Agriculture.
14. Sunil Kumar Vyas, Gunwant Sharma and Y.P.Mathur "Rainfall Runoff Regression Model for Galwa Catchment in Banas River Basin", International Journal of Advanced Technology in Engineering and Science, Volume No.3, Special Issue No.2, February 2005.

A STUDY ON INCREASED CHLOROPHYLL CONCENTRATIONS AFTER THE DUST STORM

S. Lavanya[1*], and M.Viswanadham[2]

[1*]Ph.D Scholar, Centre for Earth Atmosphere and Weather Modification Technologies (CEA&WMT), IST, Jawaharlal Nehru Technological University, Hyderabad-85, India, email : lavanya.kusa@gmail.com.

[2*]Professor of Civil, and Director of DUFR, J.N.T.U, Hyderabad-85, India, email: maviswa14@gmai.com.

ABSTRACT

Dust Storm is a meteorological event common in arid and semi-arid regions. Deserts Surrounding the Arabian Sea are the dominant source of the dust aerosols. Desert dust is rich in nutrients, which is beneficial for the growth of the Phytoplankton's over the Arabian Sea. This paper highlights the effect of dust storm on chlorophyll concentrations of phytoplankton, which occurred on 8 October 2004 over the Arabian Sea. Results from observation of satellite images revealed that due to dust storm of 8 October 2004 there was increase in chlorophyll concentrations of Phytoplankton's. This increase in chlorophyll concentrations was more prominent on 13 October 2004.

Keywords: Dust storm, Chlorophyll concentration, MODIS Ocean color data.

1. INTRODUCTION

Dust storms originating over the world's arid regions contribute a large fraction of aerosols in the atmosphere. Dust storm arise when strong wind blows loose sand or dirt lifted from a dry surface and transported from point of origin to another places. Deserts surrounding the Arabian Sea are the main sources of Dust Storms. The Arabian Sea is surrounded by arid and semi-arid areas which are dominant sources of atmospheric dust. The largest one Rub Al khali desert located in Saudi Arabia. Other sources are located in Afghanistan, Pakistan and north-west India. Phytoplankton's are autotrophic components of the Plankton community, live in the sunlit layer of the ocean. They consume carbon dioxide (CO_2) and release oxygen. Phytoplankton's growth depends on the availability of CO2 sunlight, and nutrients. When all these conditions are required sufficiently Phytoplankton's can grow explosively, a phenomenon called the Bloom. Blooms in the ocean may cover 100's of square Km's and are easily visible in the satellite images.

2. STUDY AREA

The study area covering the northern part of the Arabian sea is located between latitudes 30°N and 0^0S and longitudes 50°E and 78°E bordered by India, Pakistan, Iran, Somalia, Arabian Peninsula, Oman. Productivity is highest in the northern region of the Arabian Sea.

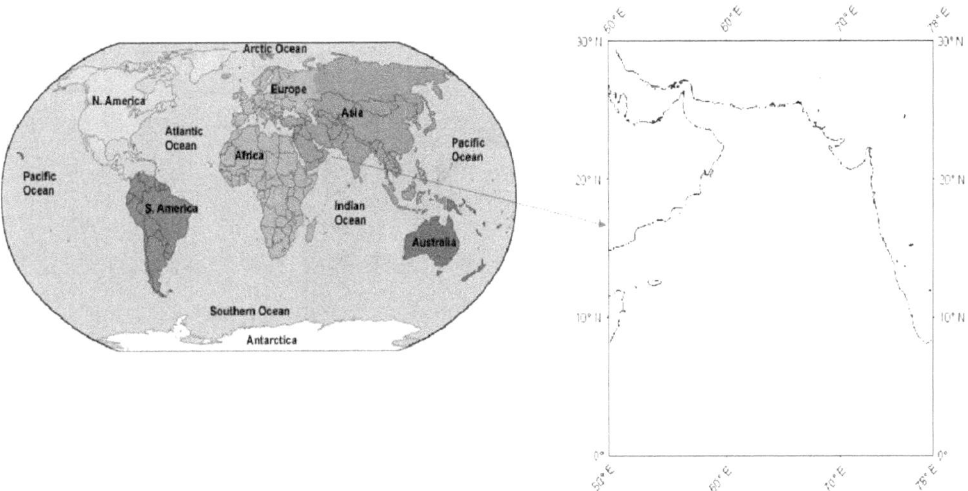

Fig. 1 Study Area of the Arabian Sea.

3. MATERIALS AND METHODS

3.1 *DATA AND METHODOLOGY*

MODIS Aqua Level-3 daily datasets of 4 km resolution were taken to study the Chlorophyll Concentration over the Arabian Sea. These data sets were downloaded from Ocean Color website. MODIS Terra and OrbView- SeaWiFS true Color images were used to study the Dust storm migration from the point of origin to Arabian Sea.

SeaDAS is a comprehensive image analysis package for the processing, display, analysis, and quality control of ocean color data. By using SeaDAS crop the datasets. After analysis Chlorophyll Concentration increased after the Dust Storm. Phytoplankton Bloom was found at Gulf of Oman.

4. RESULTS AND DISCUSSIONS

Dust Storm originated from the Sistan Basin located in South-western (SW) Afghanistan and North Eastern Iran on 7 October 2004. Sistan Basin is one of the world's largest deserts in SW Afghanistan. Because of strong winds loose silt over the Sistan Basin lifted from the surface and blew across the Sistan Basin and moved towards the Southern parts of Afghanistan and Pakistan. MODIS Terra image on 7 October 2004 showed that Dust veil over the Southern parts of the Afghanistan and Pakistan. Most of the Dust was trapped in northern Pakistan and South-western Afghanistan by Central Makran Mountains, however some of the Dust was escaped through river valleys and blew over the Arabian Sea and Gulf of Oman. Orbview–2 SeaWiFS image on 8 October 2004 showed that Dust was covered the Arabian Sea and Gulf of Oman.

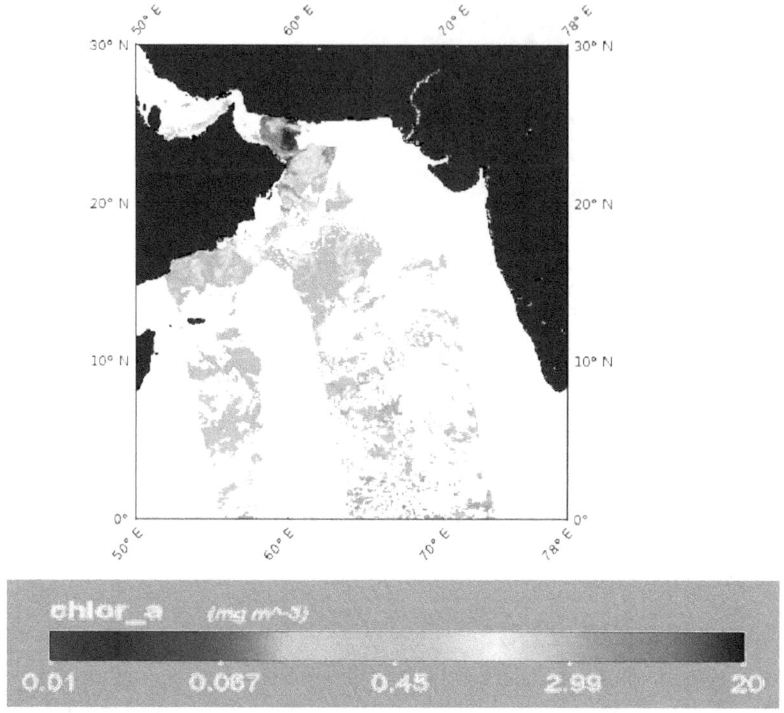

Fig. 2 Satellite image depicting Phytoplankton Bloom over the Gulf of Oman.

After observing Satellite images of Arabian Sea, it was found out that Chlorophyll Concentration increased after the Dust Storm when compared to before and during Dust Storm. Before Dust storm Chlorophyll Concentration was found to be 0 mg/m^3, during Storm about 1.36 mg/m^3 and after the Storm it was 29 mg/m^3 at 62^{0}45'E Longitude over the Arabian Sea. Phytoplankton Bloom was found at Gulf of Oman on 13 October 2004.

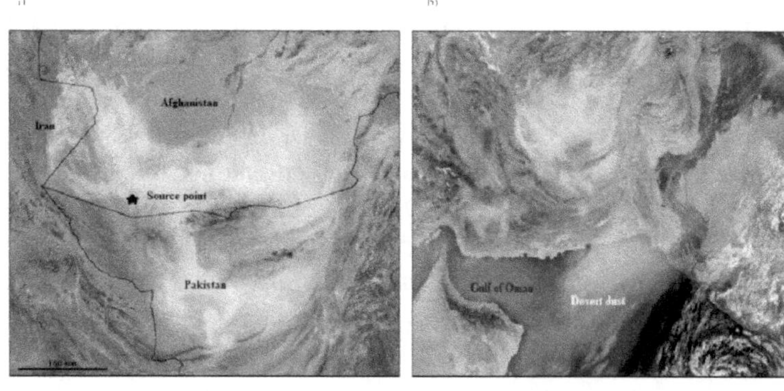

Fig. 3 Dust Storm migration from source region to Arabian Sea, a) shows that dust covered most parts of Southwestern Afghanistan and Pakistan on 7 October 2004, b) shows that dust storm over the Gulf of Oman and Arabian Sea on 8 October 2004.

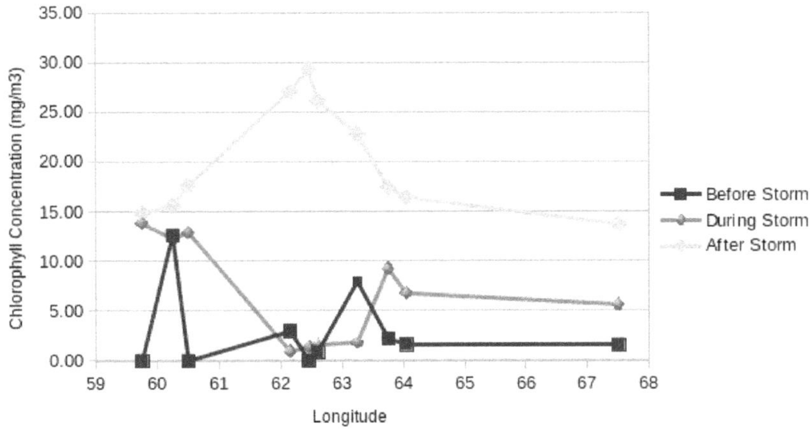

Fig. 4 Increased Chlorophyll Concentrations (mg/m^3) After the Dust Storm.

Before Storm - 29 September to 1 October 2004

During Storm – 2 October to 9 October 2004

After Storm- 10 October to 17 October 2004.

5. CONCLUSION

Arabian Sea is found to be highly productive region during dust storm. After dust storm chlorophyll concentrations (mg/m$^{3)}$ increased over the Arabian Sea because of the growth of the Phytoplankton's. Phytoplankton's using nutrients from the Dust deposition, and CO_2 (inorganic carbon) from both Atmosphere and Ocean, and sunlight to produce organic compounds (carbon).

ACKNOWLEDGEMENTS

We are thankful to the National Aeronautics and Space Administration (NASA) MODIS Team for providing Ocean Color data and SeaDAS software package developed by OBPG

REFERENCES

1. A.J.Gabric, R.A.Cropp, G.H.McTainsh, B.M.Johnston, H.Butler, B.Tilbrook, and M.Keywood., Australian dust storms in 2002-2003 and their impact on Southern Ocean biogeochemistry. Global Biogeochemical Cycles, VOL.24,GB2005,doi:10.1029/2009GB003541, 2010.
2. Dr.Tuhin Ghosh, Dr.Indrajit Pal., Dust storm and its Environmental Implications. Journal of Engineerring Computers & Applied Sciences (JECAS), ISSN No: 2319-5606, volume 3, No.4 April 2014.
3. J.H.Steele and l, Baird., Relations Between Primary Production, Chlorophyll and Particulate Carbon. Marine Laboratory, Aberdeen.

BREWERY EFFLUENT IMPACT ON CHICKPEA GROWTH, YIELD AND SOIL PROPERTIES

Rupa Salian[1], Suhas Wani[2], Ramamohan Reddy[1] and Mukund Patil[2]

[1]Centre for Water resources, Jawaharlal Nehru Technological University Hyderabad, Kukatpally, Hyderabad
[2]ICRISAT Development Centre, International Crops Research Institute for the Semi-arid Tropics, Patancheru, Medak

ABSTRACT

Use of wastewater increased largely because of water scarcity issues and also due to continuous demand on scarce fresh water sources. Several studies have been conducted on safe reuse of wastewater and its implications on soil properties and plant growth. This study is also conducted to know the effect of brewery wastewater on chickpea growth and also on soil properties. A pot scale experiment using black soil was conducted in green house of ICRISAT, Telangana. From effluent treatment plant (ETP) of SAB Miller beer factory, three types of water samples were collected i.e., 1) effluent of up-flow anaerobic sludge blanket reactor (UASBR), effluent of tertiary clarifier (TC) of ETP and reject effluent of reverse osmosis (RO) plant. Study contained five types of treatments–tap water as control, UASBR-50% (50% UASBR effluent + 50% distilled water), ETP-50%(50% TC effluent + 50% distilled water), ETP-100% (TC effluent without dilution)and RO-10% (10% RO reject + 90% distilled water) with three replications in completely randomized design. Initial soil and soil samples of respective treatment at the end of 90 days were collected and analyzed for parameters like pH, EC, NPK and Na. At the end of experiment(90 days), plants were harvested and respective treatment plant height, dry weight and yield were recorded.

Results of study revealed that root height (32.3 cm) and shoot height (46.3 cm) were highest with RO-10%, root (6.1 gm) and shoot (18.8 gm) dry weight were highest with UASBR-50%, Highest yield was achieved with ETP-100% (39.7 gm). pH, EC, N, P and Na concentrations increased and K concentration decreased considerably from Initial to end of experiment (90 days). From initial to end of experiment, 367% and 99% increase in EC and P concentrations respectively was recorded with UASBR-50% treatment. Whereas, ETP-100% showed 687% increase in Na concentration and 60% increase of N concentration achieved with ETP-50% treatment.Hence from the present study, it can be concluded that application of brewery wastewater showed positive effect on plant growth of chickpea and also enhanced the soil fertility.

Keywords: Wastewater reuse, Brewery wastewater, Chickpea, Soil properties.

INTRODUCTION

Wastewater has become valuable resource as there is continuous demand on fresh water resources which is leading to water scarcity issues all over the globe. This wastewater is produced from almost all human activities like domestic, industrial and agricultural practices. Depending on their source this wastewater may contain different kinds of pollutants which are capable of contaminating surface water bodies. Hence treatment of wastewater before releasing constitutes important aspect in protecting our environment. Due to increase in population there is continuous pressure on industrial product generation and also on food production. Industries releases large quantities of wastewater and this wastewater if used properly after treatment may solve water scarcity issues and also food production problems. Brewery industry also releases large quantities of wastewater like for production of one litre beer it was found that nearly 3 to 10 litres of wastewater is generated (Genner,1988). This brewery effluent was found to have sugars, soluble starch, ethanol and solids which are easily biodegradable and hence can be used for irrigational purposes (Driessen and Vereijeken, 2003).

Wastewater application may have both positive and negative effects as suggested by various authors. For instance, Positive effects like increase in organic matter concentration and soil fertility with brewery wastewater (Ramana et al., 2002a) and Palm mill oil effluent application (Yeop and Poop, 1983) respectively have been reported. Maize yield was increased with distillery wastewater application (Ramana et al., 2002a). Negative effects like increase sodium concentrations in soil were reported by Hati et al., (2007), which can pose negative effects on plant growth and may make the soil unfit for agriculture. Hence, the present study was carried out to assess the impact of brewery effluent on growth and yield of chickpea and on soil properties.

MATERIALS AND METHODS

From effluent treatment plant (ETP) of SAB Miller beer factory, three types of water samples were collected i.e., 1) effluent of up-flow anaerobic sludge blanket reactor (UASBR), effluent of tertiary clarifier (TC) of ETP and reject effluent of reverse osmosis (RO) plant. Study contained five types of treatments–tap water as control, UASBR-50% (50% UASBR effluent + 50% distilled water), ETP-50% (50% TC effluent + 50% distilled water), ETP-100% (TC effluent without dilution) and RO-10% (10% RO reject + 90% distilled water) with three replications in completely randomized design. Collected samples were analyzed for physico-chemical parameters using standard methods (APHA, 2005).

A pot scale experiment using black soil was conducted in green house of ICRISAT (International Crops Research Institute for the Semi-arid Tropics), Patancheru, Telangana. Each pot was planted with three chickpea plants. Pots were irrigated thrice in a week with respective treatment water source. After harvesting, shoot and root height were measured with scale. Dry weight of root and shoot was recorded using weighing balance. In each pot 10 kg of black soil was used for the study, collected soil termed as Initial soil and was analyzed for various physico-chemical parameters like pH, electrical conductivity (EC), total nitrogen, available phosphorous, exchangeable cations like potassium and sodiumusing standard procedures. All these parameters were analyzed for soil samples collected at the end of experiment (90 days).

RESULTS AND DISCUSSION

Physico-chemical parameters of wastewater collected from different units of brewery industry treatment plant are represented in Table 1 below.

Table 1 Physico-Chemical parameters of Tap water & Brewery effluent.

Parameter	Tap water	UASB Outlet	Tertiary Clarifier outlet	RO Reject Outlet
pH	7.47	7.37	7.83	6.93
Electrical Conductivity (ms)	0.62	4.37	4.04	12.07
Total Dissolved Solids (mg/l)	445.	2737.22	2556.56	7578.50
Total Suspended Solids (mg/l)	5.0	46.50	5.50	27.00
Chemical Oxygen Demand (mg/l)	49	350	202.50	365.33
Nitrate nitrogen (mg/l)	1.82	7.35	20.46	16.29
Phosphates (mg/l)	0.02	1.57	1.25	0.76
Potassium (mg/l)	14.8	50	47.20	119.05
Sodium (mg/l)	12.00	527	416.28	1126.77
Total Hardness (mg/l as CaCO3)	134.66	305.75	281.00	685.00
Total Alkalinity (mg/l)	128	731.00	551.25	1138.00

pH of brewery wastewater collected from different phases of treatment plant was almost similar to tap water (around 7-7.6), but pH of TC effluent was slightly higher (7.83). Highest value of EC was recorded with RO reject outlet (12.07 mS). Lowest EC value was observed with tap water (0.62 mS).TDS concentration was highest for RO reject outlet (7578 mg l^{-1}) and lowest TDS value was observed in Tap water (445 mg l^{-1}). Highest TSS value was recorded for UASBR outlet effluent (46.5 mg l^{-1}) and lowest value for tap water (5.06 mg l^{-1}). COD value was highest in RO reject outlet (365.33 mg l^{-1}) and lowest in TC outlet (202.50 mg l^{-1}) and tap water (49 mg l^{-1}). Nitrates was highest for TC outlet effluent (20.46 mg l^{-1}) and lowest value was observed in tap water (1.82 mg l^{-1}). Phosphate concentration was very low in tap water (0.02 mg l^{-1}) and highest value was recorded for UASBR outlet (1.57 mg l^{-1}). Potassium values were highest for RO reject outlet effluent (119.05 mg l^{-1}) and lowest value was observed for tap water (14.87 mg l^{-1}). Sodium was found to be highest for RO reject outlet (1126.77 mg l^{-1}). Low concentrations of sodium was recorded with tap water (12 mg l^{-1}). Total Hardness and Alkalinity values were highest for RO Reject outlet (685 mg/L$CaCO_3$ and 1138mg l^{-1} respectively), whereas lowest values were recorded with tap water (134.66mg/L$CaCO_3$and 128 mg l^{-1}respectively).

Root and Shoot height and dry weight of Chickpea as influenced by brewery effluent application is depicted in Figure 1.

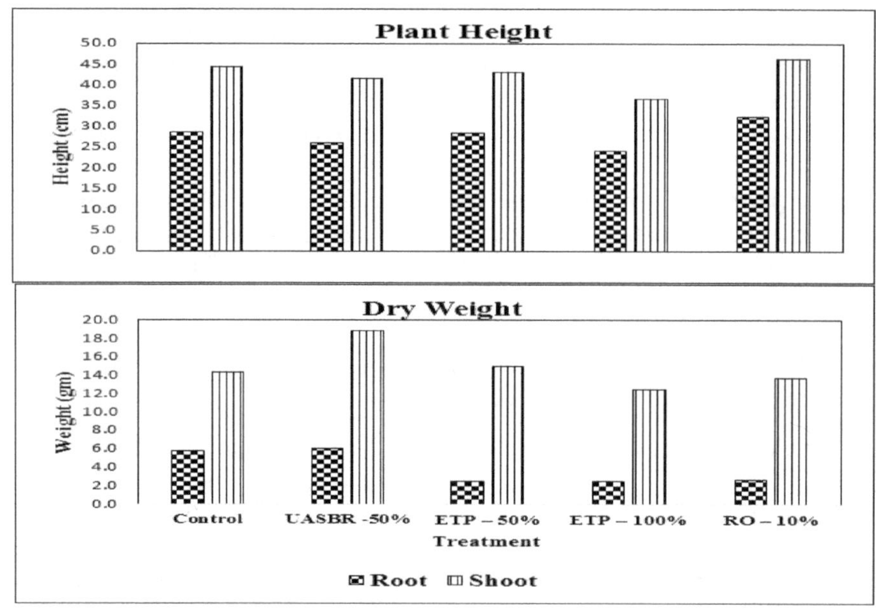

Fig. 1 Effect of Brewery Effluent application on plant height and dry weight of Chickpea.

Root height was highest with RO-10% (32.3 cm) and lowest root height was recorded with ETP-100% (24 cm). Shoot height of chickpea was also highest with RO-10% (46.3 cm) and similar to root height, shoot height was also lowest with ETP-100% (36.7 cm). Highest root and shoot dry weight was recorded with UASBR-50% (6.1 gm and 18.8 gm respectively) and lowest dry weight was recorded with ETP-100% (2.5 gm for root and 12.5 gm for shoot). This decrease in plant height and dry weight with ETP-100% might be due to highest salt concentration in it. Similar decrease of plant height was observed by Rusan et al., (2016) by applying olive mill wastewater. Promoting effect of RO-10% on plant height and UASBR-50% on plant dry weight can be attributed to nutrients present in it.Robert et al., (2005) also found out that the growth of maize was enhanced due to brewery effluent application and this might be due to high nutrient concentration of brewery effluent.

Yield of chickpea as influenced by brewery effluent application is represented in Figure 2 below.

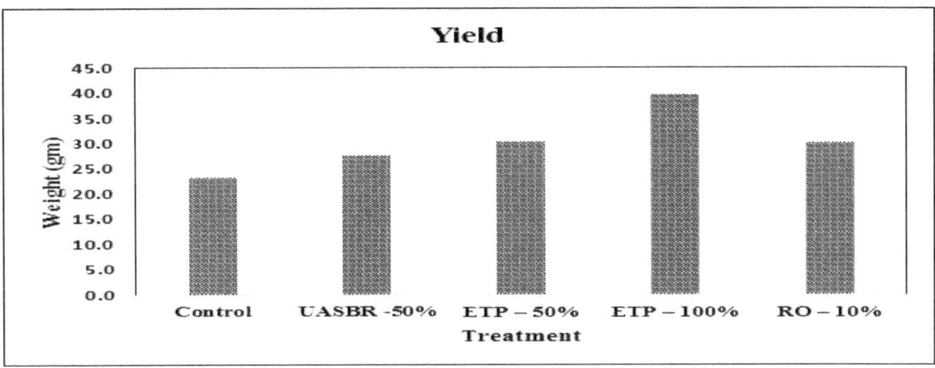

Fig. 2 Effect of Brewery Effluent Application on Yield of Chickpea.

Highest yield was recorded with ETP-100% (39.7 gm) and lowest yield was achieved with control (23.4 gm).Similar increases in growth and yield have been reported for groundnut (Ramana et al., 2002a) with distillery effluent application.

Initial soil parameters collected from ICRISAT campus were analyzed and their results are presented in Table 2. Initial soil was having normal pH, adequate OC %and good amount of nitrates and potassium levels. Sodium concentration and EC values were low and soil was found to be suitable for agriculture.

Table 2 Effect on soil properties at the end of 90 days of brewery wastewater application.

Treatment	pH	EC (mS)	N (mg/Kg)	P (mg/Kg)	K (mg/Kg)	Na (mg/Kg)
Initial Soil	**7.6**	**0.238**	**785**	**2.6**	**284**	**65**
Control	8.14	0.563	711	2.90	198	126
UASBR-50%	8.18	1.113	810	5.18	225	408
ETP-50%	8.2	0.69	831	2.18	209	354
ETP-100%	8.15	1.045	803	2.48	211	512
RO-10%	8.16	0.898	795	2.31	207	270

(Note: EC = Electrical conductivity, OC = Organic Carbon, N = Total Nitrogen, P = Available Phosphorous, K = Exchangeable Potassium, Na = Sodium) pH and EC increased in all treatments from initial to end of 90 days period. Highest increase of EC from 0.238 mS to 1.113mSwas observed withUASBR-50% and lowest increase was seen in control (0.563mS). Similar increase in pH and EC of soil was observed by application of treated Olive mill wastewater (Mekki et al., 2006 and Pierantozzi et al. 2013 respectively). Total nitrogen decreased in control from 785 mg/Kg to 711 mg/Kg, whereas highest increase of 831 mg/Kg from 785 mg/Kg was observed with ETP-50%. Garcia et al., (2010) found out that total nitrogen in alluvial soil increased with application of Brewery wastewater. Highest increase in phosphorous concentrations from 2.6 mg/Kg to 5.18 mg/Kg was recorded with UASBR-50%. Chartzoulakis et al. (2010) also reported that there was increase in phosphorous concentrations by using olive mill wastewater. Potassium values decreased from initial to end of experiment in all treatments, this might be due to uptake by plants. Sodium concentration increased from initial to end stage of experiment in all treatments. ETP100 showed highest sodium concentration (512 $mgKg^{-1}$) and lowest sodium concentration was recorded in control (126 $mgKg^{-1}$). Hati et al., (2007) reported that distillery effluent application led to increase in sodium concentrations of soil.

CONCLUSION

From the present study, it can be concluded that application of brewery wastewater showed positive effect on plant growth of chickpea and also enhanced the soil fertility.

ACKNOWLEDGEMENTS

Rupa Salian acknowledges Department of Science and Technology, India for providing fellowship under INSPIRE programme to pursue doctoral study.

REFERENCES

1. APHA, AWWA, and WEF (2005). Standard Methods for the Examination of Water and Wastewater, 21st ed. American Public Health Association, Washington
2. Chartzoulakis, K., Psarras, G., Moutsopoulou, M., &Stefanoudaki, E. (2010). Application of olive mill wastewater to a cretan olive orchard: effects on soil properties, plant performance and the environment. Agriculture, Ecosystems& Environment, 138(3–4), 293–298
3. Driessen W, Vereijeken T (2003). Recent developments in biological Treatment of brewery effluent. Paper presented to the Institute and Guide of brewery convention, Livingstone, Zambia
4. Garcia, S, Barco, M (2010). Use of brewery wastewater and its effects on soil physical and chemical properties under greenhouse conditions. Anadolu j. Agric. Sci. 25(S-1), 11-15
5. Genner, C (1988). Treatment and disposal of brewery effluents. Brewers Guardian, October. 25 – 27.
6. Hati, K.M, Biswas, A.K, Bandyopadhyay, K.K, Misra, A.K (2007). Soil properties and crop yields on a vertisol in India with application of distillery effluent. Soil and Tillage Research 92: 60-68
7. Mekki, A, Dhouib, A, Sayadi S (2006). Changes in microbial and soil properties following amendment with treated and untreated olive mill wastewater. J.Mic Res, 161, 93-101
8. Pierantozzi, P, Torres, M, Verdenelli, R, Basanta, M, Maestri, D.M, &Meriles, J.M (2013). Short-term impact of olive mill wastewater (omww) applications on the physicochemical and microbiological soil properties of an olive grove in Argentina. Journal of Environmental Science and Health, Part B, 48(5), 393–401.

9. Ramana, S, Biswas, A.K, Singh, A.B, Yadava, R.B.R (2002a). Relative efficacy of different distillery effluents on growth, nitrogen fixation and yield of groundnut. Bioresource Technology 81: 117-121.
10. Robert, O.E., Ulamen, O.A., Emuejevoke, V.D. (2005), Growth of maize (*Zea mays* L.) and changes in some chemical properties of an ultisol amended with brewery eefluent. Afr. J. Biotechnol. 4 (9), 973-978
11. Rusan M J M., Albalasmeh A A., Malkawi H I., Treated Olive Mill Wastewater Effects on Soil Properties and Plant Growth, Water Air Soil Pollut (2016) 227:135
12. Yeop K.H, Poon K.C (1983). Land application of plantation effluent. Proceedings of the Rubber Research Institute of Malaysia on oil palm by product utilization. Kuala Lumpur 1983.

ASSESSMENT OF GROUNDWATER CONTAMINATION DUE TO LANDFILL LEACHATE

C S V Subrahmanya Kumar[1] **and A. Parshuram Reddy**[2]

[1]Professor, Department of Civil Engineering, MVSR Engineering College, Hyderabad
[2]Graduate Student, Department of Civil Engineering, ACE Engineering College, Hyderabad

ABSTRACT

The objective of the study is to access the effect of leachate generated from municipal solid waste dumping yard on ground water quality by using ground water quality index in Hyderabad, Telangana. Groundwater Quality Index is one of the tools used to know the quality of groundwater. Ground water samples are collected from wells, 2 kms around the municipal solid waste dumping site and the physio- chemical analysis of water was carried out. The study revealed that municipal solid waste leachate plays a major role in contamination of the ground water.

Keywords: Ground water, leachate, water quality index.

INTRODUCTION

Water is a precious source and needs to be conserved. In several parts of the world human beings face safe drinking water problem. Waste deposited in landfills or in refuse dumps immediately becomes a part of the prevailing hydrological system. Fluids derived from rainfall, snowmelt and groundwater, along with liquids generated by the Municipal Solid Waste i.e., leachate from the dump, chiefly organic carbon largely in the form of fulvic acids migrate downward and contaminate the groundwater(Ugwu, S.A. and Nwosu, J.I., 2009).It's thus necessary that the quality of drinking water ought to be checked at regular intervals otherwise the human population suffers from a range of water borne diseases (Raman, N. and Sathiyanarayanan, D., 2011).

In unsealed landfills above an aquifer, water percolating through landfills and refuse dumps often accumulate or 'mound' within or below the landfill. This is often because of production of leachate by degradation processes operative inside the waste, additionally to the rainwater percolating down through the waste [WHO (b) 2006, Protecting Groundwater for Health]. The raised hydraulic head developed promotes downward and outward flow of leachate from the landfill or dump. Downward flow from the landfill threatens underlying groundwater resources whereas outward flow may result in leachate springs yielding water of a poor, often dangerous quality at the periphery of the waste deposit. Observation of leachate springs or poor water quality in adjacent wells/boreholes measure indicates that leachate is being produced and is moving. Leachate springs represent a major risk to public health, therefore their detection in situation assessment is vital to prevent access to such springs.

Leachate migration is also affected by the type of waste deposited. Compaction of waste before deposition reduces its permeability, whereas regular application of a topsoil cover between the loadings of waste to landfills induces layering. These characteristics bring about to preferential flow paths through landfills. The residence times for rainwater entering a landfill varied from a period of a few days to several years. This is often mirrored within the frequently temporal nature of leachate "springs", which might seem in wet seasons however afterwards; disappear in dry seasons to leave patches of discoloured soil (Johnson et al.).

2. MATERIALS AND METHODS

2.1 Study Area

The capital city of Telangana, Hyderabad is one of the most populated city. Municipal solid waste generated in the city is around 3000 metric tonnes per day. The municipal solid waste generated is disposed in open dumps, that is posing threat to quality of ground water. Municipal solid waste leachate contains variety of chemicals like inorganic and complex organic chemicals detergents and metals. The people are using the ground water for their daily use.

The part of solid waste generated from twin cities, Hyderabad and Secunderabad is been disposed at Nagole. In the present study, an attempt is made to study Ground Water Quality at Nagole.

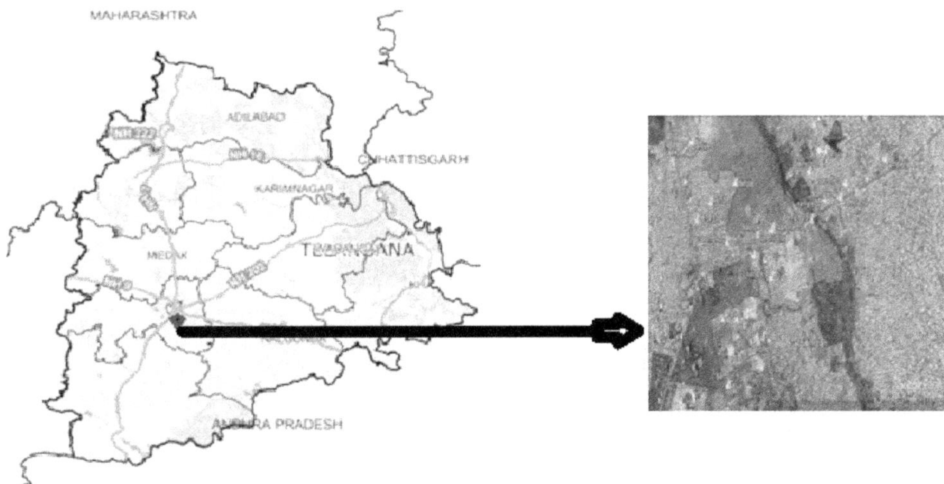

Fig. 1 Location map of the study area.

2.2 Sample Collection and Analysis

A preliminary survey was conducted to evaluate the effect of leachate on ground water quality. Sixteen bore wells were selected, at radius of 2 kilometres from the MSW site. Water samples from bore wells were collected in plastic bottles. Collected samples were placed in icebox and transported to laboratory within 3 hours for analysis.

The samples were analysed for various parameters like alkalinity, total hardness, calcium, magnesium, sodium, potassium, pH, TDS, chlorides, DO, and BOD, as per standard methods for the examination of water. The results obtained were compared with the drinking water standards as specified by Bureau of Indian Standards.

2.3 Water Quality Index

Water quality index (WQI) represents water quality assessment through the determination of physico-chemical parameters of Ground water. It can act as an indicator of water pollution because of natural inputs and anthropogenic activities.WQI is one of the most effective tools to provide feedback on the quality of water to the policy makers and environmentalists. It provides a single number expressing overall water quality status of a certain time and location.

Horton defined Water Quality Index as reflecting the composite influence on the overall quality of individual characteristics and proposed two basic steps to develop WQI namely:

(i) Selection of quality characteristics on which the index is to be based.

(ii) Establishing of a rating scale for each characteristic giving weightage to each parameter.

The simple WQI model expressed by Brown et al (1972) is given by,

$$WQI = \sum_{i=1}^{n} (Q_i W_i)$$

Where,

Q_i : quality of ith parameter

W_i : the unit weight of ith parameter

n : the number of parameters considered

Out of all the WQI models, Weighted Arithmetic Index Method (Brown et al, 1972) is popularly used and the same is applied for calculation of GWQI in this study. With the help of GWQI one can arrive at a rightful conclusion about the quality of water and can take any action that needs to be taken to improve the quality of water.

CALCULATION OF GWQI

The groundwater quality index (GWQI) was calculated using Weighted Arithmetic Index Method and the quality rating/sub index (Q) corresponding to the parameter is a number reflecting the relative value of this parameter,

The Q_i is calculated by using following expression,

$$Q_i = \{ (M_i - I_i) / (S_i - I_i) \} * 100$$

Where,

$\mathbf{M_I}$ = estimated values of the ith parameter in the laboratory

$\mathbf{I_i}$ = ideal values of the ith parameter

$\mathbf{I_i}$ = 0 for all the parameters except for DO and p^H which are 14.6 & 7.0

$\mathbf{S_i}$ = Standard values of the ith parameter.

The sign (-) indicates the numerical difference of the two values ignoring the algebraic sign. In the present study, unit weight (W_i) value is inversely proportional to the recommended standards (S_i) of the corresponding parameter.

$$\mathbf{W_i = K / S_i}$$

Where,

$$K = \frac{1}{(1/s_1) + (1/s_2) + (1/s_3) + \ldots + (1/s_i)}$$

s_1, s_2, s_3 ----------- s_i are standard values of various parameters from 1,2,3, ------ i.

The overall groundwater quality index (GWQI) was calculated by aggregating the quality rating (Q_i) with unit weight (W_i) linearly.

$$GWQI = \left\{ \sum_{i=1}^{n} (Q_i W_i) \Big/ \sum_{i=1}^{n} (W_i) \right\}$$

In this study, the permissible GWQI for drinking water is considered as 100, i.e., any value above 100 indicates groundwater contamination.

Table 1 Hydrochemical data of groundwater samples.

S.No	Parameter	Min Value	Max Value	Avg
1	Ca	24	320	111.81
2	Mg	9.7	194.5	59.18
3	Na	64	564	199.87
4	K	8	30	17.25
5	pH	7.45	9.42	7.78
6	NO_3	0	18	3.625
7	TH	34.6	1558.2	507.43
8	Alkalinity	15	125	73.12
9	TDS	394	3498	1393.43
10	Cl	104	779	351.68
11	BOD	4	6.5	6.41
12	DO	6.6	7.6	6.93

Note: All units are in ppm except pH

Table 2 Water Quality Rating.

WQI Level	Water Quality Rating
0 – 25	Excellent
26 – 50	Good
51 – 75	Poor
75 – 100	Very Poor
>100	Unfit for drinking purpose

Table 3 Ground Water Quality Index for samples.

Bore Well No.	GWQI	Water Quality Rating
1	282.356	Unfit
2	725.96	Unfit
3	357.32	Unfit
4	85.24	Very Poor
5	278.96	Unfit
6	346.28	Unfit
7	679.5	Unfit
8	407.2	Unfit
9	349.655	Unfit
10	352.09	Unfit
11	438.156	Unfit
12	722.2	Unfit
13	355.25	Unfit
14	648.45	Unfit
15	653.64	Unfit
16	352.02	Unfit

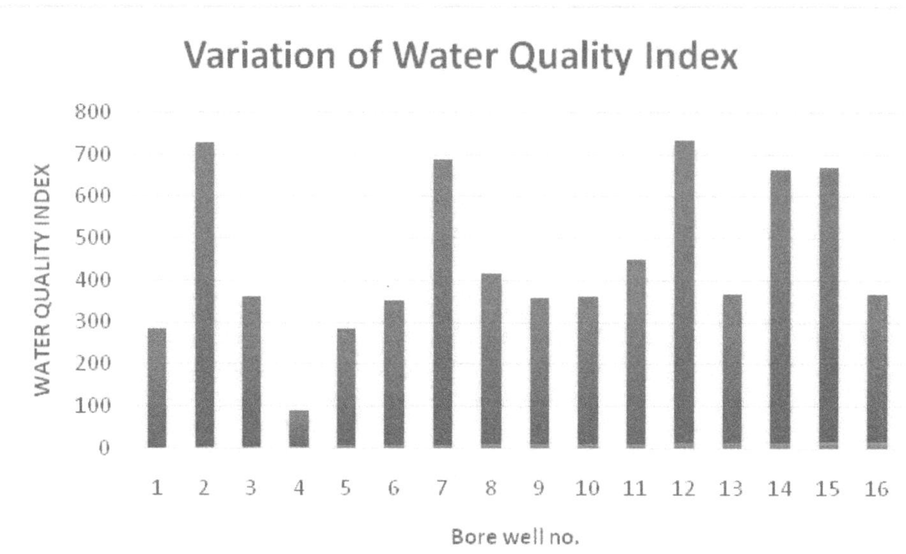

Fig. 2 Water Quality Index at sampling stations.

RESULTS

The results of hydrochemical data of ground water at various points are given in table1. The results indicate the quality of water varies considerably from location to location. The values of almost all the parameters are above the permissible limits. The wide variation is due to dissolved materials from Leachate.

P^H values are found to vary from 7.45 – 9.42. All the values are within the drinking standards. Though p^H has no direct effect on human health, all bio-chemical reactions are sensitive to the variation of p^H.

The TDS ranges between 394 ppm to 3498 ppm, greater than 500 ppm. Higher TDS value affects the palatability of food cooked and causes gastro – intestinal irritation.

Total alkalinity of water is mainly due to presence of bicarbonate in water. This is also associated with calcium and magnesium contents in water. The alkalinity of samples varies between 15 to 125 ppm.

Total hardness of water is characterized by contents of calcium or magnesium salts or both. The calcium and magnesium sulphates exert a cathartic action in human beings. It is also associated with respiratory diseases. In the study area calcium and magnesium contents of water vary from 24 to 320 ppm and 9.7 to 194.5 ppm respectively.

Nitrate concentration was found to be within the limiting value in all the wells.

DO and BOD are very important pollution parameters. The values of DO and BOD indicate degree of pollution. Generally low DO values indicate high pollution and high BOD values indicate presence of organic loadings in water sources. The observed values of DO and BOD varies from 6.6 to 7.6, 4 to 6.5 ppm respectively. DO values less than 5 ppm and BOD value greater than 5 ppm respectively indicate pollution.

Chlorides are also considered to be pollution indicating parameters. They impart salty taste to water. The range of chlorides in the study area is 104 to 779 ppm.

Water Quality Index (WQI) is one of the most effective tools to provide feedback on the quality of water to the policy makers and environmentalists. It is a number expressing overall water quality status which categorizes the ground water, based on WQI value. The water quality index of ground water is presented in Table 3.

GWQI for various wells were computed and the GWQI was found to be above 100, which indicates contamination of groundwater. The results reveal that the ground water in the entire area is contaminated.

The strategies to be adopted are,

(i) Sanitary landfill sites to be developed, which should consist of geo- synthetic liners.

(ii) The leachate should be collected and treated.

CONCLUSIONS

1. The ground water quality near the Municipal solid waste dumping site is of poor quality.
2. The Ground Water Quality Index in the entire area is above 100 except at one station, indicating that the water is unfit for drinking.
3. The sanitary landfills are to be built with liners to prevent leachate from seeping through soil into aquifers. Leachate collection systems store the liquid away from the water table. Clay caps prevent rainwater runoff from carrying pollutants from the landfill into the groundwater.
4. The municipal solid waste must be managed using composting, incineration and power generation.

REFERENCES

1. American Public Works Association, Institute for 'Solid Wastes Municipal Refuse Disposal', Chicago, III, USA 1970.
2. American Public Works Association, Institute for Solid Wastes, 'Solid Wastes Collection Practice', Chicago, III., USA,1975.
3. Baum P&ParkerC.H.,Solid Waste Disposal', Vol.I& II, Ann Arbor Science Publisher Inc., USA, 1974.
4. Bhide A.D.&Muley V.U. 'Studies on Pollution of Ground Water by Solid Wastes'. Proc.Symp. on 'Environmental Pollution', CPHERI, Nagpur, Jan.1973, p.236-243.
5. Bhide A.D. et al 'Management of Solid Wastes in Indian Cities', J. ISWA,NO.17, Dec.1975, p.9-12.
6. Bureau of Solid Waste Management,'Technical Economic Study of Solid Waste Disposal Needs& practices', Report SW-7c. US Dept. of Health, Education and Welfare, 1969.
7. Committee of San.Engg. Div. 'Sanitary Landfill', Manual No.39, ASCE,1976.
8. CPHERI Report 'Solid Waste in India', Final Report,1973.
9. APHA, 1985 -Standard methods for the examination of water and waste water by American Public Health Association, New York.
10. Akinbile C.O. and Yusoff, M.S., 2011. Environmental Impact of Leachate Pollution on Groundwater Supplies in Akure, Nigeria, International Journal of Environmental Science and Development, 2(1):81-86.

11. Amadi, A. N., P.I Olasehinde, Okosun, E.A and Yisa, J., 2010. Assessment of the Water Quality Index of Otamiri and Oramiriukwa Rivers. Physics International. 1 (2): 116-123.
12. APHA., 2012 Standard methods for the examination of water and waste water, 22ndEdn, American Public Health Association, Washington.
13. Ashwani, K.T. and Abhay, K.S., 2014. Hydro geochemical investigation and Groundwater Quality assessment of Pratapgarh District, Uttar Pradesh. Journal Geological Society of India, 83:329-343.
14. Behnke, J., 2003. A summary of the biogeochemistry of nitrogen compounds in ground water. Journal of Hydrology. 27:155-167.
15. Bureau of Indian Standards., 2012. IS 10500:2012, Drinking Water-Specification, Second Revision. Government of India, New Delhi.
16. Census., 2011. Census of India, Registrar General & Census Commissioner, India.
17. Christensen, J.B., D.L.Jensen, C.Gron, Filip, Z and Christensen, T.H., 1998. Characterisation of the dissolved organic carbon fraction in landfill leachate-polluted groundwater. Water Research. 32(1):125-135.
18. Cocchi, D and Scagliarini, M., 2005. Modelling the Effect of Salinity on the Multivariate Distribution of Water Quality Index. Journal of Mathematics and Statistics.1 (4): 268-272.
19. Craun, G.F. and Mccabe, L.J., 1975. Problems associated with metals in drinking water. Journal of the American Water Works Association. 67: 593.
20. Diodato, N., L. Esposito, G.Bellocchi, L.Vernacchia, Fiorillo, F and Guadagno F.M., 2013. Assessment of the Spatial Uncertainty of Nitrates in the Aquifers of the Campania Plain (Italy). American Journal of Climate Change. 2: 128-137.
21. Faisal, I., A.Shafaqat ,M.T.Hafiz, , B.S.Muhammad , F. Mujahid, I. Usman and Muhammad, M.N., 2013. Assessment of ground water contamination by various pollutants from sewage water in Chakera village, Faisalabad. International Journal of Environmental Monitoring and Analysis.1 (5): 182-187.
22. Hem., 1991. Study and interpretation of the chemical characteristics of natural water. US Geochemical Survey Water Supply, Scientific Publishers, India: 2254.
23. Howard, D., 2007. Chemical Speciation Analysis of Sports Drinks by Acid–Base Titrimetry and Ion Chromatography: A Challenging Beverage Formulation Project. Journal of Chemical Education. 84(1):124.
24. Huang, T., Z.Pang and Edmunds W.M., 2012. Soil profile evolution following land-use change: Implications for groundwater quantity and quality. Hydrological Process. 27(8):1238-1252.
25. Jain, C.K., A. Bandyopadhyay and Bhadra A., 2010.Assessment of ground water quality for drinking purpose, District Nainital, Uttarakhand, India. Environmental Monitoring and Assessment. 166: 663-676.
26. Jefferis, S.A., 1993. Old landfills: perceptions and remediation of problem sites. In: R.W. Sars by (Ed.) Waste Disposal by Landfill: 93-106.
27. Johnson, C.A, G.A. Richner , T. Vitvar, , N. Schittli and Eberhard M., 1998. Hydrological and geochemical factors affecting leachtae composition in municipal solid waste.
28. Muhammad S., RosliSaad& Marwan., 2013. Leachate Migration Delineation using 2-D Electrical Resistivity Imaging (2-DERI) at GampongJawa, Banda Aceh.EJGE. 18: 1505-1510.
29. Parihar, S.S., A. Kumar, R. N. Gupta, M. A. Pathak, Shrivastav and Pandey A.C., 2012. Physico Chemical and microbiological analysis of underground water in and ground Gwalior city, M.P. India. Research Journal of Recent Science. 1(6): 62-65.
30. Pocock S.J., A.G. Shaper, Cook D.G., 1980. British Regional Heart Study: geographic variation in cardiovascular mortality, and the role of water quality. Br Med J. 280: 1243-1249.

31. Rahman, S. (2002). Groundwater quality of Oman. Groundwater Quality, London. 122-128.
32. Ramakrishniah, C.R., C. Sadashivam and Ranganna, G., 2009. Assessment of Water Quality Index for the Groundwater in Tumkur, Karnataka State, Indian. Electronic Journal of Chemistry. 6(2): 523 530.
33. Raman, N. and Sathiyanarayanan, D., 2011. Quality Assessment of Ground Water in Pallavapuram Municipal Solid Waste Dumpsite Area Nearer to Pallavaram in Chennai, Tamilnadu. Journal of Chemistry. 4 (2): 481-487.
34. Riediker, S., J.F. Suter and Giger, W., (2000). Benzene- and naphthalene-sulphonates in leachates and plumes of landfills. Water Research. 34(7): 2069-2079.
35. Saadi, S., D.Khattach and Kharmouz, M., 2014. Geophysics and physico chemical coupledapproach of the groundwater contamination.Application in pollution by the landfillleachate of Oujda city (Eastern Morocco). Larhyss Journal.19: 7-17.
36. Sengupta, M. and Dalwani, R., 2008. Determination of water quality index and sustainability of an urban water body in Shimoga town, Kornataka, Proceedings of Taal 2007: The 12th World Lake Conference: 342-346
37. Ugwu, S.A. and Nwosu, J.I., 2009. Effect of Waste Dumps on Groundwater in Choba using Geophysical Method. Journal of Applied Sciences and Environmental Management. 13(1): 85-89.
38. WHO (a)., 2006. Guidelines of drinking water quality Recommendation: the 3rd edition. Geneva: World Health Organisation. 2nd ed. Geneva.
39. WHO (b)., 2006. Protecting Groundwater for Health: Managing the Quality of Drinking-water Sources. Edited by O. Schmoll, G. Howard, J. Chilton and I. Chorus. ISBN: 1843390795. Published by IWA Publishing, London, UK.
40. Yisa, J. and Jimoh, T., 2010. Analytical Studies on Water Quality Index of River Landzu, American Journal of Applied Sciences, 7(4): 453-458.

ESTIMATION OF LIFE OF RESERVOIR UNDER THE CONDITIONS OF NON AVAILABILITY OF SEDIMENT SURVEYS

M. Visweswararao[1], G.K. Viswanadh[2] and E. Saibabareddy[3]

[1]Professor of Civil Engineering, Malla Reddy Institute of Technology, Hyderabad
[2]Professor of Civil Engineering Department & OSD to Vice-chancellor, JNTUH, Hyderabad
[3]Vice-chancellor, VSS University of Technology, Burla, Odisha
gkviswanadh@gmail.com; esreddy1101@gmail.com

ABSTRACT

Pulichintala reservoir is a new reservoir constructed in between Nagarjunasagar reservoir and Prakasam barrage in Krishna basin. This reservoir will stabilize the age old ayacut of Krishna delta which is 250 years old. It is necessary and required to know for how long this reservoir will be serving Krishna delta and will become defunctional to meet its purpose of meeting the demands of Krishna delta with improved performance. Therefore it is proposed to estimate the life of this reservoir. The sediment observations at wadenepalli just upstream of Pulichintala on Krishna basin by CWC are utilized and the silt loads reaching the reservoir is estimated. Trap efficiency is estimated by Brunes trap efficiency curves. The distribution of sediment across the reservoir is done using area reduction method as specified in the code. The life of the reservoir is considered as that period when 50 % of live storage is lost due to sedimentation. The life of reservoir worked out to 138 years.

Keywords: Sedimentation, Pulichintala reservoir, Brunes trap efficiency curves, Krishnadelta, Area reduction method

1. INTRODUCTION

The River Krishna rises in the Mahadev range of the Western Ghats near Mahabaleshwar at an altitude of 1337m above sea level and flows through Maharashtra, Karnataka and Andhra Pradesh gathering water on its way from innumerable rivers, streams or tributaries and drops into the Bay of Bengal. Prakasam barrage at Vijayawada was constructed in 19 th centuary across Krishna river and serves Krishna delta which is the rice bowl of the present Andhra Pradesh state which require 152.2 TMC after modernisation. After construction of Almatti and other reservoirs upstream of this project it has become difficult to meet the demands of Krishna delta. To mitigate this problem Pulichintala reservoir was constructed in between Nagarjunasagar (which is up stream of prakasam barrage) and prakasam barrageRiver The live stotage of this reservoir is 36 TMC. As there is no storage at Prakasam Barrage it has to depend on river flows only and during flood periods lot of water flows to sea without being utilised.Pulichintala reservoir will store water in flood period and supply water to the paddy crop in needy times. Further it will also help in supplying the requirements of june and july for seed beds and transplantation of paddy when the flows are not available in the river. Hence it is proposed to study this reservoir for its life and assess the impact of sedimentation on the performance of this reservoir. The salient features of Pulichintala reservoir are given below in Table1.

Table 1 Salient features of Pulichintala reservoir.

S.No.	Description	Unit	Pulichintala
1	Sub-basin		K-7
2	Catchment area	(Sq.Km)	240732
3	Gross storage	(TMC)	45.76
4	Live storage	(TMC)	36.23
5	Dead storage	(TMC)	9.53
6	F.R.L	M	53.34
7	M.D.D.L	M	42.672
8	Crest level	M	36.54

2. LITERATURE SURVEY

A. Palman et.all provided a frame work for sustainable management of dams for economic feasibility of sediment management. This would extend the life of the reservoir for an indefinite period . Even if reduced sediment accumulation by way of catchment area treatment etc. is activated or the silt is removed by dredging the economic parameters which depend on physical, hydrological, financial are very Important. An empirical model which takes care of these parameters studies the trade off between sediment control and economic benefits short term and long term. Results show that for many reservoirs sustainable management of reservoirs is economically more desirable than the present practice of estimating the life of the reservoir by allowing silt accumulation.(Journal of environmental managementFeb-2001, A. Palman, F, Shah and others.)

The most important thing in sediment is the estimation of storage loss and the period of time at which it interferes with the reservoir performance. Fairly large number of models and procedures are available for reservoir sediment estimation which differ very widely. In the present study the rate of sedimentation and useful reservoir life have been estimated for Govindsagar reservoir adopting the trap efficiency approach and it is found that the life of the reservoir is 142 years. The trap efficiency suggested by Brunes curve is slightly modified suitable to this reservoir.(Journal of spatial hydrology 01/2008, Vaibhav garg and Vinayakan)

In this study analytical solutions were developed for calculating sediment volumes and life of the reservoir using trap efficiency relationships. Empirical relations are fitted to Brune trap efficiency curves and these equations are combined with the differential equations of the rate of sedimentation to produce final solution. The final equations which are exponential are simple to be used and found to be suitable for many reservoirs.(Journal of hydrology, Nov.1979, Mohammad Akram Gill)

The Muhammad Nur reservoir (MUR)is a majormulti purpose reservoir in Indonesia. This reservoir has high trap efficiency and there by it is envisaged that the capacity will reduce considerably causing threat to this reservoir operation inview of the inappropriate land use and large scale deforestation in the catchment. This paper uses the field data and published data from various sources. The water level variation for 22 years was studied with reference to their effect of sedimentation. The inflow patterns have changed causing changes in sedimentation volumes and trap efficiency. (Jounal of soil and sediments, March 2002, Haj Moehansyah, Basant L. Maheswari, et.al)

Aregemulu and GS Dwarakish In their research made efforts to relate the inflow and outflow of a reservoir. Forty four number of reservoirs are selected and inflow outflow relations are developed. The trap efficiency curves of Brune which is dependant on C/I ratio is replaced with C/Q where Q is the outflow. Therefore knowing the outflow and capacity the trap efficiency can be established. The outflows are generally at the reservoir and hence this method is more suitable for use when the inflow measurements are not available or not accurate due to indirect methods.(Aquatic Procedia 2015,AregaMulu, G.S. Dwarakish)

3. METHODOLOGY FOR THE PRESENT STUDY

In the present study the sediment generated up to Pulichintala is worked out considering the sediment observations at Wadenepalli CWC observation station on main Krishna river just up stream of Pulichintala and the sediment generated from free catchment between Wadenepalli and Pulichintala is added to get the total sediment reaching Pulichintala. The calculations are shown below.

3.1 Total Sediment Load

There are no hydrographic surveys in this project as this has come into partial operation in 2016 only. The original gross storage capacity is estimated as 45.77 TMC.

Average silt load from Wadenepall gauge site on main Krishna just up stream of Pulichintala = 1624286 MT/year

Adding 20 % bed load and adopting 1.4 density the average load per year from wadenapally = 1392245 cum or 1.392 Mcum

Total sediment load from Wadenepall by 2050 = 34*1.392=47.328 Mcum

Free catchment= catchment at Pulichintala- catchment at Wadenapally= 240732-235544 = 5188 Sq. Km

Total Silt load from free catchment by 2050 = 5188 *1126*34=198617392 or 198.62 Mcum

Grand total up to Pulichintala= 47.328+198.62=245.948 Mcum or 8.68 TMC

Annual sediment load= 245.948/ 34=7.2337647

minor irrigation and the small projects in the catchment between Srisailam to Nagarjunasagar form the inflows in Nagarjunasagar. These flows are again routed considering the planned utilisations of Nagarjunasagar reservoir to Prakasam Barrage. The reservoir operation is done in monthly timesteps. Success rates of meeting the annual demand at each reservoir are worked out by comparing demand planned and met for all demands.

3.2 Trap efficiency

The Trap efficiency of Pulichintala is estimated using Brunes trap efficiency curve.

Average Inflow in to reservoir = 370 TMC

Gross capacity of Pulichintala= 45.77 TMC

Capacity inflow ratio = 45.77/370=0.124

Trap efficiency from Brunes curve = 79 %

Total sediment deposited in Pulichintala = 0.79*245.948=194.29 Mcum

3.3 Sediment Distribution

The total sediment is distributed adopting empirical area reduction method as the hydrographic surveys are not available and as it is a new reservoir. The total sediment of 194.29Mcum is distributed adopting the empirical area reduction method.

3.3.1 Classification of the reservoir

The Pulichintala reservoir is classified adopting area reduction method. The reservoir comes under type II reservoir as the slope m worked out to 3.31. The graph between depth vs capacity is shown in Fig 4.1. The log values of depth and gross capacity are plotted and the slope is obtained by fitting linear relation. The slope worked out to 3.31. according to code if the slope is 2.5 to 3.5 it is classified as type II reservoir. The equation for Ap for type II is

$$Ap = 2.487 * p^{0.57} * (1-p)^{0.41}$$

Where p = relative depth.

Adopting the standard procedure of area reduction method the sediment is distributed vertically and the results are shown in table 2 The new zero elevation worked out to +27.29 m. About 70 % deposited in dead storage. The loss of live storage is 60.3 Mcum or 2.12 TMC over 34 years. The classification of reservoir is shown in FIG.1

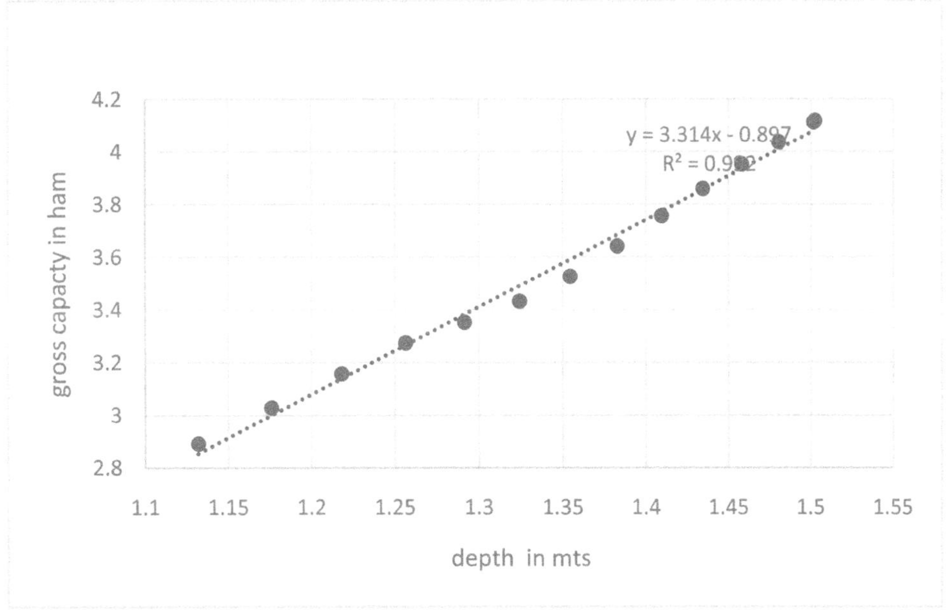

Fig. 1 Classification of Pulichintala reservoir.

Table 2 Revised area capacity table of Pulichintala Reservoir

Level in mts	original area in Mm^2	original capacity in Mm^3	sediment area in Mm^2	sediment volume in Mm^3	revised area in Mm^2	revised capacity in Mm^3
21.58	0	0	0	0	0	0
27.21	4.9	32.33	4.9	32.33	0	0
35.1282	11.79869	77.87	6.973866	79.33982	4.824824	-1.46982
36.576	14.58578	106.92	7.143445	89.55934	7.442335	17.36066
38.1	17.74448	143.79	7.26039	100.5351	10.48409	43.25493
39.624	21.2748	188.31	7.313427	111.6403	13.96137	76.66968
41.148	26.75607	224.84	7.299903	122.7757	19.45617	102.0643
42.672	32.88768	270.14	7.214904	133.836	25.67278	136.304
44.196	54.16247	335.84	7.050471	144.7062	47.112	191.1338
45.72	80.82565	438.06	6.794055	155.2557	74.03159	282.8043
47.244	93.64626	570.87	6.425369	165.3289	87.22089	405.5411
48.768	105.445	723.5	5.90923	174.7279	99.53577	548.7721
50.292	119.0088	895.66	5.175879	183.1747	113.8329	712.4853
51.816	131.3649	1086.23	4.04361	190.2	127.3213	896.03
53.34	143.9997	1296.06	1.086637	194.1092	142.9131	1101.951
53.4	145	1310	0	194.1418	145	1115.858

From the above table the percentage of sediment deposited in dead storage and live storage are separated and the life of reservoir to fill up 50 % of live storage is worked out as shown below.

3.4 Life of reservoir

Deposition by 2050=	194.14	Mcum
Deposition in dead storage=	133.84	Mcum
Percent deposition in dead storage =	68.94	%
balance dead storage available =270.14-133.84 =	136.30	Mcum
Estimated yearly silt trapped in reservoir=	5.71	Mcum
Silt trapped in dead storage/year=0.69*5.71	3.94	Mcum
Silt trapped in live storage /year=0.31*5.71	1.77	Mcum
Total time in years required to fill dead storage= 136.30/3.94	34.57	years
Available live storage as per study by 2050=1115.86-136.30	979.56	Mcum
year of impoundment	2016	year
number of years by 2050	34	years
Additional deposition for35 years=	61.95	Mcum
Net live storage for69 years= 979.56-61.95	917.61	Mcum
original live storage =1310-270.14	1039.86	Mcum
Half of live storage=1039.86/2=	519.93	Mcum
Additional deposition required for filling halfof live storage	397.68	Mcum
time to fill	69.60	years
life of Pulichintala= 34+35+70	138	years

4. RESULTS AND DISCUSSIONS

The results indicated that the annual sediment trapped in Pulichintala is 5.71 Mcum/year. The deposition in the dead storage is about 69%. The balance will be deposited in Live storage. The results indicated that the life of Pulichintala will be 138 years.

5. CONCLUSIONS

From the above study it can be concluded that the life of Pulichintala is 138 years. This is worked out assuming the trap efficiency as carry over storages will improve the system performance during deficit years. It is concluded that the carry over storages of Srisailam and Nagarjuna sagar has improved the performance of almost all projects in the system. It can also be concluding there is lot of reduction in deficits with carryover storages even in years when full demand is not met. It can further be concluded that consecutive deficit years will not be fully taken care by carryover storages. It is also clear that with Pulichintala another reservoir proposed below Nagarjunasagar the system will further improve and the demands of irrigation can be met at near 75 % the required level for irrigation.

REFERENCES

1. A.palman etc.al-econamics of a reservoir sedimentation and sustainable management of dams-journal of environmental management2/2001
2. Vaibhav garg and vinayakan-estimation of useful life of a reservoir using sediment trap efficiency-journal of spatial hydrology1/2008
3. Mohammad Akram gill-sedimentation and useful life of a reservoir-journal of hydrology.11/1979
4. Hajmoenshah et.al-impact of land use changes and sedimentation on mohammad nur reservoir-journal of soil and sediments,3/2002
5. Aregamuluet.al-different approach for using trap efficiency for estimation of reservoir sedimentation-aquatic procedia,2015
6. CWC gauge data of Krishna basin
7. Project reports of Pulichintala
8. IS 5477 part2, 1994

MAPPING OF PERMANENT SNOW COVER AREA USING SATELLITE DATA IN SUTLEJ BASIN

P.B Rakhee Sheel[1] and M.V.S.S Giridhar[2]

[1]M.Sc (WES) Student, Centre for Water Resources, Institute of Science & Technology, JNTU Hyderabad
[2]Associate Professor, Centre for Water Resources, Institute of Science & Technology, JNTU Hyderabad.

ABSTRACT

The mountains cover a large portion of the Earth surface. In these high mountains, it is estimated that the total surface area is covered from 30 to 40% of seasonal snow cover and Himalayas possess one of the largest resources of snow and ice outside the Polar Regions. Snow form a natural reservoir. Snow cover measurements are difficult and not easy because of the hostile climatic conditions and the remoteness of the areas .So remote sensing is attractive tools as a means of estimation of snow-cover properties. Keeping in view the importance of snow cover area, in this study remote sensing methods have been applied for mapping of permanent snow cover area. The study area comprises of Sutlej river basin in Western Himalayas which is snow fed and contributes to snowmelt runoff during summer month. The permanent snow cover in Sutlej basin has been mapped using NOAA/AVHRR data of 1998-1999 and IRS-P6/AWiFS data of 2007-2009 years. The permanent snow cover area (PSCA) is observed to be 1672.58 sq.km and 1703.38 sq.km. This is about 3.25% and 3.31% of the basin area. The permanent snow line is above 4500m elevation. The PSCA values derived from NOAA/AVHRR data and AWiFS data are very much comparable and majority of the PSCA is above 5500m elevation.

Keywords: Snow cover, Remote Sensing, Permanent Snow Cover Area (PSCA), Himalayas

1. INTRODUCTION

The snow cover in the Himalayas occurs and exists depending on terrain and climate conditions of the region. The Himalaya is the youngest mountain which is developed by continent-continent collision between the northward moving Indian plate and the Eurasian plate during 50-60 million years ago. The Himalaya is 2500km long mountain ranges from west to east comprised of 30 mountains rising more than 7300 meters and includes the highest mountain peak (8848 meters) of the everest. The characteristic features of the Himalaya ranges are steep sided hills and valleys, snow-capped mountains, large valley glaciers, deep river cut gorges and temperature and vegetation.

Seasonal snow cover and glacial ice are the fascinating elements of nature that has always attracted all, specially the scientists to study the apex environment. So these are an important freshwater resource and equally important resource for hydroelectric generation. However, it is also a potential cause of serious natural hazards because they are, close to the melting point and react strongly to climate change and also regional hydrology. Rama Moorthi, et al, 1991, the snow cover mapping in Himalayas using remote sensing techniques and the snowmelt runoff modeling approaches. This paper discusses in detail the techniques and operational methodologies used for remote sensing of snow cover in Himalayas. J.T Andrews et al. describes about a comparison of little ice Age glaciations levels with those based on the present distribution of permanent ice/snow bodies indicates that during the Little Ice Age the regional snowline fell between 100 and 400m, thus extensive areas of the upland plateaus of Baffin Island above 600m were mantled by a thin but extensive permanent snow cover.

2. STUDY AREA

Sutlej river basin in Western Himalayas is the study area. The river Sutlej is one of the main tributaries of Indus and has its origin near Manasarovar and Rakas lakes in Tibetan plateau at an elevation of about 4,500 m (approx.). It travels about 300km (approx.) in Tibetan plateau in North-Westerly direction and changes direction towards South-West and covers another 320km. (approx.) Up to Bhakra gorge where 225m high straight gravity dam has been constructed. This western Himalayan basin is highly rugged terrain with abundant natural water resource in the form of snow pack. It is stretched about 51,475.1 Sq. Km. in area. The Sutlej basin is geographically located between 30 ° 00' N, 76 ° 00' E and 33 ° 00'N, 82 ° 00' E. Characteristics of the basin and inaccessibility of the major part of it make remote sensing application ideal for hydrologists to monitor the snow cover information of the region and assess the resulting water resource.

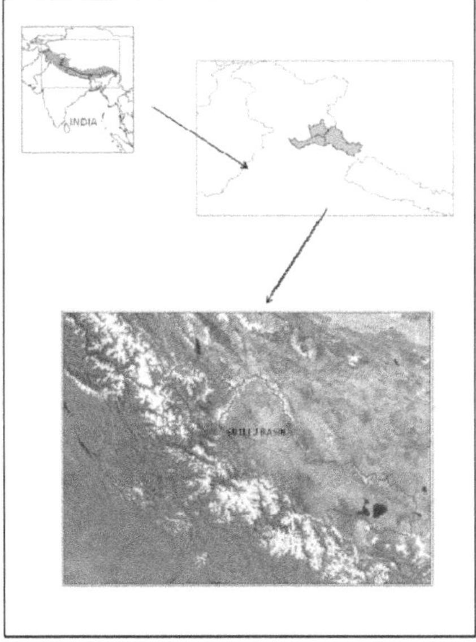

Fig. 1 Study Area.

3. DATA and SOFTWARE USED

For the permanent snow cover mapping, the satellite data from AVHRR (Advanced Very High Resolution Radiometer) sensor onboard NOAA (National Oceanic Atmospheric Administration) satellite and AWIFS (Advanced Wide Field Sensor) onboard IRS-P6 (Indian Remote Sensing-Resourcesat-1) satellite have been used. The software used for processing are ERDAS Imagine-2010, ARCGIS-9.3, Microsoft Excel, Microsoft Word and Power point.

4. METHODOLOGY

The permanent snow cover is the snow prevailing in the basin throughout the year. It is normally captured during the Sep – Dec month's period so that seasonal snowfall has already melted off and first snowfall during the winter season is yet to fall. Snow cover in Sutlej basin was mapped using ERDAS/Imagine digital image processing software. The digital analysis steps can be classified into

4.1 Pre – Processing

The NOAA/AVHRR and IRS-P6/AWIFS satellite data which is relatively cloud free on the area of interest is procured from National remote sensing centre (NRSC) Data Centre in Level 1B format and the same is imported using digital image processing steps. Satellite data is distorted by the curvature of the earth, its rotation, and the satellite platform and sensors aspects. Because of the sun-synchronous near polar orbit, the satellite data is oriented to true north during the acquisition. Hence there is a need to geometrically correct the image and so that it can be represented on a planar surface, confirm to other (Reference) images and have the integrity of a map. The standard earth curvature correction is performed to minimize systematic geometric distortions. The image is enhanced for better interpretability

4.2 Rectification

Satellite data (NOAA/AVHRR and AWiFS data) is distorted by the curvature of the earth, its rotation, and the satellite platform and sensors aspects. Because of the sun-synchronous near polar orbit, the satellite data is oriented to true north during the acquisition. Hence there is a need to geometrically correct the image and so that it can be represented on a planar surface, confirm to other images and have the integrity of a map. Each satellite data scene is registered into reference digital map base using Ground Control Points (GCPs). In the current study, a 3rd order polynomial transformation function is used in all rectifications. After selection of GCPs a polynomial transformation matrix is calculated along with RMS error which does not exceed 1 pixel for each GCP and all GCP's together. Once the transformation is over, resampling is performed to assign grey levels to the new data file. While various techniques are available in resampling to suit different situations, in the present case, Nearest Neighborhood (NN) method is employed since it has the advantage that original data values are transferred without averaging and therefore extremes and subtleties are not lost.

4.3 Classification

Before classification the first step is to create a signature file. The signature file is used to store the definition of classes to be classified in the form of thresholds in different bands of the satellite data. In the present study, snow is taken as one class or object. In this if we select some pixels of snow then the similar DN (Digital Number) value of pixels in the image will be selected and grouped as one object. This process is completed until all the pixels covered with snow are selected. The next step is to classify with this signature file. The classification of satellite image is an information extraction process that involves pattern recognition of spatial properties of various surfaces features and categorizing the similar features. The overall objective of image classification procedures is to automatically categorize all pixels in to image into specific land cover classes or themes. The set of radiance measurements obtained in various discrete wavelength bands for each land cover represent the spatial response pattern. The variability in spectral response pattern of land features from the basis for their discrimination on the satellite data.

In the classification of snow in all the images parallelepiped rule is used and unclassified pixels remain as unclassified and Maximum Likelihood parametric rule is used.The portion of Sutlej basin from the entire satellite data is extracted using basin mask image as reference. Snow cover area Statistics in the Sutlej basin are computed by generating a matrix using classified image and reference basin mask. The numbers of snow pixels in the basin are expressed as percentage of total basin area denoting snow cover area available in the basin as on that date. Thus, the snow cover area values have been computed for all the images.

5. RESULTS

The permanent snow cover image of Sutlej basin is shown in Fig. 2. The total PSCA is in Sutlej basin derived from NOAA/AVHRR data is observed to be 1672.58 sq. km. which is about 3.25% of the basin area. Using Digital elevation model the frequency distribution of the permanent snow cover area with respect to (w.r.t) elevation has been done. The Area-Elevation distribution is shown in the Fig 3.

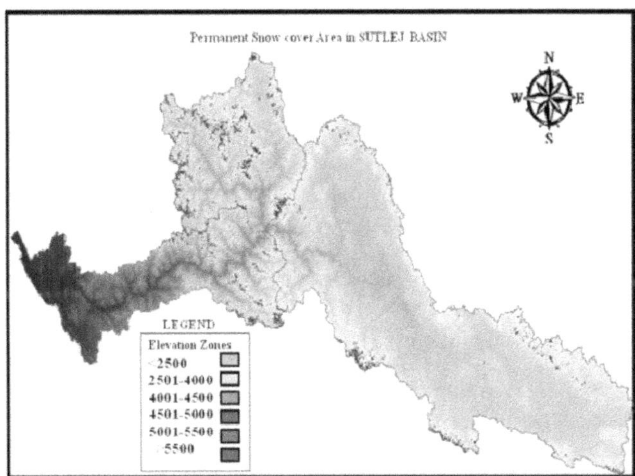

Fig. 2 Permanent Snow Cover Area in Sutlej Basin using NOAA/AVHRR data.

The basin area (sq.km) is shown on Y-axis with elevation ('m') on X-axis. The Area - Elevation curve is analyzed. The Area - Elevation curve is high at approximately 5700m elevations and slowly decreases.

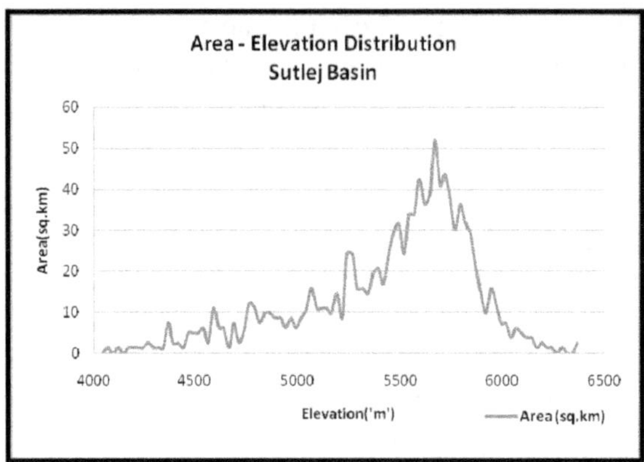

Fig. 3 The Area-Elevation Distribution of Sutlej basin.

The permanent snow cover image of Sutlej basin derived from IRS-P6/ AWiFS data is shown in Fig. 4. The total PSCA is in Sutlej basin derived from AWiFS data is observed to be 1703.38 sq. km. which is about 3.31% of the basin area.

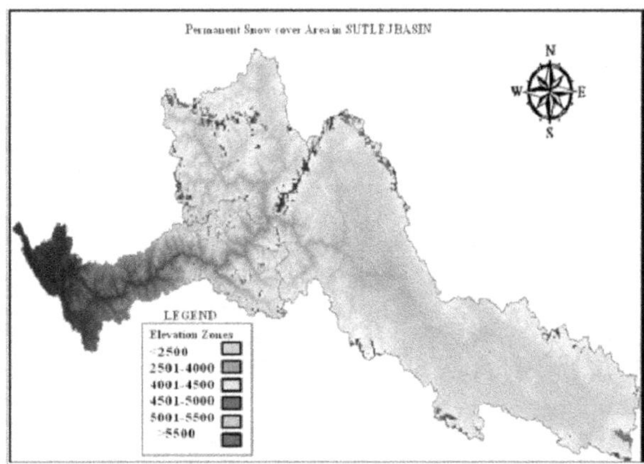

Fig. 4 Permanent Snow Cover Area in Sutlej basin using IRS-P6/ AWiFS data.

The PSCA is nil in zones 1 & 2. The PSCA in zones 3, 4, 5 are comparable. The total PSCA in entire Sutlej basin derived from NOAA/AVHRR data and AWiFS data are almost same. It is therefore, concluded that PSCA in Sutlej basin during the period 1998-1999 period and 2007-2009 period are very much comparable. Hence, no change in the PSCA over the period of about 10 years is noted.

6. CONCLUSIONS

1. The Permanent snow cover area in Sutlej basin has been mapped using NOAA/AVHRR (1998-1999) & IRS-P6/AWiFS (2007 – 2009) data. The permanent snow line in Sutlej basin is observed to be above 4500m elevation.

2. The total permanent snow cover area from Zone 1 (<2500) elevation range to Zone 6 (>5500) elevation range for NOAA/AVHRR data (1998-1999) is 1672.58 sq.km.
3. The total permanent snow cover area from Zone 1 (<2500) elevation range to Zone 6 (>5500) elevation range IRS-P6/AWiFS (2007 –2009) data is 1703.38 sq.km.
4. The total permanent snow cover area is about 3.31% of the basin area derived from AWiFS data and 3.25% of basin area derived from NOAA/AVHRR data.
5. The PSCA derived from NOAA/AVHRR data of 1998-1999 period and AWiFS data of 2007-2009 period are comparable.
6. Majority of the PSCA is observed to be above 5500m elevation in Sutlej basin.

REFERENCES

1. Dorothy K. Hall and Jaroslav Martinec(1985): Remote Sensing of ice and Snow, Chapman and Hall, London.
2. Erik Höppner & Nikolas Prechtel, Snow Cover Mapping with NOAA-AVHRR Images in the Scope of an Environmental GIS Project, Institute for Cartography of Dresden University of Technology, the Russian Altai (South Siberia).
3. H S Negi, A V Kulkarni and B S Semwal, Estimation of snow cover distribution in Beas basin, Indian Himalaya using satellite data and ground measurements, pg – 525, October 2009,H N B Garhwal University, Srinagar Garhwal, Space Applications Centre, Ahmedabad, India.
4. H S Negi , N K Thakur, Rajeev Kumar and Manoj Kumar, Monitoring and evaluation of seasonal snow cover in Kashmir valley using remote sensing and GIS, J. Earth Syst. Sci. 118, No. 6, December 2009, page - 711, Him Parisar, Sector-37A, Chandigarh, India.
5. J.T. ANDREWS, P.T. DAVIS, and C.WRIGHT, little ice age permanent snow cover in the eastern Canadian arctic extent mapped from landsat-1 satellite.
6. Klaus Seidel and Jaroslav Martinec, Hydrological applications of satellite snow cover mapping in the Swiss Alps, pg -79, Computer Vision Group, Communication Technology Laboratory, Switzerland.
7. Klaus Seidel, Jesko Schaper and Jaroslav Martinec, Computer Vision Group, Mapping of snow cover and glaciers with high resolution remote sensing data for improved runoff modeling, Communication Technology Laboratory, ETHZCH-8092 Zürich/Switzerland.
8. Majumdar,T.J. (2002) Areal snow cover estimation over the Himalayan terrain using INSAT- 1D AVHRR data. Go-cart International, v.17, No.3.
9. N. Foppa, S. Wunderle, A. Hauser, Spectral unmixing of NOAA-AVHRR data for snow cover estimation, pg-156, Remote Sensing Research Group, Department of Geography, University of Bern, Switzerland.

ESTIMATION OF REFERENCE EVAPOTRANSPIRATION USING CROPWAT

Dhondi Sindhu[1] and Giridhar M.V.S.S[2]

[1]Student M-tech Water and Environmental Technology, Center for water Resources, IST, JNTU Hyderabad, India

[2]Associate Professor in Center for Water Resources, J N T U Hyderabad, India.

Email: sindhudhondi@gmail.com

ABSTRACT

Evapotranspiration is the sum of evaporation and plant transpiration. The evapotranspiration rate from a reference surface, not short of water, called the reference crop evapotranspiration denoted as "ET_0". The Study area selected is Kaddam watershed present in the G-5 sub basin of Godavari River Basin. In this study, the normal monthly minimum, maximum temperature is analyzed and reference evapotranspiration of the study area calculated using CROPWAT for years 2000-2014. The reference evapotranspiration ET_0 can be estimated by using many methods, methods range from the complex energy balance equations to simpler equations that require limited meteorological data. For the study, the Food and Agriculture Organization (FAO) Penman-Monteith methodology was used to determine the reference evapotranspiration (ET_0). During the years 2000-2014 minimum ET_0 was observed in 2013 having value 1.42 mm/day and maximum ET_0 of 13.52 mm/day was observed in the year 2010.

Keywords: Reference Evapotranspiration ET_0, FAO Penman-Monteith method, CROPWAT.

INTRODUCTION

Evaporation is the process by which water precipitated on the earth's surface is returned to the atmosphere by vaporization, while the transpiration is a process similar to evaporation. It is a part of the water cycle, and it is the loss of water vapor from parts of plants (similar to sweating), especially in leaves but also in stems, flowers and roots. Quantitatively expressed, evaporation and transpiration are the depths of water vaporized from a unit surface in unit time (e.g. mm/day, and mm/year).

Evapotranspiration (ET) is the sum of evaporation and plant transpiration from the Earth's land and ocean surface to the atmosphere. Evaporation accounts for the movement of water to the air from sources such as the soil, canopy interception, and water bodies. Transpiration accounts for the movement of water within a plant and the subsequent loss of water as vapor through stomata in its leaves. Evapotranspiration is an important part of the water cycle. The concept of the reference evapotranspiration (ET_0) was introduced to study the evaporative demand of the atmosphere independently of crop type, crop development, and management practices. As water is abundantly available at the reference evapotranspiring surface, soil factors do not affect ET_0.

Hussein I. Ahmed & Junmin Liu (2013) Estimates of reference crop evapotranspiration (ET_0) are widely used in irrigation engineering to define crop water requirements. These estimates are used in the planning process for irrigation schemes to be developed as well as to manage water distribution in existing schemes. From the several existing ET_0 equations, the

FAO-56 application of the Penman-Monteith equation is currently widely used and can be considered as a sort of standard. The only factors affecting ET_0 are climatic parameters. Consequently, ET_0 is a climatic parameter and can be computed from weather data

The Penman–Monteith method is the best method to estimate ET_0 because of its inclusion of parameters in calculation (Patel et al. 2017). The FAO-Penman-Monteith equation is recommended as the standard method for estimating reference and crop evapotranspiration. The new method has been proved to have a global validity as a standardized reference for grass evapotranspiration and has found recognition both by the International Commission for Irrigation and Drainage and by the World Meteorological Organization (Nahla Mustafa Abdallaa et al.). In a study by Shreedhar & Sindhura (2016) Crop water requirement for each of the crops was determined using 30-year climatic data in CROPWAT. Of all the three radiation methods, Hargreaves method, Turc Method and FAO 56 Radiation method. RET estimated from Turc method resulted in 10.63% deviation when compared to FAO-56 PM method which was found to have least deviation among all the radiation methods in the present study (Giridhar and Viswanadh 2007).

Objectives of this study is to calculate daily, monthly and yearly ET_0 using FAO 56 Penman-Monteith equation for the 15 years data from 2000-2014, to calculate daily, monthly ET_0 for 15 years, to analyze estimated reference evapotranspiration and to show the variation of ET_0 over 15 years.

STUDY AREA

Kaddam watershed mostly lies in the Adilabad district. The climate of the district is characterized by hot summer and in generally dry except during the south-west monsoon season. The relative humidity is high generally during the south-west monsoon season. The air is generally dry during the rest of the year, the district part of the year being the summer season when the humidity in the afternoon is 25%. During the south-west monsoon season the sky is heavily clouded. There is rapid decrease of clouding the post-monsoon season (Giridhar et al. 2017). In the rest of the year, sky is mostly clear of light clouded. Winds are light to moderate with some strengthening in the period from May to August. During the post-monsoon and cold season, winds blow mostly from the east or northeast.

The study area selected is Kaddam watershed present in the G-5 sub basin is the 'Middle Godavari' Sub basin of Godavari River Basin. The Godavari basin extends over an area of 3,13,812 Sq.km. Godavari catchment is divided into eight sub basin in which G-5 sub basin is one of the basin, it lies between latitudes 17°04'N and 79°53'E longitude. The study area selected in the Middle Godavari sub basin is considered up to Kaddam reservoir watershed which lies between 19° 05' N and 19° 35' N latitudes and 78°10' E and 78°55' E longitudes. The watershed covers a total of twelve Mandals of which eight Mandals are taken Khanapur, Boath, Ichchoda, Narnoor, Utnoor, Indervelly, Bazarhatnoor and Kaddam all of which fall under Adilabad district. Figure 1 shows India map with state boundaries, Telangana map with district boundaries, study area Kaddam watershed, and Kaddam reservoir.

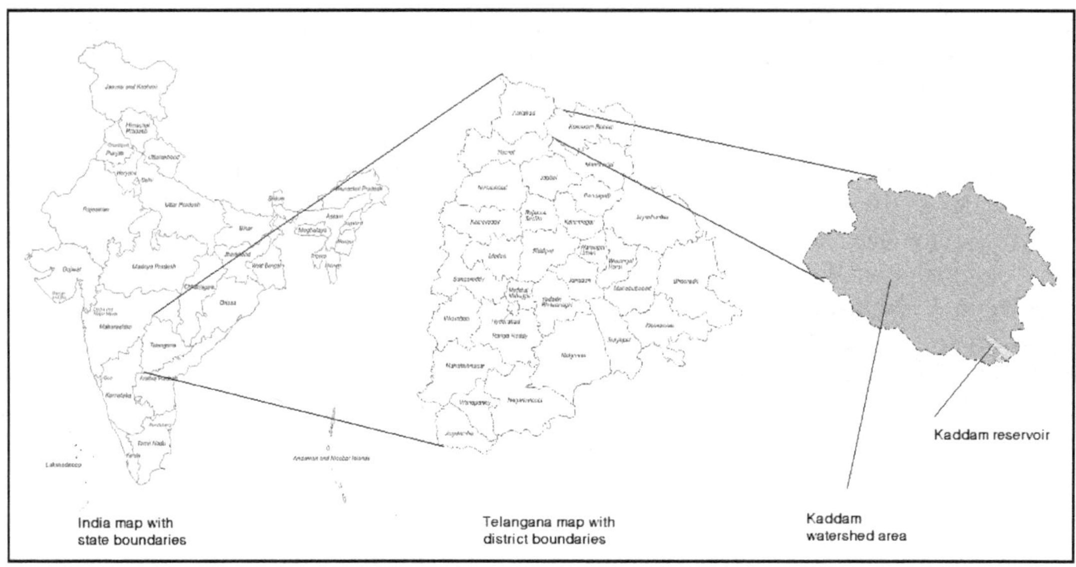

Fig. 1 India map with state boundaries, Telangana state map with district boundaries, Kaddam watershed area.

METHODOLOGY

The reference evapotranspiration "ET_0" can be estimated by many methods. Methods range from the complex energy balance equations to simpler equations that require limited meteorological data. For the study, the Food and Agriculture Organization (FAO) Penman-Monteith methodology was used to determine the reference evapotranspiration (ET_0). The Penman-Monteith method gives more consistently accurate "ET_0" than other methods. The calculation procedures allow for estimation of ET_0 using FAO Penman-Monteith method under all circumstances, even in the case of missing climatic data.

In 1948, Penman combined the energy balance with the mass transfer method and derived an equation to compute the evaporation from an open water surface from standard climatological records of sunshine, temperature, humidity, and wind speed. This so-called combination method is further developed by many researchers and extended to cropped surfaces by introducing resistance factors.

The surface resistance parameters are often combined into one parameter, the 'bulk' surface resistance parameter, which operates in series with the aerodynamic resistance. The surface resistance, r_s, describes the resistance of vapour flow through stomata openings, total leaf area and soil surface. The aerodynamic resistance, r_a, describes the resistance from the vegetation upward and involves friction from air flowing over vegetative surfaces. Although the exchange process in a vegetation layer is too complex to be fully described by the two resistance factors, good correlations can be obtained between measured and calculated evapotranspiration rates, especially for a uniform grass reference surface.

The Penman-Monteith form of the combination equation is:

$$\lambda ET = \frac{\Delta(R_n - G) + \rho_a c_p \frac{(e_s - e_a)}{r_a}}{\Delta + \gamma\left(1 + \frac{r_s}{r_a}\right)} \qquad .....(1)$$

Where

R_n - the net radiation,

G - the soil heat flux,

(e_s - e_a) - represents the vapour pressure deficit of the air,

ρ_a - the mean air density at constant pressure,

c_p - is the specific heat of the air,

Δ - represents the slop of the saturation vapour pressure temperature relationship,

γ- the psychrometric constant,

r_s - (Bulk) surface resistances,

r_a - aerodynamic resistances.

The Penman-Monteith approach as formulated above includes all parameters that govern energy exchange and corresponding latent heat flux (evapotranspiration) from uniform expanses of vegetation. Most of the parameters are measured or can be readily calculated from weather data. The equation can be utilized for the direct calculation of any crop evapotranspiration as the surface and aerodynamic resistances are crop specific.

FAO Penman-Monteith method: By defining the reference crop as a hypothetical crop with an assumed height of 0.12 m having a surface resistance of 70 s m-1 and an albedo of 0.23, closely resembling the evaporation of an extension surface of green grass of uniform height, actively growing and adequately watered, the FAO Penman-Monteith method was developed. The method overcomes shortcomings of the previous FAO Penman method and provides values more consistent with actual crop water use data worldwide. From the original Penman-Monteith equation (Equation 1) and the equations of the aerodynamic and surface resistance, the FAO Penman-Monteith method to estimate ET_0 can be derived. Apart from the site location, the FAO Penman-Monteith equation requires air temperature, humidity, radiation and wind speed data for daily, weekly, ten-day or monthly calculations.

$$ET_o = \frac{0.408\Delta(R_n - G) + \gamma \frac{900}{T + 273} u_2 (e_s - e_a)}{\Delta + \gamma(1 + 0.34u_2)} \qquad ....(2)$$

Where

ET0 - reference evapotranspiration [mm day-1],

Rn - net radiation at the crop surface [MJ m-2 day-1],

G - soil heat flux density [MJ m-2 day-1],

T - mean daily air temperature at 2 m height [°C],

u2 - wind speed at 2 m height [m s-1],

es - saturation vapour pressure [kPa],

ea - actual vapour pressure [kPa],

es - ea saturation vapour pressure deficit [kPa],

Δslope vapour pressure curve [kPa °C1]

γ psychrometric constant [kPa °C-1].

CROPWAT is a computer program developed by the Land and Water Development Division of Food and Agriculture Organization of the United Nations (FAO) to estimate irrigation requirements based on climate, crop, and soil data. CROPWAT 8.0 for Windows is a computer program for the calculation of crop water requirements and irrigation requirements. ET_0 is calculated when you enter daily/monthly climatic data (temperatures, humidity, wind speed, sunshine) in CROPWAT version 8.0 software.

This software uses the FAO (1992) Penman-Monteith methods for calculating reference crop evapotranspiration. This is a widely used tool for estimating irrigation water requirements for practical field level use and it is also used for research purposes. In the present study, this software has been used for arriving the daily, monthly, and yearly ET_0 for the duration of 2000-2014 for the study area Kaddam watershed. In order for CROPWAT to provide efficient and correct data, one need to insert data about the evapotranspiration process, i.e. the quantity of water evaporated from the soil and eliminated by plants into the atmosphere. The data can be entered manually by filling in the forms provided by the inlays from the left side of the main window. Further, the software was also validated by generating excel sheet for the FAO 56 Penman-Monteith method and the output is observed to be almost same as in the software. Daily minimum temperature, maximum temperature, humidity, wind speed, bright sunshine hours for a period of 15 years from 2000-2014 was collected from a rain gauge station located in the study area. The collected data was validated for any manual typing errors on excel sheets. The validated data is used in calculation of evapotranspiration.

RESULTS AND DISCUSSIONS

The daily data such as minimum-maximum temperature, humidity, wind speed, bright sunshine hours for a period of 2000-2014 is used in the study. The input to the software is given as daily data for each year and daily ET_0 is been calculated which is printed to extract the output from the software. The output is kept in excel sheet. The daily ET_0 for the 15 years is shown in figure 2.

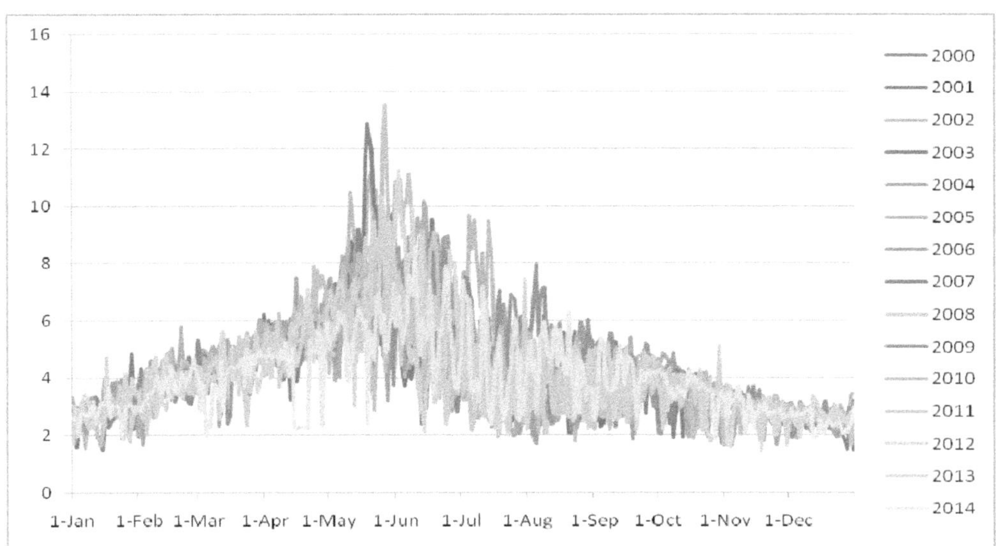

Fig. 2 Daily ET_0 variation in mm/day for the period of 2000-2014.

From the figure 2 daily ET_0 values are minimum during January, February, October, November, and December. The value of ET_0 is moderate during the months April, August and September. The ET_0 maximum can be observed during the months of May, June and July.

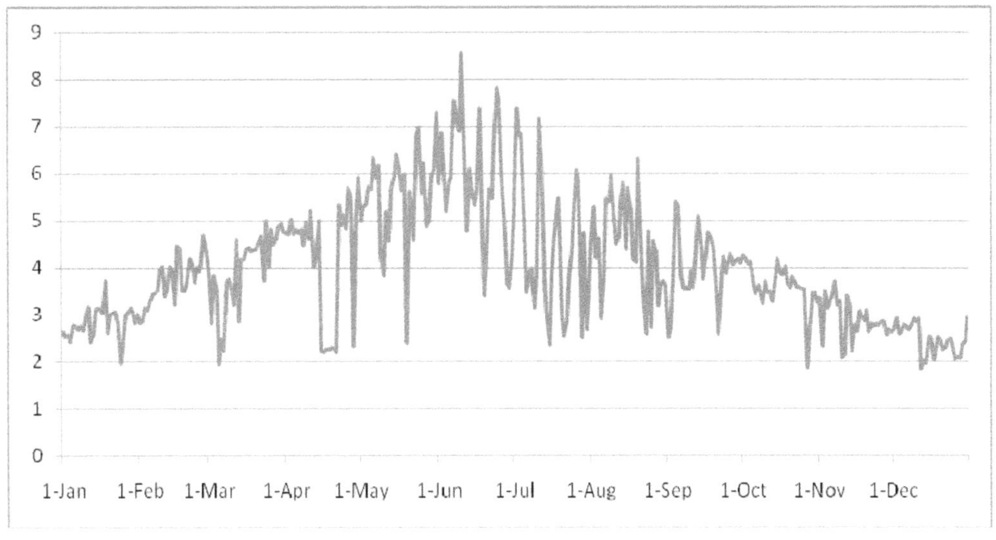

Fig. 3 Variation of ET_0 in mm/day 365days for the year 2014.

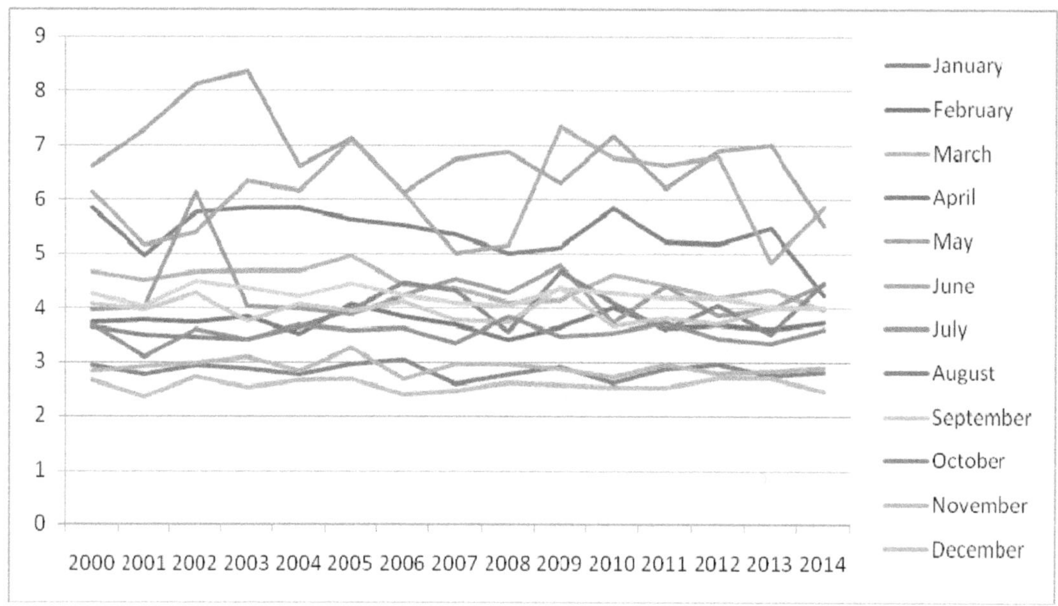

Fig. 4 Monthly ET_0 variation in mm/month for the period of 2000-20014.

As a sample, 2014 daily ET_0 is shown in figure 3. Monthly variation in the ET_0 for a period of 2000-2014 (15 years) is shown in figure 4. The values of monthly ET_0 in mm/month for 15 years are shown in the table 1.

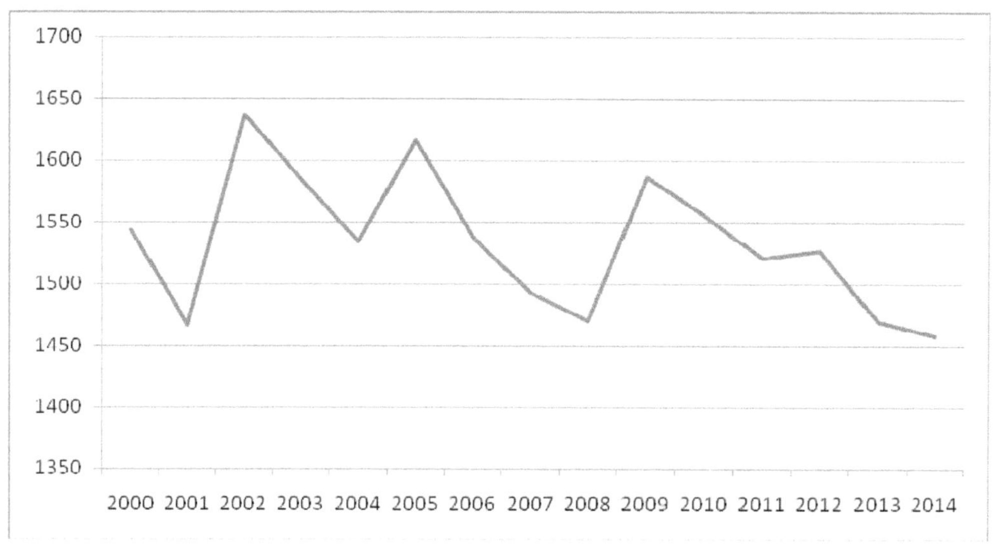

Fig. 5 Yearly ET_0 variations in mm/year for a period of 2000-2014.

Table 1 Average monthly ET_0 values in mm/month for the period of 2000-2014.

Month\ Year	2000	2001	2002	2003	2004	2005	2006	2007	2008	2009	2010	2011	2012	2013	2014
January	2.94	2.76	2.94	2.86	2.77	2.97	3.05	2.58	2.77	2.92	2.60	2.86	2.96	2.74	2.81
February	3.73	3.77	3.73	3.83	3.52	4.07	3.83	3.69	3.41	3.65	4.01	3.63	3.68	3.61	3.73
March	4.66	4.51	4.66	4.67	4.67	4.96	4.41	4.36	4.09	4.14	4.61	4.42	4.18	4.34	3.96
April	5.84	4.95	5.76	5.83	5.84	5.63	5.52	5.34	5.00	5.10	5.83	5.22	5.16	5.47	4.22
May	6.61	7.27	8.12	8.35	6.62	7.13	6.13	6.75	6.88	6.29	7.17	6.21	6.90	6.99	5.52
June	6.14	5.15	5.38	6.34	6.16	7.11	6.12	5.00	5.15	7.35	6.77	6.64	6.79	4.83	5.86
July	3.97	4.01	6.14	4.04	3.98	3.88	4.23	4.52	4.27	4.78	3.73	4.39	3.85	3.98	4.44
August	3.65	3.50	3.46	3.41	3.65	3.97	4.46	4.3	3.55	4.67	4.09	3.60	4.06	3.51	4.46
September	4.07	3.97	4.27	3.76	4.07	3.94	4.11	3.77	3.74	4.37	3.67	3.81	3.72	3.98	4.02
October	3.67	3.08	3.59	3.40	3.67	3.57	3.61	3.33	3.83	3.47	3.52	3.73	3.43	3.33	3.59
November	2.83	2.91	2.99	3.09	2.83	3.26	2.68	2.95	2.97	2.87	2.72	2.96	2.78	2.84	2.89
December	2.64	2.35	2.71	2.51	2.64	2.67	2.40	2.46	2.61	2.57	2.53	2.53	2.68	2.68	2.46
Average	4.23	4.02	4.48	4.34	4.20	4.43	4.21	4.09	4.02	4.35	4.27	4.17	4.18	4.02	4.00

The daily minimum, maximum, and average ET_0 for all the years in mm/day and the total ET_0 for the particular year in mm/year are given in the table 2. During the years 2000-2014 minimum ET_0 was observed on 18 November 2013 having value 1.42 mm/day and maximum ET_0 of 13.52 mm/day was observed on 27 May 2010. Figure 5 showing the yearly variation of ET_0. The maximum total ET_0 is 1636.64 mm/year is in the year 2002 and minimum total ET_0 is 1458.26 mm/year is in the year 2014.

Table 2 Yearly minimum, maximum and average ET_0 (in mm) values.

Year	Daily minimum ET_0 (mm/day)	Daily maximum ET_0 (mm/day)	Average daily ET_0 (mm/day)	Yearly ET_0 (mm/year)
2000	2.05	9.12	4.23	1543.60
2001	1.44	9.87	4.02	1466.54
2002	1.53	11.18	4.48	1636.64
2003	1.48	12.89	4.34	1584.04
2004	1.66	9.12	4.20	1533.84
2005	1.88	10.19	4.43	1616.14
2006	1.62	8.56	4.21	1536.97
2007	1.91	9.18	4.09	1492.20
2008	1.82	8.88	4.03	1469.85
2009	1.79	8.96	4.35	1586.55
2010	1.62	13.52	4.26	1555.76
2011	1.73	9.17	4.17	1520.34
2012	1.56	11.25	4.18	1526.59
2013	1.42	10.32	4.03	1469.25
2014	1.84	8.55	4.00	1458.26

CONCLUSIONS

The FAO-Penman-Monteith equation is recommended as the standard method for estimating reference evapotranspiration. The manual calculation evapotranspiration is a time consuming. The CROPWAT software uses the FAO Penman-Monteith method and makes it easy to estimate ET_0 by giving simple climatic input data. The calculated ET_0 can be further used in estimation of crop-water requirements in the study area. CROPWAT also calculates irrigation scheduling, cropping pattern, and scheme for water supply for the climatic data, soil data, rainfall data, and the cropping pattern. CROPWAT model appropriately estimate the ET_0, which makes this model as a best tool for irrigation planning and management.

REFERENCES

1. A. Patel, R. Sharda, S. Patel and P. Meena (2017) "Reference evapotranspiration estimation using CROPWAT model at ludhiana district (Punjab)" International Journal of Science, Environment and Technology, Vol.6, No 1, 620-629.
2. CROPWAT for windows: User guide (1998).
3. FAO, Crop Evapotranspiration - Guidelines for computing crop water requirements - FAO Irrigation and drainage paper 56 by Richard G. Allen Martin Smith, Rome, 1998.
4. Giridhar. M.V.S.S., and Viswanadh G.K. "Comparison of Radiation Based Reference Evapotranspiration Equations with FAO-56 Penman Monteith Method" International Journal of Computer Science and System Analysis 1(2) July-December 2007, PP. 149-158. (ISSN: 0973-7448).
5. Hussein I. Ahmed, Junmin Liu (2013) "Evaluating Reference Crop Evapotranspiration (ET_0) in the Centre of Guanzhong Basin -Case of Xingping & Wugong, Shaanxi, China" Scientific Research Engineering, Vol 5, No 5, 459-468.
6. M.V.S.S. Giridhar, Shyama Mohan, G. Sreenivasa Rao and P. Sowmya (2017) "Spatial and Temporal Variability of Rainfall in Kaddam Watershed Using Geomatics" National conference on Water, Environment and Society -2017, ISBN: 978-93-5230-182-9, 540-546.
7. Nahla Mustafa Abdallaa , Zhang Xiujub, Adam Ishagc, Gamareldawla Husseind "Estimating Reference Evapotranspiration Using CROPWAT model at Guixi Jiangxi Province"
8. Shreedhar R and Sindhura P (2016) "Study on Water Requirement of Selected Crops under Markandeya Command Area" International Journal for Scientific Research & Development, Vol.4, Issue 09, 54-56.

TREND ANALYSIS OF SEASONAL MAXIMUM AVERAGE RAINFALL INTENSITIES IN RAYALASEEMA REGION OF ANDHRA PRADESH

Shaik Reshma
Assistant Professor, Department of Civil Engineering,
N.B.K.R Institute of Science and Technology, Vidyanagar,
reshma7234@gmail.com

ABSTRACT

Regional seasonal rainfall analysis is essential for effective planning, designing and management of water resources such as urban water supply, drainage and irrigation system. The present study was conducted to determine the trends in 1h, 2h, 3h, 4h, 5h and 6h maximum average rainfall intensities at Ananthapur, Arogyavaram, Kadapa and Kurnool raingauge stations in Rayalaseema, a semi-arid region of Andhra Pradesh. Hourly rainfall data during pre-monsoon, monsoon, post-monsoon and winter seasons at Ananthapur (1969-2010), Arogyavaram (1969-2005), Kadapa (1972-2005) and Kurnool (1969-2010) were collected from India Meteorological Department, Pune and used in the analysis. The procedure used in the present analysis is based on the nonparametric Mann-Kendall test for the trend and the nonparametric Sen's slope estimator method for the magnitude of the trend. The analysis using these methods, has revealed no significant trend at 1h, 2h, 3h, 4h, 5h and 6h maximum average rainfall intensities during the four seasons at Arogyavaram and kadapa rain gauge stations. At Ananthapur raingauge station, significant downward trend has been noticed at 5h (-0.085 (mm/h)/year) and 6h (-0.055 (mm/h)/year) maximum average rainfall intensities during post-monsoon. At Kurnool raingauge station, significant upward trend has been identified at 1h (0.412 (mm/h)/year), 2h (0.305 (mm/h)/year) and 3h (0.147(mm/h)/year) maximum average rainfall intensities during pre-monsoon and 2h maximum average rainfall intensity (0.281(mm/h)/year) during post-monsoon.

Keywords: Maximum average rainfall intensities, Trend analysis, Mann-Kendall test, Sen's trend line method, Sen's slope estimator, Season-wise rainfall, Rayalaseema.

1. INTRODUCTION

Hydrologic phenomena, such as precipitation, floods and droughts, are inherently random by nature. These processes are not fully understood due to complexity of the hydrologic cycle. Rainfall is a physical process that transports water from the atmosphere to Earth's surface. Rainfall is characterized in terms of its frequency, duration and intensity, with intensity being most relevant to the less frequent but more damaging high-intensity events. Intensity relationships of extreme precipitation are used in the storm water drainage system and flood estimation of small watersheds. The intensity and duration of rainfall vary with pre-monsoon, monsoon, post-monsoon and winter seasons.

Convective cells during summer (pre-monsoon period) cause thunder storms and a little rain occurs in the season. The south-west monsoon (called as monsoon) is the principal rainy season and it originates in the Indian Ocean and is accompanied by high south-westerly winds and low pressure regions. The monsoon winds increase from June to July and begin to weaken in September. The weather is generally cloudy with frequent spells of rainfall. The heavy rainfalls occur in the north-eastern regions. As the south-west monsoon retreats, north-easterly flow of moist air strikes the east-coast of the southern peninsula and causes rainfall during post-monsoon. Also, tropical cyclones formed strike the coastal areas and cause intense rainfalls. During winter, disturbances of extra tropical origin travel eastwards and low-pressure causes light to moderate rainfalls in the southern parts.

In general, water resources systems are designed and operated on assumption of stationary hydrology. Existence of trends and other changes in the data invalidates this assumption, and detection of the trends in hydrological time series would help us, to revise the approaches used in assessing, designing and operating our systems. As the detailed knowledge of trends in rainfall intensities helps in water resources planning and management, present study also makes an attempt to determine trends in the maximum average rainfall intensities of different durations during various seasons.

2. COLLECTION OF DATA

Hourly rainfall data during pre-monsoon, monsoon, post-monsoon and winter seasons at Ananthapur (1969-2010), Kurnool (1969-2010), Kadapa (1972-2005) and Arogyavaram (1969-2005) were collected from India Meteorological Department, Pune and used in the analysis. Maximum average rainfall intensity for different durations (1h, 2h, 3h, 4h, 5h and 6h) were calculated for different seasons and used in the analysis. A brief description of raingauge stations is presented in Table 1 and the location of the stations is shown in Fig.1.

Table 1 Brief description of raingauge stations

Raingauge station	Latitude (°N)	Longitude (°E)	Altitude (m)	Mean rainfall (mm)			
				Pre-monsoon	Monsoon	Post-monsoon	Winter
Arogyavaram	13° 32'	78° 30'	550	85.98	383.84	211.48	48.16
Kadapa	14° 29'	78° 50'	130	48.14	369.75	163.82	16.17
Ananthapur	14° 41'	77° 39'	313	29.17	323.62	133.51	9.71
Kurnool	15° 50'	78° 04'	281	71.68	428.39	108.83	11.49

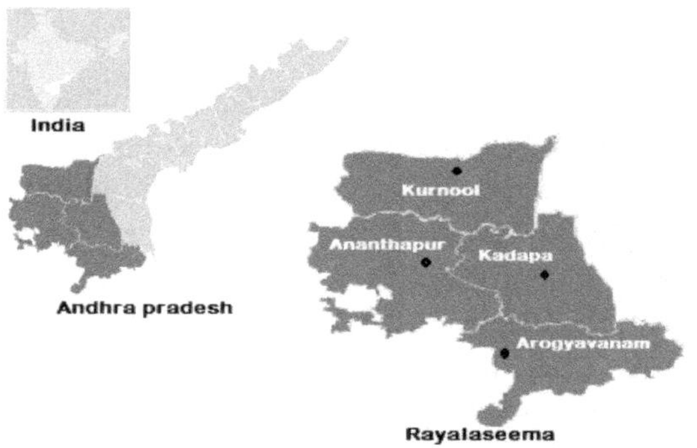

Fig. 1 Location map of raingauge stations

3. METHODOLOGY

A trend is a significant change over certain time exhibited by any variable, detectable by both statistical parametric and non-parametric procedures. Trend analysis is a method to determine the spatial variation and temporal changes for different parameters associated to climate. The climate change is too high for India compared to the global climatic variability. It has further lead to the essence of determining whether the trend is increasing or decreasing. The changes in the most important climatological parameter i.e. rainfall, may be responsible for the natural calamities like drought and flood conditions.

Due to this importance of trend analysis, the number of studies investigating trends in hydro-meteorological data has grown rapidly during the last few decades worldwide. In the present study, the trend analysis is done in two steps. The first step is to detect the presence of a monotonic increasing or decreasing trend using the non-parametric Mann-Kendall test and the second step is estimation of magnitude or slope of a linear trend with the nonparametric Sen's slope estimator. Finally, the newly proposed Sen's 1:1 trend line method is applied to series which has shown significant trend based on Mann-Kendall test.

A. MANN–KENDALL TEST

Mann (1945) presented a non-parametric test for randomness against time, which constitutes a particular application of Kendall's test for correlation commonly known as the 'Mann–Kendall(MK)' or the 'Kendall *t* test' (Kendall, 1975). Letting $x_1, x_2,...,x_n$ be a sequence of measurements over time, Mann proposed to test the null hypothesis, H_0, that the data come from a population where the random variables are independent and identically distributed. The alternative hypothesis, H_1, is that the data follow a monotonic trend over time.

In the computational procedure for the MK test, the data values are evaluated as an ordered time series. Each data value is compared with all subsequent data values. If a data value from a later time period is higher than a data value from an earlier time period, the statistic *S* is incremented by 1. On the other hand, if the data values from a later time period are lower than a data values sampled earlier, *S* is decremented by 1. The net result of all such increments and decrements yields the final value of *S*.

Under H_0, the Mann–Kendall test statistic is *S*,

$$S = \sum_{k=1}^{n-1}\sum_{j=k+1}^{n} sgn(x_j - x_k)$$

where n = number of observations; $x_j = j^{th}$ observation; and sgn(.) = sign function, which can be computed as follows:

$$sgn(x_j - x_k) = \begin{cases} +1, & if\ (x_j - x_k) > 0 \\ 0, & if\ (x_j - x_k) = 0 \\ -1, & if\ (x_j - x_k) < 0 \end{cases}$$

Under the assumption that the data are independent and identically distributed, it has been documented that when $n \geq 10$, the statistic *S* given by Kendall (1975) is approximately normally distributed with the mean and variance as

$$E(S) = 0$$

$$V(S) = \frac{n(n-1)(2n+5) - \sum_{i=1}^{m} t_i(t_i-1)(2t_i+5)}{18}$$

where m = number of groups of tied ranks, each with t_i tied observations. The original Mann-Kendall statistic, designated by Z, is computed as

$$Z = \begin{cases} \frac{S-1}{\sqrt{V(S)}} & S > 0 \\ 0 & S = 0 \\ \frac{S+1}{\sqrt{V(S)}} & S < 0 \end{cases}$$

The statistic Z here follows normal distribution. A positive (negative) value of Z signifies an upward (downward) trend. The significance level, α, is the probability of rejecting the null hypothesis, when it is true. If $-Z_{1-\alpha/2} \leq Z \leq Z_{1-\alpha/2}$, then the null hypothesis of no trend was accepted at the significance level of α. Otherwise, the null hypothesis was rejected and the alternative hypothesis was accepted at the significant level of α. A significance level α is also utilized for testing either an upward or downward monotonic trend. If absolute value of Z appears greater than $Z_{1-\alpha/2}$ then the trend is considered as significant, where α depicts the significance level .

Significance levels are normally set quite low at values of 0.01, 0.05 and 0.10. Choosing a small value for critical probability like 1% leads towards acceptance of the null hypothesis and conversely, adopting a fairly large critical probability like 10% increases the chances of rejecting the null hypothesis. Adopting 5% for the critical probability seems to be a rational choice because it is neither on the liberal nor on the conservative side (Haktanir and Hatice, 2014).

Therefore, in the present study 5% for the critical probability i.e. significance level (α) of 0.05 with confidence level (1-α) of 95% as per Mann-Kendall is used for trend analysis over the stations in Rayalaseema region.

B. SEN'S SLOPE ESTIMATOR

The magnitude of the trend in the seasonal and annual series was determined using a non parametric method known as Sen's estimator (Sen, 1968). The Sen's method can be used in cases where the trend can be assumed to be linear.

Here, the slope Q_i, of all data value pairs is first calculated using the equation

$$Q_i = \frac{x_j - x_k}{j - k}$$

where x_j and x_k are data values at time j and k $(j>k)$ respectively . If there are n values x_j in the time series there will be as many as $N = n(n-1)/2$ slope estimates Q_i. The Sen's estimator of slope is the median of these N values of Q_i. The N values of Q_i are ranked from the smallest to the largest and the Sen's estimator is

$$Q = Q_{\left[\frac{(N+1)}{2}\right]} \qquad \text{if } N \text{ is odd}$$

$$Q = \frac{1}{2}\left(Q_{\left[\frac{N}{2}\right]} + Q_{\left[\frac{(N+2)}{2}\right]}\right) \qquad \text{if } N \text{ is even}$$

Positive value of Q_i indicates an upward or increasing trend and a negative value of Q_i gives a downward or decreasing trend in the series.

C. SEN'S 1:1 TREND-LINE TEST

Sen (2012) proposed a diagrammatic method for trend detection. The basis of this approach rests on the fact that if two time series were identical to each other, their plot against each other shows scatter of points along 1:1 (45°) line on the Cartesian coordinate system. The recorded series is chronologically divided into two equal halves. The elements of the second portion are plotted versus the elements of the initial half after both segments are ordered in an ascending pattern.

If there is no trend, most of the $n/2$ plotted points fall close to the 1:1 (45°) line. For an increasing trend, the plotted points lie above the 1:1 line because the magnitudes of the elements of the second half of the recorded series (ordinate values) are greater than those of the first half (abscissa values), and conversely, for a decreasing trend, they fall below the 1:1 line. The closer the scatter points are to the 1:1 line, the weaker the trend magnitude (slope). By visual observation, no trend, or either an increasing or a decreasing trend is determined.

4. RESULTS AND DISCUSSION

The results of Mann-Kendal analysis to detect the trend for the four raingauge stations Ananthapur, Arogyavaram, Kadapa and Kurnool in the study area are presented in Tables 2 to 5 respectively.

Table 2 Mann-Kendall test results for Ananthapur during different seasons.

Season	Maximum average rainfall intensity (mm/h)	Kendall's S	Variance of S	Z_C	Trend nature	Null hypothesis	Trend significance
Pre- monsoon	i_1	112	8511.22	1.20	-	Accepted	No
	i_2	163	8514.33	1.76	-	Accepted	No
	i_3	129	8513.56	1.39	-	Accepted	No
	i_4	82	8231.22	0.89	-	Accepted	No
	i_5	24	7136.89	0.27	-	Accepted	No
	i_6	-67	6725.44	-0.80	-	Accepted	No
Monsoon	i_1	31	8514.33	0.33	-	Accepted	No
	i_2	26	8514.33	0.27	-	Accepted	No
	i_3	-29	8514.33	-0.30	-	Accepted	No
	i_4	-101	8514.33	-1.08	-	Accepted	No
	i_5	-4	8514.33	-0.03	-	Accepted	No
	i_6	-89	8514.33	-0.95	-	Accepted	No
Post-monsoon	i_1	-137	8514.33	-1.47	-	Accepted	No
	i_2	-135	8514.33	-1.45	-	Accepted	No
	i_3	-178	8514.33	-1.78	-	Accepted	No
	i_4	-207	8513.56	-1.92	-	Accepted	No
	i_5	-195	8487.11	**-2.23**	Negative	**Rejected**	**Yes**
	i_6	-165	8470.78	**-2.11**	Negative	**Rejected**	**Yes**
Winter	i_1	95	8445.11	1.02	-	Accepted	No
	i_2	31	8078.78	0.33	-	Accepted	No
	i_3	-52	6492.11	-0.63	-	Accepted	No
	i_4	-101	4270.78	-1.53	-	Accepted	No
	i_5	-109	3397.33	-1.85	-	Accepted	No
	i_6	-87	2359.00	-1.77	-	Accepted	No

* Bold values indicate existence of a trend with level of significance $\alpha = 0.05$.

From Table 2, it can be concluded that, no significant trends in maximum average rainfall intensities were found during pre-monsoon, monsoon and winter seasons. Significant falling trend was noticed in 5h and 6h maximum average rainfall intensities during post-monsoon season.

Table 3 Mann-Kendall test results for Arogyavaram during different seasons.

Season	Maximum average rainfall intensity (mm/h)	Kendall's S	Variance of S	Z_C	Trend nature	Null hypothesis	Trend significance
Pre- monsoon	i_1	35	5844.44	0.44	-	Accepted	No
	i_2	-9	5846.00	-0.10	-	Accepted	No
	i_3	-33	5846.00	-0.42	-	Accepted	No
	i_4	-113	5802.44	-1.47	-	Accepted	No
	i_5	-102	5317.11	-1.39	-	Accepted	No
	i_6	-84	4823.22	-1.20	-	Accepted	No
Monsoon	i_1	13	5846.00	0.16	-	Accepted	No
	i_2	20	5846.00	0.25	-	Accepted	No
	i_3	-26	5846.00	-0.33	-	Accepted	No
	i_4	56	5846.00	0.72	-	Accepted	No
	i_5	91	5846.00	1.18	-	Accepted	No
	i_6	110	5846.00	1.43	-	Accepted	No
Post-monsoon	i_1	-18	5845.22	-0.22	-	Accepted	No
	i_2	2	5846.00	0.01	-	Accepted	No
	i_3	-7	5846.00	-0.08	-	Accepted	No
	i_4	-36	5846.00	-0.46	-	Accepted	No
	i_5	-72	5846.00	-0.93	-	Accepted	No
	i_6	-80	5845.22	-1.03	-	Accepted	No
Winter	i_1	28	5845.22	0.35	-	Accepted	No
	i_2	44	5845.22	0.56	-	Accepted	No
	i_3	28	5818.78	0.35	-	Accepted	No
	i_4	36	5802.44	0.46	-	Accepted	No
	i_5	-40	5562.89	-0.52	-	Accepted	No
	i_6	-17	5492.11	-0.22	-	Accepted	No

From Table 3, it can be concluded that, no significant trends were identified in all the (1h, 2h, 3h, 4h, 5h and 6h) maximum average rainfall intensities during various seasons at Arogyavaram raingauge station.

Table 4 Mann-Kendall test results for Kadapa during different seasons.

Season	Maximum average rainfall intensity (mm/h)	Kendall's S	Variance of S	Z_C	Trend nature	Null hypothesis	Trend significance
Pre- monsoon	i_1	77	4550.33	1.13	-	Accepted	No
	i_2	92	4547.22	1.35	-	Accepted	No
	i_3	71	4327.89	1.06	-	Accepted	No
	i_4	29	4194.87	0.43	-	Accepted	No
	i_5	21	3795.10	0.32	-	Accepted	No
	i_6	36	3458.33	0.60	-	Accepted	No
Monsoon	i_1	70	4550.33	1.02	-	Accepted	No
	i_2	54	4550.33	0.79	-	Accepted	No
	i_3	107	4550.33	1.57	-	Accepted	No
	i_4	133	4550.33	1.96	-	Accepted	No
	i_5	121	4547.22	1.78	-	Accepted	No
	i_6	131	4523.11	1.93	-	Accepted	No
Post-monsoon	i_1	4	4549.55	0.04	-	Accepted	No
	i_2	28	4550.33	0.40	-	Accepted	No
	i_3	75	4550.33	1.10	-	Accepted	No
	i_4	55	4550.33	0.80	-	Accepted	No
	i_5	82	4547.22	1.20	-	Accepted	No
	i_6	99	4523.11	1.46	-	Accepted	No
Winter	i_1	-29	4521.55	-0.42	-	Accepted	No
	i_2	-8	4506.77	-0.10	-	Accepted	No
	i_3	26	4015.22	0.39	-	Accepted	No
	i_4	-3	3380.55	-0.03	-	Accepted	No
	i_5	10	3352.55	0.16	-	Accepted	No
	i_6	4	3172.89	0.05	-	Accepted	No

From Table 4, it can be concluded that, no significant trends were identified in all the (1h, 2h, 3h, 4h, 5h and 6h) maximum average rainfall intensities during various seasons at Kadapa raingaugestation.

Table 5 Mann-Kendall test results for Kurnool during different seasons

Season	Maximum average rainfall intensity (mm/h)	Kendall's S	Variance of S	Z_C	Trend nature	Null hypothesis	Trend significance
Pre- monsoon	i_1	253	8514.33	**2.73**	Positive	Rejected	Yes
	i_2	271	8514.33	**2.93**	Positive	Rejected	Yes
	i_3	214	8498.77	**2.31**	Positive	Rejected	Yes
	i_4	129	8231.22	1.41	-	Accepted	No
	i_5	22	7627.663	0.24	-	Accepted	No
	i_6	40	6239.333	0.49	-	Accepted	No
Monsoon	i_1	50	8513.55	0.53	-	Accepted	No
	i_2	144	8514.33	1.55	-	Accepted	No
	i_3	77	8514.33	0.82	-	Accepted	No
	i_4	120	8513.55	1.29	-	Accepted	No
	i_5	95	8514.333	1.02	-	Accepted	No
	i_6	105	8514.333	1.13	-	Accepted	No
Post-monsoon	i_1	173	8514.33	1.86	-	Accepted	No
	i_2	192	8514.33	**2.07**	Positive	Rejected	Yes
	i_3	125	8514.33	1.34	-	Accepted	No
	i_4	123	8513.55	1.32	-	Accepted	No
	i_5	144	8511.223	1.55	-	Accepted	No
	i_6	123	8407.773	1.33	-	Accepted	No
Winter	i_1	135	8078.77	1.49	-	Accepted	No
	i_2	80	6239.33	1.00	-	Accepted	No
	i_3	11	4707.89	0.15	-	Accepted	No
	i_4	15	2961.00	0.26	-	Accepted	No
	i_5	-1	2471.003	-0.04	-	Accepted	No
	i_6	44	1406.223	1.15	-	Accepted	No

* Bold values indicate existence of a trend with level of significance $\alpha = 0.05$.

From Table 5, it can be concluded that, no significant trend was identified in all the maximum average rainfall intensities (1h, 2h, 3h, 4h,5h and 6h) during monsoon and winter seasons, whereas significant increasing trend in (1h, 2h, 3h) maximum average rainfall intensities during pre-monsoon season and (2h) maximum average rainfall intensity during post-monsoon season was observed at Kurnool station.

Maximum average rainfall intensities which have shown significant trends based on Mann-Kendall test are shown in Figs. 2 and 3.

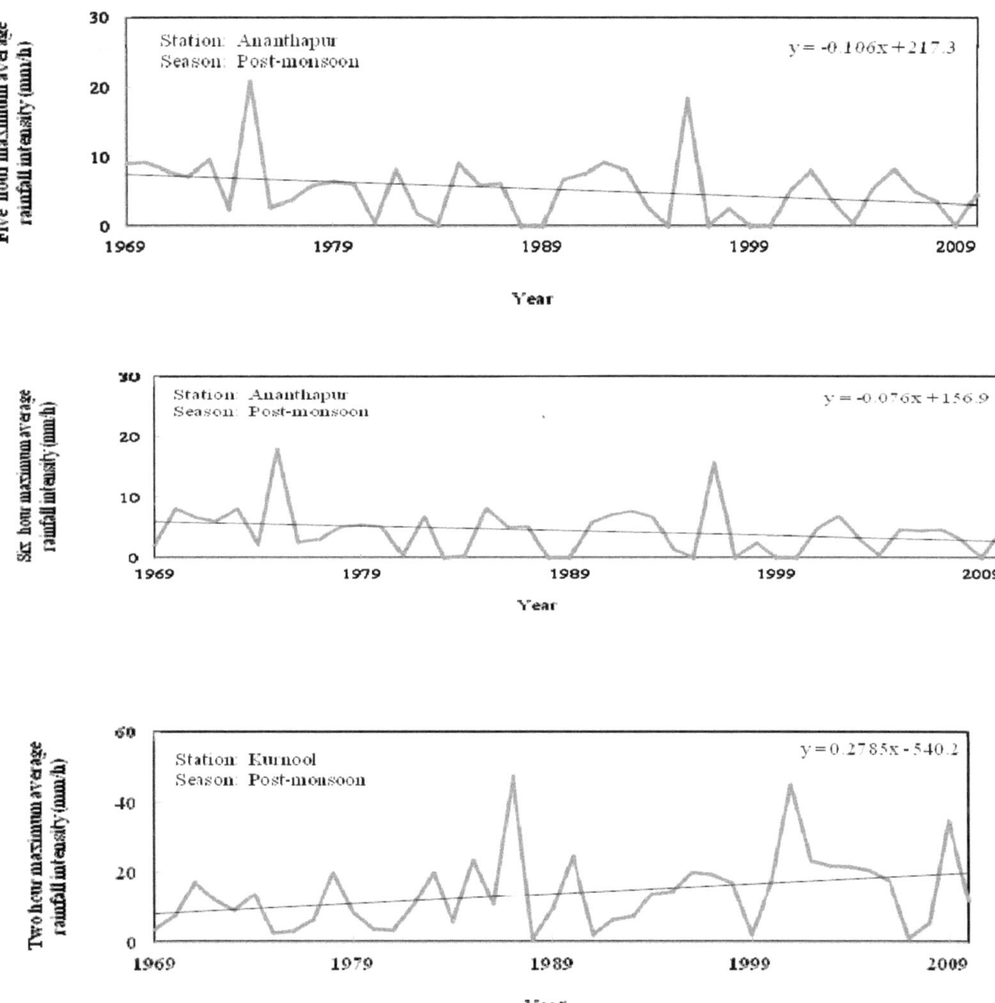

Fig. 2 Trend of hourly maximum average rainfall intensities (Season: Post-monsoon)

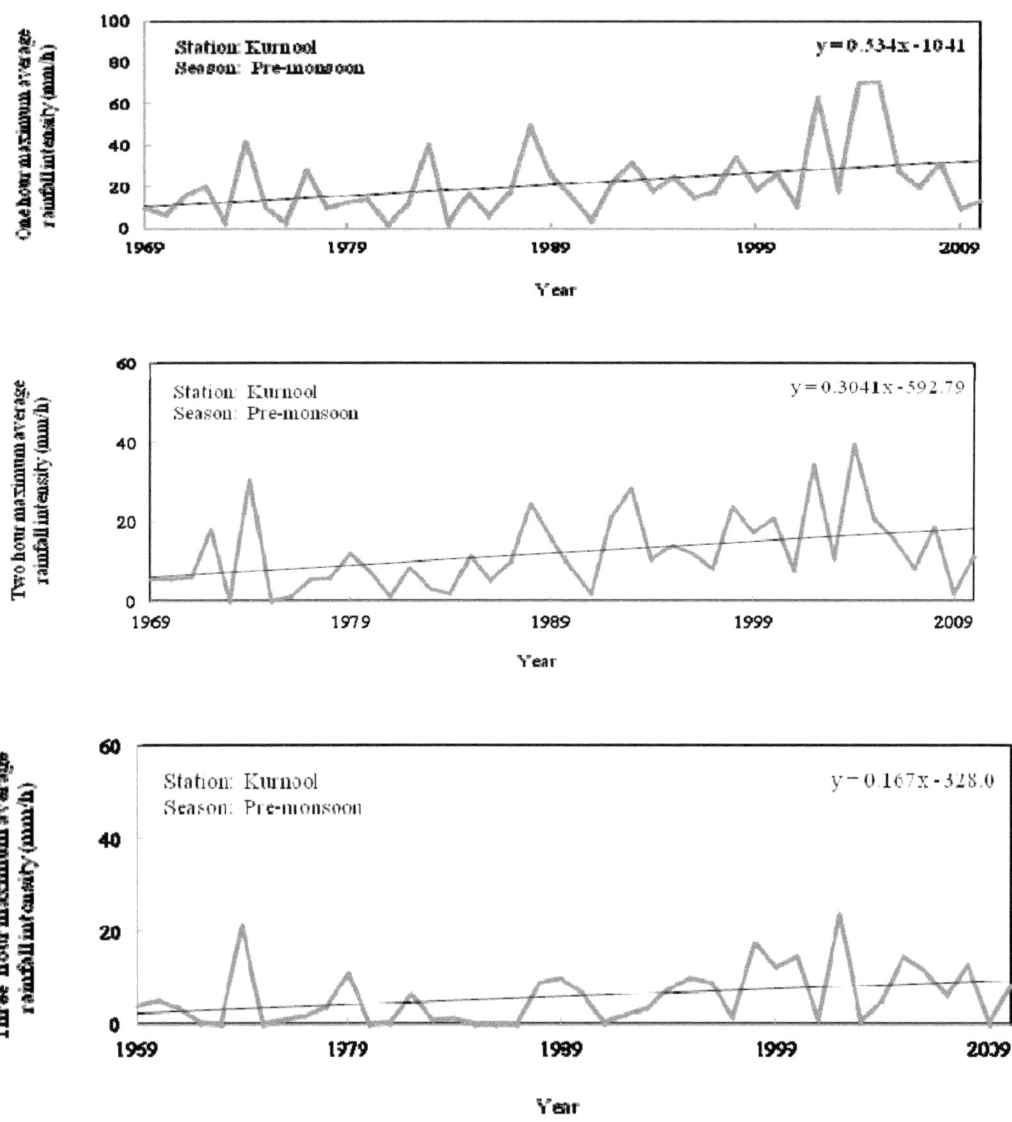

Fig. 3 Trend of hourly maximum average rainfall intensities (Season: Pre-monsoon)

The magnitude of the trend for the maximum average rainfall intensities of 1h to 6h durations during various seasons at the four raingauge stations are determined using a non parametric method known as Sen's estimator and the results are shown in Tables 6 to 9.

Table 6 Estimated Sen's slope for different seasons (Station: Ananthapur).

Maximum average rainfall intensity (mm/h)	Estimated Sen's slope for different seasons			
	Pre-monsoon	Monsoon	Post-monsoon	Winter
i_1	1.283	0.056	-0.238	0.014
i_2	0.896	0.037	-0.150	0.012
i_3	0.490	-0.047	-0.134	-0.009
i_4	0.308	-0.022	-0.108	-0.010
i_5	0.146	-0.004	**-0.085**	-0.015
i_6	-0.037	-0.007	**-0.055**	-0.014

* Bold values indicate statistical significance at 95% confidence level as per the Mann-Kendall test.

Table 7 Estimated Sen's slope for different seasons (Station: Arogyavaram).

Maximum average rainfall intensity (mm/h)	Estimated Sen's slope for different seasons			
	Pre-monsoon	Monsoon	Post-monsoon	Winter
i_1	0.069	0.063	-0.053	0.020
i_2	-0.008	0.034	0.003	0.043
i_3	-0.036	-0.053	-0.006	0.004
i_4	-0.082	0.047	-0.032	0.011
i_5	-0.010	0.072	-0.049	-0.008
i_6	-0.005	0.091	-0.040	-0.003

Table 8 Estimated Sen's slope for different seasons (Station: Kadapa).

Maximum average rainfall intensity (mm/h)	Estimated Sen's slope for different seasons			
	Pre-monsoon	Monsoon	Post-monsoon	Winter
i_1	0.242	0.300	0.000	-0.013
i_2	0.177	0.106	0.069	-0.009
i_3	0.150	0.133	0.132	0.070
i_4	0.090	0.167	0.103	-0.011
i_5	0.073	0.131	0.088	0.010
i_6	0.004	0.141	0.087	0.006

Table 9 Estimated Sen's slope for different seasons (Station: Kurnool).

Maximum average rainfall intensity (mm/h)	Estimated Sen's slope for different seasons			
	Pre-monsoon	Monsoon	Post-monsoon	Winter
i_1	**0.412**	0.077	0.333	0.080
i_2	**0.305**	0.157	**0.281**	0.052
i_3	**0.147**	0.081	0.124	0.010
i_4	0.047	0.082	0.117	0.009
i_5	0.029	0.066	0.100	-0.006
i_6	0.010	0.061	0.052	0.002

* Bold values indicate statistical significance at 95% confidence level as per the Mann-Kendall test.

From the Tables 6 to 9, it can be concluded that Sen's slope is also indicating increasing and decreasing magnitude of slope which are similar to that of Mann-Kendall's test values.

Sen's trend line method has been applied to the maximum average rainfall intensity series, which represented significant trend based on Mann-Kendall test and its results are observed to be in agreement with the Mann-Kendall test. Plots of Sen's 1:1 trend line for maximum average rainfall intensity series at Ananthapur and Kurnool are presented in Figs. 4 and 5 respectively.

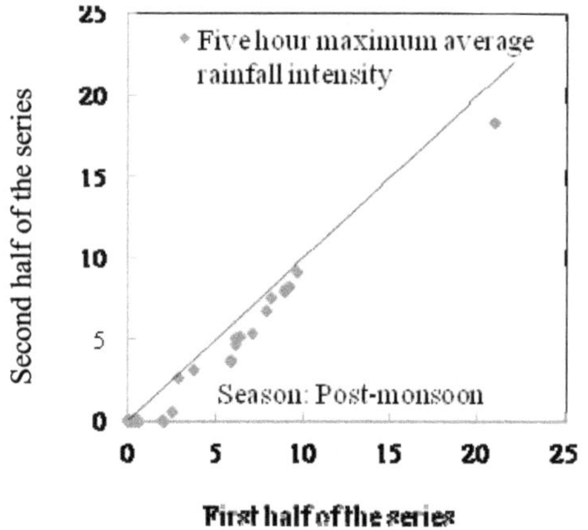

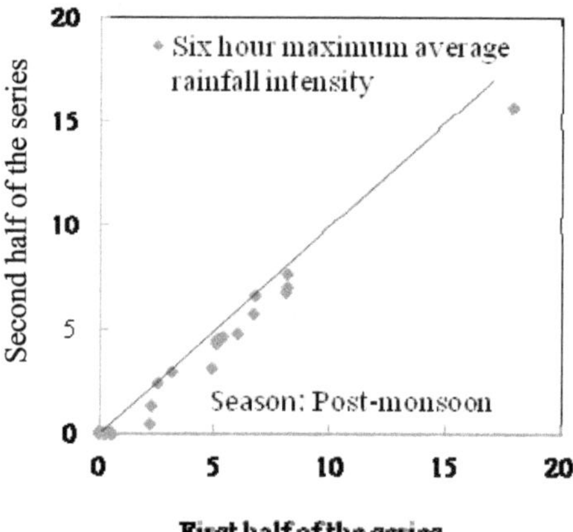

Fig. 4 Plot of Sen's 1:1 trend line method for maximum average rainfall intensity series at Ananthapur

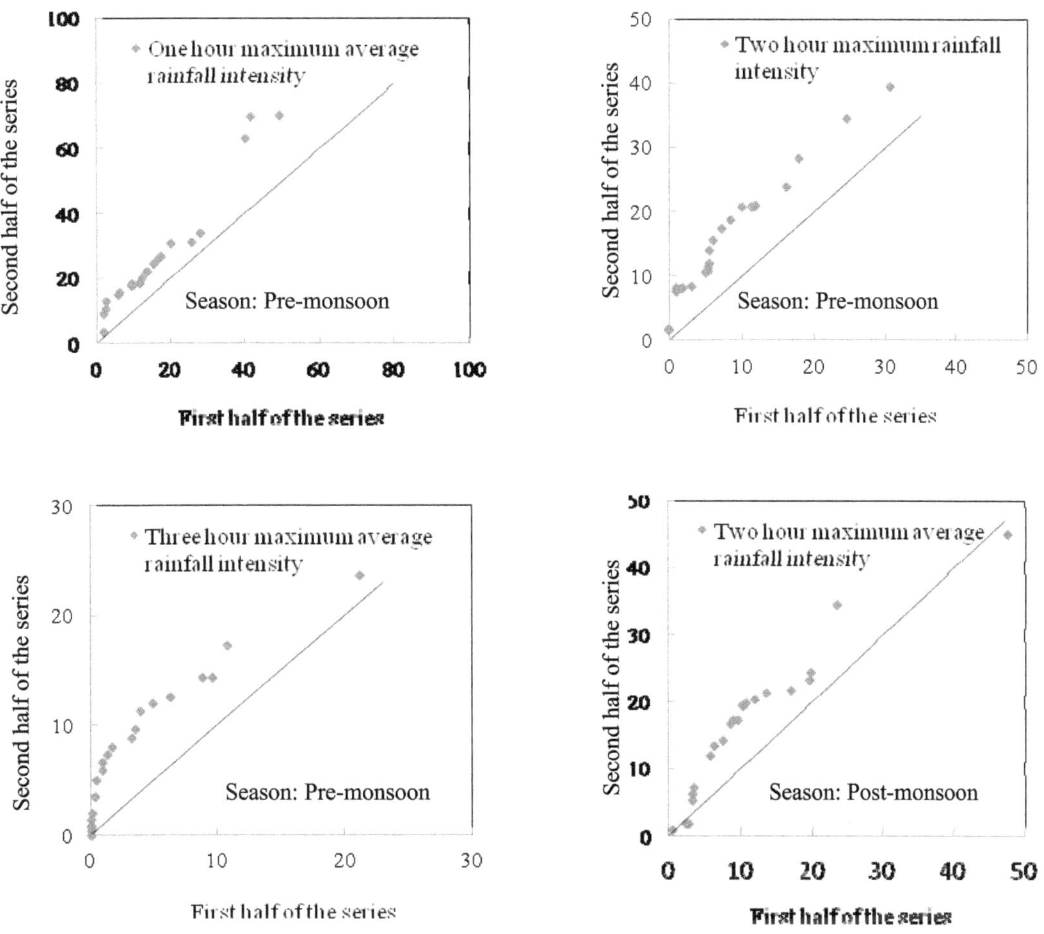

Fig. 5 Plot of Sen's 1:1 trend line method for maximum average rainfall intensity series at Kurnool.

Fig. 5 depicts that, scatter points fall below and close to 1:1 trend line indicating a weaker decreasing trend at both 5h and 6h maximum average rainfall intensities during post-monsoon season at Ananthapur whereas in Fig. 5, all the scatter points fall above 1:1 trend line indicating increasing trend at 1h, 2h and 3h maximum average rainfall intensity during pre-monsoon season and 2h maximum average rainfall intensity during post-monsoon season.

5. CONCLUDING REMARKS

From both of the statistical test results it can be concluded that, there is an evidence of no significant change in trend of hourly maximum average rainfall intensities at Arogyavaram and Kadapa raingauge stations. There appears a little change at larger durations during post-monsoon season at Ananthapur raingauge station and similar changes were appeared during pre-monsoon season at smaller durations at Kurnool raingauge station.

The following are the station-wise concluding remarks drawn based on the trend analysis of maximum average rainfall intensities:

Raingauge Station: Ananthapur

- Significant downward trend has been noticed at 5h and 6h maximum average rainfall intensities during post-monsoon.
- Sen's slope estimation method has also presented significant downward trend at 5h (-0.085 (mm/h)/year) and 6h (-0.055 (mm/h)/year) maximum average rainfall intensities during post-monsoon.

Raingauge Station: Arogyavaram

- No significant trends have been found at 1h, 2h, 3h, 4h, 5h and 6h maximum average rainfall intensities during all the four seasons.

Raingauge Station: Kadapa

- No significant trends have been identified at 1h, 2h, 3h, 4h, 5h and 6h maximum average rainfall intensities during all the four seasons.

Raingauge Station: Kurnool

- Significant upward trend has been noticed during pre-monsoon for 1h, 2h and 3h maximum average rainfall intensities.
- Significant upward trend has been appeared for 2h maximum average rainfall intensity during post-monsoon.
- Sen's slope estimation method has also revealed significant upward trend at 1h (0.412 (mm/h)/year), 2h (0.305 (mm/h)/year) and 3h (0.147(mm/h)/year) maximum average rainfall intensities during pre-monsoon and at 2h maximum average rainfall intensity (0.281(mm/h)/year) during post-monsoon.

ACKNOWLEDGEMENTS

The author thanks Smt. S. Aruna Jyothy and her supervisor Prof. P. Mallikarjuna, Department of Civil Engineering, SV University for their guidance. Author thanks the staff of India Meteorological Department (IMD), Pune, for their cooperation in providing the requisite data. Administrative support given during the period of investigation, by the Principal and Head of the Department of Civil Engineering, Sri Venkateswara University College of Engineering, Tirupati is greatly acknowledged.

REFERENCES

1. Duhan D and Pandey A (2013), "Statistical Analysis of long term spatial and temporal trends of precipitation during 1901- 2002 at Madhya Pradesh", India Atmospheric research, Elsevier, 122: 136-49.
2. Haktanir T and Hatice C (2014), "Trend, Independence, Stationarity, and Homogeneity Tests on Maximum Rainfall Series of Standard Durations Recorded in Turkey", Journal of Hydrologic Engineering, Vol. 10, No. 1, pp. 1-13.

3. Jayawardene, H.K.W.I., Sonnadara, D.U.J. and Jayewardene, D.R. (2005), "Trends of Rainfall in Sri Lanka over the Last Century", Sri Lankan Journal of Physics 6: 7-17.
4. Kendall, M.G. (1975), Rank Correlation Methods. Charles Griffin, London, UK.
5. Kumar V, Jain, S. K. and Singh, Y (2010), "Analysis of long-term rainfall trends in India", Journal of Hydrological Sciences, Vol. 55, No. 4, pp. 484–496.
6. Longobardi A and Villani P. (2010), "Trend analysis of annual and seasonal rainfall time series in the Mediterranean area", International Journal for Climatology, 30: 1538-46.
7. Manikandan M and Tamilmanin D (2012), "Statistical Analysis of Spatial Pattern of Rainfall Trends In Parambikualam Aliyar Sub Basin, Tamil Nadu", Journal of Indian Water Resources Society, Vol. 32, No. 1, pp. 40-50.
8. Mann, H.B. (1945), "Nonparametric tests against trend", Econometrica, 13: 245-259.
9. Mondal A, Kundu S and Mukhopadhyay A (2012), "Rainfall trend analysis by Mann-Kendall Test: A Case Study of North-Eastern Part of Cuttack District, Orissa", International Journal of Geology, Earth and Environmental Sciences, ISSN: 2277-2081, Vol. 2, No. 1, pp.70-78.
10. Sen PK (1968), "Estimates of the Regression Coefficient Based on Kendall,s tau", Journal of American Statistical Association, Vol. 39, No. 4, pp. 1379-1389.
11. Sen Z (2012), "Innovative Trend Analysis Methodology", Journal of Hydrologic Engineering, Vol. 17, No. 1, pp. 1042-1046.
12. Sabyasachi S, Manikant V and Verma M.K (2015), "Statistical Trend Analysis of Monthly Rainfall for Raipur District, Chhattisgarh", International Journal of Advanced Engineering Research and Studies, Vol. 4, No. 2, pp. 87-89.
13. (13) Yilmaz A.G and Perera B.J (2015), "Spatiotemporal Trend Analysis of Extreme Rainfall Events in Victoria, Australia", Journal of Water Resource Management, Vol. 10, No. 3, pp. 1070-1085.

SUSTAINABLE URBANIZATION USING REMOTE SENSING AND GIS – AN OVERVIEW

G. Padmaja[1], M.V.S.S. Giridhar[2], V. Shiva Chandra[3], G. Sreenivasa Rao[4] and R. Sandhya Rani[4]

[1,3,4]Asst. Prof., Dept. of Civil Engg., MVSR Engg. College, Hyderabad
[2]Assoc Prof, Center for Water Resources, IST, JNTU, Hyderabad
gantipadmaja6@gmail.com

ABSTRACT

Globally, the migration to urban areas is rapidly increasing leading to an unprecedented shift and significant changes relating to population distribution and spatial patterns. The change in the dominant habitat of world population makes the process of urbanization a significant global trend of the twenty-first century. Urban areas no longer represent and function as mere spaces for settlements and habitation but have significant role in the global development. These areas not only shape the present land cover patterns but also prominently influence social and economic issues. Poorly managed urban growth and development can exaggerate inequalities, increase in vulnerability among marginalized population. It is essential that urban cities need to focus more on orderly expansion of the existing land resources by implementing sustainable practices without depriving the needs and aspirations of future generations. Issues related to unsustainable development in urban areas needs to be addressed with the help of recent technologies such as Remote Sensing and GIS. Mapping of urbanization specifies locations where growth/expansion is taking place, help in identifying areas having serious environmental issues like depletion of natural resources and pollution. The integration of these new technologies with proper planning not only helps judicious utilization of the existing natural resources but also emphasize on environmental protection such that basic needs of future generations are secured.

INTRODUCTION

Urban areas occupy a small portion on the surface of the earth while devouring two thirds of the energy resources and are responsible for over 70% of CO_2 emissions. Urban population is growing at a faster rate which cities cannot accommodate. Cities are not self-sufficient and are dependent on neighboring towns and villages for their food resources and other utilities. Increased migrations from neighboring areas during the last few decades have expanded cities to such an extent that they are now struggling hard to provide basic amenities and services. Globally, major cities have crisis for power, transportation, water systems and pollution related issues which continue to increase due to overcrowding. Urban poverty associated with unemployment and inadequate housing facilities is another serious socioeconomic challenge. The circumstances that led to rapid urbanization may vary from one city to another, but they have one thing in common that is urban areas are becoming increasingly fragile and are not capable of providing needs for all its residents.

"The world is going through an unprecedented transition. The global balance of power is shifting, extreme poverty has dropped to historic lows, more people than ever before now live in cities, and new technologies are revolutionizing social behaviors and entire industries." (UNDP Strategic Plan 2014-17).

Additionally, cities require amenities like building materials, fuel, industrial and household chemicals, foodstuffs, water and land, resulting in adverse impacts on the environment. Water bodies and aquatic ecosystems are an integral part of urban landscape constituting an important source of fresh water. Degradation of water bodies in urban areas has detrimental impact on domestic water supply, industrial & commercial sector, recreational & agricultural activities. On most occasions human interaction with water bodies leads to contamination and pollution, thus it is necessary to restore, preserve and build up such water bodies. Continuous spatial and temporal monitoring of these changes is required to assess the effects of urbanization on the environment. Therefore the capabilities of RS & GIS can be effectively used to monitor the expansion of cities on different spatiotemporal scales.

Urbanization & Environmental Impact: Urbanization refers to unprecedented population shift from rural to urban areas, "the gradual increase in the proportion of people living in urban areas", and the ways in which each society acclimatizes to the change. Many people move into cities for better economic opportunities but it disturbs the ecological balance thus leading to loss of water bodies, agriculture land, increased built up area and diminished green cover. Demand for water and energy proliferates where the supply of the same may be a difficult proposition. The environmental impacts related to urbanization are as follows:

- **Poverty**: Enormous growth in urban population leads to poverty as cities cannot accommodate and also fail to provide basic amenities and services for all its residents.
- **Pollution:** Several human activities along with increased energy consumption due to urbanization lead to air, water, noise and thermal pollution with significant impact on human health.
  In 2011, EPA identified India and Russia as the fourth largest contributors to the global CO_2 emissions due to fossil fuel combustion, cement manufacturing, and gas flaring (EPA 2015). In 2003, the construction sector in India accounted for maximum CO_2 emissions (22 %), while the transportation industry (road transport, aviation, and shipping) contributed to 12.9 % of the national CO_2 emissions (Parikh et al. 2009). In 2007, 87 % of the total CO_2 emissions from the transport industry were attributed to road transport (MoEF 2010). Hyderabad witnessed a reduction in the content of black carbon, particulate matter, CO, and ozone by about 57%, 60%, 40%, and 50%, respectively, during a weeklong nationwide truck strike (Sharma et al. 2010). Rapid urbanization has also enhanced living standards leading to increased vehicle ownership pattern.
- **Solid Waste:** Urbanization is increasing all over the planet and municipal solid waste is one of the byproducts of urban living. Cities are becoming hubs for garbage production, and the amount of garbage they create is increasing even faster than its population, according to a recent report from the World Bank.
- **Flash Floods:** Changes in Land Use and Land cover due to Urban Sprawl magnifies the risk of environmental hazards such as flash flooding."As the urban sprawl of rapid urbanization expands outwards and upwards, it provides ready opportunities for hazards such as floods, storms and earthquakes to wreak havoc. Half the world's population now lives in urban areas, and that figure is estimated to rise 70% by 2050. That's a lot of

vulnerable and exposed people given that urban floods will represent the lion's share of total flood impact because of infrastructure, institutions and processes that are not yet up to the task ahead", warns Wahlström.

- **Diminishing Green Cover& Destruction of habitat:** Construction activities and other physical barriers due to urbanization triggers to loss of green cover there by destructing the natural habitat of the area. Increase in population has adversely affected the green cover in urban India—Chennai and Mumbai have a meager 0.46 m^2 (Srivathsan 2013) and 0.12 m^2 (FAO 1998) of green space per capita, respectively, as compared to the UN recommended standard of 9 m^2 of green space per capita. The shrinking of residential gardens adds to environmental degradation. The observed change in lifestyle at the expense of garden space indicates reduction of urban green cover and impact on the environment and changes in the weather patterns especially rainfall.
- **Loss of biodiversity:** urbanization has adverse effect on the animal populations which are inhibited by toxic substances leading to the loss of habitat and non-availability of food sources in those areas.
- **Urban heat islands:** Urban sprawl and unplanned growth of Indian cities pose significant threat to the local climate. Cities are becoming urban heat islands due to less vegetative cover and due to direct exposure of soil to sun's energy leading to higher surface temperatures. Additionally vehicular emissions, factories, industries and domestic heating and cooling units release even more heat as a result of which most of the cities are often 1 to 3 °C (1.8 to 5.4 °F) warmer than surrounding landscapes. Impacts also include reducing soil moisture and a reduction in absorption of carbon dioxide emissions.
- **Social Unrest:** Urban centers provide livelihood opportunities and also act as accelerators for social and economic progress. Since ancient times cities are centers of innovation in technology, commerce, social organization and ideas. The concentration of people, resources and ideas allows innovation to occur at tremendous speed, generating economic activity and wealth at unparalleled rates. Globally cities are emerging as hot spots for entrepreneurs to achieve their dreams. However, cities are also home to significant concentrations of the poor and marginalized population. Urban poverty is growing and the World Bank estimates that, by 2035, most of the world's extreme poor will be found in urban areas leading to social unrest.
- **Energy consumption:** Population growth is intrinsically linked to increased demand for energy. The compounded annual growth rate of total installed capacity for power generation in India during 1971–2012 stood at 6.58 % (Ministry of Statistics & Program Implementation 2013). The building industry consumes 40 % of the national electricity consumption and is estimated to increase to 76 % by 2040 (Centre for Science & Environment, New Delhi, India 2014). The domestic sector accounted for 22 % of the total electricity sales in 2011–2012 (Ministry of Statistics & Program Implementation 2013). Excess of domestic energy consumption may be eliminated through energy friendly practices. Around 3000–5000 kWh of energy can be saved through implementation of energy efficient measures primarily aimed at reducing building cooling needs (NHB 2015).

- **Heat stress induced health hazards**: Warmer neighborhoods increase the vulnerability of residents to heat exposure (Harlan et al. 2006). Heat waves aggravate thermal discomfort in heat islands, often culminating in health issues. Heat stroke, heat exhaustion, infectious diseases, cardiovascular and respiratory problems aggravate during summer seasons (Harlan et al. 2006). The population group most susceptible to these harmful effects comprises of poor, the physically weak, and the elderly (CDC 2015). More than 21 % of the Indian population lives below the poverty line (Reserve Bank of India 2013), and remains extremely vulnerable to health hazards.

Sustainable Urban Development

Sustainable urban development is the balance between meeting the needs of the present generation while protecting the environment to safeguard the needs of future generations. The growing human population and its demands on the earth's resources generate a need for sustainable practices. Implementing these practices often requires collaboration among diverse organizations, stakeholder's participation and appropriate decision making. RS and GIS allows users across the globe to share ideas on how to meet their resource needs, plan efficient land use, and protect the environment to guarantee the survival of future generations.

Role of RS & GIS

Satellite Remote Sensing data can be used for mapping the land use pattern in urban areas and to gather detailed up to-date information while GIS can be used to develop a database system for storing urban information. The spatial and temporal changes of urban areas can be detected efficiently by using RS and GIS. Preparation of digital database for all features of land use and recurrent changes due to urbanization are needed for better environment management of existing resources. These play a pivotal role in efficient formulation and implementation of the land based sustainable development strategies.

The spatial patterns of urban expansion over different time periods can be mapped, monitored and assessed accurately using satellite data (remotely sensed data) along with conventional ground data (Lata et al.,2001). Mapping of urban sprawl indicates the location where grow this taking place and this helps in identifying the environmental crisis like depletion of natural resources and also suggests the likely future directions and patterns of sprawling growth. Ultimately, the power to manage sprawl resides with local municipal governments that vary considerably in terms of will and ability to address sprawl issues.

Ways and Means for Sustainable Urbanization

- **Urban Rain water Harvesting:** Ironically, urban areas in India are facing acute water scarcity while most of the streets are often flooded during monsoon. This has led to serious problems with quality and quantity of ground water. This is despite the fact that all these cities receive good rainfall. However, this rainfall occurs during short spells with high intensity. (Most of the rain falls in just 100 hours out of 8,760 hours in a year). Because of such short duration of heavy rain, most of the rain falling on the surface tends to flow away rapidly leaving very little for recharge of ground water. Most of the traditional water harvesting systems in cities are neglected and have fallen into disuse, worsening the urban water scenario. One of the solutions to the urban water crisis is rainwater harvesting - capturing the runoff. GIS is an efficient and cost-effective system that helps in monitoring larger areas which are suitable for RWH and artificial recharge suitability, thus facilitating decision-making for investments in RWH.

- **Afforestation:** Planting trees and incorporation of green spaces should be considered as a fundamental component in urban planning. Improper implementation of afforestation programs is a major drawback in our country. Urbanization should be accompanied with afforestation to ensure planned proliferation of green spaces. Green belts should be raised along railway lines, canals, and streams as per the National Policy on forests to improve the green cover. Remote Sensing satellite imagery can be effectively used for close monitoring of the degree of deforestation and take immediate actions. GIS can be helpful to create more effective and efficient farming practices. It can also be used to analyze the soil type, to determine the best crop, best site / place, how to maintain nutrition levels to best benefit crop. It may be fully integrated thus helping government and other agencies to manage programs that support farmers and protect the environment thereby increasing the food production leading to sustainable development.
- **Pollution Control:** Combating all sources of pollution by adopting practices such as zero emissions, zero discharge and noise control to reduce air, water, noise pollution with stringent legislation and also by creating awareness among public about the consequences they may likely have to face in an event of non compliance. Digital mapping technologies not only offer benefits in capturing and maintaining timely and accurate information but also help in identifying areas having severe pollution levels especially with respect to air and water pollution so that necessary actions may be proposed for mitigation.
- **Transportation and Mobility Systems:** High density of Indian roadways can be utilized as green corridors for attenuating atmospheric pollution, and inducing uniform cooling. Incorporation of permeable pavements such as grassed footpaths and greening of parking lots will help to decrease the proportion of paved areas, aid in storm water retention, and reduce surface heating. Significant reduction in atmospheric pollutant levels in many areas during truck strike period highlights the need for lesser polluting modes of transport. Public transport and use of CNG-based vehicles should be promoted to reduce mobile emissions from private vehicles. Car pooling and use of public transport like metro rail needs to be encouraged. Implementation of stringent rules aimed at regulating vehicular life, such as banning of commercial vehicles older than 15 years as in Delhi (Times of India 2014), will help to overcome the lack of public support. RS and GIS can be used for improved transportation facilities by providing information regarding best possible routes, finding the shortest path, monitoring the road conditions etc. there by reducing traffic congestion, air pollution and noise pollution in urban areas.
- **Energy Systems:** Cities face a trade-off in energy generation systems therefore it is essential to increase the dependency on renewable sources such as solar, wind or hydro which are less polluting, produce fewer GHG emissions and often have lower life-cycle costs. Energy efficiency measures in buildings, businesses and industries can provide additional benefits including cost-savings and increased income. Effective management of energy systems is a complex challenge but GIS has the potential to address these issues efficiently and in a cost effective way.

- **Environmental Protection**: Urbanization at the cost of environment may lead to substantial degradation which may be irreversible. Water contamination, pollution and depletion of natural resources are side effects to rapid modernization. It is essential to protect the environment by practicing sustainable methodologies as a key element in urban planning. RS and GIS can be used to monitor the environmental conditions at regular time intervals and analyze changes such as identifying areas which are likely to be prone to natural or man-made hazards.
- **Waste Management:** Cities have enormous scope for innovative ideas related to waste management such as waste-to-energy technologies (e.g., methane from landfills), reusing and recycling as an economic opportunity and ecosystem-based sewage treatment. Solid waste management techniques such as composting, generating energy from methane combustion helps in reduction of methane emissions in landfills, increase forest carbon sequestration and contribute to overall reduction of greenhouse gases. GIS technology can be used for selection of suitable sites for waste disposal sites.
- **Participation of Stakeholders:** Involving the local community and local government in developmental and planning activities helps in restoring the environment. Public participation is a prerequisite for the success of any urban development program.
- **Job Creation:** Combat poverty by promoting economic development and job opportunities and discouraging mass migration of people from neighboring villages and towns in search of opportunities and better quality of living. Government should take initiative and help them in creation of opportunities at the local level and discourage migration towards cities.
- **Planning / Design:** Environmental Impact Assessment may also be included in the design along with the economics in every project that is likely to commence. During the design stage itself the assessment of impact helps in finding solutions to the impending problems related to environmental sustainability. Therefore it is need of the hour for reformation of urban planning and design to meet the goals of sustainable development. Human activities such as construction and operation of highways, rail roads, pipelines, airports, radioactive waste disposal have potential adverse effects on the environment. RS & GIS play a vital role in adopting efficient EIA strategies by integrating various GIS layers thus assisting in quick assessment of available natural features.

CONCLUSIONS

The aim of urban sustainability is planned urban growth so as to reduce over exploitation of natural resources and dependency on non-renewable energy. Sustainable urban development can mitigate the impacts of climate change by reducing the ecological footprint, pollution, by increasing the efficiency of land use, recycling waste, increased use of sustainable materials and by switching to renewable energy resources. In a nut shell, the march towards sustainable development in urban areas is to manage the existing natural resources without degradation and provide services through effective design and implementation of policies. Sustainable practices also help in solving issues related to air quality, water quality, waste management, effective transport, improved health and well-being thereby reducing the environmental impact on the planet earth. Thus Satellite Remote sensing and GIS offer a cost-effective solution for collecting vast amounts of data compared to resource-intensive conventional approaches such as survey and field monitoring.

REFERENCES

1. Aabshar U. K. Imam[1] and Uttam Kumar Banerjee[2],"Urbanisation and greening of Indian cities: Problems, practices, and policies", Jan2016
2. CDC. Extreme Heat Prevention Guide—Part 1. 2015. Retrieved 23 September, 2015.
3. Centre for Science &Environment, New Delhi, India. 2014. Energy and Buildings. Retrieved December 23, 2014
4. Food and Agricultural Organisation of the United Nations (FAO). 1998. Urban forestry in the Asia-Pacific region: status and prospects.
5. Harlan SL, Brazel AJ, Prashad L, Stefanov WL, Larsen L"Neighborhood microclimates and vulnerability to heat stress" , Dec 2006
6. Lata K.M [1], C. H. Sankar Rao[2], V. Krishna Prasad[3], V. Rahgavasamy[4], "Measuring urban sprawl: A case study of Hyderabad", Jan, 2001
7. Ministry of Statistics & Program Implementation, Government of India. 2013. Energy Statistics 2013. New Delhi: Central Statistics Office, NSO, Ministry of Statistics & Program Implementation, Govt. of India.
8. MOEF, India. 2010. India: Greenhouse Gas Emissions 2007. Indian Network for Climate Change Assessment.
9. Monica Machuma Katyambo[1], Moses Murimi Ngigi[2] , "Spatial Monitoring of Urban Growth Using GIS and Remote Sensing: A Case Study of Nairobi Metropolitan Area, Kenya", Jun 2017
10. Nada Kadhim[1], Monjur Mourshed[2], Michaela Bray[3]., Advances in remote sensing applications for urban sustainability", December 2016
11. Parikh J, Panda M, Ganesh-Kumar A, Singh V, "CO_2 emissions structure of Indian economy. Energy", 2009
12. Sharma AR, Kharol SK, Badarinath KVS , "Influence of vehicular traffic on urban air quality- A case study of Hyderabad. India", 2010
13. Srivathsan, A., "Where is our patch of green", 2013
14. UN DP's support to sustainable, inclusive and resilient cities in the developing world.
15. United Nations Office for Disaster Risk Reduction - Regional Office for Asia and Pacific (UNISDR AP)
16. United States Environment Protection Agency (EPA). 2015. Global greenhouse gas emissions data.
17. Reserve Bank of India, Government of India. 2013. Number & percentage of people below poverty line.

SURFACE WATER QUALITY AND POLLUTANT CONTROL

D. Pankaja, E.M. Sunitha, A. Rajani and A. Kavitha

Department of Botany R.B.V.R.R. Women's college Hyderabad

INTRODUCTION

Control of water pollution has reached primary importance in developed and a number of developing countries. The prevention of pollution at source, the precautionary principle and the prior licensing of wastewater discharges by competent authorities have become key elements of successful policies for preventing, controlling and reducing inputs of hazardous substances, nutrients and other water pollutants from point sources into aquatic ecosystems. Some water pollutants which become extremely toxic in high concentrations are, however, needed in trace amounts. Copper, zinc, manganese, boron and phosphorus, for example, can be toxic or may otherwise adversely affect aquatic life when present above certain concentrations, although their presence in low amounts is essential to support and maintain functions in aquatic ecosystems. The same is true for certain elements with respect to drinking water. Selenium, for example, is essential for humans but becomes harmful or even toxic when its concentration exceeds a certain level. The concentrations above which water pollutants adversely affect a particular water use may differ widely. Water quality requirements, expressed as water quality criteria and objectives, are use-specific or are targeted to the protection of the most sensitive water use among a number of existing or planned uses within a catchment. Water quality criteria often serve as a baseline for establishing water quality objectives in conjunction with information on water uses and site-specific factors. Water quality criteria aim at supporting and protecting designated uses of freshwater, i.e. 1] use for drinking-water supply 2] livestock watering ,3]irrigation 4]fisheries 5] recreation or other purposes 6] supporting and maintaining aquatic life , 7]functioning of aquatic ecosystems. The establishment of water quality objectives is not a scientific task but rather a political process that requires a critical assessment of national priorities. Such an assessment is based on economic considerations, present and future water uses, forecasts for industrial progress and for the development of agriculture, and many other socio-economic factors. Water quality criteria for phosphorus compounds, such as phosphates, are set at a concentration that prevents excessive growth of algae. Criteria for total ammonia (NH3) have been established, for example by the EPA, to reflect the varying toxicity of NH3 with pH (EPA, 1985). Criteria have been set for a pH range from 6.5 to 9.0 and a water temperature range from 0 to 30 °C (Table 2.2), Ammonium (NH4+) is less toxic than NH3.

1. **Drinking water quality**

   Drinking-water criteria define a quality of water that can be safely consumed by humans throughout their lifetime. Such criteria have been developed by international organisations and include the WHO Guidelines for Drinkingwater Quality (WHO, 1984, 1993) and the EU Council Directive of 15 July 1980 Relating to the Quality of Water Intended for Human Consumption (80/778/EEC), which covers some 60 quality variables. These guidelines and directives are used by countries, as appropriate, in establishing enforceable national drinking-water quality standards.

Water quality criteria for raw water used for drinking-water treatment and supply usually depend on the potential of different methods of raw water treatment to reduce the concentration of water contaminants to the level set by drinking-water criteria. Drinkingwater treatment can range from simple physical treatment and disinfection, to chemical treatment and disinfection, to intensive physical and chemical treatment. Many countries strive to ensure that the quality of raw water is such that it would only be necessary to use near-natural conditioning processes (such as bank filtration or low-speed sand filtration) and disinfection in order to meet drinking-water standards.

2. **Livestock watering**

   Livestock may be affected by poor quality water causing death, sickness or impaired growth. Variables of concern include nitrates, sulphates, total dissolved solids (salinity), a number of metals and organic micropollutants such as pesticides. In addition, bluegreen algae and pathogens in water can present problems. Some substances, or their degradation products, present in water used for livestock may occasionally be transmitted to humans. The purpose of quality criteria for water used for livestock watering is, therefore, to protect both the livestock and the consumer.

   Criteria for livestock watering usually take into account the type of livestock, the daily water requirements of each species, the chemicals added to the feed of the livestock to enhance the growth and to reduce the risk of disease, as well as information on the toxicity of specific substances to the different species

3. **Irrigation**

   Poor quality water may affect irrigated crops by causing accumulation of salts in the root zone, by causing loss of permeability of the soil due to excess sodium or calcium leaching, or by containing pathogens or contaminants which are directly toxic to plants or to those consuming them. Contaminants in irrigation water may accumulate in the soil and, after a period of years, render the soil unfit for agriculture. Even when the presence of pesticides or pathogenic organisms in irrigation water does not directly affect plant growth, it may potentially affect the acceptability of the agricultural product for sale or consumption.

4. **Recreational use**

   Recreational water quality criteria are used to assess the safety of water to be used for swimming and other water-sport activities. The primary concern is to protect human health by preventing water pollution from faecal material or from contamination by microorganisms that could cause gastro-intestinal illness, ear, eye or skin infections. Criteria are therefore usually set for indicators of faecal pollution, forms and pathogens. There has been a considerable amount of research in recent years into the development of other indicators of microbiological pollution including viruses that could affect swimmers. As a rule, recreational water quality criteria are established by government health agencies.

5. **Amenity use**

   Criteria have been established in some countries aimed at the protection of the aesthetic properties of water. These criteria are primarily orientated towards visual aspects. They are usually narrative in nature and may specify, for example, that waters must be free of floating oil or other immiscible liquids, floating debris, excessive turbidity, and objectionable odours. The criteria are mostly non-quantifiable because of the different sensory perception of individuals and because of the variability of local conditions.

6. **Protection of aquatic life**

Within aquatic ecosystems a complex interaction of physical and biochemical cycles exists. Anthropogenic stresses, particularly the introduction of chemicals into water, may adversely affect many species of aquatic flora and fauna that are dependent on both abiotic and biotic conditions. Water quality criteria for the protection of aquatic life may take into account only physico-chemical parameters which tend to define a water qualitythatprotects and maintains aquatic life, ideally in all its forms and life stages, or they may consider the whole aquatic ecosystem.

Water quality parameters of concern are traditionally dissolved oxygen (because it may cause fish kills at low concentrations) as well as phosphates, ammonium and nitrate (because they may cause significant changes in community structure if released into aquatic ecosystems in excessive amounts). Heavy metals and many synthetic chemicals can also be ingested and absorbed by organisms and, if they are not metabolised or excreted, they may bioaccumulate in the tissues of the organisms. Some pollutants can also cause carcinogenic, reproductive and developmental effects.

7. **Functioning of Aquatic Ecosystem**

More recently within the concept of the ecosystem approach to water management, attempts have been made to address criteria that indicate healthy aquatic ecosystem conditions. In addition to traditional criteria, new criteria try to describe the state of resident species and the structure and/or function of ecosystems as a whole. In developing these criteria, the assumption has been made that they should be biological in nature. In some countries, research is under way on the development of biocriteria that express water quality criteria quantitatively in terms of the resident aquatic community structure and function.

Biocriteria are defined as measures of "biological integrity" that can be used to assess cumulative ecological impact from multiple sources and stress agents. In the UK, quality criteria for the protection of aquatic ecosystems are now being based on an ecological quality index. In other countries, considerable efforts have been made to identify key species which may serve as useful integrative indicators of the functional integrity of aquatic ecosystems.

Ongoing research suggests that such criteria and indicators should include both sensitive, short-lived species and information about changes in community structure resulting from the elimination of key predators.

Biomarkers are becoming an increasingly useful approach for identifying the impact of deteriorating water quality at an early stage. A biomarker is a variation in cell structure or in a biochemical process or function that is induced by a pollutant and that can be measured, for example, by changes in the activity of enzymes. Ideally, a biomarker should respond to a pollutant with a dose-response quantitative change which is sensitive to concentrations found in the environment and which is specific to a particular class or classes of pollutants. Thus for toxic metals, delta-aminolevulinic acid dehydratase

(ALAD) inhibition provides a signal of a potential problem and is a definite indicator of metal pollution. It is also a predictive indicator of long-term adverse effects.

Commercial and sports fishing

Water quality criteria for commercial and sports fishing take into account, in particular, the bioaccumulation of contaminants through successive levels of the food chain and their possible biomagnifications in higher trophic levels, which can make fish unsuitable for human consumption. They are established at such a concentration that bioaccumulation and biomagnifications of any given substance cannot lead to concentrations exceeding fish consumption criteria, i.e. criteria indicating the maximum content of a substance in fish for human consumption that will not be harmful. The FAO European Inland Fisheries Advisory Commission (EIFAC), for example, has been investigating these issues and has published relevant guidance (Alabaster and Lloyd, 1982).

Suspended particulate matter and sediment

The attempts in some countries to develop quality criteria for suspended particulate matter and sediment aim at achieving a water quality, such that any sediment dredged from the water body could be used for soil improvement and for application to farmland. Another goal of these quality criteria is to protect organisms living on, or in, sediment, and the related food chain. Persistent pollutants in sediments have been shown to be accumulated and biomagnified through aquatic food chains leading to unacceptable concentrations in fish and fish-eating birds. Water quality objectives are being developed in many countries by water authorities in co-operation with other relevant institutions in order to set threshold values for water quality that should be maintained or achieved within a certain time period. Water quality objectives provide the basis for pollution control regulations and for carrying out specific measures for the prevention, control or reduction of water pollution and other adverse impacts on aquatic ecosystems.

Water Quality Objectives

In some countries, water quality objectives play the role of a regulatory instrument or even become legally binding. Their application may require, for example, the appropriate strengthening of emission standards and other measures for tightening control over point and diffuse pollution sources.

In some cases, water quality objectives serve as planning instruments and/or as the basis for the establishment of priorities in reducing pollution levels by substances and/or by sources.

A major advantage of the water quality objectives approach to water resources management is that it focuses on solving problems caused by conflicts between the various demands placed on water resources, particularly in relation to their ability to assimilate pollution. The water quality objectives approach is sensitive not just to the effects of an individual discharge, but to the combined effects of the whole range of different discharges into a water body. It enables an overall limit on levels of contaminants within a water body to be set according to the required uses of the water. It is generally recognised that water quality objectives, the setting of emission limits on the basis of best available technology, and the use of best environmental practice should all form part of an integrated approach to the prevention, control and reduction of pollution in inland surface waters. In most cases, water quality objectives serve as a means of assessing pollution reduction measures. For example, if emission limits are set

for a given water body on the basis of best available technology, toxic effects may, nevertheless, be experienced by aquatic communities under certain conditions. In addition, other sensitive water uses, such as drinking-water supplies, may be adversely affected. The water quality objectives help to evaluate, therefore, whether additional efforts are needed when water resources protection is based on using emission limits for point sources according to the best available technology or on best environmental practice for non-point sources. Current approaches to the elaboration and setting of water quality objectives differ between countries. These approaches may be broadly grouped as follows:

- Establishment of water quality objectives for individual water bodies (including transboundary waters) or general water quality objectives applicable to all waters within a country.
- Establishment of water quality objectives on the basis of water quality classification schemes.

The first approach takes into account the site-specific characteristics of a given water body and its application requires the identification of all current and reasonable potential water uses. Designated uses of waters or "assets" to be protected may include: direct extraction for drinking-water supply, extraction into an impoundment prior to drinkingwater supply, irrigation of crops, watering of livestock, bathing and water sports, amenities, fish and other aquatic organisms.

In adopting water quality objectives for a given water body, site-specific physical, chemical, hydrological and biological conditions are taken into consideration. Such conditions may be related to the overall chemical composition (hardness, pH, dissolved oxygen), physical characteristics (turbidity, temperature, mixing regime), type of aquatic species and biological community structure, and natural concentrations of certain substances (e.g. metals or nutrients). These site-specific factors may affect the exposure of aquatic organisms to some substances or the usability of water for human consumption, livestock watering, irrigation and recreation.

In some countries general water quality objectives are set for all surface waters in a country, irrespective of site-specific conditions. They may represent a compromise after balancing water quality requirements posed by individual water uses and economic, technological and other means available to meet these requirements at a national level.

Another approach is to select water quality criteria established for the most sensitive uses (e.g. drinking-water supply or aquatic life) as general water quality objectives.

Principles for Water Pollution Control

During recent years there has been increasing awareness of, and concern about, water pollution all over the world, and new approaches towards achieving sustainable exploitation of water resources have been developed internationally. It is widely agreed that a properly developed policy framework is a key element in the sound management of water resources

Prevent pollution rather than treating symptoms of pollution

The most logical approach is to prevent the production of wastes that require treatment. Thus, approaches to water pollution control that focus on wastewater minimisation, in-plant refinement of raw materials and production processes, recycling of waste products, etc., should be given priority over traditional end-of-pipe treatments.

Apply the polluter-pays-principle.

The polluter-pays-principle, where the costs of pollution prevention, control and reduction measures are borne by the polluter, is not a new concept but has not yet been fully implemented, despite the fact that it is widely recognized that the perception of water as a free commodity can no longer be maintained. The principle is an economic instrument that is aimed at affecting behavior, i.e. by encouraging and inducing behavior that puts less strain on the environment. Examples of attempts to apply this principle include financial charges for industrial waste-water discharges and special taxes on pesticides.

Encourage participatory approach with involvement of all relevant stakeholders.

The participatory approach involves raising awareness of the importance of water pollution control among policy-makers and the general public. Decisions should be taken with full public consultation and with the involvement of groups affected by the planning and implementation of water pollution control activities. This means, for example, that the public should be kept continuously informed, be given opportunities to express their views, knowledge and priorities, and it should be apparent that their views have been taken into account.

Non-point source pollution

Identification of sources; It is more difficult to control non-point source pollution than defined discharges. Even though stringent controls may be placed on industrial and municipal sewage discharges, environmental water quality may not improve to the extent expected. This may be due to diffuse pollution caused by agriculture or by urban run-off. The first problem lies in the identification of sources. The catchment inventory approach is recommended and is already used in a number of countries.

The input of nitrogen and phosphorus from diffuse sources, mainly agriculture, is as significant as that from sewage works. Those areas which use sewage on land, either as a disposal route or for soil conditioning, may also be contributing to diffuse pollution.

Agricultural sources

The major causes of concern associated with agricultural pollution are: organic matter (which often leads to nutrient enrichment of water bodies) including the disposal of solid organic wastes and slurries from livestock, effluents from silage clamps and, in some situations, domestic effluents from farmstead septic tanks; pesticides and fertilizers; and soil erosion.

Nutrient flow Control

The most common source of the nutrients nitrogen and phosphorus in agriculture, and this is closely followed in the industrialized world by sewage effluents. The reduction of nitrogen and phosphorus from agriculture relies upon changes to farming practices because they give rise to diffuse sources.

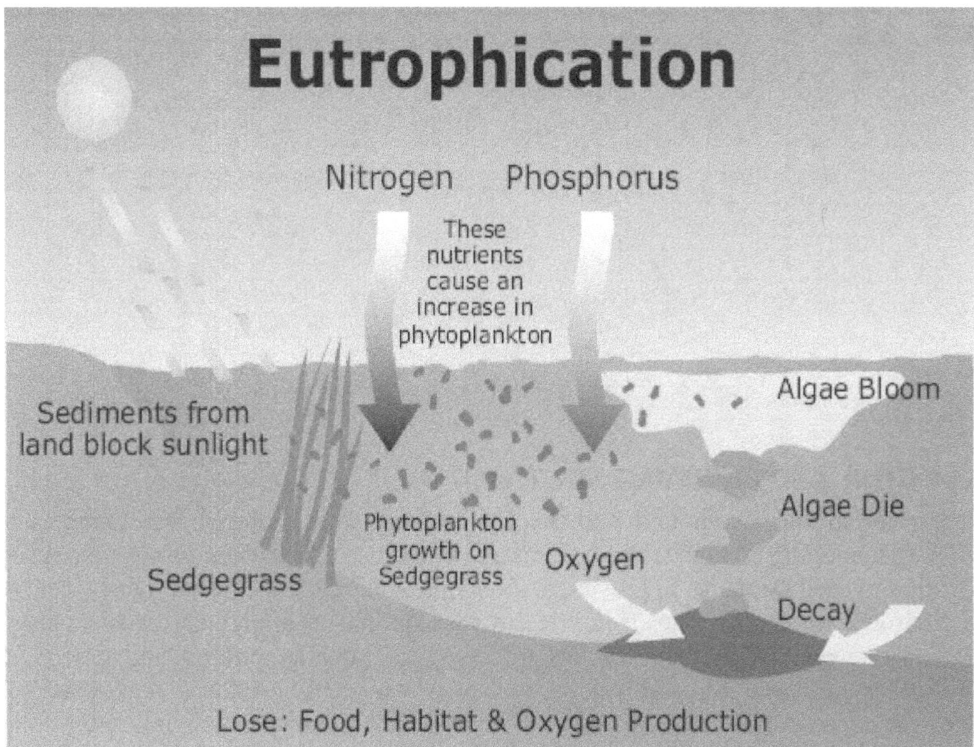

Nitrogen

Ploughing of grassland and other crops, particularly during autumn, leads to the release of large quantities of soil nitrogen and, therefore, a general move towards permanent pasture regimes assists in lowering nitrate leaching. When this is not possible, the use of short-term rotational crops to take up nitrogen, followed by their harvesting and subsequent removal from the catchment, is helpful. Animal wastes should be used carefully, avoiding over-use and direct run-off into water courses; but wherever possible they should be used in place of synthetic fertilisers. Use of all types of fertilisers should be carefully controlled and matched to crop requirements.

Phosphorus

A key issue controlling phosphorus input from agriculture is the need to prevent erosion from field surfaces. Phosphate tends to bind to soil particles which, when washed from fields into watercourses, become a source of phosphate in suspended form and in deposited sediments. Sediments act as a long-term source of phosphate by releasing it (i.e. by redissolution) under certain environmental conditions. Physical removal of the sediment layer, in order to remove the bound phosphate from the catchment, has been tried in a number of locations around the world.

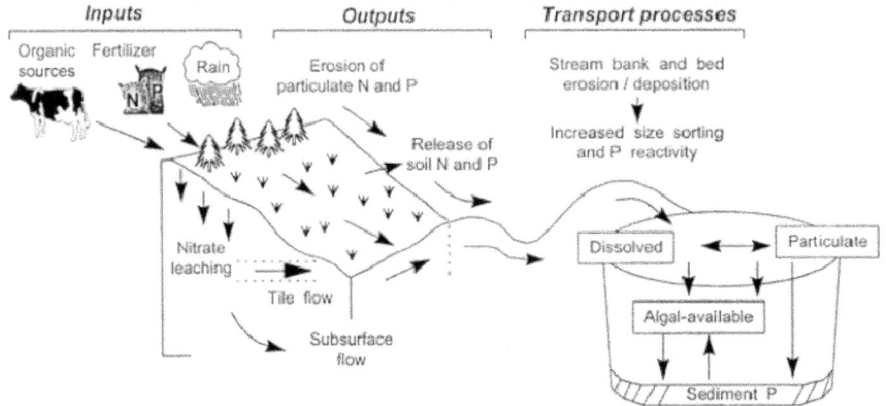

Input of Nitrogen &Phosphorus into Surface Waters.

CONCLUSIONS AND RECOMMENDATIONS

Many chemical substances emitted into the environment from anthropogenic sources pose a threat to the functioning of aquatic ecosystems and to the use of water for various purposes. The need for strengthened measures to prevent and to control the release of these substances into the aquatic environment has led many countries to develop and to implement water management policies and strategies based on, amongst others, water quality criteria and objectives. To provide further guidance for the elaboration of water quality criteria and water quality objectives for inland surface waters, and to strengthen international co-operation the following recommendations have been put forward (UNECE, 1993):

- The precautionary principle should be applied when selecting water quality parameters and establishing water quality criteria to protect and maintain individual uses of waters.
- In setting water quality criteria, particular attention should be paid to safeguarding sources of drinking-water supply. In addition, the aim should be to protect the integrity of aquatic ecosystems and to incorporate specific requirements for sensitive and specially protected waters and their associated environment, such as wetland areas and the surrounding areas of surface waters which serve as sources of food and as habitats for various species of flora and fauna.
- Water-management authorities in consultation with industries, municipalities, farmers' associations, the general public and others should agree on the water uses in a catchment area that are to be protected. Use categories, such as drinking-water supply, irrigation, livestock watering, fisheries, leisure activities, amenities, maintenance of aquatic life and the protection of the integrity of aquatic ecosystems, should be considered wherever applicable.
- Water-management authorities should be required to take appropriate advice from health authorities in order to ensure that water quality objectives are appropriate for protecting human health.

- In setting water quality objectives for a given water body, both the water quality requirements for uses of the relevant water body, as well as downstream uses, should be taken into account. In transboundary waters, water quality objectives should take into account water quality requirements in the relevant catchment area. As far as possible, water quality requirements for water uses in the whole catchment area should be considered.
- Water quality objectives for multipurpose uses of water should be set at a level that provides for the protection of the most sensitive use of a water body. Among all identified water uses, the most stringent water quality criterion for a given water quality variables should be adopted as a water quality objective.
- Established water quality objectives should be considered as the ultimate goal or target value indicating a negligible risk of adverse effects on use of the water and on the ecological functions of waters.
- The setting of water quality objectives should be accompanied by the development of a time schedule for compliance with the objectives that takes into account action which is technically and financially feasible and legally implementable. Where necessary, a stepby-step approach should be taken to attain water quality objectives, making allowance for the available technical and financial means for pollution prevention, control and reduction, as well as the urgency of control measures.
- The setting of emission limits on the basis of best available technology, the use of best environmental practices and the use of water quality objectives as integrated instruments of prevention, control and reduction of water pollution, should be applied in an action-oriented way. Action plans covering point and diffuse pollution sources should be designed, that permit a step-by-step approach to water pollution control which are both technically and financially feasible.
- Both the water quality objectives and the timetable for compliance should be subject to revision at appropriate time intervals in order to adjust them to new scientific knowledge on water quality criteria, to changes in water use in the catchment area, and to achievements in pollution control from point and non-point sources.
- The public should be kept informed about water quality objectives that have been established and about measures taken to attain these objectives.

REFERENCES

1. Alabaster, J.S. and Lloyd, R. 1982 Water Quality Criteria for Freshwater Fish. 2nd edition. Published on behalf of Food and Agriculture Organization of the United Nations by Butterworth, London, 361 pp.
2. ten Brink, B.J.E., Hosper, S.H. and Colijn, F. 1990 A Quantitative Method for Description and Assessment of Ecosystems: the AMOEBA Approach. ECE Seminar on Ecosystems Approach to Water Management, Oslo, May 1991. ENVWA/SEM.5/R.33, United Nations Economic Commission for Europe, United Nations, Geneva.
3. CCREM 1987 Canadian Water Quality Guidelines. Prepared by the Task Force on Water Quality Guidelines of the Canadian Council of Resource and Environment Ministers, Ottawa.
4. Chiaudani, G. and Premazzi, G. 1988 Water Quality Criteria in Environmental Management. Report EUR 11638 EN, Commission of the European Communities, Luxembourg.

5. Dick, R.I. 1975 Water Quality Criteria, Goals and Standards. Second WHO Regional Seminar on Environmental Pollution: Water Pollution, Manila, WPR/W.POLL/3, WHO Regional Office for the Western Pacific, Manila.
6. ECLAC 1989 The Water Resources of Latin America and the Caribbean: Water Pollution. LC/L.499, United Nations Economic Commission for Latin America and the Caribbean, United Nations, Santiago de Chile.
7. Enderlein, R.E. 1995 Protecting Europe's water resources: policy issues. Wat. Sci. Tech., 31(8), 1-8.
8. Enderlein, R.E. 1996 Protection and sustainable use of waters: agricultural policy requirements in Europe. HRVAT. VODE, 4(15), 69-76.
9. EPA 1976 Quality Criteria for Water. EPA-440/9-76-023, United States Environmental Protection Agency, Washington, D.C.
10. EPA 1985 Ambient Water Quality Criteria for Ammonia. EPA-440/5-85-001, United States Environmental Protection Agency, Washington, D.C.
11. EPA 1986 Ambient Water Quality Criteria for Dissolved Oxygen. EPA 440/5-86-003, United States Environmental Protection Agency, Washington, D.C.
12. ESCAP 1990 Water Quality Monitoring in the Asian and Pacific Region. Water Resources Series No. 67, United Nations Economic and Social Commission for Asia and the Pacific, United Nations, New York.
13. FAO 1985 Water Quality for Agriculture. Irrigation and Drainage Paper No. 29, Rev. 1. Food and Agriculture Organization of the United Nations, Rome.
14. FEPA 1991 Proposed National Water Quality Standards. Federal Environmental Protection Agency, Nigeria.
15. Hespanhol, I. 1994 WHO Guidelines and National Standards for Reuse and Water Quality. Wat. Res., 28(1), 119-124.
16. ICPR 1991 Konzept zur Ausfüllung des Punktes A.2 des APR über Zielvorgaben. Lenzburg, den 2. Juli 1991 (Methodology to implement item A.2 of the Rhine Action Programme related to water quality objectives, prepared at Lenzbourg on 2 July 1991). PLEN 3/91, International Commission for the Protection of the Rhine against Pollution, Koblenz, Germany.
17. ICPR 1994 Unpublished contribution of the secretariat of the International Commission for the Protection of the Rhine against Pollution, Koblenz (Germany), to the ECE project on policies and strategies to protect transboundary waters. United Nations Economic Commission for Europe, Geneva.

ASSESSMENT OF SPECTRAL SIGNATURES FOR DIFFERENT VARIETIES OF COLACASIA SPECIES USING CONTINUUM REMOVAL METHOD

P. Sowmya[1], M.V.S.S. Giridhar[2] and M. Prabhakar[3]

[1]Research scholar, [2]Associate Professor, [3]Principal Scientist

[1,2]Centre for water Resources, IST, JNTUH, Kukatpally, Hyderabad

[3]Entamology, CRIDA, Santhosh Nagar, Koti, Hyderabad.

ABSTRACT

Every natural and artificial object reflects and emits electromagnetic adiation over a range of wavelengths in its own characteristic manner, according to its chemical composition and physical state. A basic assumption made in remote sensing is that specific targets have an individual and characteristic manner of interacting with incident radiation that is described by the spectral response of that target. Vegetation has a unique spectral signature which enables it to be distinguished readily from other types of land cover in an optical/near-infrared image. In this present research work developed spectral libraries using SVC Spectroradiometer for the different stages of growth and different varieties of colacasia (Taro) which is a tuber crop in the Sri Konda Laxman Telangana State Horticultural University Rajendranagar, Hyderabad by measuring and analyzing their reflectance curves using SVC and ENVI software. By developing the spectral libraries for the different varieties of colacasia at different stages of growth, and the continuum removal is performed for the developed spectral signatures. The analysis is done on the continuum removed spectral libraries to distinguish the different species and growth stages using the spectral libraries.

Keywords: Spectral signatures, Spectral Library, Colacasia, Spectroradiometer.

INTRODUCTION

The amount of solar radiation that is reflected will vary with wavelength for any given material. This important property of matter allows us to separate distinct cover types based on their response values for a given wavelength. When we plot the response characteristics of a certain cover type against wavelength, we define what is termed the *spectral signature* of that cover. By comparing the response patterns of different features we may be able to distinguish between them, where we might not be able to, if we only compared them at one wavelength. Spectral response can be quite variable, even for the same target type, and can also vary with time and location. Knowing where to "look" spectrally and understanding the factors which influence the spectral response of the features of interest are critical to correctly interpreting the interaction of electromagnetic radiation with the surface.

Continuum removal and feature finding

The normalization procedure for quantification of absorption features in spectra, the overall concave shape of a spectrum should be removed referred to 'continuum removal' or 'convex-hull' transform and allows comparison of spectra that are acquired by different instruments or under different light conditions. The technique of making a continuum, or hull, is similar to fitting a rubber band over the spectrum, the spectrum is normalized by setting the value of the hull to 100% reflection.

After continuum removal, a part of the spectrum with no absorption will have a value of 1, whereas complete absorption would be 0, with most absorptions falling somewhere in between. The spectral continuum can be thought of as what the original spectrum *would* look like if there were no absorption band. The continuum-removed spectrum is the original spectrum divided by the continuum.

REVIEW OF LITERATURE

Continuum-removal analysis enables the isolation of absorption features of interest, thus increasing the coefficients of determination and facilitating the identification of more sensible absorption features. The purpose of this study was to test Kokaly and Clark's methodology with aircraft-acquired hyperspectral data of eucalypt tree canopies, which are more complex than are spectra from many coniferous canopies and much more complex than the spectra from dried ground leaves. The results of the continuum-removal analysis were most encouraging. It identified, in one experiment or another, almost all of the known nitrogen absorption features.

Although continuum-removal analysis appears superior to standard derivative analysis in estimating chemical concentrations in dried leaves, Kokaly and Clark (1999) point out that interference from leaf water shall present the greatest challenge to extending the method to the analysis of fresh whole leaves and canopies. Recently, Mutanga et al. (2003) showed that continuum-removal analysis could be used to better discriminate differences in foliar nitrogen concentrations in grass grown in the greenhouse with different fertilization treatments. Clark and Roush (1984) suggested using continuum-removal analysis to remove those absorption features of no interest and thus to isolate individual absorption features of interest. Removing the continuum standardizes isolated absorption features for comparison (Clark, 1999). In order to minimize the effect of spectral variability that is independent of the biochemical concentration, Kokaly and Clark (1999) applied a refined approach that enhances and standardizes known chemical absorption features. In their study, continuum removal was applied to broad absorption features of dry leaf spectra in the shortwave infrared region (1730, 2100 and 2300 nm) and absorption band depths relative to the continuum were calculated. Like NIRS, this approach uses stepwise regression. The risk of over fitting is minimized by concentrating on known absorption troughs that are enhanced by continuum removal (Clark & Roush, 1984). The method showed strong correlation (r2=0.95).

The amount of the chlorophyll present in the plant during the different growth stages also effect the reflectance of the plant. Due to the low chlorophyll content and the low water content in the leaf the low reflectance can be observed while studying the ground truth data (Sowmya and Giridhar, 2017)

STUDY AREA

For the purpose of study, area chosen which is belongs to Ranga Reddy district of Sri Konda Laxman Telangana State Horticultural University Rajendranagar, Hyderabad, Telangana State, India. This is enclosed with geographical regions of 17°19'17"N, 78°25'23"E. Study area boundary delineated from BHUVAN website for the study area using Arc Map 10.2 version. The study area map has been prepared and shown in Fig.1.

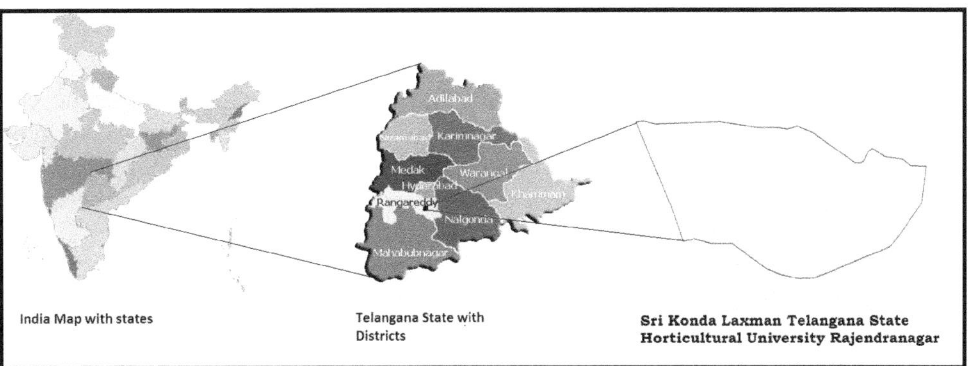

Fig. 1 Sri Konda Laxman Horticultural University boundary Study Area Map

Sri Konda Laxman Horticultural University the crops are identified in which the research is to be continued. Sri Konda Laxman Telangana State Horticultural University (SKLTSHU), a dedicated & splendid institute of horticultural learning. It is the first Sri Konda Laxman Horticultural University Horticultural University in the state of Telangana and fourth in the Country. The mission goal & primary mandate of the University is to excel as Centre of Excellence in Education, Research and Extension Education in the field of Horticulture and allied sectors. The spectral signatures of the different varieties of the colacasia and the growth stages are studied in this university.

METHODOLOGY

The spectral signatures of Colocasia Esculenta is collected using the Spectroradiometer and processed in ENVI software for the removal of continuum for the analysis of the spectral library is explain as below.

Collection of Ground truth data: The ground truth data is collected for the species of Colacasia Esculenta, using the portable SVC HR-512i Spectroradiometer which consists of the 512 bands and produces a hyperspectral signature of the object. Using this instrument the spectral signatures of the Colacasia Esculenta grown in Sri Konda Laxman horticultural university Rajendranagar Hyderabad, is collected at different stages of its growth and the libraries are developed at different stages.

Colacasia Esculenta is a perennial herbaceous tropical and subtropical regions and requires a warm humid climate, plant usually grown for its starchy but sweet flavored tuber. Under rain fed conditions, it requires a fairly well distributed rain fall around 120-150cm during the growth period. Well drained soil is suitable for uniform development of tubers. It is also called as Taro. It is also known as "dasheen", "eddo", and "Kalo". Taro plant grows from 1 meter to 2 meter tall. It belongs to family of "Araceae". The Taro root crop attains maturity about in 9 to 12 months, when leaves turn yellow colour or die down and also there is a slight lifting in tubers can be observed. As Taro root does not store for longer than a month, leaving tubers in the soil recommended until they are needed. The ground truth data collected on 29th August, 6th September, 19th September , 4th October, 14th October, 22nd October 2016 is used to analyze the spectral properties of the plant at different stages of its growth which shows the different stages of growth of the plant. The initial stage to the harvesting stage the spectral signature collected is analyzed.

Development of continuum removed spectral signatures: After re-sampling the collected spectral signatures of the plant at different stages of growth, the spectral libraries are generated, and the continuum removed spectral signatures are developed in the ENVI software. For quantification of absorption features in spectra the overall concave shape of a spectrum should be removed. This normalization procedure is referred to continuum removal or convex-hull transform and allows comparison of spectra that are acquired by different instruments or under different light conditions. The continuum is a convex hull fit over the top of a spectrum using straight-line segments that connect local spectra maxima. The first and last spectral data values are on the hull, therefore, the first and last bands in the output continuum removed data file are equal to 1.0. The formula representing the continuum equation is as shown in equation 1.

$$Scr = S / C \qquad .....(1)$$

Scr = Continuum removed spectra

S = Original spectrum

C = Continuum curve

The resulting image spectrum is equal to 1.0 where the continuum and the spectra match, and less than 1.0 where absorption features occur. The resulting spectrum is analyzed for changes in the chlorophyll at different stages of the growth. The growth stages and the varieties of the crops of the same species are identified.

RESULTS AND DISCUSSIONS

The ground truth data collected from the field is analyzed using the software of the SVC Instrument is used and the spectral signatures are developed and the collected spectral signatures are analyzed by preparing a spectral library. The spectral library developed for the different stages of the growth of the crop and the analysis is done at different wave lengths; the analysis in this paper is compared to the second stage of growth i.e., flowering stage and the budding stage. The different stage of the growth is shown in figure. 2.

Fig. 2 Growth of colacasia species at different stages of growth.

The spectral signatures collected and developed for 6th September, 4th October considered and analyzed by continuum removal and by doing the stacking of the spectral signatures. The stacking helps in the analysis and comparison of the different signatures at different wavelengths. The ups and downs of the graph i.e. spectral signature can be compared easily. At 0.75 μm and 0.95 μm the absorption can be observed in both the stages i.e. the spectral signatures collected in the flowering stage and the budding stage.

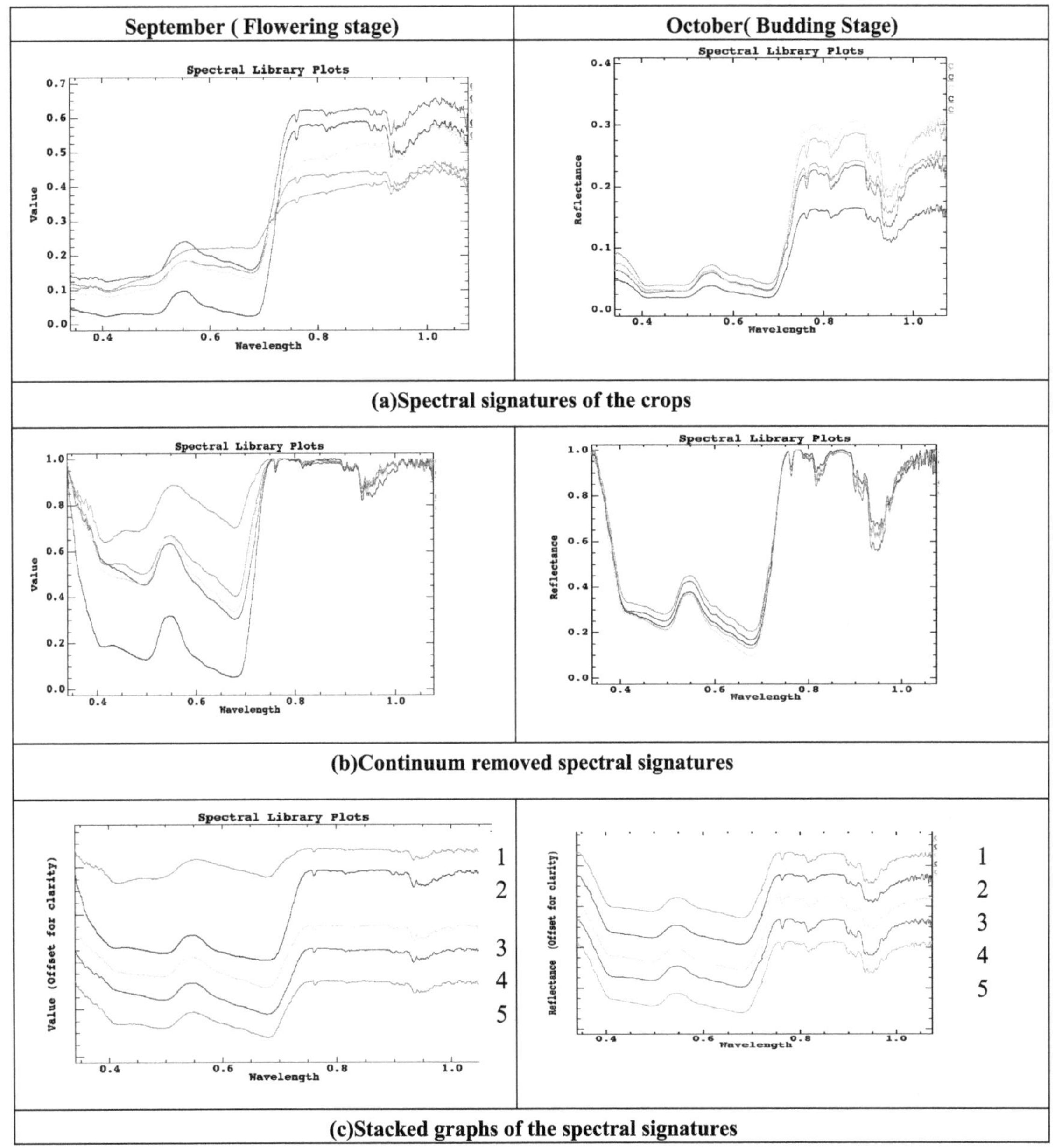

1. C. esculenta var. antiquorum, 2. C. esculenta var. esculenta, 3. C. esculenta var . kochuo, 4. C. esculenta var .nwine, 5. C. esculenta var .ogeriobosi.

Fig. 3 Spectral signatures, Continuum removed signatures and stacked graphs of spectral signatures on September 6th and 4th October

When compared the spectral library of the flowering stage, the amount of reflection is more and that of the budding stage is less. At 0.8 µm wavelength the flowering stage has the less absorption and the budding stage has the more absorption. The analysis and comparison of the signatures is done by three graphs i.e the spectral signatures library, continuum removal graph, and the stack graph. These graphs are presented in the figure 3 (a),(b), (c). In the continuum removed graph in figure 3 (b) it is observed that the red edge for the flowering stage is uncommon and meet at 0.75 µm while in budding stage the red edge is same and common for the signatures at 0.75 µm. but in common the average red edge for both the stages is observed at 0.75 µm.

In the stacked graph figure 3 (c) it is observed that the absorption is more at 0.8 µm in budding stage when compared to the flowering stage. More absorption bands are seen between at 0.8 µm – 0.9 µm in budding stage than in the flowering stage. This is because of the increased water absorption levels and the chlorophyll content in the budding stage.

REFERENCES

1. ZhiHuangaBrian J.TurneraStephen J.Duryal1Ian R.WallisbWilliam J.Foleyb , Estimating foliage nitrogen concentration from HYMAP data using continuum removal analysis Remote Sensing of Environment Volume 93, Issues 1–2, 30 October 2004, Pages 18-29 https://doi.org/10.1016/j.rse.2004.06.008.
2. R.F. Kokaly, R.N. ClarkSpectroscopic determination of leaf biochemistry using band-depth analysis of absorption features and stepwise multiple linear regression Remote Sensing of Environment, 67 (1999), pp. 267-287.
3. O. Mutanga, A.K. Skidmore, S. van Wieren Discriminating tropical grass (Cenchrus ciliaris) canopies grown under different nitrogen treatments using spectroradiometry ISPRS Journal of Photogrammetry and Remote Sensing, 57 (2003), pp. 263-272.
4. R.N. Clark, T.L. RoushReflectance spectroscopy: Quantitative analysis techniques for remote sensing applications Journal of Geophysical Research, 89 (B7) (1984), pp. 6329-6340.
5. Sowmya. P and M.V.S.S. Giridhar "Analysis of Continuum Removed Hyper Spectral Reflectance Data of Capsicum Annum of Ground Truth Data" Advances in Computational Sciences and Technology ISSN 0973-6107 Volume 10, Number 8 (2017) pp. 2233-2241 © Research India Publications

A SYSTEMATIC REVIEW ON HYPERSPECTRAL DATA PROCESSING AND CLASSIFICATION METHODS

Veeramallu Satya Sahithi[1,2], MVSS Giridhar[2] and I V Murali Krishna[3]

[1]National Remote Sensing Centre (NRSC-ISRO)
[2]Jawaharlal Nehru Technological University, Hyderabad
[3]Research Centre Imarat (RCI-DRDO), Hyderabad
sahithi.geo@gmail.com, mvssgridhar@gmailcom

ABSTRACT

Classification is a process of assigning a label to all the pixels in a digital image inorder to produce an accurate thematic map. Factors like - spectral pattern present within the data for each pixel, spatial associations, purity and size of the training samples are used as a numerical basis for the classification. Hyperspectral data contain huge number of spectral bands with spatial resolution varying between few meters to hundreds of meters (fine to coarse). The spatial resolution of the hyperspectral data plays a very important role in imparting mixed pixels in the data and thus bringing challenges like data redundancy and ambiguity in processing and classifying the datasets.There are various advanced classification methods that have come up that help in handling these challenges and accurately classifying the voluminous hyperspectral data. Kernel based Support Vector Machines, Angle Mappers, Random Forests, Artificial neural networks, sub pixel unmixing classifiers; Markov Random Fields etc. are discussed in detail along with few other techniques. The current paper gives a systematics review of initial hyperspectral data processing and various classification methods that can be used for classifying the hyperspectral datasets. Also, certain results that were obtained from classifying hyperspectral datasets using various classifiers and their comparisons are also addressed.

Keywords: Hyperspectral data, classifications, Support Vector Machines, Artificial Neural Networks, Random Forest, Unmixing.

INTRODUCTION

Imaging Spectroscopy has become an active area of research in Remote Sensing. Hyperspectral data, as the name itself suggests contains hundreds of spectral bandswith a very narrow band width. This high spectral information helps in distinguishing the materials that are spatially similar and spectrally unique. Hence, these Hyperspectral datasets (HS datasets hereafter) find applications in various fields like crop type monitoring, crop health monitoring, water quality assessment, mineral mapping, vegetation mapping, defence target identification and many others. However, the high spectral resolution leads to many challenges in processing and analysing the data.

Challenges in Hyperspectral data processing

1. Curse of dimensionality
2. The voluminous data contains many bands that contain similar information leading to redundancy in the data
3. Mixed pixels: the moderate to coarse spatial resolution and high spectral information leads to mixing of information from within the pixel. Thus, finding pure pixels from a HS image is a challenging task.
4. Adding contextual information along with spectral information during classification.

Apart from these, atmospheric corrections and dimensionality issues also need to be addressed while processing the HS data. Various atmospheric correction models are available that can convert the data from radiance to reflectance. The dimensionality issues can be addressed by transforming the data to a higher dimensional space and reducing the volume of the data. In this paper, we discuss the basic steps to be adopted for hyperspectral data processing followed by a review on various classification methods techniques that can be used for hyperspectral data classification.

The general flow of hyperspectral data classification is as shown in the flowchart. (Table 1)

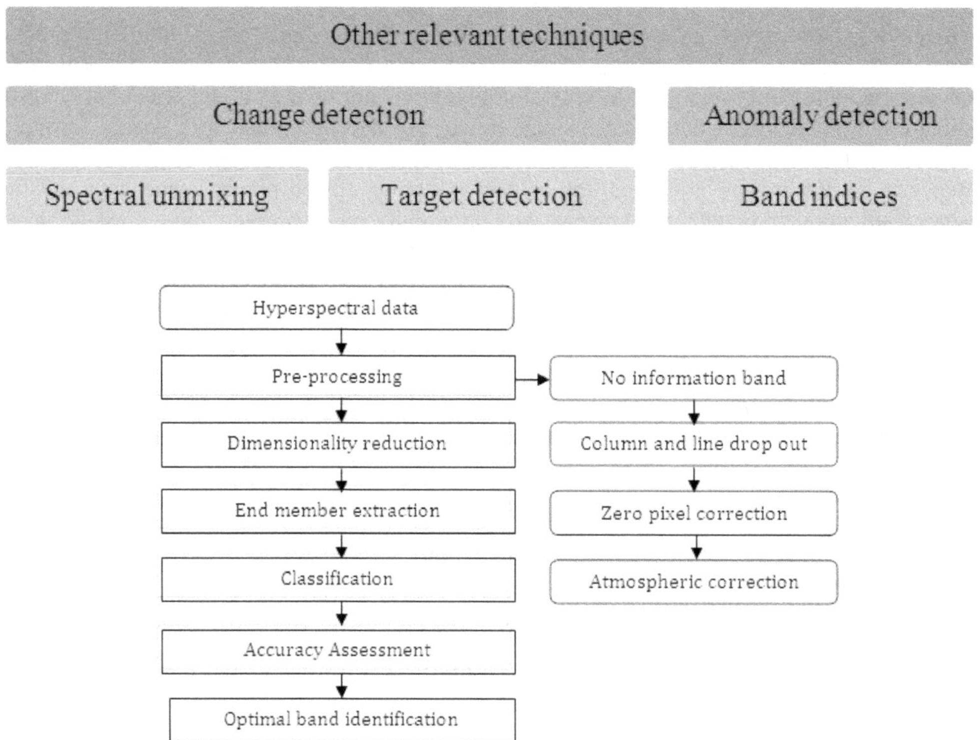

Fig. 1 Flow of hyperspectral data processing

RADIOMETRIC AND ATMOSPHERIC CORRECTIONS

Radiometric and atmospheric corrections are one of the important and prerequisite steps of hyperspectral data processing. Radiometric errors like line and column dropouts should be addressed before doing atmospheric corrections. Averaging and weighted averaging methods are used for correcting the column and line dropouts. Atmospheric correction methods can be relative methods, empirical methods, absolute methods based on radiative transfer models and combination based methods. The relative methods can be as simple as flat field method/ averaging method where the mean of a large area that is assumed to have a similar spectral reflectance is taken and the whole scene is normalized to the mean[1,2]. Whereas the absolute methods utilize various insitu parameters like solar zenith and azimuth angles, elevation, location of the scene, time of acquisition and atmospheric and topographic models to obtain a atmospherically corrected image. FLAASH, ATCOR, ACORN, ATREM[5]etc. are few absolute atmospheric correction methods that use radiative transfer models for obtaining a reflectance image.

The "log residual" technique entails taking means along the wavelength axis as well as spatially to develop pixel spectra that are normalized to remove illumination as well as albedo effects. The calculations employ the differences of algorithms instead of ratioing and hence the name[3].

A universally used technique that characterizes both the gain or transmission and the offset or path radiance is called the empirical line method8. It consists of acquiring field reflectance spectra of a bright and dark target in the field, preferably large enough to encompass several pixels. A regression equation is created for each spectral band that provides a relationship between reflectance and raw radiance data. The result is a gain factor that consolidates all the multiplicative influences such as atmospheric transmission, solar irradiance and instrument response as well as an offset that is related to the sensor and the path radiance[4].

DIMENSIONALITY REDUCTION (DR)

As the name itself suggests, dimensionality reduction techniques transform the data into a new domain where the data in each band is made uncorrelated to other band based on certain criteria[9]. DR techniques can be either linear or non-linear type. The three well known feature extraction/dimensionality reduction techniques include the principal component analysis (PCA), minimum noise fraction (MNF) and independent component analysis (ICA) techniques. PCA, also called the Karhunen-Love transform (KLT) or the Hotelling transform, is a classical statistical technique used to reduce the dimensionality of the multi-dimensional data. PCA finds a new set of orthogonal axes that have their origin at the data mean and that are rotated to maximize the data variance[9]. Minimum Noise Fraction on the other hand is a two-step process where the first step involves noise whitening and the second step is a PCA[10]. Bothe MNF and PCA consider the first and second order statistics like mean, variance, co-variance etc. One major issue for both PCA and MNF is that many subtle material substances that are uncovered by very high spectral resolution hyperspectral imaging sensors cannot be characterized by second-order statistics. This may be due to the fact that the samples of such targets are relatively small and not sufficient to constitute reliable statistics. In this case, these targets may not be captured by the second-order statistics-based PCA/MNF in its PCs. Another key issue arising in the PCA/MNF is the determination of number of dimensions to be retained. A common criterion to resolving this problem is to calculate the accumulated sum of eigenvalues that represents a certain percentage of energy needed to be preserved.[6] Unlike the PCA/MNF which prioritizes principal components in accordance with magnitude of eigenvalues or SNR, the ICA does not have such a criterion to prioritize the components based on the information available. Hence it uses a virtual dimensionality method inorder to select the components with information. IK Fodoret.al., 2002 conducted a survey on various linear and non – linear dimensionality reduction techniques like PCA and vertex components and ICA and non-linear PCA, vector quantization, neural networks etc. Vandermaten et.al., 2009 has reviewed 13 different non – linear dimensionality reduction techniques like Locally Linear Coordination, Manifold

Charting, Multi-layerautoencoders, sparse spectral techniques, convex and non convex techniques etc[14,15,16,17] and compared them with the conventional PCS method. In his review he has concluded that the diverse variants of non linear DR techniques could not outmatch the conventional vertex component , PCA or factoral analysis methods for DR. Many other techniques like self-organizing maps[19] and their probabilistic extension GTM[20] which combine a dimensionality reduction technique with clustering, Linear Discriminant Analysis[21], Generalized Discriminant Analysis[22], and Neighbourhood Components Analysis[23, 24], and recently proposed metric learners[25, 26] etc are available which possess a supervised nature of dimensionality reduction.

ENDMEMBER EXTRACTION

Hyperspectral data classification based on per pixel or sub pixel methods mainly involves two important steps- the first one is to find spectrally unique signatures of pure ground components from the image which are commonly known as end members and the second one is to assign each pixel or sub pixel portions to one of the endmembers class. The endmembers selection process can be carried out in two ways: The first one is by deriving the endmember/spectra from the image (image endmembers)and the second one is from field or laboratory spectra of known target materials (library endmembers).[26] gives a comparison between the two. The drawback of using library spectra is that these spectra are rarely acquired under the same conditions as the image. Image endmembers have the advantage of being collected at the same scale as the data and can, thus, be more easily associated with features on the scene[27]. A number of algorithms have been developed over the past decade to accomplish the task of finding appropriate image endmembers for spectral mixture analysis[28]. Antonio Plaza et.al., 2004 presented a comparative study between six standard end member extraction algorithms for linear unmixing using a widely available database of simulated and real AVIRIS images.Pablo et.al., 2006 made a short review on various endmember extraction techniques – N-FINDR, AMME (Automatic Morphological Endmember Extraction), fully constrained linear spectral unmixing methods were tested using RMSE and reference abundance maps[34].

Pixel Purity Index or PPI[36] is one the well known and successful methods of finding the pure pixels from the image. This method is based on the geometry of convex sets[37]. PPI considers spectral pixels as vectors in a N-dimensional space. A dimensionality reduction is first applied to the original data cube by using the Minimum Noise Fraction. Further, large number of random N-dimensional vectors, called as skewers[38] are generated. Every data point is projected onto each skewer, along which position it is pointed out. The data points which correspond to extreme values in the direction of a skewer are identified and placed on a list. As more skewers are generated, the number of times a given pixel is identified as the extreme pixel is also counted. The pixels with the highest count are considered the purest ones, since a pixel count provides a pixel purity index[34].

HYPERSPECTRAL DATA CLASSIFICATION

Classification of HS datasets depends on many criteria like size of the training samples, purity of the training samples, classifier used, parameter fine tuning in a classifier etc.

Table 1 Various criteria and methods of classification

Criteria	Methods of classification	
Based on availability of training samples	Supervised	Training samples are available and prior knowledge about the study area is available Eg: Maximum likelihood, minimum distance, artificial neural network, decision tree classifier.
	Unsupervised	No prior information about the study area and image is available. No training samples are given as input for classification. Eg: ISODATA, K-means clustering algorithm.
Based on assumptions on distribution of data	Parametric	Gaussian distribution of data is assumed. When landscape is complex, parametric classifiers often produce 'noisy' results. Another major drawback is that it is difficult to integrate ancillary data, spatial and contextual attributes, and non-statistical information into a classification procedure. Eg: MLC, LDA
	Non-Parametric	No assumption on distribution of data. These are close to real world scenario. Eg: ANN, SVM, DTC etc
If spatial data is included during classification	Spectral classification	Pure spectral information is used in image classification. A 'noisy' classification result is often produced due to the high variation in the spatial distribution of the same class. Eg:
	Spatio-spectral classification/Contextual classifiers.	The spatially neighbouring pixel information is used in image classification. Eg: frequency-based contextual classifier, ECHO (Extraction and Classification of Homogeneous Objects) etc[38-40]
Based on pixel information or object information	Hard classifiers	Land cover continuity is not shown. Classification is done based on the reflectance value of one full pixel. Eg: MLC, SVM, ANN, DTC etc.
	Soft classifiers	Continuity of land cover classes is seen in the classifier. It considers the within pixel information for classifying the data. Eg: Fuzzzy classifiers, Linear unmixing, MTMF etc.
	Object based classifiers	Image segmentation merges pixels into objects and classification is conducted based on the objects, instead of an individual pixel Eg: Segmentation and clustering procedures.
Whether a single classifier is used or multi classifiers are used.	Single classifier	All conventional multispectral and hyperspectral classifiers like MLC, SVM, ANN etc.
	Ensemble classifiers	More than one classifiers are integrated and used for classification [41,42,43]

The normal classification techniques available for multispectral data like maximum likelihood, minimum distance to mean etc. does not work well for the hyperspectral data due to its the higher dimensionality. The criteria considered for the multispectral datasets that –"the number of training samples required for classification should be 10 times the number of bands" is very difficult to be satisfied in the case of hyperspectral data (Hughes phenomenon). Hence several new algorithms have been developed for the classification of hyperspectral data. A brief review of various advanced hyperspectral per pixel and sub pixel classification techniques are discussed below.

SPECTRAL ANGLE MAPPER

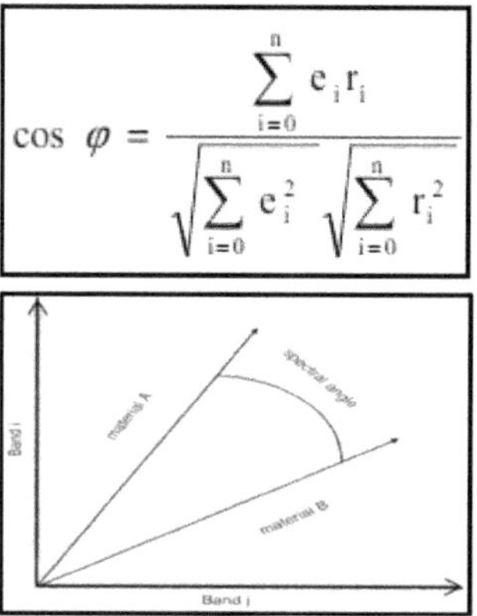

$$\cos \varphi = \frac{\sum_{i=0}^{n} e_i r_i}{\sqrt{\sum_{i=0}^{n} e_i^2}\sqrt{\sum_{i=0}^{n} r_i^2}}$$

The Spectral Angle Mapper Classification (SAM) is an automated method for directly comparing image spectra to a known spectrum (usually determined in a lab or in the field with a spectrometer) or an end member[46]. SAM assumes that the data have been reduced to apparent reflectance. This method treats both (the unknown and known) spectra as vectors in an n-D space, where n is the number of bands and calculates the spectral angle between them. Each vector has certain length and direction. The length of the vector represents brightness of the pixel while the direction of the vector represents the spectral feature of the pixel. This method is insensitive to illumination since the SAM algorithm uses only the vector direction and not the vector length. i.e., If the overall illumination increases or decreases (due to the presence of a mix of sunlight and shadows), the length of this vector will increase or decrease, but its angular orientation will remain constant [21]. The result of the SAM classification is an image showing the best match at each pixel. The spectral angle values are between 0 to $\pi/2$.

Φ = Spectral Angle,

e = given image spectra,

r = reference spectra (end member/training spectra),

n = number of classes

SAM technique fails when vector magnitude is important in providing discriminating information, which happens in many instances [30]. However, if the pixel spectra from the different classes are well distributed in feature space there is a high likelihood that angular information alone will provide good separation. This technique functions well in the face of scaling noise[31].[47] used a band max algorithm and spectral angle mapper classifier for identifying the white blood cells in vegetation. Also, many works [48] [49] [50] showcase the utility of hyperspectral data for land use land cover, vegetation, mineral and many other applications.

ARTIFICIAL NEURAL NETWORKS

Artificial Neural Networks have been developed with an inspiration from the human brain. The Neuron receives the input from the left; each of the input is multiplied by a weight factor[51]. Learning occurs by adjusting the weights in the node to minimize the difference between the output node activation and the output. This input is fed to a summing function, and the output of the summing function is fed to a transfer function, which uses some mathematical function. This can be an input to another neuron or an output multilayer feed forward ANN is used to perform a nonlinear classification. Its design consists of one input layer, at least one hidden layer and one output layer.

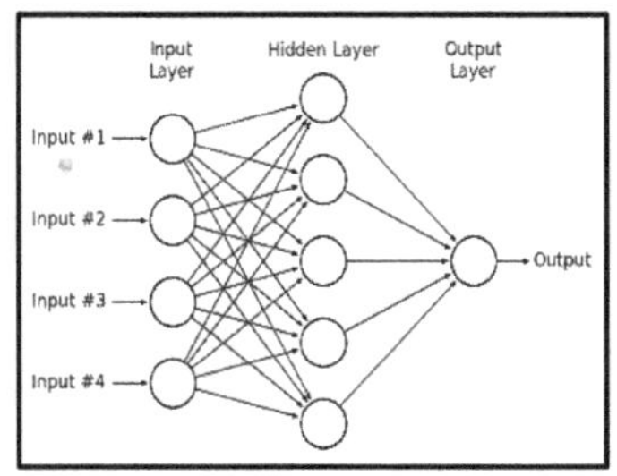

The basic representation of a feed forward neural network is represented in the figure.Applying a NN to image classification makes use of an iterative training procedure in which the network is provided with matching sets of input and output data. Each set of input data represents an example of a pattern to be learned, and each corresponding set of output data represents the desired output that should be produced in response to the input. During the training process the network autonomously modifies the weights on the linkages between each pair of nodes in such a way as to reduce the discrepancy between the desired output and the actual output [52][31]. Many experiments using ANN classifier were conducted onimages from Landsat Multispectral Scanner (MSS) [52][53],Landsat TM [54], Synthetic Aperture Radar [55], AVHRR [57], SPOT HRV [56] and aircraft scanner data [58].

SUPPORT VECTOR MACHINES

SVM classification technique is based on the statistical learning theory. This technique is said to be independent of dimensionality of feature space. The main idea behind this classification is to separate the classes with a surface that maximize the margin between them, using boundary pixels to create the decision plane[62]. “A decision plane is one that separates between a set of objects having different class memberships. The decision planes may not always be exact straight lines as it is not possible in many classification tasks. Classification tasks based on drawing separating lines to distinguish between objects of different class memberships are known as hyper plane classifiers. The data points that are close to the hyper plane are termed ‘support vectors’ [63]. The performance of the support vector mainly depends on the kernel types used to transform the data into higher dimensional feature space. SVM provides good classification results from complex and noisy data.” [65]

In practical consideration, the major advantage of support vector machine classifier is that even a single training pixel for a particular class would be enough for the SVM to classify the corresponding matching spectra in the image to that class[64].Mountrakiset.al.,2011 used SVM classifier for classifying various LULC classes from DIAS hyperspectral data.

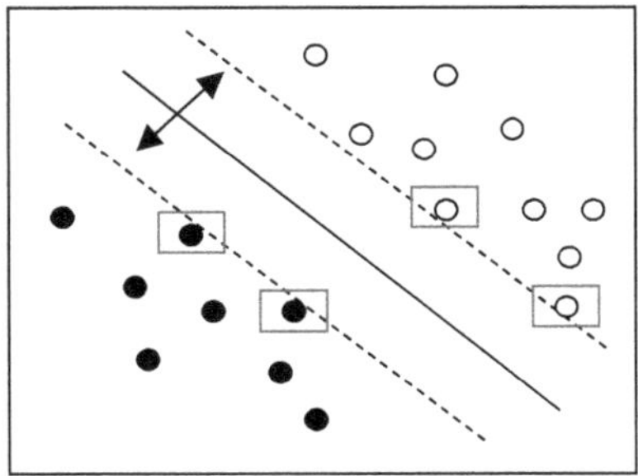

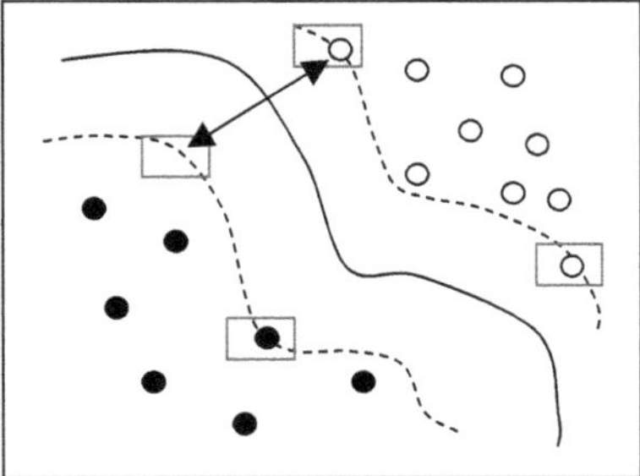

Choosing a proper kernel plays an important role in SVM classification. The most commonly used SVM kernels are linear kernel, polynomial, Radial Basis and Sigmoid kernels. Many studies were made to test the performance of each of these kernels. However, it is difficult to generalize one best kernel for various datasets. Hence a combination of more than one kernels or improvement of existing kernels is used as a method to obtain accurate classification results.

DECISION TREE

Decision tree approach is a nonparametric classifier and an example of machine learning algorithm. It involves a recursive partitioning of the feature space, based on a set of rules that are learned by an analysis of the training set [70]. A tree structure is developed where at each branching a specific decision rule is implemented, which may involve one or more combinations of the attribute inputs. A new input vector then "travels" from the root node down through successive branches until it is placed in a specific class [69]. The thresholds used for each class decision are chosen using minimum entropy or minimum error measures. It is based on using the minimum number of bits to describe each decision at a node in the tree based on the frequency of each class at the node. With minimum entropy, the stopping criterion is based on the amount of information gained by a rule (the gain ratio) [68].

The thresholds used for each class decision are chosen using minimum entropy or minimum error measures. It is based on using the minimum number of bits to describe each decision at a node in the tree based on the frequency of each class at the node. With minimum entropy, the stopping criterion is based on the amount of information gained by a rule (the gain ratio) [69]. P Kumar et.al.2015, made a comparison between SVM, SAM, ANN classifiers for classifying crop types using LISS IV data.

SPECTRAL FEATURE FITTING (SFF)

Spectral Feature Fitting is a method for analyzing hyperspectral data. It is an absorption-feature based method for matching image spectra to reference end members. Prior to analysis this method requires the data to be converted to reflectance and that a continuum be removed from the reflectance data [71]. A continuum is a mathematical function used to isolate a particular absorption feature for analysis. It corresponds to a background signal unrelated to a specific absorption feature for analysis. Spectra are normalized to a common reference using a continuum formed by defining high points of the spectrum (local maxima) and fitting straight line segments between these points[72]. The continuum is obtained by dividing it into the original spectrum.

Thus SFF requires that the reference and endmembers be selected from either the image or a spectral library, with the continuum removed, and that each reference endmember spectrum be scaled to match the unknown spectrum. A "scale" image is produced for each endmember by first subtracting the continuum removed spectra from one, thus inverting them and making the continuum zero[73]. A single multiplicative scaling factor is thus determined that makes the reference spectrum match the unknown spectrum. A large scaling factor resembles a deep spectral feature, while a small scaling factor indicates a weak spectral feature.

RANDOM FORESTS

Random forest algorithm is a supervised classification algorithm. As the name suggest, this algorithm creates the forest with a number of trees.Random forest algorithm can use both for classification and the regression kind of problems andare considered for classification of multisource remote sensing and geographic data. The most widely used ensemble methods are boosting and bagging. Boosting is based on sample re-weighting but bagging uses bootstrapping. The Random Forest is also an ensemble based classifier that uses bagging, or bootstrap aggregating, to form an ensemble of classification and regression tree (CART)-like classifiers. In addition, it searches only a random subset of the variables for a split at each CART node, in order to minimize the correlation between the classifiers in the ensemble. This method is not sensitive to noise or overtraining, as the resampling is not based on weighting. Furthermore, it is computationally much lighter than methods based on boosting and somewhat lighter than simple bagging.

Random Forest uses an unbiased method to evaluate the classification accuracy in case a separate test set is not available. For each new training set that is generated, one-third of the training samples are randomly left out, called the out-of-bag (OOB) samples. The remaining (in the-bag) samples are used for building a tree. For accuracy estimation, votes for each case are counted every time the case belongs to the OOB samples. A majority vote will determine the final label. Only approximately one-third of the trees built will vote for each case. The OOB error estimate has been shown to be unbiased in many tests.

Although the structure of a decision tree provides information concerning important features, such an interpretation becomes impossible when using hundreds of trees. One additional feature of Random Forest, however, is its ability to evaluate the importance of each feature based on the use of internal OOB estimates. Random Forest is described as a very accurate technique, with no risk of overfitting, low bias and low variance.

SPECTRAL UNMIXING – SOFT CLASSIFIER

Pixel-wise classification identifies the class to which a pixel spectrum closely resembles, but it does not yield any further insight into the other materials that exist within the mixed pixel of the hyperspectral data which occurs due to its moderate spatial resolution[67]. In such cases, sub-pixel classification techniques like spectral unmixing come into picture which give the abundance fraction of a particular material within a pixel. The results of unmixing are highly dependent on the end members used for unmixing and the results change with the changing end members. The method of unmixing a mixed pixel depends on the way in which these constituent material substances in a pixel combine to yield the composite spectrum measured at the sensor. Hence, the spectral unmixing techniques are classified into two types – i) linear unmixing, ii) non-linear unmixing (Fig 4a and 4b). The linear spectral unmixing assumes that the radiation incident on the earth interacts with only a single material before it reaches the sensor. On the other hand, the nonlinear technique assumes that the incident ray interacts with more than one material on the earth's surface, before reaching the sensor. In linear spectral unmixing, the resultant spectral signature of a pixel will be the weighted sum of all the materials within the pixel, where the weights are the abundance fractions of the corresponding material[59].

Unmixing simply solves a set of *n* linear equations for each pixel, where *n* is the number of bands in the image. The unknown variables in these equations are the fractions of each endmember in the pixel. To be able to solve the linear equations for the unknown pixel fractions it is necessary to have more equations than unknown, which means that we need more bands than the endmember materials[60]. The results of linear spectral unmixing include one abundance image for each endmember. The pixel value in each image indicates the percentage of the pixel made up of that endmember. An error image is usually calculated to help evaluate the success of the unmixing analysis.The results of spectral unmixing highly depend on the end members and the constraints used in the process of unmixing. Spectral unmixing requires all the materials in the pixel to be defined as end members without missing one[61].

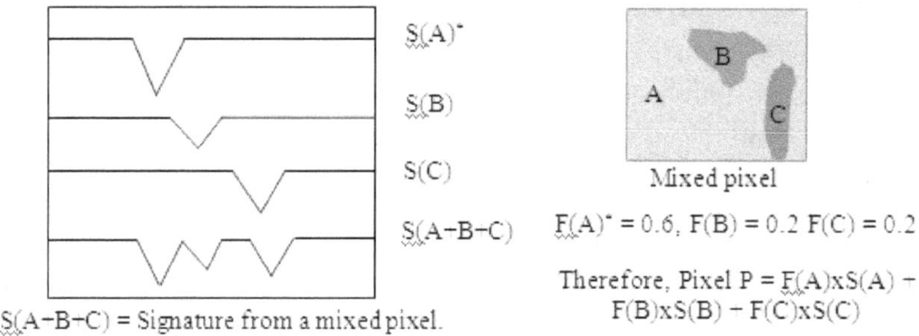

F(X) – Fraction of material X, S(X)- Spectral signature of material X

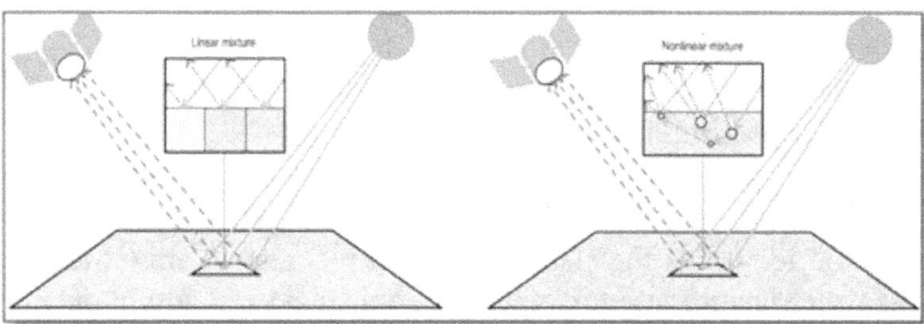

Fig. 4a Linear Unmixing **Fig. 4b** Non-Linear Unmixing

The two important constraints used in spectral unmixing are – a) Non negativity – which requires all the fractions from a pixel to be positive, b) Sum to one – where sum of fractions of all the features should be equal to one. However the constrained unmixing results do not match the real world scenario as it is not possible to identify all the end members in a pixel.

MIXTURE TUNED MATCHED FILTERING (MTMF)

MTMF is another sub-pixel unmixing type in which user chosen target spectra can be mapped. In complete unmixing to get an accurate analysis finding out the spectra of all endmembers from the data is required but this type of unmixing is called a 'partial unmixing' because the unmixing equations are only partially solved[75]. MTMF was originally developed to compute the abundances of targets that are relatively rare in the scene. If the target is not rare, special care must be taken when adapting this technique [ENVI user guide].

Matched filtering “filters” the input image for good matches to the chosen target spectrum by maximizing the response of the target spectrum within the data and suppressing the response of the target spectrum within the data and suppressing the response of everything else[74]. Any pixel with a value of 0 or less would be interpreted as background. The problem with MF is that it is possible to end up with false positive results but the solution, but the solution to this problem is to calculate an additional measure called “infeasibilty”. Infeasibility is based on both noise and image statistics and indicates the degree to which the matched filtering results are a feasible mixture of the target and the background. Pixels with high infeasibilities are likely to be false positive regardless of their matched filter value.

MTMF results are presented as two sets of images, the MF score presented as a grey scale images with values from 0 to 1.0, which provide a means of estimating relative degree of match to the reference spectrum (1.0 being a perfect match) and the infeasibilty image, where highly infeasible numbers indicate that mixing between the composite background and the target is not feasible[76]. The best match to a target is obtained when the matched filter score is high (near 1) and the infeasibilty score is low (near 0).

SUMMARY

This paper presents a summary of flow of hyperspectral data processing and a brief review on various advanced hyperspectral data classifiers.

REFERENCES

1. Gao, Bo-Cai, et al. "Atmospheric correction algorithms for hyperspectral remote sensing data of land and ocean." *Remote Sensing of Environment* 113 (2009): S17-S24.
2. Goetz, Alexander FH, et al. "Atmospheric corrections: on deriving surface reflectance from hyperspectral imagers." *Proceedings-SPIE The International Society For Optical Engineering*. SPIE International Society For Optical, 1997.
3. Green, A. A. and M. D. Craig, "Analysis of aircraft spectrometer data with logarithmic residuals", Proc. of the Airborne Imaging Spectrometer Data Analysis Workshop, JPL Pub. 85-41,pp. 1 1 1-1 19, 1985. 8.
4. Conel, J. E., R. 0. Green, G. Vane, C. J. Bruegge, R. E. Alley, and B. Curtiss, "Airborne Imaging Spectrometer-2: Radiometric spectral characteristics and comparison of ways to compensate for the atmosphere", Proc. SPIE 834, pp. 140-157, 1987.
5. Richter, Rudolf. "Atmospheric Correction Methods for Optical Remote Sensing Imagery of Land." *Advances in Environmental Remote Sensing: Sensors, Algorithms and Applications* (2011): 163-171.
6. Wang, Jing, and Chein-I. Chang. "Independent component analysis-based dimensionality reduction with applications in hyperspectral image analysis." *IEEE transactions on geoscience and remote sensing* 44.6 (2006): 1586-1600.
7. J. Richards and X. Jia, Remote Sensing Digital Image Analysis, third ed. New York: Springer-Verlag, 1999.
8. A. A. Green, M. Berman, P. Switzer, and M. D. Craig, “A transformation for ordering multispectral data in terms of image quality with implications for noise removal,” IEEE Trans. Geosci. Remote Sens., vol. 26, no. 1, pp. 65–74, Jan. 1988.
9. Sahithi, V.S., Iyyanki V Murali Krishna, “Performance evaluation of dimensionality reduction techniques on CHRIS hyperspectral data for surface discrimination”, Journal of Geomatics., Vol 10 No. 1, pp 7-11, April 2016.

10. GuangchunLuo, Guangyi Chen, Ling Tian, Ke Qin &Shen-En Qian (2016) Minimum Noise Fraction versus Principal Component Analysis as a Preprocessing Step for Hyperspectral Imagery Denoising, Canadian Journal of Remote Sensing, 42:2, 106-116, DOI: 10.1080/07038992.2016.1160772
11. Fodor, Imola K. *A survey of dimension reduction techniques*. No. UCRL-ID-148494. Lawrence Livermore National Lab., CA (US), 2002.
12. De Backer, Steve, Antoine Naud, and Paul Scheunders. "Non-linear dimensionality reduction techniques for unsupervised feature extraction." *Pattern Recognition Letters* 19.8 (1998): 711-720.
13. Van Der Maaten, Laurens, Eric Postma, and Jaap Van den Herik. "Dimensionality reduction: a comparative." *J Mach Learn Res* 10 (2009): 66-71.
14. Z. Zhang and H. Zha. Principal manifolds and nonlinear dimensionality reduction via local tangent space alignment. SIAM Journal of Scientific Computing, 26(1):313–338, 2004.
15. D.L. Donoho and C. Grimes. Hessian eigenmaps: New locally linear embedding techniques for high-dimensional data. Proceedings of the National Academy of Sciences, 102(21):7426–7431, 2005.
16. M. Belkin and P. Niyogi. LaplacianEigenmaps and spectral techniques for embedding and clustering. In Advances in Neural Information Processing Systems, volume 14, pages 585–591, Cambridge, MA, USA, 2002. The MIT Press.
17. S.T. Roweis and L.K. Saul. Nonlinear dimensionality reduction by Locally Linear Embedding. Science, 290(5500):2323–2326, 2000.
18. T. Kohonen. Self-organization and associative memory: 3 rd edition. Springer-Verlag New York, Inc., New York, NY, USA, 1989.
19. C. Bishop, M. Svensen, and C. Williams. GTM: The generative topographic mapping. Neural Computation, 10(1):215–234, 1998.
20. R.A. Fisher. The use of multiple measurements in taxonomic problems. Annals of Eugenics, 7:179–188, 1936.
21. G. Baudat and F. Anouar. Generalized discriminant analysis using a kernel approach. Neural Computation, 12(10):2385–2404, 2000.
22. J. Goldberger, S. Roweis, G.E. Hinton, and R.R. Salakhutdinov. Neighbourhood components analysis. In Advances in Neural Information Processing Systems, volume 17, pages 513–520, Cambridge, MA, 2005. MIT Press.
23. R.R. Salakhutdinov and G.E. Hinton. Learning a non-linear embedding by preserving class neighbourhood structure. In Proceedings of the 11th International Conference on Artificial Intelligence and Statistics, volume 2, pages 412–419, 2007.
24. K.Q. Weinberger and L.K. Saul. Distance metric learning for large margin nearest neighbor classification. Journal of Machine Learning Research, 10(Feb):207–244, 2009.
25. A. Bar-Hillel, T. Hertz, N. Shental, and D. Weinshall. Learning a Mahalanobis metric from equivalence constraints. Journal of Machine Learning Research, 6(1):937–965, 2006.
26. D. A. Roberts, G. T. Batista, J. L. G. Pereira, E. K. Waller, and B. W. Nelson, "Change identification using multitemporal spectral mixture analysis: Applications in eastern Amazonia," in Remote Sensing Change Detection: Environmental Monitoring Applications and Methods, R. S. Lunetta and C. D. Elvidge, Eds. Ann Arbor, MI: Ann Arbor Press, 1998, pp. 137–161
27. N. Keshava and J. F. Mustard, "Spectral unmixing," IEEE Signal Processing Mag., vol. 19, pp. 44–57, Jan. 2002.
28. Antonio Plaza, Pablo Martinez, Rosa Pérez, and Javier Plaza, "A Quantitative and Comparative Analysis of Endmember Extraction Algorithms From Hyperspectral Data", IEEE transactions on Geoscience and Remote Sensing, vol. 42, no. 3, March 2004.

29. E. Vermote, D. Tanre, J. L. Deuze, M. Herman, and J. J. Morcrette, May, 1997, IEEE Transaction on Geoscience and Remote Sensing. Second simulation of the satellite signal in the solar spectrum: An overview, vol. 35, Pp. 675 – 686.
30. Richards, J.A., and Jia, X., Remote Sensing Digital Image Analysis An Introduction, Third Edition, Springer, May 1996.
31. Kumar, Uttam. "Comparative evaluation of the algorithms for land cover mapping using hyperspectral data." ITC, 2006.
32. Lu, Dengsheng, and QihaoWeng. "A survey of image classification methods and techniques for improving classification performance." *International journal of Remote sensing* 28.5 (2007): 823-870.
33. Veganzones, Miguel A., and Manuel Grana. "Endmember extraction methods: A short review." *International Conference on Knowledge-Based and Intelligent Information and Engineering Systems*. Springer, Berlin, Heidelberg, 2008.
34. Martínez, Pablo J., et al. "Endmember extraction algorithms from hyperspectral images." *Annals of Geophysics* 49.1 (2006).
35. Boardman, J.W., F.A. Kruse And R.O. Green (1995): Mapping Target Signatures via Partial Unmixing of AVIRIS Data, in Summaries of the V JPL Airborne Earth Science Workshop.
36. Ifarraguerri, A. And C.-I. Chang (1999): Multispectral and hyperspectral image analysis with convex cones, IEEE Trans. Geosci. Remote Sensing, 37 (2), 756- 770
37. Theiler, J., D.D. Lavenier, N.R. Harvey, S.J. Perkins And J.J. Szymanski (2000): Using blocks of skewers for faster computation of Pixel Purity Index, SPIE Proc., 4132, 61-71.
38. Biehl, L. and Landgrebe, D., 2002, MultiSpec—a tool for multispectral-hyperspectral image data analysis. Computers and Geosciences, 28, pp. 1153–1159.
39. Kontoes, C.C. and Rokos, D., 1996, The integration of spatial context information in an experimental knowledge based system and the supervised relaxation algorithm: two successful approaches to improving SPOT-XS classification. International Journal of Remote Sensing, 17, pp. 3093–3106.
40. Chen, D., Stow, D.A. and Gong, P., 2004, Examining the effect of spatial resolution and texture window size on classification accuracy: an urban environment case. International Journal of Remote Sensing, 25, pp. 2177–2192.
41. Debeir, O., Van Den Steen, I., Latinne, P., Van Ham, P. and Wolff, E., 2002, Textural and contextual land-cover classification using single and multiple classifier systems. Photogrammetric Engineering and Remote Sensing, 68, pp. 597–605.
42. Tansey, K.J., Luckman, A.J., Skinner, L., Balzter, H., Strozzi, T. and Wagner, W., 2004, Classification of forest volume resources using ERS tandem coherence and JERS backscatter data. International Journal of Remote Sensing, 25, pp. 751–768.
43. Qiu, F. and Jensen, J.R., 2004, Opening the black box of neural networks for remote sensing image classification. International Journal of Remote Sensing, 25, pp. 1749–1768.
44. Thomas, N., Hendrix, C. and Congalton, R.G., 2003, A comparison of urban mapping methods using high-resolution digital imagery. Photogrammetric Engineering and Remote Sensing, 69, pp. 963–972.
45. Kumar, Pradeep, et al. "Comparison of support vector machine, artificial neural network, and spectral angle mapper algorithms for crop classification using LISS IV data." *International Journal of Remote Sensing* 36.6 (2015): 1604-1617.
46. Zhang, Xiya, and Peijun Li. "Lithological mapping from hyperspectral data by improved use of spectral angle mapper." *International Journal of Applied Earth Observation and Geoinformation* 31 (2014): 95-109.

47. Hou, Xiyue, et al. "A BandMax and spectral angle mapper based alogrithm for white blood cell segmentation." *Ninth International Conference on Digital Image Processing (ICDIP 2017)*. Vol. 10420. International Society for Optics and Photonics, 2017.
48. Krishna, Gopal, et al. "Hyperspectral satellite data analysis for pure pixels extraction and evaluation of advanced classifier algorithms for LULC classification." *Earth Science Informatics*(2017): 1-12.
49. Elatawneh, Alata, et al. "Evaluation of diverse classification approaches for land use/cover mapping in a Mediterranean region utilizing Hyperion data." *International Journal of Digital Earth* 7.3 (2014): 194-216.
50. Zhang, Xiya, and Peijun Li. "Lithological mapping from hyperspectral data by improved use of spectral angle mapper." *International Journal of Applied Earth Observation and Geoinformation* 31 (2014): 95-109.
51. Samarasinghe, Sandhya. *Neural networks for applied sciences and engineering: from fundamentals to complex pattern recognition*. CRC Press, 2016.
52. Benediktsson,J. A., Swain P. H., and Ersoy, O. K., Neural network approaches versus statistical methods in classification of multi source remote sensing data, IEEE Transactions on Geoscience and Remote Sensing, 1990, vol. 28(4), pp. 540–552.
53. Lee, J., Weger, W. C., Sengupta, S. K., and Welch, R. M., A neural network approach to cloud classification. IEEE Transactions on Geoscience and Remote Sensing, 1990, vol. 28, pp. 846 855.
54. Yoshida, T. and Omatu, S., Neural network approach to land cover mapping. IEEE Transactions on Geoscience and Remote Sensing, 1994, vol. 32, pp. 1103–1110.
55. Hara, Y., Atkins, R. G., Yueh, S. H., Shin, R. T., and Kong, J. A., Application of neural networks to radar image classification. IEEE Transactions on Geoscience and Remote Sensing, 1994, vol. 32, pp. 100 –111.
56. Tzeng, Y. C., Chen, K. S., Kao, W.L., and Fung, A. K., A dynamic learning neural network for remote sensing applications. IEEE Transactions on Geoscience and Remote Sensing, 1994, vol. 32, pp. 1096–1103.
57. Gopal, S., Sklarew, D. M. and Lambin, E., Fuzzyneural networks in multitemporal classification of land cover change in the Sahel. New Tools for Spatial Analysis, Proc. of the Workshop, Lisbon, 18 to 20 November, 1993, Office for Official Publication of European Communities, Luxembourg, pp. 69 – 81.
58. Benediktsson, J. A., Swain, P. H., and Ersoy, O. K., Conjugategradient neural networks in classification of multisource and veryhighdimensional remote sensing data. International Journal of Remote Sensing, 1993, vol. 14, pp. 2883–2903.
59. Keshava, Nirmal, and John F. Mustard. "Spectral unmixing." *IEEE signal processing magazine* 19.1 (2002): 44-57.
60. Asner, Gregory P., and Kathleen B. Heidebrecht. "Spectral unmixing of vegetation, soil and dry carbon cover in arid regions: comparing multispectral and hyperspectral observations." *International Journal of Remote Sensing* 23.19 (2002): 3939-3958.
61. Dobigeon, Nicolas, Jean-Yves Tourneret, and Chein-I. Chang. "Semi-supervised linear spectral unmixing using a hierarchical Bayesian model for hyperspectral imagery." *IEEE Transactions on Signal Processing* 56.7 (2008): 2684-2695.
62. Melgani, Farid, and Lorenzo Bruzzone. "Classification of hyperspectral remote sensing images with support vector machines." *IEEE Transactions on geoscience and remote sensing* 42.8 (2004): 1778-1790.
63. Pal, Mahesh, and Paul M. Mather. "Assessment of the effectiveness of support vector machines for hyperspectral data." *Future Generation Computer Systems* 20.7 (2004): 1215-1225.

64. Pal, Mahesh, and P. M. Mather. "Support vector machines for classification in remote sensing." *International Journal of Remote Sensing* 26.5 (2005): 1007-1011.
65. Gualtieri, J. A., and S. Chettri. "Support vector machines for classification of hyperspectral data." *Geoscience and Remote Sensing Symposium, 2000. Proceedings. IGARSS 2000. IEEE 2000 International.* Vol. 2. IEEE, 2000.
66. Mountrakis, Giorgos, Jungholm, and Caesar Ogole. "Support vector machines in remote sensing: A review." *ISPRS Journal of Photogrammetry and Remote Sensing* 66.3 (2011): 247-259.
67. Keshava, Nirmal. "A survey of spectral unmixing algorithms." *Lincoln Laboratory Journal* 14.1 (2003): 55-78.
68. Kuching, Sarawak. "The performance of maximum likelihood, spectral angle mapper, neural network and decision tree classifiers in hyperspectral image analysis." *Journal of Computer Science* 3.6 (2007): 419-423.
69. Safavian, S. Rasoul, and David Landgrebe. "A survey of decision tree classifier methodology." *IEEE transactions on systems, man, and cybernetics* 21.3 (1991): 660-674.
70. Quinlan, J. Ross. "Learning decision tree classifiers." *ACM Computing Surveys (CSUR)* 28.1 (1996): 71-72.
71. Fassnacht, Fabian E., et al. "Comparison of feature reduction algorithms for classifying tree species with hyperspectral data on three central European test sites." *IEEE Journal of Selected Topics in Applied Earth Observations and Remote Sensing* 7.6 (2014): 2547-2561.
72. Pan, Zhuokun, Jingfeng Huang, and Fumin Wang. "Multi range spectral feature fitting for hyperspectral imagery in extracting oilseed rape planting area." *International Journal of Applied Earth Observation and Geoinformation* 25 (2013): 21-29.
73. XU, Ning, and Bin LEI. "Mineral information extraction for hyperspectral image based on modified spectral feature fitting algorithm." *Spectroscopy and Spectral Analysis* 31.6 (2011): 1639-1643.
74. Williams, Amy Parker, and E. Raymond Hunt. "Estimation of leafy spurge cover from hyperspectral imagery using mixture tuned matched filtering." *Remote Sensing of Environment* 82.2 (2002): 446-456.
75. Mitchell, J. J., and Nancy F. Glenn. "Subpixel abundance estimates in mixture-tuned matched filtering classifications of leafy spurge (Euphorbia esula L.)." *International Journal of Remote Sensing* 30.23 (2009): 6099-6119.
76. Boardman, Joseph W., and Fred A. Kruse. "Analysis of Imaging Spectrometer Data Using $ N $-Dimensional Geometry and a Mixture-Tuned Matched Filtering Approach." *IEEE Transactions on Geoscience and Remote Sensing* 49.11 (2011): 4138-4152.

NUMERICAL SIMULATION OF FRANCIS TURBINE USING COMPUTATIONAL FLUID DYNAMICS – CFX

Prasanna S.V.S[1], Sankeerthana[2], Praneeth[3], Xenia Vatsalya[4] and Roshini Mendez[5]

[1]Assistant Professor, [2-5]Students, Civil Engineering Department,
University College of Engineering (A), O. U, Hyderabad, Telangana
palaparthyprasanna7@gmail.com

ABSTRACT

Turbine is the most critical component in hydropower plants because it affects the cost and as well as overall performance of the plant. Hence, for the cost effective design of any hydropower project, it is very important to predict the hydraulic behaviour and efficiency of hydro turbines before they are put in actual use. Experimental approach of predicting the performance of hydro turbine is costly and time consuming compared to CFD approach and hence the numerical simulation using CFD approach plays a vital role. The main aim of the project is to predict the performance and efficiency of Francis turbine using CFD approach and to validate the same with analytical calculations. The efficiency of draft tube is also predicted though CFD approach. The numerical simulation is carried out in CFX solver using ANSYS 17.2. The overall efficiency of the turbine is determined based on the fundamental equation. The overall efficiency when compared with the analytical values for the present case study is in proper agreement using k-epsilon model in CFX solver. This makes us understand to a large extent that, the CFD approach has once again proven to be a helpful tool in analyzing various features and performance of hydro turbines.

Keywords: Turbine, efficiency, draft tube, CFD, CFX.

INTRODUCTION

There are many components in hydropower plant but turbine is the heart of any hydropower plant because it affects the cost as well as overall performance of the whole plant. In case of high head plants the turbine cost is less compared to the cost of civil components as it is very difficult to carry out construction work in hilly areas. But for medium and low head hydropower plants, the typical turbine cost may vary from 15 to 35 percentage of the whole power project cost. Thus, for the cost-effective design of hydropower project it is very crucial to understand the flow characteristics in different parts of the turbine i.e. how energy transfer and transformation take place in the different parts, which help in predicting their performance in advance before manufacturing them. The normal practice to predict the efficiency of a hydro turbine is based on theoretical approach or experimental model testing. Theoretical approach for prediction of efficiency just gives a value; but it is unable to identify the main cause for the poor performance. Conversely, model testing is considered to be costly and time consuming process.

A hydraulic turbine is a rotary mechanical device that extracts energy from water flow and converts it into useful work. The work produced by a turbine can be used for generating electrical power when combined with a generator. A hydraulic turbine is a turbomachine with at least one moving part called a rotor assembly, which is a shaft or drum with blades attached. Moving fluid acts on the blades so that they move and impart rotational energy to the rotor.

The Francis turbine is a type of water turbine that was developed by James B. Francis. A typical model of Francis turbine is shown in Fig 1. It is an inward-flow reaction turbine that combines radial and axial flow concepts (i.e. water flow is radial into the turbine and exits the Turbine axially). It is the first hydraulic turbine with radial inflow. Water pressure decreases as it passes through the turbine imparting reaction on the turbine blades making the turbine rotate.

Francis Turbine is a reaction turbine. Reaction Turbines have some primary features which differentiate them from Impulse Turbines. The major part of pressure drop occurs in the turbine itself, unlike the impulse turbine where complete pressure drop takes place up to the entry point and the turbine passage is completely filled by the water flow during the operation[1]. The main parts of Francis turbine are **a.** Spiral casing **b.** Guide or stay vanes **c.** Runner blades **d.** Draft tube

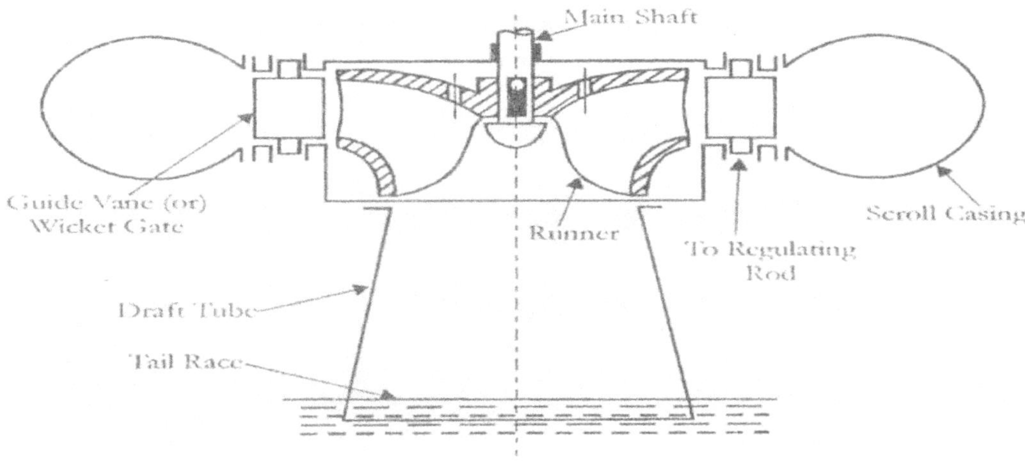

Fig. 1 Components of Francis Turbine (Source: Google images)

ANSYS can import CAD data and enables to build geometry with its "preprocessing" abilities. Similarly in the same preprocessor, finite element model (mesh) which is required for computation is generated. ANSYS CFX is a general purpose software suite that combines an advanced solver with powerful preprocessing and post-processing capabilities. It includes the following features:

- Full integration of problem definition, analysis, and results presentation.
- An intuitive and interactive setup process, using menus and advanced graphics[2].

METHODOLOGY

The first and foremost step in CFD analysis is the explanation and creation of computational geometry of the fluid flowing region. 3-D modeling of all the parts of the turbine is done separately in SOLIDWORKS and all the parts are assembled. The assembled 3-D model is downloaded from GrabCAD, an online community that helps engineering teams to view, manage and share CAD files[3]. The assembled model consists of three domains namely Spiral Casing (including stay vanes and guide vanes), Runner, Draft Tube as shown in Fig 2. The various specifications of the Francis turbine model are as shown in Table 1.

Table 1 Specifications of Francis Turbine Model (Source: GrabCAD)

S.No	Specification	Value
1	Diameter of Runner	0.646 m
2	No. of Runner blades	13
3	No. of Guide Vanes	24
4	No. of Stay Vanes	12
5	Net Head	38.34 m
6	Nominal Discharge	2.25 m^3/s
7	Hydraulic Efficiency	0.93
8	Power	786.92 kW
9	Specific Speed	1000 rpm
10	Water Density	1000 kg/m^3
11	Acceleration due to Gravity	9.81 m/s^2

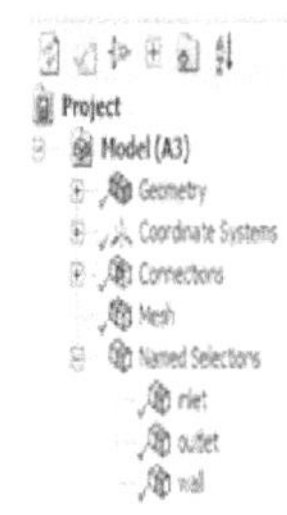

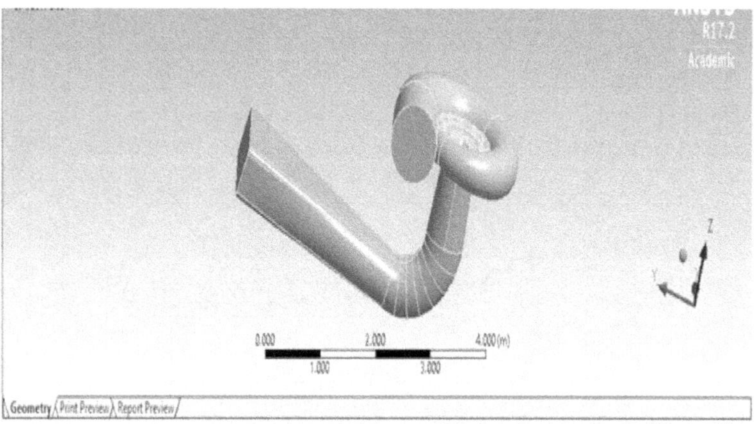

Fig. 2 3-D Model of Francis Turbine (Source: GrabCAD).

The analytical calculations proved that the diameter obtained through analytical calculations is same that of the diameter of the runner in the geometrical model. Thus we can say that the data taken from GrabCAD is accurate. The next important step in the numerical analysis is setting up the grid and mesh associated with the construction of the geometry. The simplification is made using which is called discretization technique. The construction of mesh involves discretizing or sub dividing the geometry into a number of cells or elements at which the variables will be computed numerically. By the Cartesian co-ordinate system, the fluid low governing equations are solved based on discretization of domain. Meshes consisting of too few nodes cause quick solution of simulation but not a very accurate one. However a very dense mesh of nodes causes excess computational time and memory. For CFD more nodes are required in some particular areas of interest in order to capture the large variation of fluid properties. The mesh generated consists of cells of tetrahedron shape and size 0.05. The details of Mesh data are shown in Table 2 and the meshed Francis turbine is shown in Fig 3.

The precision of boundary conditions on the solution domain plays a capital role for the accuracy of the results. The boundary conditions to be specified consist of flow inlet and outlet boundaries, which have to be defined with the flow properties such as turbulence parameters, velocity and pressure. Walls and internal faces which have a direct interaction with the flow have been defined as well. All of these boundary conditions adopted are discussed below:

1. **Inlet boundary**: The inlet section is at the entrance of the spiral casing. Velocity of magnitude 5m/s and direction are specified at the casing inlet as inlet boundary condition.
2. **Outlet boundary:** The outlet of the domain at the exit of the draft tube. Static pressure 0 Pa is specified at outlet of draft tube as outlet boundary condition.
3. **Walls:** All the faces except casing inlet and draft tube exit are specified as walls to simplify the operational conditions with no-slip conditions and to be in stationary conditions at all times.

The simulation is carried out using k-epsilon turbulence model in the CFX solver. The equations are solved iteratively as a transient condition after which the solution converges. The solution converged at 1480th iteration given the number 1500. The converged solution is shown in the Fig 4.

Table 2 Summary of Mesh data

Domain Part	No. of Elements	No. of Nodes	Type of Element
Francis Turbine	2129870	409094	Tetrahedron

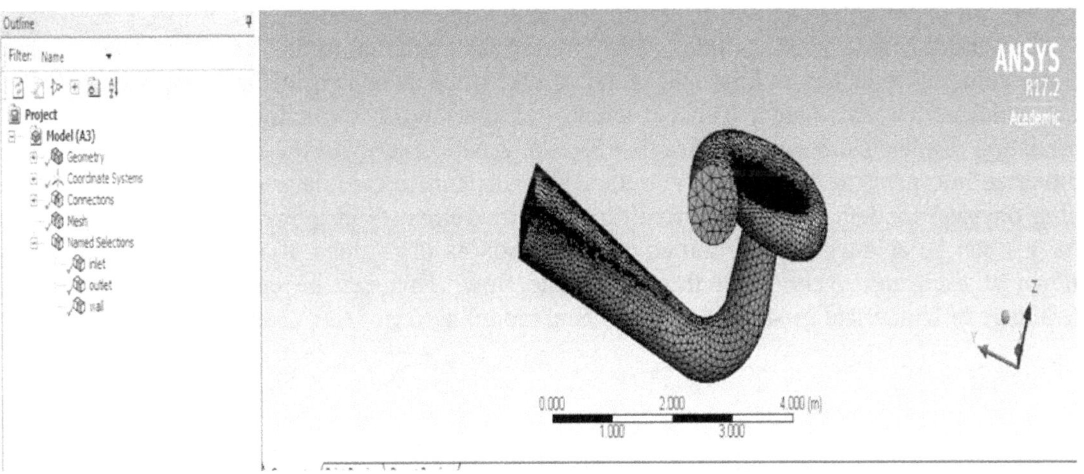

Fig. 3 Meshing of Francis Turbine

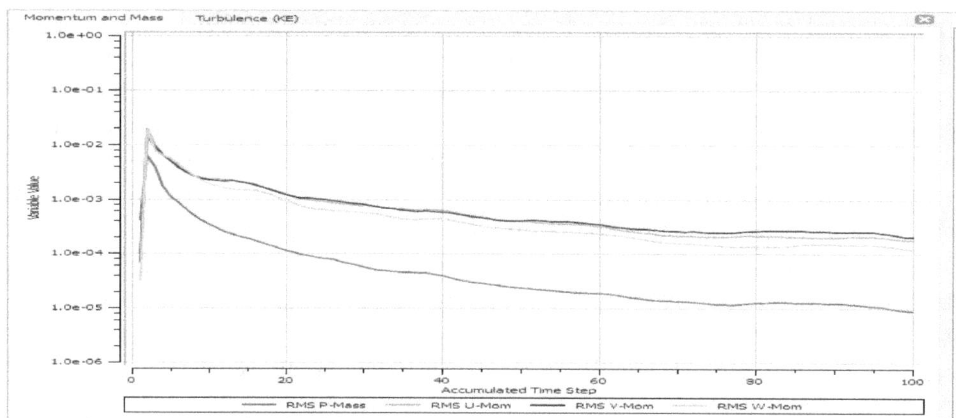

Fig. 4 Converged Solution in CFX Solver

When the equations completely satisfies the governing equations and the solution gets converged, the simulations are analyzed, interpreted and presented in a appropriate convenient way from the results like contour plots, vector plot, streamlines, data curve etc, using graphical representation and reports. Results like variation of pressure, velocity, kinetic energy, eddy viscosity, etc across the flow though the Francis turbine are obtained for clear understanding of the flow through Francis turbine. It is observed that the solution converged at a velocity of 4.73 m/s which can also be observed approximately from Fig 5. For this exit velocity i.e. V_{f2} the efficiency is calculated to be 0.88which varies with the analytical efficiency which is 0.93.

The variation of streamlines lines which are drawn tangential to the direction of flow, based on the value of velocities at inlet and outlet of draft tube as shown in Fig 6, the value of draft tube efficiency is calculated. Thus efficiency of draft tube is obtained from the numerical simulation results as 0.88. It signifies that 89% of kinetic energy at the inlet of the draft tube is converted into pressure energy at the outlet. It is difficult to calculate the draft tube efficiency using physical modeling as it is difficult to measure velocities at internal points of the turbine but this can be achieved using numerical simulation as the values of flow parameters can be known at each and every point throughout the flow. This can be considered as a greatest advantage of Numerical modeling over Physical modeling.

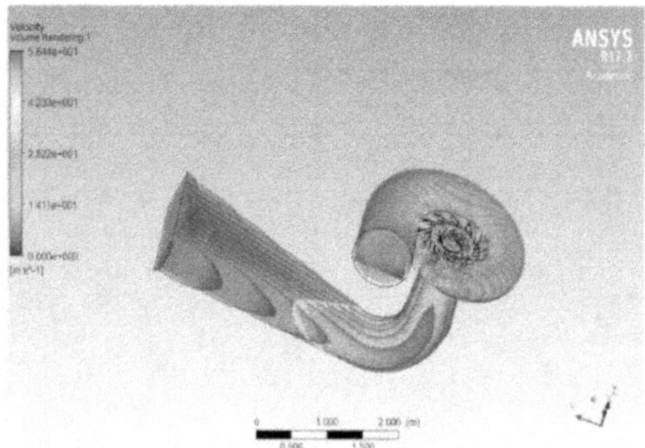

Fig. 5 Variation of Velocity in Turbine

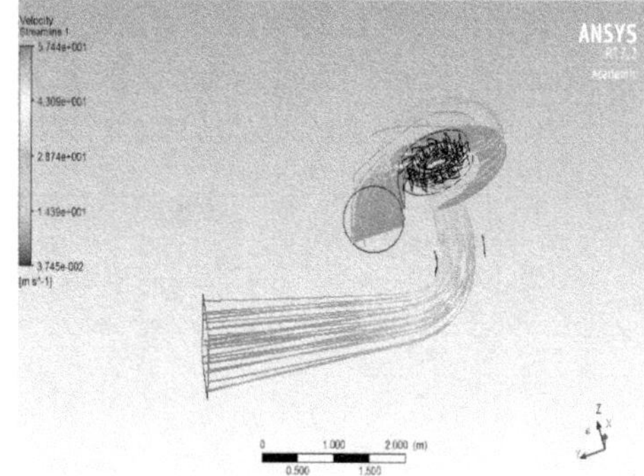

Fig. 6 Variation of Streamlines in Turbine

In the previous sections, efficiency for head of 38.34m has been has been calculated both analytically and from the results of numerically simulated model. Now, head has been varied for an interval of 2m from 30m to 50m and for these variations efficiencies have been calculated both analytically and from numerically simulated model results and a graph is potted between the varying head and obtained efficiencies as shown in Graph 1. It is clear evident from the figure that there is a very slight variation between both the efficiencies.

RESULTS

- ✓ A series of analytical calculations were performed and the values of the following parameters are determined as Inlet flow velocity is 5 m/s; Guide vane angle at inlet is 25.8^0; Runner blade angle at inlet is 168^0 and Exit velocity at runner outlet is 5 m/s.
- ✓ Based on the specified head and discharge proving study was carried out to estimate the accuracy of the model that has been downloaded from GrabCAD. The diameter of the model obtained from proving study is same as that of the CAD geometry i.e. 0.646m which clarifies that the data used is accurate. The hydraulic efficiency used in the proving studies is 0.93 (93%).
- ✓ The simulations were carried out on the downloaded 3-D model of Francis turbine in CFD using CFX solver. The mesh is generated in Workbench of ANSYS – 17.2 with a total number of nodes for the whole turbine as 409094 and the element of tetrahedron shape as 2129870. The results were analyzed and the efficiency of turbine from the simulation results is obtained as 0.88 (88%).
- ✓ The numerical simulations were also further analyzed to determine the efficiency of draft tube and is estimated to be 0.89 (89%) which signifies that 89% of kinetic energy present at the entrance of the draft tube is converted into pressure energy at the exit.

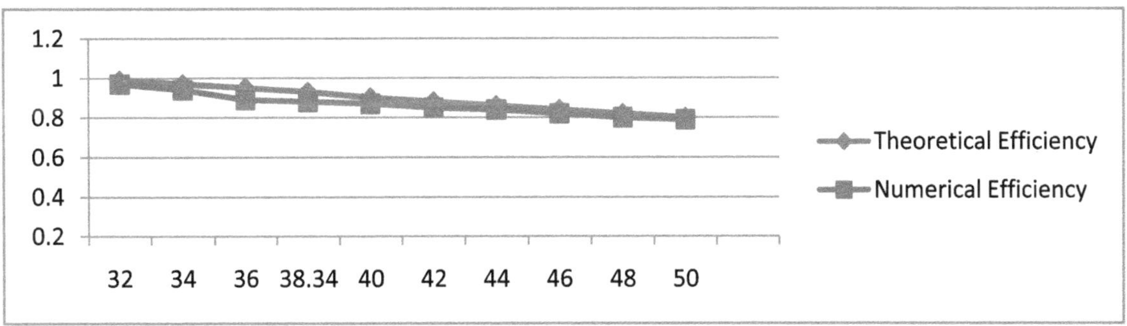

Graph 1 Efficiency vs Head

CONCLUSIONS

- The efficiency of the Francis turbine obtained through analytical calculations is in good agreement with the efficiency obtained though numerically simulations using the turbulence models.
- The calculation of the efficiency of the draft tube is very easy to estimate with the results of the numerical simulations which otherwise is difficult to calculate by using the traditional methods since the variation of velocity inside the turbine cannot be identified.

- From the present study, it is also proved again, that the numerical modeling is very user friendly to know the values of different parameters involved in the model applicable to the flow at each and every node of the element which otherwise is very difficult to know in physical modeling.

REFERENCES

1. Modi P. M., and Seth S. M, "Hydraulics and Fluid Mechanics", seventeenth edition, published by Standard Book House (2009).
2. https://www.designmasters.in/mechanical-cad/ansys.html
3. GrabCAD, an online community that helps engineering teams to view, manage and share CAD files.

GEO-SPATIAL TECHNIQUES FOR AN AUTOMATIC SHIP DETECTION ON SPACEBORNE SAR IMAGES

V. Madhavi Supriya[1*], S.K. Patra[2], and B. Asha Rani[3]

[1]JRF, NRSC, ISRO, [2]Sci/Eng.-SF Group Head (SDA&A), ADRIN, DOS, ISRO, [3]Sci/Eng.-SE, ADRIN, DOS, ISRO
venigallasupriya@gmail.com

ABSTRACT

The human interpretation of SAR images is often complicated, and time taking, but the Satellite–based Synthetic Aperture Radar (SAR) provides a powerful surveillance capability allowing the observation of broad expanses, independently from weather effects and from the day and night cycle. Since the SAR imaging technology many models for ship detection have been developed such as the K-distribution Constant False Alarm Rate (CFAR) method and two-parameter CFAR method etc. Using the advantages of both the methods a new improved two-parameter CFAR method has been proposed. Thus, the method not only has high detection rate but can improve processing speed.

Key Words: SAR, ship target detection, CFAR, K- distribution, Two-parameter distribution

1. INTRODUCTION

The detection of ships assumes greater significance from strategic as well as from commercial view point. The ship detection is, in fact, historically related to the invention of radar by Taylor and Youngman in Wisconsin, USA in 1922. Attempts have been made for automated detection of ships using optical and SAR data. Amongst various ship detection methods, namely Gamma distribution, G_o distribution, Inverse gamma distribution, Rician inverse Gaussian, Log-normal model, Weibull model including K-distribution Constant False Alarm Rate (CFAR) method developed by Vachon (1997), and Qingshan et al.(1999) and two-parameter CFAR method of Eldhuset (1996); David(1999); Wakerman et al.(2001) are noteworthy.

1.1 K- Distribution and Two-Parameter CFAR Method Implementation

As mentioned earlier, there are two most commonly used approaches for automatic ship detection i.e. K- distribution and two-parameter CFAR method. The K-distribution CFAR method doesn't need *a priori* knowledge of ship target characters but only needs the estimation of a global threshold of back scattering coefficient (sigma knot) obtained from the statistics of the whole SAR imagery. It can give good results at a fast speed when the SAR imagery background is homogeneous. As a matter of fact, ships are largely confined to water bodies especially oceans. SAR imagery of the ocean is typically inhomogeneous and a large number of false alarms can arise because of the existence of the speckle noise when the K-distribution CFAR method is used.

In order to circumvent the problem of false alarms and of poor ship detection accuracies as an adaptive threshold algorithm, two- parameter CFAR method can avoid most false alarms caused by speckle noise. Compared with the fact that K-distribution CFAR method does not need the prior knowledge, two-parameter CFAR method needs to know the size of the detected ship first, only after which the size of the detecting windows can be determined. But it uses a sliding window to detect the possible existing ship pixels and needs much time for the statistical computation of the three windows (target window, guard window, and background window). So that method is more computationally expensive. Realizing the scope for timeliness of generating the information with the improved efficiency the study was taken up (i) to automated detection of ships from SAR data, (ii) to improve the accuracy and efficiency of existing most commonly used algorithms

The K-distribution takes the added advantage of the usual shape parameter and the Two-parameter method requires the fine-tuning and analysis of the look parameter (L) and shape parameter (v). If we fix these parameters the detection probability increases. With this advantage we try to solve the problems of increasing detection rate and reducing false alarms.

2. DATABASE

2.1 ENVISAT ASAR AND SENTINEL-1 SAR:

In order to realize the objective SAR data covering two test-sites- one in coastal Galicia, Spain and another in North Sea near to Dunkirk, France, For the Spain test-site ENVISAT ASAR data of Wide Swath Mode (WSM) with a spatial resolution of 150m × 150m (table 1), For the France test-site SENTINEL-1 SAR data of Interferometric Wide Swath mode (IWSM) with a spatial resolution of 5m×20m (table 2).

3. METHODOLOGY

The process adopted involves SAR image feature extraction, generation of moving mean, k-distribution, thresholding, binary image and ship detection.

In order to be unbiased, an automated ship detection technique for extraction of ships from the remote sensing data must be scene and season independent.

Table 1 Sensor characteristics of ENVISAT ASAR.

Parameters	Characteristics
sensor scan	SAR, wide-beam SAR
Waveband	C-band
Polarization	(VV, HH,VV/HH, HV/HH, or VH/VV)
Radiometric Accuracy	Radiometric resolution in range: 1.5-3.5 dB, Radiometric accuracy: 0.65 dB
Spatial resolution	*Image, Wave and Alternating polarization modes*: approx. 30m x 30m. *Wide Swath mode*: approx. 150m x 150m. *Global Monitoring mode*: approx. 1000m x 1000m.
Swath Width	*Image and alternating polarization modes*: up to 100km, *Wave mode*: 5km, *Wide swath and global monitoring modes*: 400km or more

The steps involved in the methodology development are as follows:

- Reading the SAR imagery.
- A mean using moving window for the SAR imagery is computed.
- Two-parameter such as shape parameter (v) and number of statistically independent looks (L) were considered for K-distribution CFAR probability density function implementation.

Table 2 Sensor Characteristics for SENTINEL-1.

Parameters	Characteristics
Orbit	Near polar, sun synchronous 693 km, 6 days repeat cycle
Waveband	C-band
Polarization	VV or HH , VV+VH or HH+HV
Center frequency	5.405 GHz
Spatial resolutions	Strip map: 5 m×5 m Interferometric wide swath: 5 m×20 m Extra wide swath: 25 m×100 m Wave: 5 m×20 m
Swath width	Strip map: 80km Interferometric wide swath: 250km Extra wide swath: 400 km Wave: 20 km
Radiometric Accuracy	Radiometric accuracy: 1dB, Radiometric stability: 0.5 dB

- A global threshold values was applied to the entire image and extracted binary image. The pixels in the imagery, whose grey values lying above the threshold, are marked as possible ship targets and their values are set as 1, while the other pixels are marked as background ocean surface and their values are set as 0.
- Ship detection from the image.

In the first step: ENVISAT ASAR image of study area with the Wide Swath Mode (WSM) and spatial resolution of 150m×150m, with image size (1920×5192) near coast of Galicia, Spain and SENTINEL-1 image with the Interferometric Wide Swath mode (IWSM) and spatial resolution of 5m×20m in VV polarization with image size of (7989×5894) in North Sea near to Dunkirk, France is taken. The reading of the above two SAR imageries is done in MATLAB.

In the second step: We have used the two-parameter with the K-distribution probability density function, so for implementation of K-distribution probability density function we analyzed the mean and shape parameters by assigning different values for them.

The K-distribution of probability density function of sea clutter in SAR images is

$$f(x) = \frac{2}{x\Gamma(v)\Gamma L}\left(\frac{Lvx}{\mu}\right)^{\frac{L+v}{2}} K_{v-L}\left(2\left(\frac{Lvx}{\mu}\right)^{\frac{1}{2}}\right)$$

Where μ-mean,

L- Number of statistically independent looks,

v- Shape parameter,

K_{v-L} (.)-Modified Bessel function of (-L) th order

For the computation of mean using moving window, we used window size equal to 32, the shape parameter (v) equal to 3 and statistically independent number of looks (L) equal to 3.

4. RESULTS AND DISCUSSION

The proposed detector has been applied to test ENVISAT ASAR and SENTINEL-1 SAR real SAR images with respect to sea state for evaluating its performances in the following situations:

4.1 Strong Noise Background

Under weak noise background, most detectors, such as conventional CFAR detectors, can achieve high detection rate and low false alarm rate. However, things are different in strong noise cases. Ships will be mixed with the ocean clutters due to strong backscattering echo from them, which brings many difficulties for the detection of them.

We can observe that the in figure 1 SAR image has strong noise background. They are a total of eleven ships in the SAR image and all the eleven ships are detected by our two-parameter CFAR method. Thus ensuring that the detection performance is 99.5% by the two-parameter CFAR method and no false alarms are present in the strong noise background.

4.2 Heterogeneous Background

Performance under heterogeneous background condition is a very important evaluation criterion to any target detector. In a SAR image there could be heterogeneities such as the oil slicks, transitions between regions with different wind conditions, current boundaries, breaking waves, low wind spiral marks, effects due to land areas such as outlying rocks, shoals and islands, backscattering alteration due to bathymetry and the presence of ship wakes etc.

Generally, ships are brighter than background clutters in marine SAR images, since the scattering of ship targets can last longer than sea clutters in azimuth. This is helpful to detect ship targets. But when the size of a ship is small or the scattering of a ship is faint, it is rather difficult to separate the ship target from the non- homogenous background clutters.

The figure 2 shows they are twenty ships in the SAR image and all the twenty ships are detected by our two-parameter CFAR method. Thus ensuring that the detection performance is 99.5% by the two-parameter CFAR method and no false alarms are present in the strong noise background.

4.3 Presence of Multiple Targets

The actual task arises when there are different scales of ships in the same image. The big ships may be easily detected but the small ships detection becomes difficult and sometimes the small ships may be missed out.

The figure 3 represents they are a total of fifteen ships in the SAR image and all the fifteen ships are detected by our two-parameter CFAR method. The detection performance is 99.5% by the two-parameter CFAR method and no false alarms are present in the presence of multiple targets.

4.1 ENVISAT ASAR Image Representing the Strong Noise Background

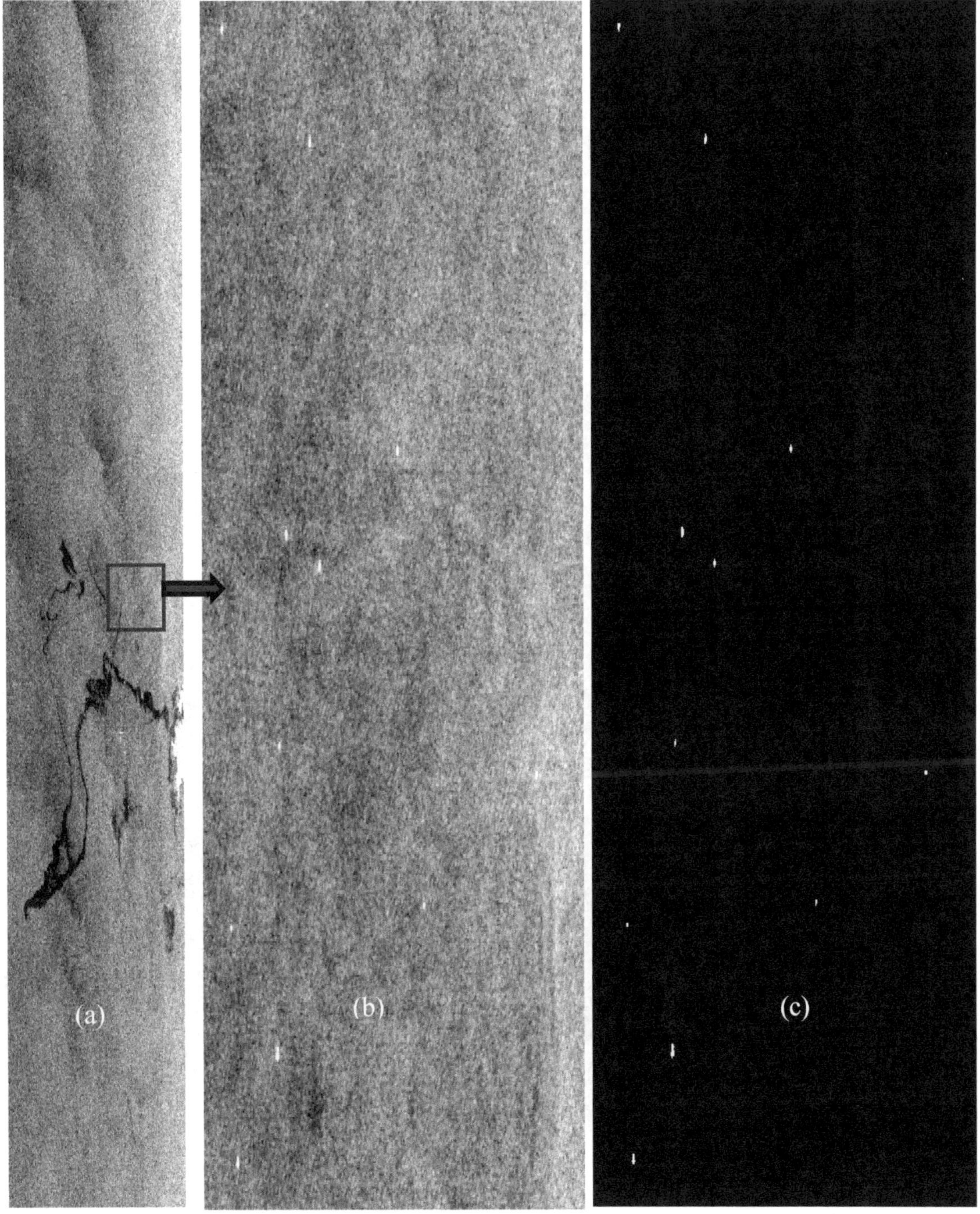

Fig. 1 a) ENVISAT ASAR image, b) Zoom portion of the ENVISAT ASAR region in red square, c) Detected output for the zoom portion for the ENVISAT ASAR image region in red square

4.2 ENVISAT ASAR Image Representing the Heterogeneous Background

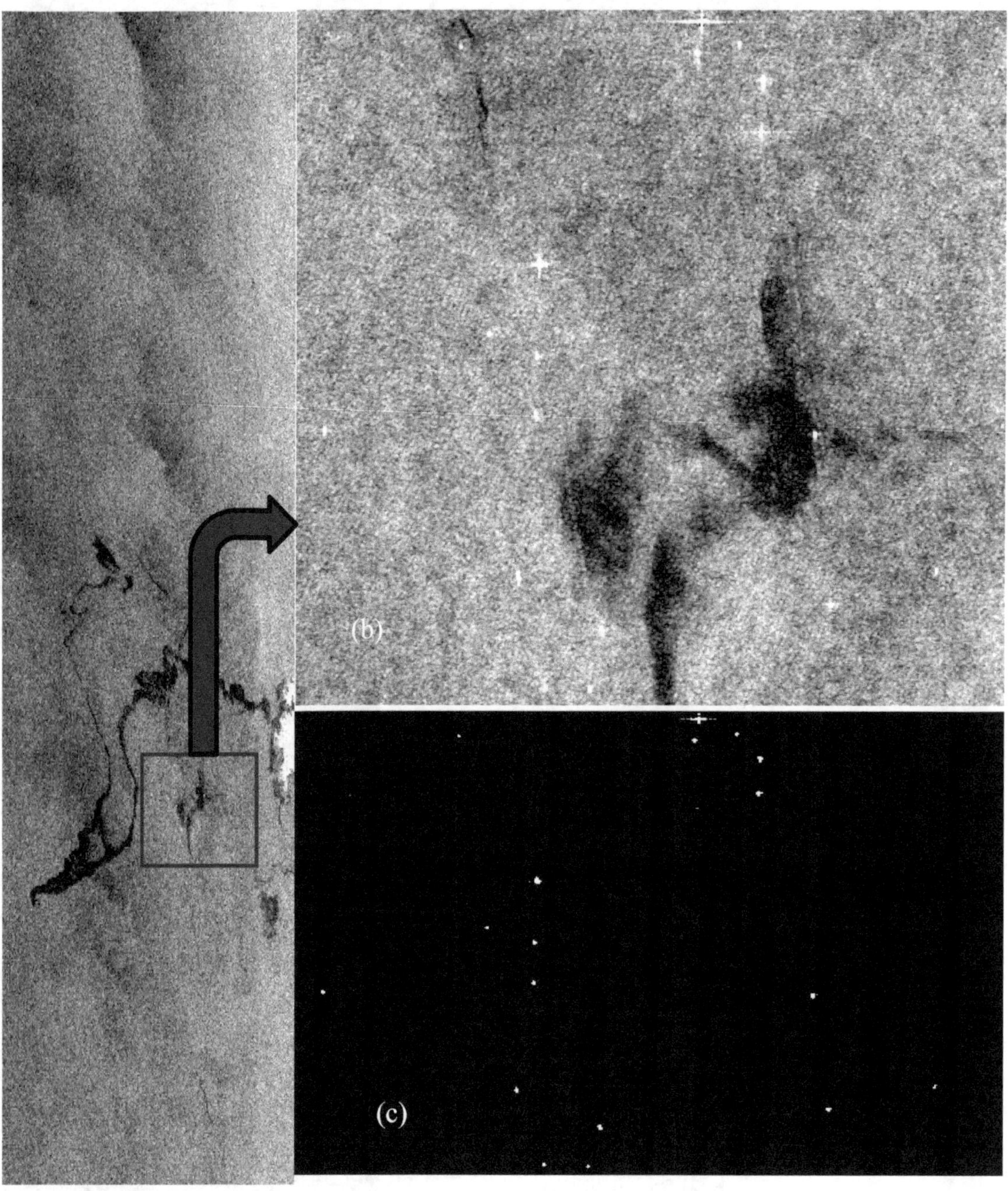

Fig. 2 a) ENVISAT ASAR image, b) Zoom portion of the ENVISAT ASAR region in red square, c) Detected output for the zoom for the ENVISAT ASAR image region in red square

4.3 SENTINEL-1 SAR Image of Different Area Representing the Presence of Multiple Targets

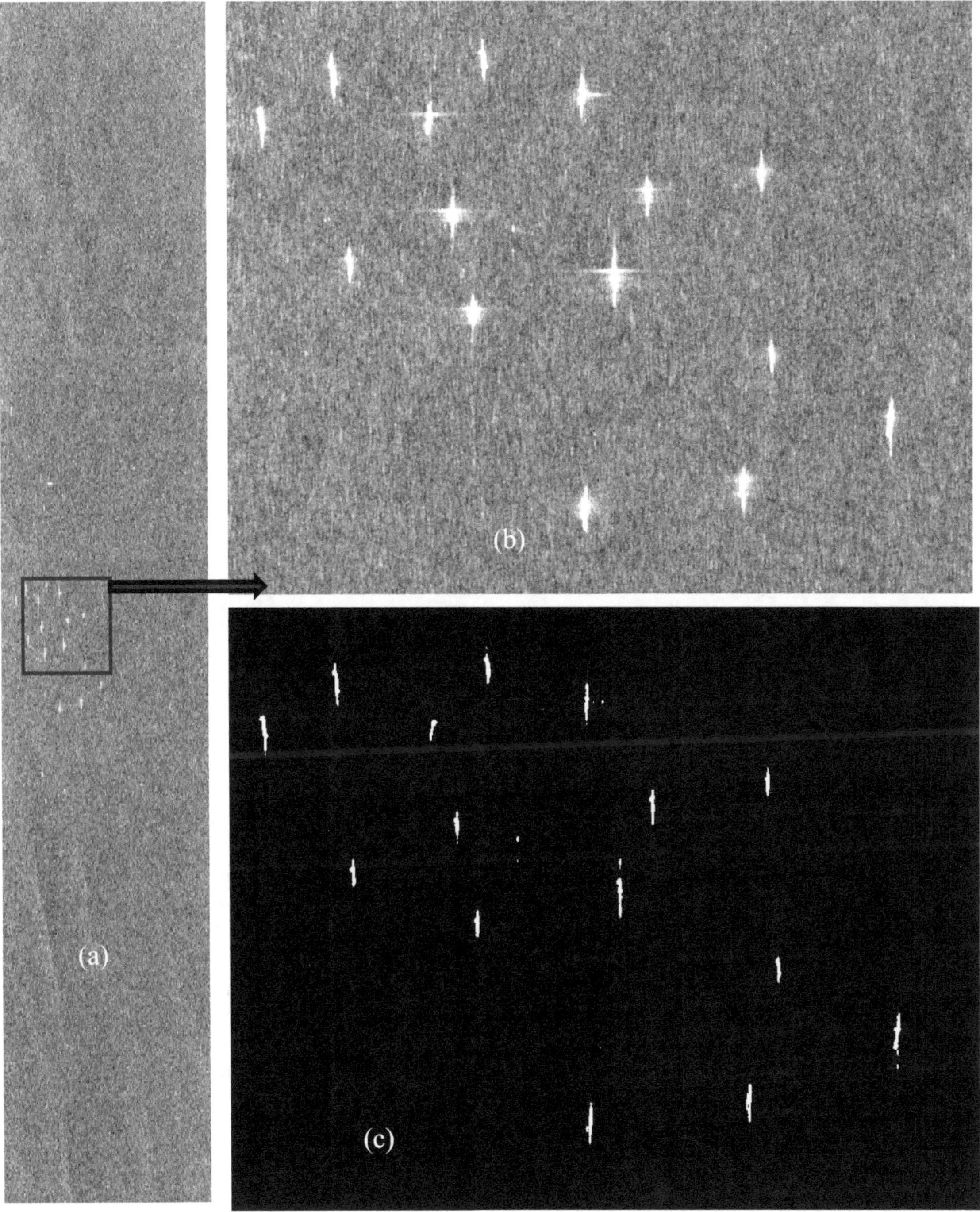

Fig. 3 a) SENTINEL-1 SAR image, b) Zoom portion of the SENTINEL-1 SAR region in red square, c) Detected output for the zoom portion for the SENTINEL-1 SAR image region in red square

CONCLUSION

An automatic ship detection model is being explored by using K-distribution and two-parameter CFAR method. This method combines the advantages of K-distribution CFAR method and two-parameter CFAR method. Compared with conventional CFAR methods, the merits of this method can be concluded as:

- This method does not require any prior knowledge therefore able to perform automatic detection.
- This can work well in different background circumstances.
- This has high detection rate.
- The method has an adequate level of immunity to interference, more suitable for applying in the multiple targets detection.

The method also provides a great benefit to the faint target detection under the higher sea state conditions. This method has given robust and reliable results when applied to the SENTINEL-1 SAR and ENVISAT images, compared with CFAR method. In this research work we dealt with both eight bits (ENVISAT) and sixteen bits (SENTINEL-1) data. Thus, this method can be applied to any type of data and is a good automatic ship detection model for any SAR image.

REFERENCES

1. Andreas Arnold-Bos, Ali Khenchaf, Arnaud Martin, "An evaluation of current ship wake detection algorithms in SAR images", IEEE transactions on geoscience and remote sensing, vol. 42, October 2006.
2. Q. Jiang, E. Aitnouri, S. Wang & D. Ziou, "Automatic detection for ship target in SAR imagery using PNN-model", Canadian Journal of Remote Sensing, Vol. 26, No. 4, pp. 297-305, August 2000.
3. F. Zhang & B. Wu , "A scheme for ship detection in inhomogeneous regions based on segmentation of SAR images", International Journal of Remote Sensing, Vol. 29, No. 19, pp. 5733–5747, 10 October 2008.
4. Biao Hou, Xingzhong Chen and Licheng Jiao, "Multilayer CFAR detection of ship targets in very high resolution SAR images", IEEE geoscience and remote sensing letters, vol. 1, no. 4, pp. 811- 815, April 2015.
5. Wang Juan , Sun Lijie , and Zhang Xuelan, "Study Evolution of Ship Target Detection and Recognition in SAR Imagery", Proceedings of the 2009 International Symposium on Information Processing (ISIP'09), Huangshan, P. R. China, ISBN 978-952-5726-02-2, pp. 147-150, August 21-23, 2009.
6. MariviTello, Carlos Lopez- Martinez and Jordi J. Mallorqui, " A novel algorithm for ship detection in SAR imagery based on the wavelet transform", IEEE geoscience and remote sensing letters, vol. 2, no. 2, pp. 201-205, April 2005.
7. JunminMeng, Jie Zhang, Changying Wang, Jungang Yang, "Field truthing experiment of ship detection by using SAR imagery", Remote sensing and GIS data processing and applications, Proceedings of SPIE, vol. 6790, 67903E, pp. 67903E-1 to 67903E-6, 2007.

A GEO SPATIAL BASED STUDY ON ARTIFICIAL RECHARGE OF GROUND WATER RESOURCES MANAGEMENT IN GVMC, VISAKHAPATNAM DISTRICT, AP

B.K. Appala Raju[1], M. Leela Priyanka[2] and D. Manaswi[3]

Assistant Professor, Assistant Professor Vignan's IIT, Assistant Professor

Raju.nv167@gmail.com, priyankamuddada09@gmail.com,manaswi.dutta@gmail.com

ABSTRACT

Water being one of the most important natural resources need to be conserved and managed properly. Owing to industrialization and urbanization Indian metropolitan cities, were constructed unscientifically leading to irrational urban planning. Unless the need for scientific planning is realized, it is sure to clear up the environment.

The Greater Visakhapatnam Municipal Corporation (GVMC) city forms the area of the study in the present investigation. The population in the city is increasing at a rapid pace due to migration from surrounding areas, with the corresponding increase in suburban areas also. The extensive growth of residential apartments along the coastal stretch resulted in the excess withdrawal of groundwater initiating seawater intrusion in these tracts. All these factors combined, leading to an acute environmental crisis in the area. Forecasting urban water demand can be of use in the management of water utilities (Salvatore Campisi-Pinto, 2012). Thus, there is need to recharge groundwater. The artificial recharge techniques enhance the sustainable yield in the area and utilize the rainfall runoff which otherwise goes to the sewer. Hence, it is necessary to manage the available water resources in an efficient manner. This is done by using advanced technologies like Remote Sensing and Geographical Information System

Keywords: Aquifer, Groundwater, Artificial recharge, Remote sensing & GIS.

1. INTRODUCTION

Water Harvesting refers to collection & storage of rainwater and other activity such as harvesting surface water extracting groundwater prevention of loss through evaporation & seepage. Rainwater harvesting has been practiced for more than 4,000 years; it is also a good option in areas where good quality fresh surface water (or) groundwater is lacking. The role of rainwater harvesting systems as sources of supplementary, backup (or) emergency water supply will become more important and the possibility of greater frequencies of droughts & floods in many areas.

The vast development of urbanization would have the major hydrological impact of peak rate of runoff, controlling rates of erosion and delivering pollutants to rivers (Goudie, 1990). Because of severe soil erosion and water paucity in many areas, conservation of natural resources is a vital issue. In urban areas, the rainfall-runoff processes are directly shown in classical hydrological cycles and are quite different from their natural conditions. The outcome of urbanization, even if it varies, affects the dimension of the flood. The measurement of fairly larger floods with increased recurrence interval decreases the effect of urbanization (Hollis, 1975). The dissimilar stages of urban development like the activities of removal of trees and vegetation, construction of houses, pavement of streets, culverts, etc., would have the impact of reducing evapotranspiration, interception, infiltration, groundwater table and increased stream sedimentation (Kibler, 1982).

Potential sites for construction of rainwater harvesting structures in the Bakhar watershed of Mirzapur District, Uttar Pradesh, India have been identified using remote sensing and GIS techniques. Various thematic maps such as Land use / Land cover, Geomorphology, and Lineaments, etc are prepared using remote sensing. These layers along with Geology and Drainage were integrated using GIS techniques to derive suitable water harvesting sites. Each theme was assigned a weightage depending on its influence on groundwater recharge (for example weightages 20,18,15,25,25 and 0 were assigned to Geomorphology, Landuse, Geology, Lineament, Drainage and Road and Villages respectively). The composite layer, obtained by multiplication of the layers weightage & rank as score, were further averaged into four classes of Excellent (> 200), Good (121 – 200), Moderate (81 –120) and poor (< 80). The suitability of Check dam, Contour bunding/trenching and Recharge pits and wells were suggested accordingly near to villages, Girish Kumar, et al., (2002).

Multi-criteria evaluation is carried out in Geographic Information system to help the decision makers in determining suitable zones for water harvesting structures based on the physical characteristics of the watershed. Different layers which were taken into account for multi-criteria evaluation are; Soil texture, slope, rainfall data (2000-2012), land use/cover, geomorphology, lithology, lineaments, drainage network. The soil conservation service model was used to estimate the runoff depth of the study area. Analytical Hierarchy Processes (AHP) is used to find suitable water harvesting structures on the basis of rainfall. The produced map will help in the selection of the suitable location of harvesting structures. Harish Chand Prasad et al., (2014)

2. STUDY AREA & DATA COLLECTION

2.1 Study Area

The study area the Greater Visakhapatnam Municipal Corporation (GVMC) is located between $17^{0}32^{1}30^{11}$ - $17^{0}52^{1}30^{11}$ northern latitude and $83^{0}04^{1}30^{11}$ - $83^{0}24^{1}30^{11}$ eastern longitude. The urban area of GVMC is divided into six zones. These six zones are further divided into 72 municipal wards covering a total area of 545km^{2}. The city is bounded by the Bay of Bengal on the eastern side, Duvvada hills, (Adavivaram hills) on the western side, Yarada konda on the southern side and Madhurawada dome on the north side. (Fig.1.1)

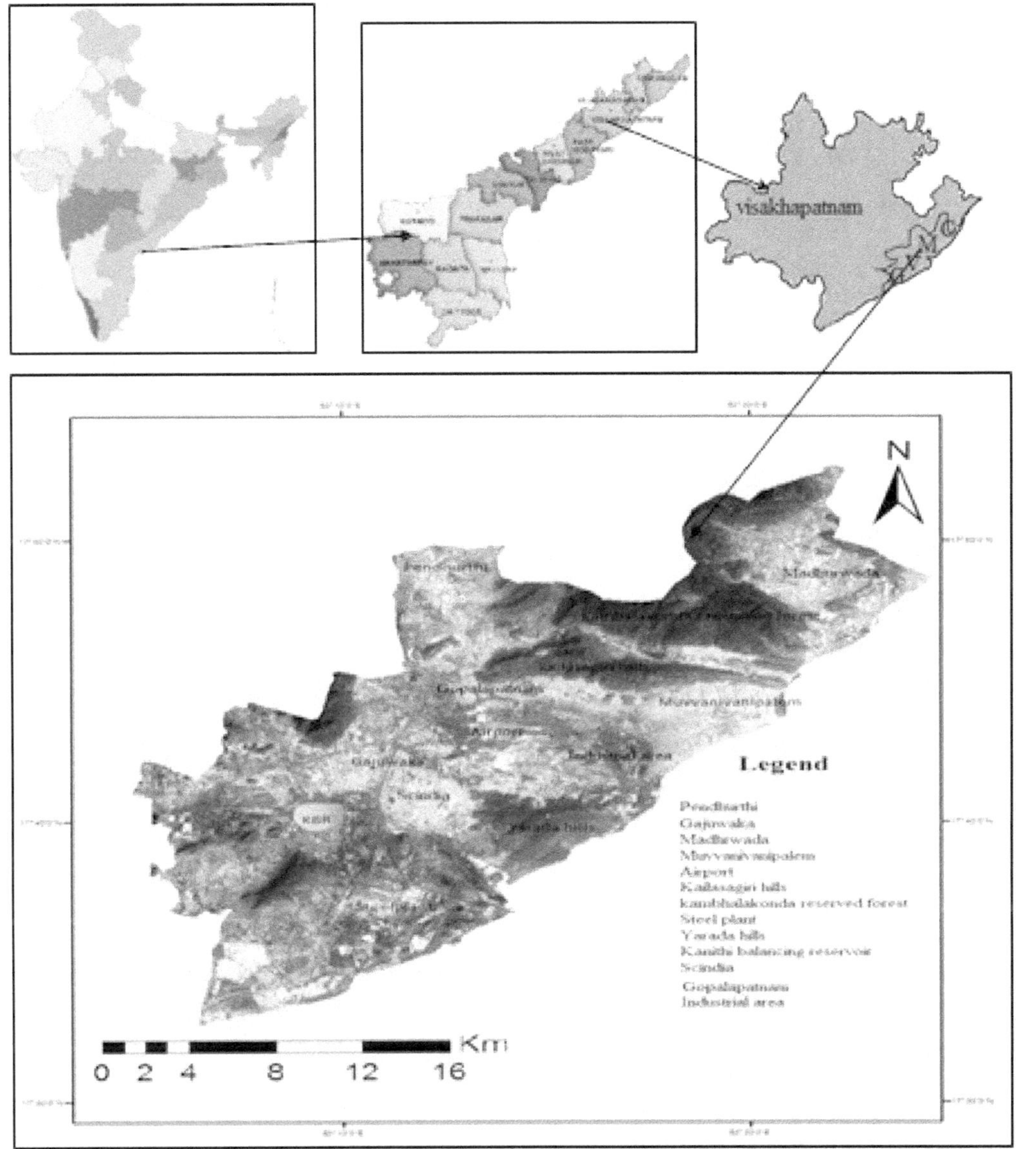

Fig. 1 Location map of the study area.

2.2 Data collection

In this study, the following data have been used for groundwater resources management

- The study area is covered in 65O / 1, 2, 4 & 6 of Survey of India toposheets on 1: 50,000 scales.
- Google Earth image of the study area is downloaded in the Elshayal Smart Web Online Solutions open source software with an elevation of 500m.

- LANDSAT 8 image of the study area is downloaded from the website the United States Geological Survey.
- Population for the years 2021 and 2031 is calculated
- Dug and bore well distribution data collected from GVMC.
- Agriculture data collected from Statistical Abstract of Visakhapatnam District, 2012-13.

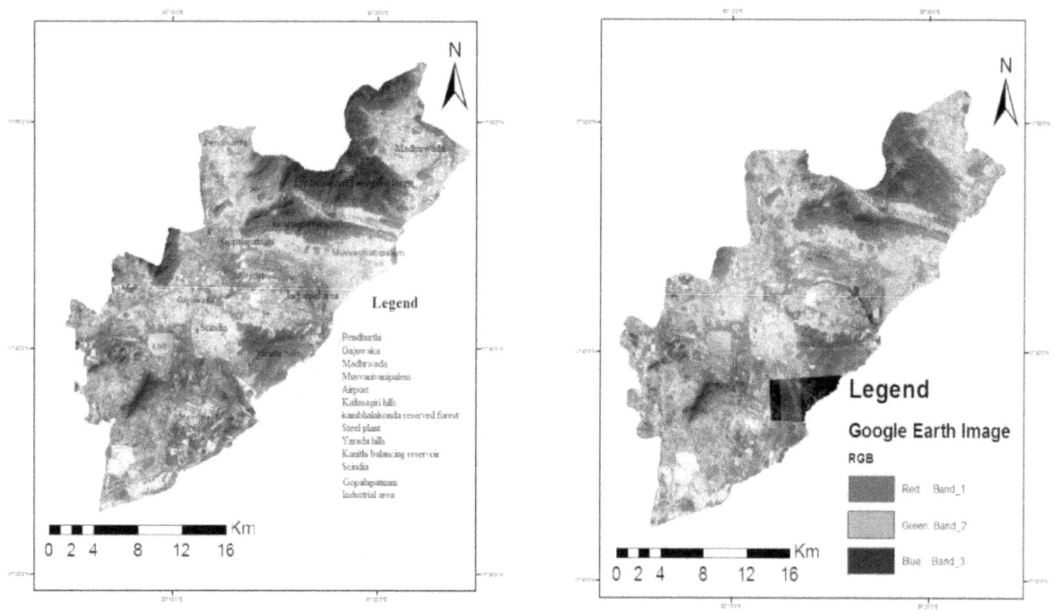

Fig. 1.2 LANDSAT-8 Image of the study area.

Fig. 1.3 Google Earth Image.

3. METHODOLOGY

Land use map, Soil map, NDBI, NDWI and Slope maps are used for identifying the suitable sites for water harvesting structures by overlaying. Overlaying of these maps are done by using "Intersect" from "Overlay" option of "Analysis Tools" in ArcGIS.

3.1 Identification of suitable sites for water harvesting structures

Normalised differential Built-up index

Normalised Difference Built-up Index, NDBI was one of the most successful of many attempts to simply and quickly identify built-up areas. And it remains the most well-known and used index to detect live built-up areas in multispectral remote sensing data. The development of the index was based on the unique spectral response of built-up lands that have a higher reflectance in SWIR wavelength range than in NIR wavelength range. NDBI is generated using the software Arc GIS 10.2. NDBI is a formula based tool used to extract the built-up area.

In this analysis, urban built-up is shown with medium gray color whereas other non-built-up is shown in dark color. The study area has reserved forests within the GVMC jurisdiction.

Similarly, there are about 6 major reservoirs are in GVMC area are shown in dark color. With this analysis built-up and non-built-up can be easily demarcated.

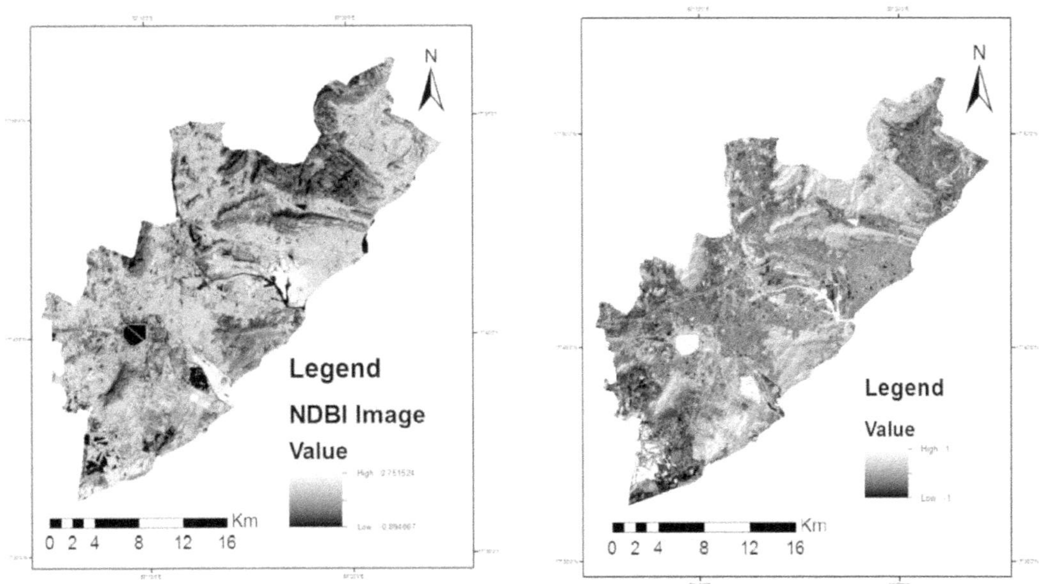

Fig. 1.4 NDBI. **Fig. 1.5** NDWI.

Modified Normalised Differential Water Index

Normalized Difference Water Index (NDWI) is the technique of extracting waterlogged (cropped areas inundated with water) areas in GVMC and it remains the most well- known and used index to detect live water bodies areas in multispectral remote sensing data. But the water features were less enhanced in the NDWI due to negative values of NDWI and, mixing of water with built up features. The water features were more enhanced with MNDWI and the values of MNDWI were positive for water features mixed with vegetation. The overall accuracy of waterlogged areas extracted from the MNDWI image was 96.9%.

In this analysis, all water bodies that include a major reservoir, sea creek and tanks were highlighted with bright white color. The extent of water in each water body can be easily assessed with this tool. The study is mainly focusing on groundwater recharging techniques therefore, the surface water bodies are the first order indicators through which groundwater recharge zones can be planned.

Land Use / Land Cover

Land Use / Land Cover information is the basic requisite for land, water and vegetation resources utilization, conservation and measurement. Information on existing land use/land cover and pattern of their spatial distribution forms the basis for any development planning. The current land use needs to be assessed for its suitability in the light of the land potential before suggestion alternate land use practices. The land cover maps derived by remote sensing are the basis to know the hydrological response units in the area. By image processing techniques, an image can be produced which depict some of the characteristics, primarily the cover types such as areas with vegetation, water bodies, areas with bare soils or outcrops and settlement areas etc. land use pattern of any watershed influences the runoff and evapotranspiration.

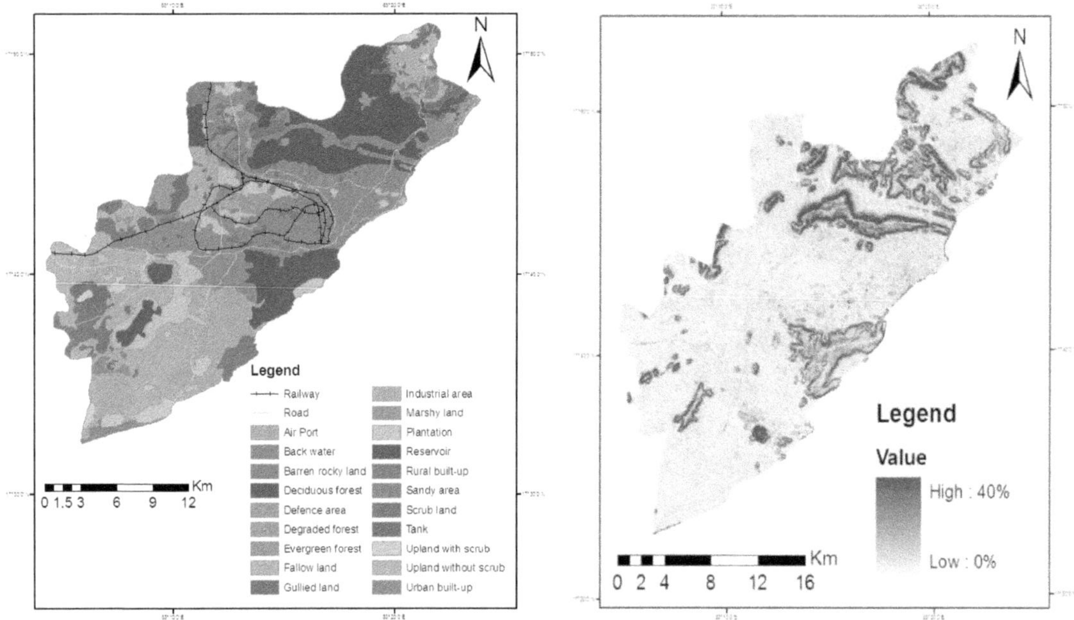

Fig. 1.6 Land Use/ Land Cover. **Fig. 1.7** Slope Map.

Slope map

Slope map shows the elevation of the particular area. The steepness value is bounded by maximum and minimum values. The slope of elevation can be depicted in two ways one is in degree and another in percent wise.

Soil map

Soil map is a geographical representation showing the diversity of soil types and soil properties. Soil maps are most commonly used for land evaluation; traditional soil maps typically show the only general distribution of soils. Generally, soil maps are used to simply identify soils and their properties, but sometimes required for more specific purposes, such as determining the suitability of a soil for particular crops, harvesting structures and land drainage capabilities of an area.

Soil map of GVMC area consists of different soils namely Brown clay soil, Brown gravel clay soil, Brown gravel loamy soil, Clay soil, Red clay soil, Red coastal clay soil, Red loamy soil and Sandy soil (fig.1.8).

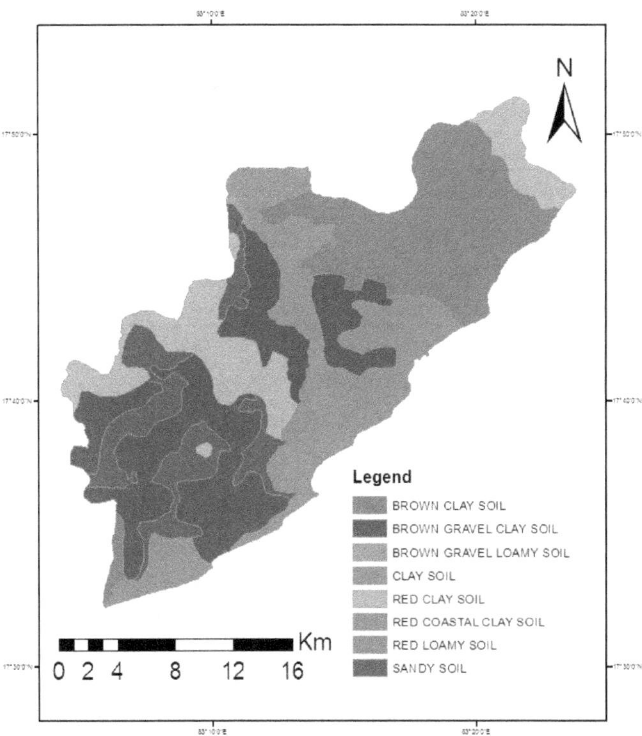

Fig. 1.8 Soil map.

3.2 Water Resource Management

With the worldwide depletion of groundwater and the intensified use around the world, particularly in many arid and semi-arid regions for irrigation and municipal use, there is no satisfactory approach to groundwater sustainability. The lack of and miss-management of this valuable resource has not only created serious groundwater pollution problems but has created present and/or future water supply problems (Larry W. Mays).

Growing urban areas are forced to increase their clean water supply in order to meet the demand from households and industries. When possible, increasing local pumping is an economically viable option (Calderhead et al. 2012). The scarcity of water is now the biggest threat in many parts of the world, especially in arid and semi-arid regions. Establishing a balance between water resources and the demands in a catchment scale basis could be one of the most important strategies to overcome this problem (Mohammad Taghi Dastorani & Samaneh Poormohammadi, 2012).

The following table shows the total requirement of water demand and total deficit in GVMC area.

Table 1 Water Requirement

S.No	Purpose of water demand	Quantity(MGD)
1.	The demand of water in study area	57.09 MGD
2.	Water supply for the domestic purpose	49.46 MGD
3.	Deficit of water for domestic purpose	7.632 MGD
4.	Water supplied for industrial purpose	17.97 MGD
5.	Demand for industrial purpose	19.16 MGD.
6.	Deficit of water for industrial purpose	1.98 MGD
7.	Total deficit	8.822MGD
8.	The demand of water for population 2784171 in the year 2021	91.86MGD
9.	The demand of water for population 4581075 in the year 2031.	151.154MGD

4. RESULTS & DISCUSSIONS

The sites for harvesting structures can also be identified by the visual interpretations of different maps like soil map, slope map, land use land cover map (supervised classification image), DEM (digital elevation model) map and drainage map. The overlapping of different raster images such as soil map, land use land cover map, DEM, supervised classification and slope give us suitable site for harvesting structures. There are different methods for water harvesting and some of these are

Contour Trenching

Contour trenching is an agricultural technique that can be easily applied in semi- arid areas to allow for water, and soil conservation, and to increase agricultural production. In GVMC, Kailasa and Yarada hill ranges covering a large area and trending linear to curvilinear shape. These hill ranges are demarcated as denudation hills which are covered by thin hydrophilic soils supporting luxuries vegetation. A part of these hill ranges Kambalakonda and surrounding hills are demarcated as reserved forests. Huge colluvium deposited at foot hills all along these hill masses. Contour trench along these foothills area may increase groundwater levels in the study area.

Afforestation

Afforestation is the transformation of wasteland into the forest or the woodland. Forests are fundamentally essential in water harvesting. The GVMC consists of fallow land, scrubland and barren land. These lands are chosen to grow up the forest which holds the soil and reduce the soil erosion. It also helps in raising the groundwater level. The area for afforestation is shown in fig.1.8.

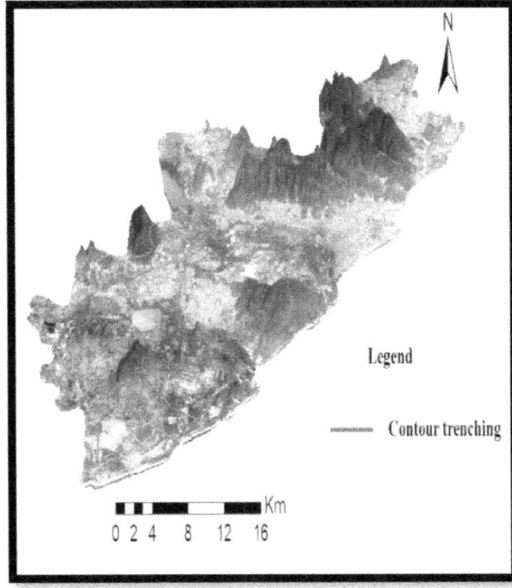

Fig. 1.9 Contour Trenching.

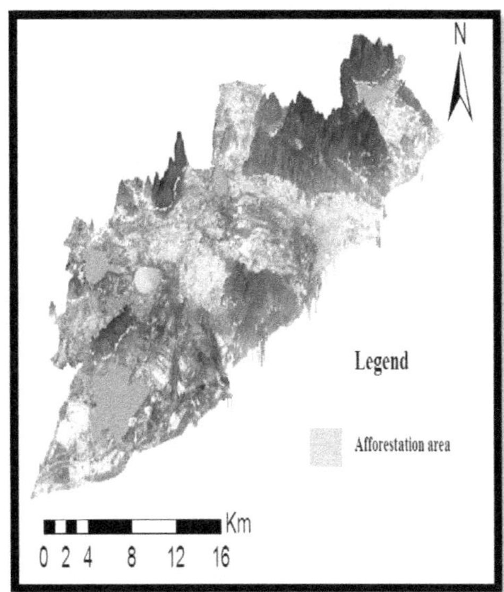

Fig. 1.10 Afforestation.

Check Dams

Check dam may be a temporary structure constructed with locally available materials. The various types are brush wood dam, loose rock dam, and woven wire dam. The main function of the check dam is to impede the soil and water removed from the watershed. A permanent check dam can be constructed using stones, bricks, and cement. Small earth work is also needed on both sides. The Check Dams will store or divert surplus water flowing to the sea at the end of monsoon.

Table 1.2 The table shows the location of Check Dams.

S.no	LULC	Name of the village
1	Fallow	Kommadhi
2	Plantation	Narava
3	Plantation	Kunmannapalem
4	Gullied	Vennela palem

Percolation tank

The percolation pond is a multipurpose conservation structure depending on its location and size. It stores water for livestock and recharges the groundwater. It is constructed by excavating a depression forming a small reservoir or by constructing an embankment in a natural ravine or gully to form an impounded type of reservoir.

Table 1.3 The table shows the location of Check Dams.

S.no	Land use/ land cover	Name of the village
1	Fallow land	Islampet
2	Gullied land	Steel Plant

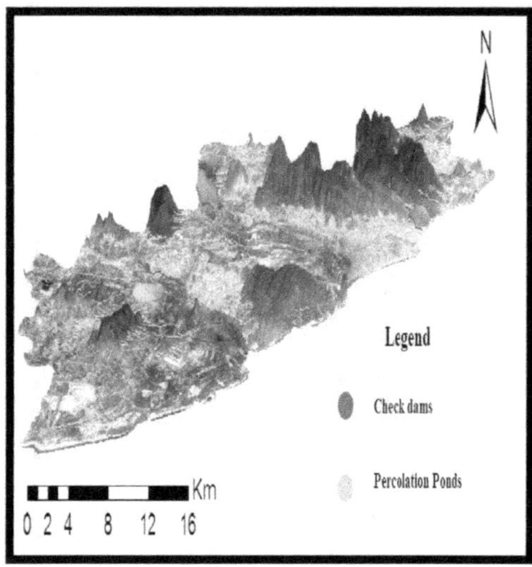

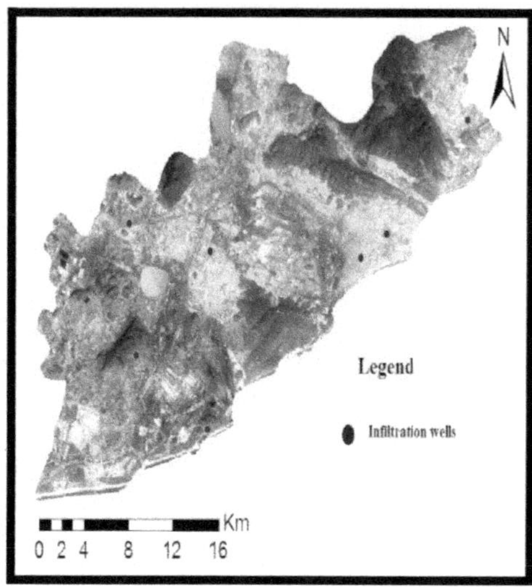

Fig. 1.11 The location of Check Dams, percolation ponds& Infiltration wells.

Infiltration well

Infiltration wells are also called as the interception wells, these are shallow wells which draw water into the subsurface. Infiltration wells are useful to intercept as large a quantity of water as possible. Infiltration wells are generally proposed in the river bank, riverbed to tap water from the unconfined aquifer. Infiltration wells are preferred where the minimum saturated thickness of the aquifer is at least 5m. This will provide a very good supply of water throughout the year. Even if the river dries up during the times of little rain, water will be available from the underground. The degree of purification will depend on the extent of contamination of the stream and on the soil type.

Table 1.4 The table shows the location of infiltration wells.

S.no	Land use/land cover	Name of the village
1	Fallow land	Madhurawada
2	Degraded forest	Radar observation centre
3	Built-up	Muvvani vani palem
4	Vacant land	Andhra University
5	Gullied land	Simhapuri colony
6	Plantation	Puttambotlapalem
7	Fallow land	Donkada
8	Upland with scrub	Islampet
9	Fallow land	Vedulla narava
10	Plantation	Vedulla narava

In the study area, there are 12 infiltration wells are recommended (Fig. 1.11). Site selection for infiltration is done on the basis of water demand and suitability of the topography and type of soil. A few wells exclusively suggested near the sea shore. This is due to arrest of sea water encroachment in the villages

5. CONCLUSIONS

Importance and need of water harvesting methods are discussed. Rainwater harvesting structures are suggested to store water and increase the water levels in GVMC area. Desiltation of tanks reis suggested for main tanks like KBR and Mudasarlova tanks. Sites for check dams are suggested.4 check dams are suggested. 2 Sites for percolation ponds are suggested. 10 infiltration wells are also suggested in all over the study area. Contour trenching technique is also suggested for the hill forests in GVMC area. Afforestation is also suggested in some areas of study area and the total area is 2479.88ha

6. REFERENCES

1. Ratnakar Dhakate & Gurunadha Rao.V. V. S, Anandagajapathi Raju.B, Mahesh.J, Mallikharjuna Rao .S. T & Sankaran. S 2012. Integrated Approach for Identifying Suitable Sites for Rainwater Harvesting Structures for Groundwater Augmentation in Basaltic Terrain.
2. Yogesh Bamne1, Patil2. Dr. K. A, Vikhe. S. D 2014. Selection of Appropriate Sites for Structures of Water Harvesting in a Watershed using Remote Sensing and Geographical Information System International Journal of Emerging Technology and Advanced Engineering. ISSN 2250-2459, ISO 9001:2008 Certified Journal, Volume 4
3. Singh.P.J, Darshdeep Singh, Litoria .P.K 2009. selection of suitable sites for water harvesting structures in soankhad watershed, Punjab using Remote sensing and Geographical information system (RS & GIS) Approach- a case study Research article
4. Harish Chand Prasad, Parul Bhalla and Sarvesh Palria 2014. Site Suitability Analysis of Water Harvesting Structures Using Remote Sensing and GIS – A Case Study of Pisangan Watershed, Ajmer District, Rajasthan The International Archives of the Photogrammetry, Remote Sensing and Spatial Information Sciences, Volume XL-8
5. Deivalatha.A, Senthikumaran.P and K.Ambujam.N 2014. Impact of testing of irrigation tanks on productivity of crop yield and profitability of farm income African journal of agricultural research
6. Jasrotia.A.S. (2008). Water Balance Approach for Rainwater Harvesting using Remote Sensing and GIS Techniques, Jammu Himalaya, India.
7. Hanqiu Xu., 2007. Extraction of Urban Built-up Land Features from Landsat Imagery Using a Thematicoriented Index Combination Technique Photogrammetric Engineering & Remote Sensing Vol. 73, No. 12, pp. 1381–1391.
8. Hordur V. Haraldsson, Harald U. Sverdrup, Salim Belyazid, Bjarni D. Sigurdsson AND Guomundur Halvorson 2007. Assessment of effects of afforestation on soil properties in Iceland, using Systems Analysis and System Dynamic methods
9. Li Xiaoyan, Zhang Ruiling, Gong Jiadong and Xie Zhongkui 2002 Effects of Rainwater Harvesting on the Regional Development and Environmental Conservation in the Semiarid Loess Region of Northwest China
10. Ranzi.R, Bochicchio.M and Bacchi.B 2002. Effects on floods of recent afforestation and urbanization in the Mella River (Italian Alps) Hydrology and Earth System Sciences

11. Girish Kumar. M, K. Agarwal. A, Rameshwar Bali 2008. Delineation of potential sites for water harvesting structures using Remote Sensing and Geographical Information System Journal of the indian society of remote sensing
12. B. Balakrishna. H, C. Jayaramu.K 2014. Identification of Potential Sites for Rainwater Harvesting using Remote Sensing and Geographical Information System
13. Lejiang Guo,Lei Xiao Lejiang Guo,Lei Xiao Hu 2010. Application of GIS and Remote Sensing Techniques for Water Resources Management, 2nd Conference on Environmental Science and Information Application Technology.

INFLUENCE OF VELOCITY ON MAJOR LIFT IRRIGATION PROJECTS

M Pratibha Satyabodha[1], A.R.N Sharma[2] and P. Raja Sekhar[3]

[1]Asst Prof, CED, Matrusri Engg College, [2]DEE,ISWR/I&ICAD, [3]Prof,CED,UCE,OU

ABSTRACT

Lift Irrigation Projects are drawing more attention in changed scenario owing to the non-feasibility of Dams and Barrages; Lift Irrigation Projects gained momentum. In the days to come ;Lift irrigation projects are going to play vital role in irrigation sector to fill the gap ayacut created due to non provision of conventional irrigation structure. There are many regions located far away from water source high elevation requiring water immediately and providing lift irrigation projects has become inevitable. With thrust and lift irrigation projects, many projects are being taken up with lifting huge quantity of water from rivers and high heads with lengthy pipes; which was not dealt before with those magnitudes. Discharge and pumping head are the parameters; which govern the planning of the project. To achieve economy, better control over them is required. Major Lift irrigation Project needs optimization in planning while fixing the pumping head and pump capacity along with length and diameter of pressure mains. For optimization, discharge if pumps can be fixed with effective usage of tanks enroute the alignment and deriving advantage of lesser demand crop during non-peak period. Pumping head can be reduced considerable by selecting the duty point of pumps with respect to water levels in the source where maximum operation period is expected. Function and efficiency of the Lift Irrigation Projects mainly depend on the performance of pumps and pressure mains of pumps act as heart of Lift Irrigation Project and pressure mains act as nerves of Lift Irrigation Project. In this study an attempt has been made to analyze influence of velocity on project cost referring a case study of Alisagar Lift Irrigation Project apart from Pumps, Pumping head, Pump Capacity, advantages of minimum number of rows, precautions taken in laying and design of pressure mains, selection of pipe materials, number of pumps for better production at lower costs and explore means to optimize the efficiency of planning ,design,construction,operation and maintenance of Major Lift Irrigation Project.

Keywords: Lift Irrigation Project, Planning and Design, Velocity, Pumping Head and Discharge, Pump Capacity and Project Cost.

Alisagar Lift Irrigation Project- A Case Study

The objective of the Alisagar Lift Irrigation Project is to irrigate 21,778 Ha (53,793 Acres) of Nizamsagar project command area situated in Nizamabad district. The project envisages lifting of 20.39 Cumecs (720 C/S) OF water from EL +321.50 m on river Godavari on right bank near Kosli village from the foreshore of Sriram Sagar Project. The project was executed on EPC basis at the cost of Rs.163.98 Cr. Alisagar Lift Irrigation Project is one among the major lift irrigation projects under taken on EPC basis by the irrigation department of Govt of Telengana.

Influence of Velocity in Pressure Mains on Pump and Project Cost

For every 0.50 m/s rise in Velocity of pipe, frictional loss rises by 75% to 100% with reduction of diameter by 11% to 13% only. Smaller diameter is economical during initial stage of construction but power consumption will be high. Higher diameter needs less power but with high initial cost. Hence, it is desirable to allow higher velocities in shorter length of pipes and lower velocities in lengthy pipes (particularly when the length of pipe is in KM) owing to the recurring power consumption annually.

Formulae

The following formulae are used in calculating discharge in each pipe and pump, different heads, head loss due to friction, hydraulic radius of pipe, velocity, pump capacities and quantity of steels etc.

Q_e(Discharge in pipe) = Q/N

Q_p(Discharge in pump) = Q/n

H_{smx}(Max Static Head) = Max water level of tank - Min water level of tank

H_{sn}(Normal Static Head) = Normal water level of tank – Normal water level in river

H_{smn}(Minimum Static Head) = Min water level of tank – Min water level in river

V(Velocity In pipe) = $Q_e/\pi/4D^2$

R(Hydraulic Radius of pipe) = D/4

h_f(Frictional loss per KM) = $L(1.1778\ V / C\ R^{0.63})^{1.853}$(William Hazen Formula)

H_f(Frictional loss in pipe) = $h_f\ L / 1000$

H_b(Total Bend Losses) = $n\ K\ V^2 /2g$

H_{max}(Max Pumping Head) = $H_{smx}+H_f+H_e+H_b$

H_{nrml}(Normal Pumping Head) = $H_s+H_f+H_e+H_b$

H_{min}(Min Pumping Head) = $H_{smn}+H_f+H_e+H_b$

Pump Capacity in KW = $9.81\ Q\ H / \eta$

Pump Capacity in hp = $9.81\ Q\ H / 0.746\ \eta$

Quantity of steel in Mtons = $\pi DLtp/1000$

Different analysis for calculating various parameters as mentioned above have been made using simple mathematical relations in **computer (Excel) program** by substituting the input parameters of Alisagar Lift Irrigation Project case study . But as far as this paper is concerned only velocity results have been tabulated which shows as velocity increases frictional losses and power requirement also increases hence project cost also increases. It is inferred that smaller diameter is economical during initial stage of construction but power consumption is very high. Higher diameter needs less power but with high initial cost.

Table

S.No	Particulars	various velocities in pipe (m/s)		
		1.5	2	2.5
1	Discharge in Cumecs	12.88	12.88	12.88
2	Dia of pipe in m	2.339	2.025	1.811
3	Length of pipe in Km	1.53	1.53	1.53
4	Thickness of pipe in mm	14	12	10
5	Hazen William Coefficient	140	140	140
6	Friction losses H_f in m	0.872	1.76	3.029
7	Quantity of steel in Mtons	247	183.4	136.7
8	Total KW for 4 pumps for normal pumping head	3392	3516	3696
9	Project cost in Millions	1959.836	1954.732	1990.292

As the velocity increases frictional losses, power consumption increases hence the project cost also increases.

RESULTS AND CONCLUSIONS

It can be noticed that increase in the velocity reduces the pressure main diameter thereby reduces capital cost, but at the same time increases power requirement as the analysis shows for 1.5 m/s project cost is 1959.836 millions where as 2.5 m/s it is 1990.292 millions, and power requirement is from 3392 KW to 3696 KW.

Future Work

As the pressure mains act as nerves of Lift Irrigation Project, care shall be taken for pipes when they are to be laid in BC soils, water logged area and at crossing of vagus/drains. Low velocity in the pipe would be economical for the projects with very lengthy pressure mains, however higher velocity may be permitted for the projects with shorter length. Larger diameter with less number of rows may be economical with respect to installation cost as well running cost. Adequate clearance shall be maintained between pipes for stability as well as maintenance purpose.

REFERENCES

1. Dracup,John.A (1996): "*The optimal use of a ground water and surface water system; A parametric linear programming approach*", Water Resources Center Contribution,No.107,University of California,Berkely,USA.
2. Dudley. N.J, (1971): "*Optimal intraseasonal Irrigation Water Allocation*", Water Resources Research; Vol.7,No. 4,Pp770-788
3. Hall. W.A.(1964):"*Optimum Design of a multiple- purpose reservoir; Journal of hydraulics division*", ASCE;Vol.90;No. HY4,Pp 141-149..
4. Hall, Warren. A and William Butcher (1968):"*Optimal Timing of Irrigation*", Journal of the Irrigation and Drainage Division, Proceedings of the ASCE, Vol.95, No.IRI Proc. Paper 6428, Pp 254-257.
5. Modi P.N(2008): *Irrigation Water Resources and Water Power Engineering*. Standard book house, India. Pp220
6. Roark.R.J(1974): *Formulas for Stress and Strain* McGraw Hill, New York, Pp851

7. Raju I S N (2008):" *Lift Irrigation Scheme a Solution to Tail end Ayacut Case Study of Alisagar and Guthpa L I Schemes"* A National Seminar on lift irrigation organized by Institution of Engineers (India), Hyderabad
8. Sharma. A.R.N (2008):"*Planning, Design and Optimization of major Lift Irrigation Schemes"* A National Seminar on lift irrigation organized by Institution of Engineers (India), Hyderabad
9. Spangler. M.G. (1956):"*Stresses in Pressure Pipelines and Protective Casing Pipes*", Proceedings of ASCE, Vol.82 Pp 1054
10. Subramanya K(1997) *Engineering Hydrology* McGraw Hill, New York,pp452.
11. Watkins. R.K., and Spangler M.G. (1958): "*Some characteristic of the Modulus of Passive Resistance of Soil –A Study in Similitude,* "Proc.37 of the Highway Research Board, HRB,Washington D C, Pp576.
12. "*Irrigation in India through ages*", Central Board of Irrigation and Power, New Delhi (1971).
13. "*A Guide for Estimation Irrigation Water requirements"*(1971): Issued by the Water Management Division, Ministry of Agriculture (Dept. of Agriculture), New Delhi-1
14. Andhra Pradesh State Irrigation Development Corporation, Hyderabad.

A REVIEW OF APPLICATION OF REMOTE SENSING AND GIS IN HYDROLOGICAL MODELING

Vangala Savinai[1] and Rathod Ravinder[2]

[1]Asst.Prof MLRIT Gandimaisamma , [2]Asst.Prof GRIET Bachupally

v.savinai@gmail.com; *rathod506ravinder@gmail.com*

ABSTRACT

In order to model the hydrological processes in a multivegetated watershed it is necessary to update the information regarding the response of these processes to various watershed parameters and acquire an in depth knowledge about the suitability of different hydrologic models for the simulation of these hydrologic processes. As most hydrologic models requires the application of Remote Sensing and GIS, it is also necessary to update information regarding the information of remotely sensed watershed information and GIS techniques by different models. Keeping this in view the present chapter deals with the review of significant contributions made by researchers in the field of hydrologic models, use of remote sensing and GIS for runoff estimation.

Application of Remote Sensing and GIS in Hydrological Modeling

The scope of hydrological applications has broadened dramatically with the advent of remote sensing and GIS. The remotely sensed data acquired from space borne platforms, owing to its wide synoptivity and multi-spectral acquisition offers unique opportunities for study of soils, land use/ land cover and other parameters required for hydrologic modeling of large areas(Schultz 1998). Remote Sensing and GIS are being widely used for solving environmental problems like degradation of land by water logging, soil erosion, contamination of surface and groundwater resources, deforestation, changes in ecological parameters and many more(Jasrotia 2002).

Tripathi, M.P. et. al. (2002) used remote sensing and GIS techniques for generation of land use, soil and contour map which were used for runoff modeling for a small watershed in Bihar.(Tripathi M. P. 2002)

Jasrotia, A.S. et. al. (2002) determined the rainfall-runoff relationship for the Tons watershed using SCS curve number technique by deriving the curve numbers through Remote Sensing and GIS techniques.(Jasrotia 2002)

Several other studies have been conducted in different parts of the world (Gupta 2001, Sharma 2001, Legesse 2003)for modeling hydrological components integrated with Remote Sensing and GIS. Kaur and Dutta (2002) highlighted the advantages of GIS based digital delineation of watersheds over conventional methods which is a pre-requisite for proper planning and development of watershed (Ravinder Kaur 2002).

Impact of Land use/ Land cover changes on hydrological response

In order to assess the impact of land use changes on hydrological response a case study was carried out by Sharma. et. al. (2001) for an area of 89.16 km^2 in Jasdan taluka (district) of Rajkot in Gujarat, India (Sharma 2001). The Curve number (CN) model was used for estimating runoff from the watershed. Satellite and other collateral data were used to derive information on land use, hydro geomorphology, soils and slope which were integrated to identify the problems and potential in the watershed and recommend measures for soil and water conservation. The impact of these conservation measures were assessed by computing runoff under alternative land use and management practices and it was observed that the runoff yield decreased by 42.88% of the pre-conservation value of the watershed.

Noorazuan (2003) evaluated the impact of urban land use- land cover change on hydrological regime for the period 1983 -1994 in Langat river basin, Malaysia, covering an area of 2271km^2.The study revealed that the landscape diversity of Langat significantly changed after 1980's and as a result, the changes also altered the Langat's streamflow response. Surface runoff increased from 20.35% in 1983-1988 to about 31.4% of the 1988-1994 events. Evidence from the research suggests that urbanization and changes in urban related land use-land cover could affect the stream flow behavior (Noorazuan 2003).

A study conducted by Ranjit Premlal De Silva et.al.(2000) to evaluate the impact land use/ land cover on hydrological regime revealed no obvious impacts of the changes of tree cover or any other land use changes on the river flow during rainy season. However obvious deviations were observed in the dry weather flow for both the sub catchments. The increase of the dry weather flow could be related to the increase of the tree cover and the reduction in canopy cover could be attributed to the decrease in dry weather flow at Kotmale. The study provided conclusive evidence that the increase in tree cover would positively contribute to the water yield in the catchments in addition to its protective role of the environment (Ranjit Premalal De Silva 2000)

Soil and Water Assessment Tool (SWAT)

Soil and Water Assessment Tool (SWAT) is a physically based distributed parameter model which have been developed to predict runoff, erosion, sediment and nutrient transport from agricultural watersheds under different management practices (Arnold J.G. 1998). SWAT is freely available which is linked to a GIS system (ArcGIS) through an interface that makes data processing and visualization easy. The model can simulate long periods, up to several years, operating with a daily time step. SWAT requires soils data, land use/management information and elevation data to drive flows and direct sub-basin routing. SWAT lumps the parameters into Hydrological Response Units (HRU) and storm runoff for each HRU is predicted with the CN equation.

SWAT is most versatile model. SWAT has been widely used in various regions and climatic conditions on daily, monthly and annual basis (Arnold J.G. 1998) and for the watershed of various sizes and scales (N. S. Kannan 2008, N. W. Kannan 2007). SWAT has been successfully used for simulating runoff, sediment yield and water quality of small watersheds for Indian conditions (Pandey 2008, Pandey V.K. 2005, M. P. Tripathi 1999, M. P. Tripathi 2003).

Application of SWAT in Hydrological Modelling

The development of SWAT model, its various components, operation, limitations has been described by Arnold. et. al. (1998) in his paper on "Large Area Hydrologic Modelling and Assessment Part-1: Model Development"(Arnold J.G. 1998). In his paper an overview has been made on SWAT model development which was developed mainly to assist water resource managers in assessing water supplies and non-point source pollution on watersheds and large river basins. The paper highlights the various components of the SWAT, methodology involved in simulating the various hydrological components, data requirement etc. The paper also gives an overview of the model limitations in simulating the various components of the hydrological cycle.

Singh et. al. (2004-08) made a comparative study for the Iroquois river watershed covering an area of 2137 sq. miles with the objectives to assess the suitability of two watershed scale hydrologic and water quality simulation model namely HSPF and AVSWAT2000. Based on the completeness of meteorological data, calibration and validation of the hydrological components were carried out for both the models. Time series plots as well as statistical measures such as Nash- Sutcliffe efficiency, coefficient of correlation and percent volume errors between observed and simulated streamflow values on both monthly and annual basis were used to verify the simulation abilities of the models. Calibration and validation results concluded that both the models could predict stream flow accurately (Singh 2004-08).

Spruill et. al. (2000) evaluated the SWAT model and parameter sensitivities were determined while modeling daily streamflow in a small central Kentucky watershed comprising an area of 5.5 km^2 over a two year period. Streamflow data from 1996 were used to calibrate the model and streamflow data from 1995 were used for evaluation. The model accurately predicted the trends in daily streamflow during this period. The Nash-Sutcliffe R^2 for monthly total flow was 0.58 for 1995 and 0.89 for 1996 whereas for daily flows it was observed to be 0.04 and 0.19. The monthly total tends to smooth the data which in turn increases the R^2 value. Overall the results indicated that SWAT model can be an effective tool for describing monthly runoff from small watersheds(Spruill 2000).

Fohrer et. al.(2002) applied three GIS based models from the field of agricultural economy (ProLand), ecology (YELL) and hydrology (SWAT-G) in a mountainous mesoscale watershed of Aar, Germany covering an area of 59.8 km2 with the objective of developing a multidisciplinary approach for integrated river basin management. For the SWAT –G model daily stream flow were predicted. The model was calibrated and validated followed by model efficiency using Nash and Sutcliffe test. In general the predicted streamflow showed a satisfying correlation for the actual land use with the observed data (Fohrer 2002).

Francos et. al. (2001) applied the SWAT model to the Kerava watershed (South of Finland), covering an area of 400 km^2.Various spatial data was used for the study. The temporal series comprised temperature and precipitation records for a number of meteorological stations, water flows and nitrogen and phosphorus loads at the river outlets. The model was adapted to the specific conditions of the catchment by adding a weather generator and a snowmelt sub model calibrated for Finland. Calibration was made against water flows, nitrate and total phosphorus concentrations at the basin outlet. Simulations were carried out and simulated results were compared with daily measured series and monthly averages. In order to measure the accuracy obtained, Nash and Suttcliffe efficiency coefficient was employed which indicated a good agreement between measured and predicted values (Francos 2001).

Eckhartd and Arnold (2001) outlined the strategy of imposing the constraints on the parameters to limit the number of interdependently calibrated values of SWAT. Subsequently an automatic calibration of the version SWAT-G of the SWAT model with a stochastic global optimization algorithm and Shuffled Complex Evolution algorithm is presented for a mesoscale catchment(Eckhartd 2001).

Tripathi et. al.(2003) applied the SWAT model for Nagwan watershed (92.46km^2) with the objective of identifying and prioritizing of critical sub-watersheds to develop an effective management plan. Daily rainfall, runoff and sediment yield data of 7 years (1992-1998) were used for the study. Apart from hydro-meteorological data, topographical map, soil map, land resources and satellite imageries for the study area were also used. The model was verified for the monsoon season on daily basis for the year 1997 and monthly basis for the years 1992-1998 for both surface runoff and sediment yield. Critical sub-watersheds were identified on the basis of average annual sediment yield and nutrient losses during the period of 3 years (1996-1998) and priorities were fixed on the basis of ranks assigned to each critical sub-watershed according to ranges of standard soil erosion classes. The study confirmed that the model could accurately simulate runoff, sediment yield and nutrient losses from small agricultural watersheds and can be successfully used for identifying and prioritizing critical sub-watersheds for management purpose(M. P. Tripathi 2003).

Daofeng et. al. (2004) made use of the SWAT model to simulate stream flow with validation and calibration of the observed yearly and monthly runoff data from the Tangnag hydrological station, and simulation results are satisfactory. Five land-cover scenario models and 24 sets of temperature and precipitation combinations were established to simulate annual runoff and runoff depth under different scenarios. The simulation shows that with the increasing of vegetation coverage annual runoff increases and evapotranspiration decreases in the basin. When temperature decreases by 2°C and precipitation increases by 20%, catchment runoff will increase by 39.69%, which is the largest situation among all scenarios(Daofeng 2004).

Chen J. et. al. (2004) used a distributed hydrological model SWAT to simulate the rainfall-runoff relationship of the Suomo Basin under different land covers in order to evaluate the impact of land-cover changes on runoff, evapotranspiration and peak flow. The results showed that if the land-cover changed from non-vegetation-cover to full-forest-cover scenarios, the runoff depth decreased, evaporation increased, while the reduced extent of runoff in dry season was less than that in rainy season, and in the first rainy season, the reduced extent of runoff was more than that in the second rainy season. With the same recurrent flood flow, the peak flow value under full-forest-cover scenario was 31.2% less than that under non-vegetation-cover scenario. The effect of land-cover between current cover and optimum cover on hydrology was small for large storm and big for small storm events (Chen J. 2004).

Chen J. et. al. (2005) conducted a study on a mesoscale river basin, the Suomo Basin that is located on the upper reaches of the Yangtze River. Land covers in the basin in the years 1970, 1986 and 1999 were mapped. A lumped hydrologic model, CHARM, and a distributed hydrologic model, SWAT, were used to model the impacts of both land-cover change and climate variation on river runoff during the past four decades. The results show that the contribution of climate variation to the change of runoff regime makes up 60%–80%, while that of land cover changes only 20% (Chen 2005).

Hua Guo et. al. (2008), used the SWAT model to examine the climate and land-use and land-cover effects on hydrology and streamflow in the Xinjiang River basin of the Poyang Lake. A major finding of this study is that the climate effect is dominant in annual streamflow. While land-cover change may have a moderate impact on annual streamflow it strongly influences seasonal streamflow and alters the annual hydrograph of the basin. Because of the vegetation and associated seasonal variations of its impact on evapotranspiration, increase of forest cover after returning agricultural lands to forest reduces wet season streamflow and raises it in dry season, thus reducing flood potentials in the wet season and drought severity in the dry season. On the other hand, losing forests increases flood potential and also enhances drought impacts. Results of this study improve our understanding of hydrological consequences of land-use and climate changes, and provide needed knowledge for effectively developing and managing land-use for sustainability and productivity in the Poyang Lake basin(Hua Guo 2008).

Faith Githui et. al. (2009), used SWAT model to investigate the impact of land-cover changes on the runoff of the River Nzoia catchment, Kenya. The model was calibrated against measured daily discharge, and land-cover changes were examined through classification of satellite images. Land-cover change scenarios were generated, namely the worst- and best-case scenarios. Historical land-cover change results showed that agricultural area increased from 39.6 to 64.3% between 1973 and 2001, while forest cover decreased from 12.3 to 7.0%. A comparison between 1970–1975 and 1980–1985 showed that land-cover changes accounted for a difference in surface runoff ranging from 55 to 68% between the two time periods. The land-cover scenarios used showed the magnitude of changes in runoff due to changes in the land covers considered. Compared to the 1980–1985 runoff, the land-cover scenarios generated changes in runoff of about −16% and 30% for the best and worst case scenarios respectively(Faith Githui 2009).

Stehr et. al. (2010), followed a multidisciplinary approach for analysis of the effect of changes in land use patterns on the hydrologic response of the Vergara watershed (4340 km^2) located in central Chile. Probable future land use scenarios were generated using heuristic rules and logistic regression models, in order to identify and represent the main pressure on the watershed, namely forestation of extensive areas used for agriculture with rapid growing exotic species. The hydrologic response of the watershed was computed with a physically based distributed precipitation-runoff model, which was calibrated and validated for the current period. Results showed that mean annual discharge increase with agricultural land use and diminish with introduced forest coverage. Thus, forestation of areas with introduced species like *Pinus radiata* and *Eucalyptus globulus* might be regulated in order to protect the water resources of the watershed(Stehr 2010).

Chesheng Zhan et. al. (2011), used Soil andWater Assessment Tool (SWAT) to simulate the impacts of LUCC on the run-off yield in the Bai River catchment—upstream of the Miyun Reservoir basin in northern China. The investigation was conducted using two 6-year historical streamflow records: from 1986 to 1991 and from 2000 to 2005. A split sample procedure was used for model calibration and validation. The data from 1986 to 1988 and from 2000 to 2002 were used for calibration, while those from 1989 to 1991 and from 2003 to 2005 for validation. The SWAT calibration was based on monthly measured discharge at Zhangjiafen station at the catchment outlet from Bai River catchment. Additionally, the influence of LUCC on the surface run-off was distinguished from that of climate change on the surface runoff through SWAT scenarios modeling, the two-way analysis of variance (ANOVA), and the rainfall–run-off double-mass analysis in the Bai River catchment. The results indicated that the SWAT model could be used successfully to accurately simulate run-off yield and different LUCC patterns affecting water quantity in this catchment. During calibration for the two periods the simulated monthly run-off satisfactorily matched the observed values, with the Nash–Sutcliffe coefficient >0.9 and 0.7 and a coefficient of determination of 0.9 and 0.65 at the outlet station (Zhangjiafen station), while during validation for the two periods the obtained values were 0.85, 0.65 and 0.9, 0.65, respectively. During the period of 1986–91, both the SWAT scenarios modeling and the analysis of the two-way ANOVA method showed that LUCC and climate change had some impact on run-off, and the impact of climate change was more significant than that of LUCC. Compared with the period during 1986–91, the run-off yield in the period during 2000–05 significantly decreased. The obtained results from the rainfall–run-off double-mass analysis indicate that since 1998 LUCC has had an increasing influence on the run-off, while the response of the run-off to rainfall has been decreasing. Since 1998, the LUCC has been a major driving force for run-off change in Bai River catchment (Chesheng Zhan 2011).

Phan D.B. et. al. (2011), implemented "Soil and Water Assessment Tool (SWAT)" model to examine the effects of land use change scenarios; associated with crop rotations and special cultivation techniques most susceptible to erosion; exert on runoff discharge and sediment yield from Song Cau catchment in Northern Viet Nam. All scenarios' simulations resulted in a decrease of soil losses and sediment yield comparing to the current land use status. SWAT successfully predicted soil losses from different HRUs that caused significant sediment yield, and it predicted explicitly the consequences of non-structural mitigation measures against erosion (Phan D.B. 2011).

Yank S.K. et. al. (2012), selected the four major streams in Jeju Island for the hydrologic analysis by the The Soil and Water Assessment Tool (SWAT) model according to the change in land use. A land use data from 1975 to 2000 from landsat satellite images provided by the Ministry of Environment and Arcview program was used. Due to the change in land-coverage in four major streams between the past and the present, the areas of impermeable land in the lower area of the streams were generally extended approximately two times higher than in the past. Accordingly, it was proved that the amount of direct runoff has been increasing by at least 1 to 6%. Especially, in the lower part of Oaedo stream, the increase in surface discharge was highest. The quantitative hydrological analysis due to land use change by SWAT model is thought to be a good approach for identifying the impact of land use in Jeju island (Yang S. K. 2012).

In a typical irrigation practice the soil water content is maintained at or near field capacity, which can affect hydrological processes such as deep percolation and runoff in different ways. Michel Rahbeh M. et. al, (2013) evaluated these effects for a partially irrigated, small watershed in the Canadian prairie using the Soil Water Assessment Tool (SWAT). The watershed was defined by upstream (inlet) and downstream (outlet) monitoring stations located along a short reach of a river. SWAT was calibrated and validated using the net flow between the upstream (inlet) and downstream (outlet) locations because the watershed was defined by this reach. Runoff contribution to the incremental streamflow was minimal, as indicated by the reduced values of the calibrated initial curve number (CNII) (CNII range 49–59). Irrigation activity increased runoff depth but the differences between irrigated and non-irrigated areas were not statistically ($\alpha \leq 0.05$) significant. The low runoff contribution was also corroborated by the streamflow record that demonstrated the low potential for runoff generation in the watershed. The only apparent runoff occurred after a major rainfall event of a cumulative depth of more than 200 mm. The modelling also showed that the shallow aquifer discharge was the main streamflow constituent. Precipitation during May to July was responsible for 70–90% of the seasonal deep percolation. By the end of the season the deep percolation from the irrigated areas exceeded that of the non-irrigated areas by up to 70%. Thus, the irrigation activity in the watershed did not change the water partitioning among the existing hydrological pathways but had temporal effects on the magnitudes of runoff and, more importantly, deep percolation and the subsequent groundwater discharge in the main reach (Rahbeh M. 2013).

Wang Xue et. al.(2013) using 3S technology, and based on the analysis and prediction of the land use change in Baimahe basin, a SWAT model was established to study the runoff response of the basin under the scenarios of different land use. The contribution coefficients of the main land use types in the basin to the runoff depth were calculated. From 1987 to 2017, the main land use types in the basin were farmland, construction land, forestland, and shrub land, occupying 96% or more of the total land area, while the grassland, waters, and unused land only had a smaller proportion. The four main land use types had different effects on the runoff depth. The contribution coefficient of forestland, shrub land and construction land to the runoff depth was 2.61, 0.38, and 0.34 mm·km-2, respectively, implying that these three land use types had positive effects on the runoff depth in this basin. On the contrary, the contribution coefficient of farmland was -0.11 mm·km-2, implying that farmland had negative effect on the runoff depth (Wang Xue 2013).

Sajikumar N. et. al. (2015), assessed the effect of land use and land cover on the runoff characteristics of two watersheds in Kerala, India. He also assessed how the change in land use and land cover in the last few decades affected the runoff characteristics of these watersheds. It is seen that the reduction in the forest area amounts to 60% and 32% in the analysed watersheds. However, the changes in the surface runoff for these watersheds are not comparable with the changes in the forest area but are within 20%. Similarly the maximum (peak) value of runoff has increased by an amount of 15% only. The lesser (aforementioned) effect than expected might be due to the fact that forest has been converted to agricultural purpose with major portion as plantations which have comparatively similar characteristics of the forest except for evapo-transpiration. The double sided action (increase in evapo-transpiration owing to species like rubber and increase percolation due to its plantation method by using terracing) might be the reason for relatively smaller effect of the land use change, not commensurate with the changes in the forest area amounting to 60% and 32% for Manali and Kurumali watersheds respectively. Water harvesting methods like rain harvesting ditches can be made mandatory where species with high evapo-transpiration are grown. This action shall enhance the groundwater percolation and shall counter act the effect due to high evapo-transpiration (Sajikumar N. 2015).

Harsh Vardhan Singh et.al. (2015) explored the performance of the soil and water assessment tool (SWAT) in predicting water quality and quantity in response to changing LULC in a coastal watershed in Alabama, USA. Using the 1992 LULC as the input, the model was calibrated and validated for flow for the period 1990–1998, and for total suspended solids (TSS), nitrate (NO_3^-), and organic phosphorus for the period 1994–1998 at several sites within the watershed. The model was then driven with the 2008 LULC data and its performance in predicting flow and TSS, NO_3^-, and total-P loads during the period 2008–2010 was evaluated (post-validation). SWAT showed good performance in predicting changes in flow and water quality during the post-validation period. The study also highlighted the importance of using the most up-to-date LULC data for effectively predicting the impacts of LULC changes on water quality (Harsh Vardhan Singh 2015).

Kim et. al. (2015) examined the potential effects of urban growth on streamflow in the Gyungan River watershed, Korea, using urban containment scenarios. First, two scenarios (conservation and development) were established, and SLEUTH model was adapted to predict urban growth into the year 2060 with 20 years interval under two scenarios in the study area. Urban growth was larger under scenario 2, focusing on development, than under scenario 1, focusing on conservation. Most urban growth was predicted to involve the conversion of farmland, forest, and grasslands to urban areas. Streamflow in future periods under these scenarios was simulated by the Soil and Water Assessment Tool (SWAT) model. Each scenario showed distinct seasonal variations in streamflow. Although urban growth had a small effect on streamflow, urban growth may heighten the problems of increased seasonal variability in streamflow caused by other factor, such as climate change. This results obtained in this study provide further insight into the availability of future water resource and can aid in urban containment planning to mitigate the negative effects of urban growth in the study area (Kim 2015).

Manisha Paul et. al.(2016) analyzed changes in hydrology between two recent decades (1980s and 2010s) with the Soil and Water Assessment Tool (SWAT) in three representative watersheds in South Dakota: Bad River, Skunk Creek, and Upper Big Sioux River watersheds. Two SWAT models were created over two discrete time periods (1981-1990 and 2005-2014) for each watershed. National Land Cover Datasets 1992 and 2011 were, respectively, ingested into 1981-1990 and 2005-2014 models, along with corresponding weather data, to enable comparison of annual and seasonal runoff, soil water content, evapotranspiration (ET), water yield, and percolation between these two decades. Simulation results based on the calibrated models showed that surface runoff, soil water content, water yield, and percolation increased in all three watersheds. Elevated ET was also apparent, except in Skunk Creek watershed. Differences in annual water balance components appeared to follow changes in land use more closely than variation in precipitation amounts, although seasonal variation in precipitation was reflected in seasonal surface runoff. Sub basin-scale spatial analyses revealed noticeable increases in water balance components mostly in downstream parts of Bad River and Skunk Creek watersheds, and the western part of Upper Big Sioux River watershed. Results presented in this study provide some insight into recent changes in hydrological processes in South Dakota watersheds (Manashi Paul 2016).

The review indicated that SWAT is capable of simulating hydrological processes with reasonable accuracy and can be applied to large ungauged basin. Therefore to assess the impact of temporal changes of land use/land cover on runoff, ArcSWAT2012 with ArcGIS interface was selected for the present study.

REFERENCES

1. Arnold J.G., R. Srinivasan, R. S. Muttiah and J. E. Williams. "Large Area Hydrological Modelling and Assessment Part-1: Model Development." Journal of the American Water Resources Association, 1998.
2. Chen J., Li Xiubin. "Simulation of hydrological response to land2cover changes." Chinese Journal of Applied Hydrology, 2004: 15 (5):833~836.

3. Chen, J., Li, X. & Zhang, M. Sci. China Ser. D. "Simulating the impacts of climate variation and land-cover changes on basin hydrology: A case study of the Suomo basin." Science in China Series D: Earth Sciences, 2005: 48: 1501.
4. Chesheng Zhan, Zongxue Xu, Aizhong Ye and Hongbo Su. "LUCC and its impact on run-off yield in the Bai River catchment—upstream of the Miyun Reservoir basin." Journal of Plant Ecology, 2011: Volume 4, Issue 1-2, Pp. 61-66.
5. Daofeng, L., Ying, T., Changming. "Impact of land-cover and climate changes on runoff of the source regions of the Yellow River." Journal of Geographical Sciences, 2004: 14: 330.
6. Eckhartd, K., J.G. Arnold. "Automatic calibration of a distributed catchment model." Journal of Hydrology, 2001: 103-109.
7. Faith Githui, Francis Mutua and Willy Bauwens. "Estimating the impacts of land-cover change on runoff using the soil and water assessment tool (SWAT): case study of Nzoia catchment, Kenya ." Hydrological Sciences Journal, 2009: Volume 54, Issue 5.
8. Fohrer, N., D. Moller and N. Steiner. "An interdisciplinary modeling approach to evaluate the effects of land use change." Physics and Chemistry of the Earth, 2002: 655-662.
9. Francos, A., G. Bidoglio, L. Galbiati, F. Bouraoui, F. J. Elorza, S. Rekolainen, K. Manni, and K. Granlund. "Hydrological and Water Quality Modelling in Medium sized coastal basin." Physics and Chemistry of the Earth, 2001: 47-52.
10. Gupta, P. K., T. Das, N.S. Raghuwanshi, S. Dutta and S. Panigrahy. "Hydrological Modelling of canal command using Remote Sensing and GIS." www.gisdevelopement.net, 2001.
11. Harsh Vardhan Singh, Latif Kalin, Andrew Morrison, Puneet Srivastava, Graeme Lockaby, Susan Pan. "Post-validation of SWAT model in a coastal watershed for predicting land use/cover change impacts." Hydrology Research, 2015: Volume 47, Issue-6.
12. Hua Guo, Qi Hu, Tong Jiang. "Annual and seasonal streamflow responses to climate and land-cover changes in the Poyang Lake basin, China." Journal of Hydrology, 2008: Volume 355, Issues 1–4, Pages 106–122.
13. Jasrotia, A.S. "Rainfall- Runoff and Soil Erosion Modeling using Remote Sensing and GIS technique- A case study of Tons watershed." Journal of Indian Society of Remote Sensing, 2002: 167-179.
14. Kannan, N., Santhi, C. and Arnold, J.G. "Development of an automated procedure for estimation of the spatial variation of runoff in large river basins." Journal of Hydrology, 2008: 1-15.
15. Kannan, N., White, S.M., Worrall, F. and Whelan, M.J. "Hydrological modelling of a small catchment using SWAT-2000 - Ensuring correct flow partitioning for contaminant modelling." Journal of Hydrology, 2007: 64-72.
16. Kim, Jinsoo, Park and Soyoung. "Potential Effects of Urban Growth under Urban Containment." Journal of the Korean Society of Surveying, Geodesy, Photogrammetry and Cartography, 2015: Vol. 33, No. 3, 163-172.
17. Legesse, D., C. Vallet-Coulomb and F. Gasse. "Hydrological response of a catchment to climate and land use changesin Tropical Africa: case study of South Central Ethiopia." Journal of Hydrology, 2003: 67-85.
18. Manashi Paul, Mohammad Adnan Rajib and Laurent Ahiablame. "Spatial and Temporal Evaluation of Hydrological Response to Climate and Land Use Change in Three South Dakota Watersheds." Journal of the American Water Resources Association, 2016.
19. Noorazuan, M. H.,Ruslan Rainis, Hafizan Juahir, Sharifuddin, M. Zain, Nazari Jaafar. "GIS application in evaluating Land use-Land cover change and its impact on hydrological regime in Langat river basin, Malaysia." Map Asia Conference 2003, www.gisdevelopment.net, 2003.

20. Pandey V.K., Panda, S.N. and Sudhakar, S. "Modelling of an Agricultural Watershed using Remote Sensing and a Geographic Information System." Biosystems Engineering, 2005: 331-347.
21. Pandey, V.K., Panda, S.N. and Pandey, A. "Evaluation of effective management plan for an agricultural watershed using AVSWAT model, remote sensing and GIS." Environmental Geology, 2008.
22. Phan D.B., C.C. Wu and S.C. Hsieh. "Land Use Change Effects on Discharge and Sediment." Journal of Environmental Science and Engineering, 2011: 92-101.
23. Rahbeh M., David Chanasyk and Jim Miller. "Modelling the effect of irrigation on the hydrological output from a small prairie watershed." Canadian Water Resources Journal, 2013: volume 38, Issue 4, Pp. 280-295.
24. Ranjit Premalal De Silva, Madusha Chandrasekhar. "Impacts of landuse changes on Hydrological Regime- A case study of Randenigala & Kotmale catchments in Sri Lanka." Map Asia 2000, 2000.
25. Ravinder Kaur, D Dutta. "GIS-based digital delineation of watershed and its advantage over conventional manual method-A Case Study on watersheds in Hazaribagh and Bankura Districts of Jharkhand and West Bengal." Ind. J. Soil Cons, 2002: 30; 1-7.
26. Sajikumar N., Remya R.S. "Impact of land cover and land use change on runoff characteristics." Journal of Environmental Management, 2015: 460-468.
27. Schultz, G. A. "Remote Sensing in Hydrology." Journal of Hydrology, 1998: 239.
28. Sharma, T., P. V. Satya Kiran, T. P. Singh , A. V. Trivedi and R. R. Navalgund. " Hydrological response of a watershed to land use changes: a remote sensing and GIS approach." International Journal of Remote Sensing, 2001: 2095-2108.
29. Singh, J., H. Verman Knapp and M. Demissie. "Hydrologic Modelling of the Iroquois River Watershed using HSPF and SWAT." Illinous State Water Survey Contract Report, 2004-08.
30. Spruill, C. A., S. R.Workman and J. R Taroba. "Simulation of daily and monthly stream discharge from small watersheds using the SWAT model." Transaction of ASAE, 2000: 1431-1439.
31. Stehr, A., Aguayo, M., Link, O., Parra, O., Romero, F., and Alcayaga, H. "Modelling the hydrologic response of a mesoscale Andean watershed to changes in land use patterns for environmental planning." Hydrol. Earth Syst. Sci., 2010: 14, 1963-1977.
32. Tripathi M. P., R. K. Panda, S. Pradhan and S. Sudhakar. "Runoff Modelling of a Small watershed using satellite data and GIS." Journal of Indian Society of Remote Sensing, 2002: 39-52.
33. Tripathi, M.P., Panda R.K. and Raghuwanshi N.S. "Identification and Prioritisation of Critical Sub-watersheds for Soil Conservation Management using the SWAT Model." Biosystems Engineering, 2003: 365-379.
34. Tripathi, M.P., Panda, R.K. and Raghuwanshi, N.S. "Runoff estimation from a small watershed using SWAT model." Hydrological Modelling, 1999.
35. Wang Xue, ZHANG Zu-lu and NING Ji-cai. "Runoff response to land use change in Baimahe basin of China based on SWAT model." Chinese Journal of Ecology, 2013: 32(1):186-194.
36. Yang S. K., W. Y. Jung, W. K. Han and I. M. Chung. "Impact of land-use changes on stream runoff in Jeju Island, Korea." African Journal of Agricultural Research, 2012: 7(46), pp. 6097-6109.

AN ASSESSMENT OF RAINFALL EROSION POTENTIAL IN AKOLA FROM DAILY RAINFALL RECORDS

P.D. Naitam and S. P. Shinde

Assistant Professor, College of Agricultural Engineering & Technology, Jalgaon (J.)

Assistant Professor, Sau.K.S.K.' KAKU' College of Agriculture, Beed

pranalinaitam2@gmail.com

ABSTRACT

This study observed rainfall graph, which are measured rainfall by automatic recording rain gauge collected from the Central Research Station, Dr. PDKV, Akola and analyzed the a rainfall data for period of eight year from 2004 to 2011. A more simple quick and time saving method for estimating Erosion Index is needed for soil conservation planner to predicts this information factor of USLE. The kinetic energy from these storm for 8 years of daily rainfall data for duration 5,10,15,30 and 60 minutes. The daily precipitation index for daily, monthly and annual index values for 5,10,15,30 and 60 minutes. Relation between the erosion index and precipitation index is obtained for 5,10,15,30 and 60 minutes for Akola Station.

Daily PI and EI values of 5,10,15,30 and 60 minutes selected time intervals were computed for the period of 8 years from 2004 to 2011 by using Raghunath's method for 144 erosive storms. It is found that, 144 erosive storm correlation co-efficient for 5 minutes is 0.73, for 10 minutes 0.42, for 15 minutes 0.50, for 30 minutes 0.77 and for 60 minutes 0.48 are resulted for the observations.

Keywords: *Erosion index, precipitatation index, USLE etc.*

1. INTRODUCTION

Increase in agriculture production is possible through modern methods. But these advances in science will not be effective unless there is enough good land for farming. Soil conservation was started around in the thirties of this century for the purpose of the future of world food production and therefore to main survived basic erosion processes and the way in which erosion research had to better understanding of the mechanics of soil erosion processes and the way in which erosion can be controlled. Land degradation is loss of land productivity, quantitively or qualitatively through various processes such as erosion, wind blowing, salinization, water logging, depletion of nutrients, deterioration of soil structure and pollution.

Total historic soil losses have been estimated at 2 billion hectare. The present arable area of the world being about 1.5 billion hectares. A figure of 5 to 7 million hectares of soil loss per year, as a result of land degradation has been put forward. If this figure is correct it would appears that 200 million hectares of additional land estimated to be required to produce sufficient food for the world population.

Universal soil loss equation is a unique tool, which is extensively used for predicting soil loss in a given set of condition. The existing method of estimating erosion index is laborious and time consuming, involved the preparation of detailed intensity tables, estimating kinetic energy there form and calculating EI values for individual storm. A more simple quick and time saving method for estimating EI value is needed for soil conservation planner to predict this information factor of USLE. This study revealed that to reduce the laborious computational process arriving at EI values without sacrificing accuracy, from the total daily precipitation and maximum intensify date for selected time intervals.

2. MATERIALS AND METHOD

The method adopted while analyzing rainfall data. Standard methods are used to compute kinetic energy, erosion index distribution.

2.1 Description of project area

Location

Dr. Panjabrao Deshmukh Krishi Vidyapeeth, Akola is entire Vidarbha region which is located in eastern Maharashtra and comprises eleven district. It lies in between $17^{0}57'$- $21^{0}46'$N Latitude and $75^{0}57'$ - $80^{0}59'$E Longitude having geographical area of 97.23 lakh ha which is 31.61% of Maharashtra. The region is agro-climatically heterogeneous and geographically which around 90% of area under rainfed farming. The region classified under agro climatic zones (NARP) viz. Central Vidarbha (AZ-97) and Eastern Vidarbha (AZ-98). Annual rainfall varies from 700 to 950 mm in western parts to more than 1250 mm in eastern parts.

2.2 Description of central research station

The region is classified as hot moist semi-arid climate with medium and deep clayey black soils (Shallow loamy to clayey black soils as inclusion). Akola centre receives an average (1971-2000) annual rainfall of 811 mm in 43 rainy days. The average rainfall during mansoon season (June to September) is 687 mm and ranges from 352 to 1155 mm. The mean daily relative humidity during mansoon, winter and summer is 73, 54 and 36 percent respectively. The rate of evaporation reaches upto 25.4 mm per day during May. The wind speed reaches to 35.3 km per hour during the same month. The major crops grown in the region are cotton, soybean, pigeon pea, green gram and black gram during kharif season and chickpea, safflower and sunflower during rabbi season.

2.3 Data Collection

Rainfall chart recorded by automatic recording rain gauge were collected from Agro ecology and Environment Centre Dr. PDKV, Akola for analysis.

2.4 Computation of Rainfall Erosion Index

Rainfall erosion potential is precisely estimated by totaling erosion index values (EI) of all the storms in a given period. The erosion index of a rain storm is one hundredth of the product of kinetic energy of rain storm and its selected durations maximum intensity (I_5, I_{10}, I_{15}, I_{30}, and I_{60}) which is the greatest average intensity experienced in any selected duration period during a storm.

There are mainly two well established methods namely rainfall erosion EI_{30} and KE > 25 mm index method for estimating the erosivity of rainfall for each storm.

2.4.1 Computation of kinetic energy and erosion Index

Kinetic energy for each storm evaluated by the equation by Wischmeier and Manning (1974). The equation in metric units is,

$$KE = 210.3 + 89 \log I \quad \ldots (2.1)$$

Where,

KE = Kinetic energy, metre tones per hectare centimeter

I = Maximum storm intensity, centimeters per hour for the selected time duration

Erosion index (EI) for each storm is computed for selected durations i.e. 5,10,15,30 and 60 minutes and EI values for each of these selected durations arrived at from the relation

$$EI_5 = \frac{KE \times I_5}{100} \quad \ldots (2.2)$$

Where,

EI_5 = Erosion index for 5 minutes duration

KE = Total Kinetic energy of storm, metre tones hectare centimeter

I_5 = Maximum storm intensity for the selected 5 minutes duration

Wischmeier suggested the relation similar to the 2.2 for the computation of erosion index R for the individual storm.

$$KE \times I_{30} = EI_{30}$$

$$R' = \frac{EI_{30}}{100}$$

Where,

R' = Erosion Index,

KE = Kinetic energy of storm, metre tones per hectare and

I_{30} = Maximum 30 minutes intensity of rain storm, Centimeter per hour

2.5 Distribution of Erosion Index

An erosion index curve can be derived from the average monthly data by plotting the mean monthly accumulation percentage values of each selected duration for each month.

2.6 Computation of Precipitation Index

Product of amount of rainfall and its intensity for the selected time intervals are derived for each storm and termed as PI_5, PI_{10}, PI_{15}, PI_{30} and PI_{60} respectively.

3. RESULTS AND DISCUSSION

Daily rain gauges charts are obtained from Agro ecology and Environment Centre, Central Research Station, Dr. P.D.K.V., Akola and were analyzed for daily precipitation index, for daily, monthly seasonal and annual erosion index values for 5, 10, 15, 30 and 60 minutes intensity durations. Relationship between the erosion index and precipitation index is obtained for 5,10,15,30 and 60 minutes durations for Akola Station.

Rainfall Analysis for Akola Station

Precipitation Index

Daily PI values for 5,10,15,30 and 60 min selected time intervals were computed for the period of eight years from 2004-2011 by using the method stated by Raghunath et.al(1971) during intensity in similar study for Chandigarh and S. M. Taley (1988) of Vidarbha region in Maharashtra State respectively.

During this period 144 erosive storms occurred which were analyses and derived daily PI values for selected time intervals and termed as PI_5, PI_{10}, PI_{15}, PI_{30} and PI_{60}.

Rainfall erosion index

Analysis of daily rainfall data for the period of eight years from 2004 to 2011 was done by adopting the method suggested by Raghunath et.al. The EI values for the period of eight year (2009-2012) are represented in Table1, Table 2 & Table 3.

Table 1 Monthwise duration of Kinetic Energy and Erosion Index (EI) of rainfall 5,10,15,30 & 60 min. duration during the year 2008-2009.

Month	Rainfall (mm)	Total K.E. (Metric unit)	EI_5	EI_{10}	EI_{15}	EI_{30}	EI_{60}
JUNE	55.00	1731.41	31.16	36.35	38.09	51.94	34.62
JULLY	140.80	3551.97	85.24	85.24	85.24	177.59	142.07
AUG	119.20	2180.69	53.33	32.71	43.61	43.61	13.73
SEPT	173.60	4918.53	442.57	22.11	177.06	137.70	7.37
OCT	13.90	313.78	1.88	2.82	2.51	4.70	0.31
Total	**502.50**	**12696.35**	**61.30**	**179.23**	**346.51**	**415.50**	**198.10**

Table 2 Monthwise duration of Kinetic Energy and Erosion Index (EI) of rainfall 5,10,15,30 & 60 min. duration during the year 2009-2010.

Month	Rainfall (mm)	Total K.E. (Metric unit)	EI_5	EI_{10}	EI_{15}	EI_{30}	EI_{60}
JUNE	157.90	4469.60	321.76	214.51	286.01	178.76	134.07
JULLY	175.70	3808.00	228.48	228.48	121.85	106.62	91.39
AUG	72.50	2190.61	210.29	197.15	87.62	78.86	43.81
SEPT	36.80	738.03	26.56	15.49	14.76	14.76	11.07
Total	**442.90**	**11205.63**	**787.09**	**655.63**	**510.24**	**379.00**	**280.34**

Table 3 Monthwise duration of Kinetic Energy and Erosion Index (EI) of rainfall 5,10,15,30 & 60 min. duration during the year 2010-2011.

Month	Rainfall (mm)	Total K.E. (Metric unit)	EI_5	EI_{10}	EI_{15}	EI_{30}	EI_{60}
JUNE	72.80	1614.56	77.49	48.43	64.56	34.87	20.18
JULLY	167.10	3714.56	378.88	356.59	267.44	148.56	79.86
AUG	121.20	2550.46	137.72	107.11	81.61	51.00	36.98
SEPT	86.30	1666.99	180.03	100.01	73.34	43.34	24.17
Total	**447.40**	**9546.57**	**774.12**	**612.14**	**486.97**	**277.73**	**161.19**

Relation between the erosion index and precipitation index

To compute EI value of 5, 10, 15, 30, 60 minute time interval from the rainfall data equation were developed between EI value and PI value in metric unit by using least square method. The equation followed the pattern

$$Y = a \pm bx$$

Duration (Min)	Estimating Equation	Correlation Coefficient (r)
5	$EI_5 = 1.83\ PI_5 + 05.23$	0.73
10	$EI_{10} = 0.39\ PI_{10} + 18.99$	0.42
15	$EI_{15} = 1.43\ PI_{15} + 11.71$	0.50
30	$EI_{30} = 1.65\ PI_{30} + 03.39$	0.77
60	$EI_{60} = 0.64\ PI_{60} + 08.24$	0.48

4. CONCLUSION

The mean annual rainfall EI values for CRS Dr. PDKV, Akola are 913.33, 549.57, 772.61, 651.31 and 409.60 vary from as low as 287.29(2005), 225.30(2008), 309.43 (2006), 174.05(2005), 139.67(2005) for a 5,10,15,30 and 60 min. duration respectively. Regarding monthly EI values on an average 98.41%, 99.45%, 99.0%, 97.61% and 99.18% of annual EI is concentrated in four months (June to September) for 5,10,15,30 and 60 min. durations, respectively. A similar order July>September>June>August for all the intensity time duration was observed in which july and September are the most erosive and critical months. As the time duration increases, the monthly erosive intensity was observed decreasing.

It is found that, 144 erosive storm correlation co-efficient for 5 minutes is 0.73, for 10 minutes 0.42, for 15 minutes 0.50, for 30 minutes 0.77 and for 60 minutes 0.48 are resulted for the observations.

REFERENCES

1. Fournier, F. (1960), Climat et Erosion, 1 a relation enter 1 erosion due precipitation atmosphere pressure, universitaries de France, Paris.
2. Hudson, (1981) Soil conservation (2nd ed), ELBS, London.
3. Kingu, P.A. (1980), World map of erosivity M.Sc. Thesis (MS/80/190), national college of Agril. Engg. Silsoe, England.

4. Ogrosky, H. O. and V. Mockus (1957), National Engineering Handbook, Section 4 Hydrology supplement A-3, 18-11 to 18-14, Soil Conserve, Service, USDA.
5. Raghunath, B.A., Khuller, R.K Thomas (1982), Rainfall energy map of India, Ind. J. Soil Conserves. 10 (2). Pp 1-5.
6. Raghunath, B. and I.I. Erasmus (1971). A method for estimating erosion potential from daily rainfall data. The India for 97(3), pp 121-125.
7. Schwab, G.O., R.K. Frevert, T.W. Edminister and K.K. Barnes (1966), Soil and water conservation engineering (2nd ed), Wiley New York.
8. Smith, D.D. (1941), Interpretation of Soil conservation data for field use. Journal of Agri. Engg., 22: pp 173-175.
9. Taley, S.M. (1988), Evaluation of erosion potential for rainfall data for Vidarbha region, Thesis submitted to the Dr. PDKV, Akola.
10. Wischmeier, W. H. (1960), Cropping management factor evaluations for a Universal Soil Loss Equation. Soil. Sci. Soc. Am. Proc. 24: pp 322-326.
11. Wischmeier, W. H. and D.D. Smith and R.E. Uhland (1958), Evaluation of factor in soil loss equation. Agricultural Engineering, 39(8), pp 458-461.
12. Wischmeier, W.H. (1959), A rainfall erosion index for Universal Soil Loss Equation. Soil Sci. Soc. Am. Proc. 23(2): 246-249.

ASSESSMENT OF GROUNDWATER QUALITY OF DOULATHABAD AND KODANGAL MANDALS,VIKARABAD DISTRICT, TELANGANA STATE, INDIA

G. Hari Krishna, A. Edukondal, M. Ramu, C. Paramesh and M. Muralidhar

Department of Geology, Osmania University, Hyderabad

krishnageology6@gmail.com

ABSTRACT

Groundwater samples are collected from parts of Doulthabad and Kodangal mandals, Vikarabad district during the Post monsoon season. Assessment of groundwater quality for suitability for drinking and domestic purposes has been carried. Groundwater quality has been assessed by examining various physico-chemical parameters. Parameters like pH, EC, TDS, F^-, Cl^-, NO_3^-, SO_4^{2-}, Na^+, Ca^{+2} and Mg^+ have been determined . Results are compared with WHO-2011 water standards. Overall view of samples reveals that out of 43 water samples concentration of Fluoride in 4, Nitrate in 26, Chloride in 09, Calcium in 08 and Magnesium in 07 Samples are exceeding the permissible limits for drinking purpose in the area.

Keywords: Groundwater, WHO, Physico-chemical.

1. INTRODUCTION

Water is the elixir of life; without it life is not possible. Although many environmental factors determine the density and distribution of vegetation, one of the most important is the amount of precipitation. Agriculture can flourish in some deserts, but only with water either pumped from the ground or imported from other areas. Civilizations have flourished with the development of reliable water supplies and then collapsed as the water supply failed (Fetter et al, 2007).

The quality required of a groundwater supply depends on its purpose; thus, needs for drinking water, domestic water and irrigation water vary widely .To establish quality criteria, measures of chemical. The aims of this study is water quality parameter of Doulthabad and Kodangal mandals so as to assess its status and suitability through the potability point of view and to compare observed levels of studied parameters with the corresponding WHO guidelines values for drinking-water quality .

2. STUDY AREA

Doulathabad and Kodangal mandals are located in the northern part of the erstwhile Mahabubngar districit. The study area covers a part of Doulatabad and Kodangal Mandals of erstwhile Mahabubnagar District and recently formed Vikarabad district, Telangana state. The present study area falls in the Survey of India toposheet numbers 56H/9 , 56G/12 and covers an area around 200 sq.km with longitude 77^0 30' 0''-77^0 37' 30" N and latitudes $17^0$7'30''-$17^0$57'30''E. The Average annual rainfall in the area is 604 mm, climate of the study area is generally hot and falls in semi-arid zone. Average temperature in summer is 40^0C, in winter is 25^0C. There are no major surface water sources in the study area, however; main sources of drinking water are bore wells and dug wells.

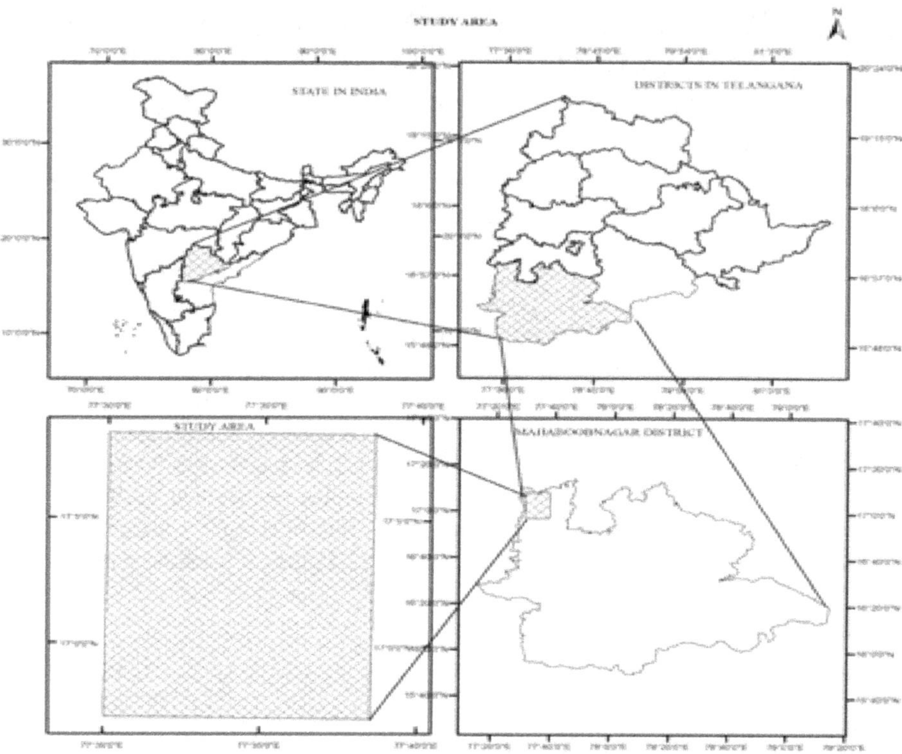

Fig. 1 Location map of the study area.

3. GEOLOGY AND HYDROGEOLOGY

Ground water occurs in all the geological formations in the district. The major rock types in the district are peninsular gneissic crystallines, limestones, conglomerates, sandstones, shales, basalts and alluvium. The district is mainly covered by three types of soils those are red sandy soil (Dubbasand Chalkas) Red earth (with loamy sub-soils and Chalkas) and black cotton soils. Red sandy soils and red earth are permeable and well drained (CGWB 2003). Geologically,the study area covers a part of the stable Dharwar Craton of South Indian shield. mostly covered by hard rocks and it comprises sandstone, green stone, purple shale, granite, gneisses, migmatites (with minor xenoliths of tonalite, trondhjemite, granodiorite, amphibolites and biotite schist) and Deccan traps. The Deccan trap formations are represented by vesicular-amygdaloidal and massive basalt. There are no major, medium or minor surface water irrigation sources in the study area and is totally dependent on groundwater.

4. MATERIALS AND METHODS

A total of forty-three Groundwater water samples were collected in cleaned polyethylene bottles from the dug wells and bore wells in the pre and post-monsoon periods. The pH was measured using the digital conductivity meter immediately after sampling. EC was estimated by the EC analyzer CM183 model of Elico-classical methods of analysis were applied for the estimation of

K^+, Na^+, Ca^{2+}, Mg^{2+} and Co_3^{2-} were analyzed by flame photometry using CL-345 flame photometer of ELICO. Sulphate was estimated by the turbidity method using the Digital Nephelo-Turbidity meter 132model of Systronics. Nitrate was analyzed applying the UV-V screen method using UV-visible spectrophotometer UV-1201model of Shimadzu. Fluoride was analyzed by the ion selective electrode method using Orion 290A+ model of Thermo-electron Corporation. The TDS were estimated by the summation of cations and anions (epm) method (Hem1991).

5. RESULTS

Ground water quality assessment; carried out to determine suitability of water samples in terms of domestic and agricultural purposes. The portability of drinking water from domestic well samples is mainly based on recommended permissible limits for certain parameters described in WHO, 2001 Standards Board limits for drinking water.

The spatial distributions of ten parameters analyzed during post-monsoon season have been represented by as per drinking water standards of WHO 2001. In the study area, pH values of groundwater samples ranged from 5.7 to 7.0 as shown in Table 1. Most of the groundwater samples had pH values below the acidic limit of 6.5. p^H values less than 6.5 are considered too acidic for human consumption and can cause health problems such as acidosis. The acidic nature of most of the samples might be due to high mineral rich rocks making up the acquifers. This is evident sample no 19, which recorded the least p^H of 5.7, showed extraordinarily high values of dissolved ions. The pH values of water from the study area were generally low. Seventy two percent (72%) of the wells had pH values lower than the safe limit for drinking water (6.5-8.5) as prescribed by WHO, 2001.The electrical conductivity in water samples is an indication of dissolved ions. Thus the higher the EC, the higher the levels of dissolved ions in the sample. The Electrical Conductivity of all the samples ranged from 192 to 1024 µS/cm. The electrical conductivity is within Permissible limit. The TDS concentrations range were between 449.38-1386.99 mg/l as shown in Table 1.Chloride ranges between 06-358 mg/l during post-monsoon season (Table 1). The chloride (Cl^-) covers an area of 16.27% in the desirable limit and 83.72% under permissible limit in post-monsoon season. Fluoride ranges between 0.28-1.93 mg/l during post-monsoon season (Table 1). Fluoride (F^-) covers 9.30% area under desirable limit, 90.69% area under permissible limit in post-monsoon season. One of the essential elements for maintaining normal development of healthy teeth and bones is Fluoride. Lower concentrations of fluoride usually below 0.6mg/l may contribute to dental caries. However, continuing consumption of higher concentrations, above 1.2mg/l however cause dental flourosis and in extreme cases even skeletal flourosis.The concentration of Na^+ in the groundwater samples ranged from 112 - 355 mg/L (Table 1). In the study area, Na^+ covers 16.27% area under desirable limit, 83.72% area under permissible limit in post-monsoon season. The nitrate concentration of the samples ranged from 02-393mg/L. Nitrate in drinking water should not exceed 45 mg/l (WHO, 2001). 26 samples recorded higher nitrate values. Excessive nitrate content in drinking water can cause health disorders such as methemoglobinemia, goiter, and hypertension.

Table 1 Analysis of groundwater quality parameters in villages of Doulatabad & Kodangal mandals, Vikarabad District.

S/No	F^-	Cl^-	NO_3^-	SO_4^{2-}	Ca^{+2}	Mg^+	Na^+	TDS	p^H	EC
1	1.00	140	140	86	72	47	242	927.38	6.1	640
2	0.97	36	34	15	68	14	136	462.59	6.8	192
3	0.71	126	46	40	69	38	198	655.15	6.1	384
4	1.18	142	12	21	42	32	172	513.64	6.4	256
5	0.82	218	75	45	76	48	198	688.32	6.5	512
6	0.28	358	48	88	65	46	155	1034.37	6.2	1024
7	0.63	245	142	66	100	48	288	935.10	5.8	576
8	1.23	8	51	5	68	43	148	543.25	6.8	256
9	0.73	8	7	4	67	47	151	536.16	6.3	256
10	0.42	100	137	2	68	43	178	599.57	6.1	256
11	1.19	10	67	12	69	41	165	577.17	6.3	320
12	1.05	12	40	2	58	17	142	472.61	6.2	256
13	0.48	32	73	12	52	47	172	571.51	6.3	320
14	0.45	51	29	31	45	49	162	540.78	6.3	256
15	0.73	56	2	28	57	44	158	554.73	6.3	384
16	1.07	73	4	37	57	32	192	599.79	6	384
17	0.54	271	187	35	142	108	236	1036.54	5.8	768
18	1.54	127	61	41	67	46	182	631.86	6.8	448
19	0.48	282	393	2	236	67	197	1386.99	5.7	1024
20	0.71	259	210	137	66	46	234	1296.26	6.5	384
21	1.07	92	43	23	30	16	182	449.38	6.5	384
22	0.80	254	152	260	201	71	342	1348.79	6.2	768
23	0.88	167	44	102	87	48	197	705.01	5.9	448
24	1.35	260	212	62	75	108	337	1119.83	5.8	1024
25	1.93	136	2	55	40	65	165	571.82	6	384
26	1.19	110	65	48	72	68	142	596.89	6	384
27	1.24	102	25	52	74	49	180	652.24	6.2	768
28	0.98	9	68	10	57	42	156	531.94	7	448
29	1.49	110	80	57	70	49	142	557.49	6.3	1024
30	1.45	76	36	32	74	47	156	590.45	6.1	384
31	0.59	6	40	10	39	41	162	501.61	6.2	384
32	0.67	97	72	36	71	48	192	678.55	6	768
33	1.88	27	48	18	52	47	140	500.17	6.4	448
34	1.93	13	20	13	58	48	128	479.90	6.2	1024
35	1.04	41	57	20	53	40	132	479.99	6.6	384
36	0.88	98	40	45	58	43	182	611.76	6.1	384
37	1.46	45	28	28	70	44	112	478.00	6.4	768
38	0.90	85	112	28	74	46	155	613.64	5.8	448
39	0.85	8	58	10	43	48	165	549.33	6.3	1024
40	0.57	85	110	35	73	68	151	629.70	5.9	320
41	1.16	252	8	262	105	49	355	1077.77	6.2	704
42	1.28	68	57	42	72	22	198	623.74	6.2	384
43	1.04	38	67	13	52	20	149	474.56	6.4	256

CONCLUSIONS

Groundwater quality studies have been carried out with the aim of determining the groundwater suitability for drinking and domestic in parts of Doulatabad and Kodangal mandals Vikarabad district, Telangana. The ground water samples analysis confirms that the p^H level of ground water was within limit. All samples were having Electrical Conductivity Permissible Limit. It is suggested that this water can be used for drinking purpose. The value of T.D.S. was more than maximum permissible limit in 35 samples, these sample *water* are not suitable for drinking purpose ,one sample is found having TDS more than 1300, This cannot be used even for irrigation purposes. In 4 samples the fluoride was found more than maximum permissible limit. Excess fluoride may lead to tooth decay and kidney disease. The concentration of Nitrate was ranging from 02-393 mg/L.Nitrate concentration was higher in 26 samples which is 60.46% of total samples. High concentration of nitrate is possibly due to more agricultural activities and more disposal of domestic and agricultural waste in to water bodies High Nitrate concentration may cause blue baby syndrome or methemoglobinemia. The concentration remains same throughout the period, irrespective of rain/flood period. Reverse Osmosis, ion exchange and biological denitrification method can be used to reduce nitrate concentration.

ACKNOWLEDGEMENTS

Authors are thankful to the Department of Science & Technology (DST), New Delhi, India, for granting Research Fellowship Innovation in Science pursuit for Inspired Research (INSPIRE). The authors are thankful to C-MET for providing the Laboratory facilities. The authors are also thankful to the Head, Department of Geology,Osmania University, Hyderabad.

REFERENCES

1. Fetter C.W. Applied Hydrogeology.
2. WHO (2011). World Health Organization. Guidelines for drinking water quality.
3. Central Ground water Board, Mahabubnagar District (CGWB-2003).

PERFORMANCE OF HARGREAVES RADIATION FORMULA IN ESTIMATION OF REFERENCE EVAPOTRANSPIRATION IN A HOT AND HUMID COASTAL LOCATION IN TAMILNADU

Manikumari N.[1©] and Murugappan A.[2]

Department of Civil Engineering, Faculty of Engg. & Technology,
Annamalai University, Annamalai nagar
[1]aumani67@yahoo.com, [2]profam@hotmail.com

ABSTRACT

Reference crop Evapotranspiration (ETo) is a major component in hydrological studies to evaluate crop water requirement in arriving the irrigation demand. Precise ETo estimates are essential in almost all water resources planning projects. There are several methods available for the estimation of ETo, the choice depends on a number of factors. FAO-56 Penman-Monteith (PM) method is the standard method for the computation of ETo adopted worldwide, but is data intensive. Many researchers have studied the consistency of the P-M method for estimating ETo. Determination of ETo involves several interacting daily meteorological factors such as air temperature, relative humidity, wind speed, bright sunshine hours as well as on the type and growth stage of the crop demanding meticulous effort and considerable time. This paper attempts to examine the efficiency of Hargreaves Radiation formula in FAO PM method for estimating ETo for a data short environment in a coastal region with a hot humid climate. The study area is Annamalainagar of Chidambaram in Tamilnadu which has observations of daily weather data with bright sunshine hours missing for certain periods.

Keywords: Evapotransipiration, PM- Method, Hargreaves Method and ETo.

INTRODUCTION

Evapotranspiration (ET) (Adeboye et. al, 2009) is one of the most vital elements in hydrologic cycle as it is a major water loss and its estimate has a significant role in the field of water resource management. It can be measured directly or estimated by calculating Reference Evapotranspiration (ETo). There have been a number of equations used to estimate ETo (Walter et al 2004; Huntington 2006; Qian et al 2007; Wild et al. 2008; Sharma and Walter 2014; Kramer et al 2015). However, the FAO-56 Penman–Monteith (FAO-56 PM) equation (Allen et al., 1998) has been confirmed to be superior in comparison with other techniques that are often used as the reference equations (Ventura et al., 1999; Lopez-Urrea et al., 2006; discussed the past and future of the FAO-56 PM as well as the accuracy and consistency of operational computations of ETo for agricultural uses.

In the FAO PM equation, the dominant factors affecting the calculation of ETo are the weather variables such as maximum and minimum air temperature; wind speed; sunshine duration; and relative humidity measured or estimated at meteorological stations for a specific location, in addition to latitude and altitude of the place that represent the geographic and climatic characteristics of the study location (Allen et al., 1998). Since 1998, the PM equation has been widely investigated and applied in many parts of the world as a reliable equation for the prediction of ETo (Son et al., 2005; Cai et al., 2007; Adeboye et al., 2009; Bakhtiari et al.,

2011; Subedi et al., 2013; Kisi, 2014). However, the main identified drawback in the application of FAO PM equation is the comprehensive weather data required in the estimation process. Also, the necessity of providing rich historical records of these weather data over a sufficient period, for each study location, to obtain reliable estimates of PM- ETo, is a real hindrance (Droogers and Allen, 2002; Murugappan et. al, 2011).

The estimation of evapotranspiration using the **Hargreaves equation** is a potential solution to data short environments. It needs only temperature and radiation data to estimate ETo. Performance of **Hargreaves equation** is compared with FAO-56 version of Penman–Monteith (PM-56). Lack of incident solar radiation is a significant impediment for most related research applications. Mathematical models have been handy in reducing challenges being posed by inability of having solar radiation instrumental sites at every point on the Earth. Hargreaves-Samani's model is one of the several empirical methods so far formulated in estimating global solar radiation from maximum and minimum temperature data. This paper examines the efficiency of Hargreaves Radiation formula in PM method in estimating ETo in a hot and humid coastal location (Annamalainagar) in Tamilnadu by adopting two different values for the radiation coefficient Krs in arriving the radiation component without using sunshine hour data. One value of Krs equal to 0.19 for coastal regions as specified by FAO and the other employed monthly calibrated values for Krs arrived for Annamalainagar in a previous study (Murugappan et al., 2011).

STUDY AREA AND DATA COLLECTION

The daily meteorological data measured at Indian Meteorological Observatory, Annamalainagar of Tamilnadu is taken for the computation of Evapotranspiration. The FAO Penman-Monteith method is the standard method for computing ETo, as recommended by FAO (FAO, 1977) since it involves many climatological parameters representative of the location. The Penman-Monteith method considers the altitude and latitude of the place factors specific of the location, in addition to the meteorological factors considered in the Modified Penman method. Annamalainagar is located at Chidambaram, Cuddalore District at 11°25' North latitude and 79°44' East longitude. It is located at an elevation of +5.79 m above Mean Sea Level. In this study, daily data on maximum air temperature, minimum air temperature, maximum relative humidity, minimum relative humidity, actual bright sunshine hours and wind speed observed at Indian Meteorological Observatory, Annamalainagar were collected for computing Daily Reference crop ET.

Twenty years of daily meteorological data 1995-2016 (except 2008 and 2009) on the parameters mentioned above were taken for study.

Hargreaves Method

The FAO Penman-Monteith method is the most desirable method universally adopted for estimating reference crop potential evapotranspiration (ETo). However, this method needs full weather data, but during some periods complete weather data does not exist in Annamalainagar due to practical reasons which arise inbetween. On the other hand, the Hargreaves equation is a more simple equation for estimating ETo. Hargreaves equation which contains monthly rainfall data was the best condition for ETo estimates. The Hargreaves equation (Hargreaves and Samani, 1985) is a simple evapotranspiration model that only requires a few easily accessible parameters: mininimum, maximum and mean temperature, and extraterrestrial radiation. Based on the knowledge that the difference between the maximum (Tmax) and minimum (Tmin) air temperature is influenced by the degree of cloud cover, humidity and solar radiation.

RESULTS AND DISCUSSION

The following Table 1 indicates the number of over estimations and under estimations in daily reference evapotranspiration when compared to PM daily ETo against each of the years taken for study. It is apparently shown from the table 1 that the number of overestimations in daily ETo have been considerably reduced in many instances when monthly calibrated K_{rs} values for Annamalainagar were adopted in Hargreaves radiation formula to compute daily ETo by FAO PM-Method as compared to Coastal K_{rs} value of 0.19 for all days in the same method. Results can be improved if the same approach is employed for a longer period like a week or month.

Table 1 No. of Overestimations in Daily ETo for the period 1995 to 2016.

Year	No. of Overestimations		No of Underestimations	
	Har C	Har A	Har C	Har A
1995	211	202	154	163
1996	139	148	227	218
1997	140	134	225	231
1998	134	123	231	242
1999	167	151	198	214
2000	139	131	227	235
2001	160	168	205	197
2002	152	121	213	244
2003	180	177	185	188
2004	182	198	184	168
2005	200	208	165	157
2006	202	213	163	152
2007	182	206	183	159
2010	206	216	159	149
2011	235	221	130	144
2012	194	200	172	166
2013	218	240	147	125
2014	204	194	161	171
2015	186	172	179	193
2016	211	216	155	150

Har C – Hargreaves Coastal Krs; Har A- Hargreaves Annamalainagar Krs

Table 2 Table of Mean Errors in ETo for different periods in 20 Years

Period	AHRF		CHRF	
	MAE	MAPE	MAE	MAPE
January	4.49	3.92	9.41	8.10
February	4.84	3.97	9.08	7.34
March	7.02	4.63	6.90	4.40
April	8.05	5.08	7.85	4.76
May	7.33	4.05	8.85	4.91
June	5.98	3.71	15.13	9.28
July	6.49	4.22	22.96	14.70
August	5.96	4.29	18.62	12.71
September	7.52	5.56	11.41	8.43
October	5.80	5.14	8.40	7.39
November	6.43	6.76	6.45	6.69
December	5.51	5.75	7.62	7.44
Winter season	8.08	3.41	18.16	7.57
Summer season	19.79	4.00	13.86	2.75
SW Monsoon	20.17	3.35	66.33	10.83
NE Monsoon	14.12	4.62	14.73	4.75
Monsoon	32.53	3.52	70.17	7.60
Annual	57.36	3.45	61.77	3.74

AHRF- Annamalainagar Krs Hargreaves Radiation Formula

CHRF- Coastal Krs Hargreaves Radiation Formula

The values shown in Table 2 were arrived by considering the cumulative daily ETo for similar period of each of the successive years from 1995 to 2016 except 2008 and 2009. The seasonal values were arrived considering the months of January and February for winter season, March to May for summer season, June to September for South West monsoon season and October to December for North East monsoon season for Annamalainagar. The last two rows of Table 2 indicate the cumulative of two monsoons as Monsoon and Annual refers to the cumulative of computed daily ETo values for each year. The values of Mean Absolute Error (MAE) and Mean Absolute Percentage Error for seven months namely January, February, June, July, August, September and December have predominantly come down when applying monthly calibrated K_{rs} coefficients for Annamalainagar as compared to Coastal K_{rs} co-efficient in computing daily ETo by FAO PM-Method.In the remaining five months of March, April, May, October and November also it has decreased but by a smaller value. It is also obvious from the table that in seasons like Winter, Summer and Southwest Monsoon ETo values converge closer to PM ETo values which is considered to be more accurate. Northeast monsoon error values do not exhibit a satisfactory improvement from coastal coefficient computations.

Figure 1 shows bar charts for the comparison of actual PM ETo, Annamalainagar using Hargreaves Method AHRF ETo and Coastal (Calibrated) CHRF ETo for the month of January as a representative graph.

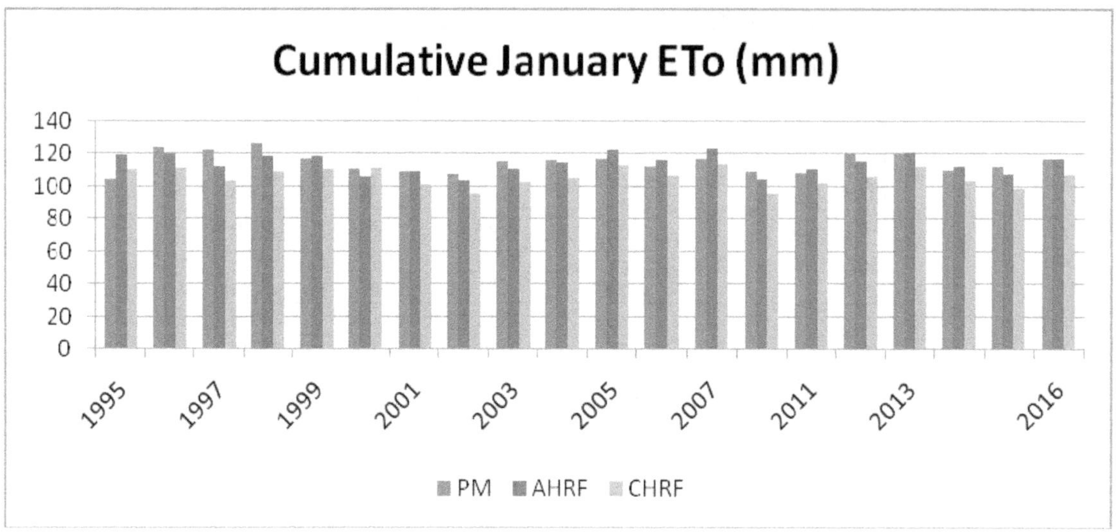

Fig. 1 Cumulative Monthly ETo for January during 1995 to 2016(except 2008 & 2009).

Figure 2 shows a line graph for the comparison of errors in cumulative monthly ETo, of Annamalainagar using monthly calibrated coefficients in Hargreaves radiation formula (AHRF ETo) with that of Coastal coefficient (CHRF ETo) in the same formula by PM method with reference to PM ETo values as basis for the month of January as representative graph with linear fit Equation and trend line.

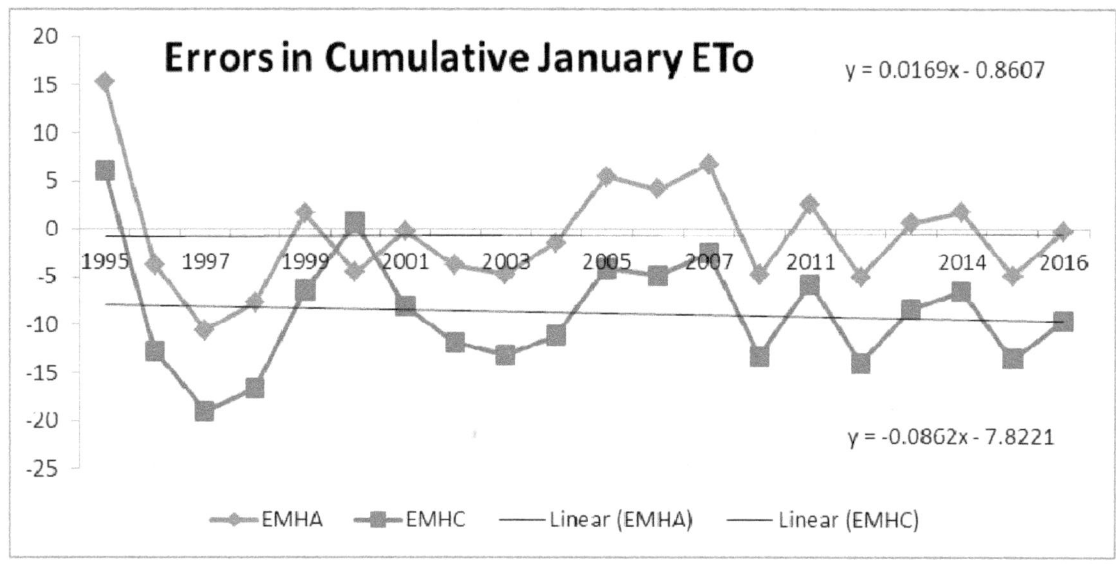

Fig. 2 Errors in Cumulative Monthly ETo for Jan (1995-2016).

CONCLUSION

The Hargreaves radiation formula using specific monthly calibrated values for Annamalainagar in PM method for arriving daily ETo has brought down the overestimation of ETo values for more than two thirds of the time for the period considered in the study from the years 1995 to 2016 excluding two years 2008 and 2009 (for which solar radiation data was not available) and overestimated less than one-third of the period, when compared to results arrived by using a general coastal coefficient of 0.19, as specified by FAO. The quantum of underestimation was found to be high during hot and humid days during summer season. So the Hargreaves radiation formula tends to work efficiently satisfactorily closer to FAO PM method which is handy in data short environments as an alternate approach in arriving daily Reference Evapotranspiration values for any location. The study can be enhanced if the approach is extended in considering the Overestimations

REFERENCES

1. Adeboye, O.B., Osunbitan, J.A., Adekalu, K.O., Okunade, D.A., 2009. Evaluation of FAO-56 Penman–Monteith and temperature based models in estimating reference evapotranspiration using complete and limited data, application to Nigeria. Agric. Eng. Int.: The CIGR J. Manuscript Number 1291, X I(1291), pp. 1–25.
2. Allen, R., Pereira, L., Raes, D., Smith, M., 1998. Crop Evapotranspiration-Guidelines for Computing Crop Water Requirements-FAO Irrigation and Drainage Paper 56. Food and Agriculture Organization, United Nations, Rome.
3. Cai, J., Liu, Y., Lei, T., Pereira, L.S., 2007. Estimating reference Evapotranspiration with the FAO Penman–Monteith equation using daily weather forecast messages. Agric. For. Meteorol. 145 (1–2), 22–35.
4. Huntington TG (2006). Evidence for intensification of the global water cycle: review and synthesis, Journal of Hydrology, 319(1-4), 83-95.
5. Huntington TG, Billmire M (2014). Trends in Precipitation, Runoff, and Evapotranspiration for Rivers Draining to the Gulf of Maine in the United States, Journal of Hydrometeorology, 15(2), 726-743.
6. Kramer R, Bounoua L, Zhang P, Wolfe R, Huntington T, Imhoff M, Thome K, Noyce G (2015). Evapotranspiration trends over the eastern United States during the 20th Century, Hydrology, 2(2), 93-111.
7. López-Urrea, R., de Santa, Martín, Olalla, F., Fabeiro, C., Moratalla, a., 2006. Testing evapotranspiration equations using lysimeter observations in a semiarid climate. Agric. Water Manage. 85, 15–26.
8. Qian T, Dai A, Trenberth KE (2007). Hydroclimatic trends in the Mississippi River Basin from 1948 to 2004, Journal of Climatology, 20, 4599- 4614.
9. Ventura, F., Spano, D., Duce, P., Snyder, R.L., 1999. An evaluation of common evapotranspiration equations. Irrig. Sci. 18 (4), 163–170.
10. Walter MT, Wilks DS, Parlange J-Y, Schneider RL (2004). Increasing evapotranspiration from the conterminous United States, Journal of Hydrometeorology, 5, 405–408.
11. Wild M, Grieser J, Schaer C (2008). Combined surface solar brightening and increasing greenhouse effect support recent intensification of the global land-based hydrological cycle, Geophysical Research Letters, 35, L17706.
12. Murugappan A, Sivaprakasam S, Mohan S, (2011). Performance evaluation of calibrated Hargreaves method for estimation of Ref-ET in hot and humid location in India, International Journal of Engineering Science & Technology, 4728 – 4743.

ACHIEVING DOUBLE DIGIT GROWTH RATE FOR A CENTURY BY INTER BASIN TRANSFER OF RIVER WATER IN INDIA

K.S. Misra[1], Neela Misra[2] and Anshuman Misra[3]

[1]University of Petroleum and Energy Studies, Dehradun
[2]Formerly Pune University, Pune
[3]Department of Geology, Kumaun University, Nainital
drksmisra@gmail.com

ABSTRACT

Long term sustained high growth rate can be achieved by proper utilization of run-off Monsoon water to the oceans. This water can be stored and transferred from surplus to deficient river basins to mitigate frequent flood and drought conditions. Cascading dams will insure perennial flow to rejuvenate river systems. Increase in agricultural land and conversion of single to double and even triple crop regions, will also advance fulfillment of national commitment to provide water to every field. Availability of surface water will reduce pressure on ground water, increase natural and artificial recharge potential. The study proposes to develop national water grid, to insure fairness to every state, on the pattern of electrical power. States will be able to charge for water provided to other provinces. Since the water supplied will be surplus rain water, it is expected to find favor from every quarter. Electricity generate from innumerable reservoirs will fulfill the present as well as future requirements at reasonable cost. This in turn will reduce nation's dependence on import of crude oil and our coal resources to mitigate effects on climate change by shifting to renewable energy.

We studied the geomorphology and geology of different geo-climatic zones of India to suggest suitable corridors for the inter-basin transfer of water. This study has brought out great potential for storage of enormous amount of rain water. Maximum utilization of existing river channels and natural gradients has been made, to avoid construction of long canals, tunnels and lifting of water. Satellite and radar imagery is interpreted for geomorphic land use details, particularly the relief features and elevation data is acquired from topographical sheets. New geological data is generated, to identify narrow gorges and competent rock section, for construction of dams, alignment of canals, tunnels, roads and availability of construction material on GIS platform.

The study identifies sites for storage of water, on the western side of Sahyadri ranges, for supply of water to Mumbai and ensures regular flow to Konkan region. Transfer of water, from the western side to the eastern side of Sahyadri mountain ranges in rain shadow regions of Maharashtra and Telangana. Furthermore, augmentation of water for supply to mega city of Mumbai and twin cities of Hyderabad and Ssecunderabad for present and future needs is proposed.

INTRODUCTION

The true potential of management of water resources has been realized centuries ago in our country. Innumerable examples of excellent and highly innovative techniques were available for every climatic zone and geomorphic/geological province. The present reporting of inter-basin transfer of water, study covering entire peninsular India. It has therefore, become imperative to give a brief account of project in totality for putting the things in proper perspective. Furthermore, Maharashtra-Karnataka sector has emerged as very important component, as it lies in between and provides significant natural links for inter-basin transfer of water from north to south. The first proposition on this subject came in late fifties from Mr. K. L. Rao, an engineer by training and a Minister in Union Cabinet. This proposition envisaged linking of river Ganga with the river Kaveri. But soon it became apparent that the project cannot be implemented because neither geomorphology and geology nor hydrology favors the idea. Firstly, it required lifting of water to higher elevations, a number of times and secondly, during non monsoon months, Ganga had hardly any water, to be transported to enormous distance to the south in Kaveri basin. The second proposition came in early seventies by Captain Dastur. He very rightly advocated, need for large scale management of water resources and compared the water potential of India, with the black gold potential of Middle East. Captain Dastur suggested two interlinked girdle canals parallel to the Himalayan mountain ranges (Fig. 1). He envisaged utilization of surplus water from northern part of the country, by the deficient peninsular part. However, this proposition also did not find support from the geomorphological and geological setting of central India. Furthermore, in the absence of specific details, cost benefit ratio could not be worked out.

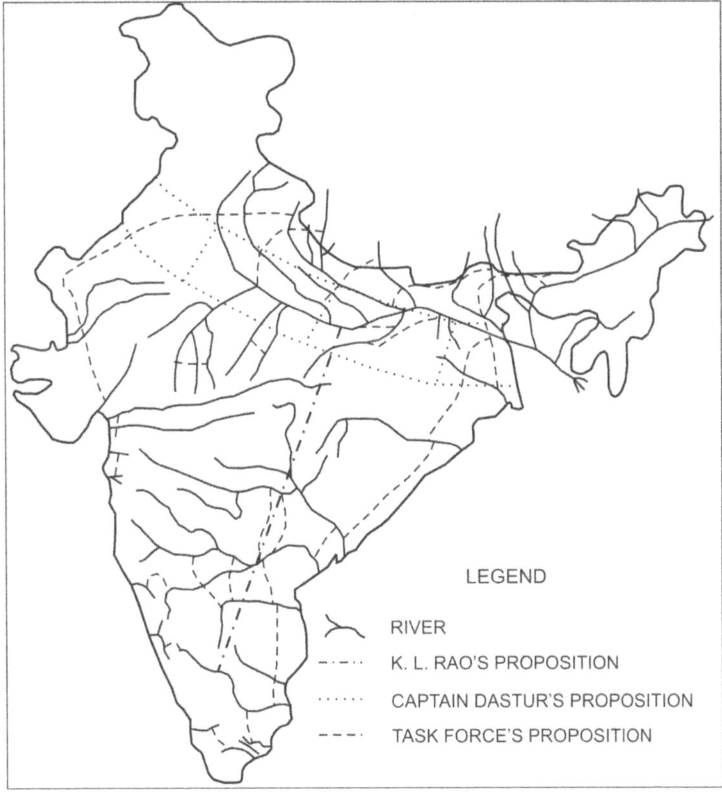

Fig. 1 Map of India showing network of major rivers with different propositions made earlier for linking of rivers and management of water resources.

More than a decade ago, the Government of India set up a task force, on this subject of linking of rivers. The recommendations of this task force had mainly two major components, along with several smaller linkages. The first eastern component, suggested bringing of water from northeast along the east coast, right up to the southern part. This component did not find favor, because in upper reaches, the delta regions of Mahanadi and Godavari have lot of fresh water. The necessity of bringing water, from northeast is thus unjustified. Moreover, this water will be available, only a few meters above the sea level and will require lifting for transportation to southern parts of peninsula. Similarly, the western component involved bringing water from Uttarakhand via Rajasthan to Gujarat. This suggestion also could not be implemented, because it involved transportation of water across Aravalli ranges, it went nearly parallel to existing Indira Gandhi canal in Rajasthan and took water to areas, where the efforts of Gujarat government have made water available from Sardar Savour project on Narmada. Among the smaller components, the linking of Ken and Betwa rivers in Madhya Pradesh, has found favor and is being implemented. Our study along Sahyadri Mountain Ranges and Peninsular India in totality, has established that formidable amount of rain water, which flows to the ocean, is available for inter basin transfer, storage to fulfill all the requirements. Scientific exploration and management of water resources can usher prosperity to the entire region.

SALIENT FEATURES OF THE PRESENT STUDY

The salient features of the present study, based on geomorphological /geological as well as hydrological data are as follows-

1. Inter-basin transfer of enormous amount of water, from surplus to deficient areas is proposed by linking of rivers, based on favorable geological and geomorphologic condition.
2. Utilization of existing river channels and natural slopes has been done, to avoid construction of long canals. Canals and tunnels wherever absolutely needed are proposed.
3. Formidable storage capacity of water is expected to be created by cascading dams and lifting of water as proposed in earlier river linking proposals is completely eliminated.
4. Geological and geomorphologic conditions favoring construction of dams, barrages and reservoirs, along main rivers and their tributaries to generate electricity, increase irrigation potential and ensure continuous flow of water throughout the year.
5. The proposed project based on scientific data will be a landmark in countrywide long term water management.
6. Availability of surface water will reduce pressure on groundwater and will also facilitate recharge and thus in turn raise the water table.
7. The implementation of the project can be started in different parts, at the same time and benefits will be available as project progresses. One does not have to wait for several decades, to reap the benefits.
8. Proposes to develop a national water grid, to insure fairness to all states, on the pattern of electrical power grid. States will be able to get cess for water provided to other state. Since the water supplied will be rain water, which usually flows to the ocean and also causes extensive floods and damage, it is expected to find favor from every quarter.

9. In many cases, it is not necessary to literally link the river channels. The dropping of water in any of its tributaries or sub basin will serve the purpose of inter-basin transfer of water.
10. This study will also contribute to national efforts to combat drought periods, which are impossible to predict and may create very difficult situation, if they are in succession.
11. The project is expected to ensure sustained double digit economic growth for very long time for the country due to accelerated activity, employment, increased potential and production.

The methodology adopted during the study, involved preparation of geological and geomorphological maps by interpretation of high resolution satellite imagery. Satellite radar imagery has provided exclusive details of relief features. Care is taken to incorporate all those details, relevant to inventory and management of water resources. Hydrological information is added and by legend organization hydro-geomorphological maps are prepared. These maps have also been utilized to identify suitable sites for small, medium and large dams. The results of several decades of working on this problem are presented here. The study also identifies sites for storage of water, on the western side of Sahyadri ranges, for supply of water to Mumbai and also ensures regular flow to Konkan region. Several corridors for transfer of water, from the high rainfall western side to the deficient eastern side from Nasik in Maharashtra to Goa are identified. In Nasik region, close to Triambakeshwar transfer can be done by a system of dams, canals and tunnels. Water from Narmada basin to Tapti basin and later to Godavari basin is very significant finding of the study have been discussed in detail by Misra et al 2017.. Furthermore, transfer of Godavari water to the easterly flowing tributaries such as Mula, Ghod, Bhama, Mula-Mutha, Nira, and finally to Krishna will ensure augmentation of water to entire peninsular India including Cauvary delta region in Tamil Nadu.

Sahyadri Mountain Ranges

Geographically north-south trending Sahyadri Mountain Ranges of western India, extend from southern Gujarat to northern part of Kerala. Geomorphologically these ranges define the most important watershed of peninsular India and detailed description of their evolution has been presented by Radhakrishna (1989). Geohydrologically these ranges form the most important ground water tower. Very high rainfall along the crest of these ranges as well as ground water located at higher reaches, seeps down to give rise to innumerable tributaries belonging to the Godavari, Krishna. Thamiraparani, Tungabhadra and Kaveri basins on the eastern side. While Kali, Bhadra, Bhavani, Malaprabha, Ghatabrabha, Hemavati Mandovi, Zuari, Netravati,Sharavati flow westward to the Arabian Sea. All these rivers carry enormous amount of Monsoon water to the sea.

Sahyadri ranges are different than the fold Mountain ranges. They form an extensive hinge zone comprising nearly vertical faults and development of great escarpment. Along these faults the successive blocks are faulted down towards west while blocks have been rising towards the east. From north to south the mountain has three distinct entities.

The northern portion comprises an overburden of very thick lava flows. The over lying flows in the succession are having gentle gradient towards east while the lower ones are moderately dipping towards west. This escarpment is controlled by either NW-SE trending faults or by curvilinear faults developed parallel to Panvel flexure (Misra, 2001). The escarpment is gradually retreating towards east and has exposed the lower sequence of felsic volcanics which form a very rugged topography on the western side. The short rivers and streams emanating from the upper reaches of the ranges cascade down rapidly. The upper sequence of volcanic rocks is tholeiitic in nature and is generally believed to be horizontally disposed flows of massive basalt. However, the present study has demonstrated that in very large part of the area, they have 1° to 3° dip, which is found to be very effective in controlling the movement, preferential accumulation and discharge of ground water. These volcanics have imperceptible dip forming the Deccan plateau on the eastern side and escarpment on the western side. This escarpment is controlled by either NW-SE trending faults or by curvilinear faults developed parallel to Panvel flexure (Misra, 2001). The escarpment is gradually retreating towards east and has exposed the lower sequence of felsic volcanics on the western side. (Fig.2) Deccan volcanic rocks are generally believed to be horizontally disposed flows of massive basalt. However, the present study has demonstrated that in very large part of the area, they have 1° to 3° dip, which is found to be very effective in controlling the movement and preferential accumulation of ground water. Exploration of ground water by mapping lineaments, Dykes and remnants of lava channel and tubes within Deccan basalts control the movement of ground water and hence provide a recharge zone on one side of these structures. . Exploration of ground water by using lineament tectonics in basaltic country is found to be extremely useful. Lineaments particularly those representing surface expression of faults due to extensional tectonics are very significant. These lineaments provide the paths of least resistance to streams and rivers, thus they largely control the hydro-dynamic aspects in these rocks. Selection of suitable drilling sites for exploration of ground water by using parameter defined by the lineament in Deccan volcanic country has considerably improved the success rate of tube wells. Dykes and remnants of lava channel and tubes within Deccan basalts control the movement of ground water and hence provide a recharge zone on one side of these structures. These zones can suitably be utilized for construction of dams, wherever a stream or river is passing through them. Vesicular, fractured and weathered flows within the Deccan volcanics are good producing zones and can be demarcated by using remotely sensed data. The Deccan volcanic country is also transacted by several structural rift and grabens. These rifts and grabens are represented by elongated valleys generally occupied by rivers and filled with thick sequence of Quaternary alluvial material. These alluvial filled valleys are occupied by Narmada, Tapti, Purna, Godavari, and Pravara rivers and are excellent structures for availability of ground water. Furthermore, they can suitably be utilized for construction of dams, wherever a stream or river is passing through them. Vesicular, fractured and weathered flows within the volcanic rocks are good producing zones and can be demarcated by using remotely sensed data.

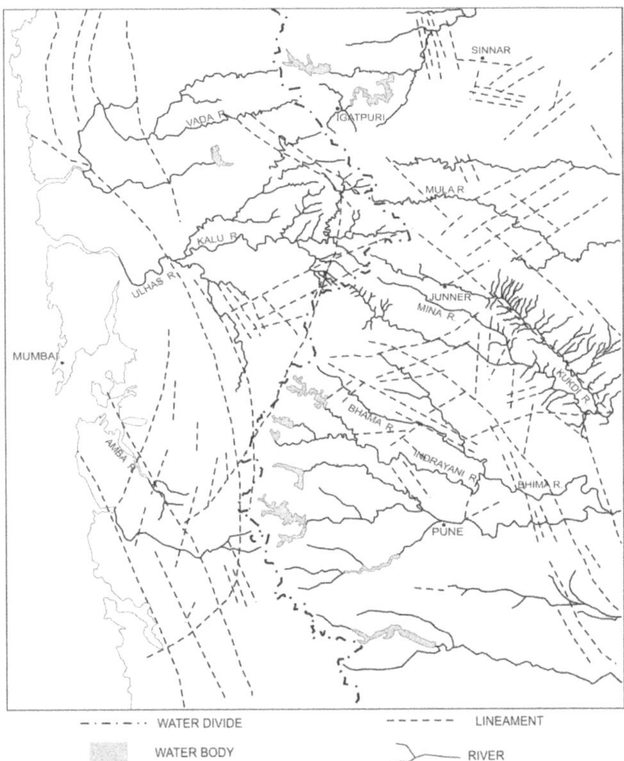

Fig. 2 Map of western Maharashtra showing distribution of major rivers and tributaries on the eastern and western side of Sahyadri mountain ranges. The disposition of major lineament sets which represent nearly vertical faults can also be seen.

Exploration of ground water by mapping lineaments, dykes and remnants of lava channel and tubes within Deccan basalts control the movement of ground water and hence provide a recharge zone on one side of these structures. . Exploration of ground water by using lineament tectonics in basaltic country is found to be extremely useful. Lineaments particularly those representing surface expression of faults due to extensional tectonics are very significant. These lineaments provide the paths of least resistance to streams and rivers, thus they largely control the hydro-dynamic aspects in these rocks. Selection of suitable drilling sites for exploration of ground water by using parameter defined by the lineament in Deccan volcanic country has considerably improved the success rate of tube wells. Dykes and remnants of lava channel and tubes within Deccan basalts control the movement of ground water and hence provide a recharge zone on one side of these structures. These zones can suitably be utilized for construction of dams, wherever a stream or river is passing through them. Vesicular, fractured and weathered flows within the Deccan volcanics are good producing zones and can be demarcated by using remotely sensed data. The Deccan volcanic country is also transacted by several structural rift and grabens. These rifts and grabens are represented by elongated valleys generally occupied by rivers and filled with thick sequence of Quaternary alluvial material. These alluvial filled valleys are occupied by Narmada, Tapti, Purna, Godavari, and Pravara rivers and are excellent structures for availability of ground water. Furthermore, they can suitably be utilized for construction of dams, wherever a stream or river is passing through them. Vesicular, fractured and weathered flows within the volcanic rocks are good producing zones and can be demarcated by using remotely sensed data.

During last three decades the Deccan volcanic region of Maharashtra, very significant developments have taken place. On mini scale check dams are made on streams which generally flow only during Monsoon period in gently undulating terrain. Water is accumulated in cascading dams and is available for agriculture and domestic use. It reduces the pressure on ground water but also facilitates recharge. Since the basaltic flows are nearly horizontal, the percolated water larger surrounding region. The hydrological properties of each flow varies due to horizontal layering, flow top breccias, ropy structures and other features associates with pahoehoe flows. Apart from horizontal connectivity, there is vertical connectivity through columnar joints and pipes.

The central portion of mountain ranges, occupying southern Maharashtra, Goa and Karnataka is made up of Achaeans comprising banded meta-sedimentary iron ore formation, quartzites and lava flows. These rocks control the geomorphology and hydro-geomorphology of this portion. Ranges formed due to differential erosion, turn around to form huge north-north westerly plunging synform. The mighty rivers such as Mandovi and Zuari originate from the top of this structure and flow in the plunge direction.

Management of Water along Sahyadri Ranges and Supply of Water to Mumbai City

Rainfall from the SW Monsoon on western side of these ranges is very heavy ranging around 400 cm. Approximately 50km to 100km wide central zone receive >400cm. However, all this rain water flows to the sea. A very meager amount of water is presently stored at a couple of places like Bhatsa, Tansa and Vihar Lakes NE of Mumbai. The water from these lakes is the main source of municipal supply to the megacity. The prosperity of our nation and particularly Maharashtra will depend in future how best our scientists and engineers can bring this water from western side to the eastern side. The present study has demonstrated that it is much easy to transfer water in the northern part of the ranges, were the ranges are narrow as well as have low elevation in the vicinity of Trimabak (Fig. 3).

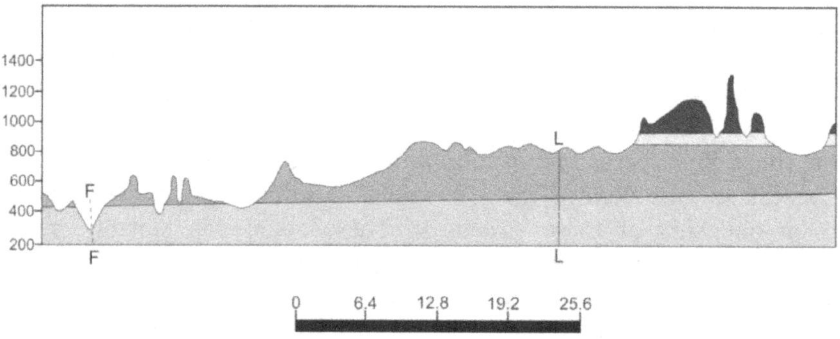

Fig. 3 Geological cross section across northern part of Sahyadri mountain ranges. Near Triambak, the ranges are less prominent and make most favorable situation for transfer of water from western side to the eastern side. Furthermore, there are many locations on western side where water can be impounded and suitably utilized. (After Misra et al 2017)

There are suitable places also where the water can be stored at high elevation and can be transferred towards the east by short tunnels. This study has also identified one such site were water from upper Vaiterna reservoir as well as a high level reservoir on river Vaiterna can be together used to transfer huge amount of water from highly surplus western side to deficient eastern side (Fig. 4).

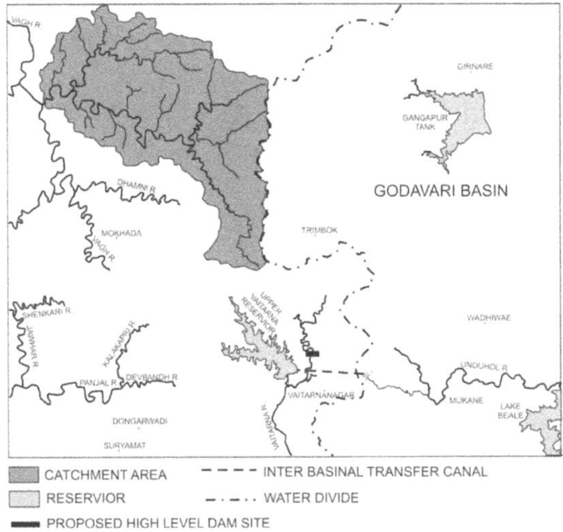

Fig. 4 Map showing upper Vaitarna reservoir located on a tributary of Vaitarna river. Through proposed dam and a tunnel the water from upper Vaitarna reservoir can be transferred to easterly flowing Undhol basin across the Sahyadri ranges.

In central and southern part Sahyadri ranges are quite broad and have very high elevation (Fig. 5). Topography along the mountain ranges is largely controlled by parallel faults. The faults are nearly vertical and are of basement origin. The north-south trending valleys along these faults have scope for accumulating rain water and proper utilization on the eastern side.

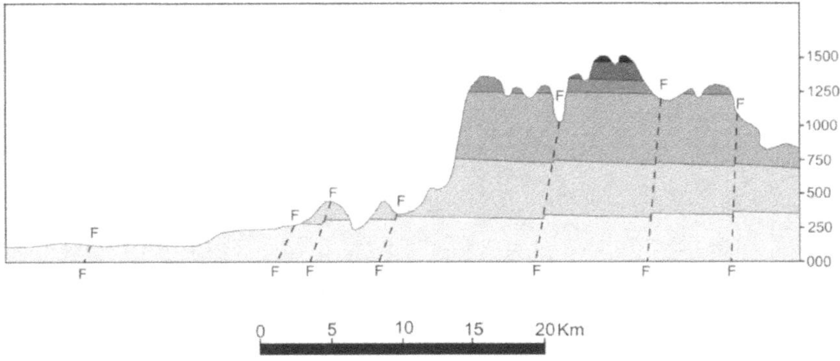

Fig. 5 Geological cross section across central part of Sahyadri mountain ranges. In the area east of Mumbai, the ranges are quite towering. Basement faults are seen controlling the topography to certain extent. Transfer of water from west to east is rather difficult. (After Misra et al 2017)

This N-S trending Konkan belt lies between the coast in the west and Sahyadri ranges in the east. The entire belt receives assured excessive (<400 cm) rainfall during June to September Monsoon months. However, all this rain water flows to the sea. A very meager amount of water is presently stored at a couple of places like Bhatsa, Tansa and Vihar lake NE of Mumbai. The water from these lakes is the main source of municipal supply to the megacity.

Geomorphologically and geologically it is a very challenging region, as far as management of water is concerned. In large parts of the region, rivers and streams emanating from the western side of the mountain, suddenly cascade down to lower levels and thus do not normally provide suitable places for construction of dams. Either the catchment areas are small or narrow gorges are not available. In such a situation available huge storage places need to be identified .Collecting canals along certain contours parallel to the ranges, at several suitable elevations can be planned. The present study has demonstrated that the conditions are very favorable on the northern part than in the southern part. Geomorphology favors storage of large amount of water and transportation to Mumbai and Konkan region. Detailed study of required parameters has shown that Vagh basin is quite suitable for storage of water. This river has extensive catchment area in high rainfall zone. Two suitable sites of Val River, a tributary of Vagh as well as one on Vagh is suggested. The areas of submergence in each of the dam are shown in (Fig. 6).

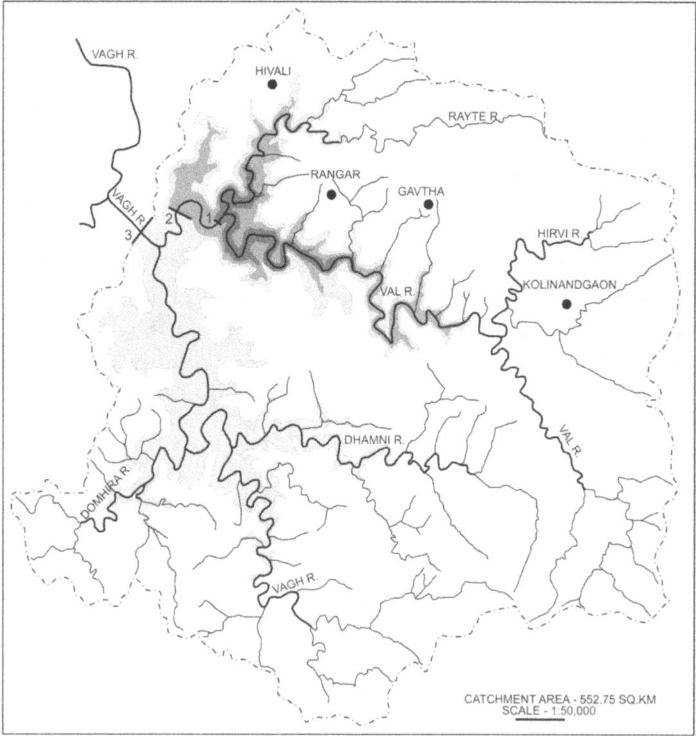

Fig. 6 Detailed drainage map of Vagh basin. This basin is located on the western side of Sahyadri ranges, receives very high rainfall and has very favorable conditions for storage of water. Three favorable sites for construction of dams are identified as 1, 2 and 3 and submerged areas related to each site are depicted.

Godavari Basin, Water Management and Water supply to the Twin cities of Hyderabad-Secundrabad

The maximum portion of Maharashtra is drained by Godavari River and its tributaries. These rivers originate from the eastern side of Sahyadri mountains and flow along nearly NW-SE trending elongated basins of river Mula, Kukdi, Mina, Ghod,Man, Nira, Bhima and Sina.The upper half of Godavari basin is extremely deficient in water, because of low rainfall, due to its location in the rain shadow region and being composed of hard basaltic terrain. To usher property in this rain deficient region of Maharashtra, two solutions have emerged from the present study. The first, suggests bringing water from the extremely high rainfall areas west of Sahaydri. The second solution envisages augmenting water from the Tapti basin in the north. The two passes identified are quite significant. The detailed study of geomorphology and geology on eastern side of Sahyadri ranges has shown that approximately 300m gradient is also available from northern fringes to the southern side and this can be utilized for power generation as well as natural gradient for the flow of water .A number of east and south easterly trending ridges emanating from the mountain ranges can be tunneled. On the northern ends of these tunnels, canals can be taken out at higher levels to irrigate upland areas and on the southern side power houses can be located for generation of electricity (Fig. 7).

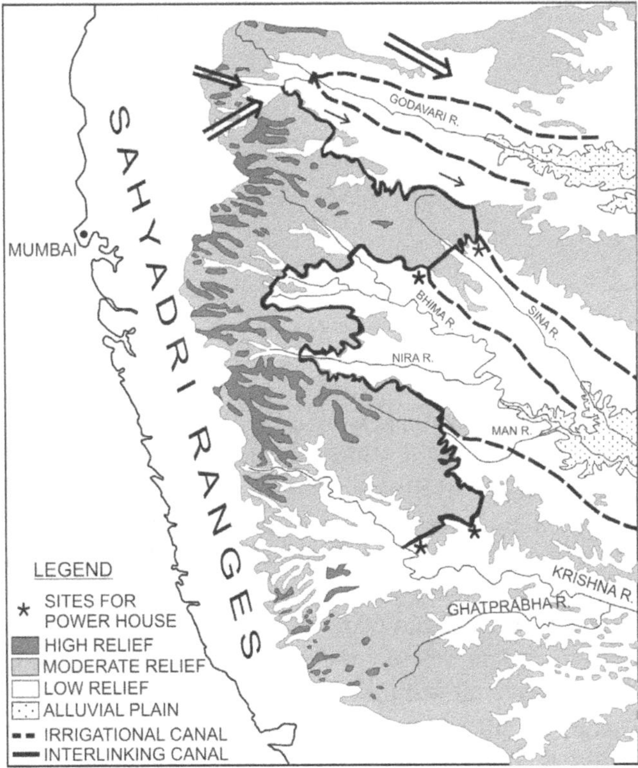

Fig. 7 Map of western Maharashtra, showing relief features on the eastern side of Sahyadri mountain ranges. Passes for inter-basin transfer from Tapti basin in the north and from western side of Sahyadri to the Godavari basin can be seen (After Misra et al 2017)

On the eastern side Godavari basin gets sufficient water after joining of two rivers Pranhita and Indravati (Fig. 8). The rainfall in the region also increases with the end of western rain shadow zone. This part of Godavari basin also gets rainfall from the eastern monsoon and cyclonic depressions approaching from Bay of Bengal. River Indravati has enormous catchment area and passes through Baster region of Madhya Pradesh.

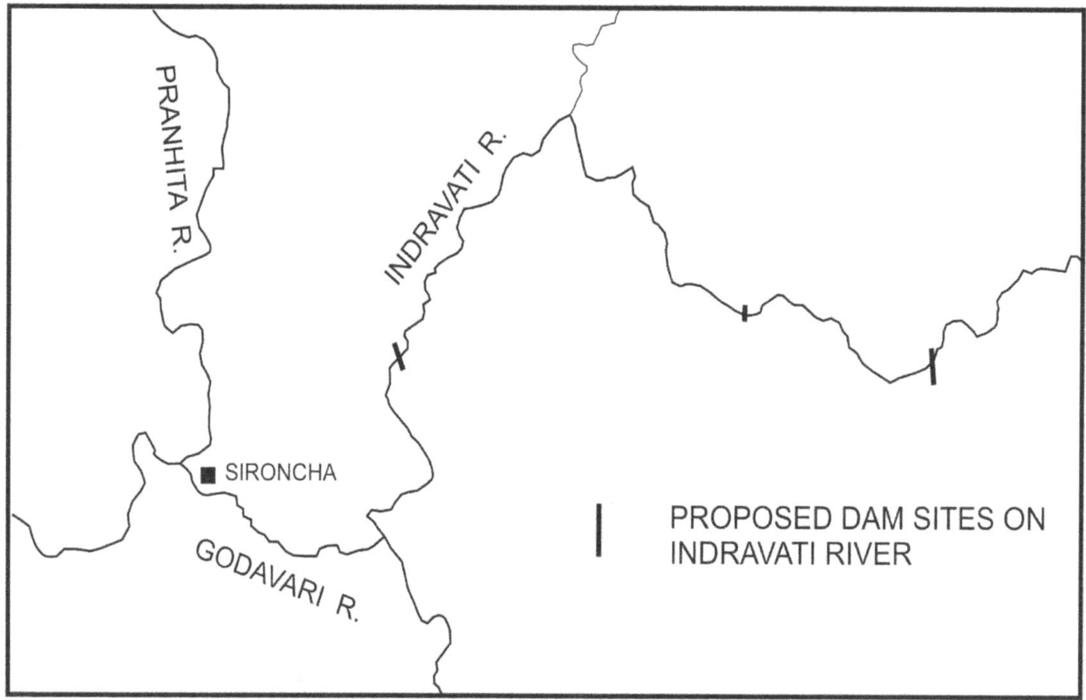

Fig 8 Map showing disposition of major rivers in eastern Maharashtra. Three sites for construction of major dams on river Indravati are marked.

During its course it flows through Achaean and Proterozoic terrains. This terrain has several strike ridges of quartzite and thus provides ideal location of large dams. The water impounded can be utilized for generation for electricity and provide regulated flow to river Godavari. Godavari is joined by river Manjira west of Nizamabad. The river Manjira like all others flows in south-westerly direction, however, north-west of twin city Hyderabad-Secunderabad, it starts flowing in north-easterly and then north-westerly direction to join Godavari river. This is very conspicuous anomaly. The geological and geomorphological map shows a number of basement faults from Achaean terrain continue below the volcanic terrain (Fig.9). Geological evidences suggest that adjacent blocks have moved up or down in relation to each other. The presence of NW-SE aligned Basavakalayan plateau appears to have risen along these faults. This fault zone is directionally parallel to the Pranhita – Godavari graben located in the north-eastern part of the area.

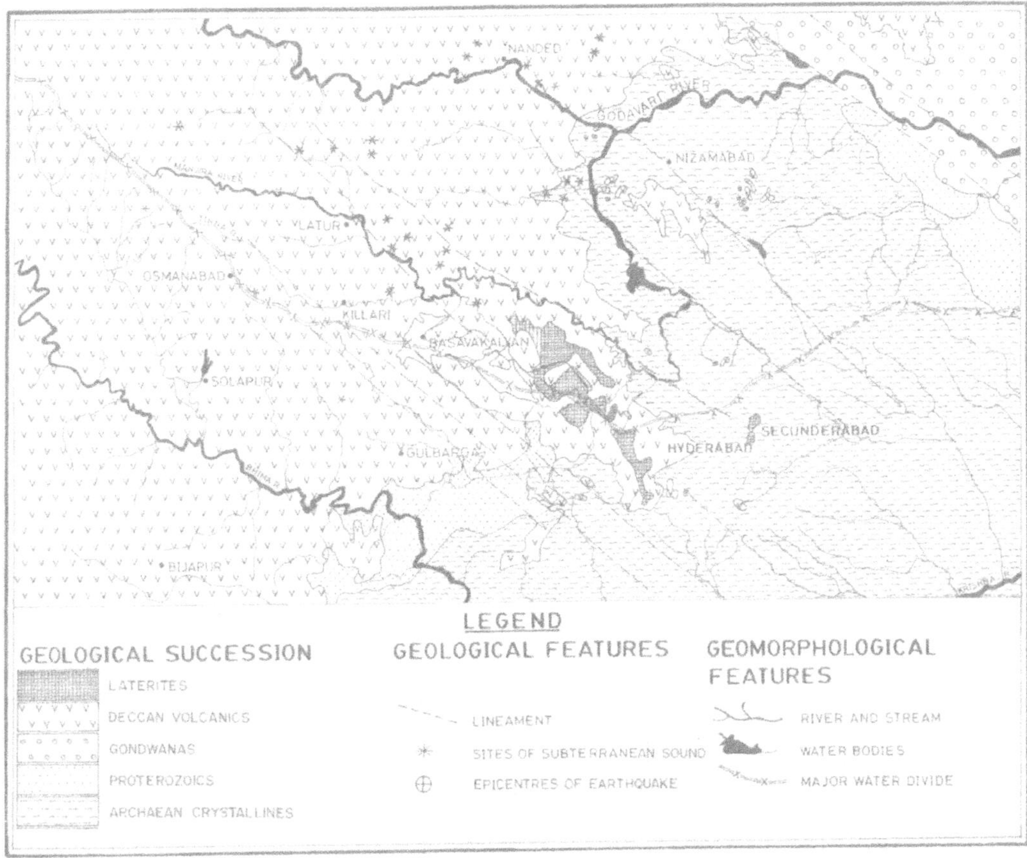

Fig. 9 Geological map of eastern part of Maharashtra and adjoining region.

The Godavari basin occupies the north-eastern part of the map area. The south-western part is drained by river Krishna and its tributaries. River Manjira flowing in south-easterly direction takes a north-easterly course and later on towards north, to join southeasterly flowing Godavari. The lineaments starting from the exposed Achaean basement can be seen continuing below the volcanics. Canal and tunnel system is envisaged for bringing the water from the bend directly to the twin cities.

CONCLUSIONS

The study has led the authors to following significant conclusions –

1. Geomorpholocical and geological study, incorporating hydrological data is required for proper planning of water management and inter-basin transfer of water.
2. Satellite radar imagery has emerged as very important tool in acquiring detailed relief features.
3. Maximum utilization of existing network of river and streams, and construction of small linking canals and tunnels is envisaged.

4. Identification of natural gradient to avoid lifting of water is one of the most significant contributions of the study. The gradient between the rivers Godavari and Krishna on the eastern side of Sahyadri ranges can be utilized as very vital link for transportation of water from central India to deficient southern Kaveri basin. The required quantity of high level water can easily be dropped in any basin of easterly flowing river and electricity can also be generated at suitable sites in both Maharashtra and Karnataka.
5. Suitable sites for construction of dams on river Narmada and Tapti have been identified for transportation of water from western side of Sahyadri to east in Godavari basin.
6. Water supply for present as well as future for the cities of Mumbai and twin cities of Hyderabad-Secundrabad through system of canals and tunnels is suggested.
7. Long term sustained high growth rate can be achieved by proper utilization of run-off Monsoon water to the oceans. This water can be stored and transferred from surplus to deficient river basins to mitigate frequent flood and drought conditions. Cascading dams will insure perennial flow to rejuvenate river systems. Increase in agricultural land and conversion of single to double and even triple crop regions, will also advance fulfillment of national commitment to provide water to every field

REFERENCES

1. Auden, J.B. (1949). Dykes of Western India: A discussion on their relationship with Deccan Traps.Trans. Nat. Inst. Sci. India, v. III.pp.123-157.
2. Dessai, A.G. and Bertrand , H.(1995). The "Panvel Flexure" along the Western Indian continental margin: an extensional fault structure related to Deccan magmatism. Tectonophysics, v. 242, pp.165-178.
3. Dessai , A.G. and Viegas, A.A.A.A. (1995). Multi-generation mafic dyke-swarm related to Deccan magmatism, south of Bombay: implications for the evolution of the western Indian continental margin. *In*: Devaraju, T.C. (Ed), Dyke Swarms of Peninsular India. Mem. Geol. Soc. India, v.33, pp.435-451.
4. GUNNEL, Y. nd Radhakrishna, B.P. (2001) Sahyadri: The Great Escarpment of the Indian subcontinent. Mem. Geol. Soc. India, no.47(1 & 2), ISBN: 81-82867-45-3 & 43-1.
5. Misra ,K.S (2001) The "Panvel Flexure" Part of a Large Structure in Western India, Member Geological Society India No.47,pp 225-232.
6. Misra K.S. (2005) Distribution Pattern, Age and Duration and Mode of Eruption of Deccan and Associated Volcanics, Gond. Geol. Mag. Spl. v. 8, pp 53-60.
7. Misra, K.S (2007) Tectonic history of major geological structures of peninsular India and development of petroliferous basins and eruption of Deccan and associated volcanics, Jour. Geophysics, v. 27, No.3, pp 3-14.
8. Misra, K.S. (2008) Dyke Swarms and Dykes within the Deccan Volcanic Province, India. Indian Dykes, Narosa Publishing House Pvt. Ltd. New Delhi, India, pp 57-72.
9. Misra, K.S. and Misra, Anshuman (2010) Tectonic Evolution of Sedimentary Basin and Development of Hydrocarbon Pools along the Offshore and Oceanic Regions of Peninsular India, Gond. Geol. Mag. Special v. no. 12, pp. 165-176.
10. Misra,K. S ,Misra, Neela, and Misra, Anshuman (2016) Geomorphological and Geological Suitability for Inter-basin Transfer of Water by Linking River Basins in Maharashtra. Geol. Soc. India, Vol., 90, pp 253-255.

11. Radhakrishna, B.P. (1989). Suspect Tectono - stratigraphic terrane elements in the Indian sub-continent. Jour. Geol. Soc. India, v.3.RADHKRISHNA, B.P. (1952) Mysore Plateau, its structural and physiographicevolution. Bull. Mysore Geologists Assoc., No.3.
12. RADHKRISHNA, B.P. (1966) Our Ground water resources and How toutilize them. Bull. Mysore Geologist's Assoc. No.20 and Groundwater Studies No. 5, of BNG.
13. RADHKRISHNA, B.P. (1993) Neogene uplift and Geomorphicrejuvenation of Indian Peninsula. Curr. Sci., v.64, pp.787-792.
14. RADHKRISHNA, B.P. (2003) Linking of Major Rivers of India: Bane orBoon. Jour. Geol. Soc. India, v.61, pp: 251-256.
15. Sawkar, R.H, Rudraiah. R. And Madhav, (2016) Transfer of surplus water from west to east flowing rivers of Karnataka: Management of surface water and groundwater in drought prone areas. Geol. Soc. Ind. Special Publication 5, pp. 76-85.
16. VAIDYANADHAN, R. (1977) Recent Advances in Geomorphic Studies in Peninsular India a review. Indian Jour. Earth Sci., Ray Volume, pp.13-35
17. VALDIYA, K.S. (2001) River response to continuing movements and scrap development in central sahyadri and adjoining coastal belt Jour. Geol. Soc. India, v.57, pp.13-20.

COMPUTATION OF RUNOFF BY SCS CN METHOD USING SPATIAL MODELING IN ERDAS IMAGINE

G. Sreenivasa Rao[1], M.V.S.S. Giridhar[2] and Shyama Mohan[3]

[1&2]Associate Professor, [3]Student,

[1]Department of Civil Engineering, MVSR Engineering College Hyderabad, India,

[2,3]Centre for Water Resources, IST, JNTUH, Hyderabad

ABSTRACT

Runoff estimation is requirement in a watershed for design of hydraulic structures, reservoir maneuver and for control measures of soil erosion. Runoff is an important parameter for water resource planning and management which is significant during draught conditions, regions of arid and semi-arid. The occurrence of runoff is affected by several factors like geo-morphological structures and type of watershed with different land use changes that effect the volume of runoff and rate of runoff significantly. Hence, a hydrological model depends on land use and type of soil to estimate runoff volume. For the computation of runoff the soil conservation service curve number (SCS-CN) method is used for a given rainfall for different Land Use/Land Cover in the study. In this study remote sensing (RS) and geographic information systems (GIS) are applied in combination with the SCS-CN method for precise and timely estimation of runoff. The remote sensing imagery used are Landsat ETM+ images and soil maps are pre-processed and maps are generated using ERDAS IMAGINE and ArcGIS softwares. The different land use land cover was generated from Landsat satellite images for the years of 2004 and 2011. The most predominant land use land cover was found to be agricultural land and water bodies. The soil groups were categorized into sol hydrological group depending on the infiltration capacity of the soil. Three types of soil hydrologic groups were mainly found namely A, B and C categories. The CN maps were prepared using the ArcGIS software with the input as land use land cover and soil map. At the final stage runoff was computed for 15 years from 2000 to 2014. The study area showed significant land use land cover changes from the year 2004 to 2011. The statistics were computed shows different runoff grid value varying across the years from 0 being the lowest when no rainfall occurred and 92.3 being the highest runoff during highest rainfall record for the year 2006.

Keywords: Runoff, SCS-CN method, ERDAS IMAGINE, Spatial modeler, ArcGIS

INTRODUCTION

In hydrological studies the most prominent problem encountered for the water balance estimation is there are records of rainfall but no records of runoff. Hence estimation of runoff is particularly important in hydrological engineering and modeling and can be used in different applications like estimation of water balance and flood design etc. There are many methods for estimating the runoff but SCS CN method is the widely used method of all due to the well established method. The integration of remote sensing and GIS has been increasingly used for the water resources planning and management. The modeler that has been developed mostly deals with the land phase data related to the topography, rainfall and other parameters of the catchments which is a necessary pre requisite. The integration of this method with GIS ensures the quality of the output with less demand on labour and human resources, which can also be less time consuming. In hydrological cycle infiltration of the basin soil is one of the most important components affecting hydrologic cycle. Hence depending on the type of the soil the infiltration rates are considered for the study area.

REVIEW OF LITERATURE

Berod et al. (1999) reported that the SCS- curve number technique has been in use widely since the tabulated curve number values provide a easy way of moving from GIS data set on soils and vegetation to a rainfall runoff model. Sarita-Sajbhive (2015)reported that_RS & GIS technique is a reliable alternative as a dependable support system to our conventional way of Planning, Surveying, Investigation, and Monitoring, Data Storing, Modeling and Decision making process. The land use/land cover parameters can be estimated accurately with the help of GIS techniques and curve number method is empirical approach to determine the runoff depth curve number method is used or measuring the runoff of un-gauged statics.

Giridhar et al. (2007) reported that all the hydrological parameters which are spatially and temporally variable were found to be more accurately estimated through RS and GIS. The average runoff was estimated for the Palleru sub basin as 37% and 39 % of rainfall from SCS-CN method and STANFORD-IV model respectively. It indicated that SCS-CN method underestimated the average yearly runoff by 2%. The yearly average runoff was estimated to be 41% and 39 % of rainfall from the ARM and STNAFORD-IV models respectively and there by indicated that ARM model overestimated runoff by 2%.

Nagarajan and Poongathai (2012) stated that Rs & GIS based SCS- CN model can be used effectively to estimate the runoff from the ungauged watersheds. The appropriate soil and water conservation measures must be planned and implemented first in the mini watersheds classified as high, followed by the Water sheds classified as moderately high , for controlling runoff and soil loss.

Orlandin and Rosso, (1996) stated the advances in remote sensing, geographic information systems, and computer technology make distributed hydrologic model approach attractive to flow simulation and prediction. GIS is useful in dealing with linkage of a distributed hydrologic model with the spatial data contained in geomorphologic maps and digital elevation models (DEM) offering the advantage given by full utilization of the spatially distributed data for analysis of hydrologic processes .

STUDY AREA

The Study area selected is Kaddam watershed (Figure 1) present in the G-5 sub basin which is the 'Middle Godavari' Sub basin of Godavari River Basin. The Godavari basin extends over an area of 3, 13,812 Sq.km. Godavari catchment is divided into eight sub basin in which G-5 sub basin is one of the basin, it lies between latitudes 17° 04' N and 79° 53' E longitude. The study area selected in the Middle Godavari sub basin is considered up to Kaddam reservoir watershed which lies between 19° 05' E and 19° 35' N latitudes and 78° 10' and 78° 55' E longitudes. The watershed covers a total of twelve mandals of which eight mandals are taken Khanapur, Boath, Ichchoda, Narnoor, Utnoor, Indervelly, Bazarhatnoor and Kaddam all of which fall under Adilabad district.

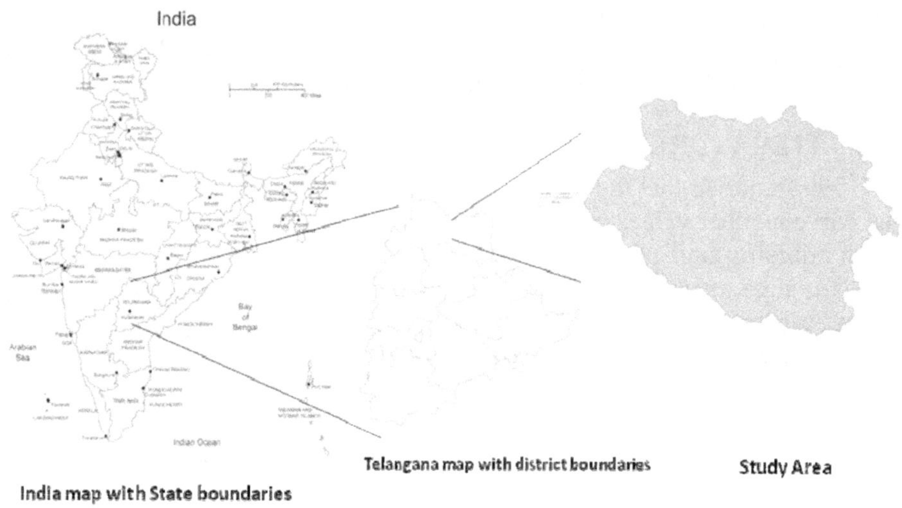

Fig. 1 Location map of Kaddam Watershed.

METHODOLOGY

Daily rainfall data have been collected for the years of 2000 to 2014 of eight rain gauge stations and are processed in the excel sheet as to requirement to obtain an interpretative spatial map. Later as the input requires the land use land cover maps are prepared using the satellite images and soil maps are processed for the study area. Arc GIS version 10.2.2 is being used for managing and generating the spatial maps and ERDAS IMAGINE software is used for the modeling of runoff computation and for further analysis.

Preparation of rainfall maps

Every day Rainfall Data for a period of 15 years from 2000 to 2014 of eight gauge stations are collected and processed on Excel sheets as indicated by the prerequisites to acquire interpretative area map. Arc GIS software version 9.3 was used for creating, managing and generating rainfall maps.

Preparation of Land Use land cover

The satellite used was georeferenced Landsat 7 ETM+ satellite image for the preparation of land use land cover images. The satellite images for two years have been taken into account i.e., 2004 and 2011. These images were preprocessed and then classified using the ERDAS IMAGINE software. The classification is done using the supervised classification were multiple training sets for each set of classes is taken. Five land use land cover classes have been identified for the classification viz., Water Body, Forest cover, Agriculture Cover, Wasteland and Fallow as shown in Fig 2. The distribution of LULC classes for different years is mentioned in Table 1.

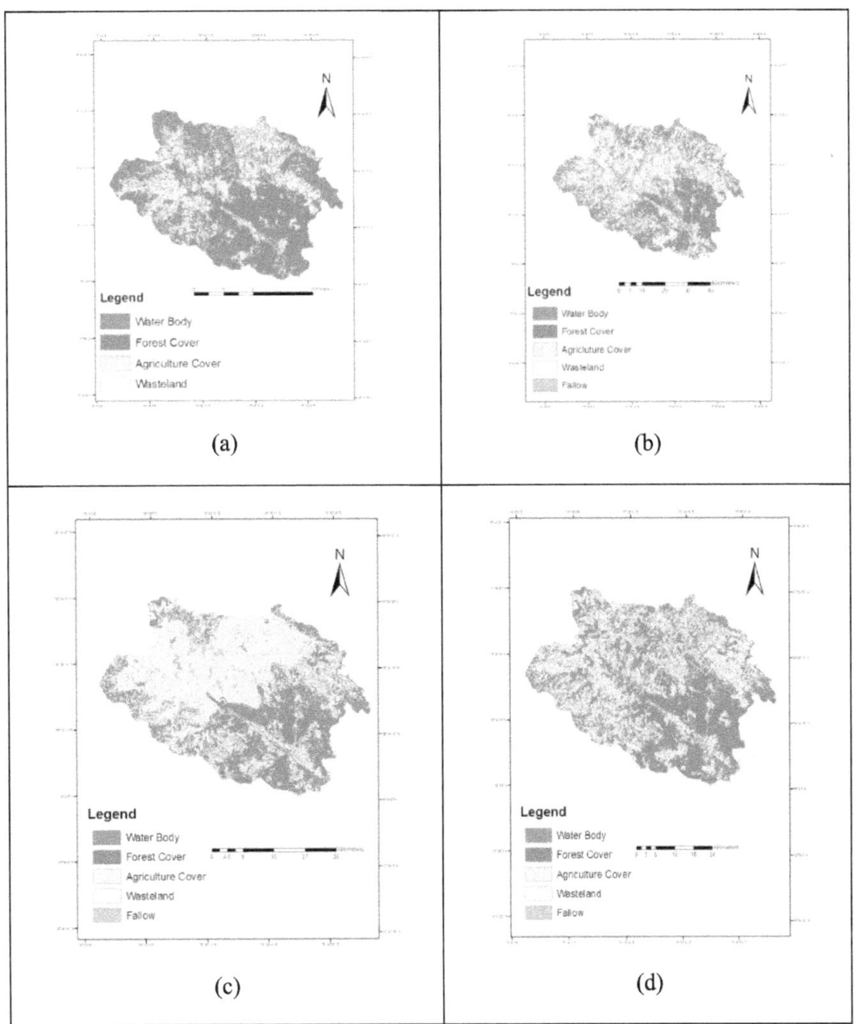

Fig. 2 (a) Land Use Land Cover image for May 2004 (b) Land Use Land Cover image for November 2004 (c) Land Use Land Cover image for February 2011 (d) Land Use Land Cover image for December 2011.

Table 1 Area in % of Land Use Land Cover for Kaddam Watershed.

S.No.	Land Use Type	Area in % for May 2004	Area in % for November 2004	Area in % for February 2011	Area in % for December 2011
1	Water Body	0.45	0.41	1.03	1.21
2	Forest Cover	69.00	52.87	47.99	55.89
3	Agriculture Cover	29.37	37.19	34.96	28.77
4	Wasteland	1.18	2.77	1.50	0.59
5	Fallow	0.0	6.76	14.53	13.54
	Total	100	100	100	100

Generating Curve Number Maps

The inputs land use land cover, soil maps and DEM are used for generating the curve number maps. In the present study DEM remote sensing images SRTM with 90 m is assessed in watershed delineation and drainage network generation. The curve number is generated using the HEC-GeoHMS extension tool of ArcGIS 10.2.2. In this study, HEC-GeoHMS is used to derive river network of the basins and to delineate sub basins of the basins from the digital elevation model (DEM) of the basins. Hydro-meteorological parameters like rainfall are collected from the different gauging station in the watershed, DEM delineated and extracted for the study area and input parameters like land-use/land-cover data of year and soil data are used for the curve number generation using the equation (1).

$$S = \frac{25400}{CN} - 254 \qquad \text{...(1)}$$

Computation of Runoff

After the generation of curve number maps the runoff is estimated. The estimation of runoff is carried out in spatial modeler in ERDAS IMAGINE. The model shown in figure 3 have been developed using the SCS-CN method equation 2, taking the input as rainfall and surface retention, that have been derived from the curve number.

$$Q = \frac{(P-0.3S)^2}{(P+0.7S)} \qquad \text{...(2)}$$

Where,

Q = Surface runoff

P = Rainfall in mm

S = Surface retention

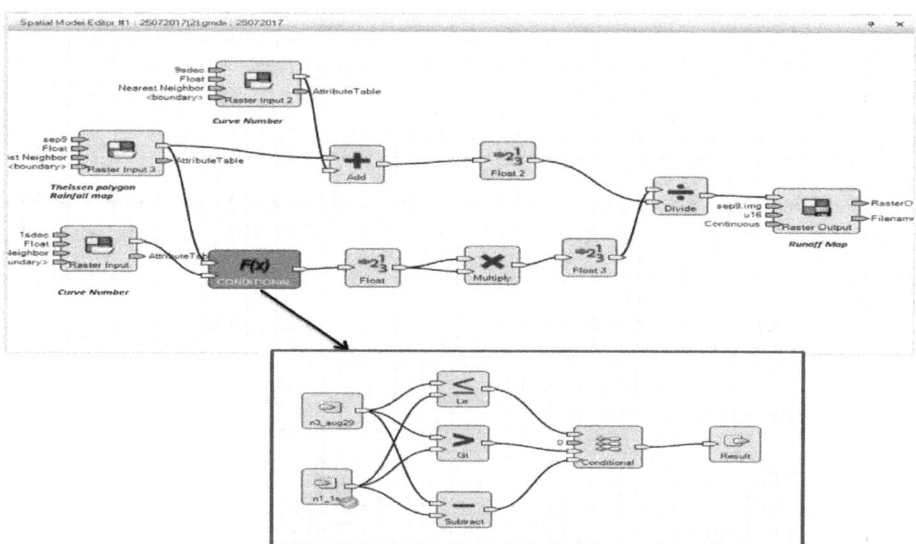

Fig. 3 Spatial Modeler of ERADS IMAGINE 2014 for estimation of Runoff.

RESULTS

For the generation of curve number, land use land cover map, hydrological soil map and runoff maps have been prepared and the results of these are explained below.

Hydrological Soil Map

The different hydrologic soil group is generated as one of the input for the computation of curve number. The hydrologic soil groups are generated depending on the type of soil the study area has. The hydrological soil group is attributed to soil mapping unit and the groups that are assigned are A, B, C and D as shown in figure 4. Each soil group has different runoff potential with different infiltration rates as shown in Table 2.

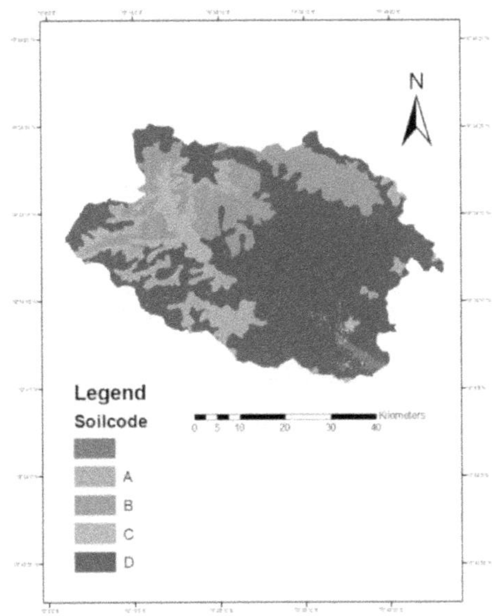

Fig. 4 Soil Map showing different Hydrologic Groups.

Table 2 Infiltration rate and runoff potential of Soil type.

S.NO.	Soil Type	Soil Code	Infiltration rates	Runoff Potential
1	Clayey soil	D	Low	High runoff
2	Cracking clay soil	C	Moderate	Moderately high runoff
3	Loamy soil	B	Moderate	Moderate runoff
4	Gravely loamy soil	A	High	Low Runoff
5	Gravelly clay soil	A	High	Low Runoff

Preparation Curve Number (CN) Maps

The curve number maps are prepared using the HEC-GeoHMS tool of ArcGIS. The Curve numbers obtained were different depending on the type of land use and soil type. The values of curve number varied from 30 to 100 (100 for water body), 30 being the least especially in the agriculture area and highest in the forest area which was found to be 85. High CN value explains that the land is exposed land and rocky outcrop that can explain the low vegetation density, with the soil showing the compactness proving the area is predominantly clay and low infiltration indicates the presence of stony surface. Curve number maps are shown in Figure 5 obtained for different years.

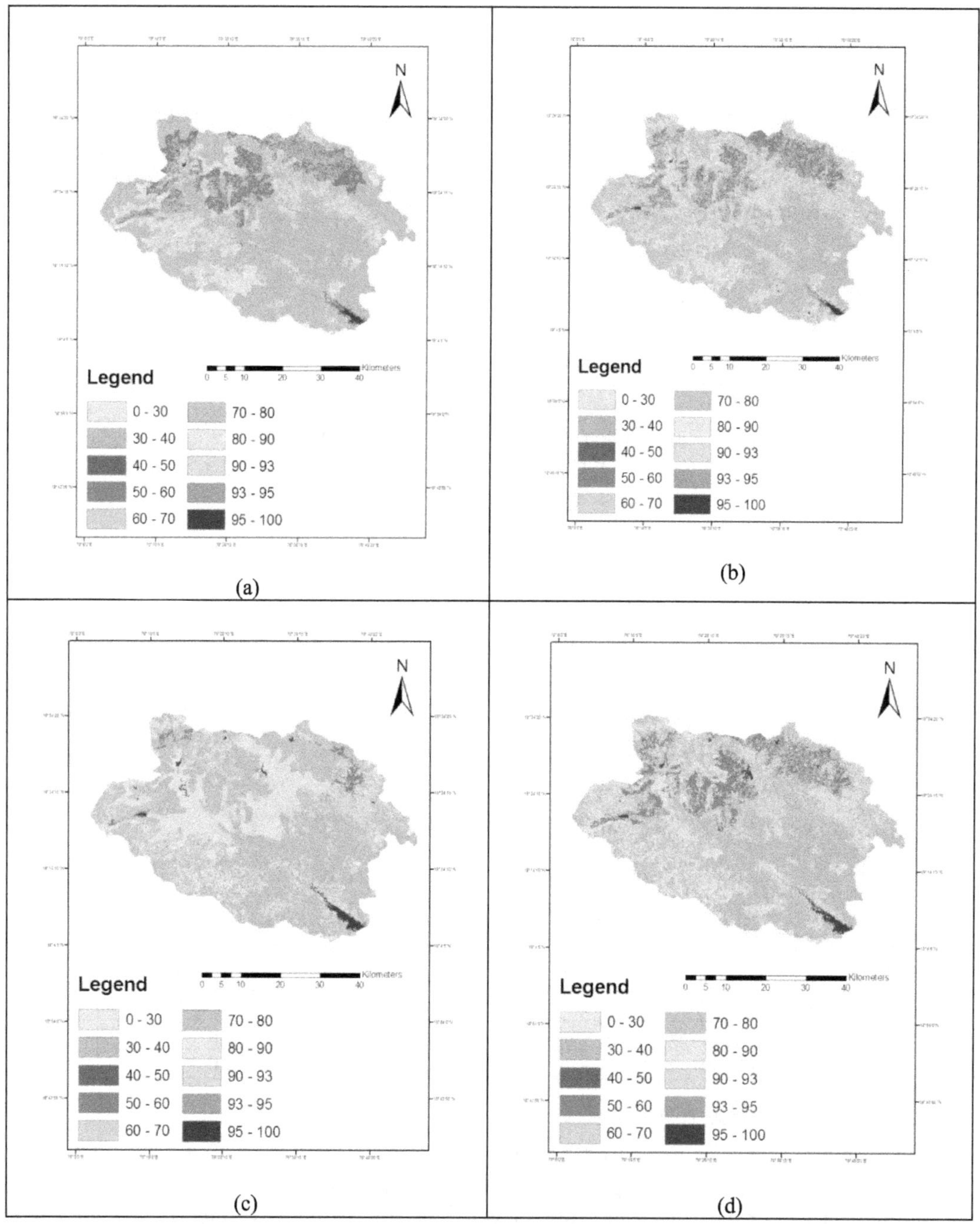

Fig. 5 (a) Curve Number image for May 2004 (b) Curve Number image for November 2004 (c) Curve Number image for February 2011 (d) Curve Number image for December 2011.

Preparation of Runoff Maps

The Curve number and rainfall maps are used as the input for the computation of runoff. The daily rainfall maps are used as the input for obtaining the daily runoff data for the total 15 years from 2000 to 2014. The curve number images are given as input to obtain the surface retention which is then used for the calculation of runoff. The runoff has been generated for the whole catchment lying in the study area. As daily runoff is calculated large data has been obtained and only for a month is shown in figure 6 generated in ERDAS IMAGINE using the spatial modeler for the year 2000. The average runoff is calculated and shown table 3 for 15 years. The statistics for the 15 years have been calculated were 0 shows the lowest value during which no rainfall has occurred and 92.3 mm being the highest runoff. The daily runoff was highest during the year 2006 which had a rainfall of 173 mm in the month of august. Mostly the recurring runoff values were found to be between 0-27, 0 was mainly seen in the water bodies. For the average rainfall taken for 14 years was 50.06 mm and the runoff was found to be 13.62 mm.

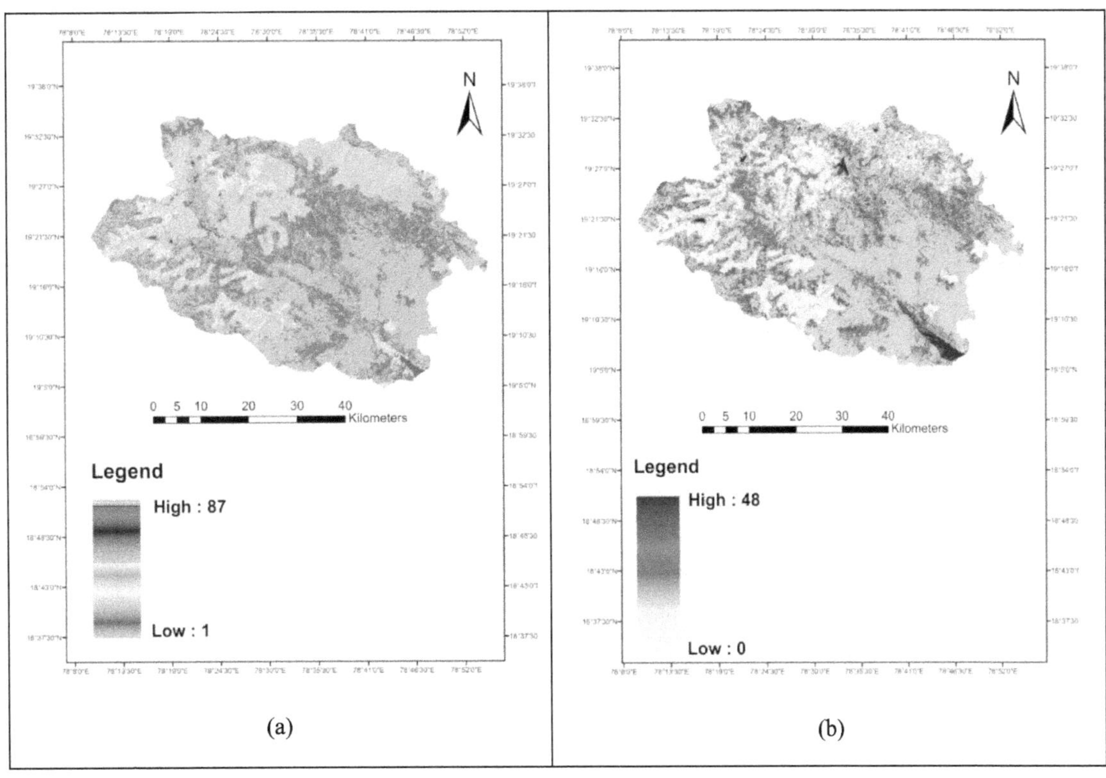

(a) (b)

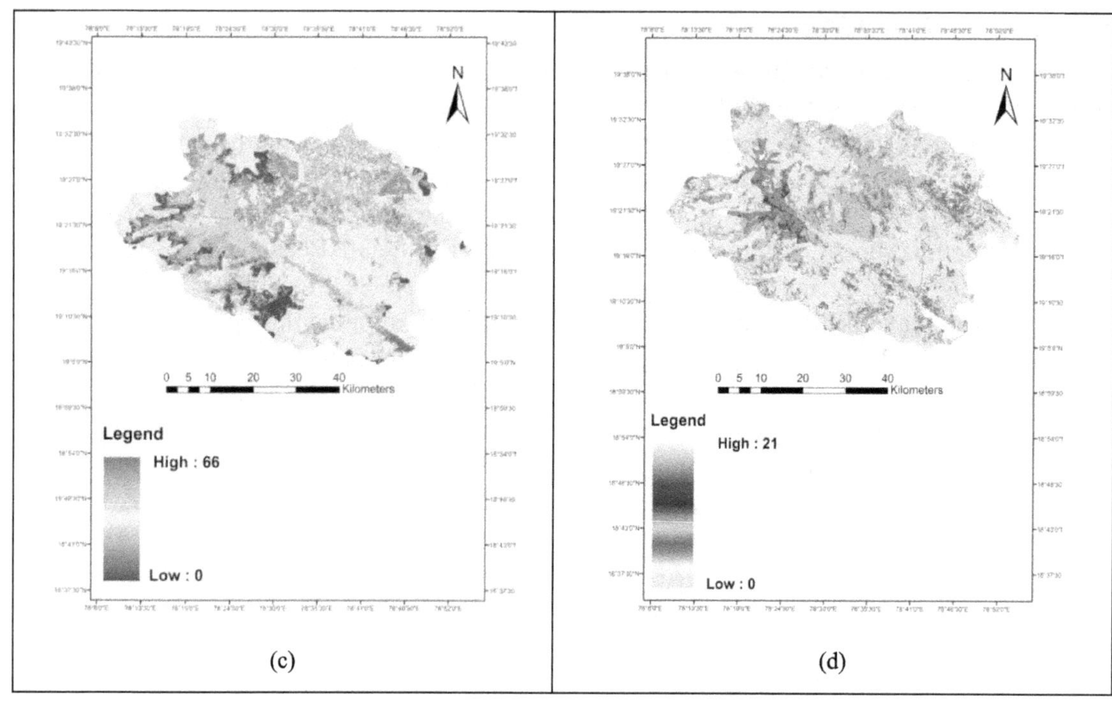

Fig. 6 (a) Daily Runoff in mm for 6th Aug for the year 2006 (b) Daily Runoff in mm for 18th June for the year 2014 (c) Daily Runoff in mm for 25th June for the year 2002 (d) Daily Runoff in mm for 2nd Sep for the year 2009.

Table 3 Runoff grid value for average rainfall 50.06mm.

S.no.	Runoff Grid Value	Pixel Count	Area in %
1	0	2459	5.72
2	4	629	1.46
3	9	870	2.02
4	12	33752	78.60
5	21	64	0.14
6	22	3172	7.38
7	33	215	0.50
8	50	1780	4.14

CONCLUSIONS

The computation of runoff can be done using many empirical formulas but for the calculation of data for many years is a cumbersome process. Hence using the modeler for the calculation of runoff was easier and less time consuming process. The calculation of runoff in GIS has found to be much uncomplicated and easy. The incorporation of SCS-CN model and GIS facilitates for runoff estimation improves the accuracy of estimated runoff. The accurate assessment of runoff for all the stream and watershed may take time and develop management problems.

Remote sensing technology has been of great value that makes the conventional method easier to a great extent in rainfall-runoff studies. The developed ArcCN-Runoff tool for the computation of curve number reduced the technical processing and the model developed in the spatial modeler in ERDAS IMAGINE for the runoff estimation made work easier from days to hours.

REFERENCES

1. Anubha Topno, Singh A.K. and Vaishya R.C (2015) "SCS CN Runoff Estimation for Vindhyachal Region using Remote Sensing and GIS", International Journal of Advanced Remote Sensing and GIS 2015, Volume 4, Issue 1, pp. 1214-1223.
2. A.S Chandra Bose, MVSS Giridhar, GK Viswanadh (2013), "GIS-based fully distributed rainfall-runoff model for suggesting alternate land use patterns", World Environmental and Water Resources Congress 2013, ASCE publisher, p-2060-2068.
3. Berod D. D., Singh V. P. and Musy A. (1999) "A geomorphologic kinematic-wave (GKW) model for estimation of floods from small alpine watersheds". Hydrological Processes 13:1391-1416.
4. Chatterjee C, Jha R, Lohani AK, Kumar R and Singh R (2001) "Runoff curve number estimation for a basin using remote sensing and GIS". Asia Pacific Remote Sensing GIS J. 4, 1-7
5. Giridhar. M.V.S.S and Viswanadh.G.K, "Development of Semi Distributed Conceptual Rainfall-Runoff Model for a Semi Arid area using STANFORD-IV Model" Procs of the international conference, organized by IASTED, at Honolulu, Hawaii, USA Aug. 2007.
6. G. Sreenivasa Rao, M.V.S.S. Giridhar and Shyama Mohan (2017) "Rainfall analysis with reference to Spatial and Temporal in Kaddam Watershed Using Geomatics", Nattional Conference on Water, Environment and Society (NCWES-2017).
7. G. Sreenivasa Rao, M.V.S.S. Giridhar, Shyama Mohan and P.Sowmya, "SCS-CN Method and Geomatics Approach for Fully distributed Runoff Modelling" International Journal of Computational Engineering Research (IJCER), ISSN (e): 2250 – 3005, Volume 07-09 September – 2017.
8. N. Nagarajan and S. Poongothai, "Spatial Mapping of Runoff from a Watershed Using SCS-CN Method with Remote Sensing and GIS" Journal of Hydrologic Engineering Volume 17 Issue 11 - November 2012.
9. S Gajbhiye (2015), "Estimation of Surface Runoff Using Remote Sensing and Geographical Information System", International Journal of U-and E-Service, Science and Technology 8 (4), 118-122

AQUIFER PERFORMANCE TESTS IN RCI RESIDENTIAL AREA SAROOR NAGAR MANDAL, RR DISTRICT, TELANGANA STATE

A. Manjunath[1], G. Sreenivasa Rao [2] and A. Samba Shiva Rao[3]

M.V.S.R. Engineering College, Hyderabad

manjunath.aluru@gmail.com

ABSTRACT

Aquifer is the water bearing geological formation, is very important in ground water yielding pump out test is the one of method for determination of aquifer performance. In the present work pump out test are conducted in Research Center Imrath (RCI), HYDEARABAD In the study area, pumping test was conducted in seven tube wells with depths ranging from 17 to 40 meters. Each well is tapping 3 to 9 meters of aquifer thickness. The aquifers of the study area, based on their litho- logy can be grouped under one category– non- leaky confined aquifer.

Keywords: Aquifer, Transmissivity (T), Storativity (S), Geomorphology.

1. INTRODUCTION

Groundwater is the water present beneath Earth's surface in soil pore spaces and in the fractures of rock formations. A unit of rock or an unconsolidated deposit is called an aquifer when it can yield a usable quantity of water. The depth at which soil pore spaces or fractures and voids in rock become completely saturated with water is called the water table.

An aquifer is a layer of porous substrate that contains and transmits groundwater. When water can flow directly between the surface and the saturated zone of an aquifer, the aquifer is unconfined. The deeper parts of unconfined aquifers are usually more saturated since gravity causes water to flow downward. The upper level of this saturated layer of an unconfined aquifer is called the water table or phareatic surface. Below the water table, where in general all pore spaces are saturated with water, is the phreatic zone. Confined aquifer is the one bounded by impermeable formations, in which ground water pressure is not equal with atmospheric pressure pump Aquifer Performance is nothing but measure the Storativity(S), Transmissivity, (T). This can be done by performing pump out tests.

2. STUDY AREA

2.1 TOPOGRAPHIC CHARACTERISTICS

The RCI Complex is located about 25 Km from Hyderabad on the Hyderabad-Srisailam State Highway and about 2 Km from Pahadisharif Village. The site is bounded by 17 degree 15’ to 17 degree 17’ N latitude and 78 degree 28’ to 78 degree 30’ E longitude. The RCI complex has undulating terrain with low rocky hillocks and boulders scattered throughout Rocky Ridges run along the northern limits of the area and also along the south-eastern corner. The ground elevation of the area varies from 570 m to 638 m above MSL, the average altitude being about 580 m above MSL. The land is generally terrain and under scrub jungle and scattered palmyra palm trees. There is general slope from south-east to north-east, and the run-off water gets drained out outside the complex area. Rock outcrops in the from of small hill mounds are scattered at the site which lend a distinctive color and texture to otherwise barren land with scrum jungle and occasional trees.

2.2 GEOMORPHOLOGY& GEOLOGY

Relict Topography is formed due to Differential Erosion. Soil Type : Sandy Silty & Gravely Soils, rock Types : Granites & Dolerites.Texture : Medium grained.

Extent of Rock Exposed : Granite Boulders are extensively exposed due West & Northwest of the site and are intruded with Dolerite Dyke trending in Northwest – Southeast direction. A quartz vein intruded into Weathered & fractured granites is Seen due Southeast of the site.

Nature and Depth of Weathering: Erratically weathered to Shallow depths. Depth of Weathering may extend from 7 to 10 m BGL.) Major Structural features : Dolerite Dyke boulders with a Width of about 10 – 15 m is well Exposed due West, at a distance of 100m, Near pump house site, trending in Northwest - Southeast direction. Hydro geological : The area towards Northeast direction.

3. METHODOLOGY

The basic concept of a pumping test is very simple , water is abstracted from a well or borehole, thus lowering the water level. The water level in the abstraction borehole and the pumping rate are monitored over time, along with various other parameters if possible (such as water levels in observation boreholes). The way in which the water levels respond to the pumping is then analyzed to derive information about the performance characteristics of the borehole and the hydraulic properties of the aquifer. In reality, the situation is much more complicated. There are many different types of test from which to choose (intermittent or continuous, short or long in duration, low or high pumping rates, etc.). What other parameters or water features should be monitored in addition to the obvious ones, i.e. the water level and pumping rate in the borehole being tested, Can conclusions about long-term behavior be extrapolated with confidence from the results of a short pumping test. The main problem with investigating groundwater in that one is effectively working blindly, because it is impossible to see into the aquifer and directly observe its behavior. One can infer information about the borehole and aquifer only by observing how the water level changes in response to pumping.

Experimental procedure:

Measure the initial measurement of water level with respect to base ground level i.e. the static water level of both test well as well as observation well in meters in the data sheet using the water level indicator before switching on the motor .Now, switch on the motor for 1 hour and within this 1 hour of pumping out, we have to note the measurements of water level depths in meters with respect to base ground level for every 15 minutes .After 1 hour of pumping, switch off the motor and from that point measure the water level depths with respect to ground level for every 2 minutes for the first 20 minutes in the recovery stage .After 20 minutes, measure the water level depth for every 5 minutes for the next 30 minutes. Again after 30 minutes of measurements, now measure the water level depths for every 10 minutes for next 30 to 40 minutes' Now measure the water level depth in observation well after taking the observations in test well .Calculate the Discharge rate of the well, i.e. rate at which water is pumped out of the well. After noting all the measurements, calculate the drawdown and residual drawdown in meters by subtracting the water level depths with the initial static water level. Draw drawdown curves between drawdown values and time in minutes Also draw a recuperation curve between residual drawdown and time in minutes·

Table 1 Litho-Log of the Station 1

0	*1.0m*	Soil.
1.0	3.2m	Disintegrated Rock.
3.2	7.3m	Weathered Rock.
7.3	70.m	Hard Rock.
70.0	75.0m	Rock with intermittent fractured zones.
75.0	120.0m	Hard Rock.
120.0	130.0m	Rock with intermittent fractured zones.

Table 2 Drawdown curve for station 1 during pumping stage

Time 10:00am	**Time(t) since pumping began(min) 15**	**Depth to water level (m) 16.7132**	**Drawdown (m) 2.7178**	**Pumping rate(Q) (m^3/Sec) 4.711×10^{-3}**
10:15am	30	16.738	2.7426	
10:30am	45	16.738	2.7426	
10:45am	60	16.738	2.7426	

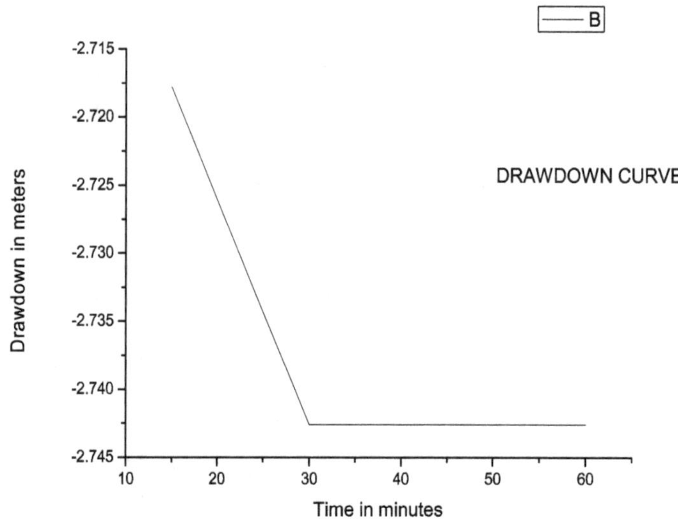

Graph 1: Drawdown curve for station 1 during pumping stage

Table 3 Data regarding time vs. residual drawdown at station 1 during recovery.

Time	Time	Depth	Residual dradown
10:45am	0	16.738	2.7426
10:45am	0	16.738	2.7426
10:47am	2	15.976	1.9806
10:49am	4	15.443	1.4476
10:51am	6	15.062	1.0666
10:53am	8	14.884	0.8886
10:55am	10	14.782	0.7866
10:57am	12	14.655	0.6596
10:59am	14	14.579	0.5836
11:01am	16	14.528	0.5326
11:03am	18	14.503	0.5076
11:05am	20	14.478	0.4826
11:10am	25	14.427	0.4316
11:15am	30	14.376	0.3806
11:20am	35	14.325	0.3296
11:25am	40	14.270	0.2746

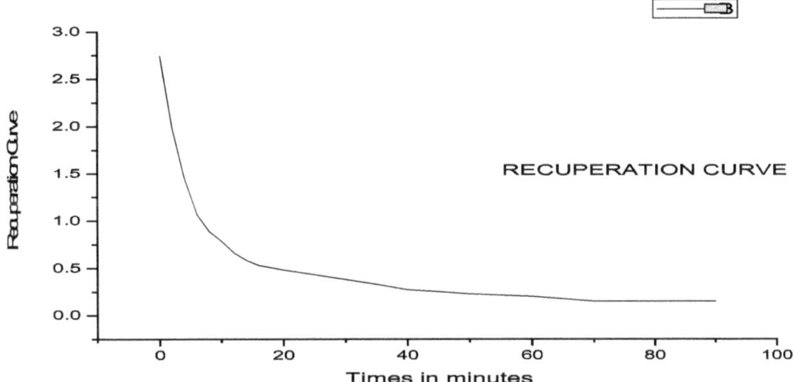

Graph 2: Regarding time vs. residual drawdown at station 1 during recovery

Table 4 Litho-log of the station No 2

0	4.4m	Soil followed by disintegrated rock
4.4	7.4m	Weathered Rock
7.4	70.0m	Hard Rock.
70.0	75.0m	Rock with intermittent fracture zones.
75.0	120.0m	Hard Rock.
120.0	130.0m	Rock with intermittent fractured zones.
0	4.4m	Soil followed by disintegrated rock

Table 5 Drawdown curve for station 2 during pumping stage

Time	Time(t) since pumping began (min)	Depth to water level (m)	Drawdown (m)	Pumping rate (Q) (m^3/Sec)
2:41pm	15	6.883	1.168	3.156×10^{-3}
2:56pm	30	6.959	1.244	
3:11pm	45	6.959	1.244	
3.26PM	60	6.959	1.244	

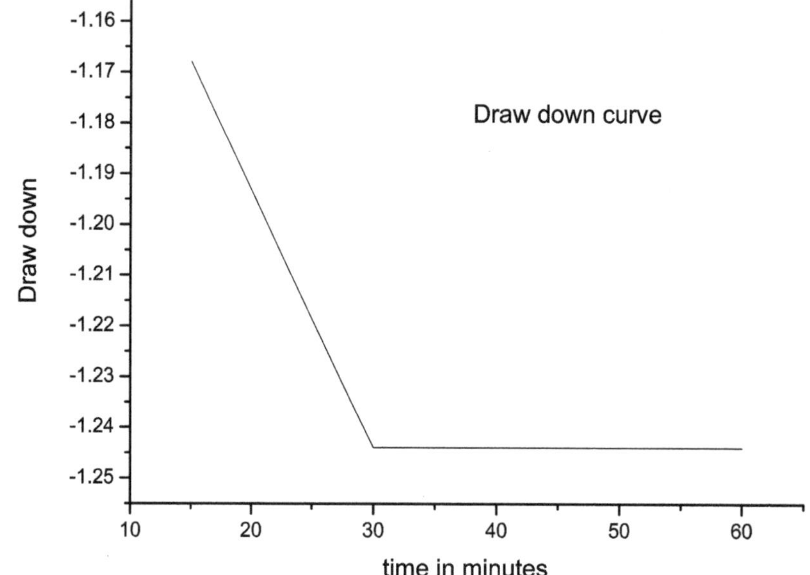

Graph 3: Draw down curve

Table 6 Data regarding time vs residual drawdown at station 2 during recovery

time	Time	Depth	Residual dradown
3:26pm	0	6.959	1.244
3:28pm	2	6.451	0.736
3:30pm	4	6.299	0.584
3:32pm	6	6.096	0.381
3:34pm	8	5.867	0.152
3:36pm	10	5.791	0.076
3:38pm	12	5.765	0.05
3:40pm	14	5.740	0.025
3:42pm	16	5.740	0.025
3:44pm	18	5.740	0.025
3:46pm	20	5.740	0.025

Table 7 Litho-log of the station no. 3

0	1.0m	Soil
1.0	10.0m	Weathered Rock
10.0	55.0m	Rock with intermittent fractured zones.
55.0	60.0m	Hard Rock
60.0	65.0m	Rock with intermittent fractured zones.
65.0	120.0m	Hard Rock
120.0	130.0m	Rock with intermittent fractured zones.

Table 8 Drawdown curve for station 3 during pumping stage

Time	Time(t) since pumping began(min)	Depth to water level (m)	Drawdown (m)	Pumping rate(Q) (m^3/Sec)
4:48pm	15	7.883	1.168	3.156×10^{-3}
4:50pm	30	7.959	1.244	
4:52pm	45	7.959	1.244	
4:54pm	60	7.959	1.244	

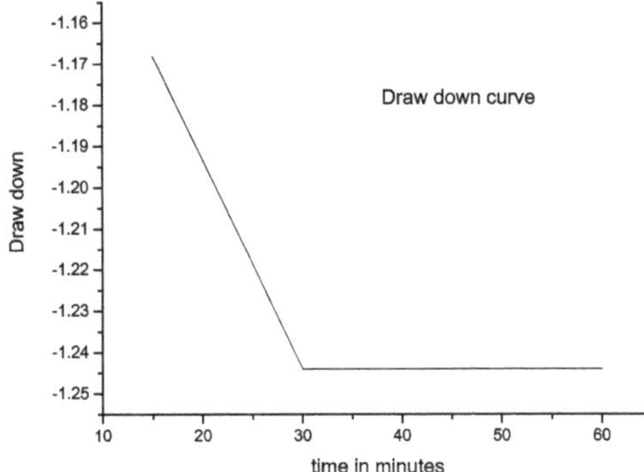

Graph 4: Draw down curve

Table 9 Data regarding time vs residual drawdown at station 3 during recovery

Time	Time(t') since	Depth	Residual
4:48pm	0	10.388	2.286
4:50pm	2	8.102	0
4:52pm	4	8.102	0
4:54pm	6	8.102	0
4:56pm	8	8.102	0
4:58pm	10	8.102	0
5:00pm	12	8.102	0
5:02pm	14	8.102	0
5:04pm	16	8.102	0
5:06pm	18	8.102	0
5:08pm	20	8.102	0
5:13pm	25	8.102	0
5:18pm	30	8.102	0

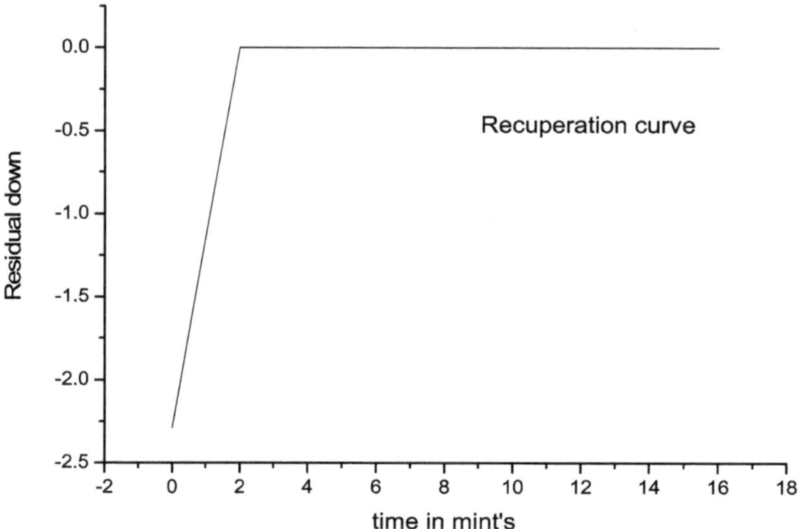

Graph 5: Recuperation curve

Table 10 Litho-log of the staton. no 4

0	1.0m	Soil
1.0	10.0m	Weathered Rock
10.0	65.0m	Rock with intermittent fractured zones.
65.0	70.0m	Hard Rock
70.0	75.0m	Rock with intermittent fractured zones.
75.0	130.0m	Hard Rock
130.0	140.0m	Rock with intermittent fractured zones.

Table 11 Drawdown curve for station 4 during pumping stage

Time	**Time(t) since pumping began(min)**	**Depth to water level (m)**	**Drawdown (m)**	**Pumping rate(Q) (m^3/Sec)**
6:30pm	15	6.883	1.168	3.156×10^{-3}
6.32PM	30	6.959	1.244	
6:36pm	45	6.959	1.244	
6.38PM	60	6.959	1.244	

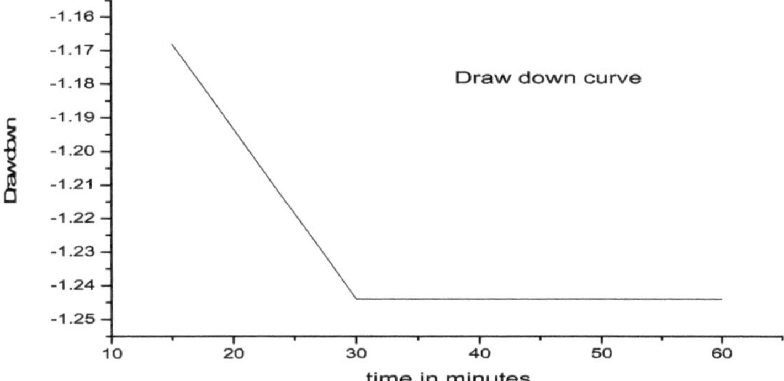

Graph 6: Draw Down Curve

Table 12 Data regarding time vs residual drawdown at station 4 during recovery

Time	Time(t') since pumping stopped(min)	Depth to water level (m)	Residual Drawdown (m)
6:30pm	0	12.750	1.422
6:32pm	2	11.557	0.229
6:36pm	6	11.480	0.152
6:38pm	8	11.455	0.127
6:40pm	10	11.404	0.076
6:42pm	12	11.379	0.051
6:44pm	14	11.379	0.051
6:46pm	16	11.354	0.026
6:48pm	18	11.354	0
6:50pm	20	11.328	0
7:00pm	30	11.328	0

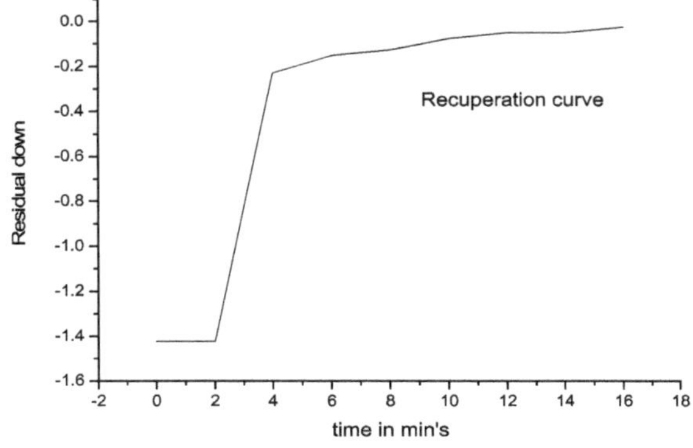

Graph 7: Recuperation Curve

Table 13 Litho-log of the staton. No5

0	1.0m	Soil
1.0	7.0m	Weathered Rock
7.0	55.0m	Hard Rock
55.0	60.0m	Rock with intermittent fractured zones.
60.0	80.0m	Hard Rock
80.0	85.0m	Rock with intermittent fractured zones.
85.0	120.0m	Hard Rock

Table 14 Data regarding time vs drawdown at station 5 during pumping stage.

Time	Time(t) since pumping began(min)	Depth to water level (m)	Drawdown (m)	Pumping rate(Q) (m^3/Sec)
7:07pm	0	5.562	0	1.777×10^{-3}
7:22pm	15	12.420	6.8586	
7:37pm	30	12.420	6.8586	
7:52pm	45	12.420	6.8586	
8:07pm	60	12.420	6.8586	

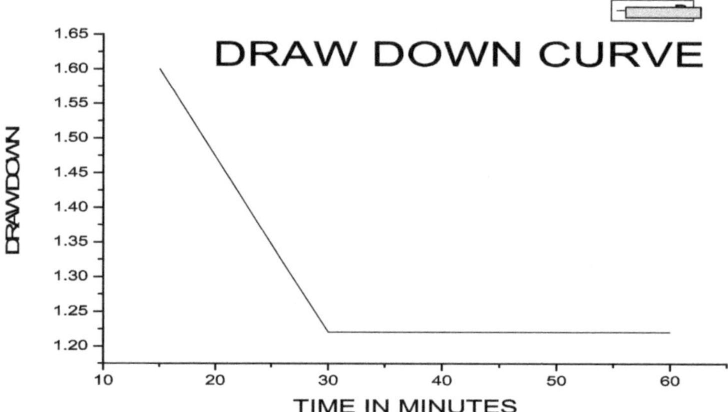

Graph 8: Drawdown Curve

Table 15 Data regarding time vs residual drawdown at station 5 during recovery

Time	Time(t') since pumping stopped(min)	Depth to water level (m)	Residual Drawdown (m)
8:07pm	0	12.420	6.8586
8:09pm	2	12.319	6.757
8:11pm	4	12.242	6.68
8:13pm	6	11.912	6.35
8:15pm	8	11.912	6.35

Contd...

Time	Time(t') since pumping stopped(min)	Depth to water level (m)	Residual Drawdown (m)
8:17pm	10	10.769	5.207
8:19pm	12	9.931	4.369
8:21pm	14	8.331	2.769
8:23pm	16	7.315	1.753
8:25pm	18	6.578	1.016
8:27pm	20	6.096	0.534
8:32pm	25	5.664	0.102
8:37pm	30	5.588	0.026
8:42pm	35	5.588	0.026
8:47pm	40	5.562	0
8:52pm	45	5.562	0
9:07pm	60	5.562	0

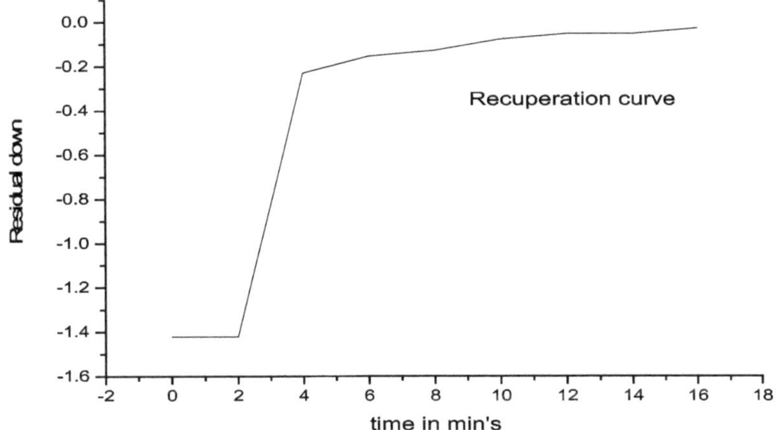

Graph 9: Recuperation Curve

Table 16 Litho-log of the staton. No 6

0	1.0m	Soil
1.0	7.0m	Weathered Rock
7.0	55.0m	Hard Rock
55.0	60.0m	Rock with intermittent fractured zones.
60.0	80.0m	Hard Rock
80.0	85.0m	Rock with intermittent fractured zones.
85.0	120.0m	Hard Rock

Table 17 Data regarding time vs drawdown at station 6 during pumping stage.

Time	Time(t) since pumping began(min)	Depth to water level (m)	Drawdown (m)	Pumping rate(Q) (m^3/Sec)
6:25pm	0	18.364	0	6.21×10^{-3}
6:40pm	15	18.542	0.178	
6:55pm	30	18.846	0.482	
7:10pm	45	18.897	0.533	
7:25pm	60	18.897	0.533	

Table 18 Data regarding time vs. residual drawdown at station 6 during recovery stage.

Time	Time(t') since pumping stopped(min)	Depth to water level (m)	Residual Drawdown (m)
7:25pm	0	18.897	0.533
7:27pm	2	18.846	0.482
7:29pm	4	18.796	0.432
7:31pm	6	18.783	0.419
7:33pm	8	18.770	0.406
7:35pm	10	18.770	0.406
7:37pm	12	18.745	0.381
7:39pm	14	18.745	0.381
7:41pm	16	18.719	0.355
7:43pm	18	18.719	0.355
7:45pm	20	18.694	0.33
7:50pm	25	18.694	0.33
7:55pm	30	18.669	0.305
8:00pm	35	18.643	0.279
8:05pm	40	18.618	0.254
8:10pm	45	18.592	0.228
8:15pm	50	18.567	0.203
8:25pm	60	18.516	0.152
8:35pm	70	18.465	0.101
8:45pm	80	18.440	0.076
8:55pm	90	18.415	0.051

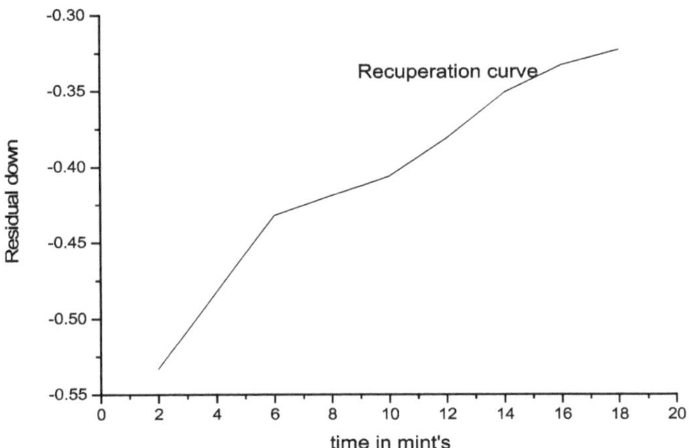

Graph 10: Recuperation Curve

Table 19 Litho-Log of the Station No 7

0	1.0m	Soil
1.0	10.0m	Weathered Rock
10.0	45.0m	Hard Rock
45.0	50.0m	Rock with intermittent fractured zones.
50.0	95.0m	Hard Rock
95.0	100.0m	Rock with intermittent fractured zones.
100.0	130.0m	Hard Rock

Table 20 Data regarding time vs drawdown at station 7 during pumping stage

Time	Time(t) since pumping began(min)	Depth to water level (m)	Drawdown (m)	Pumping rate(Q) (m^3/Sec)
10:07pm	0	18.846	0	$6.055x10^{-3}$
10:22pm	15	34.620	15.774	
10:37pm	30	34.645	15.799	
10:52pm	45	34.823	15.977	
11:07pm	60	34.823	15.977	

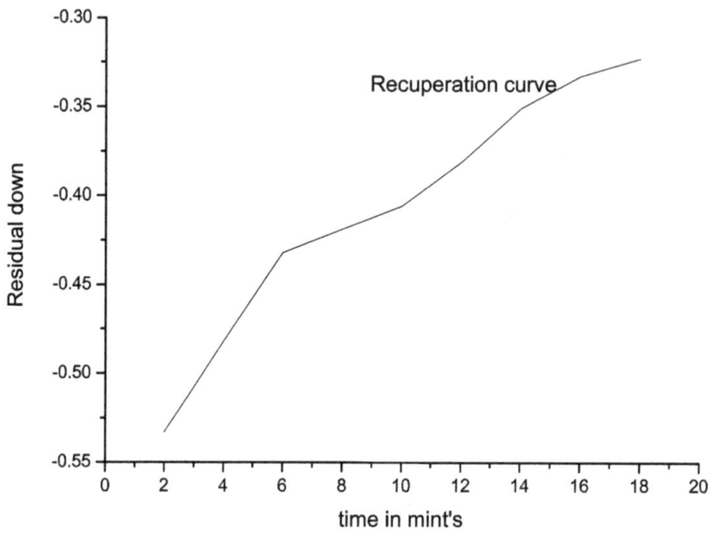

Graph 11: Recuperation Curve

RESULT AND CONCLUSIONS

The present study brought out the following results and conclusions, they are: RCI is located in S-W of Hyderabad city, which is one of the important research organization, works under the banner of DRDO..RCI extended on an area of 1800 acres which is majorly divided into 2 parts i.e. Technical area and Non-Technical Area(i.e. Residential area). Technical area is purely for the purpose of research and testing. Non-Technical area or Residential Area in which we have selected 7 bore wells at different locations along with 2 observation wells in different geomorphologic locations.

Pump out tests were conducted and the following observations are obtained. Among all the bore wells studied maximum drawdown was observed at **7th station (i.e. near Staff Institute)** with a drawdown of **15.977m** and minimum drawdown was observed at **6th station (i.e. near A-type)** with a drawdown of **0.533m** which indicates that the bore well with minimum drawdown i.e. **the 6th station bore well is connected to a subsurface stream**. The Maximum transmissivity obtained is of **6th station with transmissivity value of $2.135 x 10^{-3}$ m^2/sec** and minimum transmissivity obtained is of **7th station with transmissivity value of $6.945 x 10^{-5}$ m^2/sec.** Therefore, we can infer that the rate at which groundwater flows horizontally through the aquifer is low at Station 7.

The maximum storability obtained is of **7th station (i.e. near staff institute)** with storability value of **$6.945 x 10^{-3}$** and the minimum storability obtained is of **1st station (i.e. near C-Type)** with storability value of **$1.997 x 10^{-6}$.** Therefore, we can infer that station 7 bore well is very good to excellent in terms of yield whereas the station 1 is moderate in terms of yield.

REFERENCES

1. Aquifer Parameterization in an Alluvial Area by Poonma Dubey, MM Singh and HK Pandey-IJIRST –International Journal for Innovative Research in Science & Technology
2. Bear J., "Hydraulics of groundwater", Mc-Graw hill, 1979.
3. Cooper, H.H. and C.E. Jacob, 1946. A generalized graphical method for evaluating formation constants and summarizing well field history, vol. 27, pp. 526-534.
4. Determination of Aquifer Parameters of the Shallow Aquifers by Utpal Gogoi-international bulletin of water resources and development.
5. Groundwater hydrology by H.M.Raghunath.
6. Pumping tests for groundwater analysis-analysis and evaluation by Abdel ghafour.

URBAN WATER MANAGEMENT IN INDIAN CITIES – AN INTEGRATED APPROACH

K. Ashok[1*], K. Rakesh[2], G.K. Viswanadh[3] and Ch. Nageshwar Rao[4]

[1,*]PhD scholar, JNTU and Senior Lecturer in Civil Engg, Warangal

[2]Scientist, CSIR - National Environmental Engineering Research Institute, Nagpur

[3]Professor in Civil Engineering, JNTU, Hyderabad

[4]Professor in Civil Engineering, VNR Vignana Jyothi Institute of Engineering and Technology, Hyderabad

ashokgpw@gmail.com

ABSTRACT

Environmental Management in urban settings encompass many aspects including urban water management. Urban water management includes drinking and industrial needs of cities, the drainage system, treatment, rain water harvesting, runoff water routing to nearby tanks, flood water disposal etc. However the sewerage system is not exclusively considered in urban water management, and it is assumed to be safe in terms of leakages, collection system, treatment and final disposal. But in urban areas the sewerage system is in peril due to haphazard concrete structures, illegal occupation of drainage streams, improperly designed roads and storm water management system and disposal of solid wastes in surface drainage. Indian cities are poorly equipped with disaster mitigation, especially in the event of urban flash floods the cities pose sever environmental threat to the habitants. In regulation of concrete structures, the role of Government limits to provision of rules related to rain water harvesting, front, rear and side setbacks and limiting the total height of the building etc. There are also rules stipulated by pollution control board in safeguarding the water by getting contaminated by industrial and domestic wastes. But in reality, very less has been achieved in management of urban water systems and it will not be possible without the coordination of all stakeholders. The present paper deals with such steps essential from various aspects in the urban water management. Satellite data and GIS tools are to be better utilized for urban water management which include watershed modeling to control the drainage, run off and flood water disposal. Reliable Cross drainage arrangements at suitable points on streets, roads and at junctions are suggested for better management of water. In this theoretical paper, we propose certain solutions as part of urban water management to achieve resilient and habitable urban environment.

Keywords: Water management, Urban India, Disaster mitigation, Governance

INTRODUCTION

Water is the base for every kind of human activity. The demand on per capita consumption of water is ever increasing with the growing consumerism and greed of human beings especially in urban areas of India. Nearly 30% of population in India dwells in urban areas and the urban population is expected to be doubled by year 2050. At present most Indian cities are water stressed with hardly any city is having 24/7 water supply. Further, in cities having more than a million populations, the official per capita water supply is just 125 liters/day after considering 35% losses in leakages, which is lower than the demand of 210 liters/day. The gap between infrastructural growth and rise in urbanization is glaringly conspicuous in Indian urban cities. Further, due to growing population the share of available water resources per capita are dwindling as shown in **Table 1**. This paper provides an assessment to assist the policy makers, administrators and researchers in appreciating the multi-dimensional view of urban water management. We also discussed the possible suggestions for mitigating urban flood impact and provide a holistic view point of the problem.

Table 1 Total and per Capita Availability of Water in India

Year	Population (million)	Per capita availability (m3/year)	Utilizable water (Billion cubic meters) Annual average
1951	361	5177	1123
1955	395	4732	1123
1991	846	2209	1123
2001	1027	1820	1123
2011	1210	1545	1123
2014	1265	1463	1123
2025	1394	1341	1123
2050	1640	1140	1123

[Source: Based on Census of India, 2011 and Annual Reports (2012, 2013) Central Water Commission, MoWR, GoI.]

On another side of urban water problem, urban sanitation and sewerage is a major challenge now every municipality is facing with. Nearly 70% of urban sewerage is discharged untreated (Mara D, 2004). The gap in treatment facility is due to employing conventional treatment methods at centralized locations for a city which demands huge drainage channels and energy to gather the sewage. Further the under capacity of treatment facilities are adding owes to the challenge of sewage treatment. Discharge of partially or untreated sewage is polluting surface fresh water resources, one of the finest examples are rivers Ganges and Yamuna.

Of the many facets of urban water issues, the urban flash floods is most familiar now-a-days. Floods are commonly associated with urban setting than in the natural river banks in the haphazard concrete jungles of India. Since year 2000, the incidents of urban floods have increased in many Indian cities which include Hyderabad, Ahmadabad, Delhi, Chennai, Mumbai, Surat, Kolkata, Jamshedpur, Guwahati, Jaipur, Srinagar and Puducherry. The menace of urban flooding has become inevitable with many urban cities in India and the flood management and bringing it to normalcy will be much costlier in urban areas in comparison to rural environments (Saini et al., 2016). Flash floods is another phenomenon named recently, which is caused by heavy downpour within short duration and without any notice time, causing inconvenience to public (Ashok et al, 2017). Various reasons for high intensity rainfall in urban areas are studied which are causing flash floods. Urban Heat Island (UHI) is a well known phenomenon in cities with vast concrete structures. Big cities in India are also infamous for non compliance of air quality standards, particularly due to high concentrations of fine particulate matters and aerosols. It is reported that, UHI and certain type of aerosols cause high intensity rainfall in the urban setting which may lead to flooding.

At the same time, if a city lacks proper storm water management system, the floods will cause serious damage to the government and public. Further, poor maintenance of surface drains – chocking with solid waste, negligent to surface topology and hydrology while constructing any building, impervious road networks etc only aggravate the event of floods. Lack of vision by local governments in framing policies to make cities aesthetically appealing and at the same time making the cities resilient to floods, are need to be addressed. Poor municipal waste management in urban areas is exacerbating the problem by blocking drains and streams, while ill-planned road projects are cutting off flood flows.

LITERATURE REVIEW

Now the world population has reached 7.2 billion, and more than half of global population lives in cities and many cities are growing at a phenomenal rate. In Asia alone there is a threefold increase in number of mega-cities from 1970s to 2000. Urbanization will significantly change the land-use and land-cover of the surroundings which results in increased built-up area, reduction in vegetation, increased albedo, skyline and air pollution. All these changes will significantly interact with the micro-climate of the urban environment and result in complex feed-backs with the precipitation.

Trend analysis of long-term rainfall data suggests that there is a significant decrease in number of rainy days over the Indian urban areas (Bisht et al., 2017). The study also points that, the intensity of rainfalls has increased which was primarily attributed to urbanization and deterioration of air quality. Deforestation is another important factor affecting the variation in rainfall. Further, Ali et al. (2014) has reported that maximum rainfall at 1-3 day durations and 100 year return period are estimated to increase significantly in majority of Indian cities due to projected climate change scenarios. It can be interpreted that, increase in rainfall and decrease in rain days will result in high intensity rains over short duration on urban areas, making urban water management a challenge.

UHI is a well studied phenomenon which is resulted due to differential heating of downtown and outskirts of a city due to built-up concrete jungles (Ashok, 2007). The influence of UHI on regional precipitation was not well studied until recently with the help of numerical weather models and large simulations. Metropolitan meteorological experiments (METROMEX) were conducted in US in 1970s which estimated an increase of 5 to 20 % of urban downwind precipitation due to UHI effects (Shepard, 2005). Since then there were several studies which reported the UHI effect on precipitation. Shem and Shepard, (2009) has conducted controlled experiments in Atlanta, US on three different locations with varying degrees of urbanization and found that there is a significant increase of rainfall by 10-13% due to urbanization.

Out of the many factors affecting the precipitation in a region, aerosols play a central role in affecting atmospheric radiation, hydrological cycle, and cloud physics. Aerosol's interaction with cloud micro-physics dramatically changes the radiation and lifetime of clouds (Wu et al., 2016). Research is uncertain about the direct role of aerosols in precipitation. Long term studies reported by Yang and Li (2014), shows that aerosols produced from the mineral and carbon origin suppress the precipitation, while aerosols produced by pollution sources cause high intensity rainfall. Air pollution from various pollutants such as vehicles, residue burning and industries is concentrated heavily in and around urban areas, which will result in formation of aerosols which causes high intensity rainfalls.

DISCUSSION

Urban water systems around the world are usually managed with laying of similar series of systems for potable water, storm water and sewerage. The conventional system has lost relevance in addressing sustainability issues of future (Saraswat et al., 2016). Bahri, (2012) explains that, the conventional approach has drawbacks in addressing ecosystem-related problems in urban areas due to hydrological changes in urban catchments and quality of run-off. Rapid urbanization and its negative impacts such as increasing runoff and pollutant loads induce lot of pressure on conventional system (**Figure 1 & 2**).

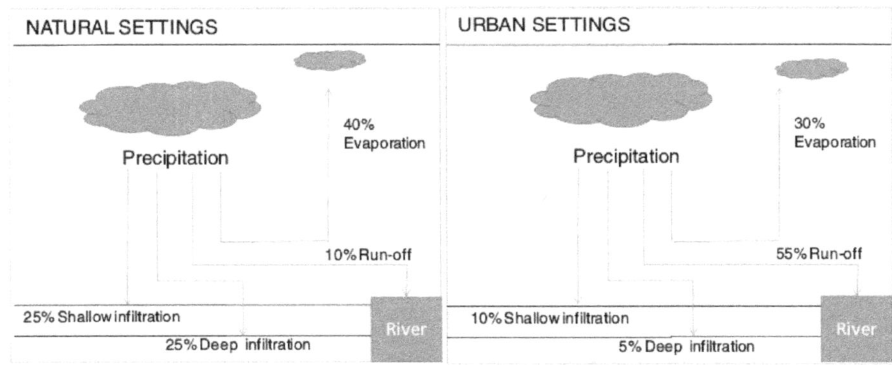

Fig. 1 Comparison of infiltration and surface runoff between natural and urban settings (Saraswat et al., 2016).

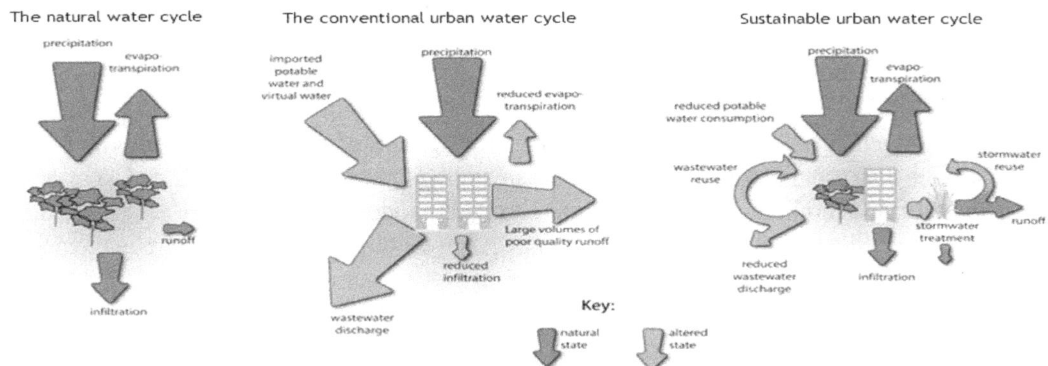

Fig. 2 The natural water cycle, the conventional urban water cycle and the *sustainable* urban water cycle. [Source: HEALTHY WATERWAYS (2011)]

The area available for recharge is increasingly covered with pavements, roads, buildings etc making very little chance for water to percolate in urban areas, leading to increased run off (**Figure 3**). Moreover the pollutants which accumulate on impervious surfaces are washed away and carried with runoff leads to non-point source pollution into streams. Studies show that the soil in urban areas is highly contaminated with heavy metals such as lead, arsenic and chromium which are toxic and carcinogenic. In case of open sewerage system, the storm water gets in direct contact with pathogens which causes water born diseases directly and indirectly.

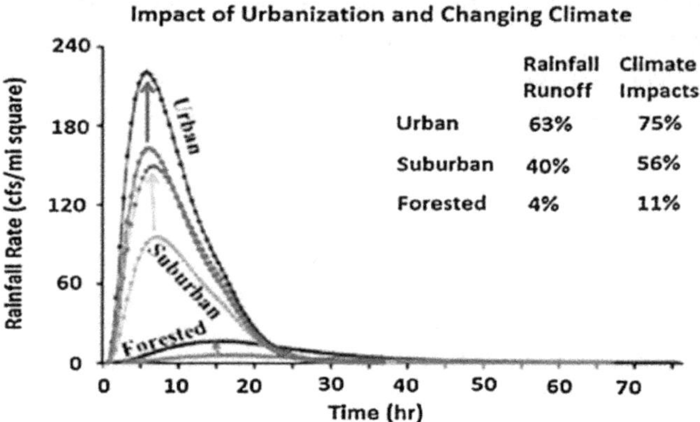

Fig. 3 Hydrographs showing runoff from urban, sub-urban and forested areas – based on the development of impervious cover. Adapted from (Saraswat et al., 2016)

Mitchell et al., (2007) calls for an alternate approach to mange urban water systems such as Integrated Urban Water Management (IUWM), where water supply, drainage and sanitation are integrated into natural landscape of the urban system. The principles of IUWM are 1) to minimize the generation of surface runoff and 2) to manage runoff to mitigate the damage caused to humans and civil structures. Proper storm water management is cruicial for any urban water management.

MANAGEMENT OF URBAN WATER

In this section we discuss various strategies for the mitigation of problems related with potable water – surface and ground water as well, in urban settings. Application of GIS and various hydrological and flood modeling, using of digital elevation models (DEM) is pertinent in planning all structures related to urban water management. Also we discuss storm water management in urban region as well as at household levels duly suggesting certain structural modifications of existing drains and roads arrangement. We also propose certain policy improvements in town planning for better management of floods in urban setting. The IUWM approach is a paradigm shift for urban water management. It is not a prescriptive model but a process that invites existing cities and emerging ones to adjust their current planning and management practices, given their own priorities, in a hydrological, environmental and socio-economic context. It is based on the following key concepts:

Structural modifications are an integral part of town planning. The plinth level of buildings should have uniform and gentle slope to follow the natural gradient of the terrain in the street. The arrangement should not clog the drains and storm water flow should be uniform without any discontinuity/obstruction. In Indian urban settings, usually the houses will be constructed first and then infrastructural facilities like, power supply, road networks and drainage water supply will be provided later. Instead, the practice of developing a layout, colony roads and drainages are to be planned in first place which should be followed by the construction of buildings. The rain water harvesting (RWH), even though insisted by certain municipalities while according construction permission, the enforcement is inadequate. The plot size of a middle income and lower income group family is very low around 100 to 180 square meters, which neither has sufficient standard setbacks and nor has enough space open to sky or space for percolation or RWH pit. Sooner, the needs change and they try to pave the entire plot area available for extended use, usually without any further permission fom municipality, which will lead to increase in runoff and reduces percolation of rain water. Since the set backs are not implemented strictly, the term slum can be applied to 60 to 80 percent of urban settings. It is also observed in the Indian urban settings, especially, the local powerful persons construct the houses deviating the rules causes inconvenience.

Application of GIS and Hydrological models

Hydrological modeling of watersheds will give us the flood discharge rate, water level, duration and direction of flow at a given point. Any new project or buildings or any developmental activities should not disturb the hydrological aspects, to avoid the flood accumulation or clogging. Sakeih et al., (2017) reported that urban morphology will significantly effects flood profile. The use of various numerical flood modeling or shallow water modeling software has become prevalent for studying water sheds and flood risk estimation. These models integrate well with GIS platform for better interpretation of the flood risk mapping and to identify locations prone to damage. HEC-HMS, HEC-RAS, HEC-Geo RAS, MIKE, Arch Hydro and ANUGA are very commonly used software for the modeling (Saini et al., 2016). Field specific data such as watershed boundary, DEM, surface roughness coefficients, land use, rainfall discharge etc will play crucial role in getting meaningful simulation results. Finally, flood modeling results are superimposed with various thematic maps of population, infrastructure, road networks, drainage lines and land-uses etc in order to appraise the vulnerability and to make mitigation measures, by determining the flood quantity at a particular point for a given amount of rainfall and duration.

The quantity can be estimated for each millimeter of rainfall using remote sensing and GIS through modelling. Road network should be such that the storm water drainage flow is unidirectional. There should be sufficient cross drainage works on the road alignment so that neither road surface will damage, nor the sub-grade of the road is damaged or eroded. The heavy load traffic should be diverted to bypass roads, exclusively laid for the purpose, since the heavy load trucks cause higher damage to the urban roads when compared with the light vehicles and two-wheelers.

Harnessing Rain Water and Floods

The context of flood control is evolving with the increased number of risks and challenges associated with it. Japan and Thailand were managing storm water until recently with only structural approaches whereas now the management has shifted to use the combination of structural and non-structural approaches (Saraswat et al., 2016). In Vietnam, the country mainly focuses on improving ground water levels in managing storm water with integrated design of surface runoff, infiltration, storage and transportation. Non-structural approach involves rain water harvesting, increasing urban forestry, building water percolation tanks at numerous places in the cities. Burden, (2006) reported that, trees will slowdown storm water runoff by obstructing 30 % of precipitation through foliage, and another 30 % through root system. Trees also filter out the pollutants carried along with runoff.

Some of the finest examples of flood control with civil structures include G-Cans project in outskirts of Tokyo, Japan. Huge underground silos (65m deep and 32m in diameter) are constructed to collect the flood water and they are interconnected. When the silos are full, the water is transported to huge storage tank (the water storage tank, popularly called the Underground Temple, is 25.4m high and 177m long) from which water is pumped into Edogawa river. While in Bangkok, storm water runoff is managed through drainage provisions, retention ponds, large scale rain water harvesting ponds, with public-private participation (Saraswat et al., 2016). Also combating urban floods with vegetation swale, road side vegetation strips and with permeable pavements. Recently Bangkok is also ambitiously pursuing huge civil structure beneath the existing eight-lane road, its double deck multipurpose tunnel with transports the storm water into collection tank from which the water is pumped to a river (**Figure 4**). Similarly other examples of huge underground tunnel infrastructure include, Chicago TARP, Hong Kong drainage system and Kuala Lumpur's SMART.

Fig. 4 Underground structure for flood water management in Bangkok.
[Source: https://www.tunneltalk.com/]

Urban foresting

Green infrastructure is another emerging concept of urban flood control and management inspired to protect ecology by inter connecting ecological spots in the urban areas such as forests, wetlands, wildlife and lakes. Water retention capacity and quality will be improved by the inter connections. India is also getting sensitive to importance of breathing spaces in the urban setting. Vast green spaces not only cleans air pollution, but also acts like buffer in storing the storm water and also helps in ground water recharge. Local municipalities should come up with green infrastructural projects in interconnecting natural spaces in the city which not only improve the flood mitigation but also dramatically changes the aesthetic beauty of the city.

Public Participation

Participation of key stakeholders coming from the public, private and social sectors representing different socio-economic activities that have an interest in water in urban areas. There can be many stakeholders involved but an agreement needs to be reached with the representatives of local government who remain the main convener. Not all have the same role and responsibility, but all need to be aware and contribute. Public have to take part of responsibility to build green roofs, rain gardens and infiltration wells in big private properties. The choking, clogging should be checked periodically, especially during pre-monsoon. The above should be implemented through civil engineering task force, without any deviations. The successful implementation leads to better urban water management. In some urban areas, provision of RWH pit is mandatory for building permission. Local government should also provide incentives to the households which take necessary actions for preventing such clogging at community level by inserting meshes in the storm water drain, which stops solids to some extent and for practicing rainwater harvesting in their houses.

Involvement and participation of various stakeholders is needed to amalgamate the specialized activities to reach a common understanding and to strengthen cooperation in management of urban water. Various government departments such as municipalities, pollution control boards etc have to work in tandem.

Cascading the usage of water of various qualities will certainly reduce the water demand of treated supply. Many households use municipality supplied potable compliant water for gardening or cleaning vehicles, for which potable quality is not required. Matching water quality with purpose of use is the better strategy in aiming for a better utilization of natural systems for water and wastewater treatment; considering storm water/rainwater catchment systems as a potential source; better managing use of water, effluents and water demand and hygiene behavior; strengthening leakage management and maintenance; strengthening resilience of urban water systems that are facing drought or floods.

Wastewater is a resource that can be used productively. Grey water can be reused for irrigation (Kadaverugu et al. 2016), urban agriculture and industrial processes, treated or partially treated depending on the purpose of its use and its legislation; nutrients in wastewater (grey and black) can be used for energy production and fertilizer production. Application of innovative technologies and optimum infrastructure design implies: selection of technology for water supply, wastewater treatment and sanitation which should be based on a multi-criteria decision support system.

Effective water governance with an IUWM perspective encompasses many aspects with the main following key elements: adopting a new mind set, a holistic and cross sectoral approach linking urban water management with overall urban planning; adjusting some of the policy and legislation concerning the use of water and reuse of waste water; analyzing aspects of centralised and decentralised management; assessing the economic and financial impact of adopting an IUWM approach; building the capacity of technical and managerial staff; and sharing information with the public and users

As we stated few examples of storm water management in cities like Japan, Hanoi, Bangkok, Hang kong and Chicago, Indian urban settings can come up with such mega infrastructural projects under public-private partnership to collect storm water beneath the landscape. Public should be made aware in rain water harvesting practices and it should be promoted through incentives.

CONCLUSIONS AND SUGGESTIONS

Our suggestions regarding urban water management are in consonance with Ashok et al., (2016, 2017), which are as follows.

1. Effort has to be made in reducing the UHI effect, air pollution in the cities. Promotion of green infrastructure – urban forestry. Public participation in decentralized rain water harvesting and urban forestry in private lands should be encouraged through incentives. While permitting house construction, the municipality should mandate certain no of plants, depending on plot size. This step invariably provides some percolation space around the plants in each house.
2. Revival and preservation of water bodies in the urban areas such as check dams, retention ponds, percolation wells and their interconnectivity of streams or canals, to absorb the intense downpour.
3. Hydrological modeling and flood risk mapping of the urban areas should be done with valid ground data. Creation of tanks and lakes for water storage, to harvest the floods and replenish depleted groundwater of urban areas as well. Permissions to a new developmental activities should also consider its impact on surrounding hydrology.
4. Creation of an integrated flood management system for each and every urban area, duly projecting the future population growth, at least for a further three decades. Sufficient Cross drainage works shall be constructed to avoid overflow of runoff on the roads for longer duration during flash floods. Control of flash floods leads to better management of urban water.
5. Government should build mega underground tunneling projects to carry the flood water out of the city. Casing of open drainage channels or enhancing the aesthetic value of streams is suggested, so that solid waste dumping can be prevented – which chokes the drainage.
6. Contamination of storm water with sewage has to be prevented, decentralized treatment of sewage at houses can be promoted through constructed wetlands.
7. Vehicle parking and Open space for parks and greenery, playing grounds etc. should be aptly provided in master plans and strictly implemented. Permission to the buildings should include plinth level fixing and pavement area and the same to be mentioned in the approved drawings. Increasing the setbacks by urban governments for new approvals. Prohibiting constructions in smaller plots, which also takes care of slum formation.
8. Slums of low lying areas may be removed to convert back to ponds, check dams and green pastures.
9. The new house constructions for slum relocation can be done in a planned way, by creating satellite towns. Draining of wetlands and their conversion into expensive real estate shall be stopped at once.

The shortcomings of existing urban setting and the suggestions for urban water management discussed in the paper can be overcome in newly developing cities with a design for resilience to the adverse impacts of floods, pollution, broadly – climate change, making the cities a better place to live.

REFERENCES

1. Ali, Haider, Vimal Mishra, and D. S. Pai. "Observed and Projected Urban Extreme Rainfall Events in India." Journal of Geophysical Research: Atmospheres 119, no. 22 (November 27, 2014): 2014JD022264. https://doi.org/10.1002/2014JD022264.
2. Ashok K, Rakesh K, Nageshwar Rao Ch, Laxmana Rao KM. "Review of impact of flash floods in urban India and its mitigation." 3rd National conference on water environment and society, Hyderabad, (June – 2016): 7-21. http://jntuhist.ac.in/web/bulletins/96_proceedings_of_3rd_national_conference_on_water_environment_and_society_NCWES2016.pdf
3. Ashok K, Kongre DN, Rohit Goyal and Gupta AB. "Urban Heat Island studies using Remote Sensing data Case study of Rajasthan cities" 11 th Asian Urbanization Conference, Hyderabad, (Dec 2011), 251-258, vol.3 of ISBN 8180699471
4. Ashok K, Rakesh K, Nageshwar Rao Ch, Viswanadh G.K. "Integrated approach to mitigate the impact of flash floods in India urban setting." 3rd Interational conference on environment and Management, Hyderabad, (November – 2017).489-498, Vol.II of Proceeds of ICEM-2017, B S Publications, ISBN: 978-93-86819-50-5
5. Bahri, A. "Integrated Urban Water Management. GWP Technical Background
6. Bisht, Deepak Singh, Chandranath Chatterjee, Narendra Singh Raghuwanshi, and Venkataramana Sridhar. "An Analysis of Precipitation Climatology over Indian Urban Agglomeration." Theoretical and Applied Climatology, June 22, 2017, 1–16. https://doi.org/10.1007/s00704-017-2200-z.
7. Burden, Dan. "Urban Street Trees: 22 Benefits Specific Applications." Glatting Jackson and Walkable Communities Inc, 2006.
8. Ducan M. "Domestic wastewater treatment in developing countries." Earth Scan in UK and USA, 2004
9. HEALTHY WATERWAYS (Editor) (2011): What is Water Sensitive Urban Design?. Brisbane: Healthy Waterways. URL: http://waterbydesign.com.au/whatiswsud/ [Accessed: 19.09.2013]
10. Kadaverugu R, Shingare RP, Raghunathan K, Juwarkar AA, Thawale PR, Singh SK. The role of Sand, Marble chips and Typha latifolia in domestic Wastewater Treatment – A Column Study on Constructed Wetlands. Environmental Technology 2016:1–26. doi:10.1080/09593330.2016.1153156.
11. Mitchell, V.G., A. Deletic, T.D. Fletcher, B.E. Hatt, and D.T. McCarthy. "Achieving Multiple Benefits from Stormwater Harvesting." Water Science and Technology 55, no. 4 (February 1, 2007): 135. https://doi.org/10.2166/wst.2007.103.
12. Papers." Global Water Partnership, Stockholm Number 16. (2012)
13. Saini, Surjit Singh, S. P. Kaushik, and Ravinder Jangra. "Flood-Risk Assessment in Urban Environment by Geospatial Approach: A Case Study of Ambala City, India." Applied Geomatics 8, no. 3–4 (December 1, 2016): 163–90. https://doi.org/10.1007/s12518-016-0174-7.
14. Saini, Surjit Singh, S. P. Kaushik, and Ravinder Jangra. "Flood-Risk Assessment in Urban Environment by Geospatial Approach: A Case Study of Ambala City, India." Applied Geomatics 8, no. 3–4 (December 1, 2016): 163–90. https://doi.org/10.1007/s12518-016-0174-7.
15. Sakieh, Yousef. "Understanding the Effect of Spatial Patterns on the Vulnerability of Urban Areas to Flooding." International Journal of Disaster Risk Reduction 25, no. Supplement C (October 1, 2017): 125–36. https://doi.org/10.1016/j.ijdrr.2017.09.004.
16. Saraswat, Chitresh, Pankaj Kumar, and Binaya Kumar Mishra. "Assessment of Stormwater Runoff Management Practices and Governance under Climate Change and Urbanization: An Analysis of Bangkok, Hanoi and Tokyo." Environmental Science & Policy 64, no. Supplement C (October 1, 2016): 101–17. https://doi.org/10.1016/j.envsci.2016.06.018.

17. Shem, Willis, and Marshall Shepherd. "On the Impact of Urbanization on Summertime Thunderstorms in Atlanta: Two Numerical Model Case Studies." Atmospheric Research 92, no. 2 (2009): 172–89. https://doi.org/https://doi.org/10.1016/j.atmosres.2008.09.013.
18. Shepherd, J. Marshall. "A Review of Current Investigations of Urban-Induced Rainfall and Recommendations for the Future." Earth Interactions 9, no. 12 (2005): 1–27.
19. Shepherd, J. Marshall. "A Review of Current Investigations of Urban-Induced Rainfall and Recommendations for the Future." Earth Interactions 9, no. 12 (2005): 1–27.
20. Wu, GuoXiong, ZhanQing Li, CongBin Fu, XiaoYe Zhang, RenYi Zhang, RenHe Zhang, TianJun Zhou, et al. "Advances in Studying Interactions between Aerosols and Monsoon in China." Science China Earth Sciences 59, no. 1 (January 2016): 1–16. https://doi.org/10.1007/s11430-015-5198-z.
21. Yang, Xin, and Zhanqing Li. "Increases in Thunderstorm Activity and Relationships with Air Pollution in Southeast China." Journal of Geophysical Research: Atmospheres 119, no. 4 (February 27, 2014): 2013JD021224. https://doi.org/10.1002/2013JD021224.

EFFECTIVE RAINWATER HARVESTING BY GIS ANALYSIS IN GRIET CAMPUS, BACHUPALLY

S. Venkat Charyulu, G.K.Viswanadh[2] and M.V.S.S Giridhar[3]

[1]Research scholort, Department of Civil Engineering, JNTUHCEH, Kukatpally, Hyderabad, Telangana.
[2]Professor of Civil Engineering and OSD to VC, JNTUH, Kukatpally, Hyderabad, Telangana.
[3]Assosiate Professor, CWR, IST, JNTUH, Kukatpally, Hyderabad, Telangana.
gorthi.kasi@gmail.com; mvssgiridhar@gmail.com

ABSTRACT

Rain water is the natural surface water resource and best useful for various purposes. It requires to store and reuse for the various human needs. The quality of rain water is especially depends on the zone and area of rain fall. The quantity of rain fall depends on the arid and semiarid region. Rain water is the best quality water which can be used for the drinking, irrigation, industrial and gardening and for the cleaning and other purposes. The rain water is very precious and good water resource freely available in seasonally and occasionally. Different countries are adopting various procedures to Harvest the rain water for the needs. In this paper discussed about the Type of harvesting methodology and selected one of the two best method to proposed to implement in the GRIET Campus .

1.0 INTRODUCTION

1.1 Rainwater harvesting:

It is simple methodology of collecting and storing of rainwater in the proposed tanks or recharge in to the ground to or rise the ground water table. Harvested water can be used for the human daily needs.

Rain fall over the catchment of hill terrain Collected and rooting in to the local rivers and streams. Rain fall over the residential building can recharge in to harvesting pits or in to existing bore wells which improve the ground water table. Harvesting of rain fall will reduce the load on water distribution system, sewer system.

1.2 Various methods of rainwater harvesting are described as.

(i) Surface runoff harvesting.
(ii) Rooftop rainwater harvesting.
  (i) Storage of Direct Use. (ii) Recharging groundwater aquifers

1.2.1 Surface runoff harvesting

In urban area rainwater flows over paved and unpaved areas will cause the surface runoff. Surface run off is depends on several factors of existing topography .This runoff could be caught and used for recharging aquifers by adopting appropriate methods.

1.2.2 Rooftop rainwater harvesting:

In this system rain water collected from the roof top. In this system the roof become the catchment and the rainwater is collected from the roof surface of existing building can be stored in a tank or diverted to artificial recharge system. This method is very less expensive and very effective and if implemented properly helps in augmenting the groundwater level of the area.

The system mainly constitutes of following sub components discussed later in this paper.

- Catchments
- Transportation
- First flush
- Filter

3.0 OBJECTIVES OF RAIN FALL HARVESTING IN THE GRIET CAMPUS

1. To meet the water requirements of the campus needs
2. Improve the ground water levels in the GRIET campus
3. Reduce the water tankers utilization and their by saving economy
4. Two methods are proposed in the campus for rainwater harvesting

4.0 STUDY AREAS AND DATA COLLECTION

Study area chosen as Griet campus Bachupally. This catchment found out as rocky area. Study area situated in the Telangana state of Rangareddy. Nearly 80 percent of the State is underlain by hard rock formations consisting of granites, gneisses, metamorphic and intrusive (Archaean's), so infiltration of rain water is very less into ground. On the other hand we are losing the ground water table day by day. Water for daily consumption of our college is not sufficient from the bore wells. Recently our college management had a two new bores to our campus to meet the daily demand. Planned to preserve the ground water table by alternate use of rain water for our daily demands. There is a another problem that, if we are not using the ground water resources frequently a phenomenon called ground water depletion will occur by preceding water bores. This situation was more in the Hyderabad. So we are planning and designing to tap rain water from roof catchment areas in the campus and to utilize them. Simultaneously recharging the ground water by letting the rain water into ground water tanks (sump)

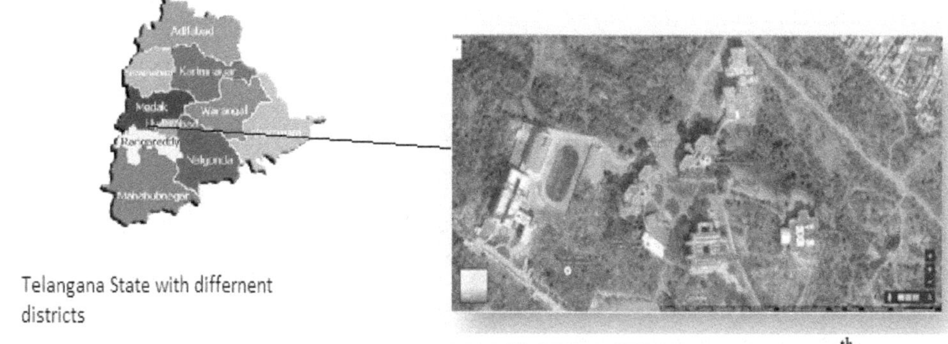

Fig. 1 Study area, GRIET campus, Bachupally.

5.0 THE MAIN COMPONENTS OF RAIN WATER HARVESTING

5.1 Catchments

The surface that receives rainfall directly is the catchment of rainwater harvesting system. It may be terrace, courtyard, or paved or unpaved open ground. The terrace may be flat RCC/stone roof or sloping roof. Therefore the catchment is the area, which actually contributes rainwater to the harvesting system.

5.2 Transportation

Rainwater from rooftop should be carried through down take water pipes or drains to storage/harvesting system. Water pipes should be UV resistant (ISI HDPE/PVC pipes) of required capacity. Water from sloping roofs could be caught through gutters and down take pipe. At terraces, mouth of the each drain should have wire mesh to restrict floating material.

5.3 First Flush

First flush is a device used to flush off the water received in first shower. The first shower of rains needs to be flushed-off to avoid contaminating storable/rechargeable water by the probable contaminants of the atmosphere and the catchment roof. It will also help in cleaning of silt and other material deposited on roof during dry seasons Provisions of first rain separator should be made at outlet of each drainpipe.

5.4 Filter

There is always some skepticism regarding Roof Top Rainwater Harvesting and doubts are raised that rainwater may contaminate groundwater. There is remote possibility of this fear coming true if proper filter mechanism is not adopted. Secondly all care must be taken to see that underground sewer drains are not over flow in close vicinity of harvesting. Filters are used for treatment of water to effectively remove turbidity, color and microorganisms. After first flushing of rainfall, water should pass through filters. A gravel, sand and 'netlon' mesh filter is designed and placed on top of the storage tank. This filter is very important in keeping the rainwater in the storage tank clean. It removes silt, dust, leaves and other organic matter from entering the storage tank. The filter media should be cleaned daily after every rainfall event. Clogged filters prevent rainwater from easily entering the storage tank and the filter may overflow. The sand or gravel media should be taken out and washed before it is replaced in the filter. A typical photograph of filter is shown in Fig 2 and 3.

Fig. 2 Photograph of Typical Gravel Filter.

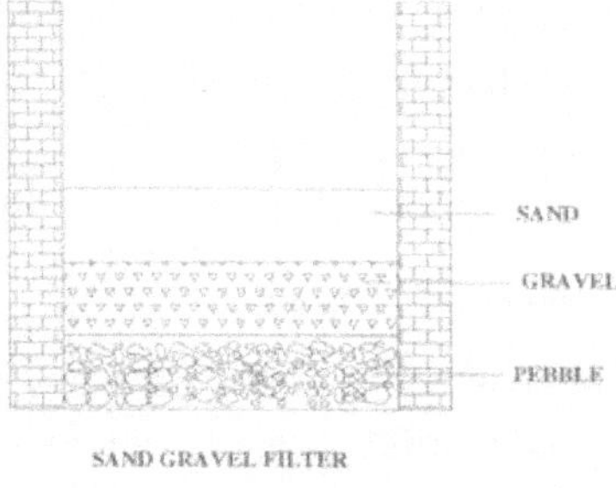

Figure 3 Sand Gravel Filter.

These are commonly used filters, constructed by brick masonry and filleted by pebbles, gravel, and sand as shown in the figure. Each layer should be separated by wire mesh. A typical figure of Sand Gravel Filter is shown in Fig 2 and 3.

7.0 METHODOLOGY ADOPTED

The surface that receives rainfall directly is called the catchment. It may be terrace, courtyard, paved or unpaved open ground. The terrace may be flat RCC/stone roof or sloping roof. Therefore the catchment is the area, which actually contributes rainwater to the harvesting system. There are two methods of harvesting system has proposed and compare them. In First method storage tank harvesting (1 Rationing method (RM) and other is Rapid Depletion Method (RDM)) other is method is ground water recharge Methods of Rooftop Rainwater Harvesting

(a) Storage of Direct Use In this method rainwater collected from the roof of the building is diverted to a storage tank. The storage tank has to be designed according to the water requirements, rainfall and catchment availability. Fig 4.

Fig. 4 A storage tank on a platform.

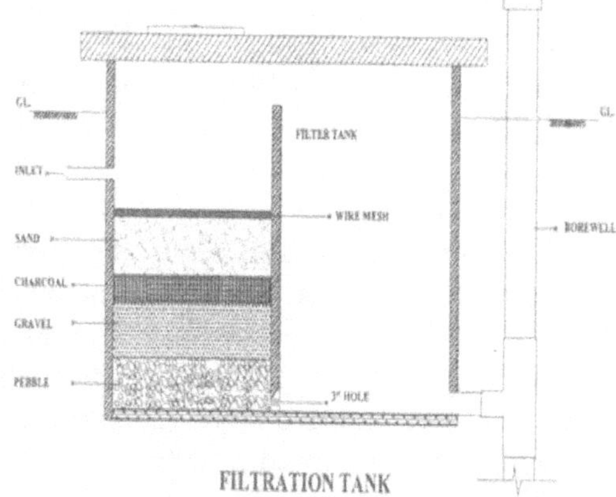

Fig. 5 Filter tank.

(b) Recharging groundwater aquifers

Groundwater aquifers can be recharged by various kinds of structures to ensure percolation of rainwater in the ground instead of draining away from the surface. Commonly used recharging methods are:-

(a) Recharging of bore wells
(b) Recharging of dug wells.
(c) Recharge pits
(d) Recharge Trenches
(e) Soakaways or Recharge Shafts
(f) Percolation Tanks

(c) Recharging of bore wells

Rainwater collected from rooftop of the building is diverted through drainpipes to settlement or filtration tank. After settlement filtered water is diverted to bore wells to recharge deep aquifers. Abandoned bore wells can also be used for recharge fig 6.

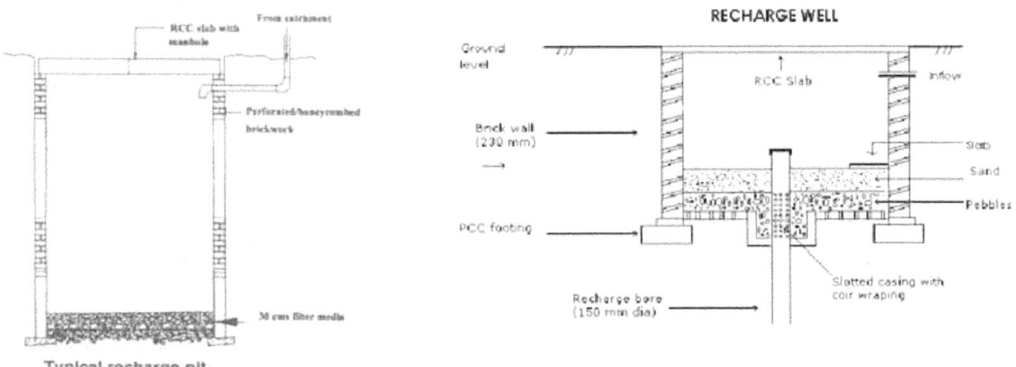

Fig. 6 Bore well recharges. **Fig. 7** Recharge pit.

(d) *Recharge pits*: Recharge pits are small pits of any shape rectangular, square or circular, contracted with brick or stone masonry wall with weep hole at regular intervals. Top of pit can be covered with perforated covers. Bottom of pit should be filled with filter media. Fig 7

(e) Soakway or Recharge shafts soak away or recharge shafts are provided where upper layer of soil is alluvial or less pervious. These are bored hole of 30 cm dia. up to 10 to 15 m deep, depending on depth of pervious layer. Bore should be lined with slotted/perforated PVC/MS pipe to prevent collapse of the vertical sides. At the top of soak away required size sump is constructed to retain runoff before the filters through soak away. Sump should be filled with filter media. A schematic diagram of recharge shaft is shown in Fig 8.

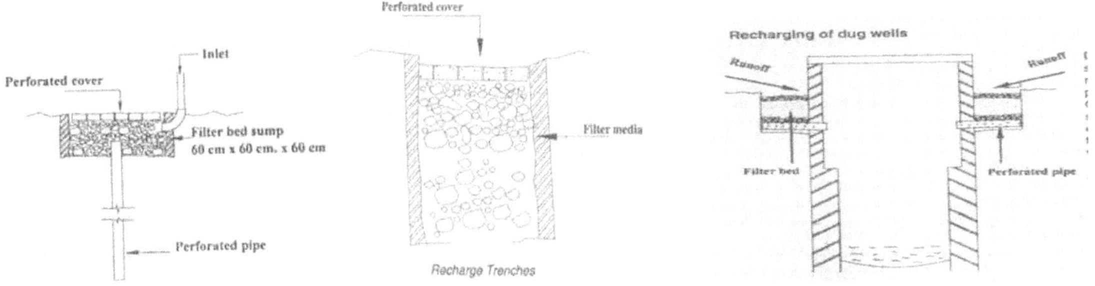

Fig. 8 Recharge shaft. **Fig. 9** Recharging to Trenches. **Fig. 10** recharging to dug well.

(f) Recharging of dug wells Dug well can be used as recharge structure. Rainwater from the rooftop is diverted to dug wells after passing it through filtration bed. Cleaning and desalting of dug well should be done regularly to enhance the recharge rate. The filtration method suggested for bore well recharging could be used. A schematic diagram of recharging into dug well is indicated in Fig 10.

(g) Recharge trenches Recharge trench in provided where upper impervious layer of soil is shallow. It is a trench excavated on the ground and refilled with porous media like pebbles, boulder or brickbats. It is usually made for harvesting the surface runoff. Fig9. It requires having rain fall details of the harvested area and run of coefficients of surface area. Table no.1 shows Hyderabad region rain fall data .and table. 2 show the run off coefficients for various surfaces. (Referred from Irrigation Engineering & Hydraulic Structure, by Garg, S.K., table 7.3)

Table 1 Monthly Average rain data of Hyderabad region for 10 Years (2001-2014)

Month	**RAIN FALL**
January	5.9
February	7.4
March	14.6
April	20.4
May	33.8
June	110.7
July	176.8
August	190.5
September	165.5
October	95.6
November	23.7
December	6.4
TOTAL	851

(h) Percolation tank percolation tanks are artificially created surface water bodies, submerging a land area with adequate permeability to facilitate sufficient percolation to recharge the groundwater. These can be built in big campuses where land is available and topography is suitable.

The stored water can be used directly for gardening and raw use. Percolation tanks should be built in gardens, open spaces and roadside greenbelts of urban

Table 2 The run off coefficients for various surfaces.

S.no	**Types of area**	**Value of K**		
		Flat land 0-5 % slope	**Rolling land 5%-10% slope**	**Hilly land 10%-30%**
1.	Urban areas	0.55	0.65	-
2.	Single family residence	0.3		
3.	Cultivated Areas	0.5	0.6	0.72
4.	Pastures	0.30	0.36	0.42
5.	Wooden land or forested areas	0.3	0.35	0.50

8.0 METHODS OF HARVESTING

Technically, there are two types of methods for distributing the harvested rainwater they are 1. Rationing method (RM) 2. Rapid depletion method (RDM). To explain these both methods, let us first apply it on any one block in the campus Block.

8.1 Civil Engineering and Mechanical Engineering block:

The details of each calculation are carried out to obtain the valuable steps. Later on, that method can be applied to all other building. Number of days for consumption of stored water is calculated using both of these methods.

8.2 Rationing Method (RM):

The Rationing method (RM) is distributes the rain water to estimated public in such away that the rainwater tank is able to service water requirement to maximum period of time. This can be done by limiting the quantity of water demand per person. For example in this method, the amount of water supplied for the student is limited let us say, 20 lit/day (per capita water demand). Number of student in block no.4 is 1128. Then, Total amount of water consumption per day = 1128 x 0.02=22.56 m^3/day. Total no. of days we can utilized preserved water = stored water/ water demand for block no.4. For example the volume of water stored in tank has taken as 2880.94 m^3. Then No of days that water can utilized = 2880.94 / 22.56 = 127.70 days (4.2 months)

8.3 Rapid Depletion Method (RDM):

There is no restriction on the use of harvested rain water by consumer. Consumer is allowed to use the preserved rain water up to the maximum requirement, In Rapid Depletion Method resulting in less number of days. The rain water tank in this method is considered to be only source of water for the consumer, and alternate source of water has to be used till next rains, if it runs dries.

For example if we assume per capita water demand = 25lit/day = 0.0 25m^3/day

(Textbook s.k garg @ etal)

Total amount of water consumption per day = 1128x0.025= 28.2 m^3/day

Total no. of days, preserved water can be utilize = stored water/ water demand = 2876.94/28.2
= 102 days (3.40 months)

Hence, finally it is observed that, if the amount of water stored is equal to 2876.94m^3, then applying

1. RDM consumer can only utilize the preserved stored water for about127 days,
2. Where as in RM, preserved stored water can be utilized for a period of 102 days.

Each block in the campus is digitized using the GIS software and the area of the roof top is calculated. The quantity of rainwater over the roof has taken monthly average rain fall for the particular month. For example it is shown in the table No.3. Entire campus has shown in fig.11

Table 3 Showing Rainfall Discharge of Block no.4 (civil & mechanical).

Sl. No	Month	Rainfall(mm)	Discharge(m^3) = (rainfall*catchment area)
1	January	5.9	17.16
2	February	7.4	21.53
3	March	14.6	42.48
4	April	20.4	59.36
5	May	33.8	98.35
6	June	110.7	322.13
7	July	176.8	514.48
8	August	190.5	554.35
9	September	165.5	481.60
10	October	95.6	278.19
11	November	23.7	68.96
12	December	6.4	18.62
	TOTAL	851	2477.21

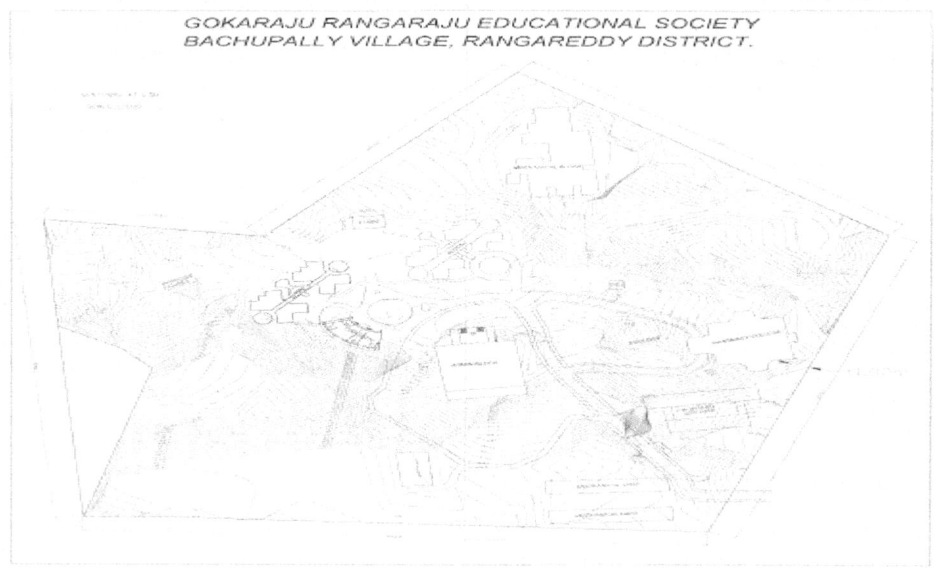

Fig. 11 Scanned contour map of G.R.I.E.T campus.

9.0 COMPUTATION OF VOLUME OF RUN OFF PER YEAR

As we know the formula for run off discharge from section 5.1. Is volume of water received (m^3) = catchment Area x Amount of rainfall Total roof area of block no.4 was calculated = 2910 m^2

The average rain fall from records for entire state = 851mm/year

Average annual rainfall at Bachupally = 909mm/year= 0.909 m/year

Total volume of surface runoff water for Block 4 to be calculated as

$$2910 \times 0.908 = 2642.28 \text{ m}^3/\text{year}$$

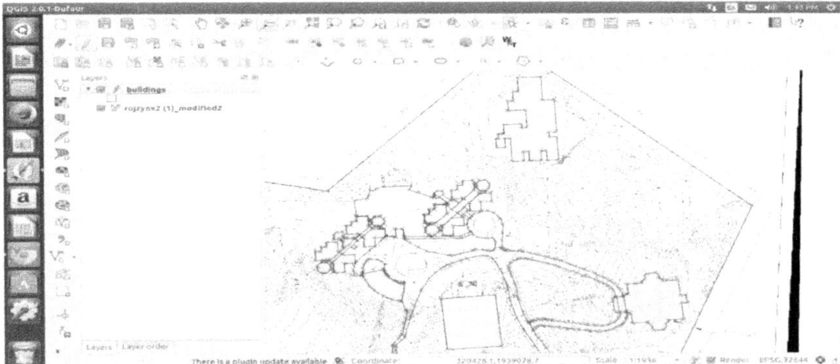

Fig. 12 Contour View with built up area of the entire campus.

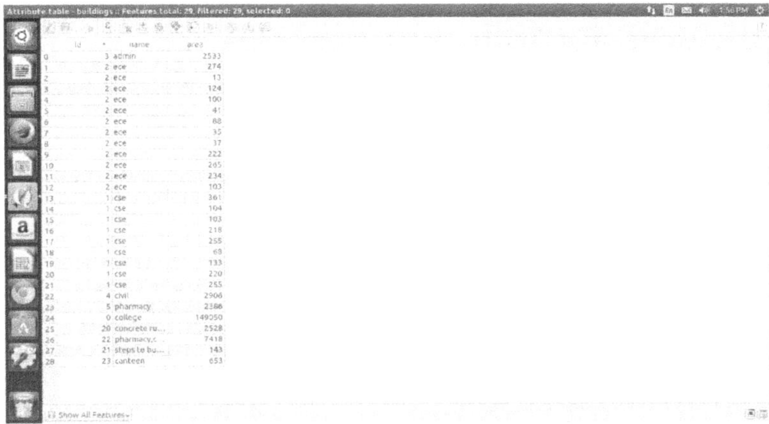

Attribute table - buildings :: Features total: 29, filtered: 29, selected: 0

	Id	name	area
0	3	admin	2533
1	2	ece	274
2	2	ece	13
3	2	ece	124
4	2	ece	100
5	2	ece	41
6	2	ece	88
7	2	ece	35
8	2	ece	37
9	2	ece	222
10	2	ece	265
11	2	ece	234
12	2	ece	103
13	1	cse	361
14	1	cse	104
15	1	cse	103
16	1	cse	218
17	1	cse	255
18	1	cse	68
19	1	cse	133
20	1	cse	220
21	1	cse	255
22	4	civil	2906
23	5	pharmacy	2386
24	0	college	149050
25	20	concrete ru...	2528
26	22	pharmacy,c	7418
27	21	steps to bu...	143
28	23	canteen	653

Show All Features

Fig. 13 Calculations of Block area ,paved area of Griet campus.

Fig. 14

Given below fig. 8 the complete GIS Referencing drawing of the G.R.I.E.T. giving clear view and dimension of the three blocks (1,2,3,4,5) and also dimensions of canteen attached to these blocks In the left hand side. The table no 3 which gives the monthly rainfall and discharge runoff obtained from the roof top area of each Blocks and corresponding graph are also plotted for rain fall collected in the year fig.15 and volume of water from the rain fall fig.16. Table No. 5 Dimension of Tank various buildings inside the campus of Griet

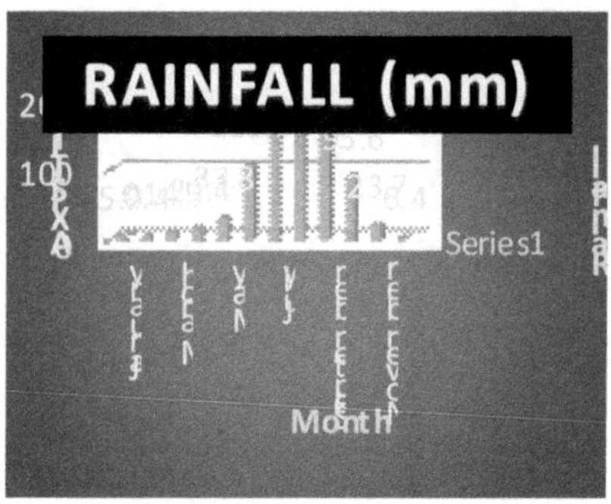

Fig. 15 Rainfall collected in the year.

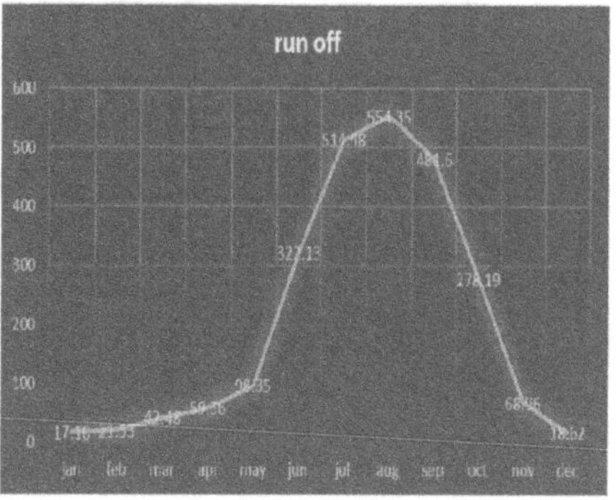

Fig. 16 Volume of water from rainfall.

Fig. 17 Optimum location of storage tank (or recharge point).

Table 4 Combined (reservoir) Storage tanks location for Different buildings.

S.No	Block Name	Storage tank(Reservoir Point)
1	Block 1	R1
2	Block 2	
3	Block 3	R2
4	Block 5	
5	Block 4	R3

Table 5 Analysis of harvested water over the individual Blocks.

Sl no.	Name of the Block	Roof top area(m2)	Runoff(m3) (rooftop areax0.908m)	Reservoir Capacity (m^3) (tank)	Dimensions Of Assume Tank	Tank name
1	Block no.1	1719	1560.852	256	8×8×7	R1
2	Block no.2	1537	1395.596	336	4×7×12	
3	Block no.3	2533	2299.964	448	7×6×8	R2
4	Block no.5	2355	2138.340	320	4×8×10	
5	Block no.4	2909	2642.280	384	4×8×12	R3

Average rain fall =0.909m

Total run off from the roof top surface =19433 x 0.909 = $(17664)m^3$ /year

Table 6 Final comparison for the entire tank capacity from all the buildings.

S.no.	Name of the Block	Roof top area(m^2)	Runoff (m^3) (roof top area x 0.908m)
1.	Block no.1	1719	1560.852
2.	Block no.2	1537	1395.596
3.	Block no.3	2533	2299.964
	Block no.4	2910	2642.280
	Block no.5	2355	2138.340
	Canteen	653	592.924
	Concrete area	10089	9160.812
	Overall college	149050	135337.4
		TOTAL=19433	TOTAL=16129.3

Average rain fall = 0.909m

Total run off from the roof top surface = 19433 × 0.909 = $(17664)m^3$ /year

Table 7 Final comparison for the entire tank capacity from all the buildings.

S.NO	Hall Name	Roof top area (m^2)	Reservoir Capacity(R)	RM =R/30 (days)	RDM= R/45 (days)
1	Block 1	1719	1324.55	59	47
2	Block 2	1537	1186.26	53	42
3	Block 3	2533	1954.96	89	71
4	Block 4	2910	2242.84	100	80
5	Block 5	2355	1840.74	108	85

Table 8 Final capabilities of tank for the individual Buildings and comparison of two methods.

Reservoir capacity or Tank capacity (m^3)	No.of days of potentially Rational Methods (RM)	No.of days of potentially Rapid depletion
17664	782.97	626.382

CONCLUSION

The average rain fall over the entire campus over the roof top is calculated and estimated the storage tank capacity . This storage tank can supply for the approximately783 and 627 day for the campus needs by the both methods.

***Alternative method*:** It can also calculated that if individual tanks of several capacities can be built as storage tanks that also survive for the number of days shown in the table No.10.

G.R.I.E.T can reduce the consumption of ground water and minimize the cost of purchasing the water from out side with the above mention utilization process. By constructing the recharge pits for the excess water after the storage tanks we can recharge the ground water table and able to use the store water seasonally. We can avoid wastage of rain water and runoff water which is entering in to gutter.

REFERENCES

1. Two books entitled1.Estimation and costing civil engineering ,by: -Dutta,B.N.2. R.C.C. Designs, By: -Punmia B.C.,Jain Ashok, & Jain Arun Kumar,was referred
2. Mr. Ranjith Kumar Sharma (2010) Titled "Rain Water Harvesting At N.I.T Rourkela" Project Report Submitted Under The Guidance Of ***Prof. K.C.Patra & Prof. Ramakar Jha*** In N.I.T Rourkela.
3. Ministry Of Water Resources Central Ground Water Board, Faridabad (2003) Tilted "*Rain Water Harvesting Techniques To Augment Ground Water" Guide Published Publically.*
4. Referred to website
5. http://www.cgwb.gov.in
6. http://www.irrigation.telangana.gov.in
7. http://irrigationap.cgg.gov.in
8. http://www.nitrkl.ac.in
9. https://theconstructor.org/water-resources/rainwater-harvesting-components/6739/

ANALYSIS OF GROUND WATER QUALITY PARAMETERS RANGA REDDY DISTRICT, TELANGANA

Sanjay Kumar Alladi[1], Ravi Sekhar Katru[2] and Sandhya Rani Regalla[3]

[1 &2]Asst. [3]Associate. Prof, CED, M.V.S.R Engg College Hyderabad, INDIA

ABSTRACT

Due to human and industrial activities the ground water is contaminated. This is the serious problem now a day. Thus the analysis of the water quality is very important to preserve and prefect the natural eco system. The assessment of the ground water quality was carried out in the different wards of Ranga Reddy District. The present work is aimed at assessing the water quality index (WQI) for the ground water of Ranga Reddy District and its industrial area .The ground water samples of all the selected stations from the wards were collected for a physiochemical analysis. For calculating present water quality status by statistical evaluation and water quality index, following 12 parameters have been considered Viz. pH, total dissolved solids, dissolved oxygen, electrical conductivity, chlorides, sodium, potassium, total alkalinity, acidity, total hardness ,temperature, turbidity . The obtained results are compared with Indian Standard Drinking Water specification IS: 10500-2012. The study of physico-chemical and biological characteristics of this ground water sample suggests that the evaluation of water quality parameters as well as water quality management practices should be carried out periodically to protect the water resources.

Keywords: Ground water, water quality standards, physico-chemical, Water Quality Index.

INTRODUCTION

Water is the most important in shaping the land and regulating the climate. It is one of the most important compounds that profoundly influence life[1].Groundwater is used for domestic and industrial water supply and also for irrigation purposes in all over the world. In the last few decades, there has been a tremendous increase in the demand for fresh water due to rapid growth of population and the accelerated pace of industrialization. According to WHO organization, about 80% of all the diseases in human beings are caused by water[2]. Once the groundwater is contaminated, its quality cannot be restored back easily and to device ways and means to protect it. Water quality index is one of the most effective tools to communicate information on the quality of water to the concerned citizens and policy makers. It, thus, becomes an important parameter for the assessment and management of groundwater. The greater part of the soluble constituents in ground water comes from soluble minerals in soils and sedimentary rocks. The more common soluble constituents include calcium, sodium, bicarbonate and sulphate ions. Another common constituent is chloride ion derived from intruded sea water, connate water, and evapotranspiration concentrating salts, and sewage wastes for example. Nitrate can be a natural constituent but high concentrations often suggest a source of pollution. Water quality standards are needed to determine whether ground water of a certain quality is suitable for its intended use. Guidelines for Drinking Water Quality have been published by IS: 10500-2012. For Drinking water, quality is commonly expressed by classes of relative Suitability, although most classification systems include units on specific conductance, sodium content and boron concentration.

Table 1 List of substances found naturally in some ground waters which can cause problems in operating wells.

Substance	Types of problems
Iron(Fe^{+2},Fe^{+3})	Encrustation, staining of laundry and toilet fixtures
Manganese (Mn^{-2})	Encrustation, staining of laundry and toilet fixtures
Silica (SiO_2)	Encrustation
Chloride (Cl^-)	Portability, Corrosiveness
Fluoride (F^-)	Fluorosis
Nitrate (NO_3^-)	Methemoglobenemia
Sulphate (So_4^{-2})	Portability
Dissolved Gases	Corrosiveness
Dissolved Oxygen	Corrosiveness
Hydrogen Sulphide (H_2S)	Corrosiveness
Carbon dioxide (CO_2)	Corrosiveness
Radio Nuclides	Portability
Miner Constituents	Portability, Health aspects
Calcium and Magnesium (Ca^{2+}, Mg^{2+})	Encrustation

WQI is an important technique for demarcating groundwater quality and its suitability for drinking purpose. It is computed to reduce the large amount of water quality data to a mere numerical value that expresses the overall water quality at a certain location and time based on several water quality parameters. In this index mathematical equation used to transform large number of water quality data into a single number which is simple and easy to understandable for decision makers about quality and possible uses of any water body. It serves as the understanding of water quality for the possible uses by integrating complex data and generating a score that describes water quality status.

LITERATURE REVIEW

General: I referred various technical research papers on assessment of ground water quality for bore wells of different cities and countries, which are presented in Dissertation Phase-1. Reported work on assessment of ground water quality index is summarized below.

Shweta Tyagi, Bhavtosh Sharma, Prashant Singh, Rajendra Dobhal[3] carried out Water quality assessment in terms of Water Quality Index at Uttarakhand (India). The study states that Water quality index (WQI) is valuable and unique rating to depict the overall water quality status in a single term that is helpful for the selection of appropriate treatment technique to meet the concerned issues. However, WQI depicts the composite influence of different water quality parameters and communicates water quality information to the public and legislative decision makers. In spite of absence of a globally accepted composite index of water quality, some countries have used and are using aggregated water quality data in the development of water quality indices. Attempts have been made to review the WQI criteria for the appropriateness of drinking water sources. Besides, the present article also highlights and draws attention towards the development of a new and globally accepted "Water Quality Index" in a simplified format, which may be used at large and could represent the reliable picture of water quality. Initially

WQI selecting 10 most commonly used water quality variables like dissolved oxygen (DO), pH, coliforms, specific conductance, alkalinity and chloride etc. and has been widely applied and accepted in European, African and Asian countries.

Manjesh Kumar and Ramesh Kumar[4] Carried out experimental work on Physico-Chemical Properties of Ground Water of U.P., (India). The study deals with evaluation of granite mines situated in jhansi (Goramachia) for their status about physicochemical contamination of ground water. Six different sites are selected for sample testing collected from mines and urban area. Three samples have been taken at various distances on the site. This location is 10Km above from Jhansi city. The physic-chemical parameters such as pH, D.O., E.C., T.D.S., alkalinity, turbidity, Ca (calcium) and Mg (magnesium) hardness, total hardness, NO_3 (nitrate), F (fluoride), Fe^{+3} (iron) and Cl^- (chloride) have been tested. It has been found that parameters are not in limit when compared with W.H.O. standards.

Shivasharanappa, Padaki Srinivas and Mallikarjun S Huggi[5] carried out research work on Bidar city (Karnataka) for their characteristics of ground water and Water quality index (W.Q.I.). This research work deals with revaluation of W.Q.I. for groundwater for the residential and industrial area of bidar. In the city there are 35 wards, samples collected from all wards and tested for 17 parameters. The parameters are pH, total hardness, Ca (Calcium), Mg (magnesium), chloride (Cl), NO_3 (Nitrate), SO_4 (sulphate), T.D.S., Fe^{+3} (Iron), F (Fluoride), sodium (Na), potassium (K), alkalinity, manganese (Mn), D.O., total solids and Zinc (Zn). Tested results were used for suggest the models for water quality analysis.

J Sirajudeen, S Arul Manikandan and V Manivel (2013)[6] Carried out the work on ground water for evaluating the W.Q.I. Samples collected an Ampikapuram area near Uyyakondan channel Tiruchirappalli district. For Evolution of water quality index following parameters are examined: pH, E.C., T.D.S., Total hardness, D.O., C.O.D., B.O.D., Cl^-, NO_3 and Mg .The WQI for these samples ranged between is 244 to 383.8.The analysis reveals that the groundwater of the area needs some degree of treatment before consumption, and it also needs to be protected from the perils of contamination.

Cristina Rosu, Ioana Pistea, Mihaela Calugar, Ildiko Martonos, A.Ozunu[7], carried out work on quality of ground water by W.Q.I. method in Tureni Village, Cluj County. The rural population from Romania is dealing even today with the absence of access to a sure drinking water source. Therefore in 2002 only 65% of the Romanian population had access to drinking water, distributed in 90% from the urban environment and 33% from the rural one. This work presents a case study referring to a 3 month (April-May-June 2011) monitoring of weekly samples of the quality of well water (10 samples) from Tureni village, Cluj County. A portable multi parameter model WTW 720 Germany was used to measure the pH, total dissolved solids (TDS), electrical conductivity (EC), temperature, oxidation-reduction potential and salinity of the collected water samples (these tests were done on site). In laboratory, using the photometric method (RQ Flex instrument, Merck) we determined: Ca2+, Mg2+, SO42-, Cl- and NO3-.

Dr. N.C. Gupta, Ms. Shikha Bisht and Mr. B.A. Patra[8] carried out Physico-Chemical analysis of drinking water quality from 32 locations in Delhi. Delhi is an old town, which has gradually grown into a popular city. It is one of the important business centers of India and thickly populated as well (Gupta et al). Since the last decade, drinking water problem has created havoc in the city. In this study, we collected 32 drinking water samples throughout Delhi. Different parameters were examined using Indian Standards to find out their suitability for drinking purposes. During this examination mainly the physic chemical parameters were taken into consideration.

G. Achuthan Nair et al[9] carried out ground water quality status by water quality index at North –East Libya The quality of groundwater was assessed to their suitability for drinking at six places of north-east Libya viz. El-Marj Albayda, Shahat, Susa, Ras al-Hilal and Derna, during November, 2003 to March, 2004, by determining their physicochemical parameters (17 parameters) and water quality index (15 parameters). Peoples of Libya are aware for ground water quality and purity level and present study will be use full for maintaining the desired levels.

ASSESSMENT OF WATER QUALITY

General: Due to increasing urbanization, surface water is getting over contaminated and more stringent treatments would be required to make surface water potable. Therefore, it is required to additional sources for fulfill the requirement of water. Because the ground water sources are safe and potable for drinking and other useful purposes of human being. Hence studies of physic-chemical characteristics of underground water to find out whether it is fit for drinking or some other beneficial uses.

Parameters to be analyzed: For the assessment of ground water quality of the bore well of the Ranga Reddy Distrit, Taking in view the following drinking water parameters are analyzed (1) pH (2)Turbidity (3) Total Dissolved Solids (4) Elec. Conductivity (5)Total hardness (6) Calcium(7) Magnesium (8) Sodium (9) Potassium (10) Acidity.(11) Total alkalinity (12) DO (13) Chlorides (14)

Parameters included in water quality assessment: Monitoring of bore wells at Ranga Reddy District requires many different parameters to be sampled. The parameters analyzed in this assessment include:

pH: pH of solution is taken as –ive logarithm of H2 ions for many practical practices. Value range of pH from 7 t0 14 is alkaline, from 0 to 7 is acidic and 7 is neutral. Mainly drinking water pH lies from 4.4 to 8.5. The pH scale commonly ranges from 0 to 14.

Turbidity: Suspension of particles in water interfering with passage of light is called turbidity. Turbidity is caused by wide variety of Suspended particles. Turbidity can be measured either by its effect on the transmission of light which is termed as Turbiditymetry or by its effect on the scattering of light which is termed as Nephelometry. As per IS: 10500-2012 the acceptable and permissible limits are 1 and 5 NTU respectively.

T.D.S: Difference of total solids and suspended solids is used to determine the filterable solids by the help of filtrate and following the procedure as above. In water sample it can also be estimated from conductivity measurement. The acceptable and permissible limits As per IS: 10500-2012 is 500 and 2000 mg/l respectively.

Elec.Conductivity: Conductivity is the capacity of water to carry an electrical current and varies both with number and types of ions the solution contains. In contrast, the conductivity of distilled water is less than 1umhos/cm. This conductivity depends on the presence of ions their total concentration, mobility, valence and relative concentration and on the temperature of the liquid. Solutions of most inorganic acids, bases, and salts are relatively good conductors.

Total hardness: As per IS: 10500-2012 Desirable limit and Permissible limit for hardness is lies between 200 to 600 mg/l respectively. The effect of hardness is Scale in utensils and hot water system in boilers etc. soap scum's Sources are Dissolved calcium and magnesium from soil and aquifer minerals containing limestone or dolomite. The Treatment of hard Water is Softener Ion Exchanger and Reverse Osmosis process. The degree of hardness of drinking water has been classified in terms of the equivalent $CaCO_3$ concentration as follows: Soft - 0-60mg/l, Medium - 60-120 mg/l, Hard - 120-180 mg/l, Very hard - >180 mg/l.

Total alkalinity: Alkalinity is the sum total of components in the water that tend to elevate the pH to the alkaline side of neutrality. It is measured by titration with standardized acid to a pH value of 4.5 and is expressed commonly as milligrams per liter as calcium carbonate (mg/l as $CaCo_3$).Commonly occurring materials in water that increase alkalinity are carbonate, phosphates and hydroxides. Limestone bedrock and thick deposits of glacial till are good sources of carbonate buffering.

Chloride: All type of natural and raw water contains chlorides. It comes from activities carried out in agricultural area, Industrial activities and from chloride stones. Its concentration is high because of human activities. As per IS: 10500-2012 Desirable limit for chloride is 250 and 1000 mg/l in Permissible limit.

Table 2 Physical and chemical properties of tube well water as per IS 10500-2012[18].

S.No	Parameter	Unit	Accept Limit	Permi. Limit
1	Colour	Hazen Unit	5	15
2	Odour		Agreeable	Agreeable
3	pH		6.5-8.5	No relaxation
4	Turbidity	NTU	1	5
5	Total Dissolved Solids	mg/l	500	2000
6	Ammonia	mg/l	0.5	No relaxation
7	Boron	mg/l	0.5	1
8	Calcium	mg/l	75	200
9	Chloride	mg/l	250	1000
10	Fluoride	mg/l	1	1.5
11	Magnesium	mg/l	30	100
12	Nitrate	mg/l	45	No relaxation
13	Total Alkalinity	mg/l	200	600
14	Sulphate	mg/l	200	400
15	Total Hardness	mg/l	200	600
16	Temperature	°C	-	
17	Sodium	mg/l	-	
18	Iron	mg/l	0.3	No relaxation
19	Cadmium	mg/l	0.003	No relaxation
20	Chromium	mg/l	0.05	No relaxation
21	Zinc	mg/l	5	15
22	Manganese	mg/l	0.1	0.3
23	Nickel	mg/l	0.02	No relaxation

Water Quality Index (WQI): There are two basically different types of Water Quality Index. Additive Water Quality Index, in the form $WQI_a = \sum_i^n w_i q_i$

Multiplicative water quality index, in the form $WQI_m = \sum_{i=1}^{m} q_i^{w^i}$

Where, q_i = quality rating for the parameter $\sum_{i=1}^{n} w_i = 1$, w_i = weight to the ith parameter, Such that, n = Number of parameter.

Canadian council has invented Canadian Water Quality Index (CWQI) which is based on W.Q.I. of british Columbia. Parameters used by CWQI are: temp, conductivity, color, turbidity, D.O., pH, alkalinity (total alkalinity), calcium (Ca). sodium (Na), magnesium (Mg), potassium (K) sulphate (SO_4^{2-}), chloride (Cl^-), fluoride (F^-), dissolved organic carbon (DOC), phosphorus (P), nitrate, nitrite (NO_3^-, NO_2^-), nitrogen (N), silicon dioxide (SiO_2), aluminum (Al), arsenic (As), barium (Ba), beryllium (Be), cadmium (Cd), cobalt (Co), chromium (Cr), copper (Cu), mercury (Hg), lithium (Li), manganese (Mn), molybdenum (Mo), nickel (Ni), lead (Pb), selenium (Se), strontium (Sr), vanadium (V), zinc (Z).

Canadian Water Quality Index is based on three attributes of water quality that relate to water quality objectives: i.Scope -F_1, ii. Frequency-F_2, iii. Amplitude-F_3

$$CCME_{WQI} = 100 - \left(\frac{\sqrt{F_1^2 + F_2^2 + F_3^2}}{1.732} \right)$$

Quality index defines ranges for each CWQI: Bed (0–44), Marginal (45–64), Good (65–79), Very Good (80–94), Excellent (95–100)

WATER QUALITY INDEX (WQI) of Ranga Reddy District

pH, alkalinity, chlorides, dissolved oxygen, biological oxygen demand, total hardness, calcium, magnesium, total dissolved solids and sulphates were the parameters considered for the calculation of Water Quality Index.

Based on their relative importance in the overall quality of water for drinking purposes and considerable effects on primary health each of the chemical parameters was assigned a weight (wi). 5 was the maximum weight assigned to the parameters which are important in maintaining water quality and have prominent effect on water quality.

The parameters which exhibit considerably low harmful effects were assigned a weight of 2. Computing the relative weight (Wi) of each parameter Eq. 1. Table 1 present the weight (wi) and calculated relative weight (Wi) values for each parameter.

The concentration of each parameter was divided by its respective standard according to the guidelines laid down by BSI (1998) for computing a quality rating scale (qi).This result was multiplied by 100 using Eq. 2. For WQI computation, the water quality sub-index (SIi) for each chemical parameter is first determined, which is then used to determine the WQI as per the Eqs. 3 and 4.

$$W = wi / \Sigma Wi \qquad (1)$$

Where Wi is the relative weight, wi is the weight of each parameter and n is the number of parameters.

$$qi = (Ci/Si|)100 \quad (2)$$

Where qi = quality rating, Ci = concentration of each chemical parameter in each water sample in mg/L, Si = Indian drinking water standard (BIS 1998) for each chemical parameter in mg/L except for conductivity (lS/cm) and pH.

$$SI = Wiqi \quad (3)$$

$$WQI = (\Sigma SIi) | \Sigma Wi \quad (4)$$

Where SIi is the sub-index of ith parameter; qi is the rating based on concentration of ith parameter and n is the number of parameters.

Table 3 The weight and relative weight of each of the physico-chemical parameters used for WQI determination

S.no	Parameters	BSI desirable limits	Qi values Saroornagar	Qi values Hasthinapuram
1	Ph	8.5	84.7	91.5
2	Alkalinity	200	104.5	102.6
3	Chlorides	250	70.2	43.56
4	DO	6	64	50.3
5	BOD	3	93	61.6
6	Total Hardness	300	91.3	100.1
7	TDS	1000	67.58	61.65
8	Na	200	11.55	16.79
9	Potassium	200	8.9	6.23

S.no	Parameters	BSI desirable limits	Weight(wi)	Relative weights(W*i*)
1	Ph	8.5	3	0.107
2	Alkalinity	200	2	0.071
3	Chlorides	250	3	0.107
4	DO	6	5	0.178
5	BOD	3	3	0.107
6	Total Hardness	300	3	0.107
7	TDS	1000	5	0.178
8	Na	200	2	0.071
9	Potassium	200	2	0.071

$$WQI = (\Sigma SIi) | \Sigma Wi$$

Where, $SI = Wiqi$

Saroor nagar:

$$WQI = 68.58$$

Hasthinapuram:

WQI = 60.59

Quality index defines ranges for each CWQI: Bad (0-44), Marginal (45-64), Good (65-79), Very Good (80-94), Excellent (95-100)

DISCUSSIONS

ALKALINITY:

SAROORNAGAR

The range of alkalinity in the area of study is 100-350mg/l and the average value is 209mg/l. Alkalinity is within the limits in this region. In P&T colony (S.No 3) the alkalinity is slightly higher than the permissible limit.

HASTHINAPURAM:

The range of alkalinity in this area of study is 144-264mg/l with 205.2mg/l as the average. All the samples in this area are in the permissible limits.

HARDNESS:

TOTAL HARDNESS

Hardness is attributed principally to Ca and Mg. Hardness is also considered as total hardness is expressed in mg/l of equivalent $CaCO_3$.

SAROORNAGAR:

All the samples in this area are under the permissible limits and are safe for drinking purpose with an average of 274 mg/l.

HASTHINAPURAM:

Although the average of hardness in this area is 300 mg/l, S.No.10 and S.No.9 are quite higher than the limit,350 & 400 mg/l respectively

CHLORIDES:

SAROORNAGAR:

The average of chlorides around Saroornagar lake is 175 mg/l which is below the desirable limit as per IS 10500-2012(which is 250mg/l). In areas of sample no.s 6 & 1, the values are as low as 100 mg/l.

HASTHINAPURAM:

With 108 mg/l as average value the chloride content is far below the desirable limit required. In all the areas the chloride content is nearly 100mg/l with an exception of sample no. 6 which has 69mg/l.

pH:

pH value represents the concentration of hydrogen ions in water and is measure of acidity and alkalinity of water. pH value below 7.0 indicate acidic character while pH greater than 7.0 is indicative of alkaline character of water.

The average value for the study area is nearly 7 in both the areas, the water is slightly alkaline. However pH values are within permissible limit.

T.D.S:

SAROORNAGAR & HASTHINAPURAM:

The average TDS of these areas are 675 &616 mg/l respectively. While the desirable limit is 500 mg/l, many of the samples of Saroornagar area contain above 700 mg/l. All the samples of Hasthinapuram are below 700 mg/l, but are slightly greater than the desirable limit.

ELECTRICAL CONDUCTIVITY:

The average EC of the samples of these areas are 1.2 & 1.1 mS respectively. They range between 0.86 to 1.6 mS for Saroornagar and .96 to 1.37 mS for Hasthinapuram. Conductivity is the capacity of water to carry an electrical current and varies both with number and ypes of ions the solution contain. The conductivity of distilled water is less than 1mS.

SODIUM (Na):

SAROORNAGAR:

Sodium presence in this area is between 18 to 32 mg/l. The average is 23.1 mg/l.

HASTHINAPURAM:

For this area it ranges between 26 to 48 mg/l with an average of 33.5 mg/l.

POTASSIUM:

SAROORNAGAR:

In majority of this area, the potassium quantity is between 23 to 84 mg/l with an average of 53.5 mg/l.

HASTHINAPURAM:

With an average of 62.5 mg/l, majority of the areas here are in the range of 38 to 83 mg/l.

DISSOLVED OXYGEN:

SAROORNAGAR:

DO value in the majority of study area is between 3.0-5.2mg/l. Areas having DO value less than 4mg/l will not be suitable for aquatic life.

HASTHINAPURAM:

DO range of this area is 2.4 to 3.8 mg/l. All the areas have low DO and are not quite suitable for aquatic life.

B.O.D:

SAROOR NAGAR

The average value of BOD in this area is 2.8 mg/l. The values range from 1.7 to 3.9 mg/l.

HASTHINAPURAM:

The values of BOD in this area are between 1.4 to 2.5 mg/l with an average of 1.85 mg/l.

CONCLUSIONS

Water quality is dependent on the type of the pollutant added and the nature of mineral found at particular zone of bore well. Monitoring of the water quality of ground water is done by collecting representative water samples and analysis of physicochemical characteristics of water samples at different locations of Indore City. Estimation of water quality index through formulation of appropriate using method and evaluation of the quality of tube well water by statistical analysis is carried out.

Overall water quality index in both areas is seems to be in good condition, ranges from 60-70.

In the study area groundwater drawn from 20 bore wells were analyzed for their chemical contents. The analytical results of physical and chemical parameters of groundwater were compared with the standard guideline values recommended by the World Health Organization (WHO, 2011) for drinking purpose. Hydrochemically the ground water contains higher concentrations of TDS, Hardness values, moderate concentrations of pH, Ca+, Na+ and K+, and lower concentrations of Cl-, and DO.

Assessment of the quality of the groundwater from 20 bore wells indicate that the groundwater belong to Hard to very Hard category and groundwater from majority of the bore wells of the study region is unfit for drinking purposes. The groundwater is laden with objectionable concentration of cations and anions which may possibly have been derived through combined sources viz., mineralization, chemical weathering of rock, mine tailings, sewage contamination and intense agricultural activities.

This preliminary study calls for continuous monitoring of the quality of the groundwater in the region as further exploitation of groundwater may increase the values of the some of the parameters viz., EC, TDS, Na+, DO and F and deteriorate the water quality in near future which ultimately will prove to be disastrous for the entire living beings in the region. Spatial distribution map of certain parameters prepared from the hydrochemical data in GIS environment is useful in assessing the best groundwater quality zone in the study area.

REFERENCES

1. S.P. Gorde and M.V. Jadhav, Assessment of Water Quality Parameters: A Review, Journal of
2. Engineering Research and Applications, 3(6), 2029-2035 (2013)
3. Ground Water Pollution South of Musi River in Saroornagar Area, Rangareddy District, Andhra Pradesh Venkateshwarlu.Ch1*, D.Vijaykumar, G.Udayalaxmi, V.Jawahar lal
4. Shweta Tyagi, Bhavtosh Sharma, Prashant Singh, Rajendra Dobhal3 carried out Water quality assessment in terms of Water Quality Index at Uttarakhand (India)
5. Manjesh Kumar and Ramesh Kumar4 Carried out experimental work on Physico-Chemical Properties of Ground Water of U.P., (India)
6. Dr. N.C. Gupta, Ms. Shikha Bisht and Mr. B.A. Patra8 carried out Physico-Chemical analysis of drinking water quality in Delhi
7. Neeraj D. Sharma, J.N. Patel carried out evaluation of ground water quality index of the Urban segments of Surat city (INDIA)
8. J Sirajudeen, S Arul Manikandan and V Manivel (2013) Carried out the work on ground water for evaluating the W.Q.I. Samples collected an Ampikapuram area near Uyyakondan channel Tiruchirappalli district

9. Cristina Rosu, Ioana Pistea, Mihaela Calugar, Ildiko Martonos, A.Ozunu7, carried out work on quality of ground water by W.Q.I. method in Tureni Village, Cluj Country
10. Shivasharanappa, Padaki Srinivas and Mallikarjun S Huggi5 carried out research work on Bidar city (Karnataka) for their characteristics of ground water and Water quality index (W.Q.I.)
11. G. Achuthan Nair et al carried out ground water quality status by water quality index at North – East Libya The quality of groundwater was assessed to their suitability for drinking at six places of north-east Libya
12. Rajankar P. N. et. al (2013)carried out evaluation of tube well water quality using W.Q.I. in Wardha (India). Using W.Q.I. Some tehsile of district Wardha were evaluated
13. K. Elangovan (2010) carried out characteristics of tube well water for district Erode (India) states that ground water quality of 60 locations in Erode district during pre- monsoon and postmonsoon seasons
14. Neeraj D. Sharma, J.N. Patel carried out evaluation of ground water quality index of the Urban segments of Surat city (INDIA) states that the development of urban regions in developing country needs the multifaceted study of qualitative and quantitative stresses n available natural resources there within.
15. Sriniwas Kushtagi and Padaki Sriniwas(2011) carried out studies on water quality index of Groundwater of Aland taluka, Gulbarga(INDIA)states
16. Padma Priya K T et al., 2015 conducted an investigation on the water quality in Saroornagar Lake
17. Amaliya N.K. and Sugirtha P. Kumar carried out ground water quality status by water quality index method at Kanyakumari (INDIA)

ANALYSIS OF WATER QUALITY IN SEWAGE TREATMENT PLANT-BUDDHA PURNIMA PROJECT

P. Sowmya

Centre for water resources, Institute of Science and technology, JNTUH, Hyderabad

sowmyapuram143@gmail.com

ABSTRACT

Sewage may be defined as the used water or liquid waste of a community, which includes human and household wastes together with street washings, industrial wastes and such ground and storm water as may be mixed with it. Sewage must ultimately be disposed into receiving waters or on the land. Treatment of the sewage is required to remove all the contaminants from it before the same is disposed off to a natural water body or on to the land, which otherwise can cause adverse affects on health and environment. The main objective of this work is to analyze the outlet samples of each chamber such as receiving chamber, filtration unit, grit chamber, aeration tank, settling tank and outlet chamber in the process of treatment of domestic waste water at Buddha purnima project site Hyderabad. The parameters analyzed are pH, EC, COD, TS, TSS, TDS, Ammonical nitrogen, Phenols and Metals like cadmium, lead, zinc, nickel, manganese, copper and chromium. The instruments such as pH meter, EC meter, desecator, hot air oven and atomic absorption spectrometry method are used in the analysis. The values obtained in this analysis are compared with CPCB standard values. From the analysis it is found that pH, TDS, Ammonical nitrogen, phenols and heavy metals such as cadmium, lead zinc, nickel, manganese, copper and chromium are within the limits of CPCB. However two parameters namely COD and TS are exceeding the limits recommended by CPCB. It is also found EC and TSS is very close to the CPCB limits. Therefore treatment for these parameters COD, TS, EC and TSS are to be improved.

Keywords: Sewage Treatment, Buddha Purnima project, Water Quality,

INTRODUCTION

Sewage may be defined as the used water or liquid waste of a community, which includes human and household wastes together with street washings, industrial wastes and such ground and storm water as may be mixes with it. Sewage or liquid waste from a community is essentially the water supply to the community after it has been fouled by a variety of uses.

The constituents of sewage are:

(i) Domestic sewage, which includes human excreta as well as discharges from kitchens, baths, lavatories etc., from public and private buildings.

(ii) Industrial and trade-wastes from manufacturing processes such as tanneries, slaughter houses, mills, distilleries, chemical plants etc.

(iii) Strom water, which is rain water from houses, roads along with surface water etc.

(iv) Ground water or sub soil water entering sewers through leaks.

The characteristics and composition of sewage mainly depends on it source. Sewage contains organic and inorganic matters, which may be in dissolved, suspension or in colloidal and settelable state. Usually large proportion of the waste matter is organic in nature, which is attacked by saprophytic organisms, which are organisms that feed upon dead organic matter.

This activity of the organisms causes decomposition of the organic matter and destroys them. Sewage also contains various types of bacteria, virus, protozoa, algae, fungi etc. Some of these are pathogens and are harmful to the human and animal life and hence must be removed from the source of generation immediately by treatment and disposal.

Sewage must ultimately be disposed into receiving waters or on the land. Treatment of the sewage is required to remove all the contaminants from it before the same is disposed off to a natural water body or on to the land, which otherwise can cause adverse affects on health and environment.

Buddha purnima project authority was originally constituted for the development of the Husain sagar lake and its environs covering an area of 902 hectares, under the provision of the A.P. urban areas (development) act 1975, vide G.O.Ms.No.575, M.A.&U.D department ,dated 12th December, 2000 and the surrounding area of the Husain Sagar lake was declared as special development area. Sewage treatment plant of 20 mld with an extended aeration process commissioned in 1998 and maintained by BPPA located on necklace road adjacent to khairatabad flyover. Treated water (satisfying approved water quality standards of CPCB) released in to HSL (Husain Sagar Lake).

NEED FOR SEWAGE TREATMENT

Sewage if allowed to accumulate without treatment leads to production of mal-odors gases due to decomposition of organic matter and leads to epidemics due to the presence of disease causing microorganisms and pathogens. It stimulates aquatic growth due to presence of nutrients and leads to ground water pollution due to percolation of polluting and toxic compounds. The land on which excessive sewage will be applied will become sewage sick and it will create dirty scenes as well as nuisance and unhygienic smells. If the excessive quantity of sewage is mixed with natural water body course the water will become septic and totally unfit for any other use. It is necessary to reduce the quantity of sewage strength and make it such that it can be applied safely to land or discharged into natural water body courses. This objective is maintained by the treatment of sewage and by reducing its strength.

Sewage treatment is required to remove organic and inorganic matter which would otherwise cause pollution and to remove pathogenic organisms. In order to protect the environment and the human health. Pathogens or disease-causing organisms are present in sewage. These need to be got rid of before the sewage can be released back into the environment. It is unhealthy for humans, pets, and wildlife to drink or come in contact with surface or ground water contaminated with wastewater. Treatment of sewage is essential to ensure that the receiving water into which the effluent is ultimately discharged is not significantly polluted. In other words the sewage treatment plants are mainly employed to supplement the natural purifying powers of land and water courses and help in maintaining their normal uses. In addition to water that we want to recycle waste water also contains pathogens (disease organisms), nutrients such as nitrogen and phosphorus, solids chemicals from cleanness and disinfectants and even hazardous substances. Obviously we need to treat waste water not only to recycle water and nutrients but also to protect human and environmental health.

Inadequate treatment of wastewater allows bacteria, viruses, and other disease-causing pathogens to enter groundwater and surface water. Hepatitis, dysentery, and other diseases may result from bacteria and viruses in drinking water. Disease-causing organisms may make lakes or streams unsafe for recreation. Flies and mosquitoes that are attracted to and breed in wet areas where wastewater reaches the surface may also spread disease. Inadequate treatment of wastewater can raise the nitrate levels in groundwater. High concentrations of nitrate in drinking water are a special risk to infants. Nitrate affects the ability of an infant's blood to carry oxygen, a condition called methemoglobinemia (blue-baby syndrome). Sewage treatment plants are essential to life. Without them we would have to dispose of tons of raw sewage each hour, and that could get very messy very fast. If the sewage treatment plants didn't exist, we would probably be up to our knees in raw sewage.

REVIEW OF LITERATURE

In 1627 the water treatment history continued as Sir Francis Bacon started experimenting with seawater desalination. He attempted to remove salt particles by means of an unsophisticated form of sand filtration. It did not exactly work, but it did paved the way for further experimentation by other scientists. In the 1700s the first water filters for domestic application were applied. These were made of wool, sponge and charcoal.

In 1804 the first actual municipal water treatment plant designed by Robert Thom, was built in Scotland. The water treatment was based on slow sand filtration, and horse and cart distributed the water. Some three years later, the first water pipes were installed. The suggestion was made that every person should have access to safe drinking water, but it would take somewhat longer before this was actually brought to practice in most countries. In 1854 it was discovered that a cholera epidemic spread through water. The outbreak seemed less severe in areas where sand filters were installed. British scientist John

Snow found that the direct cause of the outbreak was water pump contamination by sewage water. He applied chlorine to purify the water, and this paved the way for water disinfection. Since the water in the pump had tasted and smelled normal, the conclusion was finally drawn that good.

Villiers et.al (1999) reported that the treatment of sewage by lime clarification and activated carbon treatment produces good quality of water with less cost to build, but more to operate than a conventional activated sludge plant.

Anil Kumar Misra and Ajai Misra (2002) in their paper discussed the Biotechnological aspects in waste water treatment in developing countries. The wastewater treatment technologies like Membrane electrolysis, Ion exchange, reverse osmosis, vacuum evaporation, cross flow micro filtration, diffusion dialysis are very expensive when compared to biological waste treatment processes. Due to this role of biotechnology in waste water, treatment is extremely important because, it is the only technology which is low cost, low maintaince, and effective against important sewage effluents and industrial waste.

Devesh Sharma and Virendra Kumar (2002) in their paper discussed the methodology for the selection of appropriate technologies for waste water treatment using decision tree criteria. The selection of appropriate technologies and process is complex process and creates dilemma as it depends on the different factors of the waste water. This paper discusses methodologies for identifying appropriate technologies for different areas depending on the factors.

A. K. Srivastava, Deepak Rastori and Neeraj Jain (1999) studied the performance of various coagulants like alum, lime, magnesium sulphate, ferric chloride and their combination in recovering various pollution parameters of the waste water. The results indicated that lime is an economical coagulant removing 55% of COD and 79% of suspended solids. Commercial alum, and magnesium sulphate and alum in conjunction with lime produced sludge with good drain ability.

STUDY AREA

The Buddha purnima project with the sewage treatment plant is selected as the study area. Buddha purnima project authority was originally constituted for the development of the Hussain sagar lake and its environs covering an area of 902 hectares, under the provision of the A.P. urban areas (development) act 1975, vide G.O.Ms.No.575, th M.A.&U.D department ,dated 12 December , 2000 and the surrounding area of the Husain Sagar lake was declared as special development area. Sewage treatment plant of 20 mld with an extended aeration process commissioned in 1998 and maintained by BPPA located on necklace road adjacent to khairatabad flyover. Treated water is released in to Husain Sagar Lake after the treatment.

METHODOLOGY

The sewage collected from the picket nala is treated in different stages which includes the collection of the raw sewage and the screening and degriting the waste water. It further includes the aeration treatment, distribution chamber, clarifiers, filtration, chlorine contact tank and the drying yard. And the final water which are obtained after the treatment are mixed with the water in the Husain Sagar lake and some of the water is used for the fountains surrounding the Husain Sagar lake and the tank bund. The complete treatment process and the stages are explained briefly as below

SAMPLING AT BUDDHA POORNIMA PROJECT

The samples are collected at the sewage treatment plant at Buddha poornima project in necklace road. The samples atr collected in the year 2011 march month.

The samples are collected at the different stages of treatment from the following chambers

1. Receiving chamber
2. Filtration unit
3. Grit chamber
4. Aeration tank
5. Settling tank
6. Outlet chamber

The collected samples are carried to the laboratory through a container without any contamination and maintaining the moderate room temperature. The collected samples are analysed for the following parameters.

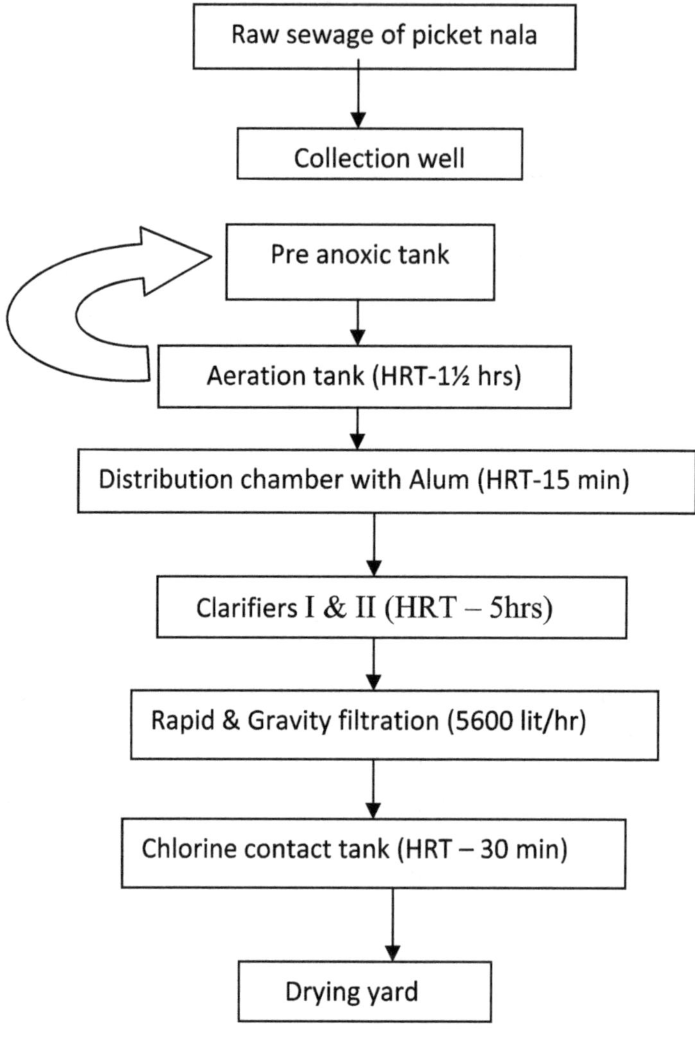

Fig. 1 Flow chart: Methodology of waste water treatment

- pH
- Electrical conductivity
- Total dissolved solids (TDS)
- Total suspended solids (TSS)
- Total solids (TS)
- Chemical oxygen demand (COD)
- Ammonical nitrogen
- Phenols
- Metals

Table 1 Measurement of are pH, EC, COD, TS, TSS, TDS, Ammonical nitrogen, phenols

Sample	pH	EC	COD (mg/l)	TS (mg/l)	TSS (mg/l)	TDS (mg/l)	Amonical nitrogen (mg/l)	Phenols (mg/l)
Receiving chamber	7.03	1142	600	848.8	297	551.8	2.80	0.15
Filtration unit	7.25	1123	680	1007.2	281	726.2	14	0.30
Grit chamber	7.37	1116	1080	1220	415	805	15.4	0.46
Aeration tank	7.39	1125	880	1502	706	1087	18.2	0.68
Settling tank	7.38	1096	600	1294	616	678	22.4	0.12
Outlet	7.6	1106	560	3085.6	160	925.6	15.4	0.1

MEASUREMENT OF METALS

Atomic absorption spectroscopy involves atomising a sample, and measuring the absorption of light passing through the vapour containing the atoms of the element. The amount of light absorbed is depending on the amount of element present in the vapour. Specific lamps emitting light of specific frequency absorbed by the element are required for each element. There are two techniques used in wastewater treatment to atomise the sample: Flame AA and Cold vapour AA.

Flame AA

This technique is used for detecting cadmium, chromium, copper, lead, silver and zinc. The sample is fed into a high temperature flame, causing it to split up into its individual atoms. The light from the specific lamp passes through the flame containing the element and a detector measures the light intensity of the light emerging from the flame. This technique determines whether the element is present, and its concentration.

In order to determine the concentrations of heavy metals such as Cd, Cr, Co, Ni, Pb, Cu, Fe and Zn the atomic absorption spectrometry method is used. The standard sample is prepared such that its concentration should include the concentration value of the unknown sample. The samples have been analyzed with the atomic absorption spectrometer with flame AVANTA GBC. This system is used for elemental analysis of a variety of samples. It has the measurement limit at 1 ppm. The water samples have been kept in polypropylene bottle, filtrated by filter paper and the level of pH has been established at 4 – 5 with HNO3. The obtained concentrations of Cd, Cr, Co, Ni, Cu, Fe and Zn are presented in Tables 4.9 - 4.15.

Cadmium (cd)

The GBC AVANTA Version 1.32 is selected and the element Cadmium (cd) is selected and the analysis is started by selecting the standards. The standard concentration and the absorbance are set to zero for the standards and the blank which are prepared. The flame is selected and optimized, and then the flame is ignited. The analysis for the sample is started by taking the sample and set for the readings for concentration and the absorbance at which the analysis is done. The results are presented in the Table 2.

Table 2 Measurement of Cadmium

Sample	Concentration mg/l	Absorption(nm)	CPCB standard for outlet
Receiving chamber	**0.00**	**0.001**	
Filtration unit	**0.00**	**0.000**	
Grit chamber	**0.00**	**0.003**	
Aeration tank	**0.00**	**0.001**	
Settling tank	**0.00**	**0.003**	
Outlet	**0.00**	**0.002**	**2 mg/l**

From the above readings it can be observed that the Cadmium metal values are as per the CPCB specifications for the disposal.

Lead (pb)

The GBC AVANTA Version 1.32 is selected and the element Lead (pb) is selected and the analysis is started by selecting the standards. The standard concentration and the absorbance are set to zero for the standards and the blank which are prepared. The flame is selected and optimized, and then the flame is ignited. The analysis for the sample is started by taking the sample and set for the readings for concentration and the absorbance at which the analysis is done. The results are presented in the Table 3

Table 3 Measurement of Lead

Sample	Concentration mg/l	Absorption(nm)	CPCB standard for out let
Receiving chamber	**0.07**	**0.003**	
Filtration unit	**0.02**	**0.001**	
Grit chamber	**0.05**	**0.002**	
Aeration tank	**0.02**	**0.001**	
Settling tank	**ND**	**0.000**	
Outlet	**0.02**	**0.001**	**0.1mg/l**

ND – Not Detected

From the above readings it can be observed that the Lead values are as per the CPCB specifications for the disposal

Zinc (Zn)

The GBC AVANTA Version 1.32 is selected and the element Zinc (Zn) is selected and the analysis is started by selecting the standards. The standard concentration and the absorbance are set to zero for the standards and the blank which are prepared. The flame is selected and optimized, then the flame is ignited. The analysis for the sample is started by taking the sample and set for the readings for concentration and the absorbance at which the analysis is done. The results are presented in the Table 4.

Table 4 Measurement of Zinc

Sample	Concentration mg/l	Absorption(nm)	CPCB standard for outlet
Receiving chamber	**1.09**	**0.115**	
Filtration unit	**0.34**	**0.036**	
Grit chamber	**0.65**	**0.069**	
Aeration tank	**0.26**	**0.028**	
Settling tank	**0.45**	**0.048**	
Outlet	**0.12**	**0.013**	**5mg/l**

From the above readings it can be observed that the Zinc values are as per the CPCB specifications for the disposal.

Nickel (Ni)

The GBC AVANTA Version 1.32 is selected and the element Nickel (Ni) is selected and the analysis is started by selecting the standards. The standard concentration and the absorbance are set to zero for the standards and the blank which are prepared. The flame is selected and optimized then the flame is ignited, then analysis for the sample is started by taking the sample and set for the readings for concentration and the absorbance at which the analysis is done and are presented in the Table 5.

Table 5 Measurement of Nickel

Sample	Concentration mg/l	Absorption(nm)	CPCB standard for out let
Receiving chamber	**0.00**	**-0.002**	
Filtration unit	**0.00**	**-0.001**	
Grit chamber	**0.00**	**-0.001**	
Aeration tank	**0.00**	**-0.002**	
Settling tank	**0.00**	**-0.001**	
Outlet	**0.00**	**-0.004**	**3mg/l**

From the above readings it can be observed that the Nickel values are as per the CPCB specifications for the disposal.

Manganese (Mn)

The GBC AVANTA Version 1.32 is selected and the element Manganese (Mn) is selected and the analysis is started by selecting the standards. The standard concentration and the absorbance are set to zero for the standards and the blank which are prepared. The flame is selected and optimized, then the flame is ignited. The analysis for the sample is started by taking the sample and set for the readings for concentration and the absorbance at which the analysis is done. The results are presented in the Table 6

Table 6 Measurement of Manganese

Sample	Concentration mg/l	Absorption(nm)	CPCB standard for outlet
Receiving chamber	0.073	0.015	
Filtration unit	0.028	0.006	
Grit chamber	0.088	0.018	
Aeration tank	0.040	0.008	
Settling tank	0.093	0.019	
Outlet	0.014	0.003	2mg/l

From the above readings it can be observed that the Manganese values are as per the CPCB specifications for the disposal

Copper (Cu)

The GBC AVANTA Version 1.32 is selected and the element Copper (Cu) is selected and the analysis is started by selecting the standards. The standard concentration and the absorbance are set to zero for the standards and the blank which are prepared. The flame is selected and optimized, and then the flame is ignited. The analysis for the sample is started by taking the sample and set for the readings for concentration and the absorbance at which the analysis is done. The results are presented in the Table 7

Table 7 Measurement of Copper

Sample	Concentration mg/l	Absorption(nm)	CPCB standards for outlet
Receiving chamber	0.016	0.002	
Filtration unit	0.011	0.002	
Grit chamber	0.026	0.004	
Aeration tank	0.019	0.003	
Settling tank	0.031	0.005	
Outlet	0.000	0.000	3mg/l

From the above readings it can be observed that the Copper values are below the CPCB specifications for the disposal

Chromium (Cr)

The GBC AVANTA Version 1.32 is selected and the element Chromium (Cr) is selected and the analysis is started by selecting the standards. The standard concentration and the absorbance are set to zero for the standards and the blank which are prepared. The flame is selected and optimized, and then the flame is ignited. The analysis for the sample is started by taking the sample and set for the readings for concentration and the absorbance at which the analysis is done. The results are presented in the Table 8.

Table 8 Measurement of Chromium

Sample	Concentration mg/l	Absorption(nm)	CPCB standards for outlet
Receiving chamber	0.00	-0.004	
Filtration unit	0.00	-0.004	
Grit chamber	0.00	-0.004	
Aeration tank	0.00	-0.005	
Settling tank	0.00	-0.005	
Outlet	0.00	-0.004	**2mg/l**

From the above readings it can be observed that the Chromium values are as per the CPCB specifications for the disposal

RESULTS AND CONCLUSIONS

Six samples are collected from the outlets of the chambers namely receiving chamber, filtration unit, grit chamber, aeration tank, settling tank and outlet chamber in the process of domestic waste water treatment at Buddha purnima project at Hyderabad. The parameters analyzed are pH, EC, COD, TS, TSS, TDS, Ammonical nitrogen, phenols and heavy metals such as cadmium, lead, zinc, nickel, manganese, copper and chromium.

The values of the analyzed parameters are compared with CPCB standards. From the comparison of results it is found that the parameters namely pH, TDS, Ammonical nitrogen, phenol and heavy metals such as cadmium, lead, zinc, nickel, manganese, copper and chromium are within the CPCB limits. However some of the parameters namely EC and TSS are very close to the CPCB limits. COD and TS are above the CPCB limits.

REFERENCES

1. Origin of sewage (2011) http://www.oppapers.com/essays/Origin-Of-Sewage/527912 accessed on 10 august 2011.
2. Wikipedia of waste water (2010) http://en.wikipedia.org/wiki/Wastewater
3. Clesceri, Leonore S., Eaton, Andrew D. and Greenberg, Arnold E. (eds); Standard Methods for the Examination of Water and Wastewater (19th edition); American Public Health Association, American Water Works Association and Water Environment Federation; 19.
4. V. Raveendra kumar (2005). "Design of Sewage Treatment System For Puda and usage of treated waste water for irrigation". Submitted to JNTU, 88 P.
5. Journal of Science and Arts Year 10, No. 1(12), pp. 113-118, 2010. "Assessment of river water quality in central and eastern parts of Romania using atomic and optical methods"
6. planet kids website (2011) http://www.planetkids.biz/documents/stpfacts.pdf accessed on 9 September 2011
7. U.S. Geological Survey(2011) URL: http://ga.water.usgs.gov/edu/wwvisit.html accessed on 5th December 2011
8. J.A.C. Broekaert (1998), "Analytical Atomic Spectrometry with Flames and Plasmas", 3rd Edition, Wiley-VCH, Weinheim, Germany.
9. Metcalf & Eddy, Inc. (1972). "Wastewater Engineering". New York: McGraw-Hill Book Company. ISBN 0-07-041675-.

10. Lenntech website (2012) http://www.lenntech.com/history-water-treatment.html accessed on 4th February 2012
11. EPA. Washington, DC (2004). "Primer for Municipal Waste water Treatment Systems." Document no. EPA 832-R-04-001.
12. T.-H. Han, E.-S. Nahm, K.-B. Woo, C. Kim, and J.-W. Ryu, "Optimization of coagulant dosing process in water purification system," in Proceedings of the 36th SICE Annual Conference, pp. 1105–1109, July 1997
13. IWA water wiki (2011) http://iwawaterwiki.org/xwiki/bin/view/Articles/Aerobic Wastewater Treatment Processes History and Development Synopsis accesed on 10 december 2011.
14. CPCB (Central Pollution Control Board) code IS 2490-1982, "Indian Standards for Industrial and Sewage Discharge".
15. Psb web waste water (2012) http://psbweb.co.kern.ca.us/eh_internet/pdfs/land/landHotTopic/WhyYouNeedGoodWastewaterTreatment.pdf accessed on 1st march 2012
16. M. N. Rao and A. K. Dutta , "Waste Water Treatment", Oxford & IBH Publishing Co.,Ltd.,1987.
17. Tunner J, Katsoulis G, Denoncourt J, Murphy S. "Design, qualification and performance of a cost-effective water purification system for a GMP pilot plant". Pharma Engg 2006; 26(4):1-8.
18. Metcalf and Eddy, "Waste water Engenering- Treatment, disposal and reuse" McGraw Hill book Co., Newyork, 1979.

TWO DIMENSIONAL INDUCED POLARIZATION IMAGING TO DELINEATE THE KAOLINIZED ZONES IN THE KHONDALITIC TERRAIN

Y. Siva Prasad[1]* and B. Venkateswara Rao[2]

[1]Research Scientist, Deltaic Regional Centre, National Institute of Hydrology, Kakinada

[2]Professor, Centre for Water Resources, IST, JNTUH

cwr_jntu@yahoo.com, sivaprasad.gw@gmail.com

ABSTRACT

An attempt is made with Two Dimensional (2D) Induced Polarization (IP) techniques for the investigation of high yielding water wells in the khondalitic terrain of northern parts of Eastern Ghats of India. This khondalitic terrain is mostly faced with the problem of identification of the extent of the depth of kaolinisation of the aquifer. The traditional One Dimensional (1D) Vertical Electrical Sounding survey could not identify the kaolinisation of the aquifer. The 2D IP Imaging surveys are attempted for the identification of kaolinised layer and the depth of kaolinisation. Number of 2D IP Imaging profiles were conducted using an ABEM SAS 1000 Terrameter near Chipurupally in Vizianagaram district along successful and failed wells located within short distances. The 2D IP Chargeability images have provided a reasonable clarity about the occurrence of the highly weathered zone (kaolinised zone) at both successful and failed wells. The layers having the high thickness obtained at greater depths with higher chargeability values below the success wells are identified as aquifer layers in the khondalitic suit of rocks. It can also be observed that below the success well, the formations having the low chargeability values with lesser thickness are obtained at the shallow depths. The layers having the high thickness obtained at greater depths with lower chargeability values below the failed well are indicating the kaolinised formations which are responsible for failure of wells.

Keywords: Khondalites, Kaolinisation, Eastern Ghats, Vertical Electrical Soundings, 2D Induced Polarization Technique

1. INTRODUCTION

The khondalitic suit of rocks (Garneti ferrous sillimanite/biotite gneiss) originally in sedimentary origin (Krishna Rao, 1952) and has become hard rock after high grade metamorphism (Krishnan, 1968). The khondalites are altered in two different manners upon the action of water, on the surface the rock changes into lateritic soil and the subsurface formation alters itself into kaoline (Mahadevan, 1929). In this terrain, higher well yields can be observed where the country rock is identified to have silicification. The main problem in the khondalitic terrain is kaolinisation of khondalite which is not properly being recognized in the vertical electrical sounding data. The khondalite becoming partly clay due to kaolinisation and is unable to support high well yield (Venkateswara Rao, 1995). With the Vertical Electrical Soundings (VES) survey, it is very difficult to distinguish between kaolinised and fractured khondalite, since they show same resistivity at both the places (Venkateswara Rao et al., 2013). Therefore, multi electrode geophysical techniques were employed in order to recognize the kaolinised layer.

Many earlier studies have used VES to delineate the anisotropy of the aquifers in the hard rock terrains. Hegde et al., (1989) have conducted radial vertical electrical soundings and computed coefficient of anisotropy to delineate the fractures at different horizons. Venkateswara Rao (2009) has developed an improved methodology for locating potential aquifers by using resistivity with VES data, hydrogeology and geomorphic parameters.

Burtman and Zhdanov (2015) have noted that Induced Polarization (IP) occurs in the bulk volume of the rock. IP surveys finds application in the study of clay minerals. IP methods can be used in the field of hydrogeology (Vacquier et al., 1957). The dependency of polarizability of rocks or soils upon their lithological composition and hydrogeological properties favors the application of the IP method for groundwater (Marshall and Madden, 1959). Polarization is a geophysical phenomenon which measures the slow decay of voltage in the ground after the cessation of an excitation current pulse (time domain method) or low frequency variation of the resistivity of the earth (frequency domain method) (Sumner, 1976). This means that the IP effect reflects the degree to which the subsurface is able to store electric charge. Hence, IP methods measure the Chargeability in 'm-sec'. Telford and Sheriff (1990) have stated that, the membrane IP effect is most pronounced in the presence of clay minerals, in which the pores are particularly small and also explained that the magnitude of polarization decreases with the clay mineral concentration. They concluded that the optimum concentration of clay is low in Montmorillonite and higher in Kaolinite. Hence, Koalinite and Shales with a high percentage of clay minerals have a relatively low polarization or lesser chargeability values. Interactions between clay minerals and groundwater can produce polarization phenomena and decreases in resistivity (Cohen, 1981). Parkhomenko (1971) stated that induced polarization increases with increasing water content until an optimum saturation is reached.

Laboratory experiments have also indicated that the Induced Polarization depends on cation exchange in the clay minerals contaminating the aquifers. It is suggested that the IP effect is due to electro-dialysis of the clay within the aquifer, which acts as a distributed electronegative membrane. The magnitude of the IP depends on the kind of clay and the kind of positive ions in the water. It is inversely proportional to the conductivity of the water and independent of the kind of negative ion dissolved therein. The rapidity with which the IP decays appears to depend only on the grain size of the aquifer (Vacquier et. al., 1957).

The IP effect in earth materials having non-metallic minerals is usually referred to as membrane polarization. This requires the presence of particles such as clays, with a negative surface charge, which in the presence of an electrolyte like water, attracts cations and forms an electrical double layer on the particle surface. The volume property of soil related to this phenomenon is the Cation Exchange Capacity (CEC) (Olhoeft, 1985) which is proportional to the surface charge density and the specific surface area of the solid phase and mostly related to clay minerals. This property supports the alternative procedure of IP in trying to detect clay minerals apart from the other conventional laboratory techniques like X ray diffraction. The CEC expresses the number of positive charges from the ions that neutralize the negative charge of soil particles. The CEC is quantified in meq/100 g i.e., the weight of ions in milliequivalents adsorbed per 100 grams of clay. The CEC is a typical property of clayish material because it relates to the clay mineral structure and is the basis for the electrical behavior of such soils.

Vacquier et. al., (1957) explained the importance of I.P. effect on different rock textures. The IP effect is characterized as a weak IP in compact clays having the low CEC in the order of 3-15 meq/100 g and low surface area of 15 m^2/g such as kaolinite and a strong IP in sediments with disseminated clay particles having the high CEC in the order of 80–150 meq/100 g and high surface area (on the surface of larger grains) of more than 100 m^2/g such as montmorillonite.

2. MATERIAL AND METHODS

Geoelectrical techniques namely One Dimensional (1D) Vertical Electrical Sounding and Two Dimensional (2D) Induced Polarization Imaging surveys have been employed at pairs of successful and failed wells covering khondalitic terrain where well failure is mostly due to kaolinisation. The pair of successful and failed well in the basin is selected in such a way that they are only few tens of meters apart and are in similar hydrogeologic setting.

2.1 One Dimensional Vertical Electrical Sounding

Vertical electrical soundings were conducted with DDR2 model resistivity meter. It has the facility to read directly the resistance offered by the formation with a voltage and current sensitivities reading up to one hundredth parts of a milli-volt and milli-ampere.

2.2Two Dimensional Inducted Polarization Imaging

Since the kaolinised material is basically a clayey material, chargeability imaging of IP surveys are attempted using an ABEM SAS 1000 Terrameter in order to delineate the kaolinisation of khondalitic formation below success and failed wells. Chargeability occurs when an electric current passes through a rock or soil. If the current is interrupted, a difference in potential which decays with time is observed. The rate of decay of this induced polarization potential depends on the lithology of the rock, its pore geometry and degree of water saturation. When direct current introduced into the ground at two points is interrupted, a small voltage, which may take several minutes to decay, appears between another pair of electrodes. This effect is called 'provoked polarization', or 'induced polarization' is found useful in prospecting for groundwater (Burtman and Zhdanov, 2015).

Unlike conventional resistivity sounding and lateral profiling surveys, 2D imaging is a fully automated technique that uses a linear array of about 48 electrodes (in the present survey) connected by multicore cable. The current and potential electrode pairs are switched automatically using a control module connected to an earth resistivity meter (that provides the output current). In this way a profile of chargeability against depth (pseudosection) is built up along the survey line. Data is collected by automatic profiling along the line at different electrode separations. In the present survey, for a 48 electrode array with an electrode spacing of 5 m this depth is approximately 43 m (Fig.1).

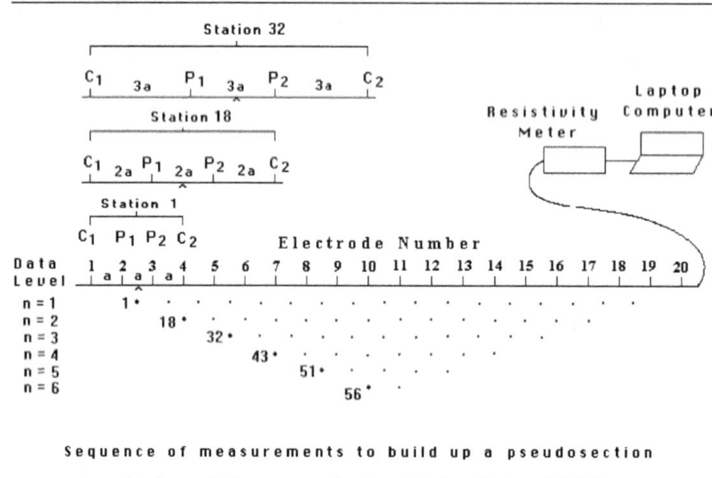

(Adopted from practical guide by Loke, 1999)

Fig. 1 Sequence of measurements in Chargeabiity Imaging to build up a pseudosection.

The imaging profiles are laid across the slope of the ground where the successful or failed wells are located and analyzed the profiles in such a way that these wells are almost in the middle of the profile. The chargeability imaging from IP surveying profiles conducted at the adjacent successful and failed wells are shown from Fig.4 to Fig.9. For the convenience of interpretation, various layers and their thicknesses have been identified from the chargeability images and are tabulated.

3. RESULTS AND DISCUSSION

3.1 Discussion of the Results from the VES

In this paper, few problematic study areas were selected pertaining to khondalitic formation of northern parts of Eastern Ghats near Cheepurapalli, Vizayanagaram district of Andhra Pradesh, India (Fig.2).

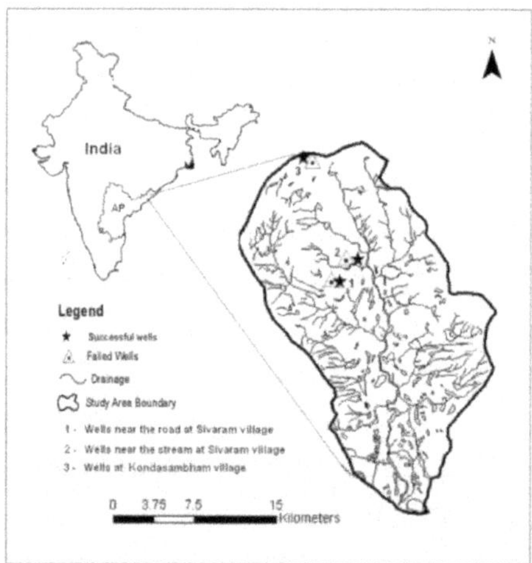

Fig. 2 Location of the Khondalitic terrain, Vizianagaram district, A.P., India

These areas are located near the road of Sivaram village, near the stream of Sivaram village and near Kondasambham village where a pair of successful and failed well has occurred at each place. These pairs of successful (> 8000 LPH) and failed wells (< 600 LPH) with their lithologs, yield logs and vertical electrical sounding curves are shown in the Fig.2 and can be observed within the pair that, the difference in well yield is very large between successful and failed well. From these VES curves, it is difficult to explain differences in their well yields but actual drilling results indicate the extent of kaolinisation that matters in deciding the well yields.

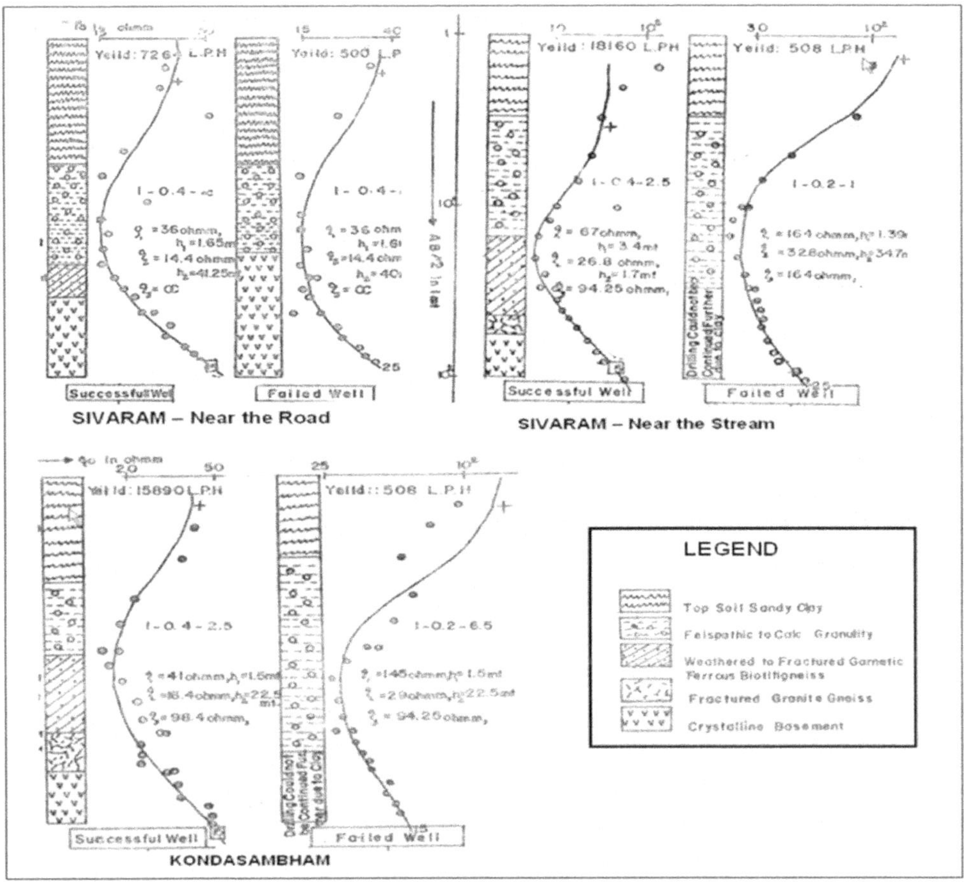

Fig. 3 Vertical electrical sounding curves and lithologs of adjacent success and failed wells

Lithologic information of 42 wells drilled in the study area have been collected and the observed kaolinized layer thicknesses were analyzed by computing percentage of success wells in various ranges of kaolinized layer thickness (Table1). From this table it can be observed that the greater the kaolinized layer thickness the lesser the success rate of the wells implying that the kaolinized layer is not supporting a good well yield. The extent of kaolinisation could not be deciphered in the VES curves interpreted with the curve matching technique.

Table 1 Resistivities of the subsurface formations with below Success and Failed wells in Khondalitic terrain

3.2 Discussion of the Results from 2D IP Surveys

Location of the well	Resistivity Data (ρ in 'Ω.m') , Thickness (h in 'm')			
	Success Well		Failed Well	
Sivaram- Near The Road	ρ_1 =34.74	h_1=2.43	ρ_1 =39.60	h_1 =1.55
	ρ_2 =14.23	h_2 =26	ρ_2 =13.96	h_2=37.95
	ρ_3=1004.2		ρ_3=1009.27	
Sivaram- Near The Stream	ρ_1 =82.91	h_1=1.93	ρ_1 =46.43	h_1 =2.79
	ρ_2 = 14.96	h_2=33.48	ρ_2 =21.76	h_2=32.60
	ρ_3=457.39		ρ_3 = 999.0	
Kondasambham	ρ_1 =41.0	h_1 =1.5	ρ_1 =145.0	h_1 =1.5
	ρ_2 =16.4	h_2 =22	ρ_2 =29.0	h_2 =22.5
	ρ_3 =98.4		ρ_3 =94.25	

Since the process of conducting Radial Vertical Electrical Sounding is cumbersome in the field and does not provide the depth section to guide the drilling process, 2D IP Imaging Surveys are conducted at successful and failed wells.

3.2.1 The Case of Sivaram Village – Near the Road

The IP Chargeability images at success well and at failed well near the road of Sivaram Village are as shown in Fig. 4 and Fig.5.

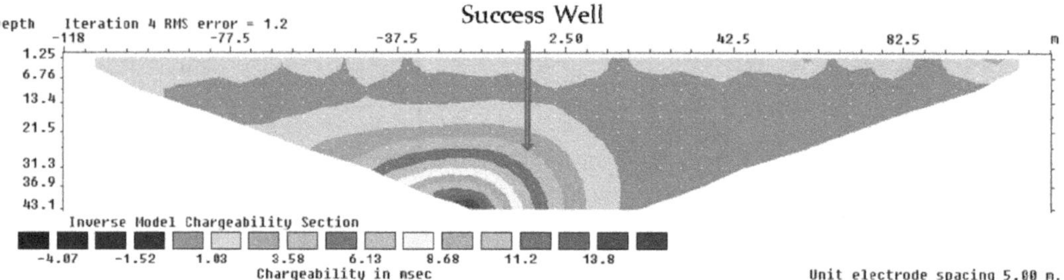

Fig. 4 Chargeability image at success well near the road of Sivaram Village

From Fig. 4 it can be observed that, below the success well the chargeability values are in the range of 1.03 m-sec, up to the depth of 21.5 m. From the depth of 21.5 m to the depth of 43.1 m, the chargeability values are gradually increasing from 1.03 m-sec to 3.58 m-sec with a thickness of 21.6 m. Table 4.5 is showing the depth ranges, the corresponding chargeability values and thicknesses below the success and failed well at Sivaram village near the road.

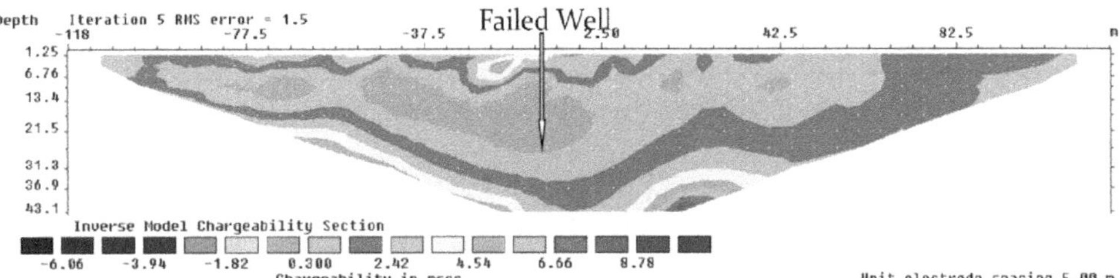

Fig. 5 Chargeability image at failed well near the road of Sivaram village

Table 2 Depth ranges and corresponding chargeability values below the success well and failed well at Sivaram village near the road

Success Well			Failed Well		
Depth in m.	**Chargeability in m-sec**	**Thickness in m**	**Depth in m.**	**Chargeability in m-sec**	**Thickness in m**
0 – 21.5	1.03	21.5	0 – 6.76	2.42	6.76
21.5 – 43.1	3.58	21.6	6.76-31.3	0.30	24.54
			31.3 – 43.1	2.42	11.8

Chargeability distribution at failed well in the Sivaram village near the road is shown in Fig. 5. It can be observed that below the failed well, a very low chargeability value of 0.3 m-sec is maintained with a thickness of nearly 25 m up to the depth of 31.3 m. From the depth of 31.3 m, the chargeability value is of the order of 2.42 m-sec. From the Table 2 it can be observed that the layers having the high thickness of 21.6 m with higher chargeability values of 3.58 m-sec are found below the success well, while the layers having the high thickness of 24.54 m with lower chargeability values of 0.30 m-sec are found below the failed well. Low chargeability values are indicating the kaolinised formations and are responsible for failure of well at the road of Sivaram village.

3.2.2 The Case of Sivaram Village – Near the Stream

The 2D IP Chargeability images at success well and at failed well near the stream of Sivaram Village are shown in Fig. 4.11 to Fig.4.12.

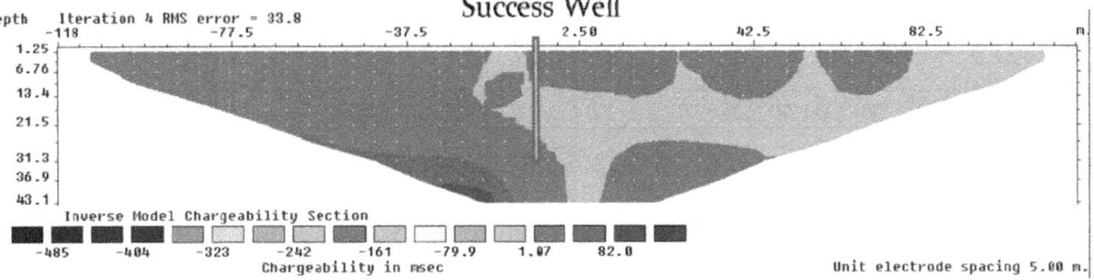

Fig. 6 Chargeability image at success well near the stream of Sivaram Village

From Fig.6 it can be observed that, below the success well the chargeability values are in the range of 1.07 m-sec, up to the depth of 43.1 m. In fact, higher side of chargeability values of 1.07 m-sec are obtained from the depth of 21.5 m. below the success well and also high chargeability values which are more than 1.07 m-sec can be observed nearer to the success well (left bottom of the image) from Fig.6.

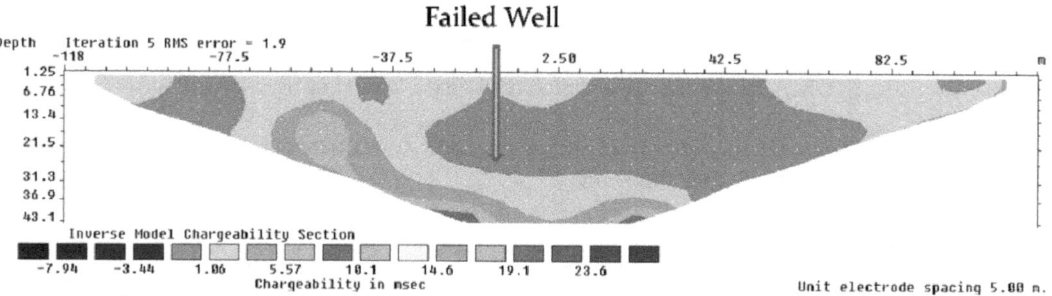

Fig. 7 Chargeability image at failed well near the stream of Sivaram village

The 2D IP Image in terms of chargeability at failed well near the stream of Sivaram Village is shown in Fig.7. Below the failed well, the chargeability values are in the range of 1.06 m-sec up to the depth of 36.9 m. In fact, lower side of chargeability values of 1.06 m-sec are observed up to the depth of 26.5 m. (thickness of 20 m) below the failed well. The kaolinised layer having the lower chargeability values with the thickness of 20 m is the responsible for failure of well at near the stream of Sivaram village. The depth ranges, the corresponding chargeability values and thicknesses below the success and failed wells in Sivaram near the stream are shown in Table 3. It can be observed from the table that differences in chargeability between success and failed well is very less.

Table 3 Depth ranges and corresponding chargeability values below the success well at Sivaram near the stream

Success Well			Failed Well		
Depth in m.	**Chargeability in m-sec**	**Thickness in m**	**Depth in m**	**Chargeability in m-sec**	**Thickness in m**
0 – 10.0	1.07	10.0	0 – 6.76	1.06	6.76
10.0 – 21.5	Lower side of 1.07	11.5	6.76-26.5	Lower side of 1.06	20.0
21.5 – 43.1	Higher side of 1.07	12.4	26.5 – 36.9	1.06	10.4
			36.9 – 43.1	5.57	6.2

3.3.3 The Case of Khondasambam Village

The 2D IP Chargeability images at success and failed wells in the Khondasambam village are shown in Fig.8 and Fig.9.

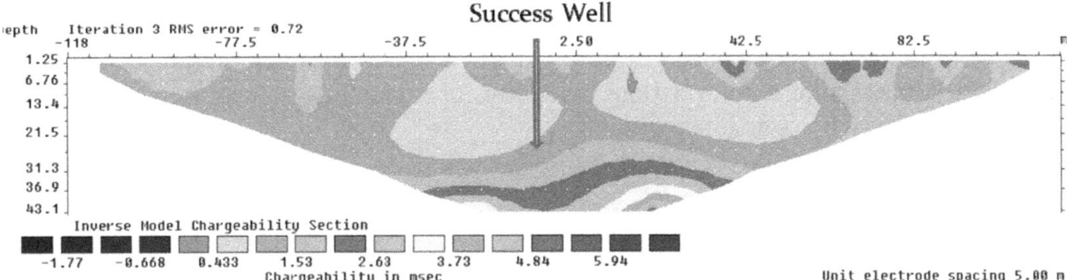

Fig.8 Chargeability image at success well at Khondasambham village

From Fig.8 it can be observed that, below the success well the chargeability values are gradually increasing from the depth of 21.5 m to 43.1 m. The higher chargeability values are associated with a formation thickness of 21.6 m.

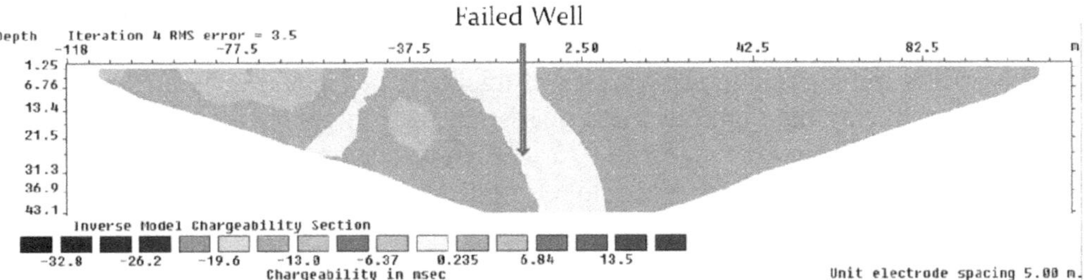

Fig.9 Chargeability at failed well at Khondasambham village

Table 4 is showing the depth ranges, the corresponding chargeability values and thicknesses below the success and failed wells in Sivaram near the road. It can be observed that high chargeability formation below the success well and low chargeability formation below the failed well.

Table 4 Depth ranges and corresponding chargeability values below the success well and failed well in the Khondasambham village

Success Well			**Failed Well**		
Depth in m.	**Chargeability in msec**	**Thickness in m**	**Depth in m.**	**Chargeability in msec**	**Thickness in m**
0 – 11	1.53	11	0 – 43.1	0.235	43.1
11 – 21.5	0.33	10.5	-	-	-
21.5 – 36.9	1.53	15.4	-	-	-
36.9 – 43.1	2.63	6.2	-	-	-

Chargeability distribution at failed well in the Khonasambham village is shown in Fig.9. It can be observed that, below the failed well, a very low chargeability value of 0.235 m-sec is maintained with a thickness of 43.1 m till the depth of 43.1 m. At this location, Fig.9 is correlating to the drilling results in which kaolinisation is extended to the deeper depths. From the Table 4 it can be observed that, the layers having the high thickness of 21.6 m with higher chargeability values of 2.63 m-sec are below the success well and the layers having the high thickness of 43.1 m with lower chargeability values of 0.235 m-sec are below the failed well indicating that the kaolinised formations with low chargeability values are responsible for failure of well in the Khonasambham village.

4. CONCLUSIONS

Since the traditional Vertical Electrical Sounding survey could not identify the kaolinisation of the aquifer, the 2D IP Imaging surveys are attempted for the identification of kaolinised layer and the depth of kaolinisation in the khondalitic terrain of Eastern Ghats of India. The IP images have provided a clear view of the thickness of the highly weathered zone (kaolinised zone) at both successful and failed wells. The IP Chargeability images have provided a clear view of the thickness of the highly weathered zone (kaolinised zone) at both successful and failed wells. The layers having the high thickness of 12 m - 21 m obtained in greater depths with higher chargeability values in the range of 1.07 - 3.58 m-sec below the success well are identified as aquifer layers which are composed of moderately weathered and fractured khondalitic suit of rocks. It can also be observed that below the success well, the formations having the low chargeability values of 0.3 – 1.0 m-sec with lesser thickness of 10 m are obtained in the shallow depths. The layers having the high thickness of 20 m – 43.1 m obtained in greater depths with lower chargeability values in the range of 0.235 – 1.0 m-sec below the failed well are indicating the kaolinised formations which are responsible for failure of wells.

REFERENCES

1. Burtman, V. and Zhdanov, M.S., (2015). “Induced Polarization Effect in Reservoir Rocks and its Modeling Based on Generalized Effective-Medium Theory Resource-Efficient Technologies”, Resource-Efficient Technologies, Vol.1, pp.34–48.
2. Cohen, M.H. (1981). “Scale Invariance of the Low–Frequency Electrical Properties of Inhomogeneous Materials”. Geophysics, Vol.46, pp.1057–1059.
3. Hegde, V.R., Gowdireddy, K. and Shetti, S.B., (1989), “Radial vertical electrical soundings-A case study on groundwater exploration”, Procs. of the International workshop on Appropriate Methodologies for Development and Management of Groundwater Resources in Developing Countries, Vol.1, pp.302-305.
4. Krishnan, M.S., (1968). “Geology of India and Burma”. Higgin Botham’s (P), Madras, India, 555p.
5. Krishna Rao, J.S.R., (1952). “The Geology of Chipurupalle Area, Visakhapatnam District with Special Reference to the Origin of Manganese Ores”, Journal of Geological Society of India, Vol. XXVI, No.1, pp. 36-45.
6. Loke, M.H. (1999). “A Practical Guide on Electrical Imaging Surveys For Environmental And Engineering Studies” 56p.
7. Mahadevan, C., (1929). “Geology of Vizag Harbour Area”. Quar. Jour. Geol. Min. Metal. Soc. India, Vol.2, No.4.
8. Marshall, D.J. and Madden, T.R. (1959). “Induced Polarization, A Study of its Causes”, Geophysics, Vol. XXIV, No.4, pp.790-816.
9. Olhoeft, G.R., (1985), “Low Frequency Electrical Properties”, Geophysics, Vol.50, pp.2492- 2503.
10. Parkhomenko, E.I. (1971). Electrification Phenomena in Rocks. Plenum Press, New York, U.S.A.
11. Sumner, J.S., (1976). “Principles of Induced Polarisation for Geophysical Exploration”, Elsevier Scientific Publishing Company, Amsterdam.

12. Telford, W.M. and Sheriff, R.F., (1990). Applied Geophysics, 2nd edition, Cambridge University Press, U.K.
13. Vacquier, V., Holmes, C.R., Kintzinger, P.R. & Lavergne, M., (1957). "Prospecting for Ground Water by Induced Electrical Polarization", Geophysics, Vol.22, pp.660-687.
14. Venkateswara Rao, B., (1995). "Integrated Studies for Evaluation of Ground Water Potential in a Typical Khondalitic Terrain" Ph.D. thesis, JNT University, India, 284p
15. Venkateswara Rao. B., (2009). "An Improved Methodology for Identification of Ground Water Potential Zones in A Typical Khondalitic Terrain" Journal of geophysics, Vol. XXX, No. 1-4, pp.73-81.
16. Venkateswara Rao. B., Siva Prasad, Y. and Srinivasa Reddy, K., (2013). "Hydrogeophysical Investigations in a Typical Khondalitic Terrain to Deliniate the Kaolinised Layer using Resistivity Imaging", Journal of Geological Society of India, Vol. 81, pp. 521-530.

OPTIMAL CONTROL OF WATER DISTRIBUTION AT DAMS & RESERVOIRS USING SCADA & TELEMETRY

V. Phani Madhav[1] and G.K.Viswanadh[2]

[1]Ph.D Scholar , JNTUH College of Engineering , Hyderabad.
[2]Professor of civil Engg. & OSD to Vice-Chancellor, JNTUH,
phanivangara@gmail.com, gkviswanadh@jntuh.ac.in

ABSTRACT

In order to meet the end customer's requirement there is a need to demand for the Supply of water that is transmitted via interconnected pipes from the source dams & reservoirs through valves and pumps by controlled pressure regulations. These supply elements are categorized as active and passive. Those elements that can be operated by controlling the flow of water and pressure viz., valves, pumps etc. are defined as active whereas the reservoirs are treated as passive. This discussion is focused primarily on overview of SCADA, Architecture of SCADA, and its Operation.The study area considered is Osmansagar & Himayatsagar Reservoirs to understand the water losses happening here using SCADA & Telemetry Units. The preliminary data of both the reservoirs were collected via the internet.

1. INTRODUCTION

As day by day we are facing a growing water shortage problem on surface and sub-surface ground, therefore water losses associated with the inefficiencies of the reservoirs & underground water distribution system is not only causing revenue loss for the water boards but also significantly negatively affecting the national water reserves. In order to avoid any pilferage, directly or indirectly, to maintain the quality , proper control and distribution of water, there is a need to implement a system which would not only reduce the operational cost but also improves the efficiency of the set-up otherwise the future generations are no more going to enjoy the resources of the nature.

The industrial control systems, which include supervisory control and data acquisition (SCADA) systems, distributed control systems, and other smaller control system configurations such as PLC's (programmable logical controllers) are often used in the industrial control sectors. The SCADA systems are generally used to control dispersed assets using centralized data acquisition and supervisory control. The study area considered is Osmansagar & Himayatsagar Reservoirs to understand the water losses happening here using SCADA & Telemetry Units. The preliminary data of both the reservoirs were collected from the internet .

2. NEED OF THE STUDY

Because of the rapid construction activities taken place in the recent times in and around the areas of Osmansagar and Himayatsagar reservoirs, were not able to render the requirements due to low rains and low intensity of water inflow and affecting the surface rainwater flow.

Osmansagar and Himayatsagar were constructed for flood control after 1908 Musi Floods submerged Hyderabad City and that is how their storage capacity was provided three times more than the expected yield of water in a year as per the analysis. The expected yield to the reservoirs was only 1 TMCFT each.

3. OBJECTIVE

The objective of the research is about an Optimal Control of water distribution network in a supervisory control system by deploying SCADA & Telemetry Units.

3.1 To monitor & control water distribution

3.2 Devices we use to perform for controlling and monitoring.

Supervisory Control and Data Acquisition (SCADA) is a computer aided system which collects, stores and analyses the data on all aspects of O&M. In a SCADA system up-to the minute real time information is gathered from remote terminal unit located at the water treatment plant, reservoir, flow meter, pumping station etc. and transmitted to a central control station where the information is updated, displayed and stored manually or automatically. SCADA systems will have probes/sensors which will sense and generate signals for the level, pressure and flow in a given unit and transmit the signals for storage and analysis in the computer. The signals are transmitted by radio, by Telephone, microwave satellite or fiberoptic transmission systems. The signals transmitted are stored as data, analysed and presented as information.

4. SCADA SYSTEM ARCHITECTURE

Broadly any of the SCADA systems consists of the following architecture:

4.1 The apparatus used by a human operator where all the processed data is presented to the operator

4.2 A supervisory system which collects all the required data about the process

4.3 Remote Terminal Units (RTUs) connected to the sensors of the process, which helps to convert the sensor signals to the digital data and send the data to supervisory stream.

4.4 Programmable Logic Controller (PLCs) used as field devices

4.5 Communication infrastructure connects the Remote Terminal Units to supervisory system.

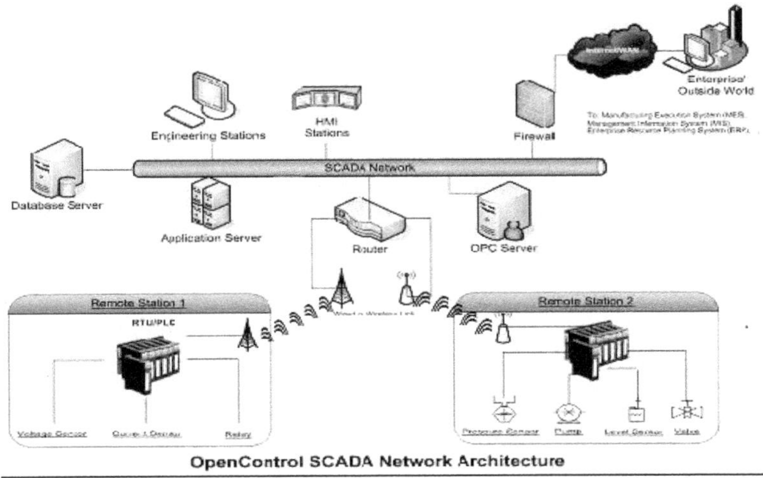

OpenControl SCADA Network Architecture

5. Advantage of SCADA Systems

5.1 Automatic Capture of Vital Hydraulic DATA where Digital data is automatically generated of all key installations and stored for record & analysis purpose. Earlier we used to depend on number of manuscript logbooks maintained at various reservoir locations for the similar analysis purpose.

5.2 The data Pertains to flows to Reservoirs, levels of water released from reservoir. Monitoring of Real Time DATA.

5.3 Online DATA is regularly monitored pertains to flows & quantities to each reservoir with reference to the allotted quantities & Short/Excess Supplies Reports.

5.4 Total quantity generated and supplied by the transmission system to each distribution system vis-à-vis short/excess supplies to regulate UFW Assessment.

5.5 UFW (Unaccounted for Water) study is taken up to assess the losses at Transmission system, Reservoir & Distribution System.

5.6 Presently, one person is generating the reports & communicating to each O&M Division for their perusal and information about the supplies of the quantities provided to them and even being made known to the customers and public representatives to improve the customer satisfaction.

6. WORKING PRINCIPLE

The SCADA base water supply system have Some important part and those are pump station, Reservoir station, main control station, sub control station , master station, server room, pipe & sub pipeline & home supply management.

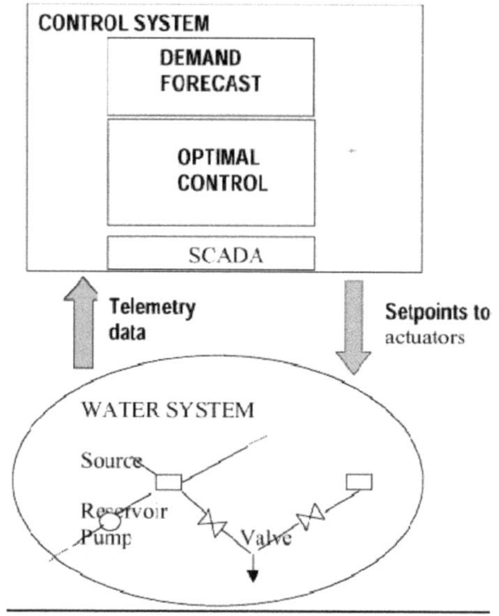

7. METHODOLOGY

The methodology adopted for conducting the study is as follows:

7.1 Collection of water inflow and outflow data at the reservoirs from various sources

7.2 Analysis on SCADA applications using MiScout Application.

8. STUDY AREA

The study area considered is **Himayath Sagar & Osman Sagar Reservoirs** which falls 17.3187° N Latitude to 78.3586° E Longitude and Osman Sagar 17.3763° N Latitude 78.2989° E Longitude and lies in Hyderabad of Telangana State in India.

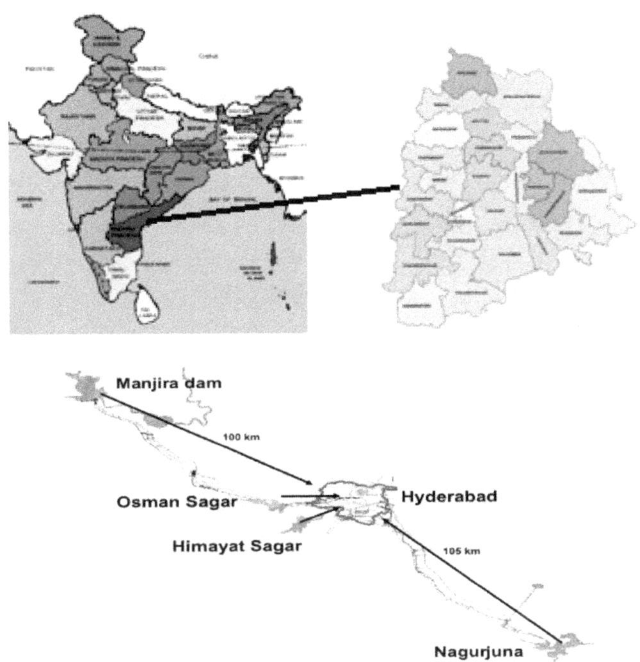

SOURCES OF WATER SUPPLY – HYDERABAD CITY

Source Name	River	Year	Distance from City *Km	Capacity		Drawls (Mgd)
				Installed (Mgd)	Storage (TMC)	
Osman Sagar	Musi	1920	15 (Gravity)	27	3.9	18
Himayat Sagar	Esi	1927	9.6 (Gravity)	18	2.967	17
Manjira –I Phase I & II	Manjira	1965 & 1981	58 (Pumping)	45 (15 + 30)	1.5	45
Manjira –II Phase III & IV	Manjira	1991 & 1994	80(Pumping)	75 (37 + 38)	30	75
Krishna Ph-I	Krishna	2004/05	116	90	150	90

			(Pumping/Gravity)			
Krishna Ph-II	Krishna	2006-08	116 (Pumping/Gravity)	90		90

Demand, Supply & Gap

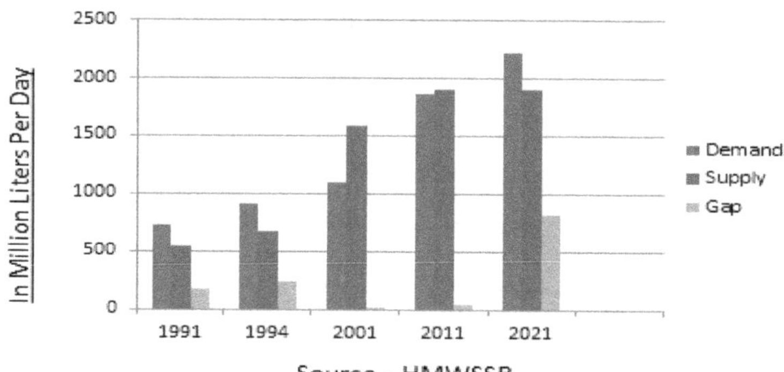

Source - HMWSSB

Year	Demand	Supply	Gap
1991	722	545	177
1994	913	680	233
2001	1100	1590	15
2011	1862	1906	44
2021	2224	1906	818

BY 2021 there will be a GAP of 818 Million Liters per day deficit which needs to be fulfilled with the existing available resource only by adopting SCADA Systems .

Total water control system using SCADA system

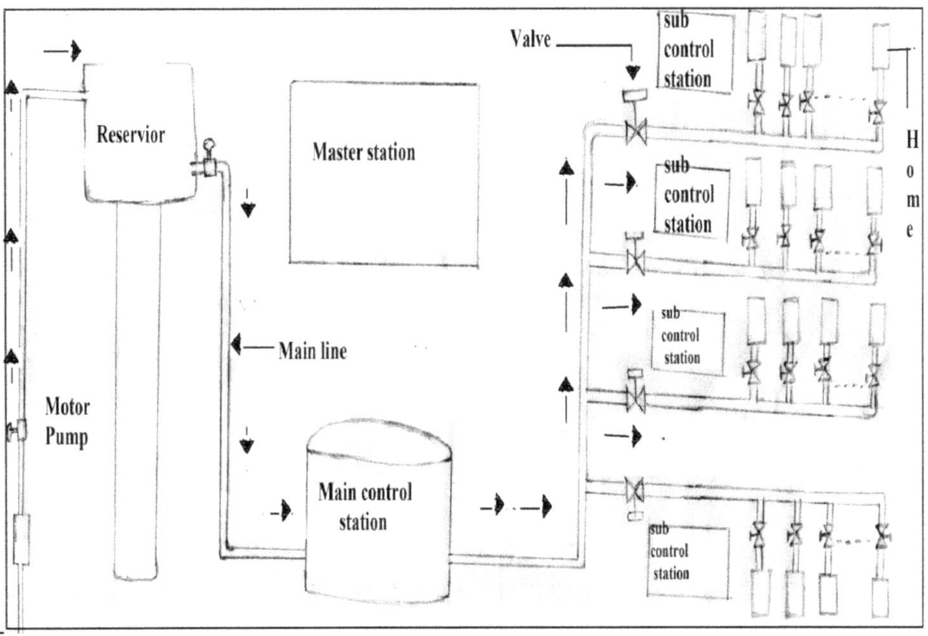

RESULTS & CONCLUSION

After working virtually for the data collected on the study area with SCADA Process Viewer the water levels capacity has drastically dropped down when compared with the maximum water levels capacity the draw down in TMC with the present situation. Nearly an average 40% gap was found between the demand vs supply since 1991 to 2011 and which is going to increase the average gap to 85% from 2011 to 2021 with an incremental of 45% when compared to the previous decade 1991-2001 loss.

The need of the hour is to immediately take steps on implementing and integrating the SCADA systems for a better outcome and also to control the UFW at catchment areas.

REFERENCES

1. Working Phases of SCADA System for Power Distribution Network by Gajendra Kumar and Prof Sunil Kumar Bhatt.
2. Butler, D., and Schütze, M. (2001). *Integrating simulation models with a view to optimal control of urban wastewater systems*
3. IT for Management of Water Supply -- "SCADA in HMWSSB" By Sri M. Satyanarayana, Director Projects, HMWSSB
4. CGWB, 2000, Ground water in Urban Environment of India, Faridabad, India
5. Irrigation water Resources and Water Power Engineering, Seventh Edition, by Dr. P.N.Modi, Published by Standard Book House, Delhi.
6. HMSSB, 2001, Musi River Conservation Project, Foundation stone laying ceremony (brochure).

Author Index

LIST OF SPONSORSHIPS

Patron

Technical Education Quality Improvement Programme-III (TEQIP-III)
JNTUH Institute of Science and Technology,
Jawaharlal Nehru Technological University Hyderabad,
Kukatpally, Hyderabad

Sponsors

National Bank for Agriculture and Rural Development
Mumbai

Science and Engineering Research Board
Department of Science and Technology
Government of India, New Delhi

Indian Space Research Organization
Department of Space
Government of India

www.ingramcontent.com/pod-product-compliance
Lightning Source LLC
LaVergne TN
LVHW080850240726
843527LV00052B/287